Trace Element Metabolism in Man and Animals

Proceedings of the Fourth International Symposium on Trace Element Metabolism in Man and Animals (TEMA-4), held in Perth, Western Australia 11–15 May 1981

Edited by

J. M. GAWTHORNE
J. McC. HOWELL
C. L. WHITE

SPRINGER-VERLAG
BERLIN HEIDELBERG NEW YORK 1982

Dr. J. M. Gawthorne
Division of Veterinary Biology
School of Veterinary Studies
Murdoch University
AUS-Murdoch, WA 6150

Prof. J. McC. Howell
Division of Veterinary Biology
School of Veterinary Studies
Murdoch University
AUS-Murdoch, WA 6150

Dr. C. L. White
CSIRO Division of Animal Production
AUS-Wembley, WA 6014

Publisher outside Australia:
Springer-Verlag Berlin Heidelberg New York

Publisher in Australia:
Australian Academy of Science
PO Box 783, Canberra City, 2601

ISBN-13: 978-3-642-68271-1 e-ISBN-13: 978-3-642-68269-8
DOI: 10.1007/978-3-642-68269-8

ISBN-13: 978-3-642-68271-1

Softcover reprint of the hardcover 1st edition 1981

2131/3140–543210

PREFACE

The Fourth International Symposium on Trace Element Metabolism in Man and Animals (TEMA-4), was held in the Sheraton Hotel, Perth, Western Australia from May 11 to 15, 1981. One of the aims of TEMA-1, which was held in Aberdeen in 1969, had been to promote a meeting at which involved scientists from a variety of disciplines would be given the opportunity to present and discuss recent research findings. It was our intention to maintain this aim. We also intended to continue the initiative taken at TEMA-3, held in Freising-Weihenstephan in 1977, which was to encourage participation by our colleagues in human medicine. We feel that a pleasing degree of success was achieved for both these aims.

The decision to hold TEMA-4 in Perth was made at the TEMA-3 meeting in 1977. The International Guiding Committee wished to acknowledge the significant body of work on trace element metabolism and disorders that had been undertaken in Australia. But in particular they wished to pay tribute to Professor Eric Underwood who for so long had been guide and mentor to so many scientists working on problems associated with trace elements. Eric was the obvious choice as Chairman of the Local Organising Committee and he set about his task with his usual vigour and good humour. We were all shocked and saddened by his death in August 1980. We feel sure that the decision to continue with arrangements for the TEMA-4 meeting is what he would have wished us to do. The members of the Local Organising Committee hope that in some measure the meeting and the Proceedings of the Fourth International Symposium on Trace Element Metabolism in Man and Animals will serve as a memorial to the work of this unique and most gifted man of science.

The Symposium was attended by 210 scientists from 24 countries, which was a most gratifying response in view of the ever increasing cost of travel. We chose six speakers for our plenary sessions and invited active workers in the field of trace element studies to submit abstracts. One hundred and forty seven contributions were presented verbally and a further twenty three were presented as posters. A wide variety of topics were covered and we were particularly pleased to have presentations and active discussions in such areas as the effect of trace elements on the development of the central nervous system and the role of trace elements in human disease. Many new ideas were presented and considerable interest centred around the series of papers given on aspects of zinc metabolism and on the role of metallothioneins.

To assist the rapid publication and to minimise the cost of the Proceedings we chose to use an offset litho process and a 'camera ready' system was used for manuscript preparation. The importance of the discussion of the contributions was recognised. We chose to record the discussions which were then typed so that they were available for correction during the meeting. In order to minimise the delay between the meeting and the publication of the Proceedings the manuscripts and discussions were lightly edited and a minimum number of changes were made. This has involved some retyping and the editors accept responsibility for the changes that may have occurred at this stage.

We wish to thank all participants and session chairpersons for their valuable contributions to the meeting. We are grateful to the Secretariat who have worked with such dedication on the manuscripts and discussion of the papers. We acknowledge the financial support of our sponsors, without their aid the Symposium could not have taken place. We are grateful to the members of the International Guiding Committee and the Local Organising Committee who worked hard and long for the success of the meeting.

J. McC. Howell

J. M. Gawthorne

C. L. White

Perth 1981

CONTENTS

INTRODUCTORY LECTURE

INTRODUCTORY LECTURE

ERIC JOHN UNDERWOOD, FRS; A COLLEAGUE, MENTOR & FRIEND

C. F. Mills, Rowett Research Institute, ABERDEEN, SCOTLAND.

My Lord Mayor, Guests, Ladies and Gentlemen;

On behalf of all my fellow participants, I would first like to express gratitude to the City of Perth for all it has already done to provide us with such a congenial environment within which to hold our meetings during the next 5 days. Experience of past International Symposia on Trace Element Metabolism in Man and Animals held in Aberdeen, Madison and Munich has taught many of us to expect that we are now at the start of a stimulating and challenging if exhausting week of discussions. The success of these past meetings was due in no small measure to the kindness and hospitality extended to us that helped to maintain the informal atmosphere so essential for frank but constructive discussion of our problems. The generous welcome given to us by you, my Lord Mayor, and the kindness already shown by members of the Local Organising Committee and their families leaves us in no doubt that this is going to be an equally successful meeting.

In all our minds at this moment are thoughts of the man who should have been delivering this Opening Address. The decision to hold this meeting in Perth, Western Australia, reflected not only our wish to acknowledge the contribution made by Australiasian Scientists to the topic that is of such interest to us but also to acknowledge the unique contributions of our late colleague, Professor Eric Underwood. In the interval between the idea and the event, we have lost from our company a man who in his lifetime did more than any other to emphasise to the world at large and to the scientific community in particular, the influence of essential and toxic trace element supply upon the health of man and his animals. For this reason, my pleasure at being invited to open the affairs of this meeting is tempered by the sadness we will all share throughout the week at the absence of the man who so often has guided our thoughts and activities.

Traditionally, the opening papers of past TEMA Symposia have taken the form of general reviews of recent progress. On this occasion my objectives will differ. Instead, before we embark upon this 4th Meeting, it is more appropriate we should pause to reflect upon some of the objectives and attitudes that led Eric Underwood to make such outstanding contributions in the field of trace element research and to consider how this approach, so successful in his hands, might continue to influence the effectiveness of our own future efforts.

TRACE ELEMENT RESEARCH AND ITS PRACTICAL IMPLICATIONS

When in the year 1920, the Underwood family moved to a 2,000 acre holding in the vicinity of Perth, Eric, then 15 years old, first became familiar with the difficulties faced by those Western Australians attempting to develop livestock farming in what at the time was a particularly hostile natural environment. This experience and, later, the emergence of evidence that the low productivity of the area might be partly attributable to mineral deficiencies, coloured almost all of his subsequent attitudes on the relationship between the practice and the science of agriculture. It was this background that led him as a young graduate to investigate variables influencing inorganic nutrient supply and pasture productivity in work for a Doctorate at the University of Cambridge and later, after his appointment as Animal Nutritionist working for the Department of Agriculture, Western Australia, led him to join Filmer in his studies of nutritional aspects of the aetiology of wasting diseases of cattle and sheep endemic in this and many other areas of Australasia.

The remit given to this 26 year old appointee would have daunted many. It was virtually to explain why earlier studies yielded many conflicting observations on the effectiveness of iron preparations in controlling such diseases and then to recommend more effective prophylactic measures. Although the subsequent history of the painstaking studies leading to the finding that cobalt was the biologically effective adventitious contaminant has often been documented, such accounts usually ignore the most important personal attributes of the worker primarily responsible for this discovery. These were the clarity of his appreciation that a successful outcome to the investigation could have important practical implications and the courage of his conviction that the approach selected, although physically and at that time technically demanding, was appropriate. The road to ultimate success was hindered by problems with which some of us are once again beginning to become familiar - shortage of funds and of equipment and conflicting advice with respect to appropriate priorities for state-financed research. That so much was achieved in establishing the essentiality of cobalt for ruminants under such circumstances is its own testimony to the enthusiasm and tenacity of purpose of the man whose work paved the way for the discoveries of others that brought him equal satisfaction and pleasure - the education of the role of cobalt in cobalamin (Vitamin B_{12}) and later, identification of the biochemical defects responsible for the clinical manifestations of the disease that first excited his interest in the trace elements.

Even this brief account is sufficient to indicate the reasons why Eric Underwood's interests in the topic of the role and metabolism of the trace elements, although based upon a wide experience of their significance for health, also extended to detailed aspects of their biochemical and metabolic role in Nature. Those of his friends with a similar outlook were often made particularly aware of his intense interest in such fundamental studies. On meeting such colleagues, even after a long interval, he would often forego many of the conventional trivia of greeting and rush directly into the topic most exciting him at the time - "Hello (very briefly!) - what did you think of that excellent piece of work on ....". His remarks on studies for which he felt less respect were equally direct and, usually, much more succinct. The voracity with which he assimilated such newly published information never diminished. It clearly reflected his conviction that progress towards the more effective diagnosis and control of practical problems would be determined largely by the extent to which studies of those biochemical and physiological processes most sensitive to trace element deficiency or excess could be encouraged and granted adequate financial support.

SCIENTIFIC INTEGRITY AND RESPONSIBILITY

To some, the role of the trace elements in health and disease is a highly emotive subject. Although, in the context of animal health, their practical significance has been established clearly and repeatedly, much less is yet known of the influence of trace element deficiency or excess upon the health of man. Indeed, many of us have become familiar with the particular difficulty of maintaining a dispassionate appraisal of the claims sometimes made that abnormalities in trace element supply or metabolism play a primary role in the aetiology of many of man's most serious diseases or that increasing pollution of man's environment must inevitably have an adverse influence upon his health. Sometimes obscured by the dust created by the often heated discussion of extravagent claims and counter-claims lies the slowly growing body of evidence that, in some circumstances, often not fully defined, anomalies in trace element supply do indeed influence the health, development or well being of man. That, despite the vagaries of popular opinion, financial support for the conduct of trace element investigations relevant to human health has been growing steadily is largely due to the fact that we have had as one of our colleagues a man with a widely acknowledged reputation as a dispassionate arbiter of progress with a gift for distilling the essentials in his writings on the subject of the trace elements. The task of maintaining such an authorative commentary was enormous, particularly in recent years as progress accelerated rapidly. The achievement was recognised widely. Eric Underwood's writings ultimately influenced not only aspects of the policies of national and international agencies concerned with nutrition and health but also the effectiveness with which many of us, approaching the subject for the first time, were able first to establish familiarity with the subject and, later, to develop lines of communication with other workers sharing similar interests.

I have often heard it asked what led Eric Underwood to undertake this challenging commitment. I have no doubts whatsoever. It was that he shared with many others the satisfaction and sheer pleasure to be gained from intellectually challenging scientific effort in a field in which the outcome, directly or indirectly, will influence man's ability to control nutritional disease. No one appreciated better than he, the impact of trace element deficiency upon animal and human health in so many parts of the world. But few have also appreciated so clearly the need to maintain and guard the standards of scientific work in a field which by its very complexity necessitates interdisciplinary effort. Indeed, it was these very attitudes and anxieties that led him to be such an enthusiastic supporter of these Symposia on Trace Element Metabolism. Initially conceived as a much more modest enterprise intended to serve European interests only, those of us in Aberdeen concerned with the organisation of the first meeting in 1968 began to develop the view that such a remit was unnecessarily restrictive. These views were not shared by all. When his guidance was solicited and freely given, it influenced not only the international and interdisciplinary character of these meetings but also their effectiveness as gatherings in which scientist and practitioner, meeting upon common ground, could assess progress and outstanding problems.

FUTURE PROSPECTS AND RESPONSIBILITIES

For many, the venue for this meeting is geographically very remote from our place of work. Some who would have made important contributions have been unable to attend. Nevertheless, the fact that by whatever means, funding has somehow been made available in these times of financial stringency for more than 200 of us to gather is one indication of the support that our work is now receiving. It is a tangible indication both of a growing recognition of the importance of its objectives and of the need for the information we are endeavouring to provide. Those, who in addition to having commitments to research, also act in consultative capacities, will be fully aware of the demands now arising for information on the roles of the trace elements in disease. While the published Proceedings of our three previous meetings clearly reflect the rapid progress made in research during the last twelve years, they do not indicate the growing volume of requests for information and help originating from nutritionists, veterinarians, clinicians, from the food industry and from government agencies. These range from enquiries relating to the detection and control of deficiency to questions concerning the likely tolerance of animals and man to elements emitted as environmental pollutants.

All too often, the provision of adequate answers demands information that is not yet available. Our immediate problem is to identify these areas of doubt and, in meetings such as this, to ensure that such uncertainties are discussed frankly. The precedent for such discussions was provided in Eric Underwood's own introductory address to TEMA-1. In this, he challenged participants to recognise that, although much was becoming known of the functional role of the essential trace elements, the nature of metabolic lesions responsible for early pathological manifestations of trace element deficiency disease was usually unknown. The comment is as valid in 1981 as when it was made in 1969. It indicates the principal reason why the biochemical and other diagnostic indices currently used for the detection of trace element-related diseases are frequently found to be inadequate in specificity or sensitivity and why, lacking adequate criteria, it is often difficult to determine minimum requirements for some essential trace elements or how these requirements are influenced by diet, by stage of physiological development or by other variables.

This example is merely one of many in which progress towards the resolution of practical problems must await the more effective deployment of scientific effort towards effective investigation of the processes influencing trace element absorption, utilisation and function. Without first identifying such gaps in knowledge and, later, more effectively recruiting to our company workers with sufficient technical expertise, stamina and enthusiasm to face the inherent technical difficulties of work in this field it is questionable whether the future will see us much better fitted to meet the requests for help and guidance that will increasingly be made.

The problem of maintaining a continually vigorous development of background research was often pre-eminent in Eric Underwood's mind in his work as a consultant to national and international agencies. He, like we, appreciated to the full that the consequences of limitations in the design, conduct or interpretation of our experimental work are increasing. They influence the effectiveness of the contributions we make to those faced with the task of establishing animal industries under unfavourable circumstances in some newly developing countries, they can affect policy decisions with respect to manufacture and control of the composition of new foods or of therapeutic diets for man and influence opinions on the potential hazards created by environmental pollutants.

Thus, however intellectually satisfying the attainment of the immediate objective of any investigation may be, we must now recognise that we are firmly denied the luxury of working in academic isolation. Sooner or later, sometimes with, but more often without our sanction, our results and experience will be drawn upon in man's efforts to identify and control nutritional diseases in which anomalies in trace element supply are believed to be involved. To ensure that his concepts of such responsibilities remain before us, it has been agreed that future meetings in this series will include a specially designated Underwood Lecture; its purpose will be to remind us of the standards of integrity, perseverance and scientific excellence set by a friend and colleague whose example we would do well to follow.

In conclusion, I would like to take this opportunity to emphasise the debt that all of us owe to the Local Organising Committee in making this meeting possible. I am certain you will appreciate the particular difficulties they suddenly had to face. The fact that we are here at all is its own testimony to their determination in the face of difficulties that could well have led to the cancellation of TEMA-4. We can best express our gratitude to them by conducting our meeting in the way their late Chairman would have wished; by maintaining the free exchange of ideas and constructive comment in the informal atmosphere for which this series of Symposia has already established an envied reputation among scientific meetings.

TRACE ELEMENT STATUS AND REQUIREMENTS

TRACE ELEMENT NEEDS OF SPECIES IN RELATION TO BODY SIZE

R.M. Smith
CSIRO Division of Human Nutrition
Kintore Avenue
ADELAIDE, SOUTH AUSTRALIA, 5000

The requirements of an animal for energy and protein depend upon body size, level of activity, rate of growth and the rate at which the animal secretes, produces or sheds its non-excretory products. Such a factorial approach to the statement of requirements is used in principle in most systems for calculating the needs of these nutrients. The factorial approach has also been applied to the estimation of mineral requirements of livestock and, in a few cases, to requirements for trace elements. Thus Weigand and Kirchgessner (1) have examined in factorial terms the requirements for zinc of the young growing rat and have derived predictive equations which account separately for maintenance needs, the increment required for growth, the inescapable excretory losses in urine and faeces and the average daily loss of zinc in shed hair. Over the limited range of growth for which the equations were derived irreducible faecal zinc excretion was found to be constant but the term for urinary zinc excretion was expressed in terms of metabolic body size ($W^{0.75}$).

There seems to be no logical reason to suppose that endogenous excretion of a trace element should relate to the rate of expenditure of energy and the analogy between the requirements for a trace element and those for energy is, in fact, a poor one. Whereas the daily increment of energy consumed for maintenance is destined to be oxidised and lost, the trace elements are in general able to be retained and recycled and in any event are incorporated into components of the tissues rather than being metabolised to meet ongoing needs. The magnitude of endogenous losses of trace elements can therefore be more logically related to body size than to metabolic rate and energy needs. This relationship became clear from an analysis of the iodine requirements for maintenance of a number of species that covered a wide range of body sizes. Expressed on a per kg body weight basis the minimum requirement of the rat for iodine was only 50% higher than that of an ox whereas total energy needs were six or sevenfold higher (2). It is the purpose of the present communication to examine the minimum requirements for zinc and copper for both maintenance and growth of a range of species with a view to establishing any relationship between these requirements and body size.

REQUIREMENTS FOR ZINC

No authentic measurements of minimum zinc requirements for adult maintenance were discovered for any species but in the upper panel of Fig. 1 are shown a number of estimates of marginally adequate intakes together with some flanking values of levels of zinc intake that are either of proven adequacy or that have led to clear symptoms of deficiency. No suitable value for the minimum requirement of the mouse for zinc could be found but this may be assumed to be less than that provided by the well established long-term mouse-maintenance diet shown. In selecting data for zinc requirements care has been taken to avoid cases where substances such as phytate, known to interfere with zinc availability, were present. Inspection of the figure shows that the maintenance requirement per kg body weight for zinc is virtually constant from the mouse to the ox although on the same basis the normal energy expenditure of the mouse is about 20 fold higher than that of the ox.

In the lower part of Fig. 1 is set out the much more abundant data on requirements and marginal requirements for zinc for growth and these are again flanked by values for adequate and deficient intakes. The requirements recorded for growth are much more scattered than those for maintenance and this is attributable in part to differences in growth rates between species. Thus at a body weight of 20 kg the pig may grow at a rate of 25g/kg body weight per day whereas the growth rate of a 20 kg child is only about 0.3 g/kg per day. Assuming a zinc content of the tissues laid down of 25 mg/kg (3) then the difference in growth rate would account for a difference of 0.62 mg Zn/kg per day in the requirements of the two species for growth. This represents only about 30% of the difference seen in Fig. 1 and despite the large uncertainties involved suggests that the requirement of the pig for zinc during growth may be intrinsically higher than that for man or, indeed, for the other species represented.

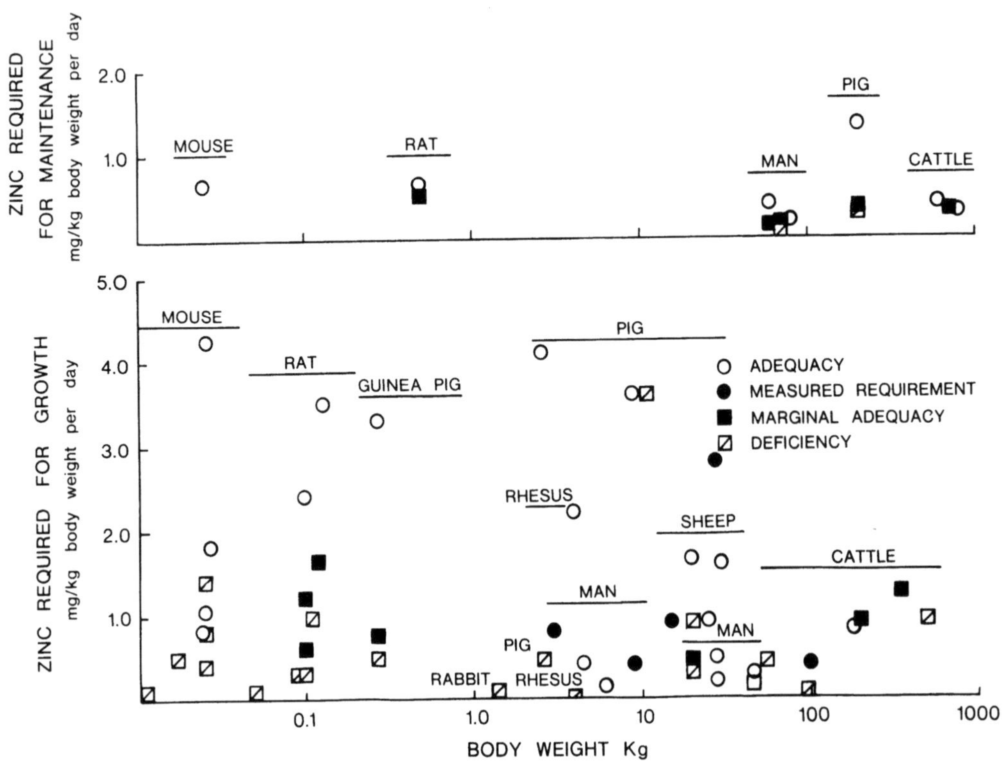

Fig. 1. The requirements for zinc for maintenance shown in the upper panel were derived from references 6-9. The zinc requirements for growth in the lower panel came from references 9-27. Where this was not provided in the primary reference information on feed requirements, rates of body weight change and representative body weights was derived from the Biology Data Book, Vol. III (Fed. Amer. Soc. Exp. Biol., Bethesda).

Apart from the special case of the growing pig it is apparent from Fig. 1 that even for growth the requirements for zinc across species cover a range that is small in comparison with the range of energy requirements that are dictated by the basal metabolic rates.

REQUIREMENTS FOR COPPER

As shown in the upper part of Fig. 2 values for the minimum requirements for copper are available for several species and, as in the case of zinc, the requirement appears to be relatively constant across species. The observation is rather surprising in view of the wide differences that exist in the availability of copper between ruminants and non-ruminants (4,5) although this trend emerges to some extent. At all events the minimum maintenance requirement for dietary copper across all species examined rarely exceeds 0.1 mg/kg body weight per day and shows no relationship to metabolic rate.

The requirements of growing animals for copper are seen in the lower panel of Fig. 2 and reveal an even wider scatter than was the case for zinc. Once again the pig shows an elevated requirement that is only partly accounted for on the basis of faster growth. Assuming a content of 1 mg Cu/kg body weight laid down then less than 10% of the additional requirement appears to stem from this cause. The requirements for copper during growth, however, may also have been affected by large differences in dietary availability (5), a factor that is less important in the case of zinc. No overall trends in copper requirements in relation to body weight can be perceived in Fig. 2.

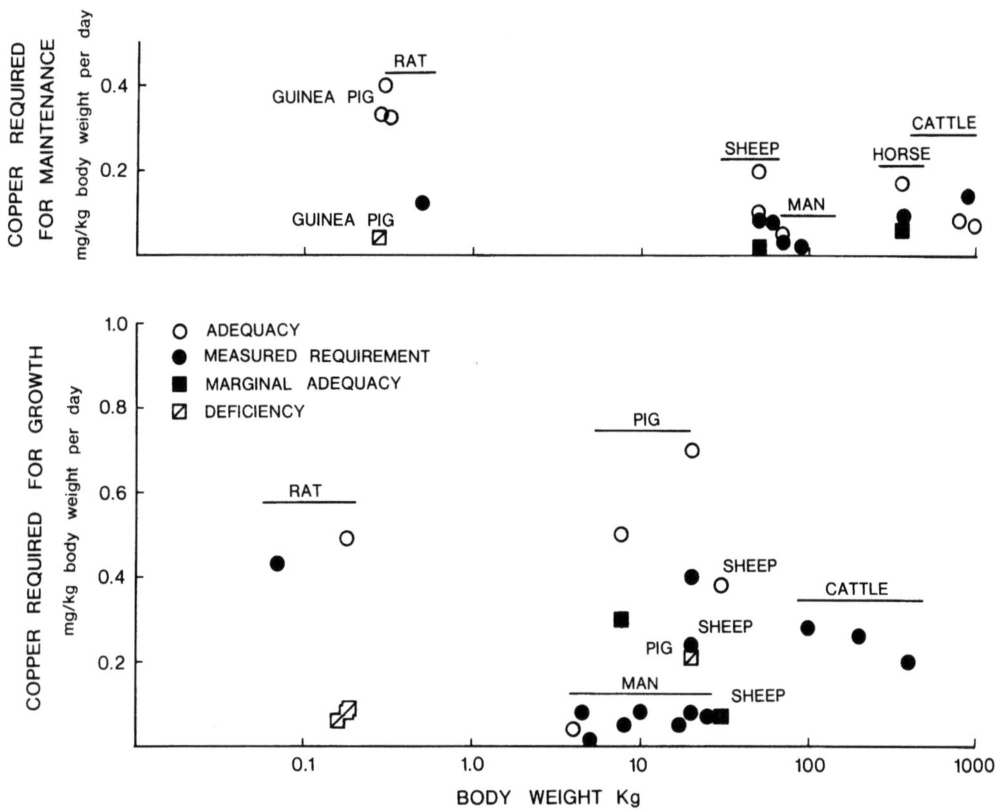

Fig. 2. In the upper panel are shown the maintenance requirements for copper derived from references 4, 9 and 28-32. The copper requirements for growth shown in the lower panel came from references 9, 28, 29 and 33-37. Additional data were obtained as described in the legend to Fig. 1.

CONCLUSIONS

The present study has revealed that as in the case of iodine (2) the maintenance requirements for dietary copper and zinc over a wide range of species fall within a relatively narrow range of values when expressed on a per kg body weight basis. Thus zinc requirements are clustered round a value of 0.4 mg/kg per day while copper requirements approach 0.1 mg/kg per day. The minimum requirements for iodine for maintenance were previously shown to be around 0.001 mg/kg body weight per day in an equally wide range of species. Requirements of zinc and copper for growth likewise show no general relationship to metabolic rate but do show some effects of the rate of growth. In the case of copper, availability for absorption from the diet is probably also a major factor.

REFERENCES

1. Weigand, E. and M. Kirchgessner, 1977, Z. Tierphysiol., Tierenahrg, u. Futtermittelkde 39:84-95.
2. Smith, R.M., 1980, Proc. Nutr. Soc. Aust. 5:36-43.
3. Sandstead, H.H., 1973, Amer. J. Clin. Nutr. 26:1251-1260.
4. Suttle, N.F., 1974, Br. J. Nutr. 32:395-405.

5. Suttle, N.F., 1976, Chem. and Ind. 3:559-562.
6. Hunt, I.F., N.J. Murphy, J. Gomez and J.C. Smith, 1979, Am. J. Clin. Nutr. 32:1511-1518.
7. Taper, L.J., M.L. Hinners and D.J. Ritchey, 1980, Am. J. Clin. Nutr. 33:1077-1082.
8. Underwood, E.J., 1978, Persp. in Prac. 72:177-179.
9. Underwood, E.J., 1977, Trace Elements in Human and Animal Nutrition, 4th Ed. (Academic Press, NY).
10. Ziegler, E.E., B.B. Edwards, R.L. Jensen, L.J. Filer, Jr. and S.J. Fomon, 1978, "Proceedings of the Third International Symposium on Trace Element Metabolism in Man and Animals" 292-295.
11. Greger, J.L., O.A. Bennett, S. Buckley and P. Baligar, 1977, "Proceedings of the Third International Symposium on Trace Element Metabolism in Man and Animals" 300-303.
12. Hambidge, K.M., M.N. Chavez, R.M. Brown and P.A. Walravens, 1979, Am. J. Clin. Nutr. 32:2532-2539.
13. Vuori, E., 1979, Br. J. Nutr. 42:407.
14. Ritchey, S.J., M.K. Korslund, L.M. Gilbert, D.C. Fay and M.F. Robinson, 1979, Am. J. Clin. Nutr. 32:799-803.
15. Hardie-Muncy, D.A. and A.I. Rasmussen, 1979, J. Nutr. 109:321-329.
16. Joseph, C.E., S.H. Ashrafi and J.P. Waterhouse, 1981, J. Nutr. 111:53-57.
17. Fang, M.M., K.Y. Lei and L.T. Kilgore, 1980, J. Nutr. 100:1032-1036.
18. Beach, R.S., M.E. Gershwin and L.S. Hurley, 1980, J. Nutr. 110:201-211.
19. Saylor, W.W. and R.M. Leach, Jr., 1980, J. Nutr. 110:448-459.
20. Weigand, E. and M. Kirchgessner, 1980, J. Nutr. 110:469-480.
21. Swenerton, H. and L.S. Hurley, 1980, J. Nutr. 110:575-583.
22. Luecke, R.W. and P.J. Fraker, 1979, J. Nutr. 109:1373-1376.
23. Mertz, W., 1977, J. Anim. Sci. 44:469-474.
24. Alberts, J.C., J.A. Lang, P.S. Reyes and G.M. Briggs, 1977, J. Nutr. 107:1517-1527.
25. Kincaid, R.L., 1979, J. Dairy Sci. 62:572-576.
26. Miller, E.R., R.W. Luecke, D.E. Ullrey, B.V. Baltzer, B.L. Bradley and J.A. Hoefer, 1968, J. Nutr. 95:278-286.
27. Miller, W.J., W.J. Pitts, C.M. Clifton and J.D. Morton, 1965, J. Dairy Sci. 48:1329-1334.
28. Schroeder, H.A., A.P. Nason, I.H. Tipton and J.J. Balassa, 1966, J. Chron. Dis. 19: 1007-1034.
29. Mason, K., 1979, J. Nutr. 109:1979-2066.
30. Klevay, L.M., S.D. inHyg, S.J. Reck, R.A. Jacob, G.M. Logan, J.M. Munoz and H.H. Sandstead, 1980, Amer. J. Clin. Nutr. 33:45-50.
31. Givens, D.I. and Hopkins, J.R., 1978, J. Agric. Sci. Camb. 91:13-16.
32. Cymbaluk, N.F., H.F. Schryver and H.F. Hintz, 1981, J. Nutr. 111:87-95.
33. Vuori, E., 1979, Br. J. Nutr. 42:407-411.
34. Cerklewski, F.L., 1979, J. Nutr. 109:1529-1533.
35. Saylor, W.W. and R.M. Leach, 1980, J. Nutr. 110:448-459
36. Paynter, D.I., R.J. Moir and E.J. Underwood, 1979, J. Nutr. 109:1570-1576.
37. Okonkwo, A.C., P.K. Ku, E.R. Miller, K.K. Keahey and D.E. Ullrey, 1979, J. Nutr. 109:939-948.

DISCUSSION

HORVATH (USA). You have stated that the larger species are more likely to develop trace element deficiencies, yet in the wild there is little evidence for this. Zinc deficiency was reported in moose on the Kenai Peninsular of Alaska, but I have not noticed other reports of mineral deficiencies in wild species. It would be interesting to know if wild species are similar or different to domestic species in mineral requirements.

SMITH (Australia). I would like to obtain extra data from any species, and it would be very valuable to have information for wild species if the data exists.

BRAY (Australia). Could you comment on how the differences in trace element requirements between breeds of a given species fits in with your overall thesis?

SMITH (Australia). For a given trace element, the minimum requirement for a particular species is a range of values that includes the variation between breeds. The copper requirements of sheep, for instance, vary as much as five fold. However, when these ranges are graphed, variations of this magnitude are small. In contrast, the differences in energy requirements between species are very much greater. Thus, when mineral requirements/kg body weight deviate markedly from the pattern that exists for most species then the reasons for the deviation warrant investigation.

AGGETT (Scotland). Have you examined your data in relation to the growth rate of the various species?

SMITH (Australia). The data as presented takes no account of differences in growth rate of the species. However, I have looked more closely at the data for pigs as the requirements for this species are outstandingly different from the others. The pig has the fastest growth rate of the species studied, but this accounts for only one third of the difference in trace element requirements relative to the other species. It is possible that additional amounts must be present in the diet to establish adequacy of retention, and in particular the pig has some unusual constraints on its absorption of zinc.

ULLREY (USA). I suspect that if you chose different data for the pig you would find a better match with the other species. Your data seems to be based on a dietary zinc requirement of 50-60 ppm for pigs. However, if estimates are made with purified diets where availability is not a problem the zinc requirement is 12-20 ppm and this would bring the values for pigs into line with the other species.

SMITH (Australia). I encountered a large range of values for pigs and tried to avoid data where phytate or the calcium:phosphorus ratio was likely to influence zinc availability. However, I did not have estimates from experiments where purified diets were used and would appreciate receiving the data you mention so that it can be included.

THE DIETARY INTAKE OF TRACE ELEMENTS IN SWEDEN

M. Abdulla, S. Svensson, Å. Nordén and P-A. Öckerman

Unit for Community Care Sciences at Dalby and Research
Department 2, University Hospital, S-221 85 Lund, Sweden

SUMMARY

In spite of the recent advances in trace element research, little is known about the dietary intake of essential trace elements from prepared foods. We have investigated the dietary intake of the important essential inorganic chemical elements in a group of healthy adults, pensioners, vegans and lactovegetarians in Sweden, employing the duplicate portion technique. The duplicate portion is a copy collected while eating normally to represent as exactly as possible the food and drinks consumed during 24 hours. Pooled material during 24 hours was homogenised and part of it was lyophilised after fat extraction with chloroform and methanol. The fat-free powder was wet-ashed with sulphuric, nitric and perchloric acids and the residue left was used for the determination of the elements. Most essential elements were determined by atomic absorption spectrophotometry. Our results showed that except for calcium and sodium, the intake of the remaining elements analysed was low as compared to the recommended dietary allowance (RDA) levels. The intake of potassium, magnesium, zinc, copper and selenium was particularly low as compared to the RDA values. The vegetarian diets by contrast had significantly higher contents of most elements than the other diets, except for selenium. Our results on the whole indicate that the intake of most essential inorganic chemical elements from prepared meals in industrialised countries like Sweden is low as compared to the RDA levels.

INTRODUCTION

Alongwith the remarkable advances in analytical technology and sophisticated instrumentation during the last few decades a number of inorganic chemical elements that were hitherto regarded as unimportant have been detected in living systems and their food chain. Recent estimates indicate that most tissues of a healthy human body are fairly rich in as many as 46 elements represented in the Periodic Table (1). In spite of the recent progress in trace element nutrition research, the biological role and the minimum requirement of many trace elements are still hypothetical. The number of elements that may be essential in human nutrition and health will probably increase in future with improvements in our knowledge of the chemical elements. A number of trace elements that have been found to be essential in human nutrition on the other hand are required only in such small concentrations that it is commonly believed that a pure nutritional deficiency of these micronutrients hardly ever occur in man. During the past few decades, the rapid development in food technology has enabled the growing food industry to offer the general public in industrialised countries an enormously great choice of various types of food products. In countries with high standards of living many of the conventional foods prepared for an average house-hold are being replaced by processed foods, food analogues and empty calories. The increased consumption of highly refined food and food analogues for prolonged periods, however, may induce marginal deficiencies of trace elements (2). Although the prevalence of clinical malnutrition is low in affluent countries, there are indications that specific sub-clinical deficiencies of trace elements may occur in certain vulnerable groups. Children, pregnant women, elderly and alcoholics are specially susceptible to marginal deficiencies of trace elements. A lack of characteristic symptoms and diagnostic techniques are the major reasons that these deficiencies are not identified at present.

Although most of the essential inorganic chemical elements can be measured accurately today, there are no internationally accepted standards for the minimum requirements of many of them. Also, the recommended dietary allowances vary considerably from country to country. Moreover, the data found in the literature regarding the dietary intake of most essential trace elements is often unsatisfactory since most results are based on conventional techniques like the dietary history and interview techniques in combination with food tables. The present study has been done to investigate the dietary intake of major and minor nutrients from prepared foods in a group of healthy adults, pensioners, vegans and lactovegetarians in Sweden and to compare the data with that of the recommended daily allowance (RDA) levels.

MATERIAL AND METHODS

Daily food samples were collected from 20 normal adults (10 men and 10 women), 37 pensioners (17 males and 20 females > 67 years old), 6 vegans and 6 lactovegetarians employing the duplicate portion technique. The duplicate portion is a copy collected while eating normally to represent as exactly as possible the food and drinks consumed during 24 hours. A total of 450 daily portions were collected. Material collected during the day was by and by pooled in one container. It was later homogenised and a definite volume of the homogenate was lyophilised after extraction of the fatty components with methanol and chloroform (3, 4). The lyophilised powder was used for the analysis of the major and minor food components. For the analysis of the inorganic chemical elements, 0.5 g of the dry powder was wet-ashed using ideal concentrations of concentrated sulphuric, nitric and perchloric acids. The residue left after the wet-ashing was later diluted with ion-free water to a constant volume. This solution was used for the determination of the chemical elements. Sodium and potassium were determined by flame photometry and calcium, magnesium, iron, zinc, copper, chromium, nickel, lead, cadmium and mercury were analysed by atomic absorption spectrophotometry. Determinations of nickel, chromium, selenium, cadmium, lead and mercury were made on 4 - 7 day pooled samples.

RESULTS AND DISCUSSION

Table I and II summarize part of the results. The losses due to fat extraction were less than 5 % in most cases. The wet-ashing procedure resulted in a further loss of 10 % in the case of sodium, potassium and magnesium. The difference between duplicate analyses was less than 2 %.

Table I. Intake of major elements (mg/day), mean ± S.D.

Subjects	Ca	Mg	Na	K
Adults	725 ± 278	220 ± 62	2691 ± 828	2106 ± 663
Pensioners	704 ± 226	204 ± 54	2024 ± 805	1911 ± 722
Vegans	625 ± 244	541 ± 123	2208 ± 713	4017 ± 1092
Lactovegetarians	904 ± 269	428 ± 138	2218 ± 876	4132 ± 1318

Table II. Intake of trace elements (mg/day), mean ± S.D.

Subjects	Fe	Zn	Cu	Se*	Iodine
Adults	13.5 ± 4.2	8.5 ± 3.2	1.4 ± 0.7	-	-
Pensioners	11.6 ± 4.3	8.3 ± 2.8	1.2 ± 0.4	0.038 (0.008 - 0.098)	0.28 ± 0.12
Vegans	16.3 ± 6.5	11.8 ± 3.5	3.6 ± 0.8	0.011 (<0.003 - 0.059)	0.07 ± 0.02
Lactovegetarians	14.7 ± 3.9	10.8 ± 2.5	2.1 ± 0.7	0.064 (0.042 - 0.082)	-

*The values for selenium are given as mean and range

As can be observed from the tables, the intake of most of the elements studied was low as compared to the recommended dietary allowance levels. The intake was particularly low for potassium, magnesium, zinc, copper and selenium. Since the soil in Sweden is rather poor in selenium, it is well-documented that meat and dairy products contain less selenium than that is found in other parts of Europe. The vegetarian diets by contrast had significantly higher intake of most trace elements than the other diets. The vegan diets were very poor in selenium and several 24-hour portions had concentrations below 5 μg/day, whereas the lactovegetarian diets contained more selenium than the other diets. One reason for the high contents of many essential trace elements in the vegetarian diets may probably be due to the limitation of refined food items in the diet. Food processing results in a considerable loss of many essential trace elements.

Food selection in relation to requirements has been the subject of continuous revision during the last few decades. Depending on a number of local factors, the dietary allowances recommended for different nutrients vary to a considerable extent in different parts of the world. One reason for this discrepancy is our limited knowledge regarding the nutritional requirements for different ethnic groups with divergent food habits. Because of these factors, there is at present a global confusion, interest and concern regarding nutritional factors affecting general health. The importance of food composition in relation to the general health of the public is the subject of many studies conducted in various parts of the world at present. As mentioned earlier, most information concerning the composition of food is usually computed from food tables specifying the concentrations of various nutrients in raw food materials. There is no guarantee that these figures apply to prepared meals served on the plate for consumption. The refining and cooking may alter the composition of the original food materials. New products marketed by food industries may not give adequate information regarding several nutrients including trace elements. Duplicate portion technique eliminates many of the problems mentioned above. This technique gives, however, lower values for total intake of energy, protein and other major nutrients and trace elements. The results in different series in our laboratory and in studies by others are, however, reproducible. Further, the intake levels of most trace elements found in Swedish populations are similar to that found in several nutrition institutes in United States (5). The concentration of toxic metals like mercury, cadmium and lead was investigated in the diets of pensioners and the data showed that the intake levels were much lower than the tolerable dietary intake proposed by FAO/WHO (6).

Although the marginal deficiencies of trace elements that may occur in the general population of affluent countries like Sweden may not be severe enough to receive medical attention, certain vulnerable groups like pregnant women and children may suffer more than the population as a whole. It may be necessary to watch such groups in future for subtle symptoms produced by marginal deficiency of inorganic micro-nutrients like zinc, copper and selenium. The pensioners who participated in the present study were followed for a number of years and so far we have not found any clinical parameter indicating deficiency of trace elements.

REFERENCES

1. Hamilton, E.I. 1979. In "The Chemical Elements and Man" (Charles C. Thomas Springfield, Illinois, USA) pp xiii and 363-410

2. Mertz, W. 1970. Fed. proc. 29:1482-1488

3. Borgström, B., Å. Nordén, B. Åkesson, and M. Jägerstad. 1975. Scand. J. Soc. Med. suppl. 10, pp 14-25

4. Borgström, B., Å. Nordén, B. Åkesson, M. Abdulla, and M. Jägerstad. 1979. Scand. J. Gastroent. suppl. 52, pp 28-41

5. Mertz, W., and H.H. Sandstead. 1980. Personal communication

6. FAO/WHO. 1972. WHO Tech. Rep. ser. 505, p 32

DISCUSSION

KRISHNAMACHARI (India). One of the difficulties in assessing the adequacy of diets for minerals is that in addition to total intakes of each element, the absorption or availability is needed also. Would you care to comment?

ABDULLA (Sweden). Yes I agree. For instance although we found relatively high levels of trace elements in vegetarian diets, the absorption from such diets may be low because the elements are probably bound to complex substances.

ULLREY (USA). One of your surprising findings was the fairly high intake of zinc by the vegetarians. Do you have any explanation for this?

ABDULLA (Sweden). One reason may be the large amount of nuts that they eat and another could be zinc contamination in imported foods.

ROBINSON (New Zealand). You might expect vegetarians in New Zealand to have low intakes of selenium but actually they have quite high intakes because of the quantities of imported food they consume.

ABDULLA (Sweden). In our study lacto-vegetarians had a higher intake of selenium than vegans. We were surprised to find such high selenium values because the soil in Sweden is low in selenium and vegetables are usually poor sources of selenium.

COPPER AND ZINC STATUS OF PARENTERAL NUTRITION PATIENTS

T.F. Hartley, L.G. Worthley*, B. Dollman** and P. Coyle

Institute of Medical and Veterinary Science, Adelaide
*Intensive Care Unit, Royal Adelaide Hospital
**Pharmacy Department, Royal Adelaide Hospital

INTRODUCTION

The techniques and procedures for total parenteral nutrition (T.P.N.) are now well documented(1). Hypertonic glucose and fat emulsions are used to meet the patient's calorie requirements and there are a variety of commercially available L-amino acid mixtures to meet the demands for protein synthesis. However these nutrients cannot be fully utilised unless a variety of additions are made and infused simultaneously. The necessary additions include Na, K, Mg, phosphate, trace metals and the fat and water soluble vitamins. Our protocol involves making some of the additions directly into 1 litre bags of a base solution prepared from a 5:4 mixture of 70% dextrose and the 10% L-amino acid solution, Synthamin 17(2). The typical stabilised T.P.N. patient receives 70-150 mmol/day of Na, 60 mmol/day of K of which 20-40 mmol is provided as KH_2PO_4, 4-8 mmol/day of Mg, as $MgSO_4$, 76-152 mmol/day of Zn as $ZnCl_2$ and 2 ml/day of a multi-vitamin preparation. This provides 50 mg of ascorbic acid, 10 mg of thiamine, 5 mg of riboflavin, 5 mg of sodium pantothenate, 50 mg of nicotinamide and 2.5 mg of pyridoxine. Weekly supplements of folate, 15 mg, ascorbic acid, 1 g, vitamin K1, 10 mg, and hydroxycobalamin, 1 mg, are given as slow intravenous injections.

SERUM COPPER AND ZINC IN T.P.N. PATIENTS RECEIVING NO TRACE METAL SUPPLEMENTS

In 1977 we were involved with a clinical trial of an L-amino acid solution, Aminofusin-L-10% and at that time no routine protocol of Cu and Zn supplementation had been devised. Mg was added to the intravenous fluids. The average supplement was 9.9 mmol/day. A total of 48 T.P.N. patients (25 males and 13 females) were involved in the study which ran for 401 patient days. The median duration of T.P.N. was 7 days with a range of 1-40 days.

Each patient underwent a strict protocol of biochemical and haematological investigations during which their serum Mg, Cu and Zn concentrations were measured using flame atomic absorption spectrophotometry. We have examined the individual patient mean serum concentrations of Cu and Zn and found that 5 patients (10%) had low serum Cu concentrations and 18 patients (38%) had low serum Zn concentrations. These abnormalities were assessed in comparison to our normal reference ranges for serum Cu of 10-22 μmol/l for males and 10-30 μmol/l for females and that for serum Zn of 9-18 μmol/l. Elevated serum Cu and Zn concentrations were observed in 6% and 8% of the patients respectively. Further data from this study are presented in Table 1. The distribution of the serum Zn data was skewed towards low concentrations whereas the Cu data had a Gaussian distribution. These data highlighted the need for routine Zn supplements for T.P.N. patients.

RESPONSE TO ZINC SUPPLEMENTS: THE 1979-1980 SURVEY

The routine protocol of Zn supplementation described in the introduction was finalised in 1979 and a prospective study of the responses was commenced. The serum Zn concentrations increased and overall rises in the mean and median of 3 and 2 μmol/l respectively were observed. In addition the shape of the distribution of the 177 serum Zn measurements was clearly Gaussian and only 8% of the results were below the lower limit of normal. There was an almost fivefold reduction in the prevalence of low Zn concentrations when compared with the unsupplemented patients. Additional data from this study are shown in Table 1.

COPPER AND ZINC STATUS OF T.P.N. PATIENTS: THE 1980-1981 SURVEY

During the past 17 months we have collected serum Cu and Zn data from patients receiving the dextrose plus Synthamin 17 mixture and the standard T.P.N. supplement already described. The 231 serum Zn estimations had an almost identical distribution to that observed in 1980-1981 survey. The prevalences of high and low serum Zn concentrations had not changed; see

Table 1.

The serum Cu data, however, revealed a deterioration in the Cu status of our T.P.N. patients. The prevalence of low Cu values had increased from 10%, in the 1977, to 18% in this most recent study. The relevant data, shown in Table 2, suggested that the mean and median Cu for males were almost identical to those observed in 1977 study. The data from the females had a lower mean and median but the spread of the data from them completely overlapped those for the males. Hence there was insufficient evidence for a sex difference effect. An alternative explanation relates to the change from Aminofusin-L-10%, used in the 1977 study, to Synthamin 17 and this is discussed later.

DISCUSSION

There have been a number of recommendations published detailing protocols for vitamin and trace metal supplementation during T.P.N. (3,4,5). The objective of such protocols is to supply sufficient Cu and Zn to maintain the total intracellular and extracellular contents of these metals and replace losses. In T.P.N. patients the losses occur via the urine, nasogastric aspirates, wound drainage and intestinal fistual drainage fluids. Apart from the urinary losses the other losses can vary considerably from patient to patient and this places limitations upon any theoretical approach to trace metal supplementation. However, we can estimate the total contents of Cu and Zn in the intracellular and extracellular compartments and thereby gain an overall perspective on a supplementation protocol. If the respective volumes of the extracellular and intracellular compartments are assumed to be 14 and 21-28 litres respectively then:-
Extracellular Cu = 14 l x (16 to 20 μmol/l) = 224-280 μmol.
Intracellular Cu = (21 to 28 l) x 14 μmol/l = 294-392 μmol.
Extracellular Zn = 14 l x 14 μmol/l = 196 μmol
Intracellular Zn = (21 to 28 l) x 210 μmol/l = 4410-5880 μmol.

These calculations illustrate some important points:

(i) Our routine Zn supplement corresponds to only 0.7-1.4% of the total Zn present in the combined intracellular and extracellular compartments but 19-38% of the extracellular compartment.

(ii) Relatively small leakages of Zn from the intracellular compartment into the extracellular compartment would lead to measurable elevations of the serum Zn concentration. We believe that this does occur because during the 1977 survey high serum Zn concentrations were associated with those patients who had extensive sepsis within their abdominal cavity. Additional evidence from our other biochemical investigations would support this opinion because alkaline phosphatase and lactic dehydrogenase are elevated in 67% and 52% of the T.P.N. patients. Both enzymes leak from damaged cells and both contain Zn.

(iii) In view of (ii) and from a consideration of the gross differences that exist between the total intracellular and extracellular contents of Zn, the serum Zn concentration is probably a poor indicator of Zn status. Therefore future work should be directed towards devising a better index of Zn status such as lymphocyte Zn content. Lymphocytes are an accessible tissue which can be separated from blood samples using density gradient centrifugation. A well controlled study of this nature is required to establish the clinical usefulness of this approach which has been described with undue scepticism in the literature (6).

(iv) Cu is distributed in approximately equal amounts between the intracellular and extracellular compartments. Hence the serum Cu concentrations should be a good index of Cu status However, the Cu occurs in caeruloplasmin. This is an acute phase reactant and hence its synthesis by the liver is stimulated by stress and infections. These conditions are common in T.P.N. patients and hence this is the most probable reason for the 4-6% prevalence of high serum Cu concentrations.

(v) The increased prevalence of low Cu concentrations in the 1980-1981 survey may have been associated with the changeover to Synthamin 17. Neither solution contained detectable concentrations of Cu. The amino acid formulations may have affected the transport of Cu across the cell membranes and hence the intracellular to extracellular ratio. For example Synthamin 17 contains 4.4 g/l of histidine whereas Aminofusin contains only 2 g/l and a histidine-Cu complex has been reported to be the labile Cu species in plasma involved in Cu transport into cells (7,8). Unfortunately our data cannot provide more than a speculative explanation of the low Cu concentrations.

(vi) The Cu results indicate that we should provide a routine Cu supplement during T.P.N.

with the Synthamin 17 plus dextrose mixture. Reports in the literature recommend Cu supplements of 5-16 μmol/day. At present our three long term home T.P.N. patients are given 16 μmol/day of Cu as $CuSO_4$ and the effects are being monitored prospectively.

CONCLUSIONS

Zn supplementation during T.P.N. has been shown to substantially reduce the prevalence of low serum Zn concentrations. However, the serum Zn concentration is probably a poor index of the Zn status and the measurement of the intracellular Zn concentration in cells such as lymphocytes has been proposed as a more sensitive and specific alternative.

Cu supplements were not given to the T.P.N. patients and an increased prevalence of low serum Cu concentrations has been documented. The reasons for this are not clearly understood, however, routine Cu supplements are now being seriously considered.

TABLE 1: SUMMARIES OF THE THREE SURVEYS

	1977 Survey - No Cu or Zn Supplements		1979-1980	1980-1981 Survey	
			With Zn Supplements		
	Copper	Zinc	Zinc	Zinc	Copper
Median, μmol/l	18.8	9.9	13.0	12.9	16.6
Mean, μmol	18.6	11.5	13.5	13.6	17.0
S.D.	6.7	5.7	4.0	4.1	6.4
Range, μmol/l	4.6-36.6	4.9-35.1	6.0-31.6	5.4-31.6	5.9-31.6
No. of observations	48	48	177	231	51
% of Low Results	10	38	8	7	18
% of High Results	6	8	14	14	4
T.P.N. Soln. Used	Aminofusin-L-10%		4:5 Mix of Synthamin 17 and 70% Dextrose		

TABLE 2: SERUM COPPER CONCENTRATIONS (μmol/l) IN MALE AND FEMALE T.P.N. PATIENTS,1980-1981

	Male	Female
Number of measurements	26	25
Mean	17.7	16.3
Median	18.0	15.0
Maximum	31.6	30.8
Minimum	6.2	5.9
S.D.	6.2	6.8

ACKNOWLEDGEMENTS

The authors would like to acknowledge the support of their respective organisations and also their colleagues in the Parenteral and Enteral Nutrition Team at the Royal Adelaide Hospital.

REFERENCES

1. "Total Parenteral Nutrition. Premises and Promises", 1975, H. Ghadimi (ed), (John Wiley).
2. Worthley, L.I.G., Hartley, T., Dollman, B., and Martin, D., 1980, Aust. Soc. for Parenteral Nutrition, 7th Annual Scientific Meeting, Melbourne.
3. Wretlind, A., 1972, J. Nutrit. 107 : 458.
4. Shils, M.E., 1972, Drug Intell. Clin. Pharm. 6 : 385.
5. Hull, R.L., 1974, Amer. J. Hosp. Pharm. 31 : 759.
6. Editorial, 1981, B.M.J. 282 : 1098.
7. Neumann, P.Z. and Silverberg, M., 1967, Nature 213 : 775.
8. Hallman, R.S., Perrin, D.D., and Watt, A.E., 1971, Biochem. J. 121 : 549.

DISCUSSION

SMITH (Australia). It may be inappropriate to supplement the infusate with inorganic copper salts. Wouldn't it be better to provide the copper in a more physiological form such as the tripeptide Cu-glycyl-histidyl-lysine? This substance would promote the passage of copper across cell membranes and into the liver where the copper would be incorporated into caeruloplasmin, the major form of copper in plasma. Would you like to comment?

COYLE (Australia). I have no knowledge of the compound you have mentioned, but successful supplementation has been reported in the literature using salts such as copper sulphate. Presumably the copper is readily incorporated into caeruloplasmin.

DANKS (Australia). Have you noticed aminoaciduria in these patients? With aminofusin a massive aminoaciduria occurs and is likely to affect the copper and zinc balance by removing these elements into the urine. In very sick patients I would expect high serum copper values because of the acute phase reaction. The fact that they were not high in your patients may be because there was insufficient copper in the liver for additional caeruloplasmin synthesis. This could be tested by giving some copper to each patient for a few days and examining the response in caeruloplasmin levels.

COYLE (Australia). Yes, aminoaciduria could be a factor in causing the low serum copper values. The data presented was from patients on short-term T.P.N. We also have several long-term parenteral nutrition patients who have a broader supplementation program including copper. The patients have been given 16 μmol Cu/day and we are currently monitoring the effects.

ELMES (Wales). Dr Keeling, at St Thomas's Hospital used white cells (lymphocytes and neutrophils) to estimate the zinc status in chronic disease. This is a long-term estimate and I wonder if the turnover of these cells would be fast enough to be useful in T.P.N. studies. Also any shift in the white cell population towards lymphocytosis or leucocytosis could be a complication and would need to be carefully studied.

COYLE (Australia). Yes, at the moment we are merely suggesting lymphocyte zinc as a measure of zinc status. Detailed studies are required to establish whether the turnover time of lymphocytes and the problems of the patients condition will enable it to be a reliable indicator.

HORVATH (USA). Could you tell us how you obtained the reference values?

COYLE (Australia). The reference range was established by analysing serum from a population of laboratory staff.

Christine D. Thomson and Annabelle Duncan

Nutrition Department, University of Otago,
Dunedin, New Zealand.

The enzyme glutathione peroxidase (GSHPx) is the only known function for selenium (Se) and since the discovery of this association, activities in whole blood and erythrocytes have been used as a measure of Se status in man.

Early work in our laboratory suggested that at least for New Zealand residents this method is satisfactory, but with some limitations. A close correlation was found between Se and GSHPx activities for Se concentrations below 140 ng/ml erythrocytes[1] which included most of the values for New Zealand residents. Above these concentrations there was no correlation and GSHPx activities were not greater with the greater Se concentrations of overseas subjects. This lack of an association for our overseas visitors agrees with work from most overseas laboratories.[2,3] These observations suggested that the Se requirement for GSHPx activity had been met and that a dietary intake that maintains this level of activity represents a dietary requirement for Se. Further, this could mean that GSHPx activity may assess suboptimal Se status only, and that no relationship exists between enzyme activity and Se status when the latter is adequate.

Results of subsequent supplementation studies are consistent with these suggestions. In long-term dosing studies with 100 μg Se as selenite or selenomethionine, GSHPx activities increased in all twelve subjects but tended to plateau at erythrocyte Se concentrations of 110 to 140 ng/ml. On the other hand Se concentrations in whole blood, erythrocytes and plasma also tended to plateau with selenite supplementation, but with selenomethionine they continued to rise as long as dosing continued. Such changes are illustrated in Fig. 1 for Subject JR who participated in both studies.

Se concentrations in whole blood erythrocytes and plasma and GSHPx activities in whole blood of subject J R during daily supplementation.

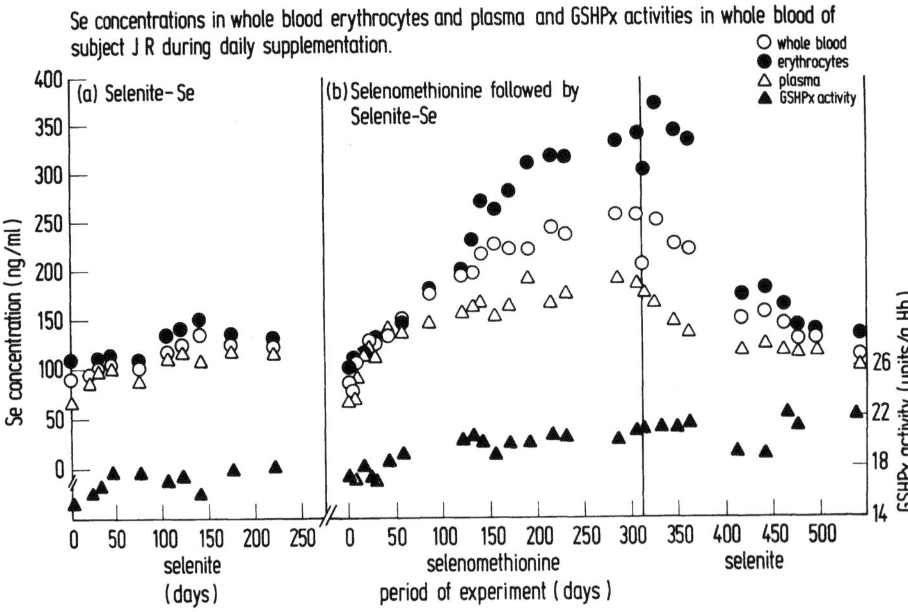

The levels at which GSHPx activities plateaued varied with the individual, suggesting that each person may have an optimal level of enzyme activity which could be determined by need or by a number of other factors. In chicks the form of supplemented Se influences GSHPx activities,[4] but we found no significant difference in the increase in GSHPx activities between selenite and selenomethionine. We have found no effect of age except for reduced GSHPx activities and Se concentrations in infants[5] and in elderly folk over the age of 60 years.[6] Also we have found no difference between males and females. Other factors which may effect the response of GSHPx to Se are oxidant stressors, other toxicants, iron deficiency anemias, heavy metals and vitamins E and A.

Hafeman and coworkers[7] have suggested that an excess of selenite itself may exert an oxidant stressor effect on erythrocyte GSHPx in rats. The small increase in enzyme activities in our subjects indicated that this did not occur in man for physiological amounts of selenite. However we have also observed extremely high GSHPx activities (41, 34 and 43 units/g Hb compared with the normal Dunedin range of 8 to 23 units/g Hb) in three persons in a group of self-dosers, two taking regular but large doses of selenite (0.46 and 0. 37 mg Se daily) and one taking large doses of selenium dioxide (1.4 mg Se) at regular intervals. It is possible that these high enzyme activities result from an oxidant stressor effect of the large doses of selenite-Se.

In another subject GA, who ingested initially 500 µg Se daily as selenite for 4 weeks and then 100 µg daily for a further 28 weeks, GSHPx activities continued to rise to a relatively high level without a corresponding increase in erythrocyte Se concentrations after the first few weeks. This subject was an athlete and training hard. Although exercise itself does not effect GSHPx activities in erythrocytes of horses[8] or rats,[9] a stressor effect of physical activity might possibly increase the requirement for the enzyme. GSHPx activity might then be raised when extra Se is given. Further, exercise has been shown to increase lipid peroxidation in man as measured by pentane production.

For lack of any other known function for Se, GSHPx has been used as a measure of biological availability of Se. However, GSHPx accounts for only 10% of erythrocyte Se in man[2,11] and we need to consider the possibility of other functional Se compounds. The situation is further complicated by the existence of a non Se-dependent GSHPx activity or glutathione transferase,[12] although this does appear to be present in human erythrocytes.

Many laboratories are now introducing measurement of platelet GSHPx activities following the observation that these cells contain relatively high levels of Se[13] and GSHPx.[14] Furthermore Bryant and Levander[15] have shown changes in platelet GSHPx activities during Se depletion/repletion studies in man[16] and these were accompanied by temporary changes in arachidonic acid metabolism.

In preliminary studies in New Zealand we have found higher platelet GSHPx activities in New Zealanders returned from overseas travel than in Dunedin residents (Table 1). A similar effect was shown earlier for erythrocytes.[1]

Table 1. GSHPx Activities in Platelets, Plasma and Whole Blood

	n	Whole blood units/g Hb	Whole blood units/ml	Plasma units/ml	Platelets units/10^{10} cells
Dunedin residents	10	18.3 $\pm$ 4.6	2.7 $\pm$ 0.7	0.15 $\pm$ 0.03	4.2 $\pm$ 0.8
NZ travellers overseas	6	27.0* $\pm$ 4.0	4.1 $\pm$ 0.5	0.18^{+} $\pm$ 0.02	5.8^{+} $\pm$ 1.6

Values are means and standard deviations
*$P<0.005$ $^{+}P<0.05$

Plasma activities, although only 5% of whole blood activities, were also higher in the travellers overseas. GSHPx activities in platelets, plasma and whole blood of two of these subjects declined during 4 months after return from overseas and previous dosing with high Se bread. On the other hand activities remained relatively constant for two subjects who were not exposed to higher Se intakes. It remains to be seen whether the lower platelet activities of New Zealanders result in functional impairment of arachidonic acid metabolism as shown by Levander and Bryant,[15] and what might be the consequences of such impairment.

This work was supported by the Medical Research Council of New Zealand. We wish to thank Professor Marion F. Robinson for her guidance and support, and Ms Gaylene Friend, Ms Dianne Campbell and Mrs Janice Sullivan for their technical assistance.

REFERENCES

1. Rea, H.M., C.D. Thomson, D.R. Campbell, and M.F. Robinson, 1979, Br. J. Nutr. 42: 201-208.

2. Schmidt, K., and W. Heller, 1979, Blut, 3: 247-251.

3. Schrauzer, G.N., and D.A. White, 1978, Bioinorg. Chem. 8: 303-318.

4. Contor, A.H., M.L. Scott, and T. Noguchi, 1975, J. Nutr. 105: 96-105.

5. McKenzie, R.L., H.M. Rea, C.D. Thomson, and M.F. Robinson, 1978, Am. J. Clin. Nutr. 31: 1413-1418.

6. Thomson, C.D., H.M. Rea, O.W. Chapman, and M.F. Robinson, 1977, Proc. Univ. Otago Med. Sch. 55: 18-19.

7. Hafeman, O.G., R.A. Sunde, and W.G. Hoekstra, 1974, J. Nutr. 104: 580-587.

8. Brady, P.S., J.E. Shelle, and D.E. Ullrey, 1977, Am. J. Vet. Res. 38: 1045-1047.

9. Brady, P.S., L.J. Brady, and D.E. Ullrey, 1979, J. Nutr. 109: 1103-1109.

10. Dillard, C.J., R.E. Litov, W.M. Savin, E.E. Dumelin, and A.L. Tappel, 1978, J. Appl. Physiol. 45: 927-932.

11. Behne, D., and W. Wolters, 1979, J. Clin. Chem. Clin. Biochem. 17: 133-135.

12. Lawrence, R.A., and R.F. Burk, 1976, Biochem. Biophys. Res. Commun. 71: 952-958.

13. Kasperek, K., G.V. Iyengar, J. Kiem, H. Borberg, and L.E. Feindegen, 1979, Clin. Chem. 25: 711-715.

14. Karpatkin, S., and Weiss, 1972, New Engl. J. Med. 287: 1062-1066.

15. Bryant, R.W., J.M. Bailey, J.M. King, and O.A. Levander, 1981, In "Proceedings of the Second International Symposium on Selenium in Biology and Medicine" (ed. J.E. Spallholz) (AVI Publishing Co., Westport, Conn.) in press.

16. Levander, O.A., B. Sutherland, V.C. Morris, and J.C. King, 1980, Fed. Proc. 39: 339.

DISCUSSION

PROHASKA (U.S.A.). Two questions, firstly with respect to those people who are self dosing with those selenium compounds where there was very poor agreement between the level of the selenium in the whole blood and enzyme activity. Do you think it is possible that the selenium was being taken up by the liver and was incorporated into serum glutathione peroxidase and not into erythrocytes and that is why the blood analysis did not agree with the enzyme activity. The second question is, if I come to Dunedin will I have to give you any blood?

THOMSON (N.Z.). In answer to the second question, yes, and in answer to the first question, that is possible. We did not look at serum glutathione peroxidase activites in those subjects.

LEVANDER (U.S.A.). In your last slide I think you showed another reason why platelets might be interesting because if I saw your scale right, it looked like the platelet activities decreased to about half the level in the same time period that the red cells decreased by about twenty per cent. That would indicate that they are turning over faster and/or perhaps that selenium turnover is a little bit more rapid in platelets than in the red cells.

THOMSON (N.Z.). Yes they did decrease to about 50% of initial levels.

SELENIUM IN PLATELETS OF CHILDREN WITH LOW SELENIUM STATE

Ingrid Lombeck*, Kasperek,K.+, Iyengar,G.V.+*, Kiem,J.+, Feinendegen,L.E.+, and Bremer,H.J.*

*University Children's Hospital, Moorenstraße 5, 4000 Düsseldorf, F.R.G.

+Institute of Medicine, Nuclear Research Centre, 5170 Jülich, F.R.G.

The selenium content of platelets exceeds that of any hitherto examined body tissue (2).

The purpose of this investigation is to show the correlation of the selenium content of platelets to the selenium content of serum and erythrocytes, furthermore to investigate whether a reduced selenium intake leads also to a decrease of the selenium content of platelets.

METHODS

10 ml of whole blood were taken from 10 healthy children, age 1.5 - 9.0 years and from 8 patients of similar age during routine control of their amino acid metabolism. Because of phenylketonuria (PKU) or maple-syrup-urine disease (MSUD) those patients were treated by approved dietary methods. Their diets are deprived of natural proteins and poor in selenium (4). The dietary treatment in these patients lasted between 2.0 and 8.7 years.

About 10 ml of citrated venous blood served to prepare two samples each of erythrocytes and plasma and to isolate and prepare platelets (3). Platelet pellet contained 9.8 ± 2.6 mg trapped plasma (wet weight) and 8.5 ± 2.4 mg platelets including approximately 0.5 mg of erythrocytes and 0.2 mg of leucocytes. Total platelet number amounted to $1.01 \pm 0.31 \times 10^9$/sample. The erythrocyte samples were not washed and included 23% of plasma as estimated by iron concentration. The plasma contained about 20% of citrate anticoagulant. The selenium content of erythrocytes, plasma and platelets were determined by instrumental neutron activation analysis (1) with the acception that the platelet samples were irradiated for 10 days and washed out from the quartz-tubes using nitric acid.

RESULTS

1. The platelet selenium content in healthy children amounted to $\bar{x} = 555 \pm 105 \times 10^{-9}$g/g, n = 10.

2. There is a significant correlation between the selenium content of plasma and platelets, r = 0.87. A similar positive correlation exists between erythrocytes and platelets, r = 0.86. A child with a high platelet selenium content e.g. 668×10^{-9}g/g platelets exhibit also high selenium values of erythrocytes and plasma (112×10^{-9}g/g erythrocytes, 88×10^{-9}g/g plasma, resp.) (fig. 1).

3. The platelet selenium content is dependent on the selenium intake. Platelets of dietetically treated patients (n = 8) with PKU or MSUD ($\bar{x} = 264 \pm 98 \times 10^{-9}$g/g) show decreased values in comparison to healthy children (n = 10) consuming normal food ($\bar{x} = 555 \pm 105 \times 10^{-9}$g/g) ($p < 0.01$) (fig. 2).

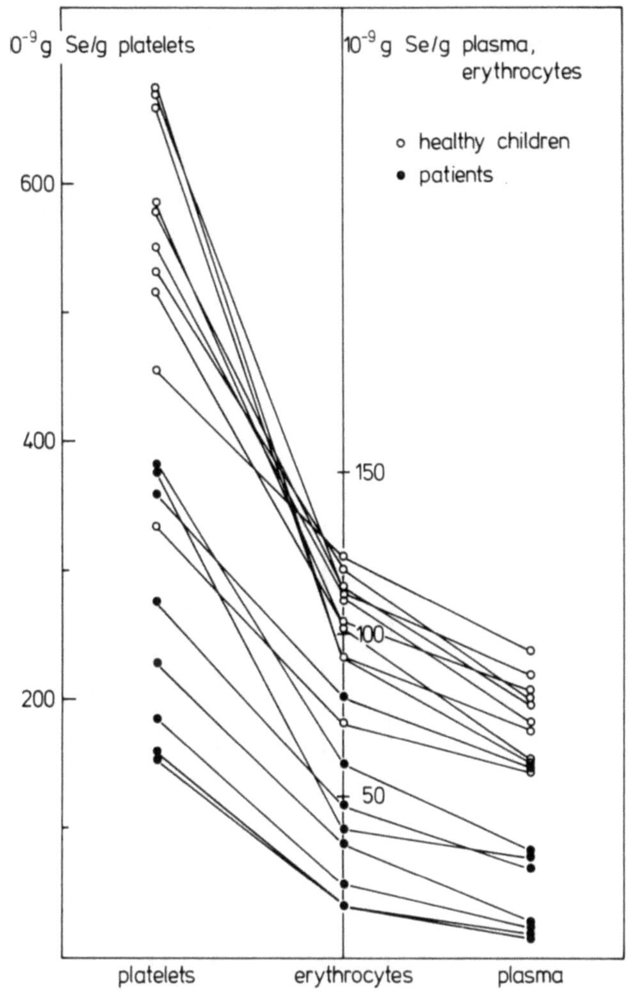

Fig. 1: The corresponding values of the selenium contents of platelets, erythrocytes and plasma in healthy children (o) and dietetically treated patients with PKU and MSUD (•).

DISCUSSION

Because of the low wet weight of the platelet samples of about 8.5 mg the absolute selenium content of the samples amounted only to 1.2 - 5.6 ng. In children the selenium content of platelets is also higher than in other tissues and dependent on the selenium intake. In dietetically treated patients it is reduced to 47% of the values of healthy children. The decrease of the platelet selenium after several years of a low selenium intake is not as pronounced as the decrease of selenium in plasma or erythrocytes. At the moment one can only speculate if platelets have a higher affinity for selenium than other tissues.

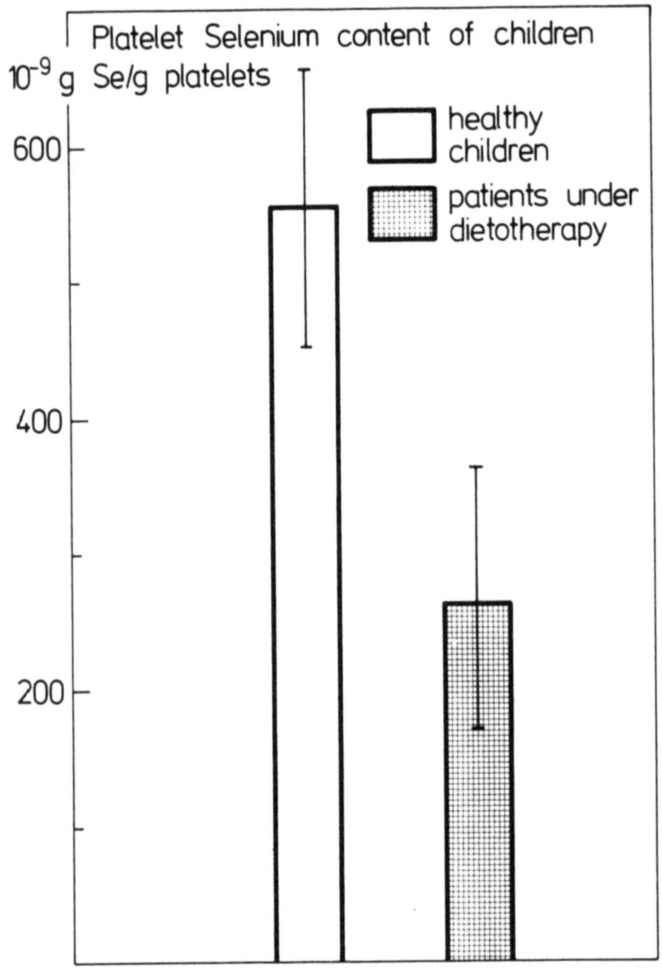

__Fig. 2:__ Platelet selenium content of healthy children and dietetically treated patients with PKU and MSUD.

REFERENCES

1. Kasperek,K., 1974. In "Proc. of the 2nd internat. conf. on nuclear methods in environmental research, U.S. Energy research and development administration", Conf. 740701.

2. Kiem,J., Borberg,H., Iyengar,G.V., Kasperek,K., Siegers,M., Feinendegen, L.E., and Gross,R., 1979, Clin. Chem. 25: 705 - 710.

3. Kiem,J., Wang,Y.X., Iyengar,G.V., Kasperek,K., and Feinendegen,L.E., 1981, Blood submitted.

4. Lombeck,I., Kasperek,K., Harbisch,H.D., Becker,K., Schumann,E., Schröter, W., Feinendegen,L.E., and Bremer,H.J., 1978, Eur. J. Pediatr. 128: 213 - 223.

*ACKNOWLEDGEMENT

This study was supported by the "Deutsche Forschungsgemeinschaft"

DISCUSSION

LEVANDER (U.S.A.). Did you say that the platelet selenium was proportional to plasma selenium over a range of approximately 6 to 95 micrograms per litre?

LOMBECK (Germany). Yes, that is correct.

LEVANDER (U.S.A.). Then is the platelet glutathione peroxidase also related to platelet selenium over that same range?

LOMBECK (Germany). I cannot answer the question at the moment because we did not analyse for glutathione peroxidase due to that very small volume of blood we collected.

PETER (Australia). What is the turnover rate of platelets as opposed to erythrocytes?

LOMBECK (Germany). It is much quicker, the erythrocyte turnover has a half lifetime of about 100 days and in platelets I think it is about 11 days. So we have an intermediate between plasma selenium turnover and red cell turnover.

PETER (Australia). If I saw your last slide correctly though, you are saying that erythrocytes and plasma are both lower than platelets. Is that correct?

LOMBECK (Germany). Yes they are lower, that is right. I cannot say why at the moment.

SUNDE (Scotland). Can you incubate platelets in vitro with selenium and amino acids and so on and increase glutathione peroxidase activity or are platelets a pre-formed package that cannot synthesize any more enzymes once they are in circulation.

LOMBECK (Germany). I think platelets synthesize the glutathione peroxidase during formation in the marrow and no further synthesis occurs in the blood. However, I do not know if anyone has done the type of incubation studies you mentioned using platelets.

SELENIUM STATUS OF SEVEN IVH PATIENTS

H.W. Lane, A.O. Barroso, S.J. Dudrick, D.A. Englert, and B.V. MacFadyen

Program in Nutrition and Dietetics and the Department of Surgery
The University of Texas Health Science Center Houston, Texas, 77025, U.S.A.

Trace element deficiences have been observed in patients receiving total nutritional needs from parenteral solutions(1). One such trace element is selenium, an essential component of glutathione-peroxidase (GSH-Px) whose level of activity is an indicator of selenium status in animals(2). New Zealand patients receiving intravenous hyperalimentation (IVH) for 10-20 days had decreases in their blood selenium levels and GSH-Px activity as well as negative selenium balances(3). One of these patients had remission of muscle pain after selenium was added to the IVH solution(3).

The purpose of this study was to determine if patients receiving long term IVH therapy developed deficiency of selenium as measured by plasma and erythrocyte selenium level and erythrocyte GSH-Px activity. The correlation between blood GSH-Px activity and blood selenium level was determined. The IVH patients in this study were followed clinically for metabolic and nutritional changes throughout the study.

METHODS

All patients in this study were ambulatory and all of their nutritional support was given parenterally. These patients have inflammatory bowel disease or short gut syndrome. Weights were kept constant by varing the calories in the IVH solution which contained protein, dextrose, electrolytes, calcium, phosphate, magnesium, zinc, copper, multi-vitamin infusate, folacin, and vitamin B_{12}. Essential fatty acids were given two to three times per week as soybean oil emulsion.

The IVH patients were compared to 275 healthy volunteers, 18-80 years old, with a mean age of 39 years. The control population had one blood analysis for selenium level and GSH-Px activity but blood samples were collected from the seven IVH patients monthly for 4 to 14 months. GSH-Px activity was determined by the coupled assay using hydrogen peroxide(4) and the selenium levels were measured fluorometerically using 2,3, diaminonaphthalene(5).

RESULTS

The overall means for plasma and erythrocyte selenium concentration and erythrocyte GSH-Px activity were significantly lower in the IVH patients than in the healthy controls (Table 1).

TABLE 1

COMPARISON OF A HEALTHY POPULATION TO IVH PATIENTS FOR SELENIUM LEVELS IN PLASMA AND ERYTHROCYTES AND GSH-PX ACTIVITY IN ERYTHROCYTES

	Healthy Population	IVH Patients
	$\bar{x} \pm$ S.D.	$\bar{x} \pm$ S.D.*
Plasma Se ug Se/g	0.097 ± 0.030	0.041 ± 0.025
RBC Se ug Se/g of Hgb	0.807 ± 0.421	0.300 ± 0.260
RBC GSH-Px activity units/g of Hgb	34.95 ± 15.32	21.30 ± 6.60

*P 0.005.

Within one month after initiation of this study, four of the IVH patients had plasma Se levels below levels found in the control healthy population and eventually, all seven patients had depressed plasma Se levels. The erythrocyte GSH-Px activity also decreased below the normal levels in five patients within one month after initiation of the study and all patients have below normal GSH-Px activity sometime during the study. Two patients and GSH-Px activity that returned to the low normal range during the study. Erythrocyte Selenium concentration varied with two patients having below control levels throughout the study and five patients had levels that were in the low normal range.

There was a negative correlation between the individual erythrocyte selenium level and erythrocyte GSH-Px activity whereas there was a positive correlation ($r=0.51$, $p<0.01$) between the plasma selenium level and erythrocyte GSH-Px activity (Fig. 1).

One of the IVH patients plasma selenium concentration was below the control levels but the erythrocyte selenium levels and GSH-Px activity remained at the low control level of just below the control level.

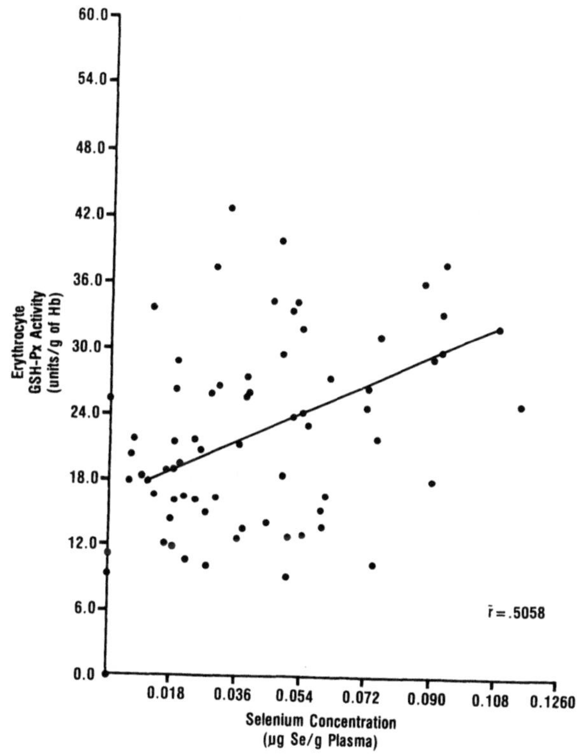

FIG. 1 Correlation of plasma Se levels to erythrocyte GSH-Px activity for all the measurements from the seven IVH patients.

CONCLUSIONS

No correlation was found between erythrocyte selenium levels and erythrocyte GSH-Px activity but a positive correlation existed between plasma selenium levels and erythrocyte GSH-Px activity. Plasma Se levels and erythrocyte GSH-Px activity may be more sensitive indicators of exogenous Se levels than erythrocyte Se levels. The New Zealand IVH patients had similar response and it was suggested that plasma Se levels may be the best indicator of Se nutritional status (3).

In general the IVH patients had lower selenium indicies than healthy controls. But, no well defined pathological condition was attributed to these low selenium indicies. Further research is needed to evaluate clinical symptoms and selenium levels in long term IVH patients.

Since selenium is toxic to animals (6), the NRC (7) has recommended a maximum of 200 ug of oral Se per day. In our study one patient had relatively normal blood selenium levels and GSH-Px activity indicating that she had adequate selenium stores and/or absorbed selenium from her gastrointestinal tract from consumption of foods. It has been assumed that dietary selenium is easily absorbed and thus this patient must have enough intestine to absorb adequate selenium from oral food stuffs. Because of patient variability and potential selenium toxicity, it is recommended that the initial assessment of selenium status should be completed prior to the addition of selenium to the IVH solution and the selenium status should be routinely evaluated while the patient receives supplemental selenium.

REFERENCES CITED

1. Kay, R.G., C.T. Jones, Pybus, J. et al, 1976, Ann. of Surg. 183:331-340.

2. Rotruck, J.T., A.L. Pope, H.E. Ganther, et al., 1973, Science 179:588-590.

3. van Rij, A.M., O.D. Thomson, J.M. McKenzie, and M.F. Robinson. 1979, Am. J. Clin. Nutr. 32:2076-2085.

4. Paglia, D.E. and Valentine, W.N. 1967, J. Lab. Clin. Med. 70:158-169.

5. Olson, O.E., 1969, J. Assoc. Off Anal Chem, 52:621-634.

6. Rosenfeld, I., O.A. Beath, 1964, In "Selenium:Geotatany, Biochemistry, Toxicity and Nutrition." (Adeamic Press, New York).

7. Food and Nutrition Board, Recommended Dietary Allowance, 7th edition, National Academy of Science, Washington, D.C.

This research was supported by #5901-0401-8-0086-0 of the United States Department of Agriculture Competitive Research Office.

DISCUSSION

BEHNE (Germany). You calculated your selenium content in plasma per gram of wet weight. Is there a difference in the ratio of water to protein in the plasma between the patients and the healthy control?

LANE (U.S.A.). No there was not.

PETER (Australia). How are you determining your erythrocyte selenium concentrations?

LANE (U.S.A.). The erythrocyte selenium concentration was determined directly by fluorometric assay using diaminonaphthalene.

PETER (Australia). This is on pure erythrocytes with 23% plasma?

LANE (U.S.A.). We try to wash out the plasma by extensive washing of erythrocytes, thus we feel that the Se was determined on pure erythrocytes.

VERSIECK (Belgium). You expressed your selenium values in red blood cells per gram of haemoglobin. Would the conclusions have been the same if you had expressed the selenium concentrations per gram red blood cells and not per gram haemoglobin?

LANE (U.S.A.). We looked at it three different ways actually, per gram haemoglobin, per gram protein and per ml of blood. The results are the same. Obviously, if the concentration of haemoglobin or protein changes, then the method of expression of selenium concentration may affect the results.

GARNYS (Australia). As a comment, we also found negative selenium balances in our parenteral nutrition patients. Did you freeze your blood samples before doing glutathione peroxidase assays?

LANE (U.S.A.). Glutathione peroxidase was determined on fresh erythrocytes. Selenium assays were done later on stored samples.

PETER (Australia). Would you care to speculate on the possible reasons behind the poor correlation between selenium and glutathione peroxidase in human erythrocytes?

LANE (U.S.A.). In healthy subjects the correlation between erythrocyte Se and GSHpx was about 0.51. In IVH and cancer patients, the correlation was virtually non existent. The patients appeared to be outwardly healthy, and so I presume the lack of correlation was due to a specific disease effect on trace element metabolism.

PETER (Australia). Do you think it could be due to a membrane effect on selenium?

LANE (U.S.A.). Possibly yes.

SOIL TYPE AND CATTLE BLOOD SELENIUM LEVELS: N.I. N.Z.

A.J. Fraser & J.A. Kirk
Ruakura Animal Health Laboratory
PRIVATE BAG, HAMILTON
NEW ZEALAND

INTRODUCTION

New Zealand has a low selenium status compared with other countries (Robinson 1975), and selenium responsive conditions in livestock are widespread (Audrews *et al.* 1968). Comprehensive trials (Robertson and During 1961) showed that liveweight gain responses in lambs to selenium supplementation were fairly well related to soil type and soil selenium level (Watkinson 1962). New Zealand lies on the margin of two active crustal plates and thus the soil-scape has been mainly determined by mountain building, volcanism and erosion processes. Selenium levels in the blood of grazing animals are dependent on the selenium levels of their diet (pasture), which in turn would be expected to depend on the nature and selenium content of the soil, and the parent rock from which the soil was formed. The degree of selenium responsiveness in young sheep and probably also cattle is related to their blood selenium concentrations, although critical levels are not yet well defined (Andrews *et al.* 1976). The development of a semi-automated method of selenium analysis (Watkinson 1980) has made it practical to carry out large numbers of blood selenium assays. The Ruakura Animal Health Laboratory has used this to provide a regular service to aid practising veterinarians in their diagnosis of problems such as poor production, ill thrift and infertility in grazing livestock, or as a "check on status" of clinically normal animals. The opportunity was thus available to use this data to further define the relationships between soil type and the blood selenium concentrations of grazing animals and hence the areas on which selenium responsiveness may be expected.

MATERIALS AND METHODS

Samples of cattle blood were collected into evacuated tubes containing EDTA, heparin or oxalate as an anti-coagulant. Selenium concentrations were determined in a minimum of three and usually five samples from each farm. A farm mean blood selenium concentration was calculated from log transformed values. Coloured pins were used to plot the farm location and mean selenium level on soil maps according to Table 1. Results from a farm were excluded if the cattle had been supplemented with selenium or shifted from another area within the preceding 6 months, or if a large within herd variation was found.

Table 1: Classification of farm mean cattle blood selenium concentrations.

Pin colour	Blood selenium conc. in the range (ug/ℓ)	Classification
red	<11	low
pink	11-19	marginal
blue	19-35	low normal
dark blue	35-65	normal
black	>65	high normal

Cattle blood selenium levels of over 1000 farms in the North Island were mapped during the period 1973-1980.

RESULTS

The map data was analysed by classifying it into soils grouped according to their parent rocks. For each soil group the median blood selenium level and the range in which 95% of the farms lay was calculated (Table 2).

Marked difference between the different soil groups were apparent. Low blood selenium levels were found in cattle grazing on the Taupo and Kaharoa pumice soils, peat soils, some coastal sand soils, and the podzols of Northland. Marginal levels were often found on soils from the alluvium of the Waikato and Hauraki basins, and on gley soils. Most other soil groups had normal selenium concentrations with the highest level occuring on basaltic ash from Rangitoto.

Table 2: Soil group and cattle blood selenium levels.

Soil Group			Cattle Blood Selenium Level (ug/ℓ) No. of farms	median	(95% range)
From volcanic ash showers - age (years) - nature					
Kaharoa (Rangitaiki)	930	rhyolitic	33	9	(5- 16)
Taupo	1900	rhyolitic	31	11	(5- 18)
Mairoa, Tirau, Waihi	5000-50 000	composite	127	45	(22- 82)
Egmont, Stratford	500-50 000	andesitic	14	35	(19- 65)
Rangitoto	600	basaltic	1	170	-
- complexed with sedimentary rock hills					
Taupo	as above		31	58	(33-120)
Tirau, Waihi	as above		15	86	(50-150)
Mairoa	as above		29	48	(24- 95)
Egmont	as above		15	44	(24- 60)
- complexed with ignimbrite					
Kaharoa	as above		87	24	(9- 70)
From sedimentary rocks - Northland					
Yellow brown earths			232	29	(13- 85)
Podzols			98	14	(7- 30)
From andesites and basaltic flows - Northland					
Brown loams			230	42	(22-110)
From alluvium, gleyed, peat					
Waikato and Hauraki basins					
rhyolitic alluvium			91	19	(10- 38)
gley			35	16	(8- 36)
peat			106	12	(5- 27)
Northland region					
gley			78	21	(9- 65)
peat			37	18	(7- 45)
Taranaki (Egmont)					
gley and peat			7	24	(13- 43)
From coastal sand					
Yellow brown sand			26	29	(8- 90)

DISCUSSION

Many of the soils of the North Island are relatively young. In many areas the flooding of rivers over alluvial plains, the drifting of sand and dust, and the fall of ash from erupting volcanoes have interrupted soil development. On these soils it may be expected that the nature and mineral content of the parent rock from which the soil was formed could have a major influence on plant and animal selenium levels (Wells 1967). In particular low blood selenium levels may be expected on soils from rocks composed mainly of silica (acidic rocks) which have a low content of iron, magnesium and associated trace elements (ferromagnesians) and higher levels on soils from rocks which have a higher ferromagnesian content (more basic).

This is illustrated in the results for soils from volcanic rocks. Recent volcanic activity has been mainly confined to the central region of the North Island where paroxymal eruptions over the last 50 000 years have deposited ash over wide areas. Low cattle blood selenium levels are found on the young rhyolitic (acidic) showers of Taupo and Kaharoa pumice, and normal levels on the older andesitic (more basic) showers from Egmont and high levels on the young basaltic (basic) ash from Rangitoto. Normal blood selenium levels are also found on the Mairoa, Tirau and Waihi ashes which are formed from several showers believed to be mainly rhyolitic. The higher blood selenium levels found on these soils may be partly because some of their constituent showers are andesitic in nature, or because they are older and more weathered than the pumice soils (Taupo and Kaharoa); Hodder and Watkinson (1976) have shown that the weathering of rhyolitic glass to allophane results in an increase in soil selenium content.

In many areas the ash showers (Taupo, Tirau, Waihi, Mairoa, Egmont) have fallen on hilly country composed of sedimentary sandstones and mudstones of mainly tertiary ages. The ash has eroded from the steeper slopes, but still covers the dissected plateaus and terrace lands. Most farms in these areas have both groups of soil. The relatively higher cattle blood selenium levels found is probably related to higher selenium levels in pastures from sedimentary rocks, but more detailed study is needed to verify this. Similarly, higher and more variable cattle blood selenium levels are found on soils from the Kaharoa shower, where it has fallen on a dissected ignimbrite plateau sloping down to the Bay of Plenty, than on the flat country of the Rangitaiki valley.

The influence of soil parent material is also seen in the Northland region, where cattle grazing soils from basic volcanic rocks (andesites and basalt flows) have on the average higher selenium levels than those on the yellow brown earth soils from sedimentary rocks. However in the warm moist climate of the region soil forming processes are more rapid and other factors such as the former vegetation become important. Under mor forming trees, particularly kauri, iron and associated minerals tend to be leached to lower soil horizons forming podzols. Lower cattle blood selenium levels are found on these soils.

Alluvial sands, silts, clays and gravels mainly of recent and late Pleistocene age are the parent rocks of the soils of flood plains and terraces in many parts of the North Island. The most extensive deposits occur in the Hamilton basin and Hauraki lowlands, where they consist mainly of rhyolitic sediments believed to be derived from the central volcanic plateau during the last ice age. At the northern end of the Hauraki lowland the sediments are partly estuarine. Gley soils occur in areas of poor drainage, and where the water table is permanently high peats have formed. Cattle grazing on the better drained soils have predominantly marginal or low normal blood selenium levels, with lower levels being found with increasing poor drainage and peatiness. This is probably due to reducing conditions in soils of high water table decreasing the availability of selenium to plants (Watkinson 1962), and also because peats have lower selenium levels than the sediments on which they formed. The same trend is apparent in the Northland region and on the Egmont ashes. However, blood selenium levels are higher on the gley and peat soils from the Egmont ashes, and in Northland, than the corresponding soils of the Waikato and Hauraki basins. The higher levels found on the Egmont soils is probably due to the andesitic nature of the parent ash. Similarly the higher (and more variable) levels on the Northland soils is probably related to their mixed sedimentary and basic volcanic origin. Variable selenium levels are also found on the soils from coastal sands. From the limited data available it seems that much of this variation can be attributed to the variable ferromagnesian content of the sands. Low selenium levels are found in cattle grazing the acidic "pumice" sands of the Bay of Plenty coast and mainly normal levels on the more basic "iron sands" of the western coast.

The findings reported here are in general agreement with those of Robertson and During. Low blood selenium levels in this study were generally in areas found to be selenium responsive by Robertson and During, and usually marginal and normal blood selenium levels were found in non-responsive areas.

ACKNOWLEDGEMENTS

The authors are indebted to J. Watkinson for the use of his automated facilities for selenium determination, and for helpful advice and criticism, and to the many veterinarians who assisted with the locations of their clients' farms.

REFERENCES

1. Andrews, E.D., W.J. Hartley, and A.B. Grant, 1968, N.Z. Vet. J. 16: 3-17.

2. Andrews, E.D., K.G. Hogan, and Sheppard, A.D., 1976, N.Z. Vet. J. 24: 111-116.

3. Hodder, A.P.W., and J.H. Watkinson, 1976, N.Z. J. Sci. 19: 397-400.

4. Gibbs, H.S., 1968. In "New Zealand Soil Bureau Bulletin 26(1)" (Govt Printer, Wellington, N.Z.) pp 47-67.

5. Robertson, T.G., and C. During, 1961, N.Z. J. Agric. 103: 306-310.

6. Robinson, M.F., 1975. In "The moonstone : more about selenium" (ed. E. Hove) (Keeling & Mundy, Palmerston North, N.Z.) pp 13-30.

7. Watkinson, J.H., 1962, Trans. Int. Soc. Soil Sci., Comm. IV and V: 149-154.

8. Watkinson, J.H., 1979, Anal. Chim Acta. 105: 319-325.

9. Wells, N., 1967, N.Z. J. Sci. 10: 142-179.

DISCUSSION

JUDSON (Australia). Do you find much seasonal variation, particularly with the marginal areas, in cattle blood selenium levels in the light of some of your later comments.

FRASER (N.Z.). Yes, in all areas there is a seasonal variation in blood selenium levels but this is a relatively minor component compared with the major factor of the soil type, but it is a significant variation none-the-less.

KUBOTA (U.S.A.). In the literature one finds that the selenium concentrations in soils vary with the kind of material that you describe, but one also gets the impression that the selenium in the soil also increases as soil organic matter increases. Now I was just wondering, as you go from well drained to poorly drained to peat, do you get an increase in soil selenium concentration with increasing organic matter? The blood selenium levels you observe reflect low availability of soil selenium to plants in the presence of soil selenium levels that increase with poor soil drainage. Your observations also show that the soil selenium that is tied up in organic matter is not readily available to plants.

FRASER (N.Z.). I am not too sure whether there is an increase in soil selenium levels with organic matter. I think the peat soils do have lower selenium levels than the more well drained soils in our country, although I know it seems different in other countries. On the last point you made, the reason for this decrease in animal blood Se levels is, I believe, due to the fact that in soils of high water table you get reducing conditions so that the selenium is present in a relatively unavailable form compared with the more well drained soils.

WEINER (U.K.). I was wondering whether the relationship between soil type and cattle blood selenium level was mediated through the animals grazing different species of grasses and legumes on these different soil types or whether through differing selenium levels in more or less the same species.

FRASER (N.Z.). In New Zealand most pastures are ryegrass and clover on the more developed areas anyway, which is most of the area sampled, so it is more due to differences in plant selenium levels within the one species than differences between species; which do also occur.

KUBOTA (U.S.A.). I'd like to comment on that point about species effect on the selenium concentrations in plants. It seems that all of the data that we show that when selenium concentration in the plant is low, species effects become almost negligible. Would you want to comment on that?

FRASER (N.Z.). Yes, I think that is the same experience in New Zealand, there is less species effect at the low levels.

JANGHORBANI (U.S.A.). In well drained areas you might find that your plants have a higher concentration of selenium simply because the selenium in the soil is selenate which is more available for uptake by the plant, and that this is really the primary effect.

FRASER (N.Z.). Yes I think that you are right, it is primarily this availability effect. Another reason may be that in peat soils a high amount of organic matter is present, and in New Zealand anyway, this organic matter has a lower content of selenium than the sediments on which they were formed so that plants growing on this would have even lower levels again.

BINNERTS (The Netherlands). Just in order to define the conditions: could you explain your sampling methods? Did you sample whole blood? Is there, apart from the seasonal variation, also a diurnal variation? Do you have to sample at a certain time of the day, and is there an effect of stress?

FRASER (N.Z.). The answer to the first question is that we sampled whole blood, we have standardised on that. As far as the other two questions go, I don't really know but I would expect they would be comparatively minor factors compared with the soil types and seasonal variations.

SOME ASPECTS OF A SURVEY INTO SELENIUM DEFICIENCY IN CATTLE IN NORTHERN CALIFORNIA

J.D. Williams*, W.H. Johnson†, B.B. Norman†, H.F. Mayland§ and J.R. Dunbar†

* NSW Department of Agriculture, PO BOX 488, LISMORE NSW 2480, AUSTRALIA
† University of California, DAVIS CALIFORNIA 95616, USA
§ United States Department of Agriculture, PO BOX 1109, SIDNEY MONTANA 59270, USA

Selenium deficiency is known to exist in Northern California (1,2) and so a survey was undertaken to provide some knowledge of the epidemiological features of this deficiency in cattle in that region. While this survey covered many aspects of selenium deficiency in cattle, this paper reports only those findings which relate clinical expression of the deficiency as observed by stock owners.

MATERIALS AND METHODS

Blood samples were collected in a convenience sample of 1465 cows from 150 ranches in 22 counties through Northern California over a five-month period from January to May 1980. Ten co-operative ranches in each county were selected and ten adult cows on each ranch were to be bled for the survey. Adult cows only were sampled as the survey formed part of a larger, nationwide survey of selenium deficiency in cows conducted by the United States Department of Agriculture.

An adult cow was defined as having produced at least one calf. All sampled cows had been on the ranch for at least 60 days and grazing the same pasture for at least 30 days prior to sampling. Nearly all cows sampled were beef cattle.

Blood samples were assayed for whole blood selenium concentrations using the method of Olson *et al.* (3), modified as follows: decahydronaphthalene was used as the extractant in lieu of cyclohexane and 1 mℓ whole blood digested in HNO_3-$HClO_4$ mixture was used in lieu of digested plant samples. Individual whole blood selenium assays were used to produce a mean blood selenium concentration for each herd. These herd values were used in subsequent analysis of results.

At the time of blood sampling, a questionnaire on various ranch and animal details was completed. Numerous variables were examined by the questionnaire, including such ranch variables as topography, altitude, rainfall, fertilizer history, feed type, soil type, disease history and selenium supplementation history, as well as individual animal variables such as production and breed data for the sample cows.

BMDP Biomedical Computer Programmes were used to perform the statistical analysis of this data (4). One way analysis of variance was made using logarithmically (base 10) transformed blood selenium data as the response variable and ranch disease history as the different grouping variables (Table I). In the presence of statistically significant results, the least significant difference multiple pairwise comparison procedure was used to compare means with an overall significance level of 0.05. This data was also analysed by the χ^2 frequency distribution test to confirm the results of the one way analysis of variance. Observing a significant difference in blood selenium levels between herds reporting white muscle disease and those not reporting the disorder (0.0308 μg Se/mℓ *vs* 0.0527 μg Se/mℓ, Table I), an arbitrary level of 0.04 μg Se/mℓ was selected to categorise herds into two groups.

Herds with mean blood selenium values below 0.04 μg Se/mℓ were classed as low selenium herds, while herds above this level were classed as high selenium herds. Contingency tables were constructed using herd selenium classification as the column variable and ranch disease history as the different row classification variable. χ^2 analysis determined significant departures between the number of herds observed in a cell and the number of herds expected in a cell (Table II).

It should be noted that responses to all ranch disease history variables refer to the past disease history of the ranch and these disorders were not necessarily apparent at the time of sampling.

RESULTS

Results of the statistical analyses are summarised in Tables I and II.

TABLE I. Results of one way analysis of variance comparing mean herd blood selenium concentrations between different groups of the ranch disease history variables

Variable	Group	No. of Herds in Group	Group Mean Herd Blood Se Conc. (μg Se/mℓ)	Analysis of Variance Level of Significance (p)
White Muscle Disease (WMD)	No history WMD	110	0.0527[a]	0.0000
	Positive history WMD	27	0.0308[b]	
	WMD history unknown	13	0.0380[a,b]	
Illthrift	No history illthrift	107	0.0518[a]	0.0055
	Positive history illthrift	26	0.0331[b]	
	Unknown illthrift history	17	0.0401[a,b]	
Weak Calf Syndrome (WCS)	No history WCS	102	0.0501[a]	0.0826
	Positive history WCS	39	0.0379[a]	
	WCS history unknown	9	0.0479[a]	
Retained Placentae (RP)	No history RP	71	0.0448[a]	0.7442
	Positive history RP	71	0.0476[a]	
	Unknown history RP	8	0.0524[a]	
Undiagnosed Abortions (UA)	No history UA	74	0.0478[a]	0.8486
	Positive history UA	61	0.0457[a]	
	Unknown history UA	15	0.0433[a]	

Variable group means with superscripts in common are not statistically different from each other at the $p = 0.05$ level.

TABLE II. Results of Chi square frequency distribution analysis of disease history variable groups versus low or high selenium herds

Variable	Degrees of Freedom	χ^2 Value	Level of Significance
White Muscle Disease	2	7.55	0.0215
Illthrift	1	4.95	0.0266
Weak Calf Syndrome	1	2.84	0.1139
Retained Placentae	2	2.613	0.2708
Undiagnosed Abortions	2	0.283	0.8683

There is good agreement of results obtained under the two different statistical analyses. From these analyses the following significant disease history variables arise:

1. White Muscle Disease (WMD): Significantly lower selenium levels were recorded in herds with past histories of WMD. This group also contained significantly more low selenium herds than statistically expected under the χ^2 frequency distribution analysis.

2. Illthrift: Significantly lower selenium levels were recorded in herds with past histories of illthrift. Similarly, this group also contained significantly more low selenium herds than expected.

DISCUSSION

Although mature cows only were bled in this survey, and clinical WMD was not necessarily present on the ranch at the time of sampling, the observed significant difference in blood selenium concentration between herds with past histories of WMD and those with no WMD history is not unexpected. Assuming accurate diagnoses of WMD by veterinarians and owners in the past, herds with past histories of this disease would be expected to have lower selenium levels because of lower selenium levels in soil and forage plants on these ranches.

Selenium responsive illthrift has been reported in both young and adult cattle (5). In this survey, herds reporting illthrift problems had significantly lower selenium values than herds with no previous history. This suggests that the illthrift problems reported on these ranches may be related to selenium deficiency and the problem may respond to effective selenium supplementation.

There was no statistically significant association between selenium levels and the problem of weak calves born on the ranch (weak calf syndrome). However, this association was close to statistical significance and may indicate that on certain ranches the problem of weak calves being produced was related to selenium deficiency. It is not possible to totally discount the role of selenium deficiency in the weak calf syndrome observed on many ranches in this survey.

The survey results do, however, indicate very little association between selenium status and the problems of retained placentae or undiagnosed abortions.

Workers in Britain and the USA have reported dramatic reductions in the incidence of retained placentae in dairy cows following selenium supplementation (6,7). In a trial in commercial dairy herds, the incidence of retained placentae was reduced from a mean of 51.2% in the control cows to 8.8% in cows injected with selenium and vitamin E (8). Such high incidence rates of retained placentae were not reported on ranches in this survey. Most ranches reporting this disorder estimated the incidence to be less than 10%. In view of these results and the reported experimental findings, it therefore appears that, when present at low incidence rates, the problem of retained placentae in cattle is not selenium dependent.

The problem of undiagnosed abortions in cattle in this region also does not appear to be selenium dependent and selenium-responsive abortions have not been reported previously in the literature.

There is little agreement in the literature concerning the blood selenium concentration considered adequate to protect sheep and cattle against selenium-responsive diseases. Suggested values include 0.01 μg Se/mℓ (5), 0.03 μg Se/mℓ (9), 0.05 μg Se/mℓ (10), 0.10 μg Se/mℓ (11) and 0.20 μg Se/mℓ (12). These are absolute values above which stock should be considered selenium adequate.

In this survey, herds were arbitrarily classified as low or high selenium herds if their mean blood selenium concentration was below or above 0.04 μg Se/mℓ. Agreement of results obtained from the analysis of variance and χ^2 distribution using this arbitrary classification indicate a mean herd blood selenium concentration of 0.04 μg Se/mℓ to be a useful dividing point in determination of herd selenium status. It would also suggest that the larger estimates of absolute blood selenium concentration are more representative of adequate selenium levels in cattle.

BIBLIOGRAPHY

1. Allaway, W.H. 1964. J. Anim. Sci. 23:271-277.
2. Johnson, W.H., Norman, B.B., and J.R. Dunbar. 1979. Calif. Agric. 33:14-16.
3. Olson, O.E., Palmer, I.S., and E.E. Cary. 1975. J. Assoc. Official Anal. Chemists 58:117-121.
4. Dixon, W.J., and M.B. Brown (eds). 1979. "BMDP Biomedical Computer Programmes - P. series" (Berkley, C.A.) 880 pp.
5. Andrews, E.D., Hartley, W.J., and A.B. Grant. 1968. N.Z. Vet J. 16:3-17.
6. Julien, W.E., Edwards, S.J.A., Forsyth, W.M. *et al.* 1978. J. Dairy Sci. 59:1954-1959.
7. Trinder, N., Hall, R.J., and C.P. Renton. 1973. Vet. Rec. 93:641-643.
8. Julien, W.E., Conrad, A.R., and A.L. Moxon. 1976. J. Dairy Sci. 59:1960-1962.
9. Jenkins, K.J., Hidiroglou, M., Wauthy, J.M., and J.E. Proulx. 1974. Canad. J. Anim. Sci. 54:49-60.
10. Hartley, W.J. 1967. In "Symposium: Selenium in Biomedicine" (ed. O.H. Muth)(Westport, Conn.) pp 76-96.
11. Anderson, P.H., Berrett, S., and D.S.P. Patterson. 1979. Vet. Rec. 104:235-238.
12. Andrews, E.D., Hogan, K.G., and A.D. Sheppard. 1976. N.Z. Vet. J. 24:111-116.

DISCUSSION

LANGLANDS (Australia). One of the problems with this sort of survey is the seasonal variability in selenium concentration. Were these samples taken throughout the year or at specific times of the year and to what extent would you expect seasonal variability to be a problem?

WILLIAMS (Australia). In an attempt to reduce the problem of seasonal variability, samples were collected from January to May 1980, with the majority of samples collected during the months of February, March and April. As a consequence of this winter sampling time, we found ranchers to be using one of three basic feeding programmes: One, access to pasture with no supplementation, Two, pasture access with hay supplementation, and Three, hay as the predominant feed source. We found that ranchers who were using hay as the dominant source of feed for their cattle through the winter period had cattle with significantly lower selenium levels than those who allowed their cattle access to pasture through the same period. Now one of the possible reasons for this was that a lot of the hay was produced on the ranch for winter supplementation. It was produced during the summer period, on irrigated and with heavy fertilized pasture where there may have been a selenium dilution effect resulting in hay of a fairly low selenium content.

KUBOTA (U.S.A.). With regard to your last comment on hay and low selenium, I just wondered if you had considered the possible role of losses of Vitamin E with storage in hay. This seems to be a fairly common observation - the longer you keep hay stored the lower the Vitamin E content. This suggests that Vitamin E may have a role; Would you comment on that?

WILLIAMS (Australia). No Vitamin E estimations were made on hay, pasture, blood or animal organs in this survey. White muscle disease in cattle has been reported to be associated with deficiencies of both selenium and Vitamin E although I doubt whether low Vitamin E levels in hay would be related to the observed low blood selenium levels in cattle predominantly fed hay.

ULLREY (U.S.A.). Relative to the relationship between the yield and selenium concentration, I cannot comment on forage but in some of our research with corn there is a very highly significant inverse correlation between yield and selenium and so your thesis may indeed have some relevance to your study.

WEINER (Scotland). In view of the fact that Langlands and his colleagues have been able to show breed variation in selenium levels, as we have done for copper, I really would like to ask two questions. Was there uniformity of breed across these different herds, particularly, of course, those with a history versus those with no history of white muscle disease? The second question concerns possible genetic variation within breed. Was there any indication of a difference in the variability within the herds with no history of white muscle disease as compared with that within herds having such a history?

WILLIAMS (Australia). I am not very confident in answering the second question. I do not think I can answer that one satisfactorily. In relation to your first question, there was a very large number of breeds and cross breed combinations observed in this survey. However, 76% of all animals in the sample were either pure bred Herefords, pure bred Angus or Hereford/Angus cross. Significantly higher blood selenium levels were observed in cross bred animals compared with pure breds. Significant differences were observed between the Hereford/Angus cross and the Angus but not between the Hereford Angus cross and the Hereford, although the mean Hereford and Angus levels were very close to each other. This breed difference perplexes me a little but I can't offer a physiological basis for the observed difference. It may possibly be a statistical error due to inadvertant sampling bias. If anybody else has information on this breed difference I would like to hear it.

EFFECTS OF HORMONAL CHANGES ON THE SELENIUM AND GLUTATHIONE PEROXIDASE LEVELS IN THE RAT

D. Behne, T. Höfer, W. Wolters, R. von Berswordt-Wallrabe[+]
W. Elger[+], P. Brätter and D. Gawlik

Hahn-Meitner-Institut für Kernforschung Berlin
[+]Schering AG, Berlin

ABSTRACT

In order to investigate the influence of hormones on the distribution of selenium and the seleno-enzyme glutathione peroxidase, rat experiments were carried out. Changes in the hormone metabolism were induced by pregnancy, hypophysectomy and adrenalectomy. The element content and the enzyme activity were determined in plasma, erythrocytes and different tissues such as liver, kidney, muscle, spleen, brain, adrenals, thymus, pancreas and testis. From the changes in the levels, which were investigated in pregnant and lactating animals, in female and male animals after hypophysectomy or adrenalectomy, and in selenium-deficient animals after hypophysectomy, information on the hormonal regulation of selenium and glutathione peroxidase metabolism was obtained.

DISCUSSION

IYENGAR (Germany). Dr Behne you showed a slide of some selenium concentrations. Could you just name the figures that you expect to come across?

BEHNE (Germany). Which tissues do you mean?

IYENGAR (Germany). Selenium concentration in liver or muscle, I just wanted to know the difference in actual figures with respect to controls.

BEHNE (Germany). Could we have the first slide please. Here you see the selenium contents in some of the tissues. Kidney, 1.45 mg Se per gram of wet tissue, plasma, .36, liver, .96, spleen, .43 erythrocytes, .42, pancreas, .29 and muscle .15.

WHITE (Australia). I would like to ask you about the time scale on which you took your measurements after you did the surgery, and whether you think there may have been some cyclical changes, or changes that were at one time different from another time.

BEHNE (Germany). The tissues were investigated 22 days after the hypophysectomy. A similar study was carried out on male animals at seven weeks after hypophysectomy. We found the same results in both studies.

WHITE (Australia). In the light of the work from the UK where cattle are turned out and show symptoms of white muscle disease due to some form of stress, would you care to speculate as to the possible hormone that may account for these stress-induced white muscle diseases?

BEHNE (Germany). At this stage it is too early to speculate about which hormones are involved. Hypophysectomy leads to many disturbances of hormone metabolism, and we do not know which ones are responsible for the changes.

THE SELENIUM STATUS OF HORSES IN WESTERN AUSTRALIA

*N.D. Costa, +H.G. Masters, +P. Steele, *B.J. Hilbert, *J.F. Hudman and *J.R. Bolton

*School of Veterinary Studies, Murdoch University
Murdoch, W.A. 6150
+Animal Health Laboratory, Department of Agriculture
South Perth, W.A. 6151

Introduction:

Western Australia contains selenium deficient soils in the high rainfall areas of the State [1]. Selenium responsive conditions in cattle, sheep and pigs have been diagnosed on farms in these areas. However, horses grazing on these farms do not generally receive selenium supplementation, despite the report of a foal which at necropsy showed lesions of nutritional myopathy [2]. Low blood selenium concentrations in horses have been associated with myocardial failure, skeletal muscle disease [3], reproduction disorders [4], unsatisfactory performance and subclinical hepatic insufficiency [5]. Also azoturia or 'tying-up syndrome', common in race horses, is reported to be responsive to selenium/ vitamin E therapy [6]. The purpose of this investigation was to survey the blood selenium concentrations of horses in Western Australia in the light of the above reports. The seleno-enzyme, glutathione peroxidase, has been shown to be a good indicator of selenium status in horses [4,7] and therefore its activity was also assayed in the latter part of the survey.

Experimental:

Animals and Assays: All samples were taken from thoroughbred or standard bred horses maintained in the higher rainfall (>410mm) areas of the south-west of Western Australia. Samples from animals with nutritional myopathy (NM) or azoturia were submitted to the diagnostic laboratory of the Department of Agriculture or Murdoch University. Livers from foals which died from causes other than myopathies were obtained at post-mortem. Heparinized blood samples were assayed for glutathione peroxidase (GSH-Px) activity by a modification of the method of Paglia and Valentine [8] and for selenium by the method of Watkinson [9] as modified by Gabbedy et al [1].

Results and Discussion:

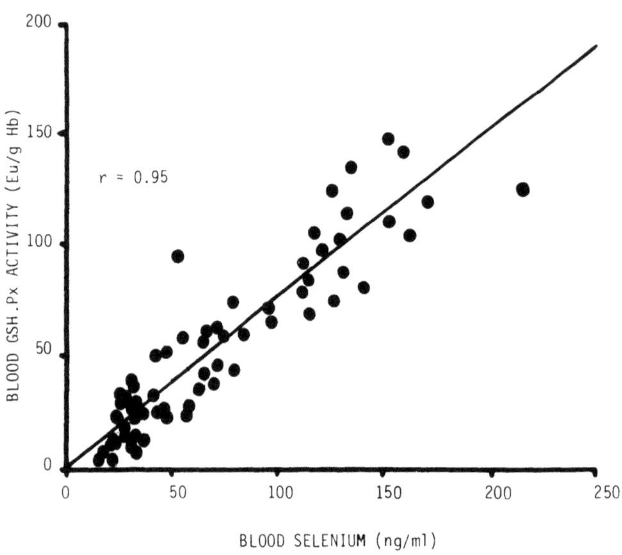

Figure 1

The relationship between blood selenium concentrations and glutathione peroxidase activity.

The regression equation is $y = 0.45 + 0.76x$ ($r = 0.95, P<.001, 69$ d.f.)
EU = µmoles NADPH oxidized/min.
Hb = haemoglobin

Figure 1 shows that GSH-Px activity (y) is significantly correlated with selenium (x) in blood samples taken from 71 horses ($r = +0.95$, $P<0.001$, 69 d.f.). The regression equation, $y = 0.45 + 0.73x$ is comparable to that reported by Caple et al. [7] in Victoria when selenium concentrations are adjusted to the same units. Thus the blood GSH-Px activity can be used as a rapid alternative method for the assessment of selenium status in horses in Western Australia.

Livers from foals with histologically confirmed nutritional myopathy were significantly lower ($t_{28} = 2.57$, $P < 0.05$) in selenium concentration than those from foals with no history of this syndrome (Table 1). Thus liver selenium could be used diagnostically to confirm equine nutritional myopathy where only post-mortem material is available.

Table 1 Selenium Content in Livers from Foals

Source of Liver	Selenium Content (ppm dry weight)
Foals with confirmed nutritional myopathy	0.18 ± 0.02 (4) (0.16 - 0.24)
Foals with no history of myopathy	0.44 ± 0.06 (26) (0.24 - 0.67)

Values shown are means ± S.E.M. with the number of foals in parentheses. The range of the selenium content is shown underneath each mean.

Racing horses that were stable-fed had the highest selenium status of those surveyed (Table 2). These animals were being fed commercial rations which contain grains grown in selenium adequate areas. There was no significant difference ($t_{26} = 0.85$, N.S.) in the selenium status of racing horses with azoturia and those with no history of this condition. Thus the reported selenium/vitamin E responsiveness of this condition cannot be accounted for on the basis of lower selenium status. Investigators at Newmarket, U.K. [5] suggest that blood selenium concentration in the range 30-90 mg/ml is associated with subclinical hepatotoxicity that can affect the racing fitness of horses. However we have insufficient history to establish any relationship between blood selenium concentration and racing performance in our groups of horses.

Table 2 The Selenium Status of Horses in Western Australia

Group Description	Whole Blood Se (ng/ml)	Whole Blood GSH-Px activity (EU/g Hb)
Stable-fed, racing horses; no azoturia	108 ± 18 (13) (34 - 214)	115.2 ± 16.3 (25) (69.5 - 177.2)
Stable-fed, racing horses; azoturia confirmed	93.2 ±24.4 (15) (21 - 184)	
Mares from farms with no history of NM	39.1 ± 6.9 (26) (21 - 77)	42.5 ± 8.9 (13) (26.4 - 95.4)
Mares of normal foals from farms where NM confirmed	22.8 ± 3.6 (14) (9 - 39)	9.4 ± 2.8 (5) (2.9 - 14.3)
Mares of foals with suspected NM	17.8 ± 2.4 (10) (10 - 26)	
Mares of foals with confirmed NM	13.2 ± 4.9 (5) (6 - 30)	10.7 (1)
Foal with confirmed NM	16.0 (1)	5.3 (1)
Unaffected foals from farms where NM confirmed	20.3 (3) (16 - 24)	13.6 (3) (12.9 - 14.4)

Each value represents the mean ± S.E.M. with the number of horses in parentheses. The range of values is shown underneath each mean. EU is μmoles NADPH oxidized/min. NM = nutritional myopathy.

Mares from farms where there was no recorded history of nutritional myopathy had a mean blood selenium concentration of 39.1 ± 4.9 ng/ml with a range of 21 to 77 ng/ml (Table 2), which is comparable to the 'normal' range of selenium concentrations reported from Sweden [10] for whole blood and from Denmark [11] for serum. However in New York State, blood selenium concentrations in horses from farms with no history of nutritional myopathy were in the range 41 to 131 ng/ml while selenium concentrations in horses grazing pastures where a foal had died from nutritional myopathy were in the range 25 to 49 ng/ml [4].

Under conditions prevailing in the high rainfall areas of Western Australia, both foals which developed nutritional myopathy and their dams had blood selenium concentrations of 30 ng/ml or less as did dams of foals with suspected nutritional myopathy (Table 2). It is important to note that there were unaffected foals from farms with a history of nutritional myopathy which had blood selenium concentrations in this range (i.e. 16 to 24 ng/ml, Table 2).

The range of glutathione peroxidase activities in groups of horses where nutritional myopathy had been confirmed compares favourably with those reported in similar groups of horses from Victoria, Australia [7]. These affected groups have significantly lower ($P <0.001$) blood glutathione peroxidase activities than those recorded from farms with no history of nutritional myopathy (Table 2).

Thus our data shows that determination of blood selenium concentrations and glutathione peroxidase activity or liver selenium concentrations must be a useful aid to the diagnosis of equine nutritional myopathy. However these parameters cannot be used to establish with certainty those horses which will develop nutritional myopathy because factors other than selenium status (e.g. vitamin E status, stress, activity) can be involved in the aetiology of this syndrome [12]. Also the variation in blood selenium concentrations and glutathione peroxidase activities reported from various countries [4,5,7,11] makes it very difficult to compare and interpret these parameters on an absolute basis in horses.

References

1. Gabbedy, B.J., Masters, H.G. and Boddington, E.B. (1977) Aust. Vet. J. 53, 482.
2. Gabbedy, B.J. and Richards, R.B. (1970) Aust. Vet. J. 46, 111-112.
3. Owen, R.R., Moore, J.M., Hopkins, J.B. and Arthur, D. (1977) J. Amer. vet. med. Assoc. 171, 343-348.
4. Maylin, G.A. Rubin, D.S., and Lein, D.H. (1980) Cornell Vet. 70, 272-289.
5. Blackmore, J.J., Wilett, K., and Agness, D. (1979) Res. vet. Sci. 26, 76-80.
6. Geiser, D.R. (1975) Vet. Med. Small Anim. Clin. 70, 710.
7. Caple, I.W., Edwards, S.J.A., Forsyth, W.M., Whiteley, P., Selth, R.H. and Fulton, L.J. (1978) Aust. Vet. J. 54, 57-60.
8. Paglia, D.E. and Valentine, W.N. (1967) J. Lab. Clin. Med. 70, 158-168.
9. Watkinson, J.H. (1966) Anal. Chem. 38, 92.
10. Bergsten, G., Holmback, R. and Lindberg, P. (1970) Acta vet. scand. 11, 571-576.
11. Schongard, H., Basse, A., Gissel-Nielsen, G., and Simesen, M.G. (1972) Nord. vet. Med. 24, 67-84.
12. Wilson, T.M., Morrison, M.A., Palmer, N.C. Finley,G.G. and van Dreumel, A.. (1976) J. Am. vet. Ass. 169, 213-217.

Cu, Mo, S, Zn Periodontitis in Brazilian Cattle

Waldemar V. de A. Camargo[1], Joao S. Veiga[2] and Joseph H. Conrad[3].

Institute Biologico, C.P. 7119, Sao Paulo, Brazil[1], Tortuga Co. Zootecnica Agraria, Sao Paulo, Brazil[2] and University of Florida, Gainesville, Fl 32611, U.S.A.[3]

"Cara inchada" (swollen face) is a periodontal disease affecting young cattle predominantly, mainly during the weaning period. It is clinically characterized as a periodontal inflammatory process with loosening and shedding of the teeth (premolar and molar) and swelling of the maxillary and, less frequently, mandibular bones.

Examination indicates that a periodontitis in calves as young as 30 days of age accompanies the oral lesions, emaciation, severe diarrhea, acromotrichia and retarded growth. Gingivitis appears to be aggravated by forage consumption and reduced integrity of the periodontal ligament permits the entrance of food particles in this damaged tissue that results in an alveolar "pyorrhea" and chronic ossifying periostitis. Normal mastication is impeded resulting in unproductive animals and death from starvation.

The disease occurs in certain areas of Central Brazil but only in purebred and crossbred zebu cattle raised in those regions.

The highest incidence is found in cattle grazing newly established pastures which have been planted on fertile soils from which tropical forest has been removed, with pH around 6,0. "Cara inchada" also occurs in acid and sandy soils of medium fertility where pastures of"Brachiaria decumbens" have been established. The incidence of the disease is very variable according to region, farm and pasture.

The disease in Brazil was first referred to by Giovine _et. al_. (1943) and more recently by Dobereiner _et al_. (1975) and other workers. The occurrence of a somewhat similar disease of sheep in New Zealand has been reviewed by Cuttress and Ludwig (1969).

"Cara inchada" is a most important health problem, responsible for the death of thousands of young cattle, with an incidence of about 10-20%.

As well as "Cara inchada", in Mato Grosso North a wasting disease of adult cattle resembling molybdenosis occurs which can largely be controlled by Cu administration, while in Mato Grosso South, in areas where "_B, decumbens_" has been established, a skin disorder of weaner steers is encountered which can be diminished by Zn supplements.

An extensive survey of minerals in forage, in parts of Amazonia Legal where "Cara inchada" also occurs (Camargo _et. al_., 1980), revealed levels of Cu, Co, Zn and P ordinarily considered too low for grazing cattle.

As a result of these findings, the possible involvement of minerals in the aetiology of "cara inchada" and related diseases has led, in all of the areas involved, to further intensive examination of minerals in soil, forage and animal tissues as well as the effects of mineral supplementation on large numbers of potentially affected animals.

Analysis of 10 soil samples from "Cara inchada" areas in Mato Grosso North revealed acceptable levels of organic matter, Al, Ca, Mg and K. The concentration of P was rather low (6.5 ppm), while that of Mo was high (8.8 ppm). In the same region the examination of 20 samples of "Panicum maximum" revealed no appreciable differences between those collected in either wet or dry seasons; several values were somewhat low (P, 0.10-0.12%; Mg, 0.13-0.15%, Cu, 5.7-6.0 ppm; Zn, 15-16 ppm).

Analyses relating to a large number of forage samples of (mainly) "P. maximum" collected from "cara inchada" areas in Mato Grasso North, Mato Grosso Central (near Rondonopolis, where the disease is endemic) and Mato Grosso South (near Paranaiba) are contrasted below with those relating to unaffected areas:-

	Mean Values (ppm DM)			
Affected areas	Cu	Zn	Co	Mo
Mato Grosso North	6.7	22	0.08	4.6
Central	8	34	0.16	3.9
South	4.5	23	0.10	2.5
Unaffected areas				
Mato Grosso Central	15	16	0.31	2.6
South	7	17	0.10	1.5

Liver analyses from affected steers and from normal steers (from unaffected areas), all from the vicinity of Paranaiba, are set out below:-

	Mean values from 10 steers (ppm DM)				
	Cu	Zn	Mn	Fe	Co
Affected	70	58	7.5	250	0.12
Normal	166	89	9.0	137	0.13

An initial experiment was conducted near Rondonopolis with 10 groups of 10 yearling steers; several in each group being affected by "cara inchada". Salt licks containing a comprehensive mixture of macro- and minor- elements were provided. In groups which consumed daily 50g of lick (60mg Cu, 4.5mg Co, 90mg Zn and 4g P) or more, the incidence of periodontitis was prevented and existing cases largely cured. In a subsequent experiment in the same area, three groups of steers were offered salt licks containing P, Ca, Co, Cu, Zn and S. The concentrations of Cu and S were varied between groups, the other elements remaining constant. A fourth group was offered salt plus Mo to exacerbate Cu deficiency. The average consumption of lick, Cu, Zn, S and Mo during 95 days (April - July) were:-

	Mean intake/day					Mean weight gain/day (kg)
	Lick (g)	Cu(mg)	Zn(mg)	S(g)	Mo(mg)	
Group A	33	51	79	0.18	-	0.31
Group B	34	277	82	0.29	-	0.43
Group C	28	42	84	1.39	-	0.27
Group D	12	-	-	-	2.4	0.13

In further experiments near Paranaiba, where P. maximum forage contained 4, 2 and 20 ppm of Cu, Mo and Zn, respectively, the provision of 75, 75 and 45 mg/day of Cu, S and Zn, respectively, to cows during gestation ensured the production of healthy calves, while one-third of the calves from cows not given trace elements developed periodontitis within 50 days.

It is clear that periodontitis, a major deterrent to cattle production in vast areas of Brazil (and neighbouring Bolivia), is related to low levels of Cu and possibly Zn in pastures, that a Cu-Mo-S inter-relationship is involved and that the condition can be effectively controlled by appropriate trace element supplementation. Several aspects of the disease and its treatment are being studied.

REFERENCES

Camargo, W.V. de A., N.S. Fernandes and A.M.H. Santiago. 1980. Arq. Inst. Biol, Sao Paulo, Brasil, 47 (4):83-111.

Cuttress, T.W. and T.G. Ludwig. 1969. Periodontal Disease in Sheep 1. Review of Literature. J. Periont. 40:529-534.

Dobereiner, J., J.A. Chaves, I.V. Rosa and R.H. Houser. 1975. Efeito da Transferencia de Bovinos com "Cara Inchada" (Doenca Peridentaria) para Pastos de Regiao Indene. Pesq. agrop. bras. Ser. Vet. 10:99-1103.

Giovine, N., N. Rangel, A.V. Machado, R.D. Lamonnier and A. Wilwert. 1943. Necrobacilose. Sumula nosologica, a necrobaciolose em Minas Gerais. 1:35-65.

DISCUSSION

GARTNER (Australia). I wonder whether the macro elements may not also be involved in the etiology of this disease. I say this because you mentioned that it occurred when tropical pastures replaced your native pastures. Green panic contains about 1% total oxalate and I am convinced that all the calcium is present as calcium oxalate crystals. In digestability trials with cattle we found that only 20 - 40% of that calcium was absorbed by the animal. When you talk of facial swellings you presumably have a depletion of the skeletal minerals in that area and replacement by organic matter. Thus I am just wondering whether calcium and phosphorus may not be involved and you have in fact quite a complicated situation here.

CAMARGO (Brazil). I have carried out experiments with calcium supplementation as calcium carbonate ($CaCO_3$) and we had no response to calcium. But we do get response when we use dicalcium phosphate as a supplement. Analysis of bones from the affected animals revealed a normal calcium content but, low levels of phosphorus. Calcium phosphate will not prevent the swollen face syndrome unless copper and zinc are added to the salt mixture.

KRISHNAMACHARI (India). Two questions, one is whether you have any idea about the fluoride content of drinking water for these animals and second question is whether you have looked into the bones of these animals histologically to see if they were showing any osteoporosis.

CAMARGO (Brazil). In answer to the first question the drinking water in these areas is low in fluorine. The addition of fluorine to the mineral mixture supplement does not precipitate "Cara inchada". In answer to the second question we have evidence of osteoporosis. Histologically "Cara inchada" is a osteoporotic dystrophy.

MINERAL STATUS OF CATTLE IN TROPICAL LOWLANDS OF BOLIVIA

B. Bauer, E. Galdo, L. R. McDowell, M. Koger, J. K. Loosli and J. H. Conrad

Department of Animal Science
University of Florida
Gainesville, Florida 32611
U. S. A.

In tropical regions of the world, mineral deficiencies and imbalances are severely inhibiting the grazing cattle industry (1, 5). Only rarely can forages completely satisfy each of the mineral requirements. Borderline or deficient levels of certain elements were noted for many forages in the 1974 Latin American Tables of Feed Composition (7). These minerals and the percentage of the forage samples deficient were: Co, 43%; Cu, 47%; Mg, 35%; P, 73%; Na, 60%; and Zn, 75%. Many supplementation studies in Latin America have noted dramatic increases in calving percentages by merely providing minerals to grazing cattle (2, 6). This present study was initiated to evaluate the mineral status of beef cattle in the Beni of Bolivia, South America, and to study the influence of mineral supplementation on growth and reproductive efficiency.

Materials and Methods

A three-year experiment was carried out on a ranch in the Beni lowlands which are part of the Amazon river basin, with a carrying capacity of 3 to 8 hectares per animal unit. The average annual temperature is 27 C (9 to 40) and the mean rainfall, 1800mm, of which 85% falls within a 7 month period, causing low pastures to be regularly flooded. Soil is a clay loam type with a mean pH of 5.0.

For Year 1, 872 steers and 894 female 12-15 month old Zebu-Criollo crossbred cattle were assigned to 3 treatments: 1) control - no minerals (C), 2) common salt only (S), and 3) salt (60%) + bone phosphate (40%) (S + BP). Injections of Se-vitamin E, Cu, I and Co bullets were administered singly and in all possible combinations to 160 animals within each major treatment. Number of females on experimentation for years 2 and 3 were 655 and 385, respectively, for treatments S and BP. Analyses were made of nonfertilized, native forages, blood and liver (biopsy) samples taken during June (dry season) and November (wet season) for each year. Methods of sample collection, preparation and mineral analysis for animal tissue and forage analyses have been previously described by Fick *et al*. (3). Statistical analyses of weight gains, animal tissue parameters, forages and soils were conducted using Harvey's Least-Squares Program (4).

Results and Discussion

At 338 days, cattle receiving S + BP had higher gains of 96.4 vs 82.7 and 79.4 kg ($P < .001$) for C and S treatments, respectively (Table 1). Mineral consumption per animal was low with only 11.5 and 16g/day for the S and S + BP treatments, respectively. There was a nonsignificant advantage for trace mineral treated over nontreated animals for each of the trace elements. Weight gains (kg) for the minor element treatments administered singly on in combination were as follows: I, 86.8; Co, 86.7; Cu, 86.1; and Se, 87.3, *vs* 82.1 for animals not receiving trace minerals. Weight gain advantages from Se were in evidence, since only the trace element treatments of Se + Cu, Se + I and Se + Cu + Co + I were significantly higher ($P < .05$) than nontreated controls. Males had higher ($P < .001$) gains than females of 88.8±1.2 *vs* 83.5±1.2, respectively. A sex x treatment interaction ($P < .01$) existed, with only control animals not expressing a gain difference due to sex.

Year 2 pregnancy rates for 655 animals for the S and S + BP treatments were high, 89.5 and 88.5%, respectively. Second pregnancy rate comparisons (year 3) for 385 animals were 12.6% higher for S + SP vs S treatments (86.4 vs 73.8%). For year 3, females consuming the S + BP supplement gained 12.8 kg while the S treatment lost 9.2 kg, resulting in a total difference of 22 kg in favor of the S + BP group. Mineral consumption for the last two years and treatments were approximately 16-17 g/day.

Table 1. Weight gains (kg) influenced by Mineral Supplementation and Sex at 338 days[a]

	Control		Control + Salt		Control + Bone phosphate		Sex Mean	
	Mean	S.E.	Mean	S.E.	Mean	S.E.	Mean	S.E.
Steers	82.9	2.2	83.6	1.9	100.1	1.7+	88.8	1.2++
Heifers	82.5	2.0	75.2	1.9	92.7	1.7+	83.5	1.2
Steers and Heifers	82.7	1.8	79.4	1.7	96.4	1.6+		

[a] Animal numbers were control (333), + salt (445) and + bone phosphate (988)

[b] Salt + bone phosphate mixture analyzed: Ca (8.2%), P (8.9%), protein (.16%), Mg, (150ppm), Co (.18 ppm), Cu 4.8 ppm), Fe (182 ppm), Mn (3.6 ppm) and Zn (36 ppm)

[+] Significantly greater than other treatments ($P < .001$)

[++] Steers greater gains than heifers ($P < .001$)

Mean analysis of forages are presented in Table 2, with season and year differences found ($P < .05$). The majority of individual forage samples were below critical levels in P (<.25%), Na (<.06%), Cu (<10 ppm) and Zn (<30 ppm). Protein (<7.0%) and K (<.60%) forage concentrations were at critical levels for the majority of forages only during the dry season of 1977. The majority of forages were low during both seasons of 1978 in Ca (<.30%).

Table 2. Nutrient Composition of Bolivian Forages (Dry Basis)

Analysis	Critical Level[a]	Dry Season 1977 (12)[b]	Dry Season 1978 (10)	Wet Season 1977 (18)	Wet Season 1978 (14)
Protein, %	7.0	6.5^{c}	11.9^{d}	11.4^{d}	12.5^{d}
Calcium, %	.30	$.49^{d}$	$.16^{c}$	$.44^{d}$	$.20^{c}$
Potassium, %	.60	$.58^{c}$	1.55^{e}	1.02^{d}	2.59^{f}
Magnesium, %	.20	.19	.24	.18	.23
Sodium, %	.06	.03	.04	.03	.02
Phosphorus, %	.25	.20	.30	.18	.19
Cobalt, ppm	.10	.38	.36	.60	.20
Copper, ppm	10	4.6	7.8	5.8	6.1
Iron, ppm	30	162^{d}	249^{d}	178^{d}	96^{c}
Manganese	20	416	434	365	326
Molybdenum, ppm	>3.0	1.4	.32	1.6	.72
Selenium, ppm	.10	.11	.18	.14	.16
Zinc, ppm	30	25	17	24	22

[a] McDowell and Conrad (6)

[b] Numbers in parentheses are numbers of forages

[c-f] Means not followed by the same superscripts are different ($P < .05$)

Mean liver and blood serum concentrations are presented in Table 3. Mean serum Ca, P, Cu and Zn concentrations were above critical levels for each collection period. Serum Mg was lower ($P < .05$) during the 1978 wet season with 83% of samples less than 2.0 mg/100 ml. The majority of liver Cu samples were deficient (<75 ppm) while 1/3 to 1/2 of the liver Zn values were low (<84 ppm) during the 4 collection periods.

Table 3. Mineral concentrations of cattle tissues

Item	Critical level	Dry Season 1977 (48)	Dry Season 1978 (39)	Wet Season 1977 (48)	Wet Season 1978 (30)
Blood Serum					
Calcium, mg/100ml	8.0	10.9^{e}	8.1^{c}	11.6^{f}	9.1^{d}
Magnesium, mg/100ml	2.0	2.0^{d}	2.2^{d}	1.7^{c}	3.0^{e}
Phosphorus, mg/100ml	4.5	6.4	7.7	7.5	7.5
Copper, µg/ml	.65	.99	1.06	.79	1.20
Zinc, µg/ml	.80	1.07^{c}	1.23^{d}	$.96^{c}$	1.47^{e}
Liver (ppm, dry basis)					
Cobalt	.07	$.48^{d}$	$.55^{d}$	$.37^{c}$	1.2^{e}
Copper	75	51^{c}	106^{d}	49^{c}	44^{c}
Iron	180	239^{c}	203^{c}	197^{c}	344^{d}
Molybdenum	>5	2.8^{d}	3.8^{e}	1.8^{c}	2.5^{d}
Selenium	.25	-----	.39	------	.48
Zinc	84	115	99	93	97

[a] McDowell and Conrad (6)

[b] Numbers in parentheses are sample numbers

[c-f] Means not followed by the same superscripts are different ($P < .05$)

Summary

A 3-year study was designed to determine the mineral status and benefits of mineral supplementation for grazing beef cattle in tropical Bolivia. Cattle receiving bone phosphate had greater gains of 96.7 vs 82.7 and 79.4 kg for control and salt treatments, respectively. Males had greater gains than females (88.8 vs 83.5 kg). For bone phosphate vs salt treatment second pregnancy rate was higher by 12.6% and gains for year 3 were 22 kg higher. On the basis of forage, liver and blood serum analyses, deficiencies of Fe, Co and Mn, or toxicity of Mo, are unlikely.

References

1. Conrad, J. H. and L. R. McDowell (1978). Latin American Symposium on Mineral Research with Grazing Ruminants. University of Florida, Gainesville.

2. Fick, K. R., L. R. McDowell and R. H. Houser (1978). Latin American Symposium on Mineral Nutrition Research with Grazing Ruminants, University of Florida, Gainesville.

3. Fick, K. R., L. R. McDowell, P. H. Miles, N. S. Wilkinson, J. D. Funk and J. H. Conrad (1978). Methods of Mineral Analysis for Plant and Animal Tissues, Second Edition, University of Florida, Gainesville.

4. Harvey, W. R. (1977). Harvey's Least-Squares and Maximum Likelihood Computer Program, Ohio State University, Wooster.

5. McDowell, L. R. (1976). Beef Cattle Production in Developing Countries, University of Edinburgh, Scotland.

6. McDowell, L. R. and J. H. Conrad (1977). World Animal Review 24:24.

7. McDowell, L. R., J. H. Conrad, J. E. Thomas, L. E. Harris and K. R. Fick (1977). Trop. Anim. Prod. 2:273.

DISCUSSION

KUBOTA (USA). Could you cover some of the methods you used to determine your mineral elements, particularly elements like selenium, molybdenum, cobalt.

McDOWELL (USA). Selenium was the fluorometric procedure. Molybdenum and cobalt were determined using a Perkin-Elmer with a Graphite Furnace attachment atomic absorption. Phosphorus was determined by a coloropetric procedure and all other elements by flame atomic absorption.

GARTNER (Australia). I don't agree with your critical levels of sodium in forages to be 0.06% for non lactating animals with the weight gains that you encountered and I would suggest that a figure of .03% or less might be more realistic. I think the proof of it is that you did not get any response to sodium chloride supplementation.

McDOWELL (USA). This could be true. The o.06% figure is from 1976 beef cattle National Research Council requirement.

LLOYD DAVIES (Australia). I think we could argue about the critical levels quite a bit but I think what must be remembered in this situation is that Dr. McDowell is working in very difficult situations and has shown a clear and unequivocal response to a mineral supplementation. Dr. Gartner and I would disagree on your liver copper level but I don't think we want to get onto minuté like that.

McDOWELL (USA). Well I think probably 25 parts per million would be a more critical level. I think that values of less than 75 ppm are borderline and suggest we maybe approaching the more definite critical level of 25 ppm.

ARORA (India). What critical forage level did you keep in your mind about the zinc status.

McDOWELL (USA). I am relying on requirements (or allowances) suggested by the 1976 NRC (30 ppm), 1978 NRC (40 ppm) and 1965 ARC (50 ppm). To me 30 ppm Zn would be a reasonable critical level to which to be concerned about a deficiency.

KUBOTA (USA). I wonder a little about your forage analyses as we have many areas in the United States, particularly in the arid West, where we have values about as low or much lower than you have reported. I have difficulty evaluating some of your values when I compare them to those from our Western states, particularly our rangeland. The cattle seem to do fairly well in the West.

McDOWELL (USA). Referring to Latin America specifically I guess one of the first reports was in 1960 from Guyana (Legg and Sears) where they had forage values of in about the 30 - 40 parts per million with very definite skin lesions. By administering zinc oxide they virtually eliminated the lesions. In an earlier paper today Dr. Carmago noted skin lesions from cattle grazing Panicum Maximum in Brazil with zinc levels at 20 ppm. We also see various reports where skin ailments are not present but growth responses are made by giving a zinc supplement.

ARORA (India). I have a comment about the zinc level in the feed. In India we are finding that as and when the zinc level is less than 30 parts per million, either subclinical syndromes of deficiency or even sometimes clinical syndromes of deficiency have been recorded.

McDOWELL (USA). There is of course some data that has shown responses in zinc supplementation when there is no clinical signs whatsoever. I think a very important consideration to keep in mind is which nutrient is the most limiting. Perhaps energy-protein or phosphorus are the most limiting and therefore zinc deficiency is not expressed. When we first solve the other limiting nutrients then we would expect a zinc deficiency to be expressed or response derived from zinc supplement.

LLOYD DAVIES (Australia). Closing on that comment I would agree with you entirely Dr. McDowell. Our experience with sheep on zinc in Scotland is that they seem to be able to get along on very small quantities indeed.

MINERAL STATUS OF GRAZING FARM ANIMALS IN MALAYSIA

R. I. Hutagalung, P. Y. Hew, M. R. Jainudeen,
T. A. Bongso and Syed Jalaludin

Departments of Animal Sciences and Clinical Studies,
Faculty of Veterinary Medicine and Animal Science,
Universiti Pertanian Malaysia,
SERDANG, SELANGOR, MALAYSIA

Low ruminant productivity in the humid tropics is due to many limitations, the most apparent being the low feed intake and poor nutrition, which consequently result in poor growth and reproductive rates. Mineral insufficiencies, particularly phosphorus, cobalt and copper, in cattle grazing pastures in Malaysia and other tropical regions have been reported to be responsible for this low productivity (1, 2, 8, 9).

MINERAL COMPOSITION OF SOIL AND PASTURES. Pasture is the major source of feed for cattle throughout the year. Any mineral deficiencies in pasture and indirectly in soil will be reflected by their deficiencies in animal tissues. Soil and forage samples were collected at the same time from similar spots in paddocks where the experimental animals grazed. These collections were made periodically during rainy and dry periods. The soils had pH values ranging from 4 to 7 in the dry season and 5 to 6.9 in the wet season. Soil extractable macro minerals were adequate to meet the requirements of forage plants, except

TABLE 1. Mineral composition of soil and forage at a commercial farm

	pH	N (%)	Ca (ppm)	P (ppm)	K (ppm)	Mg (ppm)	S (ppm)
Soil	5.62	0.10	455.2	20.1	207.2	98.0	-
Forage	Dry	1.62	150	1800	12100	1800	600
	Rainy	1.43	200	900	8900	2400	600
	Season	Fe (ppm)	Mn (ppm)	Zn (ppm)	Cu (ppm)	Co (ppm)	Mo (ppm)
Forage	Dry	98	140	63	6.0	0.10	2.0
	Rainy	87	98	54	5.4	0.07	1.7

phosphorus (P) which is considered low (Table 1). Forage P was inadequate for grazing animals, but it was adequate in the legume (0.32 ppm). Forage K, Mg, Mn, Fe, Cu and Mo was adequate at both seasons. Forage cobalt is considered barely adequate to deficiency for grazing cattle.

MINERAL SUPPLEMENTATION IN CATTLE. Two experiments were conducted to study the effect of mineral and concentrate supplementation on the performance of a commercial beef herd in Malaysia. This study was prompted by low and declining calf branding percentage in 1977 (66%) and 1978 (51.5%). The farm maintained 4,300 Droughtmaster cattle on 2,000 ha established grass-legume pasture. The average annual rainfall was 1400 mm with higher concentration in November and December.

EXPERIMENT 1. A total of 175 cows, 2 to 6 years old, nursing calves and weighing from 350 to 450 kg were randomly allotted to 4 treatments: (1) mineral lick + salt + concentrate (14% protein; 3 kcal ME/g); (2) mineral lick + salt; (3) dical. phosphate ($CaHPO_4$) + salt + concentrate; (4) salt + concentrate. Mineral lick (10% P; 8.5% Ca; 50% salt; 0.5% Mg; 0.3% Fe; 0.25% Mn; 300 ppm Zn; 50 ppm Co; 300 ppm I; 0.04% S) or $CaHPO_4$ was offered free-choice with salt throughout the trial. Concentrate was group fed to cows (except group 2) and bulls for the first 60 days of the 4-month mating period at the rate of 1.82 kg/cow/day. Data on cow performance showed that for mating occurring during the first 3 to 4 months postpartum, the pregnancy rates ranged from 71 to 91% for cows in group 2 and 3, respectively (Table 2). These were apparently higher than those reported for the herd during 1978 (60%) or for control cows in the herd for 1979 (60.3%). Pregnancy rate was improved by concentrate during 8 weeks postpartum as reflected by lower rate (71%) for cows in group 2 compared to those in group 1 (86%). Supplementation of $CaHPO_4$ gave higher pregnancy rate (91%) than those receiving the lick (86%).

TABLE 2. Effect of mineral supplementation on cattle performance - Experiment 1

Item	Group 1 (n=41) Mineral lick + Salt + Concentr.	Group 2 (n=42) Mineral lick + Salt	Group 3 (n=48) CaHPO4 + Salt + Concentr.	Group 4 (n=44) Salt + Concentr.
Pregnancy (%)	86.0	71.0	91.0	84.0
% calf crop *	85.4	69.0	87.5	77.3
% calf crop weaned	82.9	69.0	83.3	72.7
Avg calf birth wt (kg)	40.5	39.5	31.5	30.4
Avg calf weaning wt (kg)	237.1	235.2	220.8	192.8

*Calf crop percentage for the years 1977 and 1978 were 66 and 51.5%, respectively.

Percent calf crop followed somewhat the rate of pregnancy, having 69% for cows in group 2 against 87.5% for those in group 3. An overall calf crop improvement of about 14% from 1977 (66%) and 28% from 1978 (51.5%) was obtained from combined mineral-concentrate supplementation. Birthweights were heavier in groups 1 and 2 than those in groups 3 and 4. Similarly, heavier weaning weights were noticed in calves receiving the lick than those in groups 3 and 4. These findings suggest that concentrate tends to affect the rate of pregnancy whereas mineral supplementation affects weights at birth and weaning. Composition of blood mineral revealed an inconsistent trend among the treatments but the values tended to be higher at initial phase than at pregnancy phase.

EXPERIMENT 2. This study was undertaken to compare the effectiveness of mineral, salt and concentrate supplements on performance of cattle. Mineral meal (8.8% P; 12% Ca; 50% salt; 0.5% Fe; 0.25% Mn; 0.5% Zn; 100 ppm Cu; 20 ppm Co; 160 ppm I; 10 ppm Se) was used instead of mineral lick. A total of 193 cows were randomly assigned to 4 treatments: (1) mineral meal (MM) + 0.68 kg concentrate (C); (2) MM + 1.36 kg C; (3) salt; and (4) MM.

The rates of pregnancy were 84.2, 87.2, 88.7 and 94.6% for cows receiving treatments 1, 2, 3 and 4, respectively. Concentrate has less effect compared to MM on percent calf crop. Cows on MM had the highest calf crop (89.1%) while those receiving salt showed the lowest calf crop (79.2%). Calves receiving MM or MM + C were heavier (32.9 kg) than those given salt only (24.9 kg).

TABLE 3. Effect of mineral supplementation on blood minerals of cattle - Experiment 2

Mineral (ppm)	MM + 0.68 kg C		MM + 1.36 kg C		Salt		MM	
	Start	Final	Start	Final	Start	Final	Start	Final
Ca	79.3	101.2	89.4	105.4	80.4	112.4	90.4	116.8
Mg	9.82	20.27	9.92	19.2	9.31	19.93	9.45	15.5
P	71.0	108.3	69.0	108.2	60.0	100.4	77.0	95.0
K	181.4	267.2	190.1	250.5	196.0	281.0	212.8	271.6
Fe	1.07	1.01	1.05	0.85	1.35	1.03	0.98	0.66
Zn	0.93	1.11	0.93	0.97	0.89	0.90	0.78	1.08
Cu	0.46	0.53	0.46	0.51	0.46	0.48	0.42	0.49

Cows receiving MM or MM + C had slightly higher plasma P than those on salt only (Table 3). Plasma Cu levels are low compared to reported values (6). Plasma mineral values at final (pregnancy) phase are relatively higher than those collected at initial phase.

MINERAL SUPPLEMENTATION IN BUFFALO. Swamp buffalo (Bubalus bubalis) have been used for ploughing rice fields and for meat production in Malaysia. However, poor reproduction (± 529 days calving interval and heat detection problems) has been the major constraint for improvement in its productivity (3, 4). Supplementation of phosphorus-rich mineral mixture has been reported to improve the performance of cattle (6). This work was undertaken to study the performance of buffalo cows and calves to mineral supplementation. Suckled swamp buffaloes (4 to 9 years old and calving normally between Oct. 1979 and Oct. 1980) and calves were subjected to: (1) no mineral lick; and (2) with mineral lick. All animals were maintained on a rotational grazing system on established pasture with stocking rate of about 2 adults/ha. Animals were weighed at biweekly intervals from calving to weaning period. Blood was collected at weekly intervals for mineral determination.

Data on the performance of calves from birth to weaning showed that calves receiving mineral lick were heavier at birth (27.2 vs 26.4 kg) and at weaning (132 vs 112 kg), and exhibited faster rate of gain (0.543 vs 0.498 kg) compared to those without the lick. Cows receiving mineral lick had higher plasma P than those without the lick (5.10 vs 4.18 mg %) (Table 4). It is interesting to note that slightly higher P level was observed in cows given mineral lick at 2 to 3 month postpartum. Plasma P of cows without mineral lick remained relatively constant. Calves given mineral lick maintained higher P level than those without the lick. Plasma P level of calves receiving the lick was high at birth and it declined at 6 month of age from 7.66 to 5.23 mg %, respectively. Calves receiving no mineral lick followed a similar declining pattern from 6.89 mg % P at calving to 4.21 mg % P at weaning.

TABLE 4. Effect of mineral supplementation on blood phosphorus (mg %) of buffalo - Experiment 3

Postpartum period (month)	With mineral		No mineral	
	Cow	Calf	Cow	Calf
1	4.55 ± 0.27	7.66 ± 0.33	3.85 ± 0.19	6.89 ± 0.26
2	6.09 ± 0.15	7.22 ± 0.24	4.80 ± 0.39	6.38 ± 0.11
3	5.37 ± 0.31	6.56 ± 0.13	4.28 ± 0.14	5.62 ± 0.23
4	5.16 ± 0.18	5.87 ± 0.20	4.04 ± 0.15	5.14 ± 0.32
5	4.56 ± 0.22	5.52 ± 0.29	4.17 ± 0.08	4.78 ± 0.08
6	4.88 ± 0.14	5.23 ± 0.17	3.95 ± 0.13	4.21 ± 0.21

SUPPLEMENTATION IN SHEEP. The major plantation crops in Malaysia which have potentials for integration with small ruminants represent about 3 million ha or 84% of the areas under plantation crops (7). Mineral supplementation trial was carried out for 100 days in sheep reared under rubber plantation. A total of 60 sheep averaging about 11 kg liveweight were randomly assigned to 4 treatments: (1) control; (2) mineral lick (ML); (3) concentrate (C) and (4) ML + C. Sheep receiving either ML (22.6 kg) or ML + C (22.1 kg) were slightly heavier than those on control (19.5 kg) and C (20.1 kg). Sheep receiving ML grew slightly faster (118 g) compared to those on control (83 g), C (91 g) and ML + C (109 g), but the difference was not significant. Sheep appeared to have less apparent response to mineral lick than large ruminants.

SUMMARY

The data show that supplementation of mineral mixture to grazing animals improved pregnancy rate, percent calf crop, birthweight, weaning weight and growth.

REFERENCES

1. Hew, P. Y. and R. I. Hutagalung. 1980. First Asian-Australasian Animal Science Congress. Abstr. No. 75, p. 24.

2. Hill, R. R. Thambyah, S. P. Wan and C. S. Shanta. 1962. J. Agric. Sci. 59:409-415.

3. Jainudeen, M. R. 1977. Proc. Conf. Health & Prod. of Local & Australian Cattle in Southeast Asia. Ministry of Agric. Malaysia. Bull. No. 146, p. 162-169.

4. Jainudeen, M. R., R. I. Hutagalung and S. Dass. 1980. First Asian-Australasian Animal Science Congress. Abstr. No. 125, p. 38.

5. Lebdosoekojo, S., C. B. Ammerman, N. S. Raun, J. Gomez and R. C. Littell. 1980. J. Animal Sci. 51:1249-1260.

6. Little, D. A. 1975. J. Exp. Agr. Anim. Husb. 15:25-32.

7. Mahyuddin, M. D., R. I. Hutagalung and Wan Mohamed, W. E. 1978. In: Proc. Sem. Integr. Anim. Pl. Crops, MSAP, p. 70-82.

8. Mannetje, L. T., S. Ajit Singh and M. Murugaiah. 1976. MARDI Res. Bull. 4(1):90-94.

9. McDowell, L. R. 1976. In: Beef Cattle Production in Developing Countries (A. J. Smith ed.). Edinburgh, Scotland, p. 216-241.

DISCUSSION

SCHWARZ (Germany). In your different treatments did you find that the animals preferred to eat mineral lick plus concentrate or mineral lick plus salt and so on. If so, this would influence your result.

HUTAGALUNG (Malaysia). The objective of combining the mineral lick and concentrate was based on the observation that animals in that farm were loosing weight. We thought that the animals might respond both to the concentrate and mineral supplementations. We found that the mineral lick gave a more favourable response compared with a concentrate. As far as the appetite is concerned, mineral lick alone appeared to be consumed well by the animals.

PALLAUF (Germany). Did you distinguish between the organic phosphorus blood level and the inorganic phosphorus blood level because it is known that the inorganic phosphorus blood fraction gives a better indication of the phosphorus supply.

HUTAGALUNG (Malaysia). We did not make any differentiation between organic and inorganic phosphorus. We measured only inorganic phosphorus.

DEFINING A DIAGNOSTIC PLASMA VITAMIN B_{12} CONCENTRATION FOR CATTLE

A. MacPherson

The West of Scotland Agricultural College,
Auchincruive, Ayr, KA6 5HW, U.K.

Earlier work (1) with growing calves had revealed the severe weight loss that can occur in advanced Co deficiency. Unfortunately, at that time it had proved impossible to monitor the changes in plasma vitamin B_{12} concentration during the development of the deficiency condition. Very low concentrations around 50 ng/l were recorded in the advanced stages while following supplementation these were raised to 400-600 ng/l. Since that time the Veterinary Investigation Laboratory at Auchincruive has been determining vitamin B_{12} concentrations of bovine sera from all over Scotland as a diagnostic service for veterinary practitioners. One result of this has been the difficulty encountered in establishing criteria of normality and in generally interpreting the diagnostic significance of the results. Consequently it was decided to set up an experiment to try and determine the critical plasma vitamin B_{12} concentration below which an animal's productive performance was impaired.

Twelve 5-week old castrated male Friesian calves were put on to a Co-deficient diet of timothy hay (0.05 mg Co/kg DM), flaked maize (0.01-0.03 mg Co/kg DM) and fishmeal (0.10 mg Co/kg DM). Four of the calves were given one cobalt bullet (60% w/w cobalt oxide) in order to maintain adequate Co status; the remaining 8 calves were left unsupplemented. The calves were weighed and blood sampled weekly. Vitamin B_{12} concentrations were determined by microbiological assay using *Lactobacillus leichmanii*.

The changes in the mean plasma vitamin B_{12} concentrations of the supplemented and control calves are detailed in Table 1.

Table 1. Changes in Mean Plasma Vitamin B_{12} Concentrations (ng/l)

Date	7/1/80	12/2	17/3	15/4	19/5	16/6	23/7	11/8	8/9
Co-Supplemented	230	201	199	195	156	196	240	249	203
Control	259	145	141	133	99	63	91	86	38
SED	51.9	22.4	38.3	14.7	25.5	17.7	13.7	9.6	10.0
Significance	N.S.	P <0.05	N.S.	<0.05	<0.05	<0.001	<0.001	<0.001	<0.001

Concentrations for both groups were normal (2) (>200 ng/l) at the start of the experiment but within 3 months consistently and often highly significantly higher values were obtained for the supplemented group than for the controls. Over the last 4 months of the experiment the mean plasma vitamin B_{12} concentration of the unsupplemented animals was always <100 ng/l. The fall in the concentration of the supplemented group to 156 ng/l on 19th May indicated that the cobalt bullet treatment was not being effective in maintaining adequate Co status. Consequently further supplementation of this group was given by fortnightly injection of 1000 µg vitamin B_{12}. These proved to be effective in restoring and maintaining normal concentrations thereafter.

Table 2 details the changes in the mean rate of daily liveweight gain (DLWG) over the course of the experiment. The Co-supplemented animals grew somewhat more slowly than the controls over the first 8 weeks of the experiment but thereafter both groups grew at a similar rate. Over the whole experiment there was no significant difference in growth rate between the 2 groups. The experiment had to be terminated at this point due to the facilities being required for other work and the animals were turned out onto autumn grass in very wet cold and windy conditions. This caused a rapid loss of liveweight in both groups being 16.8 kg and 27.0 kg for the Co-supplemented and control groups respectively over the first week at grass.

Table 2. Changes in Mean Rate of DLWG (kg)

Day	0-36	0-70	0-100	0-135	0-160	0-200	0-220	0-245
Co-Supplemented	0.42	0.54	0.58	0.69	0.70	0.71	0.72	0.76
Control	0.62	0.67	0.65	0.73	0.76	0.75	0.76	0.79

A number of other parameters were monitored over the course of the experiment including haemoglobin, plasma glucose, plasma alkaline phosphatase, aspartate transferase and both pyruvate and pyruvate kinase concentrations. Previous work (1,3) had shown abnormal values for all these parameters in Co-deficiency but no abnormalities nor significant differences between supplemented and control animals were seen in the present experiment.

Despite the long time over which the unsupplemented animals were in a nominally, severely vitamin B_{12}-deficient state (i.e. <100 ng/l) there was no indication of any impairment in productive performance or indeed in any other parameter measured. Individual animals were as long as 20 weeks at concentrations of less than 100 ng/l plasma vitamin B_{12} with concentrations down to less than 30 ng/l in some cases without any detectable effects. It is interesting to compare these values with those reported earlier at this meeting (4) where growth responses to Co supplementation were obtained with cattle at grass when the plasma vitamin B_{12} concentration of the untreated animals was around 130 ng/l.

The explanation of this anomaly does not lie in the breed of cattle involved as these were all Friesians in this experiment and predominantly so in the other. It may lie in the different type of diet consumed as the one was essentially a dry diet fed indoors with a highly concentrated portion (flaked maize and fishmeal) although still about 40% roughage content (timothy hay) while the other was a completely forage diet consumed outside and often in quite wet conditions. Again the absence of stress factors in this experiment where the animals were housed and fed a ration of very adequate nutritional content and free from the effects of any parasitic assault or other environmental stresses may have enabled them to continue to grow normally at sub-optimal cobalt status. The severe loss of condition and liveweight particularly in the unsupplemented animals following turnout onto wet pasture in severe climatic conditions suggests that this may indeed be the case. This possibility is being pursued in an experiment currently in progress where it is intended to monitor the effect of a measured amount of stress superimposed on animals of low vitamin B_{12} status.

Another possible explanation of the lack of any detrimental effects of maintaining animals at apparently low vitamin B_{12} concentrations for long periods is that our method of assaying vitamin B_{12} status may not be sufficiently specific for the active analogue particularly in the presence of greater than normal amounts of inactive analogues whose production is reputedly favoured by high concentrate diets (5,6). This again, however, is unlikely to be the cause as all samples were analysed not only microbiologically but also by a radioassay method. Results of both methods correlated well and although the radioassay method gave higher results, which incidentally confirms the results of Millar & Penrose (7), the difference was only of the order of 10% and therefore did not markedly alter the absolute vitamin B_{12} concentration and thus the cobalt status of the animals. Thus, although there was some indication of analogue production this appears to have been fairly limited. A full report of this comparative assay is due to appear shortly (8).

CONCLUSIONS

1. Very low vitamin B_{12} status, as assessed by current standards, was maintained for a long period (16-20 weeks) in growing cattle without otherwise affecting their biochemical status or physical performance.

2. Current mimimum standards of normality for vitamin B_{12} status in cattle may have to be revised and/or more specific assay methods developed.

REFERENCES

1. MacPherson, A., F.E. Moon and R.C. Voss, 1973. Brit. vet. J. 129:414-426.

2. Underwood, E.J. 1977. In "Trace Elements in Human and Animal Nutrition". Academic Press, New York. 4th ed. pp 132-158.

3. MacPherson, A., F.E. Moon and R.C. Voss, 1976. Brit. vet. J. 132:294-308.

4. MacPherson, A. 1981. In "Proceedings of the fourth international symposium on trace element metabolism in man and animals" (eds. J.McC. Howell, J.M. Gawthorne and C.L. White), (Springer-verlag, Berlin, Heidelberg, New York).

5. Sutton, A.L. and J.M. Elliot, 1972. J. Nutr. 102:1341-1346.

6. Elliot, J.M. 1980. In "Digestive Physiology and Metabolism in Ruminants" (ed. Y. Rauckebusch and P. Thivend), (M.T.P. Press Ltd., Lancaster, England). pp 485-503.

7. Millar, K.R. and M.E. Penrose, 1980. N.Z. vet. J. 28:97-99.

8. Taylor, C.N. and C.L. Wright, 1981. In Press.

DISCUSSION

MILLS (Scotland). One of the things that Miller and Penrose have shown, is the absence of a clear correlation between serum B_{12} and liver B_{12} as assayed I think originally by *L. leichmanii* and by radio-dilution. The data that they produced from animals on pasture is in complete contrast with the Marston results where animals were fed a dry diet. I am not sure whether we are dealing in this situation with an analogue problem or something which is influencing the equilibrium distribution between serum and liver with respect to B_{12}. The differences between diets possibly do not end here because of the results of some of the studies on FIGLU and on methylmalonate. I wonder if perhaps we are expecting too much of a close correlation between methylmalonate and FIGLU and plasma B_{12} or serum B_{12}. Does the problem lie in the relationship between serum and liver?

MacPHERSON (Scotland). This is an area that we haven't worked in. I think workers at the Hill Farm Research Institute have worked in FIGLU and methylmalonate. We are interested in trying to relate B_{12} in the plasma or the serum with that in the liver, but so far the amount of data that we have is not convincing and we could do with a lot more of it.

GARTON (Scotland). I wonder if you could reveal the secret of your radio-dilution assay for serum B_{12} in cattle, because other people, ourselves included, have found that when we try the radio-dilution procedure on cattle serum our results are very unreliable.

MacPHERSON (Scotland). That work was going to be reported in the press shortly by Taylor and White.

IODINE STATUS AND THYROID ACTIVITY OF WHITE-TAILED DEER

B.E. Watkins, D.E. Ullrey, S.M. Schmitt
and R.F. Nachreiner

Michigan State University, EAST LANSING, MI
UNITED STATES OF AMERICA

INTRODUCTION

Thyroxine (T4) and triiodothyronine (T3) affect a variety of physiological processes including lipid, carbohydrate and nitrogen metabolism, calorigenesis, growth and development, neurological functions and reproductive performance. Since production of these hormones is dependent upon an adequate supply of iodine, and the Great Lakes region is an I-deficient area with a history of endemic goiter in humans and domestic animals, this study was conducted to explore the influence of dietary I level, moderate feed restriction and fasting on thyroid activity and related parameters in captive and wild Michigan deer. In addition, I concentrations of several important deer foods were determined.

METHODS

Eighteen weanling female fawns were fed diets (Table 1) containing approximately 0.26, 0.46 or 0.96 ppm I for over 2 years. They were bred after one year, and their offspring were also included in the study. Blood samples and weights were taken at approximately 6 to 8 week intervals, and serum concentrations of total T4, free T4 (FT4), total T3 and reverse T3 determined. Milk from lactating does and thyroids removed at necropsy were analyzed for I. Like determinations were made on trapped, wild deer and on captive deer subjected to 50% feed restriction for 4 months or to fasting for 16 to 20 days. Deer foods were collected in February, May, August and November for I analysis by spectrophotometric quantification of the iodide-catalyzed reduction of Ce IV by As III. Data were analyzed, where appropriate, by analysis of variance, chi-square or Tukey's test.

RESULTS AND DISCUSSION

Dietary I levels of 0.26, 0.46 or 0.96 ppm produced no significant differences in growth, serum total T4 or FT4, reproductive performance, serum T4 and T3 levels of offspring, milk I concentration, or weight and I concentration of the thyroid gland. These data indicated that 0.26 ppm I in a dry diet consumed ad libitum was sufficient for normal reproduction and lactation in white-tailed does. During the 2 year study, serum T4 and FT4 followed a consistent circannual pattern with high levels during early winter and spring, and low levels during late winter, summer and fall. These changes may have been related to cyclic changes in body weight, feed consumption and ambient temperature. Serum T3 did not follow a consistent seasonal pattern. Serum T3 and T4 were significantly higher in fawns than in adults, and in nonlactating versus lactating does. Approximately 50% feed restriction for 4 months reduced weight gain and thyroid weight but did not affect serum T3, T4, FT4 or thyroid I concentration in weaned fawns. Fasting of fawns for 16 to 20 days markedly reduced serum T3, T4 and, to a lesser extent, FT4. Fractional turnover rate, distribution volume and metabolic clearance rate of ^{131}I-T4 were unaffected, but T4 secretion rate was greatly reduced. Starvation-induced hypothroidism may play a critical role in energy and protein conservation and enhanced survivability during winter periods of food deprivation. Wild deer during late winter were found to have low T3 and T4 levels, similar to those of fasted fawns. Michigan is an I-deficient area and twigs of browse species were typically low in I (13-338 ppb); leaves (30-481 ppb), terrestrial herbaceous species (17-836 ppb) and aquatic species (85-3,100 ppb) tended to be higher. Deer could avoid incipient I deficiency by appropriate selection of the higher I foods.

TABLE 1. Composition of the basal diet used for iodine studies with white-tailed deer.

Ingredient	Percent
Corn cob product[a]	35
Soybean meal (44% crude protein)	24.5
Shelled corn	18 (23)[b]
Wheat middlings	10
Alfalfa meal (17% crude protein)	5
Cane molasses[c]	5
Soybean oil	1
Limestone (38% calcium)	0.5
Trace mineral salt[d]	0.5
Vitamin A, D, E and selenium premix[e]	0.32
Mold check[c]	0.2
Calcium propionate	(0.2)[b]

[a]Bracts and pith (soft parenchyma without vascular bundles).

[b]Values in parentheses are for the reformulated diet used after May 1978.

[c]Excluded from the reformulated diet.

[d]Contained 94.6% noniodized NaCl, 1% ZnO, 0.6% MnO, 2.5% $FeSO_4 \cdot 7H_2O$, 0.2% $CuSO_4 \cdot 5H_2O$, 0.044% $CoCO_3$, and 1% corn oil.

[e]Supplied 3300 IU vitamin A, 220 IU vitamin D, 88 IU vitamin E, and 0.2 mg Se (as sodium selenite) per kg of diet.

COPPER STATUS OF CATTLE IN THE SOUTH EAST OF SOUTH AUSTRALIA

J.D. McFarlane* and G.J. Judson+

* Department of Agriculture, Box 618, Naracoorte, South Australia
\+ Institute of Medical and Veterinary Science, Frome Road, Adelaide, SOUTH AUSTRALIA

INTRODUCTION:

The successful development of pastures in the South East of South Australia relied on the use of trace elements copper and zinc (Tiver 1955; Tiver and Webber 1957). Riceman (1961) concludes that only superphosphate is needed to maintain copper and zinc levels for pasture production after they have been spread at development. Reuter (1975) reports that Riceman based this conclusion on observations that: (1) pasture and livestock production were satisfactory for at least seven years after the initial application; (2) applied copper is not leached to any great extent (Jones and Belling 1967); and (3) the copper removed in farm produce approximately equalled their addition as impurities in superphosphate.

Investigations by Hannam (1979) have shown that, for mineral soils, an application of 2 kg/ha of copper at pasture development has alleviated copper deficiency in both pastures and sheep and could be expected to be effective for a large number of years. However, copper deficiency has occurred in some cattle herds grazing similarly fertilized pastures (S.A. Department of Agriculture, Animal Health Branch Records). For this reason the nature of copper deficiency in cattle was investigated.

METHODS:

The systematic blood sampling of cattle herds under the Brucellosis and Tuberculosis Eradication Scheme offered the opportunity to survey the copper status of herds in the region. A random sample of serum (pooled from 5 - 10 animals) was taken from every herd tested and analysed for copper. The copper status of the herds were then related geographically to the Hundred and Section from which they came.

Having identified areas: - (1) where copper deficiency was indicated, and (2) where copper serum levels were adequate, pasture samples were collected for comparisons from 89 paddocks. Ten locations were sampled per paddock. Subterranean clover (*Trifolium subterranean)* and strawberry clover (*Trifolium fragiferum*) were selected as indicator species for sampling. Copper concentration was determined using the technique of analysis on the youngest open trifoliate. Reuter (1980) showed that the critical concentration for copper in the youngest open trifoliate to be 3 µg/g. Copper and molybdenum concentrations were determined from the clover whole tops, so that an estimate could be made of pasture concentration available to the grazing animal.

RESULTS:

Initially only properties where reliable fertilizer records had been kept were considered since a relationship between the copper concentration in clover and copper fertilizer applied was expected. No such relationship emerged. On all mineral soils where copper fertilizer had been applied the copper concentration was adequate to sustain optimum pasture production.

In the period April 1979 to October 1980, approximately 11 per cent of the 3000 herds sampled showed deficient levels of copper in sera (less than 8 µ mol/L). The variable that explained the existance of copper deficiency in cattle was the corresponding relatively high levels of molybdenum in the clover samples analysed. An interesting feature being that this occurred even where soil pH had declined under pasture to be around 5.0 (1:5 in water), i.e. in a strongly acid environment. A high proportion of cattle grazing alkaline peats showed indications of deficiency despite acknowledgement of the problem and wide use of copper supplements. Table 1 shows typical results from the analysis of pasture samples.

TABLE I Typical results of pasture concentrations from areas identified in the survey of serum copper levels

Description of Area	Cu conc. in clover	Mo conc. in clover
1. Areas of deficiency		
On a mineral soil, sand over clay Cu fertilizer history 2 kg Cu/ha	8.5	4.5
On a mineral soil, loam, pH 4.9 Cu fertilizer history 1 kg Cu/ha	8.10	3.6
On an alkaline peat, 0.25 kg Cu/ha an annual requirement for pasture production.	7.4	8.7
2. Area of sufficiency		
On mineral soils Cu fertilizer history 2 kg Cu/ha	6.4	0.4
3. Farm developed in two halves showed one herd with low copper serum levels (a), and second herd with adequate copper serum levels (b).		
(a) Mineral soil, sand over clay and deeper sand. Cu fertilizer history 1.25 kg Cu/ha	5.2	2.5
(b) Mineral soil, sand over clay and deeper sand, Cu fertilizer 4 kg Cu/ha	8.1	2.1

CONCLUSIONS:

On all mineral soils where copper fertilizer had been applied, the pasture copper concentration was adequate for optimum pasture growth.

Assuming the sulphur levels in pasture average 0.2% over the region irrespective of seasonal variation and soil type (D.C. Lewis, unpublished data), the low serum copper levels were, in every case, explained using the following criteria to calculate the pasture copper available to cattle.

A copper level of 5 µg/g in clover is needed to maintain copper sufficiency in grazing animals (Underwood 1966).

Increments of 4 µg/g molybdenum in the diet reduce the availability of dietary copper in ruminants by 50% (Suttle 1975).

REFERENCES:

1. Hannam, R.J., 1979. In "Soil Conservation Branch Report S11/79" Department of Agriculture, South Australia

2. Hannam, R.J. and P.L. Marrett, 1976. In "Soil Conservation Branch Report S3/76" Department of Agriculture, South Australia

3. Jones, G.B. and G.B. Belling, 1967, Aust. J. Agric. Res., 18:733-740

4. Reuter, D.J. 1975. In "Trace Elements in Soil-Plant-Animal Systems". (ed. D.J.D. Nicholas and A.R. Egan) (Academic, N.Y.) pp 291-324

5. Reuter, D.J., 1980. In "Distribution of copper and zinc in Subterranean Clover in Relation to Deficiency Diagnosis". (PhD Thesis). (Murdoch University, West Australia.

6. Riceman, D.S. 1961, J. Technol. Sheep Breed. 8:111

7. Suttle, N.F. L975. In "Trace Elements in Soil-Plant-Animal Systems" (ed. D.J.D. Nicholas and A.R. Egan) (Academic, N.Y.) pp 271-289

8. Tiver, N.S. 1955, J. Agric. S.A. 59:100

9. Tiver, N.S. and G.D. Webber, 1957, J. Agric. S.A. 60:442

10. Underwood, E.J. 1966. In "The Mineral Nutrition of Livestock" (Central, G.B.)

TRACE ELEMENT BALANCE STUDIES AND HOMEOSTASIS

METHOD OF TRACE METAL BALANCE IN CHILDREN

M. H. N. GOLDEN, B. E. GOLDEN & A. A. JACKSON

Tropical Metabolism Research Unit
University of West Indies, Kingston 7, Jamaica

Quantitative collection of urine and stool from uncooperative active children presents many technical problems. Although various methods have been devised to overcome these problems none are satisfactory. The constraints of technique are exaggerated with trace element balances where, unlike nitrogen balance, contamination of the specimens is as serious a problem as losses.

Harnessing the child into a chair, with a condom or rigid device (1) strapped over the penis, and an open bowl underneath for faecal collection, leads to local skin irritation and breakdown, ischial sores, losses from and contamination of the specimens, severe emotional stress and an enforced change in activity. The restraint itself leads to catabolism, with a negative calcium (2) and nitrogen balance and decreased protein turnover rate (3). These metabolic changes are thus an integral part of many balance studies and render the results of technically perfect balances open to serious criticism if they are to be extrapolated to normal children. The use of specially padded beds (4) and wide funnels for urine collection (5) prevents much of the skin irritation but often results in even tighter straight-jacketting, sensory deprivation and emotional stress. To overcome these problems Hepner & Lubchenco (6) developed a hammock technique for neonates, later adapted for young infants (7), which allows considerable freedom of movement. The patient is suspended on a nylon mesh net with a large collecting pan underneath: the stool sticks on or in the mesh. The large surface areas make the prevention of contamination and losses difficult. Urine and stool cannot be confidently separated, indeed the same technique has been advocated for diarrhoeal stool collection (8). Active children are quite capable of 'painting' stool almost everywhere.

The alternative evolution of the method was to replace the strap-on devices with sealed stick-on bags for urine collection (9,10) with intermittent aspiration or gravity drainage. This allows much more freedom for the child and confident separation of urine and stool. With these techniques stools may be collected on polythene nappy liners (9) brushed nylon napkins, or by gravity in the neonate (10). In the inactive child with solid stools these methods may be satisfactory, but in the active child or where the stools are semi-formed or liquid, the specimen often squeezes into view through the nappy's leg hole.

Alexander and Delves (11) adopted a different approach. They washed disposable nappies with EDTA and kept everything in with plastic tie on pants. This allows complete freedom of movement, but for trace element balances the nappies still give considerable contamination after EDTA washing. There is no confident separation of stool and urine, and the technique is difficult with semi-liquid stool.

To overcome these problems we have adapted the stick-on urine bag technique described by Liu & Anderson (12) and the stool-bag technique of Shenai (13) for trace metal balances in young active children.

TECHNIQUE

Urine collection.

None of the commercially available paediatric urine bags tested have trace metal free adhesive. Bags are quite easily made from sheet polythene and a domestic polythene bag sealer. A wedge of clear silicone 'bath-tub' caulking is molded around two size 10 nasogastric tubes (N-G tubes), allowed to cure and wrapped with PTFE tape. The bag is sealed along edges a,b, c and d (Fig. 1) and reinforced at point, x, with a second layer of polythene. The N-G tubes are placed against 'd' and seam 'e' made so as to create a cone of polythene. The N-G tubes are now pulled so that the wedge jams firmly in the cone to form an air tight seal. The ends of the two tubes are cut to size and the final seam 'f' is made. The bag is now inflated to check for leaks.

A hole is cut in the bag so that it will just accommodate the shaft of the child's penis. His foreskin is pushed through the hole and held through the outer layer of polythene, so that when the bag is inflated the inner layer of the bag is forced to the root of his penis.

The top of the bag is then bent over and spray-on silicone colostomy cement (Dow Corning, Medical Adhesive B), which is trace metal free, is applied to the skin and allowed to become tacky. The bag is now reinflated and 'rolled' around with the palm of the hand, while an assistant holds the skin stretched. The bottom is now stuck in a similar way. The rugal folds of the scrotum must be filled with adhesive and the scrotal skin slightly stretched. The bag is attached to the anterior scrotum only, not to the perineum. Micropore tape may now be applied from the posterior scrotum to the underside of the bag protruding below the scrotum. If there are any skin folds or creases below the bag, the bag should be removed and a new bag applied : attempts to reattach bags almost always result in leakage. The N-G tubes are taped along the crease of the groin, around the buttock and down the back of the leg. If the tubes do not lie along the groin they will kink when the sitting child bends forward.

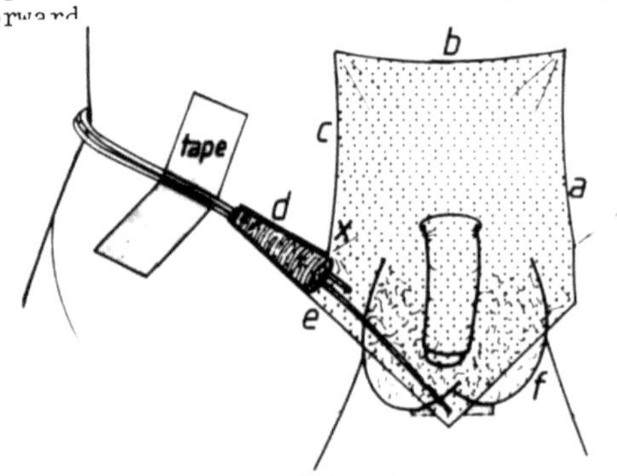

Fig. 1

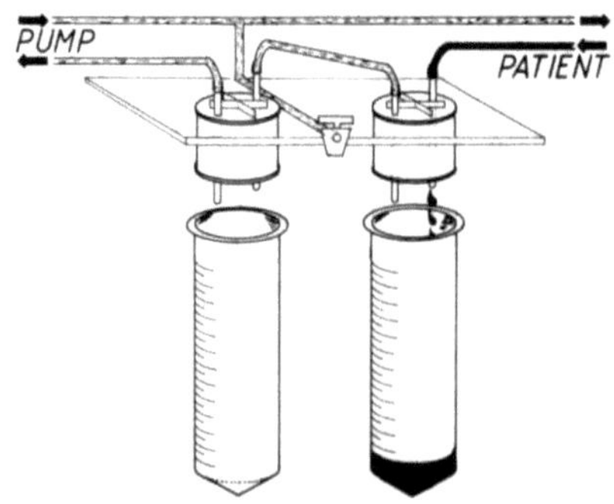

Fig. 2

Two long vinyl tubes are attached to the N-G tubes at the toes and strapped together to form an 'umbilical cord', the tubes are of sufficient length (2M) to allow the child to roam his cot. The other ends of the tubes are connected to an air pump which continuously circulates air around the closed system. Either a small diaphragm or piston fish-tank pump which generates 10-20 cms Hg pressure, is ideal.

A urine trap and overflow are incorporated into the outlet side to collect the aspirated urine. This trap can be modified to suit various applications. Thus urine can be taken into containers cooled on ice or dry-ice. We have modified a pair of 50 ml syringes by heat sealing the nozzles, setting the plunger handles in a perspex block, wrapping the plunger end in PTFE tape and perforating each end with two PTFE tubes (Fig. 2). To enable accurate timing, of at least the first and last specimen of a balance, we have incorporated two gold electrodes into the PTFE tube so that an alarm is tripped when urine is passed. The circuitry is similar to that of Dosa et al (14). The 50 ml syringes form a good seal and are quickly and easily removed, emptied and replaced. As the system is closed, the quantity of air circulating should be fixed. The input line requires to be opened when specimens are collected to prevent the bag blowing up. Following the balance, when the bag is removed from the child, complete collection can be ensured by washing acid solution from the bag, through the tubes and into the specimen collection trap.

Stool collection.

The stool bag is also simply made from sheet polythene (Fig. 3). A 1.5 x 2.5 elliptical hole is cut on a flat surface about 3 cms from the end. The pre-weighed bag is stuck over the anus with the silicone colostomy cement. The child is held in the knee-elbow position by an attendent. An ellipse of ashless filter paper, slightly larger than the hole in the bag is placed over the anus to shield it from the adhesive. The adhesive is then sprayed over the buttocks, natal cleft and perineun. One side of the bag is applied first and then the other, making sure that there is close attachment to the perineal body and natal cleft. Again there must be no skin creases or folds under the bag or it should be reapplied. A pinhole is now made in the corner of the bag away from the anus. This allows

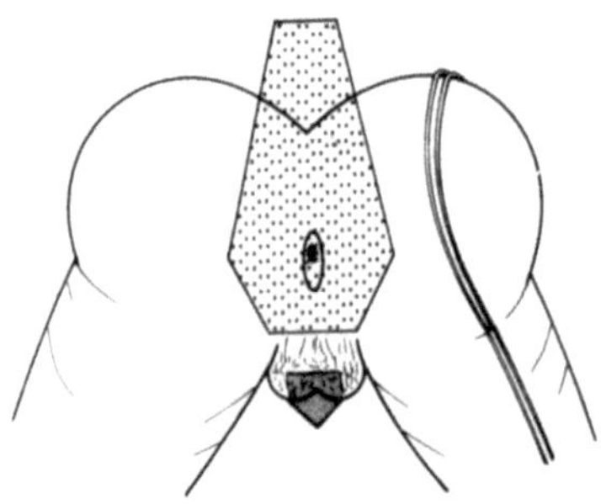

Fig. 3

flatus to escape. If the pinhole is not made the bag may act as an 'air' cushion when the child sits, either bursting the bag or loosening it by peeling when a side to side stress is applied. When stool is passed, the child is again placed in the knee-elbow position and the bag peeled from the sacral end; any stool adherent to the anus is scraped into the bag. A new bag can be applied immediately, without having to wash the child's bottom. The specimen bag is now dropped into a preweighed 'container' bag, sealed and the whole lot weighed before freezing. The balance is performed between two carmine markers, the stool being collected from (and including) the first marker, until (and excluding) the second marker. Often the marked stools require to be divided, this is best done on the fresh specimen; with semiliquid stools this division is not usually possible and is the most inaccurate part of the balance. With experience, these problems can sometimes be overcome by the judicious use of glycerine suppositories.

After the balance the stools are defrosted, the stool-bags removed from the container-bags and the polythene peeled from the solid stool into a tared glass mortar. The loss of stool, which is accurately measured, is between 0.5 and 2% of the total stool passed. The stools are then blended with a glass pestle and deionised water slowly added to make a roux in much the same way as a cook makes a sauce. When the stool is homogenous and liquid the final weight is taken, the mixture aliquoted and the aliquots freeze dried.

We have successfully conducted 60 balances, using this technique, on children with malnutrition and on normal active children; some had thin atrophic skin, skin sores, diarrhoea or severe infection, others were particularly boisterous.

DISCUSSION

Our technique allows complete urine and stool separation, full mobility of the child and because both urine and stool are passed into isolated, closed, acid-washed, plastic-only systems, chemical contamination is obviated and losses minimised. A sick child can have full nursing procedures carried out (except rectal temperature), and, by disconnecting and spigotting the N-G tubes for short periods after a urine is passed spontaneously, the child can be weighed, clinical room precedures carried out, and he may even go to the X-ray department, without danger of interupting the balance. This feature is important in studying critically ill children. A well child can be allowed time to run around the ward. The emotional and psychological trauma, as well as metabolic change due to the procedure itself are minimised.

The otherwise constant mechanical suction eliminates the possibility of a full bag of urine being forgotten and loosening the adhesive to cause leakage; it also empties the bag more completely than other methods, permits analysis of labile metabolites and prevents bacterial multiplication and metabolism. The circulating air serves to prevent penile maceration. The scrotal skin is protected from excreta and the techniques can be used therapeutically to protect an inflamed buttock, bed sore, scrotum or groin.

Quantitative recovery of liquid stool is possible and accurate measurements of faecal weight and hydration easily obtained.

ACKNOWLEDGEMENTS

We are grateful to the Wellcome Trust and MRC (UK) for support. MHNG & BEG are on secondment from the London School of Hygiene & Tropical Medicine.

REFERENCES

1. McCance, R.A., Rutishauser, I.H.E. & Boozer, C.N. 1970. Arch Dis. Child 45, 410-416.
2. Rose, G.A. 1966. Br. J. Surg. 53, 769-774.
3. Schonheyder, F., Heilskov, N.C.S. & Olsen, K. 1954. Scand. J. Clin. Lab Invest 6, 178
4. Foman, J.S. 1974. 'Infant Nutrition', Saunders, pp 548-556.
5. Newberry, E. & Van Wyk, J.J. 1955. Pediatrics, 16, 667-672.
6. Hepner, R. & Lubchenco, L.O. 1960. Pediatrics, 26, 828-831.
7. Winter, J.S.D., Baker, L. & Eberlein, W.R. 1967. Am. J. Dis. Child, 114, 150-151.
8. Gust, I.D., Seeger, V.J., Sandland, A.M. & Pringle, R.C. 1975. Lancet, 2, 933.
9. Baldwin, E.M., Clayton, B.E. et al, 1962. Arch. Dis. Child. 37, 488-490.
10. Tarlow, M.J. 1974. Arch. Dis. Child. 49, 490-492.
11. Alexander, F.W. & Delves, H.T. 1972. J. Clin. Path, 25, 927-932.
12. Liu, H-Y & Anderson, G.J. 1967. J. Pediat., 70, 276-279.
13. Shenai, J.P. 1978. J. Pediat, 93, 533-534.
14. Dosa, S., Bojtos, A. & Houston, I.B. 1971. Lancet, 2, 530.

DISCUSSION

LOMBECK (Germany). Can you collect several stools from a small baby so that the perinum skin is not shed after the first watery stool?

GOLDEN (Jamaica). Yes, we have collected up to 20 stools from the one child. We find that this technique assists healing of groin skin in children with such problems because the skin is not exposed to urine and faeces.

DIETARY ZINC AND THE OUTPUT AND COMPOSITION OF FAECES

B. E. GOLDEN, M. H. N. GOLDEN

TROPICAL METABOLISM RESEARCH UNIT
UNIVERSITY OF WEST INDIES, KINGSTON 7, JAMAICA

Children with chronic diarrhoea are often treated with soya formulae. In children recovering from protein energy malnutrition (PEM) in the Tropical Metabolism Research Unit, Jamaica, it was noted that the stool frequency was greater in those fed soya based formulae (SB) than in those fed cow's milk based formulae (CM)(1). The formulae had equal high energy and protein concentrations but, in contrast to CM, SB contained phytate (1402 μmol/kg) and less zinc (SB 52 μmol/kg, CM 73 μmol/kg)(2). The children fed SB also had very low plasma zinc concentrations during recovery(2). It was suggested that the zinc content and availability from SB were deficient relative to energy content and that the greater stool frequency in children fed SB was an early sign of zinc deficiency.

In order to define the effect of dietary zinc intake on faecal output, stool frequency, daily output and composition were measured in children during recovery from PEM on soya based formulae differing only in their zinc concentrations.

METHODS

Nine children, 6-31 months old, were studied using the balance methodology described by Golden et al (3). They were oedema free and severely wasted (< 70% wt for ht)(4) at the time of study 1 (0-10% recovery of initial deficit in wt for ht). Study 2 was conducted at 40-60% recovery and study 3 at 80-100% recovery. They were fed throughout, ad libitum, the high energy soya based formula, SB, containing 5.98 MJ and 32.7 gms protein/kg. 3 children were given unsupplemented SB (52 μmol Zn^{++}/kg, Group 1); 3 were given SB supplemented with zinc acetate to contain 132 μmol Zn^{++}/kg (Group 2). The remaining 3 were given SB supplemented to contain 194 μmol Zn^{++}/kg (Group 3).

During each study, the feed was infused continuously via a nasogastric tube, and the total faeces between carnine markers, administered 72 hours apart, was collected in zinc free polythene bags. The faeces and feeds were homogenized; samples were freeze dried to constant wt and then acid extracted and analyzed for zinc, according to the method of Hinners (5). Weekly plasma samples were also analyzed for zinc by atomic absorption spectrophotometry (Pye Unicam 2900).

The parents gave their consent to the studies which were approved by the Ethics Committee of the University Hospital of the West Indies. The results were analyzed using Student's unpaired t tests.

RESULTS

These are expressed as means ± S.E.M.

	Group 1	Group 2	Group 3
Admission height age (mths)	6.6 ± 0.5	7.2 ± 1.5	7.2 ± 2.7
Minimum weight for height %	66.0 ± 2.5	62.0 ± 6.7	63.0 ± 2.0
Rate of wt. gain (g/kg initial wt/d)	14.6 ± 0.7	14.8 ± 2.7	14.4 ± 1.7
Stool frequency (No./d)	3.4 ± 0.6	2.7 ± 0.4	2.6 ± 0.5
Mean plasma zinc conc. (μmol./l)	5.7 ± 0.3	9.9 ± 0.4	10.3 ± 0.9

Table 1 Clinical characteristics of the children

The rates of weight gain were similar in all groups. The stool frequency tended to decrease as the dietary zinc increased but the change was insignificant. The plasma zinc concentrations of the unsupplemented children were much lower ($p < 0.001$) than those of the supplemented children.

The faecal characteristics during each study are shown in Tables 2 and 3.

gm wet faeces/day

Study	Group 1	Group 2	Group 3
1	90 ± 21	101 ± 42	91 ± 28
2	103 ± 11	90 ± 11	82 ± 15
3	159 ± 24	69 ± 5	83 ± 15

Table 2 Total faeces output,
Study 3 Gp1 v Gps 2,3 p = .0035,

faeces % H_2O

Study	Group 1	Group 2	Group 3
1	84.0 ± 1.4	83.9 ± 3.1	82.7 ± 3.6
2	78.2 ± 1.2	78.3 ± 2.7	76.8 ± 3.9
3	81.9 ± 0.5	73.9 ± 1.3	74.5 ± 3.1

Table 3 Faeces % H_2O
Study 3 Gp1 v Gps 2,3 p = .0109
Gps 2,3 Study 1 v Study 3 p = .0064

In the unsupplemented group, the faeces output increased with recovery, whereas in the supplemented groups, it decreased. By late recovery, there were large differences between the unsupplemented and supplemented groups. The % water in the faeces changed little in the unsupplemented children but decreased steadily with recovery in the supplemented children. These features were obvious clinically.

The results of zinc analyses are shown graphically in Figures 1 and 2.

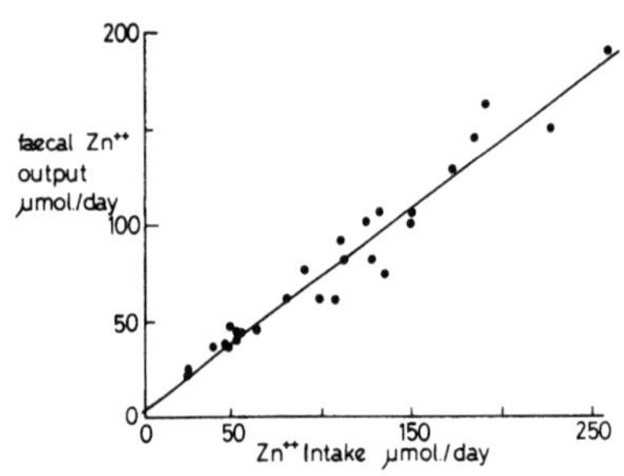

Figure 1 Effect of dietary zinc intake on faecal zinc output

Faecal zinc output = 4.20 + 0.70 zinc intake
r = 0.97, p < 0.001

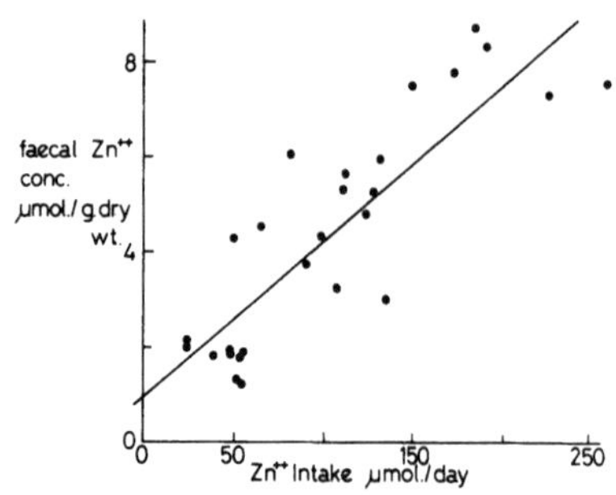

Figure 2 Effect of dietary zinc intake on zinc conc. of dry faeces

Faecal zinc conc. = 0.99 + 0.03 zinc intake
r = 0.85, p < 0.001

The zinc intake correlated positively with the faecal zinc output, giving an output at zero intake of 4.2 µmoles Zn^{++}/d. Zinc intake also correlated positively with the faecal zinc conc. expressed per gm dry weight.

DISCUSSION

We have demonstrated that dietary zinc intake is a factor influencing faecal hydration, zinc output and zinc concentration. The linear relationship between zinc intake and zinc concentration of dry faeces on this formula might prove useful in dietary zinc estimations in that concentrations are easier and more accurate measures than intakes.

Nearing recovery in weight for height, the three children on unsupplemented SB had biochemical and clinical evidence of mild zinc deficiency. They had very low plasma zinc concentrations and large daily outputs of watery stools. The latter finding appears to be a direct effect of zinc deficiency, similar to the diarrhoea characteristic of chronic severe zinc deficiency, as in acrodematitis enteropathica.

The children entered this study with severe wasting and stunting, features associated with low plasma zinc concentrations in PEM (6). Thymus hypoplasia also occurs in severe PEM and is not altered until the high energy formula is supplemented with zinc (7). Thus, these children were probably zinc deficient initially. The demand for zinc for the tissue synthesis inherent in recovery from PEM is vast. From balance studies on the supplemented children, about 0.85 μmoles zinc were retained per gm weight gained (8). The rates of weight gain were many times normal for the weights or ages of the children. That their plasma zinc concentrations fell to low levels is therefore not surprising; they reflected the gap between supply and demand, this being greater in the unsupplemented group.

Whether the watery stools are a result of gut mucosal dysfunction or infection remains in debate. Zinc deficiency is associated with reduced epithelial turnover, which in the intestine may give rise directly to mucosal dysfunction. It is also associated with depressed cell mediated immunity (6,8), and thus, less resistance to intestinal infection. That the unsupplemented children did not have other clinical signs of infection may only be a reflection of depressed inflammatory response.

It remains to point out that children with chronic diarrhoea, from whatever cause, usually fail to thrive and are often prescribed soya formulae. At least some already have low plasma zinc concentrations associated with their basic pathology. Those who achieve "catch up" gain in weight run a risk of developing overt zinc deficiency, particularly if the formula is supplemented with energy but not zinc.

ACKNOWLEDGEMENTS

B.E.G. and M.H.N.G. are seconded from the London School of Hygiene and Tropical Medicine. This work is supported by the M.R.C. (BEG) and the Wellcome Trust (MHNG).

REFERENCES

1. Golden, M.H.N., Golden, B.E. and Jackson, A.A., November 1979, Proc. Intern. Conference on Infant Nutrition & Diarrhoeal Diseases. Kuala Lumpur.

2. Golden, B.E. and Golden, M.H.N., May 1981, Amer. J. Clin. Nutr. 34 : 00-00.

3. Golden, M.H.N., Golden, B.E. and Jackson, A.A., May 1981, T.E.M.A.-4, Perth.

4. Waterlow, J.C., 1973, Lancet 2 : 87-89.

5. Hinners, T.A., 1975, Z. Anal. Chem., 277 : 377-378.

6. Golden, B.E. and Golden, M.H.N., 1979, Amer. J. Clin. Nutr., 32 : 2490-2494.

7. Golden, M.H.N., Jackson, A.A. and Golden, B.E., 1977, Lancet, 2 : 1057-1059.

8. Golden, B.E. and Golden, M.H.N., unpublished.

9. Golden, M.H.N., Golden, B.E., Harland, P.S.E.G. and Jackson, A.A., 1978, Lancet, 1 : 1226-1228.

DISCUSSION

HARLAND (U.S.A.). Could you please elaborate on the fibre content of your diet. We have done some work on fibre at Food and Drug and I appreciate how complex it is, but it would appear to be important to know about it in your experiments. Could you also tell us the moisture content of faeces in infants.

B. GOLDEN (Jamaica). We are using infant formulas and the level of fibre is not significant. In answer to your second question, the children are aged between one and two years, and they are rapidly growing. Their stools are about 80% water normally, but drop to 74% on recovery. Faeces are mainly bacteria, and faecal hydration is controlled mainly in the colon. We have done sodium, potassium and nitrogen balance studies on these children and have seen some association between faecal ion concentration and zinc concentration. These may be related to the bacterial flora of the colon, but we don't know.

OBERLEAS (U.S.A.). You mentioned malnourished children when you started the studies. How do you discern that these malnourished children are not zinc deficient in the beginning?

B. GOLDEN (Jamaica). We acknowledge that it is very difficult to measure zinc status. The malnourished children generally have lower plasma zinc concentrations than control children of the same age. Children with oedema, skin ulceration, severe wasting or stunting have low plasma zinc concentrations. We think probably all of these children are zinc deficient. What we are saying is that the demand for rapid weight gain is so high that their soya formula containing a reasonable amount of zinc per kilogram of formula is not sufficient when you look at it per calorie of formula or per gram of weight gain per day. So these children are not only zinc deficient to begin with but become progressively more zinc deficient until they actually show clinical signs of zinc deficiency.

OBERLEAS (U.S.A.). I think you have confirmed what we have found in rats. By increasing the zinc in the diet you improve utilisation of soya bean protein so what we may be seeing here is that you get an increased hydrolysis, increased digestion of protein as it passes throught the GI tract, therefore there is less of that particular protein for the bacteria to process when it gets down to the colon and therefore it has a different affect on osmotic equilibrium.

B. GOLDEN (Jamaica). I think that is at the bottom of the list for possible factors but we agree with your findings in that in these children, zinc retention is correlated with nitrogen retention. The added zinc improved protein utilization.

INTERPRETATION OF TRACE ELEMENT BALANCE IN ADULT TPN

Garry D. Phillips and Vyt P. Garnys

Department of Anaesthesia & Intensive Care, Flinders Medical Centre, and
Australian Mineral Development Laboratories, Adelaide, South Australia

ABSTRACT

Balance studies of zinc, copper, manganese, chromium, molybdenum and selenium were carried out in 8 critically ill patients receiving total parenteral nutrition (TPN). In 6 of these, nickel balance was also measured and the amount lost in urine and gastrointestinal fluid was estimated on samples taken from 24 hour fluid volumes on each of 7 consecutive days. Atomic absorption spectrometry was employed - flame for zinc and copper, filament in furnace for manganese, and graphite furnace for chromium, molybdenum and nickel. Selenium was estimated using hydride evolution.

Mean balances over 7 days indicated that an input of 1.8 mg zinc, 0.5 mg copper, 0.2 mg manganese, and 0.04 mg chromium resulted in acceptable balances. No molybdenum, selenium or nickel was intentionally administered and small negative balances of all 3 elements were observed.

The interpretation of balance studies in these patients is made difficult by factors such as impaired renal function, use of dialysis, and pooling of gastrointestinal secretions, even if problems of sample collection, quantitation, and analytical difficulties are overcome.

INTRODUCTION

There have been many reports of zinc deficiency developing in patients receiving parenteral nutrition. Deficiencies of copper, chromium, and manganese have been documented, and a recent paper has reported selenium deficiency. Molybdenum and nickel are known to be essential trace elements in animals, but not in man. Various recommendations for trace element requirements during parenteral nutrition have been made, based either on accepted oral requirements, or on balance studies. Because of the large variations in recommended doses, because locally made solutions may not contain trace elements as contaminants in similar quantities to those reported in other studies, and because most reported studies have not been carried out in critically ill patients, balance studies were carried out in eight critically ill patients receiving parenteral nutrition to determine their trace element requirements.

METHODS

Chemical Analyses

Flame atomisation atomic absorption spectrometry (AAS) with a Varian-Techtron AA-775 instrument was used to measure zinc and copper. Filament in furnace atomisation (FIFA) AAS employing a semi-automated UNSW FIFA prototype mounted in a microprocessor controlled Varian-Techtron AA-775 AAS was used to measure manganese, and also zinc and copper if their levels were below the limits of detection of flame AAS. Conventional graphite furnace atomisation AAS using a Varian-Techtron AA-575 with automated ASD-53 sampler into a CRA-90 graphite furnace was used to measure chromium and molybdenum. Selenium was determined by hydride evolution and atomisation by an AA-5 Varian-Techtron AAS instrument with a flame heated quartz absorption tube.

Pre-screened containers gave negligible contribution to trace element content of fluids. Polypropylene, polyethylene and polystyrene containers and syringes were used. Taking care to avoid contamination, all samples were digested with equal amounts of concentrated nitric acid heated to 80°C, and the oxidatively hydrolysed solutions diluted five-fold prior to measurement. The selenium solutions were further digested with perchloric acid. The standard addition technique was used for all determinations. Accuracy was established for all elements by inclusion of NBS (US National Bureau of Standards) and IMVS (Institute of Medical and Veterinary Science, South Australia) bovine liver standards.

SUMMARY OF METHODOLOGY

Element	Range ($\mu g\ L^{-1}$)	Method	Wave-Length (nm)	Conditions
Zinc	50-5000	Flame AAS	213.8	Air-acetylene
Copper	250-5000	Flame AAS	324.7	Air-acetylene
Manganese	0.5-500	FIFA-AAS	279.5	2300°/4sec
Zinc	0.05-500	FIFA-AAS	213.8	1800°/2sec
Copper	0.5-500	FIFA-AAS	324.7	2300°/4sec
Chromium	5-500	Furnace-AAS	357.9	2500°/4sec
Molybdenum	5-500	Furnace-AAS	313.3	2700°/4sec
Nickel	5-500	Furnace-AAS	232.0	2500°/4sec
Selenium	5-500	Hydride-AAS	196.0	$HClO_4/HCl/Na/BH_4$

Patients

The eight patients studied all required parenteral nutrition as part of their care. A trace element solution containing zinc, copper, manganese, iodine and chromium was added to a 3 litre bag of parenteral nutrition fluid, and a 50 ml sample taken for analysis. Samples of 24 hour urine and gastrointestinal fluid were also analysed, and daily balances were calculated for each of 7 days. Serum was collected at the beginning and end of each study. The Amino Acid solution used was Synthamin (Travenol). No fat was given. Vitamins of the B group, C, folic acid and K were given.

RESULTS

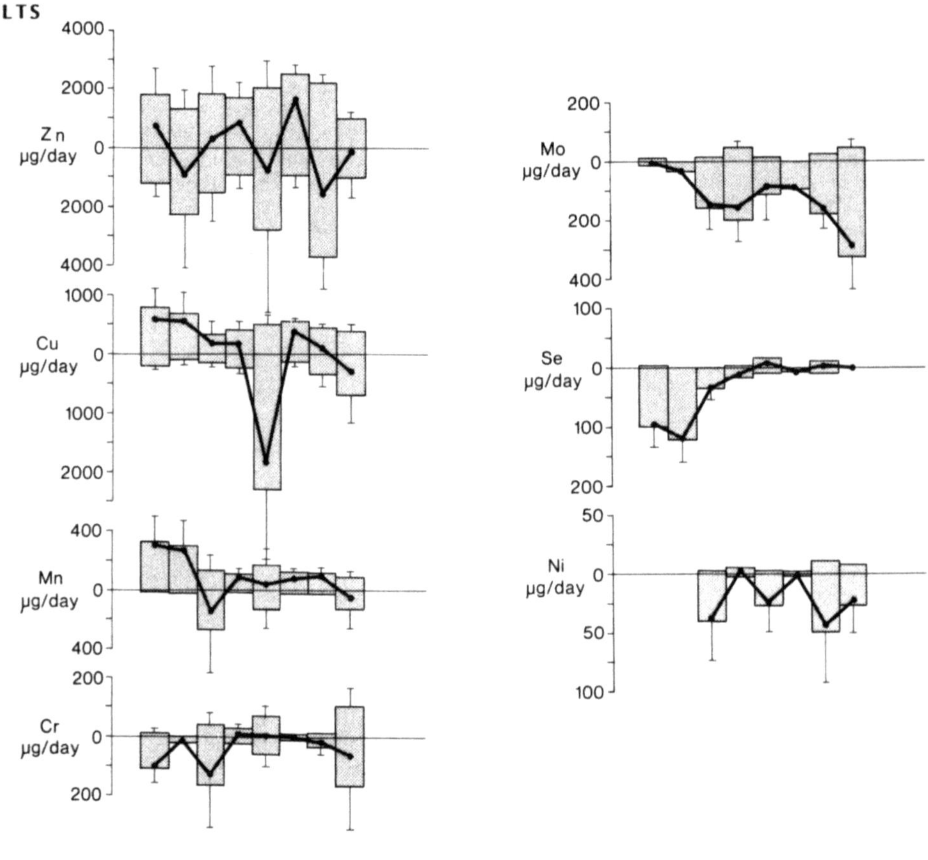

Mean of 7 day balances for each of 8 patients.
Input above zero, output below. Solid line joins balance points.
Standard deviation: I

% LOSS IN GASTROINTESTINAL FLUID

PT	MEAN DAILY VOLUME (ml)	Zn	Cu	Mn	Cr	Mo	Se	Ni
1	250	34	48	100	3	69	19	-
2	50	15	20	100	2	5	7	-
3	355	35	69	100	0	15	70	0
4	350	19	33	1	1	3	6	14
5	1606	25	22	50	15	11	0	0
6	1525	70	99	87	61	26	25	96
7	985	9	3	2	1	5	0	1
8	316	74	74	100	0	1	80	35

The percentage loss of each trace element in gastrointestinal fluid is shown. There was no consistent pattern of loss.

DISCUSSION

Zinc

An input of 1-2.5 mg per day resulted in positive balance in 4 patients, negative in 4, confirming the need for higher doses (American Medical Association [AMA] recommendation 2.5-4 mg per day).

Copper

300-800 µg/day resulted in positive balances in all but 2 patients, suggesting that the lower limit of the AMA recommended dosage is reasonable (500 µg/day).

Manganese

Positive balances were achieved in all but 2 patients with an input of 90-320 µg/day (AMA recommendation 150-800 µg/day).

Chromium

Input of 0-100 µg/day resulted in zero or negative balances in all patients. The AMA recommendation of 10-15 µg/day seems conservative.

Molybdenum, Selenium, Nickel

Small input from contaminants resulted in zero or negative balances in all patients.

Considerable day to day variation in requirements was seen. There was no consistent pattern of element loss in gastrointestinal fluid. Interpretation of trace element balance is difficult in the presence of abnormal renal and gastrointestinal function.

CONCLUSIONS

This study supports the need for zinc administration in all patients receiving parenteral nutrition. Administration of copper, manganese, chromium and selenium can be reserved for patients receiving long term parenteral nutrition. Renal failure may reduce the need for trace element administration. Excessive gastrointestinal fluid loss increases loss of trace elements.

REFERENCE

Expert Panel, AMA Department of Foods and Nutrition, Guidelines for essential trace element preparations for parenteral use. JAMA, **241**: 2051-2054, 1979.

DISCUSSION

VERSIECK (Belgium). What were the levels of manganese and chromium in the serum of your patients, and what were the normal levels?

PHILLIPS (Australia). We attempted to measure these elements in the serum of patients before and after the study, but the concentrations were at the limit of detection of the methods used (0.5 μg/l for manganese). We have not determined normal values.

VAN DEN HAMER (The Netherlands). Did you measure the loss of copper and zinc in sweat?

PHILLIPS (Australia). We did not determine losses by routes other than urine and gastrointestinal fluid.

LANE (USA). Were your patients losing weight? If so, wouldn't this explain the negative trace element balances?

PHILLIPS (Australia). Yes all patients were hypercatabolic. They were in intensive care, all had gastrointestinal problems, most had respiratory failure and six had septicaemia. The negative nitrogen balances could be attributed to their unhealthy condition. It is all very well to have daily allowances and balances for healthy volunteers but do these apply to sick patients - especially when no two sick patients are the same? For instance there are varying recommendations for zinc and it is difficult to know which to adopt. A large study on critically ill patients has been carried out in the USA and recommendations for zinc are within the range 2 - 20 mg/d for adults. Some of our patients appeared to be extremely catabolic yet 2.5 mg Zn/d gave reasonable balances.

DREOSTI (Australia). I would like to ask Dr Garnys about his FIFA method. Can biological fluids be read directly, or are there matrix effects.

GARNYS (Australia). There are matrix effects with biological material. We always use nitric acid digestion and by doing so the matrix effect is consistent for a given type of biological material. Calibration is done by standard addition, and one calibration suffices for one set of acid-digested material e.g. for urine. It is not essential to do standard additions for each sample.

THE INTAKE AND URINARY EXCRETION OF ELECTROLYTES AND TRACE ELEMENTS

M. Abdulla, Å. Nordén, B. Scherstén, S. Svensson, T. Thulin and P-A. Öckerman

Unit for Community Care Sciences at Dalby and Research
Department 2, University Hospital, S-221 85 Lund, Sweden

SUMMARY

Results of several recent investigations regarding the dietary intake of several essential inorganic chemical elements from prepared meals in industrialised countries indicate that specific sub-clinical deficiencies may occur in certain vulnerable groups like children, pregnant women, elderly and alcoholics. We have investigated the dietary intake and urinary excretion of electrolytes and a few other essential elements in a group of 60 women, 6 vegans and 6 lactovegetarians. The daily portions were collected employing the duplicate portion technique. The material collected during the day was homogenised, fat-extracted and lyophilised. The lyophilised powder was wet-ashed with concentrated acids and the residue left was used for the determination of the elements. Twenty-four hour urine samples were also collected during the days corresponding to the collection of the dietary samples. A total of 408 daily dietary portions and 228 urine samples were included in the study. The elements were analysed in most cases by atomic absorption spectrophotometry. Our results showed the following intake levels in the group of women: calcium 577 ± 256 mg, magnesium 257 ± 75 mg, zinc 8.1 ± 2.8 mg, copper 1.5 ± 0.58 mg and selenium 27 ± 6.1 µg. The 24-hour urinary excretion for the corresponding intake levels were: calcium 138 ± 58 mg, magnesium 63 ± 23 mg, zinc 0.31 ± 0.12 mg and copper 0.05 ± 0.015 mg. The selenium levels in the urine samples were below the detection limits. The vegetarian diets had a much higher concentration of most trace elements than the non-vegetarian diets. The present data concerning the women did not show any correlation between salt intake and blood pressure levels.

INTRODUCTION

In spite of the remarkable progress in trace element nutrition research during the last few decades, we still do not have a satisfactory answer to the fundamental question concerning the minimum requirement of several essential elements. Most studies in the past have been concerned with the biological functions of inorganic chemical elements. Rapid improvements in analytical technology and instrumentation during the last few years have helped to recognize the presence of several elements in living systems that were hitherto considered as unimportant. Although an average human body is but 3 % metals, recent estimates indicate that most tissues of healthy adults contain as many as 46 elements represented in the Periodic Table. With improvements in our knowledge of the chemical elements, the number of trace elements that may be essential in animal and human nutrition is likely to be increased in future. With this background, it is not surprising that most basic research in the field of trace element nutrition in the past had been concerned with their biological functions rather than the requirements.

In affluent countries, it is generally believed that a pure nutritional deficiency of trace elements hardly ever occur in man since many of them are required only in small concentrations. On the other hand, the common assumption that an average all-round western-diet will furnish an adequate intake of all trace elements is being debated at present. Recent development of food technology has enabled the growing food industry to offer the general public of privileged countries an enormously great choice of food products meeting all gustatory preferences and demands. Snack foods, convenient foods and low-calory diets have replaced much of the conventional food prepared for the common house-hold. It is being recognized today that the increased consumption of refined food materials for prolonged periods may induce marginal deficiencies of several trace elements. It has also been found that sub-clinical deficiencies of many trace elements do occur in parts of the population of industrialised countries (1).

In order to assure an optimal intake of essential elements, it is necessary to have reliable data showing minimal requirements. Such data can only be attained through controlled balance studies together with tissue concentrations. This is not often practicable. With the advent of nuclear era, most elements represented in the Periodic Table can be measured accurately in food materials and biological fluids. And yet, there is only limited data in the literature regarding the intake of essential elements from prepared meals.

Moreover, the information that is available is often unsatisfactory, since most results are based on conventional techniques like diet history and interview techniques in combination with food tables. Food tables cannot provide accurate data concerning the concentrations of several minor trace elements. Thus, there is a great need for reliable data regarding the intake and excretion of trace elements that are found to be essential in human nutrition. The present study has been done to investigate the dietary intake and urinary excretion of electrolytes and a few important trace elements in a group of women, vegans and vegetarians living in various parts of Sweden.

MATERIAL AND METHODS

The material consisted of 60 women, 6 vegans (3 men and 3 women) and 6 lactovegetarians (3 men and 3 women). The women were selected from a group who participated in an ophthalmological survey in the south of Sweden during 1969 - 1970. The subjects at the time of the present investigation were 52 ± 13.5 years old. The vegetarians were investigated to relate the influence of vegetarian diets on general health. None of the subjects participating in the present study were on medical treatment. The dietary habits were studied by employing the duplicate portion technique (2, 3). Participants were informed about the duplicate portion technique and precise instructions were given regarding the collection procedure by a trained dietician. Food collections were made within a period of 1 - 4 weeks. Each subject in the group of women collected six duplicate portions and three 24-hour urinary samples corresponding to the days of food collection. The vegans and lactovegetarians collected 24 daily dietary samples and the same number of 24-hour urine samples each. Written protocols regarding the type and approximate quantities of food consumed were also registered by the participants. Food collected during the day and night was by and by pooled in one container and later homogenised. The homogenised samples were later lyophilised after fat extraction with chloroform and methanol (2). The fat-free lyophilised powder was used for the determination of the elements. The electrolytes were analysed by flame photometry and the other elements by atomic absorption spectrophotometry after suitable pre-treatment of the samples.

RESULTS AND DISCUSSION

A summary of the major findings is shown in figure 1 and table I. As can be observed from figure 1, the dietary intake of sodium and potassium in the 60 women was lower than the corresponding values for excretion. In the vegetarians, the intake and excretion were almost the same. The intakes of calcium, magnesium, iron, zinc, copper and selenium in the women studied were almost the same as that was found in non-vegetarian diets of other groups in Sweden. The vegetarian diets on the other hand had higher contents of most trace elements analysed as compared to the non-vegetarian diets. The mean values of urinary excretion of calcium, magnesium, zinc and copper were much lower than the intake levels as expected.

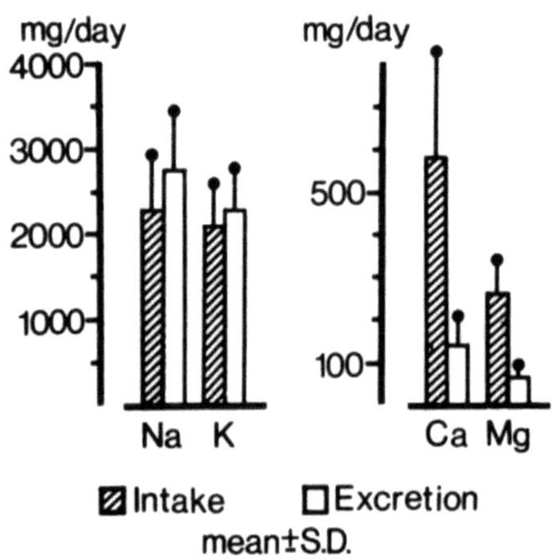

Fig. 1. Dietary intake and urinary excretion of electrolytes in women

Table I. Intake and excretion of essential elements

	Intake mg/day ± S.D.			Excretion mg/day ± S.D.		
Element	Vegans	Lactoveg.	Women	Vegans	Lactoveg.	Women
Zn	11.8 ± 3.5	10.8 ± 2.5	8.1 ± 2.8	0.32 ± 0.11	0.37 ± 0.24	0.31 ± 0.12
Cu	3.6 ± 0.8	2.1 ± 0.7	1.5 ± 0.6	0.04 ± 0.02	0.04 ± 0.02	0.05 ± 0.02
Fe	16.3 ± 6.5	14.7 ± 3.9	15.1 ± 6.1	-	-	-
Mg	541 ± 123	428 ± 138	257 ± 75	81 ± 25	67 ± 20	63 ± 23
Ca	625 ± 244	904 ± 269	577 ± 256	66 ± 34	99 ± 61	138 ± 58
Na	2208 ± 713	2218 ± 876	2300 ± 644	2220 ± 1047	2123 ± 738	2750 ± 700
K	4017 ± 1092	4132 ± 1318	2106 ± 508	3791 ± 1178	3135 ± 1112	2300 ± 525
Se*	< 3 - 58	42 - 82	17 - 39	-	-	-

*The selenium values are given as range (µg/day)

The urinary excretion of the electrolytes is generally influenced by the intake levels as well as the functional capacity of the kidneys, suprarenal glands and the gastrointestinal tract. All subjects who participated in the present study were healthy and we expected a balanced intake and excretion. Our findings that the intakes are lower than the excretion levels obviously suggest an underestimation of the food intake. Do the participants continue their normal food habits during the sampling period? This aspect had been investigated in detail. The 24-hour urinary excretion of electrolytes in the group of 60 women participating in the study were estimated twice earlier, without the simultaneous collection of food samples. The results showed no significant difference between the estimations. Thus, it appears that the food collection did not disturb the dietary habits to any significant extent.

Although the metabolism of calcium and magnesium is alike in many respects, the mechanism of absorption and urinary excretion differs to some extent. Calcium absorption is influenced by a number of constituents in the diet like fat, protein, lactose, phytates and vitamin D. In affluent countries, dairy products are good sources of calcium and it is estimated that 75 % of the dietary calcium originates from this source. Magnesium on the other hand is rich in foods of vegetable origin. Unlike calcium, magnesium absorption is not influenced by vitamin D. Intestinal absorption of magnesium appears to be an active process that in part is affected by an independent mechanism with regard to calcium uptake. The excretion of calcium in urine is dependent on the glomerular filtration rate as well as the concentration of diffusible calcium in plasma on the one hand and the activity of parathormone on the other hand. In this respect the amount of calcium absorbed is not of much significance. Magnesium excretion in urine on the other hand is related to the dietary load. When the dietary intake is low, the urinary excretion is rapidly lowered. The renal mechanism is very efficient in conserving magnesium. The excretion levels of calcium and magnesium in the women were around 25 % of the intake levels.

Very little data is available regarding the excretion of zinc, copper and selenium. For zinc and copper, we found excretion levels upto 3 - 4 % of the intake levels. Selenium contents in the diets were so low that we could not measure any selenium levels in urine when using the available techniques in our laboratory.

REFERENCES

1. Mertz, W. 1970. Fed. Proc. 29:4, pp 1482-1488

2. Borgström, B., Å. Nordén, B. Åkesson, and M. Jägerstad. 1975. Scand. J. Soc. Med. suppl. 10, pp 14-25

3. Borgström, B., Å. Nordén, B. Åkesson, M. Abdulla, and M. Jägerstad. 1979. Scand. J. Gastroent. suppl. 52, pp 28-41

DISCUSSION

LANE (USA). It has been reported in the literature that selenium can be lost during lyophilization. This may account for the low selenium levels you obtained.

ABDULLA (Sweden). We had 5% losses for some of the elements during lyophilization.

BINNERTS (The Netherlands). Could you comment on the low excretion of magnesium? Did you measure the amount of magnesium lost in sweat? I recently had a student who found that he lost a lot of magnesium in sweat, particularly during sporting activities.

ABDULLA (Sweden). I am aware that several trace elements can be lost in sweat. We did not measure sweat losses. It is unlikely that these women sweated very much.

CHANGES IN THE METABOLISM (RETENTION, ABSORPTION, EXCRETION) OF COPPER, ZINC, AND MANGANESE IN GRAVIDITY AND LACTATION

M. Kirchgeßner, F. J. Schwarz and Dora A. Roth-Maier

Institut für Ernährungsphysiologie
Technische Universität München
D-8050 FREISING-WEIHENSTEPHAN
FED. REP. GERMANY

For the trace elements copper, zinc, and manganese homeostatic regulatory mechanisms can be established in metabolism, as well as in absorption and excretion. Thereby the organism tries to maintain constant concentrations of these trace elements in the body, and to adjust total balance despite a deficient or high nutritive supply. Therefore experiments were conducted to examine, to what extent changes in the metabolic situations as pregnancy and lactation may influence homeostasis of retention, absorption, and excretion of trace elements. The experiments were carried out with farm animals (sows) and laboratory animals (rats).

In an experiment (Kirchgeßner et al. 1980, 1981a) with 16 sows of the German Landrace eight sows were mated and eight sows served as nongravid control animals. From the 21st day of gravidity four gravid and nongravid sows, each, were fed a complete feed, providing a low and high trace element supply, respectively. The balances were carried out by quantitative collection periods of the feces of all sows from 30th - 40th, 70th - 80th and from 105th day of gravidity until farrowing. At birth, number and weight of the piglets and weight of placenta and amniotic fluid were ascertained. By means of representative samples trace element content of these gestational products was analysed. The evaluation of trace element intake, apparent absorption, and retention from 21st day of gravidity until parturition is presented in Table 1. In nonpregnant control sows homeostatic regulation of absorption and excretion brought about that at a high dietary supply of zinc and manganese, respectively, retention of

TABLE 1. Total supply, apparent absorption, and retention of copper, zinc, and manganese of gravid and nongravid sows in the experimental period (21st day of gravidity until farrowing), per animal

Physiological stage	Nongravid		Gravid	
Trace element supply	low (n = 4)	high (n = 4)	low (n = 4)	high (n=4)
1. Copper				
Intake, mg	905	6,120	983	6,475
Apparent absorption, %	-3	27	1	34
Total retention, mg	-50	1,622	-10	2,197
Gestational products, mg	-	-	23	44
Anabolism, mg	-	-	-33	2,153
2. Zinc				
Intake, mg	2,903	18,931	3,172	20,129
Apparent absorption, %	-14	-1	-12	13
Total retention, mg	-429	-193	-449	2,605
Gestational products, mg	-	-	94	256
Anabolism, mg	-	-	-543	2,349
3. Manganese				
Intake, mg	2,465	16,178	2,649	17,163
Apparent absorption, %	-25	-3	-24	12
Total retention, mg	-639	-580	-674	2,069
Gestational products, mg	-	-	8	19
Anabolism, mg	-	-	-682	2,057

these elements did not increase. In contrast, homeostasis of copper was not maintained at this elevated supply. During gravidity at the low supply all elements remain in a negative balance. By contrast, at the high supply status a significant pregnancy anabolism occurred due to improved apparent absorption rates. The excretion in the gestational products is 1.7 % copper, 2.0 % zinc and 0.2 % manganese compared with the excretion in feces and urine. Thus, this excretion via the conception products is almost negligible in comparison to anabolism. When the apparent absorption rates between the three measuring points are compared, it is evident that apparent absorption of gravid sows at the high trace element supply during the first two thirds of gravidity is many times higher than shortly before parturition (Table 2).

TABLE 2. Apparent absorption of copper, zinc, and manganese of gravid sows at high dietary trace element supply during different periods of gravidity, in %

	Copper	Zinc	Manganese
30th - 40th day of gravidity	47	16	16
70th - 80th day of gravidity	37	16	12
105th day - farrowing	10	3	3

Changes in apparent absorption may be evoked by variation in true absorption, and/or endogenous excretion. For clarifying this, model studies were conducted using isolated everted sacs of small intestine of rats to test the intestinal transfer to the serosal solution and the uptake into the intestinal wall of zinc (Schwarz et al. 1981 a) and of manganese (Kirchgeßner et al. 1981 b). Both criteria agree and deliver a comparable pattern of the absorption. The relative level of absorption was continuously measured during gravidity (6th, 12th, 18th, and 20th day) and lactation (1st, 3rd, 5th, 8th, and 21st day). At four different times absorption of nongravid, nonlactating control animals was measured simultaneously. These results did not differ significantly and were therefore summarized to an average control mean. The rats were fed a commercial diet providing a high trace element supply. Fig. 1 and 2 show the time course of intestinal transfer of zinc and manganese during the experimental periods. Both elements indicate, especially at the beginning of the last third of gravidity, a highly significant increase in absorption in comparison to the control rats. Shortly before parturition absorption decreased again. During lactation Zn absorption fell continuously, so that only in the first days of lactation Zn absorption was higher, compared with control rats. Mn absorption increased in the first days of lactation, while it decreased after the 8th day. The level of absorption during lactation might be influenced significantly by the extent of pregnancy anabolism during gravidity.

Differences in apparent absorption, as it was demonstrated during gravidity of sows, may also be due to changes in endogenous excretion. Therefore endogenous excretion of copper was measured besides intestinal absorption. In Table 3 preliminary results are presented (Schwarz et al. 1981 b). Endogenous excretion was determined by

TABLE 3. Relative measured values of absorption and endogenous excretion of copper during gravidity and lactation (nongravid, nonlactating controls = 100 %)

	Control	18th day of gravidity	1st-6th day of lactation
Cu absorption in vitro (transfer + uptake)	100	123	86
Endogenous Cu excretion via intestine	100	93	67

the excretion of an i.v. ^{64}Cu-$CuSO_4$ dose. Cu absorption increases on the 18th day of gravidity and endogenous excretion is slightly reduced. In contrast, the lactating rats absorb slightly less copper, enteral endogenous excretion, however, is significantly reduced during lactation. Distribution studies of the i.v.-injected copper simultaneously indicate a high proportion of excreted copper via rat milk.

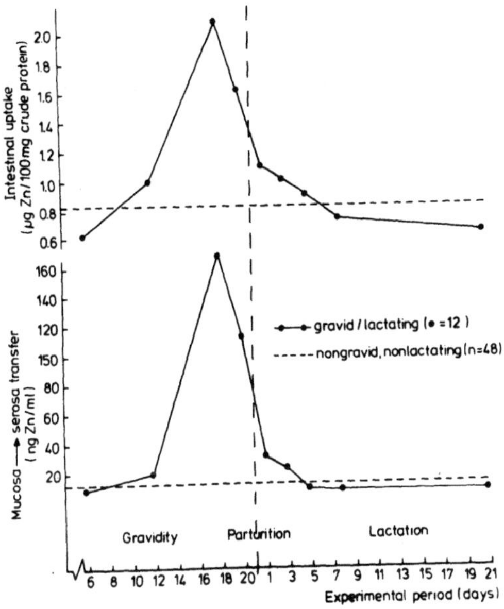

Fig 1 In vitro absorption of zinc during gravidity and lactation

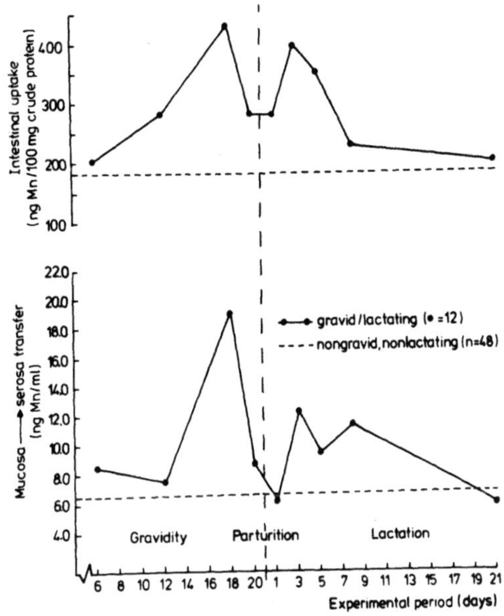

Fig 2 In vitro absorption of manganese during gravidity and lactation

SUMMARY

During gravidity and lactation considerable changes occur in absorption, excretion, and consequently retention of copper, zinc, and manganese in the maternal body compared with a nongravid, nonlactating organism. The extent of the changes is significantly influenced by the level of trace element supply. Homeostasis of trace elements in the organism, however, is abolished by pregnancy anabolism.

REFERENCES

1. Kirchgeßner, M., R. Spörl, and Dora A. Roth-Maier, 1980, Z. Tierphysiol., Tierernährg. u. Futtermittelkde. 44: 98 - 111

2. Kirchgeßner, M., Dora A. Roth-Maier, and R. Spörl, 1981 a, Arch. Tierernährg. 31: 21 - 34

3. Kirchgeßner, M., S. Y. Sherif, and F. J. Schwarz, 1981 b, Nutr. Metabol. (in press)

4. Schwarz, F. J., M. Kirchgeßner, and S. Y. Sherif, 1981 a, Res. Exp. Med. (in press)

5. Schwarz, F. J., M. Kirchgeßner, and Maria A. Spengler, 1981 b, Z. Tierphysiol., Tierernährg. u. Futtermittelkde. (in press)

DISCUSSION

HURLEY (USA). I just want to comment that the fall in the absorption at the end of pregnancy, before parturition, may be related to the fall in placental transfer. The rate of placental transfer also declines just before parturition and these two events may be functionally related.

SCHWARZ (Germany). The super-retention outside the reproductive tissues may have an even greater influence.

MILLS (Scotland). The results you showed with respect to zinc are similar to those for copper that Drs Davis and Williams found in Aberdeen using an in-vivo technique of radio copper absorption. Their results also indicate that copper retention is promoted in tissues other than the uterus and foetus during pregnancy. One point on which their results differ from yours is that they observed that the enhanced absorption continued into lactation and only began to decline during the first third of the lactation period. Your isolated 100p technique revealed an earlier decline in absorption. Would you care to comment on this difference?

SCHWARZ (Germany). I think that the level of supply of the trace element is the key point. When the supply is high during pregnancy, absorption will be high and anabolism will occur. During the subsequent lactation absorption will fall to a greater extent compared with situations where the supply is low during pregnancy. In the latter case there is less anabolism during pregnancy and thus a higher requirement during lactation.

MILLS (Scotland). It would seem to be time that the results of these sorts of experiments penetrated to those who are formulating human zinc requirements. There is at present no correction made for changes in absorption during pregnancy.

SCHWARZ (Germany). The requirements also depend on the extent of anabolism that is to be allowed, to provide for periods of temporarily low intakes.

GOLDEN M (Jamaica). Did you measure plasma concentrations at the same time as you did the absorption measurements?

SCHWARZ (Germany). Not in this study, but we have done so in previous work. Plasma zinc concentration decreases during pregnancy but the extent of the decrease depends on the level of supply. For example when 12 ppm Zn is supplied in the diet the decrease is more pronounced than with 100 ppm Zn.

THE ROLES OF THE LIVER AND GUT IN MN HOMEOSTASIS IN CATTLE

Eveline D. Hall, H. W. Symonds and W. M. Allen

Agricultural Research Council,
Institute for Research on Animal Diseases,
Compton, Newbury, Berkshire, RG16 ONN, U.K.

INTRODUCTION

There is evidence that over a wide range of dietary Mn concentrations a small, nearly constant proportion ($\simeq$ 0.01) of Mn is absorbed from the gut (Sansom et al., 1973). After absorption, at least 95% of the Mn is removed during its first passage through the liver and excreted in bile (Gibbons et al., 1976; Hall & Symonds, 1981a). These two processes maintain the concentration of Mn in systemic blood plasma between 5 and 10 ng/ml.

It is also known that, for periods of up to 6 hours, the liver of an adult cow can remove up to 4 mg of Mn per minute at the first pass (Hall & Symonds, 1981b) and in steers weighing 200-300 kg the liver can excrete Mn into bile at concentrations up to approximately 200 µg/ml, of 1-2 mg/min. This short term capacity of the liver to extract Mn from portal plasma and excrete it in bile is much greater than would be required to deal with the Mn absorbed from diets containing even 1000 mg Mn/kg DM - much more than normal diets (50-200 mg/kg DM). Thus an animal consuming 10 kg of a diet containing 1000 mg Mn/kg would absorb approximately 100 mg Mn/day or less than 0.1 mg/min. However, Mn toxicity can occur when diets containing 1000 mg Mn/kg and over are eaten (Cunningham et al., 1966). This toxicity is difficult to explain unless there is either a change in the proportion of dietary Mn absorbed when high concentrations of Mn are fed or the capacity of the liver to deal with large quantities of Mn is short-lived.

To study these aspects of homeostasis during toxicity, the excretion of Mn in bile was measured when concentrations in the duodenum were increased for long periods by means of continuous infusions. Until toxicity occurs and the liver's capacity to extract Mn from portal plasma has been exceeded, the amounts of Mn excreted in the bile should be approximately equal to the amount absorbed and there should be no increase in systemic plasma Mn concentration.

MATERIALS AND METHODS

Animals. Three Friesian x Aberdeen Angus steers weighing between 180-350 kg were used. Each had its duodenum surgically modified so that bile flow rate and trace mineral excretion could be measured with minimal disturbance to the animal. (Symonds et al., 1981).

Infusion procedure. The concentration of Mn in the duodenum of each steer was increased by 3 intraduodenal infusions of $MnCl_2$ at approximately 5, 30 and 45 mg Mn/min for 30 h, or until the animal showed an adverse response. The volume infused was always 1 ml/min and the possibility that the high toxicity of the solution could be responsible for some of the clinical effects was investigated by infusing at 1 ml/min for 6 h a solution of $CaCl_2$ containing 50 mg Ca/ml. One steer showed signs of toxicity during the infusion of Mn at 29 mg Mn/min and was later given an intramesenteric infusion of 311 µg Mn/min, equivalent to the absorption of approximately 1% of the intraduodenal infusion rate.

Bile collection. Bile flow (ml/min) was measured by collecting the bile secreted during 30 min periods into a bag attached by harness to the side of the animal and measuring the volume. A 5 ml sample was removed and the remainder returned to the duodenum. Flow was measured for consecutive 30 min periods during 1 h before starting each infusion and then during the first and last 6 h of that infusion except when signs of toxicity caused the infusion period to be shortened. The normal bile flow rates for the 3 steers ranged from 6 to 10 ml/min.

Four 20 ml samples of jugular blood were taken at 2 h intervals during the first and last 6 h periods and plasma obtained after centrifugation at 2000 rpm.

Analytical procedures. The Mn concentrations in plasma and bile were determined by atomic absorption spectrophotometry. Total cholate concentrations in bile were determined by the method of Irvin, Johnston & Kopala (1944).

RESULTS

1. Rates of infusion not producing signs of toxicity. Except in the third steer infusion rates of up to 30 mg Mn/min produced no signs of toxicity and no significant increases in plasma Mn concentrations. During the last 6 h of each infusion there was an apparent absorption rate of between 0.6 and 1.2% of the infused Mn, which was independent of the rate of infusion. At the infusion rates of 30 mg/min the biliary concentrations of Mn during the last 6 h were between 30 and 40 μg/min suggesting that at first there was an absorption rate of approximately 3-4%.

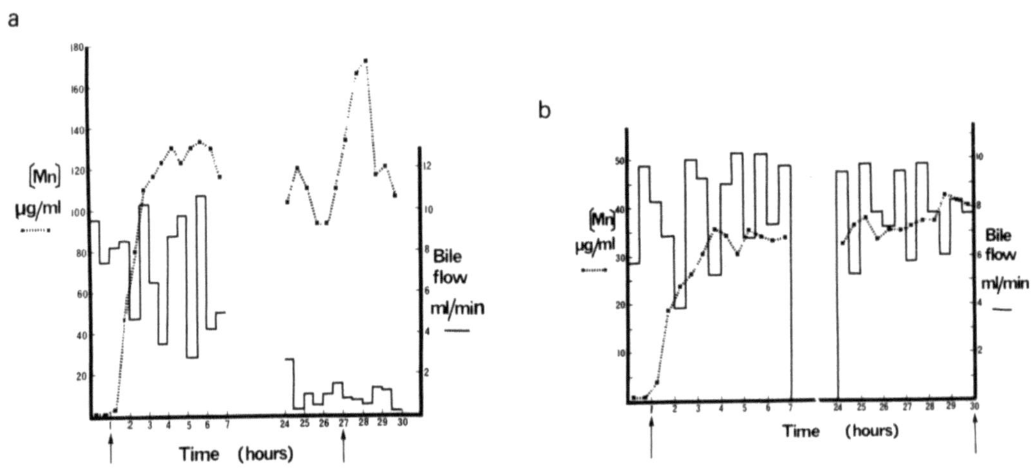

FIGURE 1. The effect on bile flow rate (——) (ml/min) and Mn concentration in bile (•---•) of an infusion of: a) 29 mg Mn/min into the duodenum and b) 311 μg Mn/min into a mesenteric vein in steer 3.

2. Rates of infusion producing signs of toxicity. The infusion of 45 mg Mn/min into steers 1 and 2 and 29 mg/min into steer 3 (Fig. 1a) caused abdominal discomfort and the infusions were stopped after 3, 6 and 27 h respectively. Bile flow rate decreased and became irregular, the steers went off their food, and the systemic plasma Mn concentration increased from normal values to 190-1060 ng/ml. The reduction in rate of bile flow was accompanied by a reduction in the concentrations of total solids and total cholate in the bile. The rate of flow of bile had returned to normal by 48 h after starting the infusions. During the first 6 h biliary Mn concentrations increased to between 130 and 160 μg/ml.

The infusion of 311 μg Mn/min for 30 h into a mesenteric vein of the third steer produced no adverse effects (Fig. 1b). Biliary Mn concentrations were between 35-40 μg/ml and the amount excreted during the last 6 h was 91% of the amount infused during this period. There were no significant changes in plasma Mn concentrations.

The 6 h infusion of 50 mg Ca/min produced no significant changes in bile flow or total solids, nor was there any abdominal discomfort.

DISCUSSION

During the last 6 h of the non-toxic infusions the absorption of Mn, as represented by the amount of Mn excreted in bile, was 0.6 to 1.2% of the amount infused and during the infusion at 30 mg/min the biliary concentration of Mn remained between 30 and 40 μg/ml. A similar concentration was observed during the intramesenteric infusion of 311 μg Mn/min into steer 3 when again no toxic effects were observed. However, the intraduodenal infusion of Mn at 29 mg/min into this steer increased plasma Mn concentration to 890 ng/ml (indicating that the capacity of the liver to remove all Mn at the first pass had been exceeded), increased bile Mn concentration to 130 μg/ml and severely decreased bile flow. The toxicity of this infusion and of those at 45 mg/min to steers 1 and 2 may be due to the fact that during the first period of the infusions more than 1% of the Mn was absorbed. During the first 6 h of the non-toxic 30 mg/min infusions biliary Mn concentrations increased to 90 and 130 μg/ml and during the toxic infusions of 29 and 45 mg/min biliary Mn concentrations increased to between 130 and 160 μg/ml. The amounts absorbed during this period must have been at least 4% of the amount infused.

It is proposed that the toxicity observed during these infusions was due to an initial high rate of absorption when the duodenal Mn concentrations were first increased. The short-term capacity of the liver to extract and excrete Mn was exceeded, plasma Mn concentrations increased and bile flow decreased further reducing the ability of the steer to excrete absorbed Mn. Toxicity ensued even though the absorption rate may have subsequently decreased to approximately 1%.

REFERENCES

1. Cunningham, G.N., Wise, M.B. & Barrick, E.R. 1966. J. Anim. Sci. 25: 532-538.

2. Gibbons, R.A., Dixon, S.N., Hallis, K., Russell, A.M., Sansom, B.F. & Symonds, H.W. 1976. Biochim. Biophys. Acta. 444: 1-10.

3. Hall, E.D. & Symonds, H.W. 1981a. Brit. J. Nutr. In press.

4. Hall, E.D. & Symonds, H.W. 1981b. Proc. Nutr. Soc. 40: In press.

5. Irvin, J.L., Johnston, C.G. & Kopala, J. 1944. J. Biol. Chem. 153: 439-457.

6. Sansom, B.F. Vagg, M.J. & Taylor, P.J. 1973. In "Production Disease in Farm Animals" (ed. J.M. Payne, K.G. Hibbitt & B.F. Sansom), Bailliere Tindall, London, pp.122-131.

7. Symonds, H.W., Mather,D.L., & Hall, E.D. 1981. Res. Vet. Sci. In press.

THE QUANTITIES OF ZN, CU, MN AND FE SECRETED IN THE PAROTID AND MANDIBULAR SALIVAS OF SHEEP FED LUCERNE CHAFF OR PERENNIAL RYEGRASS

N.D. Grace, D.H. Carr* and C.S.W. Reid

Applied Biochemistry Division, DSIR, PALMERSTON NORTH, NEW ZEALAND
*Department of Physiology, Massey University, PALMERSTON NORTH, NEW ZEALAND

Considerable quantities of Na, K and P (1), as well as smaller amounts of Mg and Ca (3), enter the ruminant stomach in the saliva. Little is known, however, as to the extent to which trace elements are recycled by this route. Recently, from studies on the flows of Zn, Cu, Mn and Fe along the digestive tract of sheep (4) and cattle (2), it has been concluded that a significant net secretion of Zn and very small net secretions of Cu, Mn and Fe can occur between the mouth and the reticulo-omasal orifice. Recycling of Zn appears to have been established: Zn occurs in the saliva of ruminants (7), while studies using ^{65}Zn have shown that Zn can be transferred from the blood to the reticulo-rumen both through the walls of that organ and via the saliva (10). The status of the other trace elements is unknown. This paper reports on the total outputs (μg/24h) and the circadian variation in output of Zn, Cu, Mn and Fe in parotid and mandibular saliva of sheep fed lucerne chaff or Ruanui perennial ryegrass.

EXPERIMENTAL

Sheep. Observations were made on 4 New Zealand Romney wethers, weighing 38-40 kg and aged 2 to 2½ years. The animals were housed indoors in metabolism crates. The surgical preparation involved unilateral cannulation of a parotid (re-entrant) and mandibular duct on the left side of each animal(8).

Diet. The animals were fed twice daily, at 08.00 and 17.00 h, a total of 900 g DM, first as lucerne chaff (*Medicago sativa* L.) [DM digestibility 58% and containing 22, 8, 50 and 200 mg/kg Zn, Cu, Mn and Fe respectively] and then as 'Grasslands Ruanui' perennial ryegrass (*Lolium perenne* L.) [DM digestibility 78% and containing 17, 6, 45 and 150 mg/kg Zn, Cu, Mn and Fe respectively]. The Ruanui ryegrass was cut daily and fed fresh (9); the very small amounts of feed refused were weighed daily.

Experimental Design. At least 21 days elapsed between surgery and the start of the collections. Each diet was fed for at least 7 days before collections began, and 2 collections a week apart, were made from each diet. A collection lasted 27 h (starting at 07.00 h and including 3 feeding periods) during which parotid and mandibular salivas secreted from the cannulae, at successive 5 minute intervals were collected and weighed to determine the flow rates. Losses of Na in the salivas were replaced by infusing 3 l of 0.9% saline by mouth during the collection.

Sampling. The 5 minute secretions for each saliva were bulked, such that samples of parotid saliva secreted over 110 to 120 minute intervals, and the mandibular saliva secreted over 30 minute intervals, were obtained for trace element analysis.

Analytical. Samples of diet, plasma and saliva were wet-ashed and the Zn, Cu, Mn and Fe determined by atomic absorption spectrophotometry.

RESULTS

The mean ($\pm$ standard errors) total weights of saliva (4 sheep x 2 collections/diet) secreted from one parotid and one mandibular gland over 24 h was 2419 $\pm$ 269 g and 271 $\pm$ 22.2 g when the sheep were fed lucerne chaff and 3001 $\pm$ 301 g and 203 $\pm$ 26.7 g when they were fed Ruanui perennial ryegrass. On either diet the flow of parotid saliva increased markedly with the start of eating but then fell to a low level from which it slowly recovered over several hours with intermittent periods of elevated secretion related to rumination. High flows of mandibular saliva occurred only during eating thereafter little or no secretion occurred until the next feed.

The mean total 24 h salivary output and concentration of Zn, Cu, Mn and Fe from single parotid and mandibular glands are shown in Table 1 together with the plasma concentrations.

Table 1. The daily quantities and concentrations of Zn, Cu, Mn and Fe secreted in the saliva by single parotid and mandibular glands, and plasma concentrations of these trace elements in sheep given lucerne chaff and Ruanui perennial ryegrass. (n=4).

Parotid	Zn	Cu	Mn	Fe
Lucerne chaff				
μg/24h	186 ± 29.1	110 ± 14.8	26.1 ± 3.92	207 ± 26.9
μg/100 ml	7.6	4.5	1.1	8.6
Ruanui ryegrass				
μg/24h	588 ± 70.6	86.2 ± 15.51	45.2 ± 6.75	336 ± 47.1
μg/100 ml	19.5	2.9	1.5	11.2
Mandibular				
Lucerne chaff				
μg/24h	8.3 ± 0.51	2.3 ± 0.13	1.40 ± 0.08	11.18 ± 1.08
μg/100 ml	3.06	0.85	0.52	4.12
Ruanui ryegrass				
μg/24h	32.6 ± 5.23	2.3 ± 0.26	1.83 ± 0.22	9.0 ± 0.82
μg/100 ml	15.3	1.07	0.85	4.22
Plasma				
Lucerne chaff				
μg/100 ml	67 ± 3	96 ± 6	2.85 ± 0.16	203 ± 15
Ruanui ryegrass				
μg/100 ml	69 ± 4	100 ± 7	2.89 ± 1.21	200 ± 11

The diurnal variation in the mean output of the four elements from a parotid gland is presented in Fig. 1.

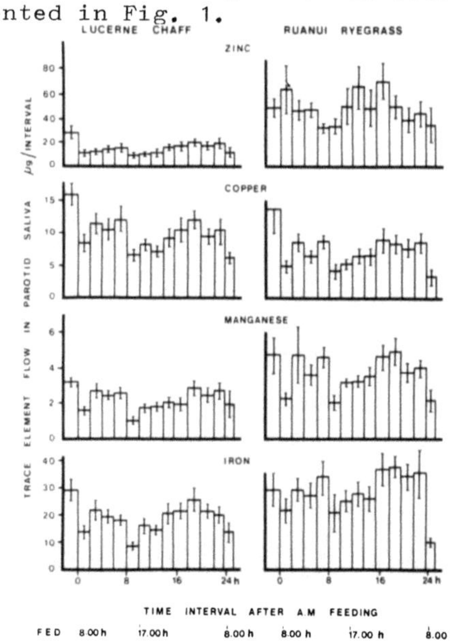

Fig. 1. Diurnal variation in the parotid secretion of Zn, Cu, Mn and Fe in sheep fed lucerne chaff or perennial ryegrass.

The circadian variation in the parotid output of Zn, Cu, Mn and Fe was not as great as that exhibited by the total salivary flow. The highest output occurred in the periods 11.00 to 14.00 h and 21.00 to 06.00 h, and the lowest in the period 16.00 to 19.00 h. Neither parotid nor mandibular glands appeared to concentrate Zn, Cu, Mn or Fe in their secretions. The level of these trace elements in plasma were 2-100 times the levels found in saliva.

DISCUSSION

As the losses of Zn, Cu, Mn and Fe via the hair, wool and skin are negligible, and only small amounts have been shown to be excreted in the urine (5), the digestive tract secretions such as the pancreatic juice and bile must be important routes of loss from the body pool. A comparison of the amounts of Zn, Cu and Mn secreted via the bile and pancreatic juice (5) with those secreted via the saliva from both parotid glands is made in Table 2.

Table 2. The quantities of Zn, Cu, Mn and Fe secretion via the digestive tract secretions in sheep.

Diet	Secretion	Zn	Cu	Mn	Fe
			μg/24h		
Lucerne pellets	Bile	1000	330	230	-
	Pancreatic juice	2100	62	90	-
Lucerne chaff	Parotid saliva	370	220	52	420
Ruanui ryegrass	Parotid saliva	1180	172	90	680

As in the case of the bile and pancreatic secretions considerable amounts of the four trace elements are secreted in the saliva. Furthermore, the saliva values are underestimates because parotid secretion only accounts for approximately half the total salivary secretion (6).

CONCLUSION

We conclude that salivary secretion is an important means of recycling Zn, Cu, Mn and Fe : it also represents a significant route of transfer of these elements between the body pool and digestive tract.

ACKNOWLEDGMENTS

To Messrs A.J. Craven, S. Dodson, M. Eggles and P.L. Martinson for technical assistance.

REFERENCES

1. Bailey, C.B. 1961, Br. J. Nutr. 15: 489-498.
2. Bertoni, G., M.J. Watson, G.P. Savage and D.G. Armstrong, 1976, Zoot. Nutr. Anim. 2: 185-91.
3. Care, A.D. 1963, Nature, Lond. 199: 818-19.
4. Grace, N.D. 1975, Br. J. Nutr. 34: 73-82.
5. Grace, N.D. and J.M. Gooden, 1980, N.Z. J. Ag. Res. 23: 293-98.
6. Kay, R.N.B. 1960, J. Physiol. 150: 515-537.
7. Little, D.A. 1975, Aust. J. Exp. Agric. Anim. Husb. 15: 437-39.
8. Titchen, D.A., and D.H. Carr, 1978, Q. J. Exp. Physiol. 63: 1-21.
9. Ulyatt, M.J., and J.C. MacRae, 1974, J. agric. Sci., Camb. 82: 295-307
10. Weston, R.H., and J. Kastelic, 1967, Aust. J. biol. Sci. 20: 975-81.

DISCUSSION

WHITE (Australia). Your conclusion that parotid saliva plays a significant part in the recycling of trace elements is based on an important assumption that salivary minerals enter a different pool in the gut than dietary sources of the elements. Do you have any evidence that this is the case? If not, then saliva contributes only a very insignificant proportion of the total mineral pool in the gut.

ZINC METABOLISM IN SHEEP GIVEN A HAY DIET

H. Lloyd Davies,* N.F. Suttle and A.C. Field

Moredun Research Institute,
408 Gilmerton Road, EDINBURGH, EH17 7JH, SCOTLAND

* On Special Studies Programme from The University of New South Wales

SUMMARY

Four pairs of Scottish Blackface ewes were given a basal diet of hay providing 8 mg Zn/d and one of four levels of supplementary Zn (0, 75, 150 or 225 mg/d) continuously by intra-ruminal infusion. 59 µCi of ^{65}Zn as zinc chloride was given intravenously after 14 d and changes in the specific radioactivity (SR) of plasma and faeces were monitored for a further 10 and 13 d respectively. It was found from the ratios of the SR in plasma and faeces that faecal endogenous losses were 5.4, 7.8, 7.7 and 8.0 $\pm$ 0.86 mg/d and these losses were largely unaffected by Zn intake. Zn retention did not increase with Zn intake and homeostasis was achieved primarily by control of Zn absorption which fell from 0.96 to 0.05 of intake as Zn intake increased.

INTRODUCTION

There are conflicting views on the mechanisms involved in Zn homeostasis. Evans, Johnson and Johnson (1979) have suggested that Zn homeostasis in the rat is maintained primarily by the control of Zn secretion into the intestine; Smith and Cousins (1980) have suggested that control over absorption is the important factor in the regulation of Zn metabolism. Previous studies using ^{65}Zn in ruminants have measured the excretion of the isotope and its distribution in the tissues (e.g. Miller, Blackmon, Howell, Gentry and Hiers (1966)) but have not used the principle of isotope dilution to indicate quantitative flows of stable Zn: we have applied these principles in our experiment.

MATERIALS AND METHODS

Eight Scottish Blackface ewes weighing 45 to 63 kg and equipped with rumen cannulae, were housed in the individual metabolism crates equipped with automatic feeders for delivering the feed allowance every two hours as described by Grace and Suttle (1979). They were allocated in pairs to one of four supplementary Zn intakes nominally 0, 75, 150 and 225 mg/d and were fed a basic diet of chaffed pasture hay 0.7 kg/d which provided 5 - 11 mg Zn/d. The supplementary Zn was given by infusing two litres of an appropriate $ZnSO_4$ solution into the rumen using double channel peristaltic pumps. On average 2 litrees of solution were infused into each sheep per day. Because the hay had a low crude protein content (6%) urea (4 g/d) and sodium sulphate (0.12 g/d) were added to the infusion solution to correct the nitrogen and sulphur deficiencies. The treatments were given for 14 days and then on day 15 59 µCi of ^{65}Zn together with 67 µg carrier Zn as zinc chloride were injected via a temporary catheter into the jugular vein in 10 ml of saline solution.

Venous blood samples for plasma Zn and ^{65}Zn determinations were obtained at 0.1, 0.5, 1, 2, 3, 4, 5, 6, 7, 8, 10, 12, 24, 28, 72, 96, 114, 192 and 240 hours after injection. ^{65}Zn was determined on plasma, urine, freeze-dried faeces, rumen contents and diet.

Duplicate samples of approximately 15 g of fresh faeces or rumen contents or 1 - 20 mls of plasma were counted in an automatic gamma-well scintillation counter fitted with a 250 mm sodium iodide crystal. Each day's total urine collection was made to constant volume and counted in a whole body counter.

The faecal endogenous excretion of Zn and the irreversible loss rate of Zn from the transport compartment were both estimated using the principle of isotope dilution, assuming that when equilibrium had been reached SR of secreted and transported Zn was the same as that of plasma Zn.

The absorption rate for Zn was assumed to equal the irreversible loss rate under steady state conditions and irreversible loss was estimated by compartmental analysis of the plasma SR curve using curve peeling techniques to obtain approximate intercept and rate constant values for the four compartments (White, Steel, Leng and Luick (1969)).

RESULTS

There were no effects of level of infused Zn upon the rate of disappearance of ^{65}Zn from the plasma. The levels of non-radioactive Zn in the plasma were also unaffected by Zn intake.

There was no effect of Zn infusion upon radioactive Zn excretion in the faeces. Faecal Zn increased linearly with increases in Zn intake and the regression coefficient was not significantly different from unity. There was therefore no detectable increase in Zn retention as intake increased. The total recoveries of ^{65}Zn in faeces for the pairs of ewes after 13 d were 49.5 and 40.0; 52.4 and 53.1; 60.2 and 61.8; 51.6 and 50.4% in order of increasing Zn intake. Maximal excretion occurred on the second day for all but one sheep when values ranged from 7.7 - 13% dose and declined exponentially thereafter.

The mean irreversible losses were estimated by the "part area" method to be 8.8, 11.3, 11.6 and 10.6 $\pm$ 2.36 mg/d with no significant treatment effect as shown in Table 1.

Table 1. Estimates by 'whole or part area' methods of the faecal endogenous secretion and irreversible loss of Zn in individual ewes on various Zn intakes

Sheep No.	Zn intake (mg/d)	Faecal excretion* of Zn (mg/d)	Fraction of faecal Zn of endogenous origin		Irreversible loss (mg/d)	
			Whole	Part	Whole	Part
1	10.4	19.7	0.35	0.325	8.3	10.5
2	7.8	14.6	0.30	0.385	5.4	7.1
3	92.4	112.7	0.057	0.072	7.4	10.0
4	89.4	111.7	0.047	0.071	8.4	12.7
5	129.0	124.3	0.029	0.0502	7.3	9.6
6	171.0	194.2	0.026	0.0427	7.7	11.6
7	249.8	238.0	0.020	0.0348	9.0	11.8
8	221.2	250.9	0.021	0.0287	6.9	9.4

* Mean for 13 days after administration of ^{65}Zn

The efficiency of Zn absorption, calculated by dividing irreversible loss by Zn intake, was the only parameter affected by Zn intake: values for the absorption coefficient were 0.963, 0.125, 0.071, 0.045 $\pm$ 0.027 in order of increasing Zn intakes. The percentage of the Zn dose excreted in urine over the entire 10 day collection period was less than 0.21% and was unaffected by Zn intake.

The mean secretion rates for Zn into the rumen based on levels of radioactive and stable Zn found in rumen contents on days 7 and 10 were estimated to be 0.8, 1.5, 1.3, and 1.7 mg/d for the 4 groups on increasing Zn intakes. Although the level of Zn secreted into the rumen was apparently higher when Zn was infused into the rumen, the increase was small relative to the increment of Zn intake.

DISCUSSION

The data provide clear qualitative evidence that the sheep in this experiment controlled their retention of Zn by changing absorption rather than endogenous loss as Zn intakes increased. Supplementary Zn was not retained but quantitatively recovered in the faeces. If absorptive control had not been the major homeostatic factor then the flux of absorbed stable Zn through the plasma into urine or faeces would have increased the rate of clearance of ^{65}Zn from plasma: this did not occur. The low urinary Zn excretion and constant ^{65}Zn excretion in faeces confirmed that secretory pathways were of minimal importance in achieving homeostasis. Our sheep therefore differed from the growing rats studied by Weigand and Kirchgessner (1980) and Evans et. al. (1979) which showed marked increases in faecal endogenous secretion over a narrower range of Zn inputs. In several studies with ruminants the apparent absorption of ^{65}Zn has been shown to decrease when Zn intakes were increased (e.g. Neathery et. al. (1973) Miller et. al. (1967)).

Some assumptions were made when we attempted to generate quantitative data for Zn absorption and endogenous losses. The crucial assumption in computing endogenous loss is the value attributed to the SR of secreted Zn. Initially we intended to use urine SR to represent the SR of Zn in the transport compartment and Zn secreted from it. Unfortunately we found that there was some stable Zn contamination in the urine. We therefore used plasma SR but this use has certain problems. There was a low SR of protein-bound Zn relative to whole plasma Zn in the first 24 hours after dosing indicating that plasma Zn consisted initially of at least two compartments. The relatively high SR of that same fraction after 72 hours probably reflects the flux of unlabelled dietary Zn through a more labile protein-free compartment. Errors in using whole plasma SR to indicate the SR of secreted Zn in faeces would therefore arise if secreted Zn came predominantly from one compartment and would be greatest early in the experiment. We have assumed that the more appropriate value for endogenous loss was the one calculated by comparing areas under the plasma and faeces SR/time curves for the time period 3 to 10 d. The values obtained were remarkably uniform when expressed on a liveweight basis and not significantly affected by Zn input. Irreversible loss should exceed faecal endogenous loss to the extent that there are additional secretory losses via urine and skin, accumulation of Zn in wool and slowly exchangeable Zn pools which feed predominantly unlabelled Zn into the transport compartment during the isotope study. We found that the difference between faecal endogenous and irreversible loss was however only 3.1 ± 1.2 mg/d and of this 1.2 mg/d would be accounted for if wool was growing at 10 g/d and contained 115 mg/kg of Zn.

It would appear that there was either a low limit to the sheep's capacity to absorb Zn or that by humoral control mechanisms the sheep absorbed Zn according to requirement. If the sheep absorbed Zn according to requirement absorbability values become progressively an animal rather than a dietary attribute as Zn intakes increase above the required level.

The relatively small changes in faecal endogenous losses in the presence of divergent absorptive efficiencies can only be explained in one of two ways: either Zn is secreted largely posterior to the principal sites of absorption or it is secreted anteriorily and remains unaffected by the processes influencing the absorption of dietary Zn.

These observations on Zn metabolism were made on a hay diet. They indicate clearly that on such a diet absorptive control rather than differences in the control of Zn secretion into the intestine is responsible for Zn homeostasis in the sheep. There is a need to repeat this work on a cereal diet.

REFERENCES

1. Evans, G.W., E.C. Johnson, and P.E. Johnson, 1979, J. Nutr. 109:1258 - 1264.

2. Grace, N.D. and N.F. Suttle, 1979, Br. J. Nutr. 41:125 - 136.

3. Miller, W.J., D.M. Blackmon, G.W. Powell, R.P. Gentry, and J.M. Hiers, 1966, J. Nutr. 90:335 - 341.

4. Miller, W.J., D.M. Blackmon, R.P. Gentry, W.J. Pitts and G.W. Powell, 1967, J. Nutr. 92:71 - 78.

5. Neathery, M.W., W.J. Miller, D.M. Blackmon and R.P. Gentry, 1973, 1973, J. Dairy Sci. 56:1526 - 1530.

6. Smith, K.T. and R.J. Cousins, 1980, J. Nutr. 110:316 - 323.

7. Weigand, E., and M. Kirchgessner, 1980, J. Nutr. 110:469 - 480.

8. White, R.J., J.W. Steel, R.A. Leng, and J.R. Luick, 1969, Biochem. J. 114:203 -214.

DISCUSSION

WEIGAND (Germany). In the introduction to your presentation you mentioned the controversy regarding the importance of absorption and endogenous excretion for zinc homeostasis. At the TEMA-3 meeting we reported work in this regard, using the laboratory rat as a model animal, work which we have published in the meantime in more detail in the March 1980 issue of the Journal of Nutrition. We have shown that, in absolute terms, zinc absorption increases with zinc intake, not at a linear rate but rather at a logarithmic rate so that eventually a sort of a plateau will be reached at a very high zinc intake. Accordingly, percent true absorption decreased greatly with increasing zinc supply. We concluded that absorption would be the most important factor for zinc homeostasis, and endogenous excretion would just serve as a fine control, so that an animal having a specific need for zinc would retain just about as much zinc as it needs. You have used sheep in your experiment. I am not familiar with the breed of sheep you used. I assume they were mature animals.

LLOYD DAVIES (Australia). Yes castrated adults.

WEIGAND (Germany). So their requirement for zinc was not materially higher than that for maintenance and the zinc levels you gave were rather high compared with the requirement of the animals. According to our findings with the fast growing rat, I would gather that zinc status in your sheep had reached a level where you would not expect a great response in absorption or in endogenous faecal excretion. What kind of conclusion would you draw from your work? Would you put the emphasis on absorption, and how would you regard the role of endogenous excretion in homeostatic regulation of zinc utilization?

LLOYD DAVIES (Australia). These were old sheep with no growth requirements whatsoever. I certainly agree with you that our work shows that absorption is the broad control. The nature of this experiment was probably insufficient to show whether there was any fine control by endogenous secretion. There is no doubt that you will see diferent requirements if you compare species at different ages. Perhaps Dr. Suttle would like to expand a little more on this.

SUTTLE (Scotland). There is no discrepancy at all; the difference, if any is that in the rat work you are starting with very deficient diets of the order of 5 ppm zinc. We are starting at a fairly respectable level and I think control of endogenous loss will become more important the further you go down the scale of zinc intake. Under the range we are working with absorption is the primary control mechanism and I think it is just worth confirming that it can go virtually up to 100%.

M. GOLDEN (Jamaica). Have you any data relating to the the validity of the assumption that the specific activity of the plasma zinc is a valid measure for the excretion of zinc into the intestine?

LLOYD DAVIES (Australia). We had hoped to use urine zinc but had to use the plasma data because among other things we suspected we had urinary zinc contamination.

M. GOLDEN (Jamaica). I was wondering that if you extrapollated your results down to the zero zinc intake, would the faecal zinc and the plasma zinc have the same specific activity?

SUTTLE (Scotland). We have not done that.

WHITE (Australia). Something that concerns me about these sorts of studies is the extrapollation from injected radio-isotopes to patterns of excretion of endogenous minerals, in ruminants especially. Would Hadyn care to discuss the metabolic relationship between injected zinc and the recirculation of intestinally absorbed Zn through the portal and hepatic system. Does he think its possible that sheep may not handle injected zinc in a way that resembles absorbed zinc?

LLOYD DAVIES (Australia). I think that the assumptions that have to be made are less with injected isotopes than with isotope infused into the rumen.

SUTTLE (Scotland). Dr. White has a reasonable question, but I think we are talking about quantative rather than qualitative differences and it seems highly unlikely to me that we are going to be so far off the mark - the only way you could check or at least get one stage nearer the physiological situation would be to infuse zinc into the portal vein rather than the jugular vein. But the results make sense, they tie in with the sort of response you get with dietary treatments, the sort of thing that Masters was reporting earlier, the ability of the sheep to cope with exceeding low concentrations of zinc in its diet. How else does the animal cope other than increasing its efficiency of absorption when intakes go down?

EFFECT OF DIETARY TIN ON HUMAN MINERAL UTILIZATION

J.L. Greger, M.A. Johnson, M.J. Baier
Department of Nutritional Sciences
University of Wisconsin, Madison, WI 53706

Several investigators have noted depressed growth, hematocrits, hemoglobin levels and serum iron levels among rats fed high (3000-6000 μg/g) levels of tin (1-3). These same symptoms have been observed in rats fed only 150 μg/g tin in the diet for one month (1). Recently, Greger and Johnson (4) have demonstrated that rats fed 200 μg/g tin in the diet for three weeks retained less zinc and copper in their tissues and lost more zinc in their feces than control animals. No investigator has published data on the effects of moderate levels of tin on mineral metabolism in human subjects.

The purpose of this study was to assess whether the amount of tin, that potentially could be present in canned foods, was sufficient to affect the utilization of zinc, copper, iron, manganese, and calcium by human subjects. The level of tin fed in the control diet in this study was typical of the amount of tin present in diets composed of only fresh, frozen, and bottled foods. The amount of tin in the test diet was typical of the amount of tin that would be present in four one-half cup servings of certain canned fruits and juices (5).

EXPERIMENTAL

Subjects were fed a mixed diet for 40 days. During the first 20 days of the study, four subjects were fed the control diet, that contained 0.1 mg tin daily, and four subjects were fed the test diet, that contained an additional 50 mg tin as stannous chloride. The stannous chloride was dissolved in the juices served to subjects at lunch and dinner. During the second 20 day period the dietary treatments were reversed so each subject served as his own control.

Eight adult males between 21.8 and 28.8 years of age participated in the study. Their average height and weight were 179 ± 7 cm (mean ±S.D.) and 76 ± 5 kg, respectively. Each gave informed consent in a manner approved by a University of Wisconsin committee on the use of human subjects.

A limited 2-day cycle menu which consisted of common foods was served throughout the study. The diets were calculated to provide 3150 kcalories and 75 g protein daily and to provide at least 100% of the Recommended Dietary Allowances for thiamin, riboflavin, niacin, and vitamins A and C. Eight composites of a complete day's menu were prepared during the study. These composites were determined by analysis to contain daily on the average 4538 mg sodium, 1201 mg phosphorus, 813 mg calcium, 371 mg magnesium, 18.79 mg iron, 13.45 mg zinc, 3.28 mg manganese, and 1.63 mg copper.

Subjects collected all their fecal and urinary excretions during the study. Venous blood samples were collected from subjects before breakfast on days 12 and 18 of each period. Care was taken to prevent trace element contamination of all samples. The zinc, copper, iron, manganese, and calcium content of food, fecal, urine, plasma, and serum samples were determined by atomic absorption spectroscopy. Radial diffusion plates were used to assess plasma retinol binding protein and ceruloplasmin levels; a radio-immuno-assay was used to determine serum ferritin levels. Food, fecal, and urine samples were wet ashed and their tin contents were determined with a colorimetric procedure that utilized pyrocatechol violet (6,7).

RESULTS AND DISCUSSION

Subjects lost 2.0 mg more zinc in their feces daily when fed the test diet rather than the control diet (Table 1). The difference was significant ($p < 0.01$). Subjects lost significantly ($p < 0.05$) less zinc in the urine but still retained significantly ($p < 0.01$)

less zinc daily when fed the test diet with 50 mg tin than when fed the control diet. Similarly apparent absorption of zinc and the retention of zinc in bone and kidney were less in rats fed 200 μg tin/g diet than in rats fed a control diet (4).

The level of tin in the diets did not affect significantly fecal or urinary excretion or apparent retention of copper, iron, manganese, or calcium by subjects (Table 1). Serum zinc, copper, iron and ferritin levels and plasma zinc, retinol binding protein and ceruloplasmin levels were also unaffected by the dietary treatments (Table 2).

Table 1: Excretion and apparent retention of zinc, copper, iron, manganese and calcium by subjects fed test (50 mg tin daily) and control (0.1 mg tin daily) diets.

	Treatments	
	Test	Control
	(mg/day)	
Zinc		
fecal loss*	15.61 ± 2.27[δ]	13.64 ± 1.77
urinary loss†	0.63 ± 0.13	0.71 ± 0.12
apparent retention*	-2.81 ± 2.23	-0.90 ± 1.72
Copper		
fecal loss	1.28 ± 0.17	1.27 ± 0.10
urinary loss	0.04 ± 0.01	0.04 ± 0.01
apparent retention	0.30 ± 0.16	0.33 ± 0.07
Iron		
fecal loss	16.82 ± 0.83	16.42 ± 1.09
urinary loss	0.14 ± 0.03	0.13 ± 0.02
apparent retention	1.79 ± 0.81	2.27 ± 1.08
Manganese		
fecal loss	3.43 ± 0.17	3.39 ± 0.15
apparent retention#	-0.16 ± 0.17	-0.10 ± 0.10
Calcium		
fecal loss	612 ± 75	631 ± 44
urinary loss	258 ± 44	248 ± 52
apparent retention	-57 ± 84	-62 ± 70

*Difference significant at $p < 0.01$ level

†Difference significant at $p < 0.05$ level

δMean ±S.D. (n=8)

#Urinary losses were less than 0.02 mg/day. Thus apparent retention = intake - fecal loss.

When subjects consumed the test diet, they apparently absorbed on the average only 2.9% of their tin intake. Although they excreted more tin in the urine when they consumed the test diet rather than the control diet (122 ± 52 μg tin daily versus 29 ± 13 μg tin daily), they still apparently retained more tin when they consumed the test diet (1.31 ± 1.54 mg tin daily versus 0.03 ± 0.03 mg tin daily).

In summary, the daily consumption of low levels of tin, 50 mg daily, is sufficient to affect zinc utilization by human subjects. This effect may have practical significance among individuals who consume a number of servings of canned fruits and juices daily and marginal levels of zinc.

Table 2: Serum and plasma levels of zinc, iron, copper and proteins in subjects fed test (50 mg tin daily) and control (0.1 mg tin daily) diets.

	Treatments	
	test	control
serum zinc (μg/dl)	105 ± 9‡	106 ± 10
plasma zinc (μg/dl)	103 ± 11	102 ± 12
plasma retinol binding protein (mg/dl)	5.0 ± 0.9	4.9 ± 1.0
serum copper (μg/dl)	87 ± 21	84 ± 16
plasma ceruloplasmin (mg/dl)	25.9 ± 4.5	24.0 ± 3.5
serum iron (μg/dl)	141 ± 17	148 ± 16
serum ferritin (ng/ml)	79 ± 65	77 ± 68

‡ Mean ± S.D.

ACKNOWLEDGEMENT

This research was supported by the College of Agricultural and Life Sciences, University of Wisconsin, Madison Project #2521 and U.S.P.H.S. Grant #5R01-AM 25666.

REFERENCES

1. deGroot, A.P., 1973. Fd. Cosmet. Toxicol 11: 955-962.

2. deGroot, A.P., V.J. Feron and H.P. Til, 1973. Fd. Cosmet. Toxicol. 11: 19-30.

3. Fritsch, P., G. de Saint Blanquat, and R. Derache, 1977. Toxicology 8: 165-175.

4. Greger, J.L. and M. A. Johnson, 1981. Fd. Cosmet. Toxicol (In press).

5. Greger, J.L. and M.J. Baier, 1981. J. Food Sci. (In press).

6. Corbin, H.B., 1970. J. Assoc. Off. Agr. Chem. 53: 140-146.

7. Corbin, H.B., 1973. Anal. Chem. 45: 534-537.

DISCUSSION

WEIGAND (Germany). What would be the tin intake of people who eat a lot of canned food?

GREGER (USA). Most canned food is in lacquered cans so the intake of tin would be only about 1-5mg/d. However, tomato paste, pineapple, orange and grapefruit juice are packaged in unlacquered cans and if modest quantities of these are consumed the tin intake could be 20-100mg/d. If opened cans are stored in a refrigerator for a few days the tin concentration in the contents may rise several fold.

SCHWARZ (Germany). Do you think that the interaction between tin and zinc occurs in the intestine to decrease the absorption, or does it occur in the metabolism and raise endogenous excretion?

GREGER (USA). We are currently investigating the site of the interaction.

ABDULLA (Sweden). In Sweden we have groups of elderly people who eat quite a lot of canned food, mainly fish, but we haven't noticed any signs of zinc deficiency.

GREGER (USA). Yes, from the experimental results one would expect to see indications of low zinc status in such cases. We have also examined patients who have a high intake of canned foods and although their zinc intake was low there seldom was any indication of poor zinc status.

THE AVAILABILITY, ABSORPTION AND RETENTION OF TRACE ELEMENTS

IMPROVEMENT OF ZINC BIOAVAILABILITY IN CEREAL GRAINS

H.-J.Lantzsch[1],S.E.Scheuermann[1] and H.Marschner[2]

Institute of Animal Nutrition[1] and
Institute of Plant Nutrition[2]
University of Hohenheim
PO BOX 700562,7000 STUTTGART 70
FEDERAL REPUBLIC OF GERMANY

The molar ratio phytic acid/zinc is a reliable indication of zinc availability from phytate-containing diets. This was first demonstrated by Oberleas (5) and later on by Davies and Olpin (1) in semisynthetic diets to which sodium phytate was added and by Lantzsch et al. (2) in natural feedstuffs containing native phytate.
From their experimental results Morris and Ellis (4) concluded that phytate in whole wheat bran affects zinc availability in nearly the same manner as does sodium phytate at a similar phytic acid/zinc molar ratio. So it seems reasonable to assume that the solubility characteristics of phytates in the gastrointestinal tract are a very important factor for the phytate effect.
Normally, phytic acid and zinc concentrations in cereal grains are found to be within the range of 13 to 15 mmol/kg (9 to 10 g/kg) and 0.3 to 0.5 mmol/kg (20 to 30 mg/kg), respectively, yielding molar ratios between 40 and 50. At these ratios insufficient zinc is available for growth and body retention of immature rats.
Zinc availability for absorption may be improved by narrowing the molar ratio phytic acid/zinc by elevating the zinc concentrations by either direct supplementation or indirectly by application of zinc fertilizers to the plants. However, particularly after soil application the amounts of zinc fertilizers nescessary to raise the availability of zinc from cereal grains on the level of $ZnSO_4$ in phytate-free diets are relatively high. Therefore, increasing the zinc concentrations of cereal grains by direct supplementation may represent the more effective way to improve the zinc availability.
In experiments carried out the biological value of zinc taken up by wheat and barley grains after soil treatment with $ZnSO_4$ was compared with that of zinc from $ZnSO_4$ supplemented directly to the grains.

MATERIAL AND METHODS

Spring wheat (cv.Jubilar) and spring barley (cv.Carina) were cultured in 1979. Soil treatment with $ZnSO_4$ (50 kg/ha) alone or in combination with foliar sprays (6.4 kg/ha), using a 0.2 % $ZnSO_4$ solution, were applied (see table 1). Grain samples from unfertilized wheat and barley plants with molar ratios of 43 and 34, respectively, were supplemented with $ZnSO_4$ so as to obtain the same molar ratios of 26 or 19 and 28 or 22, as were found in the fertilized wheat and barley samples, respectively. The overall reduction in molar ratios amounted 24 units for wheat and 12 units for barley.
The availability of zinc for intestinal absorption was determined in growing male rats by means of the EDTA-test (3), which is based on a highly significant linear relationship between the intake of available zinc and the increase in the 24 h urinary zinc excretion after an i.p. injection of a single non-toxic dose of 250 mg/kg body weight of $CaNa_2EDTA$. The suboptimal range of supply between 5 and 10 ppm dietary zinc is used for the measurements to attain maximal absorption rates and measure availability as a characteristic of the feedstuff, uninfluenced by homeostatic adaption of the animals.
The grain samples assayed were incorporated in a semipurified glucose/ovalbumin basal diet at a level to provide 10 ppm total zinc in substitution for corresponding nutrients of the basal diet. Two control diets with 5 and 10 ppm zinc from $ZnSO_4$ were run with each experiment. All diets had the same composition with respect to crude nutrients, minerals and trace elements. A calcium level of 0.75% (dry matter base) was used.
Crude nutrients were analyzed by conventional methods, minerals and trace elements by atomic absorption spectrophotometry after dry ashing, and phytic acid after the method of Schormüller et al. (6).

RESULTS AND DISCUSSION

The intake of available zinc, which depends upon the molar ratio phytic acid/zinc and the normal daily urinary zinc excretion do not show any correlationship. After EDTA-injec-

Table 1 Phytic acid and zinc concentrations in whole spring wheat and whole spring barley after soil treatment and foliar sprays with $ZnSO_4$ (experiment 1979; values expressed per kg of dry matter)

treatments	phytic acid g	zinc mg	molar ratio
	wheat		
unfertilized	9.2	21.3	43
soil treatment, 50 kg/ha	8.8	34.2	26
soil treatment, 50 kg/ha + foliar sprays, 6,4 kg/ha	8.7	46.3	19
	barley		
unfertilized	8.9	24.8	34
soil treatment, 50 kg/ha	8.7	30.4	28
soil treatment, 50 kg/ha + foliar sprays, 6,4 kg/ha	9.0	39.9	22

tion the daily urinary zinc excretion increased roughly twenty-fold and responded to the amount of the available zinc ingested (figure 1).
Lowering the molar ratio in wheat from 43 to 26 or 19 resulted in a significant increase in the urinary zinc output, both by fertilization and supplementation. However, after supplementation the increase tended to be higher. Similar results were obtained with barley, although the increase in the urinary zinc excretion after lowering the molar ratio did not reach significance in most cases. The differences in the molar ratios found between the unfertilized and fertilized barley samples are smaller than in the wheat samples.

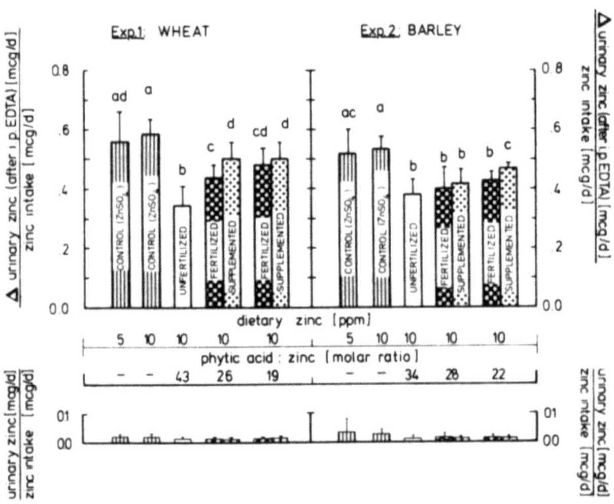

Figure 1 Effect of molar ratios phytic acid/zinc in whole wheat and barley grains, changed by $ZnSO_4$ fertilization or supplementation, on 24 h urinary zinc excretion of growing male rats. (n = 6). Lower part: untreated rats. Upper part: rats, i.p. injected with 250 mg/kg body weight of $CaNa_2EDTA$ (mean ± s.e.).

Relative availabilities, calculated from urinary zinc excretion values, using the $ZnSO_4$-control groups as reference, are shown in table 2. The tendency for a higher availability of the supplemented zinc may be attributed to faster solubility of the added metal salt as compared to that of the phytic acid, which first have to be delivered from the plant cell. In that way some zinc is protected against binding to the ligand. On the other hand, zinc incorporated in the grains by fertilization may be bound to phytic acid and thus becomes less available to the animals.

Table 2 Relative availability of zinc for absorption ($ZnSO_4$ = 100) from whole wheat and whole barley after zinc fertilization or supplementation (6 animals/group)

treatments	wheat		barley	
	molar ratio	zinc availability %	molar ratio	zinc availability %
unfertilized	43	59[a]	34	72[a]
fertilized	26	75[b]	28	76[a]
supplemented		87[c]		78[a]
fertilized	19	82[bc]	22	81[a]
supplemented		83[bc]		88[b]

values not sharing a common superscript are significantly different ($p < 0.05$)

SUMMARY

In growing male rats availability of zinc taken up by wheat and barley grains after soil treatment with $ZnSO_4$ was compared with that of zinc from $ZnSO_4$ supplemented directly to the grains. Supplemented zinc tended to be higher available.

REFERENCES

1. Davies, N.T. and S.E.Olpin, 1979, Brit.J.Nutr. 41:591 - 603.-

2. Lantzsch, H.-J., H.Schenkel, and Ingrid Nickerl, 1978. In "Trace Element Metabolism in Man and Animals" (ed. M.Kirchgessner)(ATW, Freising-Weihenstephan) pp 460-463.-

3. Menke, K.H., H.-J.Lantzsch, and H.Schenkel, 1978. In "Trace Element Metabolism in Man and Animals" (ed.M.Kirchgessner)(ATW, Freising-Weihenstephan) pp 456-459.-

4. Morris, E.R., and R. Ellis, 1980, J. Nutr. 110:2000 - 2010.-

5. Oberleas, D., 1975, Proc. Western Hemisphere Nutr. Congr. IV, 156-161.-

6. Schormüller, H., R.Höhne und G. Würdig, 1956, Dt. Lebensmittel-Rundschau 52:213 - 224.-

ACKNOWLEDGEMENT

This investigation was supported by Deutsche Forschungsgemeinschaft, SFB-142, Teilprojekt C.2 and C.6

DISCUSSION

WELCH (USA). Couldn't you invoke a different interpretation of your results - that increases in zinc fertilisation increased the form of available zinc in the seed and subsequently the zinc available to the animals? With this hypothesis you don't have to include a phytate effect.

LANTZSCH (Germany). I cannot exclude this possibility. The molar ratio phytic acid/ zinc is a calculated parameter based on the chemical analysis of phytic acid and zinc. Experiments are on the way by Dr Marschner in which the distribution and binding of the zinc taken up by the wheat grains after zinc fertilisation is investigated in more detail. In other experiments with wheat in which the molar ratios changed following phosphorus fertilisation by elevating the phytic acid concentrations without changing the zinc concentrations we got similar results with respect to the zinc availability.

NEWNHAM (Australia). Dr Shute in his book says that phosphate added to the soil tends to inhibit the availability of zinc. In your experiments was the soil fertilised by phosphate?

LANTZSCH (Germany). I have no information on whether or not phosphate fertilisers were used. The plant experiments were done by my colleagues from the Institute of Plant Nutrition.

HURLEY (USA). I would like to ask if it is likely that the fertilisers and river sludge put on the crops might have influenced the levels of other elements as well. Perhaps cadmium or copper might have been affected and these in turn affected zinc.

LANTZSCH (Germany). Only in the wheat and barley samples grown on sludge-treated soils was there a remarkable increase in essential elements, including cadmium at nearly 1 ppm. The experimental diets containing the grains were adjusted to the same level of essential elements as controls so an interaction seems unlikely. A negative effect of cadmium may exist though zinc availability was higher than 90%.

INFLUENCE OF DIETARY PROTEIN ON ZINC AVAILABILITY FROM BREAD IN RATS

Joan M. McKenzie* and Neill T. Davies

Department of Nutritional Biochemistry,
Rowett Research Institute, Bucksburn,
Aberdeen AB2 9SB, Scotland.

INTRODUCTION

The composition of a diet, especially its content of phytate, fibre, calcium and total zinc, appears to be of great importance for the ability to utilize zinc. The studies of Reinhold and his co-workers (reviewed by Reinhold et al, 1976) have shown that, in the Iranian situation, adverse effects on zinc utilization are due to both the phytate and the fibre contents of the diet, with fibre probably being of greater significance. Other studies, using the rather lower fibre contents more likely to be found in "fibre-rich" Western diets, have provided little evidence that the fibre would adversely affect mineral availability (Sandstead et al, 1978). Recent work (Sandstrom et al, 1980) has indicated a positive effect on the availability of zinc by the amount of protein in the diet, irrespective of the calcium content, and this might help to explain the apparent differences among the earlier findings. The study to be described was conducted to determine whether dietary protein concentration influenced the availability of zinc from wholemeal bread, and whether any observed influence was an effect of protein concentration on the phytate, the fibre, or phytate and fibre together.

METHODS

Weanling Hooded Lister male rats of the Rowett Institute strain were allocated randomly into groups of six for the dietary treatments. Three concentrations of protein were fed: 10%, 20% or 30%. At each protein concentration there was a control diet (egg albumin); a diet with 40% wholemeal bread; a diet with an equivalent concentration (0.3%) of phytate (Na phytate) to that containing bread; a diet with an equivalent concentration (3.6%) of neutral detergent fibre to that containing bread; and a diet with equivalent concentrations of phytate and fibre. All diets contained 15 ppm zinc, 1.2% calcium, and were made isocaloric by adjusting the amounts of arachis oil and wheat starch. The phytate: Zn ratio in the bread and phytate diets was 20:1. The animals were housed individually and were maintained on the dietary treatments for 28 days with daily measurement of body weight. The animals were then killed, hair and plasma samples were taken for analysis, and the whole carcass minus the intestinal contents was analysed for estimation of whole body retention of zinc. Zinc analyses were performed by flame atomic absorption spectrophotometry.

The data were analysed statistically using joint analysis of variance separately at each protein concentration, and also on the data combined to determine whether there was an overall effect of bread, phytate, or fibre. Interactions between phytate and fibre, and the effect of protein on any interaction were also investigated.

RESULTS

At each concentration of protein, bread and phytate significantly reduced weight gain, while fibre had no effect (Table 1). At 10% protein there also appeared to be a marginally significant interaction between phytate and fibre such that accumulative weight gain was even further reduced compared to the effect of phytate or fibre alone. Overall, the improvement in weight gain with increasing dietary protein was highly significant ($P<0.001$); furthermore the effect of bread and phytate on weight gain was partially mitigated at a high protein concentration. Plasma zinc concentration was significantly reduced by bread or by phytate (Table 1); overall there was a linear relationship ($P<0.001$) between dietary protein concentration and plasma zinc concentration. There was no interaction between phytate and fibre, or between protein concentration and the bread or phytate effect. Hair zinc concentration did not differ among the dietary treatments at 10% protein (Table 1), but it was reduced by bread and phytate at 20% protein, and by phytate at 30% protein. On the combined data the relationship between hair zinc and protein

*Present address: Nutrition Department, University of Otago, Box 56, Dunedin, New Zealand.

TABLE 1. CUMULATIVE WEIGHT GAIN, PLASMA ZINC, HAIR ZINC AND ZINC RETENTION FOR EACH DIETARY TREATMENT

	10% PROTEIN	20% PROTEIN	30% PROTEIN
(a) Cumulative weight gain (g)			
Control	89.6 ± 6.6	156.0 ± 3.8	140.0 ± 4.6
Bread	77.6 ± 3.6*	94.4 ± 1.4***	110.0 ± 3.4***
Phytate	55.0 ± 1.8***	70.4 ± 4.5***	73.5 ± 2.3***
Fibre	99.7 ± 4.7	149.0 ± 4.1	152.0 ± 5.7
Phytate + Fibre	45.4 ± 2.1$^{\phi}$	74.1 ± 3.7	83.5 ± 4.3
(b) Plasma zinc concentration (μg/ml)			
Control	1.17 ± 0.04	1.56 ± 0.08	1.55 ± 0.07
Bread	0.91 ± 0.07*	0.84 ± 0.06***	1.03 ± 0.05***
Phytate	0.65 ± 0.03***	0.80 ± 0.09***	0.90 ± 0.08***
Fibre	1.27 ± 0.05	1.56 ± 0.09	1.61 ± 0.10
Phytate + Fibre	0.77 ± 0.04	0.88 ± 0.10	0.99 ± 0.10
(c) Hair zinc concentration (μg/g)			
Control	163 ± 3	188 ± 3	188 ± 2
Bread	171 ± 4	161 ± 5**	161 ± 6**
Phytate	170 ± 2	169 ± 4**	184 ± 6
Fibre	173 ± 4	184 ± 7	182 ± 5
Phytate + Fibre	171 ± 5	175 ± 5	176 ± 3
(d) Zinc retention (% of dietary intake)			
Control	38.8 ± 2.2	65.7 ± 2.0	63.0 ± 1.6
Bread	28.9 ± 1.8**	35.5 ± 2.6***	41.0 ± 3.2***
Phytate	26.1 ± 1.2***	32.1 ± 2.2***	38.6 ± 2.2***
Fibre	49.8 ± 1.2	62.7 ± 1.0	65.9 ± 1.8
Phytate + Fibre	19.7 ± 2.7$^{\phi\phi\phi}$	38.4 ± 1.3$^{\phi}$	46.3 ± 4.8

Values are mean ± SEM for six rats per dietary treatment.
*, **, *** significantly different from controls, $P<0.05$, $P<0.01$, and $P<0.001$, respectively.
ϕ, $\phi\phi\phi$ significant fibre-phytate interaction, $P<0.05$, $P<0.001$, respectively.

concentration was linear.

Whole body retention of zinc was significantly reduced by bread and by phytate at each protein concentration, and the same was true for the data combined. As well, there was a significant interaction between phytate and fibre, especially at 10% protein (Table 1). With all protein levels combined there was a significant effect of bread and of phytate ($P<0.001$), and a marginal effect of fibre ($P<0.05$) on zinc retention. The interactions between protein concentration and the bread effect were such that relative to the controls there was a greater reduction in zinc retention at a higher protein concentration. The phytate-fibre interaction was influenced by protein concentration in an identical way.

DISCUSSION

Both bread and phytate adversely affected all measures of zinc availability at all concentrations of dietary protein except for hair zinc concentration at 10% protein. Fibre did not affect zinc availability at individual protein concentrations by any of the parameters used. In other words, the effect of bread in this experiment appeared to be due mainly to the phytate but not the fibre in the diet, confirming earlier work (Davies, 1979). However the effects of bread and phytate were influenced by dietary protein content: on the one hand the higher protein content appeared to afford some alleviation (ie on weight gain), while for zinc retention the low protein content per se (10%) might have diminished the relative difference between the dietary treatments, so that the full expression of the effect of bread or phytate was observed only at 20% or 30% protein. Furthermore there was an interaction between phytate and fibre especially at 10% protein. Such an interaction has not been reported previously, and indicates that there could be potentially important implications for zinc availability from diets in which total protein content is not high and in which a considerable proportion of energy is derived from cereals.

REFERENCES

1. Reinhold, J.G., Faradji, B., Abadi, P., and Ismail-Beigi, F. 1976. In "Trace Elements in Human Health and Disease, Vol 1, Zinc and Copper". (Academic Press, N.Y.) pp 163-180.

2. Sandstead, H.H., Munoz, J.M., Jacob, R.M., Klevay, L.M., Reck, S.J., Logan, G.M., Dintzis, F.R., Inglett, G.E., and Shuey, W.C., 1978, Am. J. Clin. Nutr. 31: S180-S184.

3. Sandstrom, B., Arvidsson, B., Cederblad, A., and Bjorn-Rasmussen, E., 1980, Am. J. Clin. Nutr. 33: 739-745.

4. Davies, N.T., 1979, In Proc. Symp. "Dietary Fibre". (Kellogg Co. of Great Britain, London) pp 113-123.

DISCUSSION

MOMČILOVIĆ (Yugoslavia). How did you decide on a phytate level of 0.3% and a fibre level of 3.6%?

McKENZIE (New Zealand). These were the concentrations contained in the diet that had 40% wholemeal bread.

GOLDEN, M (Jamaica). Coprophagia could make a difference in such experiments. Did you take any precautions to prevent it?

McKENZIE (New Zealand). The animals were housed in cages that had grids on the bottom.

HILL (England). I don't know what proportion of the iron in your diets was from added ferrous sulphate, but if it was small it maybe important to consider the possibility of a reduced availability of iron.

McKENZIE (New Zealand). There were very slight adjustments in iron intake to make sure that all diets had exactly the same iron content and in one or two diets some iron sulphate was added, but I don't think reduced iron availability is likely.

MILLS (Scotland). Iron retention has been monitored in experiments with the diets that Dr. McKenzie was using and with others that we have used previously in phytate studies. The results do show a decrease in iron retention, but no evidence of a decline in haemoglobin concentration or other adverse effects. One other point that I think supports this is that in Dr. McKenzie's work when additional zinc alone was included in the diet containing bread (25 ppm in drinking water), growth and blood zinc concentrations were restored to normal.

EFFECT OF DEHULLING ON ZINC AVAILABILITY OF FIELD BEANS

H.-J.Lantzsch and S.E.Scheuermann

Institute of Animal Nutrition
University of Hohenheim
PO BOX 700562,7000 STUTTGART 70
FEDERAL REPUBLIC OF GERMANY

Evidence has been obtained from the literature, that polyphenols, like tannins, which are widely distributed throughout the vegetable kingdom, form complexes with metal ions, i.e. iron, zinc and aluminium (3,4,9,10). In vivo it was demonstrated by Disler et al. (2, 3) and Roy and Mukherjee (6, 7, 8) that tannins decreased iron availability for intestinal absorption in normal humans and laboratory rats. Data are not available as to their influence on zinc availability in vivo.

The testa of colored-flowering field beans (Vicia faba L.) are known to contain high amounts of tannins while that of white-flowering beans are free. The objective of the experiment carried out was to elucidate the effect of field bean tannins on zinc availability.

MATERIAL AND METHODS

A white-flowering cultivar (Rowena) and a colored-flowering cultivar (Deiniol) of field beans was used in this experiment. The whole beans and their dehulled kernels were incorporated in semipurified glucose/ovalbumin basal diets at a level to provide 10 ppm total zinc in substitution for corresponding nutrients of the basal diet. The ingredients of the basal diet contribute 0.6 ppm zinc. The poor hulls, obtained by hand procedure, were added to 10 ppm zinc ($ZnSO_4$) containing basal diets at the expense of glucose and cellulose. Two semipurified control diets with 5 and 10 ppm zinc from $ZnSO_4$ were included in the experiment. All diets had the same composition with respect to crude nutrients, minerals (e.g. 0,75% Ca on dry matter base) and trace elements.

Relative availability of zinc for intestinal absorption ($ZnSO_4$=100) was determined in growing male rats (6 animals/group) of 90 g of body weight at the beginning by means of the EDTA-test (5), after a preceding 7 day balance period, in which the utilization of the dietary zinc for body retention was measured. The animals received a daily food amount of 13.0 g (air dry base) and had free access to double distilled water.

Crude nutrients were analyzed by conventional methods, minerals and trace elements by atomic absorption spectrophotometry after dry ashing, condensed tannins after the vanillin-hydrochloric acid method of Burns (1) and phytic acid after the method of Schormüller et al. (11).

Table 1 Concentrations of condensed tannins, phytic acid and zinc, and molar ratios phytic acid/zinc in white- and colored-flowering cultivars of field beans (values expressed per kg of dry matter)

	tannin[1] g	phytic acid g	zinc mg	molar ratio
cv."Rowena" (white-flowering)				
whole bean	-	18.4	36.1	50.4
kernel	-	20.3	38.1	52.7
testa	n.d.[2]	n.d.	6.0	-
cv."Deiniol" (colored-flowering)				
whole bean	5.2[3]	16.1	55.3	28.9
kernels	-	18.0	57.4	30.9
testa	43.6	n.d.	9.0	-

[1] values based on catechin standards; [2] not detectable; [3] calculated from testa analysis and testa portion (12.0%)

RESULTS AND DISCUSSION

Some analytical results of the two field bean types used are presented in table 1. In the hulls of the white-flowering beans no condensed tannins could be detected, whereas in those of the colored-flowering beans 4.4% were found. The whole beans of both cultivars had similar phytic acid concentrations. The considerable lower zinc concentration of the tannin-free Rowena caused a much wider molar ratio phytic acid/zinc of 50 compared to 29 of the tannin-containing Deiniol. The hulls of both cultivars contained no measurable amounts of phytic acid.
Tannin-containing hulls added to semipurified diets had no detrimental effect on availability (figure 1) or utilization (figure 2) of zinc from $ZnSO_4$ as compared to diets supplemented with tannin-free hulls or to the unsupplemented control diets. Consequently, avai-

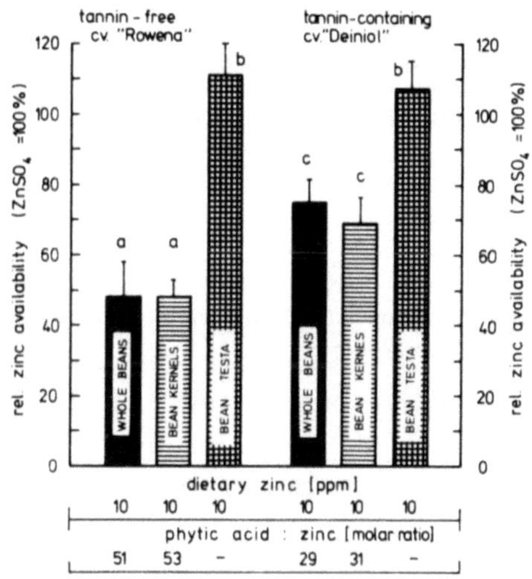

Figure 1 Availability of zinc for intestinal absorption from tannin-free and tannin-containing field beans in growing male rats, fed on a 10 ppm dietary zinc level (mean ± s.e.; n = 6).

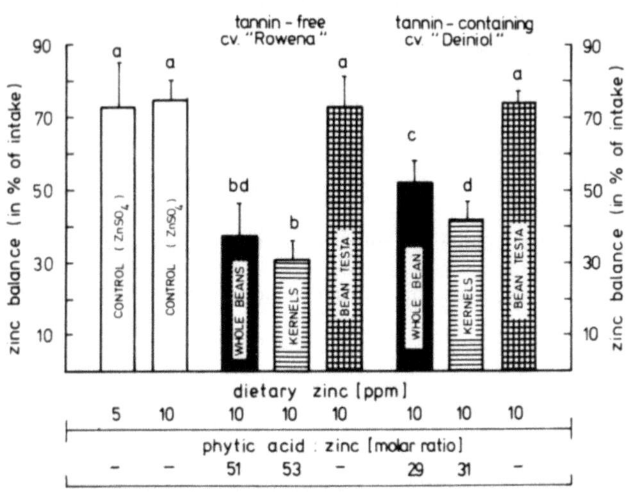

Figure 2 Utilization of zinc for body retention from tannin-free and tannin-containing field beans in growing male rats, fed on a 10 ppm dietary zinc level (mean ± s.e.; n = 6).

lability and utilization of zinc from dehulled beans was not improved as compared to whole beans. Therefore, it must be concluded that condensed tannins of faba beans, applied in the form of whole beans or isolated hulls and in dietary concentrations corresponding to those likely to occur under practical feeding conditions (20 to 30% whole bean portion) did not affect availability for intestinal absorption or utilization for body retention of zinc, neither from beans itself nor from $ZnSO_4$.
The significantly higher availability (75% and 69% vs. 48 and 48%; $p < 0.001$) and utilization (52 and 42% vs.38 and 31%; $p < 0.01$) of zinc from both, the whole and dehulled beans of the tannin-containing than from tannin-free cultivars, respectively, may be explained by the different phytic acid/zinc molar ratios (29 and 31 vs. 51 and 53).

SUMMARY

In growing male rats availability for intestinal absorption and utilization for body retention of zinc from tannin-free and tannin-containing whole or dehulled field beans (Vicia faba L.) or from $ZnSO_4$ after supplementation of isolated hulls of both bean types to semipurified diets were measured. Under the experimental conditions used no influence of condensed tannins could be stated. Availability and utilization of bean zinc depended on phytic acid/zinc molar ratio.

REFERENCES

1. Burns, R.E., 1971, Agron. J. 63:511 - 512.-

2. Disler, P.B., S.R. Lynch, R.W. Charlton, J.D. Torrance, T.H. Bothwell, R.B. Walker, and Fatima Mayet, 1975, Gut 16:193 - 200.-

3. Disler, P.B., S.R. Lynch, J.D. Torrance, M.H. Sayers, T.H. Bothwell, and R.W. Charlton, 1975, S. African J. Med. Sci. 40:109 - 116.-

4. Jurd, L., and T.A. Geissman, 1956, J. Org. Chem. 21:1395 - 1401.-

5. Menke, K.H., H.-J. Lantzsch, and H. Schenkel, 1978. In "Trace Element Metabolism in Man and Animals" (ed. M.Kirchgessner)(ATW, Freising-Weihenstephan) pp 456-459.-

6. Roy, S.N., and S. Mukherjee, 1979, Indian J. Biochem. Biophys. 16:93 - 98.-

7. Roy, S.N., and S. Mukherjee, 1979, Indian J. Biochem. Biophys. 16:99 - 104.-

8. Roy, S.N., and S. Mukherjee, 1979, Indian J. Biochem. Biophys. 16:151 - 157.-

9. Seth, P.C.C., D.R. Clandinin, and R.T. Hardin, 1975, Poultry Sci. 54:626 - 629.-

10. West, L.G., and J.L. Greger, 1978, Nutr. Rep. Int. 18:117 - 124.-

11. Schormüller, H., R. Höhne, and G. Würdig, 1956, Dt. Lebensmittel-Rundschau 52:213-224.-

ACKNOWLEDGEMENT

The authors wish to thank the Landessaatzuchtanstalt of Stuttgart-Hohenheim which supplied the bean seeds.
This investigation was supported by Deutsche Forschungsgemeinschaft, SFB-142, Teilprojekt C.2.

A VASCULAR PERFUSION TECHNIQUE FOR THE STUDY OF ZINC ABSORPTION

Linda Steel, Dorothy McMaster and A.H.G. Love

Department of Medicine, The Queen's University of Belfast,
NORTHERN IRELAND.

Vascular perfusion can be used to investigate transport of zinc by the rat small intestine (1,2). The present technique utilizes washed bovine erythrocytes and bovine serum albumin in the vascular medium.

METHODS AND MATERIALS

Animals: Male wistar rats (200-300 g) were used in the experiments. The rats were divided into two groups, one was fed a standard diet adequate in zinc (Morton Ltd; zinc content 160 µg/g) and the other was maintained for 6 weeks on a semi-synthetic diet (3); zinc content 0.67 µg/g. Group 2 animals developed symptoms of zinc deficiency. Before perfusion all animals were fasted for 24 hours.

Perfusion: The isolated vascularly perfused intestine preparation was similar to that of Hanson and Parsons (4). The perfusion apparatus consisted of two circuits. One supplied the lumen of the intestine and the other the vascular bed. The luminal reservoir contained 200 ml of a Krebs' bicarbonate solution with 11 mmol/l glucose and a nonabsorbable marker bromsulphalthein sodium (Hynson, Wescott and Dunning, Baltimore). In experiments where zinc was to be included the pH of the medium was first adjusted to pH 6.9 by the addition of glycine. The zinc was added in the form of zinc aspartate (gift from Kohler Group, Alsbach). The lumen medium was pumped via a Gilson minipuls pump through a water manometer to a cannula situated in the duodenum. The effluent was collected from a cannula in the ileum.

The vascular reservoir contained 200 ml of Krebs' bicarbonate medium with 2% w/v bovine serum albumin Fraction V (Miles Laboratories Ltd.), 11 mmol/l glucose and 20% v/v washed bovine erythrocytes. This flowed through a Watson-Marlow pump to a filter and an oxygenator. Some of the medium was returned to the reservoir through a side arm while some was pumped via a Gilson minipuls pump, an anaeroid dial gauge and a bubble trap, to a cannula situated in the aorta near the origin of the superior mesenteric artery. The vascular effluent was collected from a cannula in the portal vein. The vascular and luminal flow rates were set at 1 ml/min.

ZINC MEASUREMENTS

The zinc disappearance from the lumen was measured in µg/ml every 10 minutes for 2 hours and calculated in µg/g dry weight intestine/hr. The vascular appearance of zinc was measured not in the whole medium but in the vascular supernatant after the removal of the bovine erythrocytes and the results were calculated in µg/g dry weight/hr. The possibility that the red blood cells had removed zinc from the supernatant was investigated *in vitro*. Aliquots of bovine erythrocytes were incubated with continuous inversion at 37°C for 1 hour in oxygenated medium containing different concentrations of zinc from 0 to 2 µg/ml. Analysis of the red cells and the supernatant was carried out. The percentage recovery of zinc in the supernatant was 99.6 ± 1.9% (n = 6). Samples of tissue were dried overnight at 110°C and ashed in a 100W plasma asher (Nanotech, Manchester) for 5-7 days. The ash was taken up in 2% v/v HCl, and the zinc content calculated as µg/g dry weight.

RESULTS

The viability of the vascular perfusion preparation was tested as follows:
a) motility, vascularity and secretory activity; (b) glucose consumption and (c) the post-perfusion histology of the intestine.

In satisfactory preparations no peristalsis was detected while the gut was perfused. This may have been helped by the presence of a positive distension pressure in the lumen. The water manometer in the luminal circuit had pressure readings of between 18-26 cm H_2O.

On disconnecting the intestine from the vascular circuit at the end of the experiment peristalsis was observed. There was no intraluminal bleeding or damage of the blood vessels in the intestine. The inflow and outflow concentration of bromsulphalthein was measured for each 10 min period and there was no significant change in the concentration of this nonabsorbable marker in the two groups of rats (P>0.10) indicating that there was no hypersecretion of water. Glucose was removed from the lumen and vascular circuits of both groups of rats.

The final test of viability was the post-perfusion histology of the intestine. This was as normal after the two hours perfusion. By electronmicroscopy the intestinal epithelium has a well preserved ultrastructure with the brushborder intact although some mitochondria appeared swollen.

The zinc disappearance from the lumen and the zinc appearance in the vascular supernatant were measured and calculated in µg/g dry weight of intestine/h. The results of experiments with 0, 1, 5, 20 and 50 µg/ml of zinc in the lumen are shown in the accompanying figure.

When no zinc was added to the luminal medium none was detected in the luminal effluent but there was a small amount of zinc (2±0.2 µg/g/h Group 1 rats, n = 8 and 2.5±0.4 µg/g/h Group 2 rats, n = 4) appearing in the vascular supernatant. When zinc was added to the lumen of both groups of rats there was a loss of zinc from the luminal medium and an increase in zinc in the vascular effluent. When the values of zinc appearing in the vascular circuit at luminal concentrations of 0, 1, 5 and 50 µg/ml were compared with the value at 20 µg/ml there was a significant difference in the zinc content in the vascular effluents for both groups of rats. The luminal disappearance rates of zinc at 1, 5 and 50 µg/ml were also significantly different from the uptake at 20 µg/ml (P<0.001) for the Group 1 rats and the Group 2 rats.

The zinc content of the intestines in both groups of rats perfused with zinc concentrations of 0, 20 and 50 µg/ml were measured and more zinc was present in the intestinal ash when zinc was added to the luminal medium (P<0.001).

To assess the zinc status of the rats used in each group one testis per rat was analysed for zinc content. This organ was not included in the intestinal vascular perfusion circuit. In Group 1 with 0, 20, 50 µg/ml in the luminal perfusate the testicular zinc content was 176±3, 175±6 and 179±7 µg/g dry weight respectively. In Group 2 at similar luminal concentrations the tissue content was 155±10, 109±14 and 80±13 µg/g dry weight.

The vascular perfusion technique has advantages over other gut preparations in that changes in the vascular and luminal circuits may be determined simultaneously. The oxygen availability to the tissue is improved by oxygenation of the mucosal solution and the presence of red blood cells to supply oxygen to the vascular bed.

The preparation remained viable as judged by the absence of visible and histological damage to the intestine due to the perfusion procedure and glucose disappearance remained at a steady state for at least two hours. The preparation demonstrated that zinc disappeared from the lumen, some of this accumulating in the tissue and some appearing in the vascular circuit.

The zinc status of the two groups of rats was assessed by analysis of the zinc content of the testes. The zinc content of the testes of Group 1 and Group 2 rats with no zinc added was not different (P>0.10) indicating that the Group 2 rats were not sufficiently zinc deprived although the zinc content of their intestine was significantly less than the normal fed rats (P<0.05). However, the zinc content of testes at 20 µg/ml and 60 µg/ml was significantly less in the Group 2 rats (P<0.005 20 µg/ml and P<0.001 50 µg/ml).

REFERENCES

1. Smith, K.T., R.J. Cousins, B.L. Silbon and M.L. Failla, 1978, J. Nutr. 108:1849 - 1857.

2. Smith, K.T. and R.J. Cousins, 1980, J. Nutr. 110:316 - 323.

3. Elmes, M.E., 1974, Ph.D. Thesis, The Queen's University of Belfast.

4. Hanson, P.J. and D.S. Parsons, 1976, J. Physiol. 255:775 - 795.

ACKNOWLEDGEMENTS

I am indebted to I. Banks, Department of Pathology, The Queen's University of Belfast, for the histological analysis of tissues.

ZINC UPTAKE IN THE VASCULARLY PERFUSED SMALL INTESTINE PREPARATION

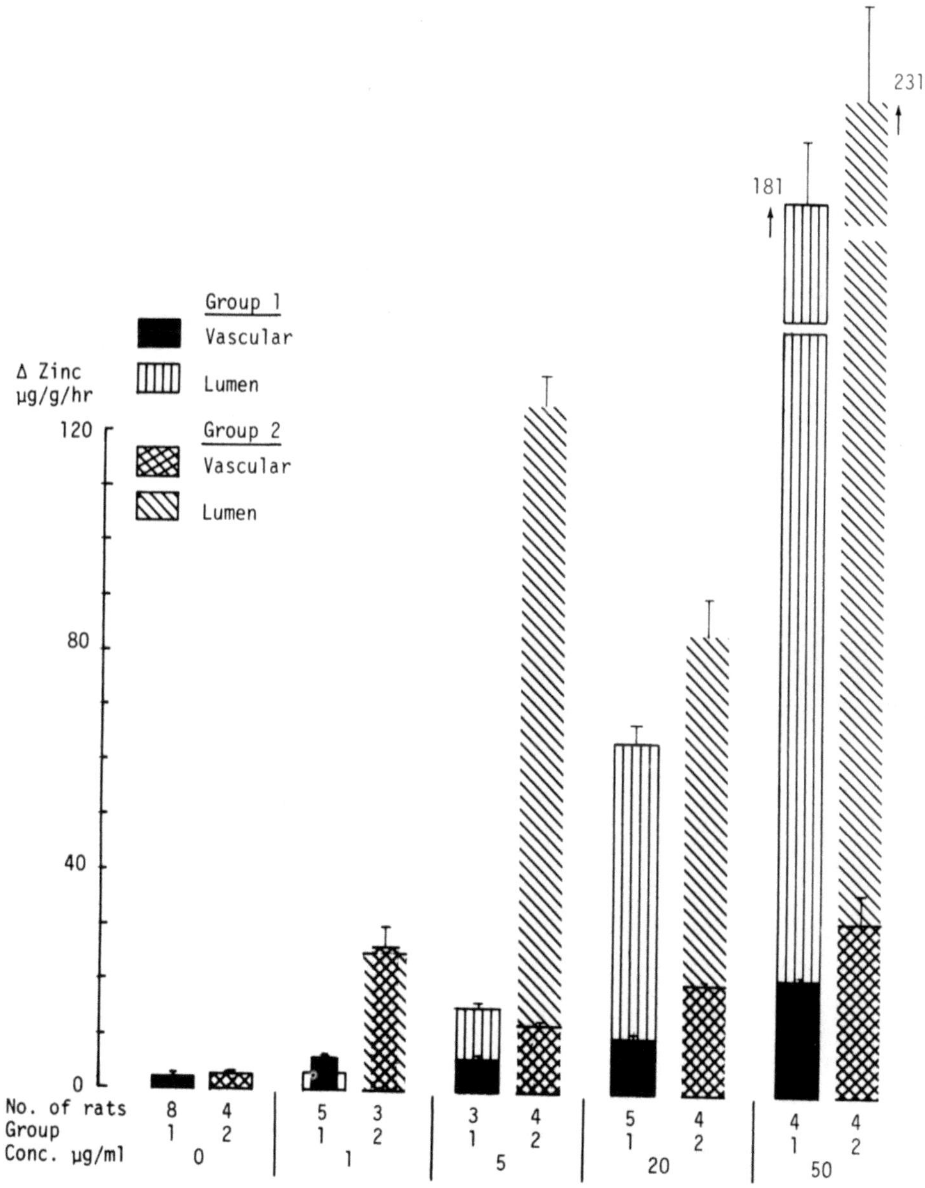

DISCUSSION

SCHWARZ (Germany). Did you examine the effect of different rates of flow of vascular effluent?

STEEL (N. Ireland). The lumen and vascular media were pumped at 1 ml per minute. We did initially try different flow rates and found that 1 ml per minute caused the least mechanical damage to the intestine.

ANON. Windmueller and Spathe found that glucose is not the physiological fuel for the small bowel in either fed or fasted rats. Glutamine and ketone bodies were used. Have you tried these substrates in your system?

STEEL (N. Ireland). No, we have always used glucose.

ZINC ABSORPTION BY VASCULARLY PERFUSED SMALL INTESTINE

Dorothy McMaster, Linda Steel and A.H.G. Love

Department of Medicine, The Queen's University of Belfast, NORTHERN IRELAND.

The vascular perfusion technique described previously (Steel et al, this symposium) has been used to study the influence of the combined endogenous secretions of the pancreas and intestine on the glucose consumption and zinc uptake by the small intestine of rats fed on a diet adequate in zinc and of rats fed on a diet depleted of zinc. In all perfusion experiments zinc aspartate was added to the luminal medium at pH 6.9 to give 20 µg Zn/ml. The zinc content of the testis was used to confirm the zinc depletion of the animals.

The description of the diets, the perfusion media and the testicular zinc content of the rats used in the study are shown in Table 1. The 11 animals of the zinc deprived reference group maintained for 6 weeks on their zinc depleted diet (Group 2) had less than 50% of zinc in the testis when compared with rats on a normal diet (Group 1). Supplementation of 8 animals for 3 days with 40 µg/g dry diet (maximum intake 1,200 µg) was not sufficient to raise the level of zinc in the testis to normal. The level of zinc in unperfused intestines from Group 2 (73±4 µg/g) and Group 3 (111±4 µg/g) animals was also considerably lower than Group 1 (161±24 µg/g). Care must be taken in the interpretation of results found with animals which are not of the same nutritional status.

In all experiments the intestine is fed with 11 mmol/l glucose from both circuits. In the absence of the influence of the liver and pancreas we have found that the perfused intestine removes glucose from either or both circuits. To study the influence of endogenous pancreatic hormones on the basal uptake of the glucose and on the absorption of zinc from the lumen the vascular effluent, which contains secretions from the head of the pancreas, was circulated to the vascular bed via the vascular reservoir. Table 2 shows the glucose consumption of the intestines in this series. The removal rates are given as µmol/g/h. Because the zinc deprived rats of Groups 2 and 3 have intestines which weigh less than those of Group 1, a direct comparison is not possible and, therefore, the % increase in the total glucose consumption is used as a measure of the stimulus from the pancreatic secretions. With the rats fed on a normal diet (Group 1) the inclusion of the pancreatic secretions in the vascular medium stimulated the glucose consumption of the intestines by 65%. In contrast, the zinc depleted rats in Groups 2 and 3 showed lower increases of 16% and 11% respectively. This may be a result of the failure of the Zn deficient pancreas to secrete insulin and/or of the intestine to respond to the stimulus.

Figure I shows the influence of the pancreas on the uptake of zinc from the lumen by the perfused intestine and the appearance in the vascular medium. With the group of 8 normally fed rats (Group 1) the presence of the pancreatic secretions resulted in an increased removal of zinc from the lumen (171%) and a concomitant 70% increase in the appearance in the vascular medium. In comparison, Groups 2 and 3 rats gave no increase in zinc removal from the lumen. The vascular increase with Group 2 rats was 45% and with Group 3 rats 24%. Analysis of the testes (Table 1) showed no difference between Groups 2 and 3. It is of interest that endogenous pancreatic secretions stimulated the removal of both glucose and zinc from the lumen of normal rats and not from zinc depleted rats.

In the absence of endogenous pancreatic influence the effects of glucagon and picolinic acid on the uptake of zinc by the intestines from rats fed on a normal diet have also been studied. The addition of 1.0 µg/ml pork glucagon (Novo Industries Ltd.) to the vascular medium led to an increase of 105% in the removal of zinc from the lumen (Figure 2). The vascular appearance rate was not altered. Thus, when compared to the endogenous pancreatic secretions, pork glucagon appears to have a similar effect on the removal of zinc from the lumen of the rat perfused small intestine.

With a series of 4 intestines perfused with 20 µg Zn/ml of luminal medium the addition of 5.0 mmol/l picolinic acid (Sigma Chemical Co.) to the vascular medium resulted in a release of 49±17 µg/g/h of zinc to the luminal perfusate (Figure 2). When picolinic acid was perfused with zinc through the lumen itself a secretion of zinc (32±3 μg/g/h) to the luminal perfusate was again seen. This finding was not a result of the movement of water. The implication of this result is that caution should be taken in the interpretation of results obtained in studies using picolinic acid and in the use of this agent in the treatment of zinc deficiency.

Table 1

Description of diets, the perfusion media and the testicular zinc of the rats used in the study

GROUP	PERFUSION MEDIUM	Testis Zinc μg/g
GROUP 1 Normal Diet Zinc Content 167 μg/g	Reference animals	163± 4 n=17
	Basal medium	172±8 n=3
	Pancreatic and intestinal secretions in vascular medium	174±2 n=3
	Glucagon in vascular medium	172±2 n=3
	Picolinic acid in vascular medium	176 n=1
	Picolinic acid in luminal medium	170±4 n=4
GROUP 2 Deficient Diet Zinc 0.67 μg/g 6 weeks	Reference animals	75±3 n=11
	Basal medium	109±14 n=4
	Pancreatic and intestinal secretions in vascular medium	105±11 n=4
GROUP 3 Deficient Diet Zinc 0.67 μg/g 6 weeks Prestudy Zn 400 μg/day 3 days	Reference animals	87±4 n=4
	Basal medium	97±19 n=3
	Pancreatic and intestinal secretions in vascular medium	118±30 n=3

Table 2

Glucose consumption by the vascularly perfused small intestine

		Vascular Circuit µmol/g/h	Lumen Circuit µmol/g/h	Total µmol/g/h	% Total Increase
GROUP 1	Basal uptake	102±4 (5)	82±8 (4)	184	65
	Stimulated uptake	168±9 (8)	136±7 (4)	304	
GROUP 2	Basal uptake	160±4 (4)	144±14 (4)	304	16
	Stimulated uptake	250±19 (4)	104±9 (4)	354	
GROUP 3	Basal uptake	131±11 (4)	115±4 (4)	246	11
	Stimulated uptake	161±10 (4)	112±8 (4)	273	

FIGURE 1

THE EFFECT OF ENDOGENOUS SECRETIONS FROM THE SMALL INTESTINE AND PANCREAS ON ZINC ABSORPTION

FIGURE 2

THE EFFECT OF GLUCAGON AND PICOLINIC ACID ON ZINC ABSORPTION

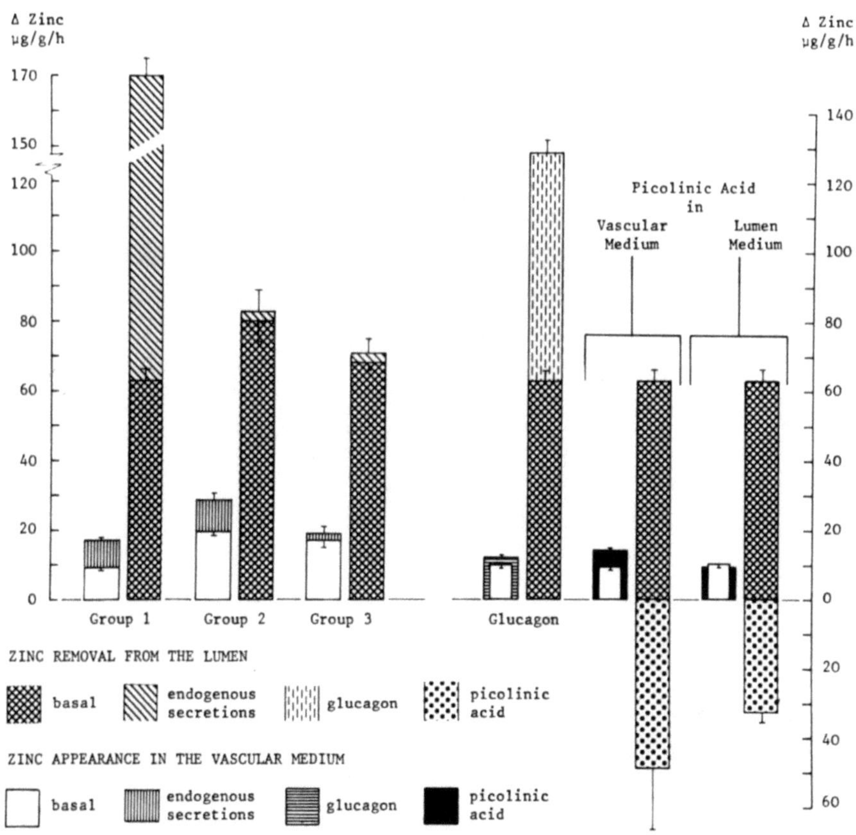

DISCUSSION

ELMES (Wales). Have you any suggestions on why glucagon had such a dramatic effect on zinc absorption?

McMASTER (N.Ireland). It may be 'opening the door' to let the zinc into the gut but it does not seem to let the zinc out of the gut - in zinc depleted rats, at least.

WEIGAND (Germany). Did you also measure the movement of picolinic acid from the vascular bed to the lumen, and vice-versa?

McMASTER (N.Ireland). Not quantitatively - we think that it goes from the lumen to the vascular side because the colour of the medium seems to change slightly.

ABDULLA (Sweden). I am a bit surprised that glucagon increases the absorption of zinc because patients with glucagonoma tend to have low serum zinc values and skin lesions.

McMASTER (N.Ireland). Perhaps we have to consider whether the glucagonoma itself has high levels of glucagon that complexes the zinc and makes it unavailable.

ABSORPTION OF ZINC AND MANGANESE IN RELATION TO AGE

M. Kirchgeßner, E. Weigand and F. J. Schwarz

Institut für Ernährungsphysiologie
Technische Universität München
D-8050 FREISING-WEIHENSTEPHAN
FED. REP. GERMANY

Trace element absorption in man and animals is influenced by a number of factors, especially by the physiological stage of the body, composition of the diet, and physico-chemical nature of the element. Studying absorption by the simple conventional balance technique is often prevented, because in the case of many trace elements, including zinc and manganese, fecal excretion may include a variable amount of the element from endogenous pools. Both in vitro and in vivo studies, using young laboratory rats as experimental animals, were designed to specifically investigate zinc and manganese absorption in relation to age during the growing stage.

1. IN VITRO ABSORPTION

In vitro absorption of zinc and manganese was determined with everted jejunal sacs by measuring the uptake of radioactively labeled zinc and manganese of known specific activity into the intestinal tissue and their net transfer into the serosal medium during one-hour incubations. A detailed description of experimental procedures and results is given by Sherif, Schwarz and Kirchgeßner (1). Overall, the series of in vitro studies included both male and female rats ranging in age from 17 days, i.e. several days before weaning, up to 83 days. Figures 1 and 2 illustrate the change in zinc and manganese absorption up to the age of 50 days. There was a sharp decline both in the mucosa-to-serosa transfer and in the tissue uptake between 17 and about 35 days. Thereafter, these absorption parameters indicate just a slight decrease with age. As checked in the case of zinc, sex of the animals had no significant effect on this age response.

Results shown in the Figures 1 and 2 were obtained with young rats that had free access to a commercial diet (55 ppm Zn; 100 ppm Mn) during weaning and later. The initially steep decline in zinc and manganese absorption suggests that this marked response was, perhaps, related to the fact that the young rats started to consume increasing amounts of solid food during this time. Therefore, two additional experiments were conducted in which the progeny had no opportunity to consume either the feed of their does or bedding. Near weaning age, the pups from each litter were divided into two groups. One group was kept with their does and could consume only their mother's milk, while the other group was weaned on a casein-starch diet (88 ppm Zn; 67 ppm Mn) after the age of 18 days. Table 1 shows the intestinal transfer of zinc and manganese. This transfer decreased with age of the animals to a similar extent in either dietary group. The intestinal uptake of zinc and manganese was overall somewhat lower in the case of the rats on the dry diet as compared to their litter mates consuming mother's milk (1). In summary, these studies did not provide clear evidence that the noted decrease in zinc and manganese absorption with age of the animals (Figures 1 and 2) could be related to the nutritional change from sucking milk to consuming dry food during weaning age.

TABLE 1. Mucosa-to-serosa transfer of zinc and manganese (ng/ml · h) by jejunal sacs from young rats consuming either mother's milk (MM) or a dry diet (DD) after 18 days of age

Age	Zinc		Manganese	
(days)	MM	DD	MM	DD
18	35.0		29.4	
21	18.0	15.3	15.5	15.9
24	10.6	6.6	7.3	10.4
29 - 36	6.7	5.9	7.1	10.4

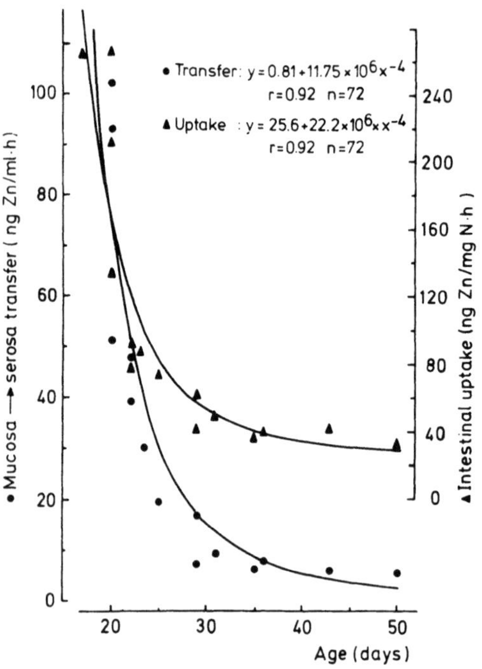

FIGURE 1. Absorption of zinc by isolated jejunal segments from growing rats of different age

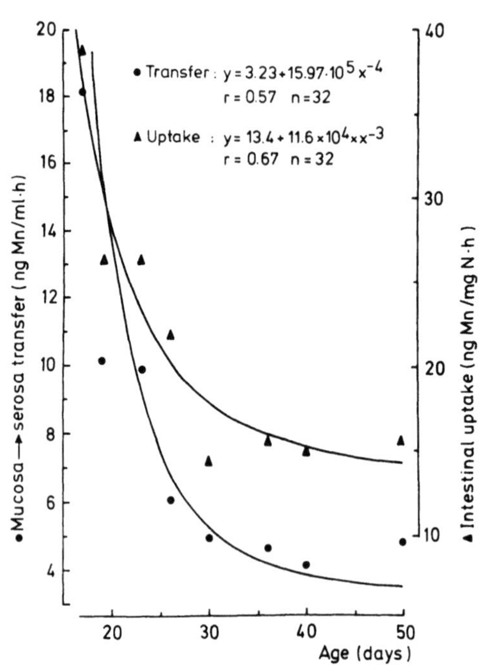

FIGURE 2. Manganese absorption by isolated jejunal segments from growing rats of different age

2. IN VIVO ABSORPTION

Zinc was chosen to further study the relationship of absorption to age under in vivo conditions, since a suitable radioisotope dilution technique (2, 3) was available for assessing true absorption. Two groups of male rats with an age difference of about four weeks between groups were fed the same diet with 12.9 ppm Zn. Details on experimental methods and results have been described by Weigand and Kirchgeßner (4). Table 2 shows the efficiency of zinc utilization by these two groups. True absorption, expressed as percent of zinc intake, was markedly lower in the older rats. These data substantiate our in vitro studies showing that intestinal uptake and transfer of zinc decline with age of the animals. It is also evident from Table 2 that the age difference in percent apparent absorption and net retention was actually larger than that in true absorption. This is explained by the much higher fecal excretion of endogenous zinc in the case of the older age group.

In order to assess the extent to which differences in zinc supply status affect true absorption and endogenous excretion, another closely comparable in vivo experiment was conducted. Six groups of rats of the same age of about five weeks were fed the same basal diet, however, with the zinc concentration varying between groups from 5.6 up to 140 ppm as described in detail elsewhere (5). True absorption, as percent of zinc intake, markedly decreased as the dietary supply increased. However, the actual amounts of zinc absorbed greatly increased as more zinc was supplied. Concomitantly to this rise in the daily rate of true zinc absorption, the rate of fecal excretion of zinc from endogenous pools also substantially increased and eventually compensated for any further increase in actual absorption (5). Accordingly, percent apparent absorption and net retention declined to a greater extent with the increasing zinc supply than percent true absorption. In summary, the efficiency of zinc absorption and retention varied in close homeostatic response to an altered supply-requirement status.

Accordingly, the age differences in the efficiency of zinc absorption and retention shown in Table 2 can be fully attributed to an age-related difference in the supply-demand status. The 9-week old group, given the very same diet as the 5-week old group, consumed about 42 % more food per gram of weight gain (4), bringing about a corresponding difference in zinc intake. Generally, food expenditure per unit mass of liveweight gain substantially increases with age and liveweight of growing animals because of rising energy needs for body maintenance and weight gain. In line with this general improvement in the trace element supply status with age, the older group of rats did actually absorb and retain a greater amount of zinc than the younger group. Thus, the physiological ability to absorb zinc did not decrease with age, but rather adapted to the particular supply status. For the younger rats, the dietary supply of 12.9 ppm Zn was clearly suboptimal. In response to this, the true absorptive and metabolic efficiency and the total utilization of zinc approached their upper limits of 100 % in this age group (Table 2). Total utilization signifies the overall true efficiency of trace element utilization and is the product of the fractional efficiencies, true absorption and metabolic efficiency (6, 7).

TABLE 2. In vivo efficiency of dietary zinc utilization (12.9 μg Zn/g dry diet) by rats of different age (percent values)

Age (weeks)	5	9
Growth rate (g/day)	5.1	5.2
True absorption	98	91
Endogenous fecal zinc	5	17
Apparent absorption	93	74
Net retention	83	69
Metabolic efficiency	100	87
Total utilization	98	80

SUMMARY

In vitro absorption of zinc and manganese assessed by everted intestinal sacs from young rats declined with increasing age, with a particularly marked decline during weaning age. This age response, which occurred alike in both male and female rats, could not be clearly attributed to the nutritional change with weaning. In vivo, percent true and apparent zinc absorption was also markedly lower in older rats as compared to younger animals. These findings are interpreted to reflect the homeostatic adaptation to the age-related shift in the trace element supply status.

REFERENCES

1. Sherif, Y.S., F.J. Schwarz, and M.Kirchgeßner, 1981, Arch. Tierernährg., in press.

2. Weigand, E., and M.Kirchgeßner, 1976, Nutr. Metabol. 20: 314-320.

3. Weigand, E., and M. Kirchgeßner, 1979, Z. Tierphysiol., Tierernährg. u. Futtermittelkde. 42: 44-56.

4. Weigand, E., and M.Kirchgeßner, 1979, Biol. Trace Element Res. 1: 347-358.

5. Weigand, E., and M.Kirchgeßner, 1980, J. Nutr. 110: 469-480.

6. Kirchgeßner, M., H.L. Müller, E.Weigand, E.Graßmann, F.J.Schwarz, J.Pallauf, and H.-P.Roth, 1974, Z. Tierphysiol., Tierernährg. u. Futtermittelkde. 34: 3-17.

7. Weigand, E., and M. Kirchgeßner, 1977, Z. Tierphysiol., Tierernährg. u. Futtermittelkde. 39: 325-341.

DISCUSSION

SALTMAN (USA). Do you use ionic zinc in the everted jejunal sacs or do you complex it with some ligand?

WEIGAND (Germany). In these experiments zinc and manganese were supplied as the chlorides. It was not the objective of this work to investigate the effects of different chelating agents. My colleague Dr Schwarz was responsible for the in vitro work and he has studied this aspect previously.

SALTMAN (USA). The reason I raised the question is that it is possible that age influences the production of endogenous chelates. The mechanism of zinc homeostasis, for instance, may be by regulation of the synthesis and secretion of a chelator that promotes zinc absorption.

WEIGAND (Germany). We know of course that chelating agents can affect absorption. Phytate inhibits it and histidine stimulates it as Dr Schwarz has shown. It would be very rewarding to pursue this aspect further to see what is happening at the cellular and subcellular level as animals get older, especially during the nursing phase.

WELCH (USA). How did you define metabolic utilisation, and how did you determine it?

WEIGAND (Germany). We consider total utilisation to be the product of true absorption and metabolic efficiency. This metabolic efficiency equals total utilisation divided by true absorption. We measured total utilisation as the true retention - the retention after corrections had been made for obligatory endogenous excretion via faeces and urine. The latter losses would occur even if intake was zero. True absorption was determined by the radio-isotope dilution technique. For more detail I refer you to our paper in the Journal of Nutrition, March 1980.

MOMCILOVIĆ (Yugoslavia). A high percentage of absorption of trace elements is observed in vivo in animals given a milk diet, but apparently with your in vitro technique this is not observed. Do you have any explanation for this?

WEIGAND (Germany). Our in vitro technique only provides relative absorption rates. Several studies have shown that trace elements contained in or added to milk are absorbed with high efficiency in animals during the phase of development when milk is consumed. We have shown that for zinc absorption in vivo in rats weaned for several weeks the percentage can vary from 100% to much lower values, depending on the level of dietary supply.

IN VITRO INTERACTION OF PHYTATE WITH TRACE ELEMENTS

D. Oberleas and N. Moody

Department of Nutrition and Food Science
College of Home Economics
University of Kentucky
LEXINGTON, KENTUCKY 40506
UNITED STATES OF AMERICA

I. INTRODUCTION

In 1960 it was reported by O'Dell and Savage that phytate decreased the bioavailability of zinc and was an effective agent in precipitating zinc deficiency. Even earlier, it was recognized that calcium also had a role in the etiology of zinc deficiency under both practical and experimental conditions (Forbes, 1960). The mechanism demonstrated later (Oberleas, 1964, 1966) indicated that phytate reacted with minerals as if in free solution. Phytate formed an insoluble complex with zinc or calcium with maximal precipitation occurring at pH 6.0. More importantly, when both zinc and calcium were added together in solution a synergism existed which altered the reaction kinetics toward complex formation to a far greater extent than would be expected simply by doubling the concentration of the cationic reactants. This same phenomenon had been observed in vivo studies with animals and thus describes the mechanism by which phytate and calcium affects zinc bioavailability.

The effect of phytate on the bioavailability of minerals such as manganese, copper, and magnesium has been suspected, however, no systematic studies have been conducted nor have any overt deficiencies of these minerals been demonstrated with phytate containing diets. Maddaiah et al. (1964) and Vohra et al. (1966) have titrated phytate with several minerals by a pH-drop method and have shown that all these minerals combine with phytate at different kinetic rates with Zn^{++}>Cu^{++}>Co^{+++}>Mn^{++}>Fe^{+++}>Ca^{++}. This methodology provided an indication that these minerals may complex with phytate but demonstration of calcium synergism or precise pH sensitivity was not apparent.

The purpose of the current research is to study the in vitro complexation of manganese, copper, and magnesium with phytate, with and without calcium, in a model used previously to study zinc:phytate and zinc:calcium:phytate interactions (Oberleas, 1964; and Oberleas et al., 1966).

II. MATERIALS AND METHODS

Dilute solutions of approximately 0.0001 molar were prepared from analytical reagent grade mineral salts; manganese chloride $4H_2O$, copper chloride $2H_2O$, magnesium chloride $6H_2O$, calcium chloride and sodium phytate (>99%, Sigma Chemical Co, St. Louis) (40 g/l). One ml of concentrated hydrochloric acid was added and the solutions made to volume with deionized water. The concentration of the mineral solutions was verified by atomic absorption spectrophotometric analysis and an aliquot of the phytate solution was wet ashed with nitric and sulfuric acid (3.0 and 0.5 ml respectively), the phosphate was measured by the Fiske-Subbarow (1925) method and phytate calculated as a hexaphosphate equivalent.

Into each of a series of 7 calibrated 50 ml polypropylene centrifuge tubes was added the 10^{-4} millimoles of each test component and the tubes diluted to 25 ml with water. Dilute hydrochloric acid was added dropwise until a pH less than 3.0 was attained. Successive tubes were adjusted to a final pH between 3 and 9 as appropriate and diluted to 30 ml. The tubes were capped, centrifuged at 1220 x g for 10 minutes and the supernatant decanted. The precipitate was washed once with 5 ml of deionized water, centrifuged as above and added to the supernatants. The supernatants were made to a final 50 ml volume for analysis. The precipitates were dissolved in concentrated hydrochloric acid and made to a 50 ml volume for analysis.

Both the supernatants and precipitates were analyzed for the test elements by atomic absorption spectrophotometry using an air-acetylene flame. Serial dilutions were made when analytically necessary. Dilutions were made with lanthanum to a final concentration of 0.5% prior to analyses for calcium and magnesium. An 1 ml aliquot of each supernatant and precipitate solution were wet ashed with concentrated nitric-sulfuric acids (3.0-0.5 ml) and analyzed for phosphate by the Fiske-Subbarow method.

III. RESULTS

The results with manganese are shown in Figure 1. With manganese and phytate (1:1 molar ratios) alone, maximum precipitate formed at pH 7 which contained 7.3% of the manganese and 2.1% of the phytate were present in the precipitate. In the presence of calcium (1:1:1) the maximum precipitate was formed at pH 6 and contained 72% of the manganese, 63% of the calcium and 25% of the phytate.

The results with copper are shown in Figure 2. With copper and phytate (1:1) only minimal amounts of precipitate formed. In the presence of calcium (1:1:1), maximum precipitate formed at pH 6 which contained 74% of the copper, 38% of the calcium and 26% of the phytate in the precipitate.

The results with magnesium are shown in Figure 3. With magnesium and phytate (1:1) the precipitate formed at all pH's was unremarkable. In the presence of calcium (1:1:1) the maximum precipitate formed at pH 9 which contained 60% of the magnesium, 60% of the calcium and 31% of the phytate in the precipitate. At pH 7 the precipitate contained 42% of the magnesium.

The average recoveries for each component was manganese 100%, copper 98%, magnesium 100%, calcium 95% and phytate 92%.

IV. SUMMARY

These results provide evidence that phytate may affect the bioavailability of several elements, particularily those classified as trace elements. Though it is well known that phytate decreases calcium availability, it appears that the kinetic synergism contributed by calcium in the presence of trace elements such as zinc, manganese, copper, and magnesium with phytate may have greater nutritional consequences than the effect of phytate on calcium bioavailability alone. However, it should also be pointed out that these other trace elements have a concommitant effect on the precipitation of and subsequently the bioavailability of calcium. The relative differences in dietary content would indicate that trace elements may become limiting nutrients before calcium.

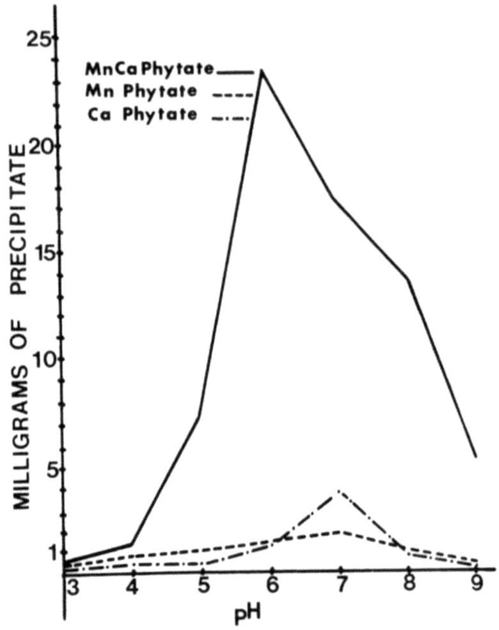

Figure 1. Precipitation curves with phytate, manganese, and/or calcium.

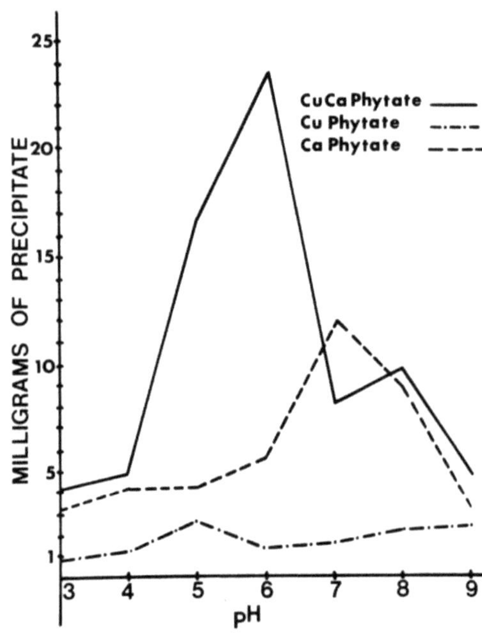

Figure 2. Precipitation curves with phytate, copper, and/or calcium.

This research was supported by the Science and Education Administration of the U.S. Department of Agriculture under Grant No. 5901-0410-8-0076-0 from the Competitive Research Grants Office.

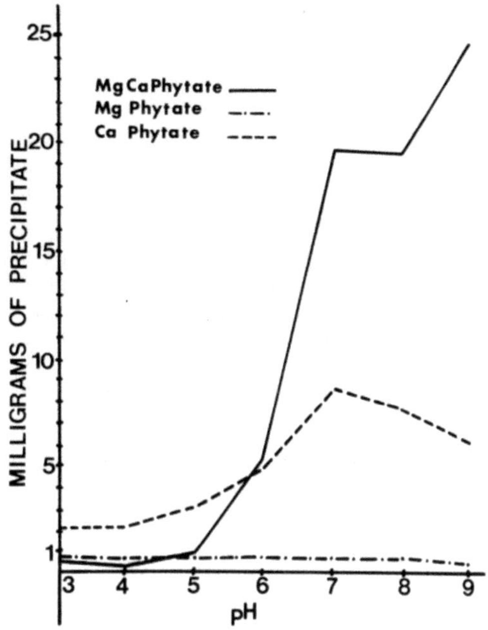

Figure 3. Precipitation curves with phytate, magnesium, and/or calcium.

V. REFERENCES

1. O'Dell, B.L. and J.E. Savage, 1960, Proc. Soc. Exp. Biol. Med. 103:304 - 306.

2. Forbes, R.M., 1960, Fed. Proc. 19:643 - 647.

3. Oberleas, D., 1964, Ph.D. Dissertation, U. of Missouri, Columbia, U.S.A. pp. 143 - 161.

4. Oberleas, D., M.E. Muhrer and B.L. O'Dell, 1966, J. Nutr. 90:56 - 62.

5. Maddaiah, V.T., A.A. Kurnick and B.L. Reid, 1964, Proc. Soc. Exp. Biol. Med. 115:391 - 393.

6. Vohra, P., G.A. Gray and F.H. Kratzer, 1965, Proc. Soc. Exp. Biol. Med. 120:447 - 449.

7. Fiske, C.H. and Y. Subbarow, 1925, J. Biol. Chem. 66:375 - 400.

DISCUSSION

ABDULLA (Sweden). I have been told that the intake of zinc by humans in the USA is around 8 - 10 mg, which is much the same as we find in Sweden. At the same time the calcium intake is 1000 - 1200 mg. Do you have any comment on the zinc availability at these high calcium intakes?

OBERLEAS (USA). I think that the actual intakes of calcium and zinc are not important in themselves. The important thing is the molar ratio of phytate : zinc at a given calcium level. At average calcium intakes there are indications that a ratio of about 10 or less would provide adequate available zinc.

MILLS (Scotland). The problem in extrapolating from this type of study with rats to humans is the large difference in calcium levels. As pointed out by Dr Chesters at a meeting in Detroit last December, studies with rats, pigs and chickens have all been done at much higher calcium levels than occur in diets for humans. Before we can say what the impact of phytate is on human nutrition we must now re-look at all the animal trials - do you agree?

OBERLEAS (USA). Yes.

SMITH (Australia). It seems to me that formation of a precipitate with phytate is one thing but the solubility product of the precipitate is another. Do you have any information on solubility products?

OBERLEAS (USA). The only work on solubility products that I know of is that done by Hoff-Jorgenson in the 1940's. He reported a solubility product between 10^{-19} and 10^{-22} for calcium phytate. The solubility products for copper and manganese are probably much lower, indicating these elements are very tightly bound.

TISSUE COPPER RETENTION OR LOSS IN YOUNG GROWING CATTLE

Anne M. Simpson, C. F. Mills and I. McDonald

Rowett Research Institute, Bucksburn, Aberdeen AB2 9SB, Scotland

INTRODUCTION

Despite the continuing economic losses caused by Cu deficiency in cattle, their requirements for Cu remain poorly defined. Data from which to estimate net requirements are scanty and take little account of the probability that endogenous losses are influenced by Cu status. Estimates of net Cu deposition differ markedly (0.47 to 1.06 mg/kg live weight gain (1-3)) and assume that there is a constant rate of gain of Cu in body tissues regardless of liver copper levels. The validity of this assumption has been investigated during a study of relationships between dietary Mo and S content and retention of dietary Cu by growing cattle.

MATERIALS AND METHODS

Four separate experiments were carried out; initial treatments were identical. Friesian bull calves (ca. 5 days old) were maintained until weaning on a low Cu (1 mg/kg DM) milk substitute (Volac, Ltd.). After castration at ca. 3 months of age, they were offered a semi-synthetic ration based on oat-husks, maize starch and dried skimmed milk, which contained 1 mg Cu/kg DM as an adventitious contaminant. Deionised water was available ad libitum.

Liver biopsy samples were taken every 4 weeks. Initially liver Cu ranged between 150 and 1000 mg/kg DM. By 9-12 weeks after initiation of Cu depletion, liver Cu had declined to ca. 100 mg/kg DM. In expt. 4 nine calves were slaughtered at this stage (LW 96-175 kg) for determination of carcass Cu content and distribution. In expts. 1-3, the animals were randomised into 4 groups of between 4 and 6 individuals. One group in each expt. continued on the same basal diet but supplemented with $CuSO_4.5H_2O$ to give 12 mg Cu/kg DM. The other groups received the same Cu-supplemented basal diet to which was added sodium molybdate and sodium sulphate to provide Mo and S concentrations comparable with concentrations normally encountered in herbage. (In the text the abbreviation (x Mo yS) is used to designate dietary Mo (mg/kg DM) and S (g/kg DM) contents.) Such diets were offered ad libitum for about 4 months during which time liver and plasma Cu were monitored every 4 weeks. Cattle were slaughtered at ca. 10 months of age. Mean LW at slaughter was 296, 255 and 315 kg for expts. 1, 2 and 3 respectively. The carcass Cu content of 10 animals from expt. 2, groups (0Mo1.8S) and (2Mo3.2S), was determined. Selected tissues from all other animals were analysed for Cu.

RESULTS

Cu-depletion phase

During each pre-experimental phase, when the weaned calves received the low Cu diet, hepatic Cu concentration declined at a rate significantly related to initial liver Cu concentration despite marked differences in the latter at the start of treatment (4). This relationship (Fig. 1) indicates an exponential decline which can be described by the equation:

$$Cu_{t_1} = Cu_{t_0} e^{-k(t_1-t_0)} \qquad \text{(Eq. 1)}$$

where Cu_t is hepatic Cu concentration on day t and k is the daily fractional decline in liver Cu. The mean values of k did not differ significantly between experiments being 0.0274, 0.0287, 0.0263 and 0.0290 for expts. 1-4 (overall k = 0.0277, S.E. ± 0.00059).

Effects of dietary Mo and S on tissue Cu concentration

Mean extra-hepatic Cu for the calves slaughtered before dietary Mo and S were varied (expt. 4) was 0.52 ± 0.01 mg/kg LW. It increased to 0.78 ± 0.03 mg/kg LW in steers given control diets (0Mo1.8S) for 4 months but only to 0.59 ± 0.01 mg/kg LW for those of group (2Mo3.2S) ($P<0.001$).

Liver Cu in the control groups (0Mo1.8S) of expts. 1-3 increased at between 0.33 and 0.42 mg/kg DM/day in response to the dietary Cu of 12 mg/kg DM. In contrast that of groups given additional Mo and S declined at 0.24, 0.46 and 0.70 mg/kg DM/day for groups (1Mo4.5S), (2Mo3.2S) and (5Mo3.2S) respectively.

Estimation of total hepatic and extra-hepatic Cu content

Although the liver contains the primary store of Cu in ruminants, data on changes in the Cu content of other body tissues are needed before effects of dietary composition upon retention or dietary "availability" of Cu can be estimated. We have investigated the practicability of predicting total hepatic and extra-hepatic Cu content from data obtained from liver biopsy samples and live weights.

Hepatic Cu content was found to be predictable from knowledge of hepatic Cu concentration and the use of the following equation relating liver dry matter (DM, kg) to live weight (W, kg) derived from carcass analysis of 73 Friesian cattle (4).

$$DM = \underset{(\pm 0.0003)}{0.0033}W + 0.30 \qquad [RSD \pm 0.073] \qquad (Eq. 2)$$

Complete carcass analysis of 19 animals in the present study indicated that hepatic Cu concentration was related to that of the remainder of the body (Fig. 2) by the expression:

$$Cu_{EH} = 0.73-0.41e^{-0.00504\ Cu_L} + 0.00072W \qquad [RSD \pm 0.055] \qquad (Eq. 3)$$

where Cu_{EH} = extra-hepatic tissue Cu (mg/kg), Cu_L = liver Cu (mg/kg DM) and W = live weight (kg). Equations 2 and 3 have been used to estimate changes in total body copper during expts. 1-3.

Effects of dietary Mo and S on body Cu content

a) Liver: Total liver Cu increased at rates of 0.86, 0.91 and 1.22 mg/day in cattle of the control groups of expts. 1-3. Increasing dietary S alone (0Mo3.2S) had no significant effect on hepatic Cu retention. Combined supplementation with Mo and S inhibited Cu retention and despite the Cu content of the basal diet (12 mg/kg DM), treatment (5Mo3.2S) promoted a mean Cu loss of 0.46 mg/day. Effects on retention were even apparent at 1 mg Mo/kg dry diet. Daily changes in hepatic Cu (Cu_L, mg/day) were related to dietary Mo (mg/kg DM) and S (g/kg DM) by the expression:

$$\Delta Cu_L = 0.98-\underset{(\pm 0.014)}{0.093}\ (MoxS) \qquad [RSD \pm 0.20] \qquad (Eq. 4)$$

b) Extra-hepatic tissues: The increments of Cu retained by tissues other than liver ranged from 0.67 to 1.16 mg/day. Mean retentions of 0.67 and 0.69 mg Cu/day by cattle of groups (2Mo3.2S) and (5Mo3.2S) were significantly lower than those of all other groups ($P<0.05$). The effects of dietary Mo and S on daily changes in extra-hepatic Cu (Cu_{EH}, mg/day) were described by the expression:

$$\Delta Cu_{EH} = 1.09-\underset{(\pm 0.007)}{0.027}\ (MoxS) \qquad [RSD \pm 0.11] \qquad (Eq. 5)$$

c) Fractional (%) retention of dietary Cu was calculated from

$$Cu_R = \frac{\Delta Cu_L + \Delta Cu_{EH}}{Cu_I} \times 100 \text{ where } Cu_I = \text{Cu intake during experimental period}$$

The fractional retention of dietary Cu by animals receiving the unsupplemented control diet was ca. 3%. Retention was reduced by all levels of Mo and by S when given with Mo (see Fig. 3). Relationships between Cu_R(%) and dietary Mo and S cnntent were described by the equation:

$$Cu_R = 3.0-\underset{(\pm 0.03)}{0.17}\ (MoxS) \qquad [RSD \pm 0.44] \qquad (Eq. 6)$$

Treatments (1Mo4.5S), (2Mo3.2S), (5Mo1.8S) and (5Mo3.2S) caused highly significant decreases in Cu retention.

INFLUENCE OF INITIAL HEPATIC Cu CONCENTRATION UPON RATE OF Cu LOSS FROM LIVER IN 17 CALVES GIVEN A LOW-Cu DIET: Expt. 2.

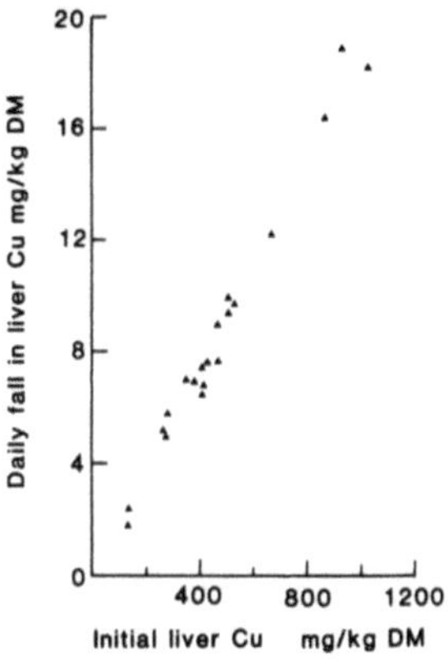

Fig. 1

RELATIONSHIP BETWEEN HEPATIC Cu (mg/kg DM) AND THE Cu CONCENTRATION OF EXTRA-HEPATIC TISSUE (mg/kg FW)

Exp 2 (N=10) WEIGHT RANGE 230-300 kg
Exp 4 (N=9) WEIGHT RANGE 94-175 kg.

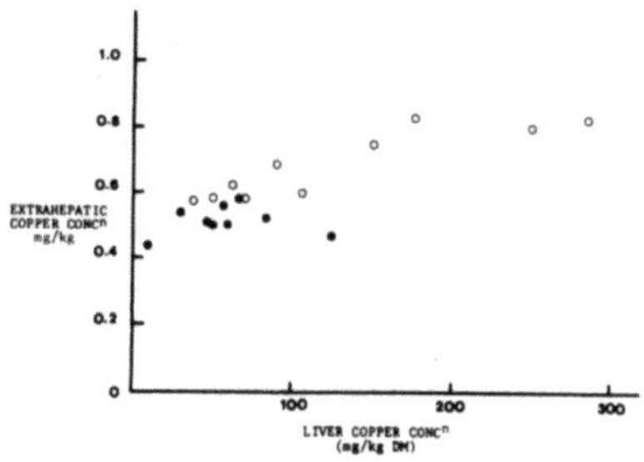

Fig. 2

INFLUENCE OF DIETARY Mo AND S ON THE FRACTIONAL RETENTION (%) OF DIETARY Cu BY GROWING CATTLE GIVEN SEMI-SYNTHETIC DIETS CONTAINING 12 mg Cu/kg DM.

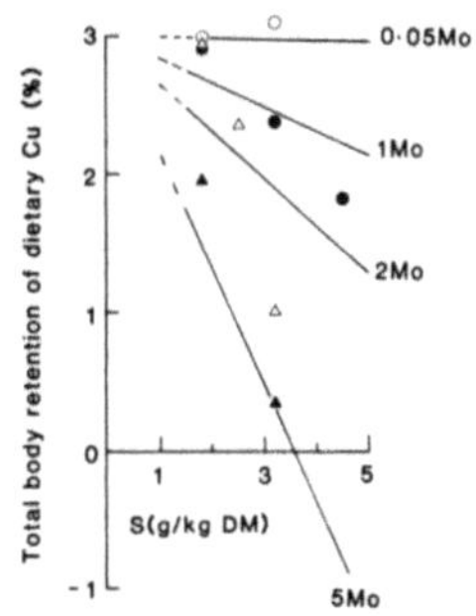

Fig. 3

DISCUSSION

Studies of the rates of hepatic Cu loss of animals gaining weight at approximately constant rates while being prepared for expts. 1-3 and during expt. 4 indicate that Cu economy is markedly influenced by existing Cu reserves. As indicated by Eq. 1, low hepatic reserves of Cu are conserved, presumably by homeostatic regulation of endogenous losses. This implies that estimates of the fractional retention of dietary Cu derived from depletion/repletion studies conducted with "marginally" deficient animals (1) will be higher than those obtained by use of animals with adequate Cu reserves. We adopted a procedure for estimation of Cu_R based on the use of cattle with hepatic Cu initially ca. 100 mg/kg DM because of the conviction that such information will usually be used to adjust dietary Cu supply to prevent depletion of hepatic Cu to marginal concentrations. Such depletion appears inevitable if data for Cu_R derived from the depletion/repletion technique are used to estimate gross requirements for Cu.

We were surprised to find that increasing dietary S from 1.8 to 3.2 g/kg DM had no influence on Cu retention unless Mo was also increased. As far as can be ascertained the Mo content of our basal diet (<0.1 mg/kg DM) was lower than any previously used in similar studies. In view of indications that Mo sources influence the metabolic fate of S^{2-} generated from dietary sources of sulphur, the latter observation may well be relevant. Other data from our study indicate that as little as 1 mg Mo/kg diet can decrease Cu retention. Thus the threshold concentrations of Mo that influence Cu utilisation are lower than suspected previously.

REFERENCES

1. Suttle, N.F., 1978. In: Trace Element Metabolism in Man and Animals-3 p.473 [M. Kirchgessner, editor] Weihenstephan: Arbeitskr. für Tierernährungsforsch.
2. Kirchgessner, M. & K.R. Neesse, 1976, Z. Lebensmittelunters U-Forsch, 161:1.
3. Agricultural Research Council, 1980. Nutrient requirements of farm livestock, 3. Ruminants [2nd edition] London A.R.C.
4. McDonald, I., C. F. Mills, A. C. Dalgarno & Anne M. Simpson, 1978, Proc. Nutr. Soc. 38:59A.

A.M.S. acknowledges the financial support of the International Copper Research Association.

DISCUSSION

SUTTLE (Scotland). Comparisons were made between different sets of data. I think it is important to realise the different situations that were in operation. In Dr Mills study he is measuring amongst other things, the sparing effect of the supply of absorbed copper on copper that was already stored in the liver before the experiment started. I think you are bound to get different results from the situation in which I was working where there was very little reserve of copper.

MILLS (Scotland). This is true, what was stored initially does not have a great deal of effect - we are both looking at animals with continual fluxes but with different equilibriua. You are working to maintain an equilibrium of 20 - 30 parts per million in liver, we are working to an equilibrium of about 100. I think the only practical point that emerges from this is that the decision as to how you are to use such information depends on your objectives. If you are defining minimal nutrient requirements so as not to maintain any tissue reserves, the repletion approach is appropriate. If you wish to prevent depletion and maintain an adequate reserve an equally arbitary decision has to be taken as to what is an adequate reserve.

SUTTLE (Scotland). There is another marked contrast in that whereas you emphasised the uniformity with which the liver copper declines, all our studies indicate that absorption, at least in the sheep, is a very variable commodity. This might arise because the endogenous component of your studies is the uniform part and that is what you are reflecting as much as absorption.

MILLS (Scotland). Probably true.

LEWIS (England). In your relationship between live weights and liver dry matter were all your calves the same breed? One can produce the same sort of relationship in sheep but it varies with breed.

MILLS (Scotland). Yes, they were all animals of the same breed, they were all Friesan, and the lightweight range was 100 to 350 kilos. We cannot apply the data to pregnant animals nor to animals over 350 kilos.

LEWIS (England). We found with our Swaledale sheep the slope of the line was quite different from the one Australian workers reported for Merinos.

MILLS (Scotland). I would emphasise too, that its full body weight not starved body weight.

GENETIC AND DIETARY FACTORS IN CU ACCUMULATION BY SHEEP

J.A. Woolliams, N.F. Suttle,[+] G. Wiener and A.C. Field[+]

ARC Animal Breeding Research Organisation, West Mains Road, Edinburgh EH9 3JQ U.K.
[+]Moredun Research Institute, Gilmerton Road, Edinburgh EH17 7JH U.K.

Only two studies of the response of sheep to a wide range of Cu intakes have been made (1, 2) and it has been widely held that hepatic retention of Cu is linearly related to intake (3). Neither of these studies included any genetic component but genetic variation has since been shown to be important in the Cu metabolism of sheep (4).

MATERIALS AND METHODS

32 mature ewes of each of 2 breeds, the Scottish Blackface (B) and Welsh Mountain (W) and their cross (B x W) were individually penned and fed a complete pelleted diet at a rate of 12.75 gDM/kg Bdy.Wt. The basal ration contained 4 mgCu, 0.5 mgMo and 3.5 gS/kgDM. After settling in the ewes were put onto one of 4 treatments; the basal diet unsupplemented or with one of 3 Cu supplements to give concentrations of 4, 9, 17 and 29 mgCu/kgDM (Part 1). After 28 weeks all ewes reverted to the basal diet for a further 18 weeks (Part 2). Finally all ewes were given the basal diet supplemented with Mo to give a concentration of 4.5 mgMo/kgDM for another 18 weeks (Part 3). The animals were then slaughtered and samples of liver and kidney were taken. During the experiment liver biopsies were taken at 6-8 week intervals. A sample of blood was taken at frequent intervals (6 weeks or less) and a sample of wool was taken after 28 weeks from over the liver biopsy site. All tissue samples and plasma were analysed for Cu concentration.

RESULTS

At the start of the experiment liver Cu concentrations were very low and breed differences were significant although mean levels were only 11, 15 and 21 mg/kgDM for the Blackface, cross and Welsh respectively. Plasma Cu values also differed between breeds with means of 0.51, 0.90 and 1.04 mg/l for Blackface, cross and Welsh.

The graphs of liver Cu concentration against time over all 64 weeks for Blackface and Welsh ewes are shown in Fig. 1.

Part 1 (weeks 1-28) : 4 levels of dietary Cu

On the basal diet ewes maintained low liver Cu concentrations. Those ewes receiving the Cu supplements all increased their liver Cu concentration, but the initial rates of Cu accumulation were not linearly related to the level of Cu in the diet and showed diminishing returns as dietary Cu increased. Breeds also differed in their rate of accumulation; Blackfaces accumulated least rapidly and Welsh the most rapidly. Further, Welsh ewes increased accumulation most over each increment of dietary Cu and Blackface the least and this gave rise to a significant breed by diet interaction. Fig. 2 shows the initial accumulation rates by breed and diet. The ewes did not maintain these initial rates of accumulation but as time progressed showed reduced accumulation, most evident in the 17 and 29 ppm diets (Fig. 1).

Plasma Cu showed a rapid increase but after 4 weeks attained a steady level. Nevertheless the levels attained showed significant breed differences, the rankings as in liver Cu concentration. Ewes on higher Cu diets had higher plasma Cu values although again a convex relationship with dietary Cu was evident.

The analysis of wool Cu showed a significant breed by diet interaction. All breeds showed increases from diet to diet but this increase was most evident in the Blackface which had the lowest concentration of the 3 breed types on 4 and the highest on 29 mgCu/kgDM - rising from 2.9 to 4.2 mgCu/kgDM. The overall mean was 3.5 mgCu/kgDM.

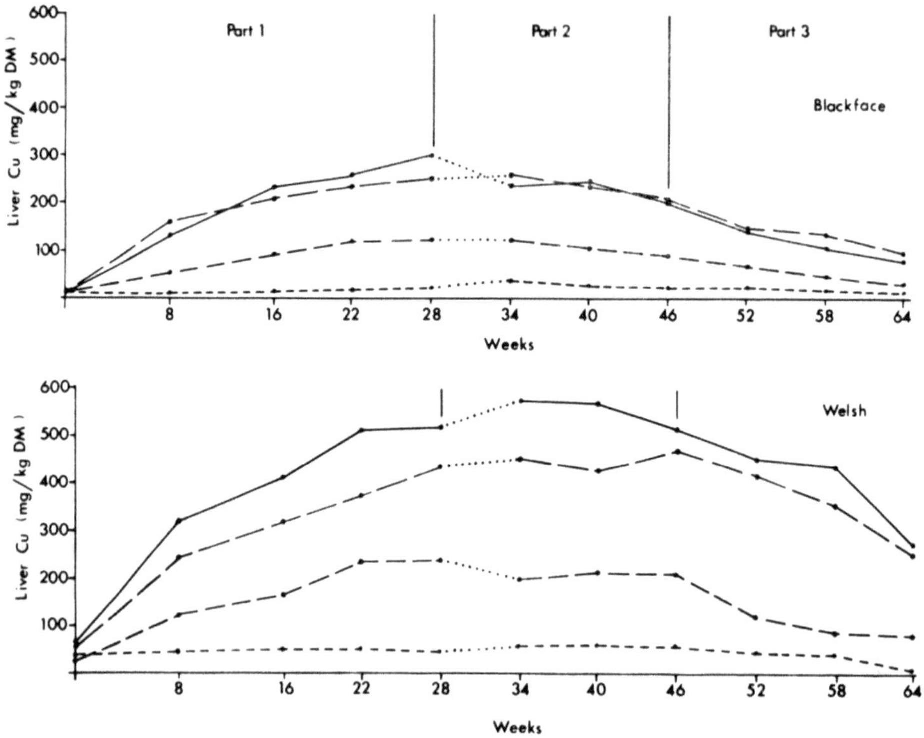

Fig. 1. Liver Cu concentration during parts 1, 2 and 3 according to breed and diet during part 1. (- - - , 4; — — , 9; —— ——, 17; ———— , 29 mgCu/kgDM)

Part 2 (weeks 29-46) : Basal diet fed to all ewes

The rates of decline in liver Cu during this period for each of the 3 high Cu treatments for each breed are shown in Fig. 3 plotted against their mean liver Cu concentration over the same period. A maximum likelihood test showed significant association between the rate of decrease and the mean liver Cu concentration. Further, the rate of decline at a given liver Cu concentration appeared to differ between breeds with Blackface showing the greatest decrease and Welsh the least. Plasma Cu maintained a steady level.

Part 3 (weeks 47-64) : Basal diet + 4.0 mgMo/kgDM fed to all ewes

The rates of decrease greatly increased and Fig. 3 shows the rates of decrease plotted against mean liver Cu concentration during the period for the same groups as previously. The association between rate of decrease and mean liver Cu concentration was of greater significance but breed differences were no longer apparent in the rate of decrease at a given liver Cu concentration. Plasma Cu showed a steady decline for all ewes but differences established during the Part 1 were maintained.

At slaughter kidney Cu concentration showed breed and diet differences; breed order was Blackface (15.1 mgCu/kgDM), cross (16.4) and Welsh (16.5) and kidney Cu increased from 14.2 to 17.1 mgCu/kgDM with dietary Cu but not linearly.

DISCUSSION

The proportion of Cu initially retained in the liver markedly decreased as dietary Cu increased (Fig. 2). This is probably due to saturation of the absorptive process because it was manifest when liver Cu concentrations were uniformly low. The further reduction in accumulation rate with time could be due to a change in partition of absorbed Cu between liver and other tissues (although on the evidence this seems unlikely); further reductions in the efficiency of absorption of dietary Cu; or to increased secretion of Cu

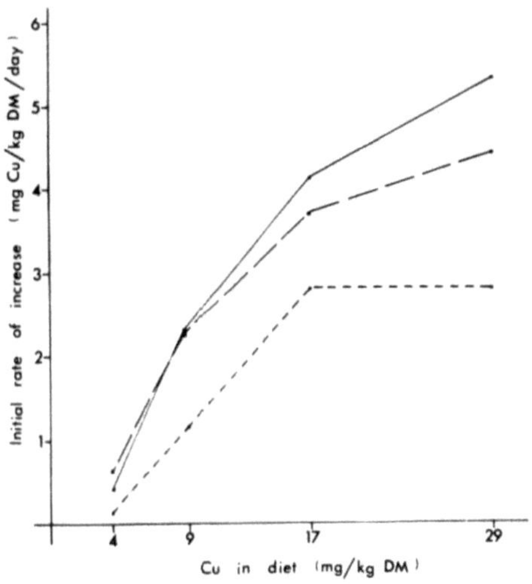

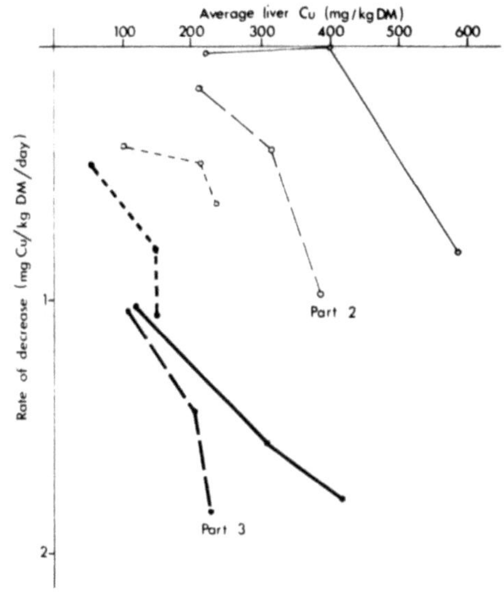

Fig. 2. Estimated initial rate of increase of liver Cu concentration according to breed and diet during part 1. (- - -, B; — —, BxW; ——, W)

Fig. 3. Rate of decrease of liver Cu concentration during parts 2 and 3 according to breed and average liver Cu concentration. (- - -, B; — —, BxW; ——, W)

into the digesta and its decreased reabsorption. Such changes could be associated with liver Cu concentration and account for the dependence of the rates of decrease during depletion on liver Cu concentration (Fig. 3). Further experiments are needed to test these possibilities as none are completely ruled out by these results.

In part 1 the Welsh showed greater increases in liver Cu accumulation from one diet to the next compared to the Blackface with the crossbred generally intermediate. In part 2 on the same diet and at similar Cu levels the Welsh showed a lower rate of decline in liver Cu than the Blackface; but in part 3 where the addition of Mo severely reduced the available dietary Cu, no such differences were apparent. These observations point to breed differences in the efficiency of absorption and support the conclusion by Wiener et al (4). When food intakes, liver weights and initial liver Cu accumulation rates for the 4 and 9 mgCu/kgDM diets are used to estimate hepatic Cu retention, the efficiencies for Blackface, Welsh and the cross are calculated to be 4.2%, 7.3% and 6.8% respectively. The value of 4.2% for the Blackface agrees well with the estimate of absorption obtained by Suttle (5) using a plasma repletion technique and a similar basal diet.

REFERENCES

1. Dick, A.T. 1954, Aust. J. agric. Res. 5:511-544.

2. Hill, R. and H.W. Williams, 1965, Vet. Rec. 77:1043-1045.

3. Underwood, E.J. 1977, In "Trace Elements in Human and Animal Nutrition" (Academic Press) pp 56-108.

4. Wiener, G., N.F. Suttle, A.C. Field, J.G. Herbert and J.A. Woolliams, 1978, J. agric. Sci., Camb. 91:433-441.

5. Suttle, N.F. 1974, Br. J. Nutr. 32:395-405.

DISCUSSION

LEWIS (England). Do you have any data for the Texel that would fit in with these patterns for Blackface and Welsh?

WOOLLIAMS (Scotland). We recently did an experiment with Blackface and Blackface Cross Texel. The liver copper retention for the Blackface was 6%, but for the Blackface Cross Texel it was 11%.

MASTERS H (Australia). Were there any differences in liveweight gain or wool production between the breeds?

WOOLLIAMS (Scotland). There were no differences in liveweight changes but there were differences in wool production between the breeds. There were no differences due to dietary treatment.

DANKS (Australia). Have similar studies been carried out with other species?

CHRISTENSEN (Canada). We have found lower concentrations of copper in liver and plasma from the Simmenthal breed of cattle, compared with Angus or Hereford, when a diet containing 5 ppm copper is fed.

WIENER (Scotland). We have also found differences in the blood or plasma concentration of copper between various breeds of cattle but as yet we do not know if they are correlated with differences in copper absorption.

COPPER DEPENDENT ENZYMES IN MOTTLED MOUSE MUTANTS

M. Phillips, J. Camakaris, J. Mann*, D.M. Danks*

University of Melbourne, Parkville 3052 AUSTRALIA
* Royal Children's Hospital, Research Foundation, Parkville 3052 AUSTRALIA

The brindled (Mo^{br}) and blotchy (Mo^{blo}) show defects in copper metabolism (Hunt, (1)). The pleiotropic effects seen in affected hemizygous males are consistent with deficiencies of a number of copper dependent enzymes consequent to the primary defect (1, 2).

Our hypothesis is that more copper is available for copper enzymes in the blotchy mutant than in the brindled mutant. Hence, if the mutations are allelic, within any one particular tissue, and at the same copper concentration, the more severely affected brindled mutant will show greater enzyme deficiency than the blotchy mutant.

Our current study involves an assay of copper dependent enzyme activity in tissues of the mutant mice both before and after treatment with copper. Changes in enzyme activity during development are also being investigated.

The enzymes chosen for study were cytochrome oxidase and superoxide dismutase. Work to date has been concentrated on cytochrome oxidase.

MATERIALS AND METHODS

The stocks of mice used, the copper content of their diet, and the copper treatment given, have been described elsewhere (Mann et al, (4)). Litters of mice were selected at random from laboratory stocks at 7 and 13 days of age. 7 day old mice where either assayed untreated or treated and assayed at 13, 23 and 60 days of age. Mice were injected with Nembutal and perfused with ice cold isotonic saline prior to dissection. Tissues were weighed and homogenized in 0.2% Triton X100 Tris HCl pH 7.5 (1:5 w/v).

Cytochrome oxidase activity was assayed by the method of Smith (3) as modified by Errede et al (8).

Chloroform-ethanol extracts of homogenates were assayed for Cu-Zn SOD by the method of McCord and Fridovich (7), using acetylated cytochrome C (Azzi, (6)).

RESULTS

Cytochrome oxidase activities in 13 day old mutant and normal mice are shown in Table I. Significant differences were found between normal and untreated mutant mice for heart, brain and skeletal muscle in both brindled and blotchy strains. Comparisons were between littermates by paired t test with two tailed distribution. An Analysis of Variance is planned for the completed experiment. The deficiencies in all three tissues were significantly greater in the brindled when compared with deficiencies in the blotchy mice. Lung and kidney activities in mutant mice were not significantly different from normal.

When mice, treated with a sub-cutaneous injection of 50 μg copper at 7 days of age, were assayed at 13 days age, the brain and heart showed significant increases in activity - ($P < 0.01$, $P < 0.05$ respectively) - compared with untreated mutants and muscle cytochrome oxidase had reached normal levels (see Table I).

TABLE I. Cytochrome oxidase activity in tissues of normal and mutant mice at 13 days.

Mouse	Liver	Brain	Heart	Skeletal Muscle
Brindled Strain				
normal ♂	80.5 ± 8.6 (6)	35.0 ± 2.2 (5)	121 ± 9.0 (4)	27.7 ± 3.0 (5)
untreated mutant ♂	83.0 ± 7.3 (6)	10.9 ± 0.33(5) (a)	47 ± 3.9 (4) (a)	16.3 ± 1.4 (4) (c)
treated mutant ♂	85.33 ± 6.93(6)	19.36 ± 1.16(6) (d)	76.9 ± 5.7 (6) (e)	34.3 ± 0.33(3)
Blotchy Strain				
normal ♂	61.8 ± 1.53(3)	39.2 ± 0.58(3)	102 ± 5.77(4)	26.7 ± 1.19(3)
untreated mutant ♂	56.2 ± 3.46(3)	16.2 ± 0.58(3) (b)	58.19 ± 3.71(4) (b)	19.9 ± 0 (3) (c)

Cytochrome oxidase activity in μ mole/min/g wet weight of tissue, Mean ± SE; number of observations (). Values significantly different from normal (a) $P < .001$, (b) $P < .01$, (c) $P < .05$. Values significantly different from untreated mutant (d) $P < .01$, (e) $P < .05$. Paired two tailed t test.

Changes in cytochrome oxidase activity during development in heart, brain and muscle of normal and untreated mutant mice are shown in Figs I → III.

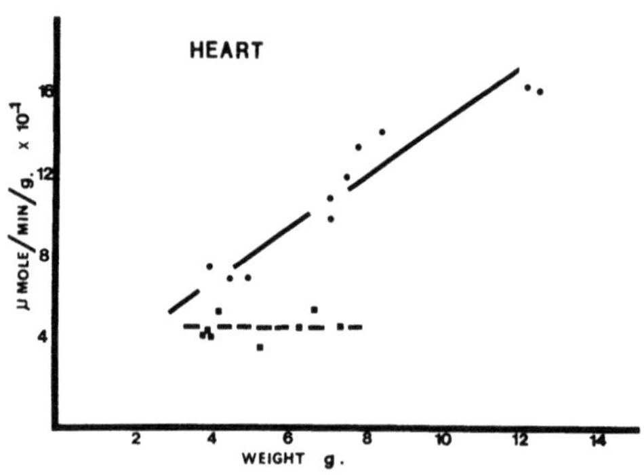

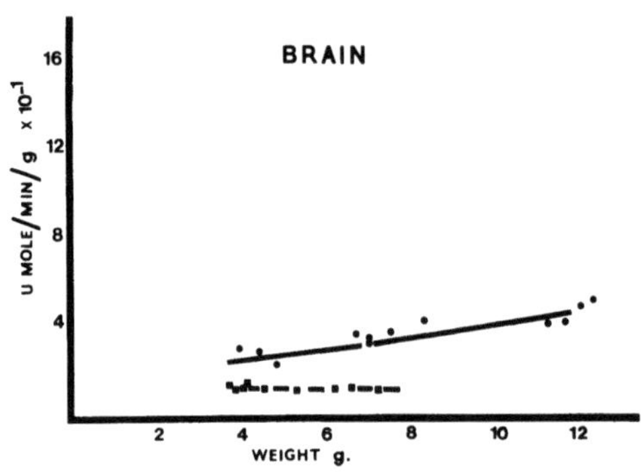

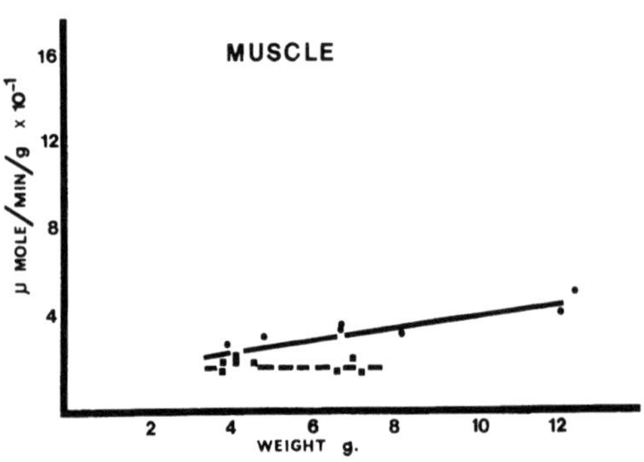

Figs I → III

Cytochrome oxidase activity μ mole/min/g wet weight tissue versus total body weight.

Brindled strain

Normal ♂

Mutant ♂

Cytochrome oxidase levels in normal brain, heart and muscle tissue increase with development. No such increase is seen in the mutant mouse.

Treated mutant values lie between normal and mutant regression lines. (Regression curves have not yet been calculated for treated mutants).

The blotchy strain shows a similar cytochrome oxidase response during development in these tissues. Blotchy mutant heart cytochrome oxidase does increase but at ¼ rate of that seen in the normal mouse.

Preliminary studies have shown deficiencies in Cu-Zn superoxide dismutase in the 13 day old brindled mutant brain (-30%), liver (-40%) and 7 day old brindled liver (-40%). Lung and kidney have normal levels of activity. No deficiencies were found in these organs in blotchy mutant males at the above ages.

DISCUSSION

The results support the hypothesis that widespread deficiencies of copper dependent enzymes are responsible for the phenotypic changes in these mice. Heart and skeletal muscle cytochrome oxidase having been added to the growing list of deficiencies.

The response to copper therapy seen in the brindled mutant suggests that manipulation of cytochrome oxidase levels in the heart and brain of both brindled and blotchy mutants, by different copper therapies, may provide valuable data on copper pathways in the cell, and on the nature of the genetic defect.

Indirect evidence for or against allelism of the mutations would also be gained. The above results are consistent with two mutations being allelic.

The increasing activity of cytochrome oxidase with development (at least between ages 7 to 23 days and weights 4 to 12 gms) in the normal mouse, and the failure to respond, or diminished response in the mutant mouse, cast doubt on data obtained by relating the two mutations at different ages (see Rowe et al, (5)).

The results suggest that cytochrome oxidase levels change in response to metabolic demands during development. Production of apo-enzyme subunits in the mutant may be normal and lack of response a reflection of the available copper level in the cells.

A similar study will now be made of Cu-Zn superoxide dismutase. Investigation of delivery of copper to these two enzymes following treatment of mutants with various copper carriers should prove informative.

ACKNOWLEDGMENTS

I would like to thank Mr R. Holt for his advice and Mr B. Phillips for practical assistance with the statistical calculations.

REFERENCES

1. Hunt, D.M., 1974, Nature. 249:852-853.
2. Danks, D.M., 1977, In: Inorganic Perspectives in Biology and Medicine. I. (Elsevier/ North Holland Biomedical Press) pp 73-100.
3. Smith, L., 1954, Arch. Biochem. Biophys. 50:285-298.
4. Mann, J.R., J. Camakaris, D.M. Danks and E.G. Walliczek, 1979, Biochem. J. 180:605-612.
5. Rowe, D.W., E.B. McGoodwin, G.R. Martin and D. Grahn, 1977, J. Biol. Chem. 252:939-942.
6. Azzi, A., C. Montecucco and C. Rechter, 1975, Biochem. Biophys. Res. Commn. 65:597-603.
7. McCord, J. and I. Fridovich, 1969, J. Biol. Chem. 244:6049-6055.
8. Errede, B., M.D. Kamen and Y. Hatefi, 1978, In: Methods in Enzymology Vol. LIII. (ed. S. Fleischer, I. Packer)(Academic Press), pp 40-48.

DISCUSSION

PROHASKA (USA). I'd like to corroborate your data on the brindled mice. We find exactly the same thing in brain, liver and kidney for cytochrome oxidase activity. When we produced dietary copper deficiency we got exactly the same thing, that is the liver seems to be spared in terms of cytochrome oxidase activity where the brain is decreased and the kidney is normal. Can you suggest why the liver superoxide dismutase activity is decreased and yet the liver cytochrome oxidase activity is not decreased in the brindled mutations?

PHILLIPS M (Australia). It is possible that there are different carriers in the liver for cytochrome oxidase which are not present for superoxide dismutase, and the carriers for cytochrome oxidase are not affected by the genetic defect. But I hope that when we use some of the different copper carriers particularly if we can radioactively label them and treat these mice we might be able to follow the passage of copper to the enzymes and gain some information on these pathways.

SALTMAN (USA). I would like to extend Dr Danks' classical admonition about one gene one enzyme one step further and say one gene one enzyme one chelating agent. This corroborates work Dr Hurley and Dr Keen and I did with crinkled mice showing that it, indeed, is a function not just of the amounts of copper there but in what form that copper is there and how it can be mobilized and moved in and out of various enzyme systems. We are going to see increasingly a series of metabolic lesions which relate to specific chelating agents being absent by mutation. We have already seen that with zinc metabolism. We are now seeing it here with copper. Hemachromatosis may ultimately be revealed as an excess of a low molecular chelating agent which mobilizes iron from the diet.

O'DELL (USA). Have you looked at any enzymes in the intestinal mucosa. I understand the intestinal mucosa accumulates copper. Is any of that copper available for copper dependent enzymes.

PHILLIPS M (Australia). We have not yet looked at intestinal enzymes. The kidney also accumulates copper and has 300% of the normal copper and in this tissue the cytochrome oxidase and superoxidase dismutase levels are normal. When we do a weight gain curve against cytochrome oxidase activities of the kidney we find that the response is normal in the mutant.

PAYNTER (Australia). In copper deficiency I found hypertrophy in the heart of rats. There were large changes in cytochrome oxidase activity but less change in Cu-superoxide dismutase. If manganese deficiency was superimposed on the copper deficiency the hypertrophy was removed, indicating that superoxide dismutases may not be causing the hypertrophy. Did you observe heart hypertrophy in your mutant mice, or did you measure superoxide dismutase activity in heart tissue?

PHILLIPS M (Australia). I have looked for cardiac hypertrophy but have not seen it in brindled mice. They only live for about 15 days so there may be insufficient time for hypertrophy to develop. I have not measured the activity of superoxide dismutase in heart tissue.

MECHANISM AND REGULATION OF INTESTINAL COPPER ABSORPTION

T. Sonsma, P. Hixon, K. McWilliams, and M.C. Linder

Department of Chemistry, California State University
FULLERTON, CALIFORNIA 92634 USA

INTRODUCTION

The mechanism by which copper is absorbed through the intestine by animals and man has not been well studied. One of the reasons has been a scarcity of suitable isotopes, like ^{67}Cu, now available from Atomic Energy of Canada, Ltd. Our work on rats with implanted malignant tumors or estrogen treatment (1) indicated that, at least in this species, copper absorption is a regulated process. Using $^{64}Cu(NO)_3)_2$ (New England Nuclear, Boston, MA) (3 - 150 μg Cu per rat) given intragastrically, we showed that rats bearing tumors absorbed about twice as much radioactivity into the blood and tissues over 2 hours than normal controls. Conversely, there was a marked reduction in overall copper absorption in animals treated for two weeks with large doses of estradiol. Enhanced absorption was seen even in rats with very small tumors (1) and preceded the elevation of serum ceruloplasmin. Ceruloplasmin is the main copper-containing component of plasma, and its concentration is greatly increased in rats and humans with many forms of cancer (2). Indeed, the presence of malignant tumors causes a variety of changes in the way copper is handled by the animal and human organism. These include: a marked elevation of serum copper concentrations (2,3); an increased rate of synthesis of ceruloplasmin (3); a change in the activation state of ceruloplasmin enzyme (3); decreased copper concentrations in liver and kidney (4); and no change in the copper content of the intestinal mucosa (1). Also, the tumors themselves contain high concentrations of copper (4) and avidly absorb it, especially from ceruloplasmin (5).

In our studies with tumor bearing and estrogen treated rats, we also noted that a more rapid rate of copper absorption was accompanied by a decreased retention of radioactive copper by the intestinal mucosa, and vice versa. Thus, rats with tumors showed the highest absorption of radioactivity into the body (excluding the intestine) and retained the least radioactivity in the mucosa, while estrogen treated rats showed the highest mucosal retention and lowest rate of transfer of radioactive copper to the blood (1). This reciprocal relationship is reminiscent of the studies of Hall *et al.* (6) with zinc fed rats, and studies on humans with Menkes' disease (7) in which there is a decreased transfer of copper from the mucosa to the blood. Similar observations have been made for mutant brindled or mottled mice (8). All of these findings prompted us to examine the nature of the copper binding components within the intestinal mucosa and their relation to the rate of copper absorption.

RESULTS AND DISCUSSION

When the cytosol of the intestinal mucosa of rats is examined after administering radioactive copper to the stomach or intestinal lumen, all of the radioactivity is found in one peak which elutes with metallothioneins on columns of Sephadex G-75 (Fig. 1). This is so whether larger doses (up to 150 Cu μg/rat), or tracer amounts are administered, and represents about 65% of the total radioactivity in the whole mucosa. A similar result is obtained on Sephacryl S-200. Analysis of the fractions for copper, by flameless atomic absorption, shows 2 major peaks, the second coinciding with the one labeled during copper absorption.

When the material in this peak is applied to polyacrylamide gel electrophoresis in the standard disc system of Davis (9), with 5% acrylamide in the separating gel, radioactivity is associated with three protein bands with characteristic migrations (Rfs) (Fig. 2). With tracer doses of copper (carrier-free $^{67}Cu(NO_3)_2$) administered by intragastric intubation, the majority of the label appeared in the first two peaks, with only about 15% in the peak with Rf 0.6 (10). With large copper doses (150 μg Cu as $^{64}Cu(NO_3)_2$), there was very little labeling of the first component (Rf about 0.35) and most of the label appeared in the central band. This suggests that the central band has the greatest capacity for copper uptake. It should be noted that the prominent central band seen on stained gels consists of at least 2 components, and that radioactive copper accumulates in the fastest migrating portion of the band (10).

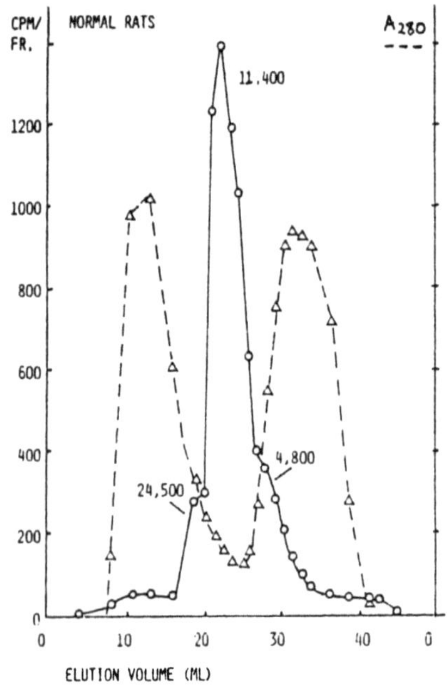

Fig. 1 - Gel filtration of 100,000 x supernatant of intestinal mucosa of rats intubated with $^{64}Cu(NO_3)_2$.

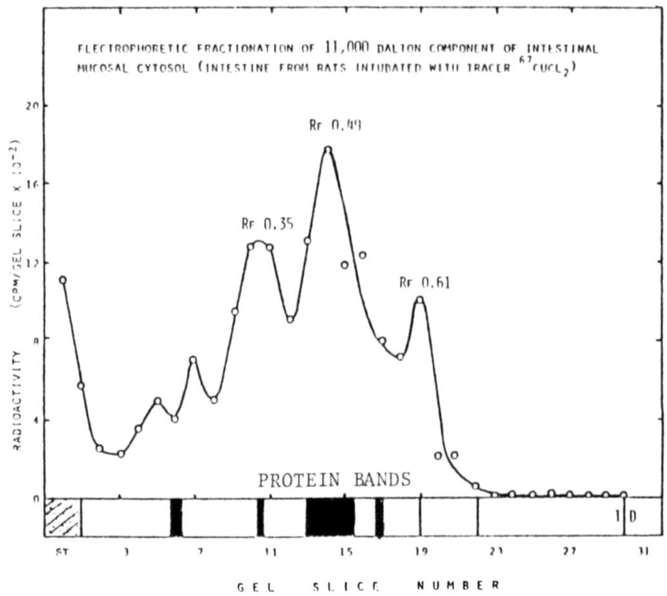

Fig. 2 - Distribution of radioactive copper among the three small molecular weight binding components of intestinal mucosa after electrophoretic separation. ST = stacking gel; TD = tracking dye (Bromphenol blue).

The mean Rf values obtained for the three proteins in numerous experiments (71 gels) are summarized in Table I, along with data obtained by labeling with ^{65}Zn, ^{109}Cd or ^{35}S-cys (11). The non-copper isotopes were administered by intraperitoneal injection 1 h before death, and intestines fractionated in the same way as for the copper studies. This was done to identify the more commonly studied forms of metallothionein. As shown, the main, central band incorporated considerable cysteine, as well as all three of the transition metal isotopes. The component with the highest Rf (0.60-0.67) also incorporated all the isotopes, implying these two were metallothioneins. On the other hand, the slowest migrating component labeled with copper did not correspond to ones labeled with the other isotopes and may not be a classic metallothionein. Current studies on amino acid composition are underway. The results of analysing these proteins by isoelectric focusing (Table I) showed further differences in their behavior. However, in contrast to the electrophoretic procedure in which more than 90% of the radioactivity was recovered in the gels, there were considerable losses of radioactivity of all metal isotopes during isoelectric focusing. None the less the isoelectric points obtained in several runs were consistent.

During our studies with ^{64}Cu, we found that the 3 mucosal copper-binding components took up copper in increasing amounts with time, and roughly in parallel (10). This implied they had a similar affinity for this metal ion. In order to determine whether the concentrations of these proteins influenced the overall rate at which copper was absorbed into the body, we now faced the problem of finding a specific assay. This was no easy matter in that assays for metallothionein are usually based upon analyses of the Zn, Cd and Cu in the peak eluting in gel filtration and are not specific for the individual proteins present. After many disappointments, we have finally evolved the method which follows. This is based upon observations of Waalkes and Bell (12) that excess Cd(II), added to the mucosal cytosol and then fractionated on Sephacryl S-200, displaces almost all the Zn from the metallothionein peak to one eluting later. We have now shown that excess copper will do the same, completely displacing copper and other metal ions from the peak eluting with metallothioneins. The displacement is accomplished as follows: 50% tissue homogenates in 0.25 M sucrose-50 mM Tris (pH

Table I: Comparison of Small Molecular Weight Copper, Zinc and Cadmium Binding Proteins by Electrophoresis and Isoelectric Focusing

	Label Administered			
	$^{67}Cu/^{64}Cu$	^{65}Zn	^{109}Cd	^{35}S-cys
Electrophoretic Bands (Rf)	0.32	-	-	-
	0.47	0.47	0.47	0.47
	0.67	0.60	0.60	0.60
Isoelectric Focusing (pI)	4.7	-	4.7	
	-	-	5.0	
	5.3	5.2	5.3	
	-	5.8	5.7*	
	6.5	-	-	

*differs from 5.8 seen for Zn

8.6), containing 10 mM mercaptoethanol, are centrifuged for 1 h at 100,000 x g. Supernatants (aliquots of 1.0 ml) are treated with 1-1.5 mM $CuSO_4$ labeled with ^{67}Cu, in the presence of 40 mM mercaptoethanol (13), and then fractionated on 50 or 100 ml columns of Sephadex G-75 or Sephacryl S-200. The peak of radioactivity with the elution volume characteristic of the copper-binding proteins is pooled, and from the total radioactivity present and the specific activity of the radioisotope used, the total copper binding capacity is determined, per g. Known volumes of this material are then fractionated further in polyacrylamide gel electrophoresis to separate the individual copper binding components. The gels are sliced and the radioactivity found associated with each peak is used to calculate the tissue concentration of each binding component in terms of copper binding capacity.

As another approach, we have examined the possibility that profiles of stained gels could be used to determine the concentration of the copper binding proteins present in these extracts. Initial studies with cytosol fractions from the intestinal mucosa of normal and tumor bearing rats indicate that the components with Rf 0.35 and 0.46 are present in lower amounts in the rats with tumors, while the most anodic component is more abundant. The application of both this and the previous assay method to the quantitation of small molecular weight copper binding proteins of intestinal mucosa, in rats with and without malignant tumors or estrogen treatment, is now proceeding. It is expected the results will indicate whether our hypothesis is correct that the presence of large amounts of one or more of these proteins acts as a "stumbling block" to the overall transfer of copper from the intestinal lumen to the blood, and that in rats with tumors, the increased absorption of copper is explained by a reduced concentrations of these binding proteins.

REFERENCES

1. Cohen, D.I., B. Illowsky, and M.C. Linder, 1979, Am. J. Physiol. 236:E309-E315.
2. Linder, M.C., J.R. Moor, and K. Wright, 1981, J. Nat. Cancer Inst., in press.
3. Linder, M.C., R.R. Bryant, S. Lim, L.E. Scott, and J.R. Moor, 19, Enzyme 24:85-95.
4. Linder, M.C. in "Morris Hepatomas:Mechanism of Regulation" (ed. H.P. Morris)(Plenum, New York) pp 643-664.
5. Campbell,C.H., R. Brown, and M.C. Linder, 1981, manuscript submitted.
6. Hall A.C., B.W. Young, and I. Bremmer, 1979, J.Inorg.Bioch. 11:57-66.
7. Danks, D.M., B.J. Stevens, and R.R.W. Townley, 1973, Science 179:1140-1142.
8. Hunt, D.M., 1974, Nature 249:852-854.
9. Davis, B.J., 1964, Ann. N.Y.Acad.Sci. 121:404-427.
10. Sonsma, T. and M.C. Linder, 1981, manuscript in preparation.
11. McWilliams, K and M.C. Linder, 1981, manuscript in preparation.
12. Waalkes, M.P. and J.V. Bell, 1980, Life Sci. 27:585-593.
13. Li,T-Y, A.J. Kraker, C.F. Shaw III, and D.H. Petering, 1980, Proc.Nat.Acad.Sci. (US) 77:6334-6338.

ACKNOWLEDGEMENTS

Supported by US PHS Grant No. CA22410

DISCUSSION

BRAY (Australia). Is there any difference in the absorption of copper along the small intestine, say the ileum compared with the jejunem?

LINDER (USA). That is something that has not been adequately looked at. We normally just use the duodenal area. The stomach by the way, in the rat, happens to be a site of absorption for copper.

BREMNER (Scotland). I think that it may be worth noting that in the work that we did on isolation of copper metallothionein from the liver of copper-injected rats or of kidneys, that we quite often got the isomers of the protein using DEAE cellulose. I would like to make another comment on the unsuitability of isoelectric focusing for the purification of metallothionein. We once tried using isoelectric focusing and obtained apparently nice separations for the zinc protein but then discovered we were looking at the zinc-ampholine complex and there was no protein there at all. We had completely separated the zinc from the protein.

LINDER (USA). We find at least two forms of zinc in the isoelectric focusing. One of them has very little protein associated with it, the other definitely has a lot.

COMPLEX SPECIATION FOR ABSORPTION AND EXCRETION

P.M.May and D.R.Williams

Department of Chemistry,
UNIVERSITY OF WALES INSTITUTE OF SCIENCE AND
TECHNOLOGY, CARDIFF CF1 3NU, UK

The optimisation of trace element diets and therapies requires a knowledge of preferred complexes for absorption and excretion. Our current researches involve the absorption of copper, iron and zinc, the excretion of these metals and cadmium, nickel and plutonium, and the *modus operandi* of pharmaceuticals dependent upon trace elements [1,2,3].

In the absence of active transport mechanisms, passive diffusion rates are dictated by equilibrium considerations. The metal involved is usually present in four forms, *viz.* inert metalloproteins from which it cannot be released from the metal ion without destroying the complete protein (for example, the iron in haemoglobin), exchangeable protein metal complexes from which it is possible to withdraw the metal providing a sufficiently powerful influence from an appropriate ligand is present (for example - transferrin), low molecular weight metal-ligand complexes, and, in keeping with strict thermodynamic principles, a very low concentration of aquated metal ion. The low molecular weight complexes are the "exchangeable currency" *in vivo* and these transition metal complexes pass through membranes on to the next stage of the biomechanism and become assimilated into larger units such as proteins [2-4].

Our results - from ultrafiltration, Stricker membrane absorption equipment, and computer simulations - for passive diffusion may be summarised as demanding low-molecular-weight complexes of low or zero charge density and a prevailing concentration gradient [5,6]. Insolubility limits the concentration of uncharged complexes in aqueous solution and we believe that above this concentration micro-particles of solids can also cross membranes or pass between cells.

Excretion may be encouraged to occur through the bile duct or through the kidneys; the former requires low-molecular-weight neutral or low-charge-density complexes to be encouraged in plasma whereas the latter requires charged complexes. The figure shows an example using cadmium: the low-molecular-weight complexes in plasma and the new additional cadmium-drug complexes formed when chelating agents are administered are depicted. Clearly, BAL promotes biliary excretion whereas DMPS is likely to cause kidney excretion. This is in agreement with animal screening data [7,8].

Neutral low-molecular-weight complexes can be absorbed into tissues or passed out in the bile. The former is desirable for treating rheumatoid arthritic inflammation using copper complexes. Our analyses suggest that EDA may be a good treatment since at 10^{-6}mol dm^{-3} it produces up to 66% of $Cu.EDA^{o}$ in plasma. This may be contrasted with TETA which, for the same dose, gives mainly $CuTETA^{2+}$ complexes which conduce cupruresis.

Although passive diffusion involves low-molecular-weight complexes, whereas high-molecular-weight species such as metalloproteins, are non-permeable, the division in molecular weight terms is not a sharp one. The intermediate area of particulate metal-ligand species is attracting increasing attention. For example, particulate lead on vegetables grown besides motorways now features in some diets. Such particles can greatly upset analyses of soluble or bio-available species. Similarly, the removal of plutonium dioxide dust from lungs and the small microparticles of plutonium hydroxide polymer that circulate in the blood-stream is not conducive to simple solution chemistry but rather one of developing an overview and a strategy to rationalise screening experiments. For the example quoted, chelation therapy should be primarily based upon the established approach to dissolving gallstones and cataracts.

Such model studies are useful in translating inorganic biochemistry research into successful therapies by producing tactical leads for animal screening.

References

1. Fiabane,A.M. and Williams,D.R. 1977, The Principles of Bio-inorganic Chemistry, The Royal Society of Chemistry, London.

2. May,P.M. and Williams,D.R. 1980. Metal Ions in Biological Systems (Sigel,H. ed.) Vol.12, Chap.7, Marcel Dekker, New York.

3. May,P.M. and Williams,D.R. 1981. Iron in Biochemistry and Medicine (Jacobs,A. and Worwood,M. eds.), Vol.2, Chap.1, Academic, London.

4. Williams,D.R. 1980. Coordination Chemistry, 20 (Banerjea,D. ed.), 129, Pergamon, Oxford.

5. Stricker,H. 1971. Pharm.Ind., 33:157.

6. May,P.M. and Williams,D.R. 1978. Trace Elements in Man and Animals (Kirschgessner,M. ed.), 3:199, ATW Press, Munich.

7. Planas-Bohne,F. 1980. Experientia, 36:1001.

8. Cherian,M.G. 1980. J.Toxicol.Envir.Health, 6:379, 393.

Figure A model depicting cadmium *in vivo*. The top portion shows the predominant low-molecular-weight complexes of cadmium present in plasma under untreated conditions. The lower portion shows the new complexes formed when different ligand drugs are added to plasma at 10^{-6}mol dm^{-3}. Charged complexes are stippled.

Abbreviations used in this figure and in the paper are BAL = 2,3-dimercapto-1-propanol, CIS = cystine, CYS = cysteine, DMPS = Unithiol = sodium dimercaptopropane-sulphonate, DTPA = diaminotriaminepentaacetic acid, EDA = ethylenediaminediacetic acid, EDTA = ethylenediaminetetraacetic acid, TETA = triethylenetetrammine.

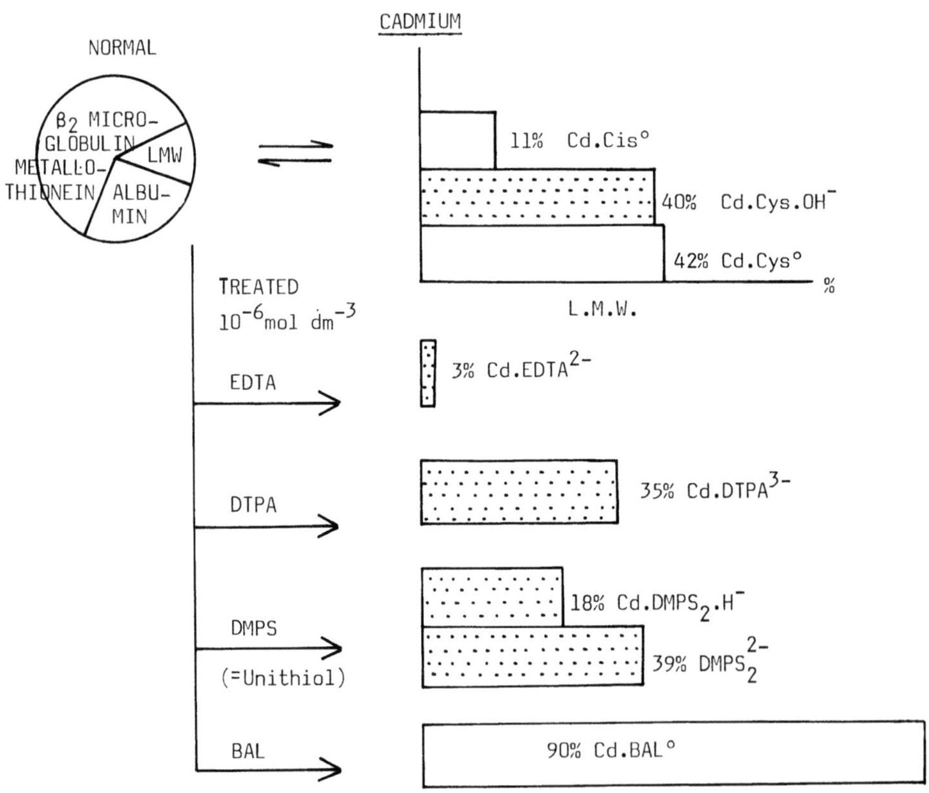

DISCUSSION

SCHWARZ (Germany). I would like to make a comment. We have published data (Schwarz, F.J., and M. Kirchgessner, Z. Tierphysiol. Tierenähr. Futtermittelk. 39, 68-83, 1977) that shows synthetic ligands such as EDTA give a completely different pattern than naturally occuring ligands such as histidine or other amino acids. In zinc-deficient rats there is a much higher absorption in the presence of histidine compared with normal-fed animals, whereas with EDTA as the ligand there is no significant difference. Moreover, the effect of an inhibitor like 2,4 Dinitropherol is to suppress this difference. In other words, the inhibitor significantly alters the transfer of zinc with histidine but has little or no effect on zinc with EDTA.

WILLIAMS (Wales). Yes of course I think one has to consider each ligand and each process step by step. We assessed some animal studies concerning the uptake of zinc from soya bean protein as encouraged, and suppressed, by various ligands, in broad-breasted bronze turkey chicks, (see J.J. Inorg. Nucl. Chem.). We suggest you consider each stage of this process as we did. I did preface my remarks by saying that we were considering passive diffusion for these ligands. When one considers amino-acids, of course, there may be active transfer processes. However, you should note that zinc forms a neutral complex with histidinate whereas it gives a charged complex with EDTA. The former is more membrane permeable and may account for your results without invoking active transport processes.

SUNDE (Scotland). From your computer models can you speculate on the oxidation state and perhaps the form of the copper/D-penicillamine complex in blood plasma?

WILLIAMS (Wales). We have established several negative features concerning the modus operandi of D-penicillamine. It certainly does not release copper, in our estimation, from serum albumin. We have also done some studies which show that it does not release the copper from caeruloplasmin. We acknowledge of course, the fact there is a great cupruresis when you administer D-penicillamine and one of our ambitions is to see whether this copper is released from metallothioneins but we have not got round to this yet. The process of "reductive chelation" for removing copper from serum albumin, I think, has now been dismissed once and for all. Copper is present in blood plasma at the +II oxidation state on both serum albumin and low molecular weight species.

ELMES (Wales). In the use of more than one ligand in therapy, do you administer the ligands simultaneously or sequentially and if sequentially how do you work out the time interval necessary?

WILLIAMS (Wales). They have to be administered sequentially. Ill-timed administration can often be counter-productive and indeed if you have a patient that has been recently polluted with lead and if this lead is still in the plasma, it can be dangerous to administer a ligand like D-penicillamine prematurely as you can form a neutral lead-penicillamine complex and actually move the lead from the plasma into tissues. So I think you have to consider each patient and their analyses step by step, naturally you analyse as far as you can using the various techniques I have referred to.

METAL BINDING BY INFUSED CHELATES

J.R.Duffield, P.M.May and D.R.Williams

Department of Chemistry
UNIVERSITY OF WALES INSTITUTE OF SCIENCE AND TECHNOLOGY,
CARDIFF CF1 3NU, U.K.

Computer simulation models of the low-molecular-weight equilibria in human blood plasma have previously identified the predominant complexes of Calcium(II), Magnesium(II), Manganese(II), Iron(III), Copper(II), Zinc(II) and Lead(II) with naturally-occurring ligands [1] and a variety of administered chelating agents [2]. In addition, a parameter called the Plasma Mobilizing Index (PMI) has been defined to quantify how well exogenous agents can compete for protein-bound metal ions [2]. Such information is not readily available from experimental techniques because the concentrations of low-molecular-weight complexes in biofluids containing proteins are so low that they cannot be reliably measured by any known analytical technique.

The computer simulation models have now been extended to examine the metal binding effects of both natural and synthetic chelating agents which are administered by intravenous infusion. It has been found that the doses of each trace metal which should routinely be added to the intravenous feeding fluids used in Total Parenteral Nutrition can be quantitatively assessed [3]. This paper describes how the approach may be applied to chelating agents used for the treatment of heavy metal intoxication. In particular, the infusion of the polyaminopolycarboxylic acids, EDTA and DTPA, is considered.

EDTA is currently the drug of choice in cases of lead poisoning [4,5]. DTPA is used in the decorporation of actinides, especially plutonium [6], and also for the occasional treatment of siderosis. Both agents are administered by infusion because they are very poorly absorbed when given orally [7].

It has long been recognised that, except in the treatment of hypercalcaemia, EDTA should not be injected either as the free acid or as its sodium salt. This is because it significantly depletes the blood calcium levels, sufficient to induce tetany. The problem can be avoided by the use of the calcium disodium salt. However, both polyaminopolycarboxylic acids bind a wide range of essential metals with great avidity so it is pertinent to ask whether other elements may also be depleted during their administration.

Results and Discussion

The simulated species distribution of EDTA and DTPA in blood plasma have been calculated using formation constants determined under biological conditions of temperature and background electrolyte (NaCl). The results show that EDTA exists in plasma mainly as its calcium complex but that some 15% of the ligand occurs as the zinc species. On the other hand, DTPA is found almost exclusively as its zinc complex. These conclusions are in essential agreement with the calculations reported previously on the basis of estimated formation constants [8]. They are also in accord with the experimental observations that (i) EDTA is an effective treatment for hypercalcaemia and the dissolution of renal calculi [9] whereas DTPA is not and (ii) both agents can significantly enhance urinary zinc excretion although DTPA is about six times more effective in this respect than EDTA [10].

Computer simulations can also quantitatively assess the amount of each trace metal which should be added to administered solutions to compensate for losses induced by infusion. As previously mentioned, it has been shown that such doses can be calculated for the fluids used in Total Parenteral

Nutrition which contain amino acids and other naturally-occurring ligands [3]. This is done on the basis that the free (uncomplexed) metal ion should occur at the same concentration in both the exogenous fluid and blood plasma. Accordingly, it can be assumed that the infused ligands will not disturb the metal-protein binding equilibria in blood plasma and, hence, will favour neither excretion nor retention of the metal ion.

The results of the calculations when EDTA and DTPA are assessed by computer simulation in this way are summarised in Tables I and II. These refer to a total ligand concentration of 1.0 mmol dm^{-3}. The free metal ion concentrations have been taken from reference [1]. It is noteworthy that, at realistic concentrations, the percentage of the ligand bound to each metal ion is independent of the total ligand concentration used in the simulations to a precision of 1%.

TABLE I Simulation of metal complexes formed in EDTA and DTPA infusates

Metal ion	Free metal concentration ($mol\ dm^{-3}$)	Total metal concentrations ($mol\ dm^{-3}$)	
		EDTA	DTPA
Ca(II)	1.13×10^{-3}	1.98×10^{-3}	1.16×10^{-3}
Cu(II)	1.00×10^{-18}	1.31×10^{-10}	6.8×10^{-10}
Mg(II)	5.20×10^{-4}	5.29×10^{-4}	5.20×10^{-4}
Mn(II)	1.00×10^{-12}	8.47×10^{-10}	7.33×10^{-10}
Zn(II)	1.00×10^{-9}	1.37×10^{-4}	9.69×10^{-4}

TABLE II Predominant metal complexes formed in EDTA and DTPA infusates

Metal ion	Percentage of total ligand bound as 1:1 complex	
	EDTA	DTPA
Ca(II)	85	3
Cu(II)	0	0
Mg(II)	0	0
Mn(II)	0	0
Zn(II)	14	96

The results suggest that in order to compensate for the loss of essential metals that would otherwise occur, the agents should be infused as the calcium disodium salt and the zinc trisodium salt respectively. Furthermore, the EDTA solution should be supplemented with zinc equivalent to 15% of the drug's concentration in the infusate. The copper and manganese requirements are minimal.

Relatively high total concentrations are calculated for magnesium with EDTA and DTPA and for calcium with DTPA, reflecting the free concentrations of these metal ions in plasma. These are the values required to prevent dilution. In practice, this factor can probably be neglected.

A rather more important matter concerns the effect of the zinc supplements on the ability of EDTA and DTPA to remove toxic metal ions. As outlined at the previous TEMA Symposium, the administration of DTPA as a calcium complex is more therapeutically effective because it results in a higher free ligand concentration in plasma [8]. This is due to the limited amount of labile zinc which can be obtained from plasma protein and which is likely to be exceeded in the presence of relatively high concentrations of the agent.

Accordingly, the computer simulations suggest that whilst it would be beneficial to supplement EDTA infusions with some zinc, DTPA should be administered as a solution of its calcium trisodium salt. The latter treatment will be most effective if the agent is injected periodically rather than in a single dose. However, the zinc status of the patient will need to be carefully monitored and when replenishment becomes necessary, this should be interspersed with the above regimen rather than given simultaneously.

References

1. May,P.M., P.W.Linder and D.R.Williams, 1977, J.Chem.Soc.Dalton, 588-595.

2. May,P.M. and D.R.Williams, 1977, FEBS Letters, 78:134-138.

3. Berthon,G., C.Matuchansky and P.M.May, 1980, J.Inorg.Biochem., 13:63-73.

4. Rieders,F., W.G.Dunnington and H.Breiger, 1955, Ind.Med.Surg., 24:195-202.

5. Hammond,P.B., A.L.Aronson and W.C.Olson, 1967, Pharm.Exp.Ther., 157:192-206.

6. Norwood,W.D., 1960, J.Occup.Med., 2:371-379.

7. Foreman,H. and T.T.Trujillo, 1954, 43:566-571.

8. May,P.M. and D.R.Williams, 1978. In Proc. 3rd Int.Symp. on "Trace Element Metabolism in Man and Animals" (ed. M.Kirchgessner) (ATW, Freising, F.R.G.) pp.179-181.

9. Perrin,D.D. and A.R.P.Agarwal, 1976. In "An Introduction to Bio-Inorganic Chemistry" (ed. D.R.Williams) (Thomas, Springfield, Illinois) pp.361-389.

10. Spencer,H. and B.Rosoff, 1966, Health Phys., 12:475-480.

DISCUSSION

SUTTLE (Scotland). Does your model assume there is no flow of the ionic element into the plasma and that the chelate does not affect that flow.

MAY (Wales). Yes. Initially when this proposal was made by Dr G Berthon my French colleague, I didn't think that there was any chance that it would give the right kind of answers. It was only after we had done the calculations that we saw it did. Apparently, the most important factor is simply the binding ability of the infused fluid.

LOW MOLECULAR WEIGHT METAL COMPOUNDS IN EDIBLE PLANTS

Colin D. Walker and Ross M. Welch,

Division of Nutritional Sciences,
Cornell University and US Department of Agriculture,
Science and Education Administration,
US Plant, Soil and Nutrition Laboratory,
Tower Road, Ithaca, New York, 14853, U.S.A.

Plants are important intermediaries in the movement of trace metals through the food chain. Plants can form organic ligands with trace metals they absorb. These compounds may be important in determining the bioavailability of trace metals in plant foods and feeds. These metal compounds may also function in the transfer of essential metals to their active sites within plants.

Several studies have revealed low molecular weight compounds in higher plants which bind trace metals(1). These compounds have been poorly characterized and little is known of their abundance in edible plant tissues. Evidence suggests that these compounds may be important in determining the bioavailability of Cd and Zn in plant foods(2), as well as being involved in trace metal transport within the plant.

Our studies reported here examine some low molecular weight (MW) trace metal compounds from soybean seeds (cv. Provar) and their interrelations with other plant minerals.

Method

Soybeans were grown in nutrient solutions labelled with ^{65}Zn. Seeds were blended with hexane to obtain a free-flowing powder (SBP). This was blended with NH_4HCO_3 buffer and the 1,000-10,000 MW fraction isolated by ultrafiltration. The extract and ultrafiltrate fractions were examined for low MW metal compounds. Fractionation methods used included gel filtration chromatography (GFC), ion-exchange chromatography (IEC) on DEAE cellulose eluted with a concave NH_4HCO_3 gradient. Metal-binding compounds were examined by paper electrophoresis, amino acid analyser, and by GC-MS of their trimethylsilyl (TMS) derivatives, the latter being prepared by reaction with hexamethyldisilazane in pyridine at room temperature. Nutrients in fractions were determined by gamma spectrophotometry, inductively coupled plasma emission spectrophotometry (ICP), or by heated graphite atomiser-atomic absorption spectrophotometry (HGA-AAS).

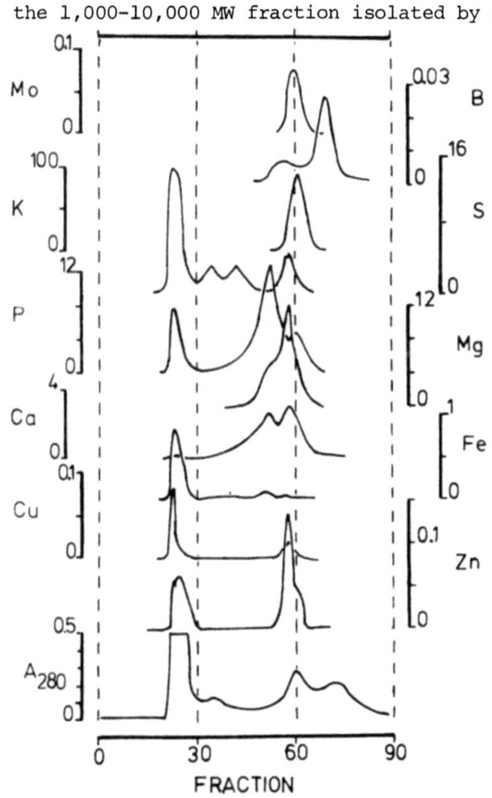

Fig. 1. Chromatogram of Soybean Extract on Sephadex G-75

Results

GFC of SBP extract revealed a number of mineral nutrient compounds (Fig. 1). A substantial proportion of these are of low MW: in particular about 70% of the extracted Zn was in the low MW peak.

IEC of the SBP 1,000-10,000 MW filtrate shows two similar anionic Zn and Cu peaks (Fig. 2). A number of other anionic species are apparent. The peaks eluting at the front of the chromatogram are presumably cationic or neutral species. The fractions containing the second Zn/Cu peak were combined and rechromatographed to yield a single UV peak containing Zn and Cu. GFC of the Zn/Cu unknown on Sephadex G-50 (Fig. 3) gave three UV peaks with Zn and Cu associated with the first of these (MW 1400). No Fe was detected by HGA-AAS. Fractions containing the Zn/Cu compound had an absorbance peak at 251 nm, but no 280 nm peak was apparent. The second peak of the chromatogram did have a 280 nm absorbance peak. All chromatogram concentrations, except C, are in ug.g^{-1}

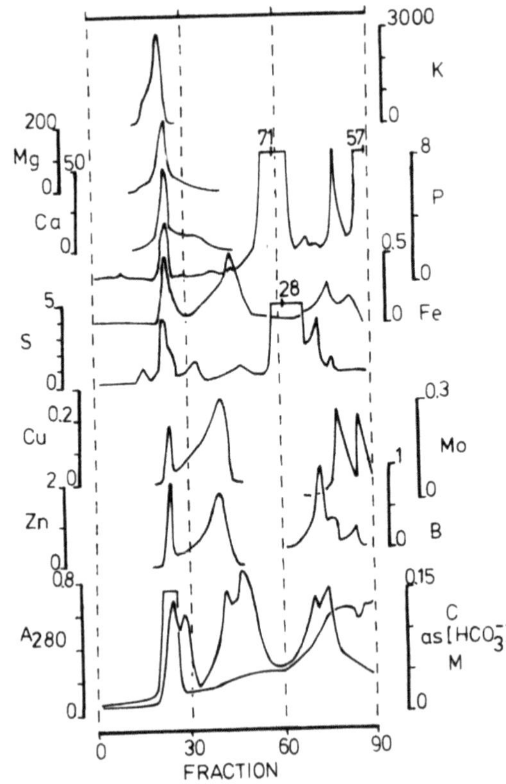

Ninhydrin reactive material in the Zn/Cu fraction was very low. Acid hydrolysis and amino acid analysis on fractions from either side of the Zn/Cu peak fractions from either side of the Zn/Cu peak fraction gave equivalent chromatograms showing four small peaks. The largest of these was definitely not a common protein amino acid.

TMS derivatives were prepared from peak fractions of the Zn/Cu unknown. Analysis by GC-MS indicated three major derivatives, two of these had very similar spectra. Computer comparison of the mass spectra of the derivatives with extensive library spectra indicated similarities of the unknowns to several small MW carbohydrate derivatives. These comparisons also eliminated several possibilities such as citric acid, malonic acid, and many other common small organic ligands.

Summary

A low MW Zn- and Cu-binding compound has been isolated from soybean seeds. It does not appear to contain S or P (Fig. 2), nor is it any of the common plant metabolites, as shown by the mass spectra of its TMS derivatives. The anionic Zn/Cu unknown has an absorption maximum at 251 nm and a MW of 1400 daltons.

Anionic low MW trace metal ligands have been reported in plants. Some of the ligand compounds observed or proposed by other researchers (e.g. citric acid(3) have been eliminated as possibilities while others (e.g. avenic acid(4) remain to be considered.

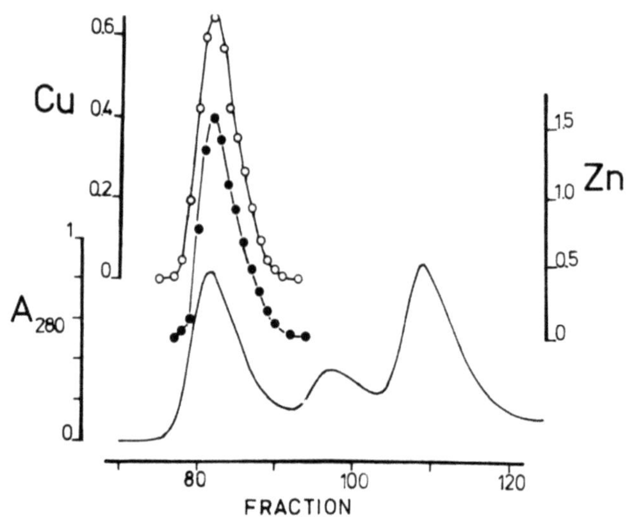

Fig. 3. Chromatogram of Second Zinc Containing Peak on Sephadex G-50.

References

1) Tiffin, L.O., and J.F. Thompson, 1975, Pl. Physiol. 56 : 19; Timperley, M.H., Brooks, R.R., and P.J. Peterson, 1973, J. Exp. Bot. 24 : 889-895; Bremner, I., and A.H. Knight, 1970, Br. J. Nutr. 24 : 279-289

2) Welch, R.M., House, W.A. and D. Van Campen, 1978. Nutr.Rep.Inter. 17 : 35-42

3) White, M.C., Chaney, R., and A.M. Decker, 1981. Pl. Physiol. 67 (In Press)

4) Fushiya, S., Sato, Y., and S. Nozoe, 1980. Tetr. Lett. 21 : 3071-3072.

DISCUSSION

HILL (England). Do you know if these compounds also bind Fe?

WALKER (USA). I found no detectable Fe by heated graphite atomiser - atomic absorption spectrophotometry.

MODIFIED FOOD STARCH INGESTION IN HAMSTERS AND RATS : EFFECTS OF MAGNESIUM AND CALCIUM

M. Lynn Buttolph and Paul M. Newberne

Laboratory of Nutritional Pathology
Massachusetts Institute of Technology
Cambridge, Massachusetts 02139

Modified food starch (MFS), was introduced into foods in 1952 as a thickening and emulsifying agent and has been used widely by the food industry since that time. Despite relatively long use there have been few studies on the metabolism of the various starches. The reports have ranged from no observed adverse effects (1,2), to growth depression, cecal enlargement and kidney mineralization (3). Because of the paucity of data about the metabolism of these interesting compounds, studies were designed in hamsters and rats to examine the effects of MFS ingestion and their possible interactions with magnesium and calcium metabolism.

Previous observations in the Syrian Golden hamster had demonstrated a relationship between the type of MFS in the diet and the incidence of renal tubular dilatation and cortical scarring (4). Hamsters fed the hydroxypropyl distarch phosphate MFS had the most severe and the highest incidence of renal lesions. This observation suggested another study which varied the magnesium content of the diet from the recommended level to 3.5 x recommended level. In this study, increasing the dietary magnesium concentration prevented the lesion (5). Hamsters fed hydroxypropyl distarch phosphate MFS required 3.5 x recommended level of magnesium while animals fed acetylated distarch adipate MFS required 2.0 x recommended level of magnesium.

Other investigators have described the relationship between kidney calcification and magnesium deficiency (6-9). In this report, a point of interest and significance is that the different types of MFS required varied amounts of dietary magnesium to inhibit development of the lesion, suggesting that the magnesium level recommended by the National Academy of Sciences (10) is too low or that MFS ingestion increases the magnesium requirement of hamsters.

Other investigators have described the relationship between kidney damage and magnesium status (9-12). In these reports a point of interest and significance is that different types of MFS required varied amounts of dietary magnesium to inhibit development of the lesion.

Several MFS have now been studied in our laboratory using rats fed various calcium/magnesium/phosphorus ratios. The basic lesion associated with imbalanced ratios of these three minerals is mineralization at the corticomedullary junction. This lesion was more dependent on the calcium/phosphorous ratio than on the magnesium dietary content or the type of MFS in the diet(5).

Further investigations designed to examine the relationship of dietary vitamin D to the renal mineralization have revealed interesting results. Previous reports have associated lactose with renal mineralization (11) and increased calcium absorption (12). Based on these observations we designed a study to compare MFS with lactose and other carbohydrate sources [sucrose and unmodified starch (UnMFS)]. A feature of these studies was to examine the interrelationships among the carbohydrate source, vitamin D status, serum calcium levels and kidney mineralization.

EXPERIMENTAL DESIGN

Seventy female Sprague-Dawley rats were divided into one of 7 groups and fed the respective diets for four weeks as shown in Table 1. The MFS and UnMFS diets were cooked in a steam jacketed cooker; lactose and sucrose diets were made with an agar base, but contained the same vitamin and mineral mixes as the starch diets. After four weeks the animals were killed and a serum sample collected for calcium analysis by atomic absorption. The kidneys were removed and prepared for routine hematoxylin and eosin staining and then examined by light microscopy.

Table 1. Serum Calcium (mg%) and Lesion Incidence

Diet	Serum Calcium + SE	Lesion Incidence
MFS: D def	5.06 ± 0.23	0%
UnMFS: D def	4.98 ± 0.07	0%
Sucrose: D def	4.37 ± 0.12	0%
Lactose: D def	8.94 ± 0.23	50%
MFS: D rec	9.82 ± 0.16	90%
UnMFS: D rec	8.92 ± 0.26	56%
Sucrose: D rec	9.68 ± 0.26	90%

RESULTS AND DISCUSSION

The serum calcium concentration and incidence of lesions are shown in Table 1.

Table 2. Diet Composition (%)

Ingredient	MFS	UnMFS	Sucrose	Lactose
Casein	20.0	20.0	20.0	20.0
Sucrose	31.0	31.0	65.0	----
Starch[1]	30.0	30.0	----	----
Lactose	----	----	----	65.0
Mineral Mix[2]	3.5	3.5	3.5	3.5
Alpha cell, fiber	5.0	5.0	5.0	5.0
Agar	----	----	4.0	4.0
Water	4850 ml	4850 ml	2500 ml	2500 ml
Vitamin Mix[3]	1.0	1.0	1.0	1.0
Corn Oil + K_1[3]	5.0	5.0	5.0	5.0
Vitamin D[4]	4.0	4.0	4.0	4.0
Choline Chloride	0.2	0.2	0.2	0.2
Methionine	0.3	0.3	0.3	0.3

[1]Starch, MFS is hydroxy propyl distarch phosphate, E1442A

[2]The mineral mix contained (in g) : $CaHPO_4$, 500; $K_3C_6H_5O_7.H_2O$, 220; $MgSO_4.7H_2O$, 116; NaCl, 74; K_2SO_4, 52; ferric citrate, 6.0; $MnCO_3$, 3.5; zinc carbonate, 1.6; chromium potassium sulphate, 0.55; cupric carbonate, 0.3; KIO_3, 0.01; Na_2SeO_3, 0.01. The mineral mix was made up to 1 kg with 26 g sucrose.

[3]Vitamin mix. Vitamin K_1 was added to the corn oil (10 mg vitamin K_1/kg corn oil) supplied in the diet (50g corn oil/kg diet). The vitamin mix contained (in g) : tocopheryl acetate, 20 g (500 IU/g); cyanocobalamin (0·1% trituration with mannitol), 5 g; retinyl palmitate, 3.2 g (250,000 IU/g); nicotinic acid, 3g; calcium pantothenate, 1.6 g; pyridoxine, 0.7 g; riboflavin, 0.6 g; thiamine HCl, 0.6 g; vitamin D_3 (cholecalciferol), 0.25 g (500 IU/g); folic acid, 0.2 g; D-biotin, 0.02 g. The vitamin mix was made up to 1 kg with 964.8 g of sucrose.

[4] Vitamin D: Add 60 mg Vitamin D (500,000 units/g) to 1000 g of sucrose, add 100 g of mixture to diet or add 100 g of sucrose to Vitamin D deficient diets.

The data demonstrate that lactose, but not MFS, UnMFS or sucrose, increased serum calcium values to normal without added vitamin D. Serum calcium concentrations correlated well with the incidence of lesion. Lesions were not observed when the serum calcium values were low. The lesion was similar in appearance in all groups and consisted of mineralization at the corticomedullary junction.

Although the mechanism of the lactose mediated enhancement of calcium absorption is not known (13), it appears that a normal serum calcium level is necessary for the development of kidney mineralization in all groups.

REFERENCES

1. Anderson, T.A., L.J. Filer, Jr., S.J. Fomon, D.W. Anderson, R.L. Jenson, and R.R. Rogers, 1973, Fd. Cosmet. Toxicol. 11:747-754.

2. Truhaut, R., B. Coquet, X. Fouillet, D. Galland, D. Guyot, D. Long, and J.L. Rouaud, 1979, Fd. Cos. Toxicol. 17:11-17.

3. DeGroot, A.P., H.P. Til, V.J. Feron, Dreef-vander, H.C. Mullen, and M.I. Willems, 1974, Fd. Cosmet. Toxicol. 12:651-663.

4. Buttolph, M.L., P.M. Newberne, and M. Janghorbani, 1979. In "Trace Substances in Environmental Health X-III". (ed. D.D. Hemphill) (University of Missouri, Columbia) pp. 217-223.

5. Buttolph, M.L., and P.M. Newberne, 1980. In "Trace Substances in Environmental Health XIV" (ed. D.D. Hemphill) (University of Missouri, Columbia) in press.

6. Greenberg, D.M., S.P. Lucia, and E.V. Tufts, 1937, Amer. J. Physiol. 121:424-430.

7. Cowey, C.B., D. Knox, J.W. Adron, S. George, and B. Pierie, 1977, Br. J. Nutr. 38:127-135.

8. Forbes, R.M., 1971, J. Nutr. 101:35-44.

9. Goulding, A., and R.S. Malthus, 1969, J. Nutr. 97:353-358.

10. National Academy of Sciences/National Research Council, 1978. In "Nutrient Requirements of Laboratory Animals," 3rd Ed. (NAS/NRC) (Washington, D.C.) p.7.

11. DeGroot, A.P., and V.J. Feron, 1975/1976. Report No. R 4812, Central Institute for Nutrition and Food Research, TAO, Zeist.

12. Lengemann, F.W., 1959, J. Nutr. 69:23-27.

13. Bushnell, P.J., and H.F. DeLuca, 1981, Science 211:61-63.

DISCUSSION

HURLEY (USA). Could you clarify the relationship between the increasing calcium phosphorus ratio and the low magnesium, because magnesium deficiency perse is exacerbated by increasing the calcium concentration of the diet.

NEWBERNE (USA). In the rat I don't think I could clarify that very well because we have used several different levels of magnesium without it seeming to have much effect. It seems that the calcium to phosphorus ratio is more important, that is, the lower the phosphorus content the less likely you are to have a hypercalcemia and renal damage. In the hamster there is certainly the magnesium effect but in the rat we have not seen very much effect of magnesium, but a great deal of effect when the calcium/phosphorus ratios are modified.

SALTMAN (USA). Would you tell us a little bit about what happens to the modified starch in terms of its digestibility and its breakdown products. The reason I raise this is because of your last comment that lactose, modified starch, and even caramel seem to enhance the calcium availability. What are those products and do they have the tendency to complex calcium?

NEWBERNE (USA). The products are now being examined and identified. The common factor seems to be caecal enlargement - a fall in caecal pH probably as a result of a change in the microflora. We believe that these changes increase calcium absorption from the hindgut.

BRUCE (Sweden). In Sweden we are concerned about the possible adverse effects of extruded starches. The high temperatures and pressures in extrusion alters the starch and decreases its digestibility. There are indications that the absorption of trace elements may also be affected. Have you studied extruded starches?

NEWBERNE (USA). No we have not, but I would not be surprised if the results were similar to those for modified starches.

MOIR (Australia). You would probably get a substantially reduced pH in the caecum-colon with this fermentative activity - we struck a similar sort of thing with whole uncooked potato starch. A great deal of it went through into the caecum and was fermented there. It may also have a secondary effect - we observed some degree of caecal re-gurgitation under those conditions. Did you observe this in your situation - it could offer an explanation for enhanced reabsorption through the ileum.

NEWBERNE (USA). We observed soft stools, but no diarrhoea. The caecum was greatly enlarged so there was opportunity for re-gurgitation. Effects on mineral absorption may well be secondary to these changes in the hindgut.

METABOLIC PATTERNS OF ARSENIC AS STUDIED BY NEUTRON ACTIVATION ANALYSIS AND RADIOTRACER METHODS

E. Marafante, J. Edel-Rade, R. Pietra, E. Sabbioni and F. Bertolero
Radiochemistry and Nuclear Chemistry Division
Commission of the European Communities
Joint Research Centre, Ispra Establishment
21020 Ispra (Varese) - Italy

In mammals absorbed inorganic arsenic is excreted mainly via urine in the form of inorganic arsenic (I As) monomethylated arsenic (MMA) and dimethyl-arsinic acid (DMMA) at different ratios depending on the animal species involved (1). The aim of this study was to investigate the association of arsenic with intracellular constituents and to evaluate its role in the process of biotransformation of arsenite in laboratory animals.

MATERIALS AND METHODS

Sprague Dawley male rats and male albino rabbits were used for the experimentation. Neutron activation analysis was performed on diet, water and tissues as described previously (2). The value stage of arsenic in the solutions of $^{74}AsO_2^-$, pH = 7.5, used for injections, was controlled by solvent extraction and ion exchange chromatography. Animals were injected i.p. with 50 μg As/Kg b.w. as sodium arsenite labelled with ^{74}As and sacrificed after the injection by cardiac puncture. The blood was collected with heparinized syringe. Tissues were removed and homogenized in buffer (0.25 M sucrose and 0.1 M Tris Cl, pH = 8.0). Ultrafiltration of the cytosols and plasma was performed on Centriflo membrane cut-off of 25000 (AMICON, B.V., Holland), by centrifuging for 20 minutes at 5000 x g. The separation of arsenic metabolites in urine was performed by the method of Tam et al. (3) on columns of AG 50 W X8 resin. Inorganic arsenic, mono-methylated arsenic and dimethyl-arsinic acid were eluted by successive elutions with 0.5 M HCl, H_2O and 20% (V/V) NH_4OH solution. The radioactivity of ^{74}As in the samples was measured with a Packard A 5320 Autogamma comparing the radioactivity with that of standard reference solutions of ^{74}As of known specific activity.

RESULTS AND DISCUSSION

Table 1 reports the levels of arsenic found by NAA in selected tissues of 1 and 14 month old rats and rabbits.

Table 1 - Concentration of arsenic in tissues (ppb ± S.D., w.w.) of rats and rabbits sacrificed at the end of nursing and as adult subjects

Age of animals	Blood	Liver	Kidney	Lung	Brain	Spleen
1 month						
rats	103 ± 27	38 ± 15	46 ± 14	96 ± 21	39 ± 18	170 ± 45
rabbit	0.8 ± 0.4	4.1 ± 2.1	6.4 ± 1.8	6.5 ± 3.1	–	2.15 ± 1.1
14 months						
rat	7270 ± 640	120 ± 50	310 ± 95	680 ± 120	59 ± 13	1610 ± 250
rabbit	9.4 ± 6.9	4.5 ± 2	11 ± 3.2	5.2 ± 0.4	3.2 ± 2.6	3.95 ± 2.00

The animals were fed ad libitum with non-supplemented diets containing 110 ± 30 and 90 ± 35 ppb, respectively, plus drinking water with 1.5 ± 0.1 ppb of total arsenic. During their lifetime, the rats accumulate remarkable amounts of arsenic, particularly in the blood and the spleen, as compared to the rabbit tissues. In the rabbits, significant accumulation of arsenic was observed only in the blood, and that to a much lower extent than was found in the rats. These findings agree with other evidence of exceptionally high levels of arsenic found in rat tissues, especially in the blood, compared with other mammals (4). The concentration of arsenic found by NAA in the blood of rabbits being of 1 to 10 ppb lies within the same range as that found in the blood of humans not excessively exposed to arsenic (5). Similarly the levels of arsenic in the rabbit tissues are of the same order, and correspond to the 10 ppb found by the same authors in the liver and kidney of unexposed persons.

There is clear evidence of the high affinity of rat tissues for arsenic, if we compare the retention of ^{74}As in the tissues of rats and rabbits after the injection of a single dose of arsenite, as reported in Table 2. Of the tissues tested, the liver, kidney and lungs of rabbits showed the highest concentration of ^{74}As at 16 and 48 hours. All these tissues showed a rapid clearance of arsenic in the first 48 hours after treatment. On the contrary, the levels of ^{74}As in the

Table 2 - Distribution of arsenic in tissues of rats and rabbits at 16 and 48 hours after the i.p. injection of 50 μg/Kg of ^{74}As as sodium arsenite (ng/g tissue w.w.)

	16 h		48 h	
	Rat	Rabbit	Rat	Rabbit
Liver	32.4 ± 0.2	59.7 ± 27.4	23.5 ± 2.7	0.98 ± 0.11
Kidney	36.5 ± 3.1	36.3 ± 15.5	30.1 ± 7.1	1.1 ± 0.01
Lung	45.7 ± 5.2	54.7 ± 19.2	52.3 ± 6.3	0.53 ± 0.11
Heart	44.1 ± 4.9	9.1 ± 3.9	49.2 ± 13.5	0.18 ± 0.03
Spleen	161.4 ± 18.3	11.4 ± 6.4	91.5 ± 7.3	0.25 ± 0.05
Testicles	3.7 ± 0.1	14.5 ± 14.6	6.9 ± 1.4	0.36 ± 0.09
Blood	295.2 ± 30.3	6.5 ± 2.7	405.3 ± 61.8	0.18 ± 0.05

rat increase from the 16th to the 48th hour, especially in the blood, and the rate of decrease in the liver and kidney was much slower than in the rabbit.

Previous studies have indicated that different binding sites occur for the intracellular arsenic in rat and rabbit tissues (6). This may explain the marked differences observed in the retention of arsenic between rat and rabbit organs.

As shown in Table 3, the ^{74}As found in the ultrafiltrable fraction (< 25000 MW) of the plasma and the soluble cytoplasmic fraction of tissues, was much less in the rat than in the rabbit. This can account for the low urinary excretion of arsenic already observed in the rat compared to other mammals (6). Considering the biotransformation of inorganic arsenic, its inorganic mono- and dimethylated forms are known to be the urinary metabolites of the absorbed inorganic arsenite (7,8,9). Investigating the biotransformation of arsenite in the body compartments of rats and rabbits, all these three chemical forms of arsenic were found as reported in Table 4. A high percentage of DMAA was found in the plasma and the lungs of rabbits, while in the corresponding tissues of rats this chemical species of ^{74}As was not detectable due to the negligible fraction of ultrafiltered ^{74}As.

Table 3 - ^{74}As content in the ultrafiltrable fraction (membrane cut-off < 25000) of plasma and the soluble fraction of liver and kidney of rats and rabbits at 1, 16 and 48 hours post i.p. injection of 50 μg of ^{74}As/Kg b.w. as sodium arsenite (percent of the total content)

	1 h		16 h		48 h	
	rat	rabbit	rat	rabbit	rat	rabbit
Plasma	92.2	95.0	0.1	66.1	N.D.	69.6
Liver	40.2	19.8	4.4	36.3	2.9	12.3
Kidney	49.9	19.1	12.1	32.1	7.3	13.3

Table 4 - Determination of the arsenic metabolites in the ultrafiltrate fraction (< 25000 M.W.) of plasma, liver, kidney and lung of rats and rabbits at 16 hours post i.p. injection of 50 μg of ^{74}As/Kg b.w. as sodium arsenite

	% of the total As in the ultrafiltrate fraction							
	plasma		Liver		Kidney		Lung	
	rat	rabbit	rat	rabbit	rat	rabbit	rat	rabbit
I As	N.D.	11.2	45.3	66.0	32.7	54.8	N.D.	4.6
MMAS	N.D.	2.4	2.1	3.4	3.5	3.6	N.D.	0.1
DMAA	N.D.	87.2	52.6	30.6	63.8	41.6	N.D.	95.3

N.D. Not detectable

In the liver and kidneys of rabbits, a higher concentration of inorganic arsenic was found than in the organs of rats. This may be a consequence of differences in the methylation processes and due also to the high affinity of the intracellular constituents of rat tissues for inorganic arsenic present in the non-filtrable fraction, thus decreasing the arsenic available to be methylated.

REFERENCES

1. Vather, M., 1981, Environ. Research, 25, in press.
2. Pietra, R., Sabbioni, E. and Marafante, E., 1981, J. Radioanal. Chemistry, in press.
3. Tam, G.K.H., Charbonneau, S.M., Bryce, F. and Lacroix, G., 1978, Anal. Biochem., 86:505-514.
4. Hunter, F.T., Kip, A.K. and Irvine Jr., J.W., 1942, J. Pharmacol. Exp. Ther. 76:207-220.
5. Liebscher, K. and Smith, H., 1980, Arch. Environm. Health, 17:881-889.
6. Marafante, E., Rade, J., Pietra, R., Sabbioni, E. and Bertolero, F., 1980, Proceeding of Arsen and Nickel Symposium, Jena, DDR, pp.49-55.
7. Bertolero, F., Marafante, E., Edel-Rade, J., Pietra, R. and Sabbioni, E., 1981, Toxicology, in press.
8. Charbonneau, S.M., Tam, G.K.H., Bryce, F., Zawidzka, Z. and Sandi, E., 1979, Toxicol. Lett., 3:107-113.
9. Crecelius, E.A., 1977, Environ. Health Perspect., 19:147-150.

BIOPOTENCY OF SELENIUM FOR GLUTATHIONE PEROXIDASE SYNTHESIS[1]

R. A. Sunde,[2] W. K. Sonnenburg, G. E. Gutzke and W. G. Hoekstra

Department of Biochemistry, University of Wisconsin-Madison, Madison, WI 53706 U.S.A.

The metabolism of selenium (Se) in animals and man has received increased attention following the discovery that Se is an essential component of glutathione peroxidase (GSH-Px) ($GSH:H_2O_2$ oxidoreductase, EC 1.11.1.9) (1). The measurement of tissue GSH-Px can be used as a sensitive method to determine the biopotency of Se compounds, and to study Se metabolism. While selenite and selenomethionine (Se-Met) have been reported to have equal biopotency, methionine (Met) and vitamin B_6 deficiencies may reduce the biopotency of Se-Met and perhaps selenocystine (Se-Cys) (2,3). This paper reports the results of experiments testing (i) the effect of dietary Met on Se-Met and selenite biopotency, and (ii) the effect of vitamin B_6 on Se-Met, Se-Cys and selenite biopotency.

METHODS. Se biopotency was evaluated by resupplementing Se-deficient rats with various Se compounds for 7 days and measuring the resulting increase in tissue GSH-Px activity. In the Met experiments, rats were fed a 30% torula yeast-based diet (containing 0.013 ppm Se) for at least 28 days to deplete them of Se. This was followed by a 7-day Met adaptation period with Met supplementation at either 0 or 0.4% DL-Met. The rats were then resupplemented with Se at 0, 0.05, 0.1, 0.2, 0.5 or 1.0 ppm Se as either Na_2SeO_3 or DL-Se-Met for 7 days. The diets and methods have been described previously in detail (4). The B_6 experiments were patterned after those of Yasumoto et al. (2), but used a 7-day B_6 adaptation period followed by a 7-day Se repletion period, as in the Met experiments. The rats were fed a 20% casein-based diet similar to that of Yasumoto et al. (2) using the mineral mix of Wagner et al. (5). In experiment B_6-1, the basal diet (containing 0.015 ppm Se) was supplemented with 0 or 2.5 mg pyridoxine•HCl/kg diet, and with 0, 0.2 or 2.0 ppm Se as DL-Se-Met, DL-Se-Cys or selenite. In experiment B_6-2, the basal diet (containing 0.035 ppm Se) was supplemented with 0.25% DL-Met. The rats were supplemented with 0 or 10 mg pyridoxine•HCl/kg diet, and with 0, 0.2 or 1.0 ppm Se. In experiment B_6-2 the feed intake of each rat was restricted to the quantity consumed by B_6-deficient rats. The 105,000 g x 60 min liver supernatant was used for all enzyme analyses. Cystathionase (EC 4.2.1.15), aspartate aminotransferase (EC 2.6.1.1), and protein were determined as described (6-8, respectively). GSH-Px was assayed using H_2O_2 (9). The results were analyzed using Duncan's multiple-range test (10).

RESULTS. The effect of dietary Met on the biopotency of Se-Met and selenite was studied in rats supplemented with 0 or 0.4% Met. These levels of Met provide roughly 50 and 100% of the sulfur amino acid requirement of the rat. Se supplementation at levels up to 1.0 ppm Se resulted in liver GSH-Px activity dose-response curves that appeared sigmoidal and that essentially reached a plateau by 0.5 ppm Se (fig. 1). Dietary Met had little or no effect on selenite biopotency below 1.0 ppm Se. Se-Met biopotency, however, was significantly increased at 0.2 and 0.5 ppm Se when the basal diet was supplemented with Met. At 0.1 ppm Se, Se-Met was 25% as biopotent as selenite, irrespective of Met supplementation, but at 0.2 ppm Se, Se-Met was less biopotent than selenite only when the diet was sub-

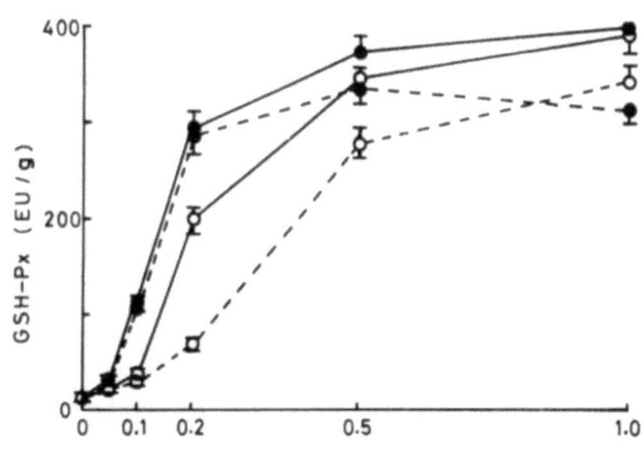

Fig. 1. Effect of Met on liver GSH-Px response to Se repletion. Rats were supplemented with selenite (•) or Se-Met (o) for 7 days. Se was supplemented as indicated and Met at 0 (– – –) or 0.4% (——). GSH-Px is expressed as EU/g protein. Each value is the mean of 6 rats. Error bars represent the SEM.

[1] Research supported by the College of Agricultural and Life Sciences, University of Wisconsin-Madison, and by United States Public Health Service Program Grant no. AM 14881.

[2] Present address: Rowett Research Institute, Aberdeen AB2 9SB, Scotland

optimal in Met. In a separate experiment, 0.9% Met supplementation increased Se-Met biopotency above that observed with 0 or 0.4% Met, but not above the biopotency of selenite (4).

Experiment B_6-1. The biopotency of Se-Met, Se-Cys and selenite was investigated in rats supplemented with 0 or 2.5 mg B_6/kg diet. In contrast to the results of Yasumoto et al. (2) our B_6-supplemented rats gained more weight (final weights 223 vrs. 136 g) and consumed more feed (37% more) than the unsupplemented rats. Se supplementation had no significant effect on growth or feed consumption. B_6 supplementation for 14 days more than tripled liver cystathionase activity (0.30 vrs. 1.17 EU/g protein), and nearly doubled aspartate aminotransferase activity (270 vrs. 470 EU/g). The decreased growth and pyridoxal-dependent enzyme activities clearly show that the unsupplemented rats were B_6 deficient. Se supplementation had no consistent effect on pyridoxal-dependent enzyme activity. Liver GSH-Px activity was not affected by B_6 supplementation or form of Se at 2.0 ppm Se (fig. 2). At 0.2 ppm Se, both Se-Met and Se-Cys were significantly less biopotent than selenite in the B_6-deficient rats; only Se-Met was significantly less biopotent than selenite in B_6-supplemented rats.

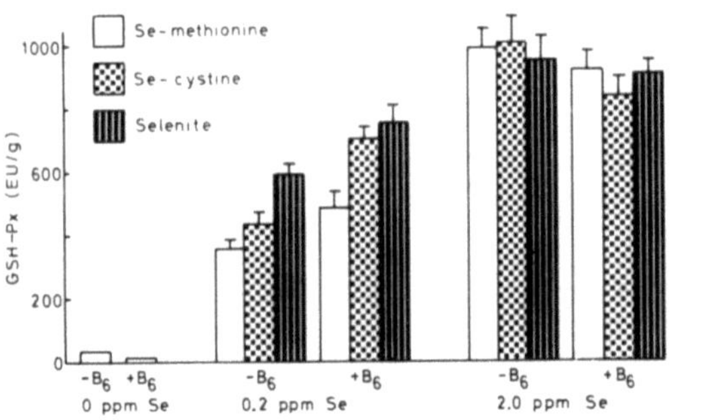

Fig. 2. Effect of B_6 on liver GSH-Px activity (exp B_6-1). Rats were supplemented with Se at 0, 0.2 or 2.0 ppm for 7 days as Se-Met, Se-Cys or selenite. B_6 was supplemented at 0 or 2.5 mg/kg diet. GSH-Px is expressed as EU/g protein. Each value is the mean of 6 rats. Error bars represent the SEM.

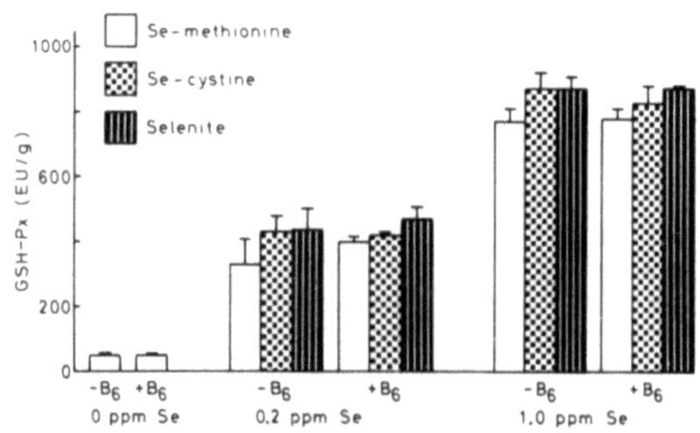

Fig. 3. Effect of B_6 on liver GSH-Px activity in restricted-fed rats (exp B_6-2). Rats were supplemented with Se at 0, 0.2 or 1.0 ppm for 7 days. B_6 was supplemented at 0 or 10 mg mg/kg. The results are shown as described in fig. 2.

Experiment B_6-2. Se biopotency was next determined in restricted-fed rats. Restricted-feeding reduced the growth of the B_6-supplemented rats so that they were not significantly larger at the end of the experiment (127 vrs. 118 g). Se supplementation again had no effect on growth. Liver aspartate aminotransferase activity was doubled by B_6-supplementation (450 vrs. 1160 EU/g) and not affected by Se supplementation. In this experiment, the liver aspartate aminotransferase activity in B_6-deficient rats was somewhat higher than in experiment B_6-1, indicating that these rats may not have been as B_6 deficient. Liver GSH-Px activity indicated that Se biopotency was not affected by B_6 status for either Se-Met, Se-Cys or selenite at either 0.2 or 1.0 ppm Se (fig. 3). These experiments clearly demonstrate that under our dietary conditions Se-Cys and selenite have equal biopotency for GSH-Px synthesis.

DISCUSSION. These experiments have demonstrated that Se-Met biopotency was reduced by suboptimal dietary Met when the level of Se supplementation was below 0.5 ppm Se. Adequate dietary Met was able to increase the biopotency of Se-Met up to that of selenite at 0.2 ppm Se but not at 0.1 ppm Se (Fig. 1). In experiment B_6-1, Se-Met was less biopotent than selenite at 0.2 ppm Se, but the addition of 0.25% Met to the 20% casein diet in experiment B_6-2 apparently raised Se-Met biopotency up to that of selenite. Se-Met and Met compete for the same intestinal transport system in the hampster (11) and presumably other animals. Se-Met readily acylates Met-tRNA (12) and is incorporated into protein in place of Met (13). Because the level of dietary Met affects the rate of Met oxidation (14), Se-Met oxidation would presumably be affected in a similar manner. The metabolism of Met and Se-Met by the same enzymes can apparently account for the impairment of Se-Met biopotency by suboptimal dietary Met. Furthermore, these experiments suggest that adequate dietary Met is essential for the utilization of Se in feedstuffs of plant origin as Se-Met is presumably a major form of plant Se.

Our experiments testing the effect of B_6 deficiency on Se-Met biopotency did not confirm the observations of Yasumoto et al. (2). The pyridoxal-dependent enzyme activities in the deficient rats, however, were only reduced by approximately 50% in either set of experiments, suggesting that the extent of B_6 deficiency was similar and not very severe. The lack of differences in growth in their experiment suggests that their rats may have been less B_6 deficient. It seems unlikely that a 50% reduction in the pyridoxal-dependent enzymes of sulfur/selenium amino acid metabolism could cause Se to be rate limiting for GSH-Px synthesis, especially when Se was fed at 2.0 ppm (see fig. 1). Because of differences in growth and feed consumption, the apparent increase in Se-Cys biopotency with B_6 supplementation in experiment B_6-1 must remain tentative.

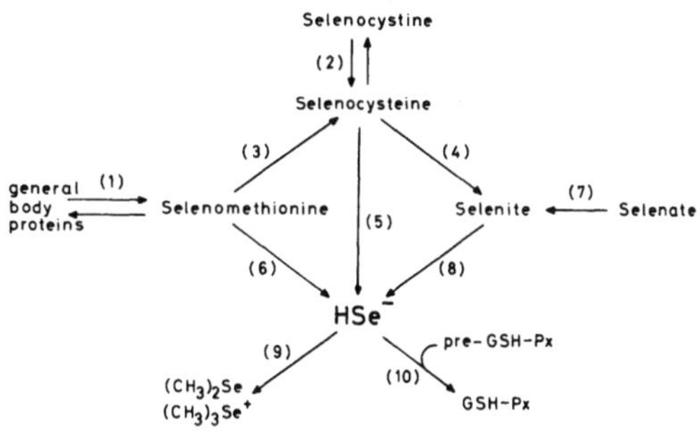

Fig. 4. Proposed diagram of Se metabolism. Reaction 1: incorporation of Se-Met into general proteins in place of Met; rxns 2-4: analogus to sulfur metabolism; rxn 5: selenocysteine may be degraded with loss of HSe^- similar to cystein desulfhydrase (16); rxn 6: Se-Met may be degraded with loss of HSe^-, analogus to sulfur metabolism (17); rxn 7: analogus to sulfur metabolism, perhaps involving APSe or PAPSe; rxns 8,9: selenite reduction pathway and subsequent methylation (18); rxn 10: hypothetical GSH-Px selenocysteine synthase (15).

The conclusions from our biopotency experiments can be used to extend the diagram of Se metabolism published previously by Sunde and Hoekstra (15). These proposed pathways (fig. 4) are consistent with the ability of Se-Met, Se-Cys, selenite and selenide to provide Se for GSH-Px synthesis (3,15). Selenide (HSe^-) is the most central compound. The release of selenide directly from Se-Met and selenocysteine is not analogus to the usual textbook sulfur metabolism, yet these pathways may be important as Se metabolism tends toward reduction. As shown here, the incorporation of Se into GSH-Px, using selenide, is postulated to be a posttranslational reaction (15). This diagram helps to stress that Se-Met has two fates: incorporation into general body proteins or degradation. The level of dietary Met apparently alters the flux of Se-Met as well as Met through these competing pathways.

SUMMARY. These experiments have demonstrated that suboptimal dietary Met reduces the biopotency of Se-Met, but not of selenite, below 0.5 ppm Se. B_6 deficiency was not able to decrease Se-Met biopotency in these experiments. These results do show that Se-Cys and selenite biopotencies are equivalent for GSH-Px synthesis.

REFERENCES

1. Rotruck, J.T., A.L. Pope, H.E. Ganther, A.B. Swanson, D.G. Hafeman and W.G. Hoekstra, 1973, Science 179:588-590.
2. Yasumoto, K., K. Iwami and M. Yoshida, 1979, J. Nutr. 109:760-766.
3. Sunde, R.A. and W.G. Hoekstra, 1980, Nutr. Rev. 38:265-273.
4. Sunde, R.A., G.E. Gutzke and W.G. Hoekstra, 1981, J. Nutr. 111:76-86.
5. Wagner, P.A., W.G. Hoekstra and H.E. Ganther, 1975, Proc. Soc. Exp. Biol. Med. 148: 1106-1110.
6. Flavin, M. and C. Slaughter, 1971, Methods in Enzymology 17B:433-439.
7. Furuno, M. and A. Sheena, 1965, Clin. Chem. 2:23-28.
8. Lowry, O.H., N.J. Rosebrough, A.L. Farr and R.J. Randall, 1951, JBC 193:265-275.
9. Lawrence, R.A., R.A. Sunde, G.L. Schwartz and W.G. Hoekstra, 1974, Exp. Eye Res. 18:563-569.
10. Steele, R.G.D. and J.H. Torrie, 1960. In "Principles and Procedures of Statistics" (McGraw-Hill, N.Y.) pp 107-109.
11. McConnell, K.P. and G.J. Cho, 1967, Am. J. Physiol. 213:150-156.
12. Hoffman, J.L., K.P. McConnell and D.R. Carpenter, 1970, Biochim. Biophys. Acta 199: 531-534.
13. McConnell, K.P. and J.L. Hoffman, 1972, FEBS Lett. 24:60-62.
14. Aguilar, T.S., A.E. Harper and N.J. Benevenga, 1972, J. Nutr. 102:1199-1208.
15. Sunde, R.A. and W.G. Hoekstra, 1980, Biochem. Biophys. Res. Commun. 93:1181-1188.
16. Matsuo, Y. and D.M. Greenburg, 1959, J. Biol. Chem. 234:516-519.
17. Steele, R.D. and N.J. Benevenga, 1979, J. Biol. Chem. 254:8885-8890.
18. Hsieh, H.S. and H.E. Ganther, 1977, Biochim. Biophys. Acta 497:205-217.

DISCUSSION

OLDFIELD (U.S.A.). How long was the initial depletion period for selenium for your animals?

SUNDE (Scotland). In the methionine experiments the rats were selenium depleted for at least 32 days before the 7 day Met adaptation period, and in the B6 experiments the rats were only depeleted for 14 days before B6 supplementation, as were those of the Japanese group. GSH-Px activites were roughly 10% of the supplemented animals, so the rats were in fact selenium deficient.

LANE (U.S.A.). When you gave your seleno-methionine was it as a DL?

SUNDE (Scotland). It was a DL seleno-methionine from Sigma, as we have published.

LANE (U.S.A.). Did you then measure how much selenium actually was in the tissue to show that DL seleno-methionine was equally available?

SUNDE (Scotland). We have not done that.

LANE (U.S.A.). Would you then predict that?

SUNDE (Scotland). I would predict one could use automated amino acid analysis and demonstrate increased selenium in the methionine peak.

LANE (U.S.A.). So you think that the amount of selenium, the total amount of selenium in the body would be the same whether it was given as L seleno-methionine or a mixture of DL seleno-methionine?

SUNDE (Scotland). In rats there is reportedly no difference between L and DL methionine when supplemented in diets. I think that seleno-methionine would be treated similarly.

DIETARY SELENIUM & RAT PLATELET GLUTATHIONE PEROXIDASE

O. A. Levander and V. C. Morris

U.S.D.A. Human Nutrition Research Center
Beltsville, Maryland U.S.A. 20705

INTRODUCTION

Platelets have a life span of 8 to 14 days in humans and 4 to 5 days in rats (1). Human platelets contain more selenium (Se) than other human tissues (2) and platelet glutathione peroxidase (GSH-Px) activity is markedly depressed in rats fed diets deficient in Se (3). Because of their relatively short biological life span, high Se content, and responsiveness to changes in dietary Se intake, platelets appear to be a promising biopsy material for estimating Se status. Therefore, a series of experiments was carried out in rats to determine the effect of dietary Se intake on the activity of platelet GSH-Px.

RESULTS AND DISCUSSION

The first experiment (4) was a dose-response study in which weanling male Fischer 344 rats were fed a Se-deficient Torula yeast basal diet (same as that described in Ref. 5 except for the addition of 100 ppm dl-α-tocopheryl acetate and 0.3% DL-methionine added at the expense of sucrose) for a 4 week depletion period and then either continued on the basal depletion diet or repleted with the basal diet supplemented with 0.1, 0.25, or 0.5 ppm Se as Na_2SeO_3. The different dietary Se levels resulted in a stepwise response in the platelet and liver GSH-Px activities (determined by the coupled assay with t-butyl hydroperoxide as the substrate) and in the total Se content of plasma and liver homogenates (measured by the fluorometric technique with diaminonaphthalene). In the groups fed 0.1 and 0.25 ppm Se, the platelet GSH-Px activity was 7 and 13 times greater and the liver GSH-Px activity 20 and 30 times greater than that of the depleted controls. There was no difference in either the platelet or liver GSH-Px activity between the 0.25 and 0.5 ppm Se groups. A second depletion/repletion study of similar experimental design showed that 0.1 ppm dietary Se fed as selenomethionine was as effective as a similar level of dietary Se fed as selenite in inducing platelet and liver GSH-Px activity in Se-deficient rats.

A third study (6) of somewhat more complex experimental design was conducted to determine the usefulness of platelet GSH-Px activity in estimating the bioavailability to rats of dietary Se fed either as selenite or as a high-selenium yeast preparation. The latter material is of interest because it is available commercially in the U.S.A. as a nutritional supplement for human consumption. The weanling Fischer 344 male rats were taken through a 14-week feeding cycle that consisted of 3 phases: 4 weeks of depletion, 4 weeks of repletion, and then 6 additional weeks of depletion. That is, the weanling male rats were fed the Se-deficient Torula yeast basal diet for 4 weeks (depletion phase). Then they were fed either the same basal diet for another 4 weeks (continued depletion) or the basal diet supplemented with 0.1 or 0.25 ppm Se as selenite or high-Se yeast (repletion). Then all remaining rats were fed the basal diet alone for another 6 weeks (second depletion phase). Liver and plasma GSH-Px activities and Se levels rose and fell quickly during repletion and depletion and platelet GSH-Px activities followed these changes closely. Smaller increases in the Se levels of heart and skeletal muscle took place during repletion and these tissues tended to hold on to their Se more tightly during depletion. Se levels in brain changed little during repletion and depletion. In all tissues studied, the retention of Se was somewhat greater when the rats were repleted with the high-Se yeast than with selenite.

Assessment of Se status by determination of GSH-Px activity with organic hydroperoxide substrates is confounded by the so-called "selenium-independent" GSH-Px activity which persists or even increases in selenium deficiency (7). The selenium-independent GSH-Px activity is now known to be due to one or more of the glutathione S-transferases (8), a family of enzymes that detoxify a wide variety of xenobiotics by catalyzing their conjugation with glutathione (9). It had been suggested that the elevation in glutathione S-transferase activity observed in selenium deficiency might be a compensatory mechanism for the loss of the selenium-dependent GSH-Px (8). However, subsequent experiments showed that glutathione S-transferase activities are not increased in selenium-deficient female, castrated male, or testosterone-treated female rats even though selenium-dependent GSH-Px was decreased in all these animals (10). Thus, the hypothesis that increased glutathione S-transferase activity might be an adaptive response to selenium deficiency remains in doubt. In agreement with the results of others, glutathione S-transferase activity (determined with 1-chloro-2,4-dinitrobenzene as substrate) was higher in the livers of our Se-deficient rats than in livers of supplemented rats (all of our rats were males). Platelets from control rats, however, had only 4% of the corresponding hepatic glutathione S-transferase activity and no increase in activity was observed in platelets from Se-deficient rats. Therefore, fluctuations in glutathione S-transferase activity due to differences in Se intake apparently will not interfere with the assessment of Se status by determination of platelet GSH-Px activity. Moreover, an ancillary study demonstrated that vitamin E deficiency had no effect on platelet or hepatic GSH-Px or glutathione S-transferase activity.

These results indicate that GSH-Px activity in rat platelets responds in a stepwise manner to graded intakes of dietary Se, that different forms of Se (selenite, selenomethionine, and high-Se yeast) are effective in inducing platelet GSH-Px activity in depleted rats, and that platelet GSH-Px activity closely follows hepatic GSH-Px activity and Se level in rats taken through a 3-phase Se depletion/repletion/depletion cycle. Since the liver is the organ that undergoes the greatest transients in GSH-Px activity and Se content as a result of fluctuations in Se intake, platelet GSH-Px appears to be a promising means for determining Se status.

Aside from its potential usefulness as a tool for monitoring Se nutriture, changes in platelet GSH-Px activity may also have important consequences for human health. Based on the high Se concentration in human platelets, it was suggested that Se-deficient platelets may be hyperaggregable and therefore may represent a risk factor for death from cardiovascular disease (2). This hypothesis derives from the fact that the aggregability state of platelets in vivo is determined by a delicate balance between the production of two peroxy-metabolites of arachidonic acid, thromboxane and prostacyclin (11,12). The former metabolite, thromboxane, is produced mainly in the platelets and is pro-aggregatory, whereas the latter metabolite, prostacyclin, is produced mainly in the blood vessels and lungs and is anti-aggregatory. Low concentrations of fatty acid peroxides strongly inhibit prostacyclin synthetase, the enzyme that converts cyclic endoperoxides to prostacyclin (13). Since the only well-documented role of selenium in mammals is in the peroxide-destroying enzyme glutathione peroxidase (14), it seems possible that a low Se status would lead to platelet hyperaggregability as a result of increased tissue levels of hydroperoxides inhibiting prostacyclin production. Platelets from Se-deficient rats have decreased GSH-Px activities and a shift in the metabolic products of arachidonic acid (3) and a preliminary human study suggested that repletion of subjects depleted of selenium resulted in platelets with transient declines in GSH-Px activity and an altered pattern of arachidonic acid metabolites (15). If verified, these metabolic changes could provide a biochemical rationale for the statistical associations that have been noted between low blood Se levels and an increased incidence of human heart disease (16).

SUMMARY

Platelet GSH-Px activity appears to be a promising tool for assessing Se status since it responds readily to changes in Se intake in rats and reflects changes in the Se state of the liver, the organ that appears to have the largest pool of labile body Se reserves. Furthermore, platelet GSH-Px activity may affect platelet aggregability in vivo and thereby play a broader role in influencing human cardiovascular disease.

REFERENCES

1. Altman, P.L., and Dittmer, D.S. 1972. "Biology Data Book" Vol. I Second edition (Fed. Amer. Soc. Exp. Biol., Bethesda, MD) pp. 101-102.

2. Kasperek, K., Iyengar, G.V., Kiem, J., Borberg, H., and Feinendegen, L.E., 1979, Clin. Chem. 25: 711-715.

3. Bryant, R.W., and Bailey, J.M., 1980, Biochem. Biophys. Res. Commun. 92: 268-276.

4. DeLoach, D.P., Morris, V.C., Moser, P.B., and Levander, O.A., 1981, Federation Proc. 40: 902.

5. Levander, O.A., Morris, V.C., Higgs, D.J., and Varma, R.N., 1973, J. Nutr. 103: 536-542.

6. Morris, V.C., and Levander, O.A., 1981, Federation Proc. 40: 902.

7. Lawrence, R.A., and Burk, R.F., 1976, Biochem. Biophys. Res. Commun. 71: 952-958.

8. Lawrence, R.A., Parkhill, L.K., and Burk, R.F., 1978. J. Nutr. 108: 981-987.

9. Jakoby, W.B., and Keen, J.H., 1977, Trends in Biochem. Sci. 3: 299-301.

10. Burk, R.F., Lawrence, R.A., Correia, M.A., 1980, Biochem. Pharmacol. 29: 39-42.

11. Bailey, J.M., 1979, Trends in Biochem. Sci. 4: 68-71.

12. Gryglewski, R.J., 1980, Trends in Pharmacol. Sci. 1: 164-166.

13. Salmon, J.A., Smith, D.R., Flower, R.J., Moncada, S., and Vane, J.R., 1978, Biochim. Biophys. Acta 523: 250-262.

14. Sunde, R.A., and Hoekstra, W.G., 1980, Nutr. Rev. 38: 265-273.

15. Bryant, R.W., Bailey, J.M., King, J.C., and Levander, O.A. 1981. In "Selenium in Biology and Medicine (eds. J.E. Spallholz, J.L. Martin, and H.E. Ganther) (AVI Publ. Co., Westport CT) in press.

16. Shamberger, R.J., Willis, C.E., and McCormak, L.J. 1979. In "Trace Substances in Environmental Health-XII" (ed. D.D. Hemphill) (Univ. Missouri Press, Columbia MO) pp. 59-63.

DISCUSSION

BEHNE (Germany). I have got a question concerning the use of platelet glutathione peroxidase as a monitor of the selenium status. We tested several indicator parameters such as the selenium content and the glutathione peroxidase activity in plasma and erythrocytes, and the selenium content in hair. We found that most of them can be used to recognise selenium deficiency but that it is very difficult to determine the selenium status in the normal selenium range. The plasma glutathione peroxidase, for instance is correlated with the liver glutathione peroxidase. However, the ratio of liver glutathione peroxidase activity to liver selenium content varies at higher selenium levels and so from the glutathione peroxidase activity in the plasma you cannot determine the liver selenium status in the normal selenium range. Now it might be different with the platelet glutathione peroxidase. Therefore my question: Is there a good correlation between platelet glutathione peroxidase activity and selenium content of the liver, if you exclude the selenium deficient animals?

LEVANDER (U.S.A). Of course these experiment are somewhat artificial and they are done with selenite or something like selenium yeast and it is possible to load up the liver under certain circumstances with selenium and not see increases in glutathione peroxidase activity. I think you saw from that one graph that there is a good correlation between liver selenium and liver glutathione peroxidase activity and platelet glutathione peroxidase activity, so at least in this type of experiment this works. I think there will be situations especially in higher selenium intakes when this relationship may not hold.

ULLREY (U.S.A.). Do you think that one could conclude that the quantitative requirement for selenium for the rat is .25 ppm or higher, or do you think this is a function of working with depleted rats which are being repleted for a short period of time?

LEVANDER (U.S.A.). I think it is exactly what you said. You can get any sort of requirement depending on how you set up the experiment.

WHITE (Australia). Would you care to elaborate a little on the reasons for using platelet glutathione peroxidase as opposed to say, plasma glutathione peroxidase, and what advantages this might have in clinical medicine or any other use.

LEVANDER (U.S.A.). I guess I am just a little bit uneasy about measuring an enzyme that is floating around in plasma for a diagnostic tool. I think that there is a good correlation at least in our experiments between plasma and platelet glutathione peroxidase activity so on the surface there would not appear to be any reason necessary to go through platelets. I feel more comfortable about having an enzyme which is inside a cell rather than just floating around freely in the pool of plasma which could be leaking from tissues for reasons other than just nutritional status considerations. Also, quite frankly this is the preliminary stepping stone for us to look at other aspects of platelet function and how they may be influenced by selenium.

WHITE (Australia). Is the glutathione peroxidase in platelets synthesised by the circulating platelets or does it have to go through the equivalent red cell system?

LEVANDER (U.S.A.). I think this is something we still have to investigate but I think these experiments indicate that it is synthesised as the platelets are coming in.

TRACE ELEMENT SUPPLEMENTATION

FIELD STUDIES WITH TRACE ELEMENT METERING DEVICES

A. MacPherson

The West of Scotland Agricultural College,
Auchincruive, AYR, KA6 5HW, U.K.

Supplementation of grazing animals has always posed the problem of the best method of achieving the desired end. Free access methods cannot guarantee complete control, application of an element to the soil or pasture runs into 'availability' problems both to the plant and animal while direct treatment of the animal by injection or bullet therapy has to cope with repeated handling for injection and possible loss or diminishing effectiveness of bullets. Consequently, the development of a metering device whereby trace elements could be provided constantly through the drinking water without any of the above difficulties seemed to provide an attractive alternative worthy of investigation and evaluation. Two metering devices, developed by the Rowett Research Institute and marketed by Elbar Ltd., Craigshaw Works, Aberdeen, were obtained and used in separate trials on one farm in Dumfriesshire.

Cu-SUPPLEMENTATION TRIAL

Fifty-one 6-month old dairy young stock (Friesian and Ayrshire) were allocated at random to each of the following treatments: Nil, Cu by injection and Cu via the drinking water. Cu injections were given at 5-weekly intervals initially providing 80 mg Cu until 26th June and thereafter 40 mg Cu. Addition of Cu to the water supply was calculated to give 2 mg/Cu/kg water. The animals which were grazed on a paddock system and moved at roughly fortnightly intervals were weighed and blood sampled initially and thereafter at 5-weekly intervals throughout the course of the experiment.

The mineral composition of the pasture is given in Table 1 for the areas grazed in both trials. The available Cu content of the pasture for the Cu-supplementation trial provided only approximately half the calculated requirements of growing cattle (1). The Co level was also below normal and the animals were consequently given an injection of vitamin B_{12} (2000 µg) at each sampling occasion in order to maintain their Co status.

Table 1. Mineral Composition of Pasture

	Co (mg/kg)	Cu (mg/kg)	Mo (mg/kg)	S (g/kg)	Available Cu (mg/kg)
1. Cu Trial	0.06	6.98	1.52	3.09	0.24
	±0.006	±0.38	±0.17	±0.25	±0.008
2. Co Trial	0.13*	8.60	1.49	2.78	0.32
	±0.02	±0.68	±0.29	±0.24	±0.01

* High Ti concentrations (15-46 mg/kg) indicating soil contamination.

The effect of the treatments on mean plasma Cu concentration is seen in Table 2. Highly significant effects of treatment were evident in the Cu-treated groups which both maintained normal plasma Cu status whereas the Nil group declined over the season to a final deficient concentration of 0.51 mg/l.

During the first 5-week period there was a marked fall in the Se status of the cattle as measured by glutathione peroxidase (GSH-Px) concentrations. It was decided therefore to supplement half the animals in each group with one Se pellet weighing approximately 30.4 g and containing 10% W/W elemental Se. The effectiveness of this treatment in halting the decline in Se status and in restoring it to normal is detailed in Table 3. Untreated animals continued to decline in Se status with the difference between them and the treated animals being highly significant at each of the three sampling occasions thereafter.

Table 2. Changes in Mean Plasma Cu Concentrations (mg/l)

Date	Nil	Cu Injected	Cu via Meter	S.E.D.	Significance
23/4	0.91	0.93	0.91	0.06	N.S.
22/5	0.84	0.89	0.88	0.05	N.S.
26/6	0.81	0.87	0.91	0.04	N.S.
31/7	0.77	1.00	0.96	0.06	$P < 0.001$
9/9	0.68	1.06	1.01	0.08	$P < 0.0001$
9/10	0.51	1.07	0.81	0.07	$P < 0.0001$

Table 3. Changes in Mean GSH-Px Concentrations (U/cm^3)

Date	Nil		Cu Injected		Cu via Meter		S.E.D.
	No Se	+Se*	No Se	+Se	No Se	+Se	
23/4	46.0		42.4		47.5		3.06
22/5	22.7		22.3		23.6		2.72
26/6	14.3		12.9		14.4		2.50
31/7	9.8	22.9	10.6	25.0	10.0	25.1	2.63***
9/9	8.3	25.6	7.7	26.9	7.9	30.2	3.18***
9/10	9.3	29.1	9.9	26.5	8.0	32.4	3.36***

* Se pellets given on 26th June.

*** Significant at $P < 0.001$.

None of the treatments (either Cu or Se) had any significant effect on daily liveweight gain (D.L.W.G.) of the animals.

Co-SUPPLEMENTATION TRIAL

Forty-six dairy heifers (mostly Friesian) were allocated randomly to the following treatments: Nil, 2 Co bullets/animal and Co via the water supply. The Co bullets were administered at the start of the trial while the Co concentration in the water supply was to be maintained at 0.2 mg Co/kg water. These animals were set stocked and the different groups were interchanged at fortnightly intervals. Weighing and blood sampling were carried out as for the Cu-supplementation trial.

Because of the set stocking system it was not possible to get pasture samples free of soil contamination (see Table 1) and consequently the Co concentration recorded is above what it should be. The lowest values found (0.08 and 0.09 mg/kg) occurred in the first half of the grazing period and higher values later on as the pasture became shorter and more contaminated with soil. Available Cu levels were somewhat higher than those found for the Cu-supplementation trial pasture.

Table 4 details the changes in plasma vitamin B_{12} concentrations over the course of the trial. All animals had normal vitamin B_{12} concentrations initially but that of the Nil group declined quickly to 125 ng/l. Both Co treatments slowed down this rate of decline over the first 5 weeks at grass and thereafter boosted mean plasma vitamin B_{12} concentrations to normal (i.e. >200 ng/l) although the bulleted group was somewhat slower in achieving this target. The fall in plasma vitamin B_{12} concentration of the metered group between 26/6 and 31/7 coincided with a blockage in the supply of Co from the concentrated tank to the trough and a consequent drop in the Co concentration of the water to 0.08 mg/kg. Restoration of the supply line was quickly followed by a return to normal plasma vitamin B_{12} concentrations.

Table 4. Changes in Mean Plasma Vitamin B_{12} Concentrations (ng/l)

Date	Nil	Co Bullets	Co via Meter	S.E.D.	Significance
23/4	225	234	234	18.0	N.S.
22/5	129	177	158	13.4	$P < 0.01$
26/6	136	165	198	18.1	$P < 0.01$
31/7	159	207	176	32.7	N.S.
9/9	189	225	226	18.9	N.S.

Over the latter half of the season the plasma vitamin B_{12} concentration of the untreated group increased gradually until it reached a near normal level of 189 ng/l at the end of the trial.

The changes in D.L.W.G. of the three groups are detailed in Table 5. A significantly ($P < 0.05$) increased rate of D.L.W.G. was achieved by the group supplemented with Co via the water supply over that of the untreated group during the second 5-week period. No significant effects on D.L.W.G. occurred in the latter half of the grazing period when the plasma vitamin B_{12} concentrations of the untreated group were approaching normality.

Table 5. Changes in the Mean Rate of D.L.W.G. (kg/day)

Date	Nil	Co Bullets	Co via Meter	S.E.D.	Significance
23/4-22/5	0.797	0.936	0.945	0.135	N.S.
22/5-26/6	0.782	0.888	1.034	0.101	$P < 0.05$
26/6-9/9	0.776	0.780	0.733	0.052	N.S.
23/4-26/6	0.789	0.909	0.994	0.092	$P < 0.1$
23/4-9/9	0.787	0.840	0.862	0.052	N.S.

The cost of providing the required amount of Co for the 15 animals was only 27p while the metering device cost £470. A response of the order of that obtained (0.075 kg increased L.W.G./day) from 50 growing young stock would more than repay the costs in one grazing season.

Despite a very wet grazing season with the rainfall being more than 25% above normal the water intake of the animals in both trials was sufficient to give them an adequate intake of the Cu or Co supplied in low concentration by this means.

CONCLUSIONS

1. The Cu and Co status of grazing cattle can be adequately maintained by supplying small amounts of these elements through the water supply even in very wet grazing seasons provided there is no other source of water available.

2. A significant growth response to Co was obtained but not to Cu. This lack of response to Cu was due to the fact that the untreated animals remained in adequate Cu status for most of the grazing period. There were some indications of a Cu:Se interaction which, although non-significant in the present trial, suggest that it would be worthy of further study.

3. The cost of providing the required amount of the necessary elements is small and the capital outlay on the metering device quickly recovered.

REFERENCE

1. Givens, D.L. and J.R. Hopkins, 1978, J. agric. Sci., Camb. 91: 13-16.

DISCUSSION

McDOWELL (USA). Are metering devices possible for the other 4 or 5 trace elements that would be required for cattle or are there solubility or other problems with the different chemical forms of the various trace elements?

MacPHERSON (Scotland). We have not tried it for any other trace elements. But it has been tried at other centres and works reasonably well for magnesium. Obviously the concentration dilution has to be greatly increased so that the reservoir tank has to be refilled pretty regularly. So from that point of view it is not so convenient. I presume it would also work reasonably satisfactorily for selenium.

SUTTLE (Scotland). One of the interesting questions over this route of supplementation is whether the element is absorbed as well as it would be were it added to the diet. We have evidence that copper added to the drinking water is absorbed with the same efficiency as copper added in the same form to the diet.

MacPHERSON (Scotland). We have no direct evidence on that from this experiment.

EFFECT OF COPPER SUPPLEMENTATION ON FERTILITY IN HEIFERS

B.E. Ruksan; A. Casaro
Departamento de Patología Animal. C.I.C.V.
Instituto Nacional de Tecnología Agropecuaria
Castelar Buenos Aires - 1712

J. Jaeschke; F. Lagos; D.G. Pondal
Departamento de Investigaciones Agropecuarias
Fundación José María Aragón - Buenos Aires - 1067
ARGENTINA

The incidence of copper deficiency of cattle in Argentina was first described by Bingley and Carrillo (2, 3) in Buenos Aires province. It was demonstrated that copper levels in cattle livers, obtained by biopsy, were below 20 ppm on dry weight basis and in the blood less than 0.50 μg/ml, levels recognized as the limits of concentration below which symptoms of copper deficiency including poor growth, diarrhoea, bone fractures, low fertility, rough coat and depigmentation of the hair specially around the eyes, were manifested.

Such symptoms were commonly observed in different regions and investigations were conducted to determine the chemical composition of the pastures which the animals grazed, particularly with respect to the copper, molybdenum and inorganic sulphate content. The soil conditions in such areas were also investigated.

The data showed that not only were the copper concentrations often below the acceptable minimum nutritional requirement for cattle and sheep (5-8 ppm dry weight) but that high concentrations of molybdenum to levels of 20 ppm dry weight were also present in pastures (14). This was not surprising because the soil studies showed that the dominant types of soils were Solonets and Solodized-Solonets, alkaline in nature and in which the molybdenum was in a state of high availability to the plant (7).

In previous studies with Fundación José María Aragón in the northern part of Santa Fe, in the submeridional low area, it was demonstrated that the animals which presented spontaneous bone fractures, had a severe copper deficiency with values less than 0.10 μg/ml in blood and 10 ppm in liver. The condition was corrected by copper administration.

This paper describes the results of copper administration on the copper status in blood, liveweight gain and fertility in heifers.

MATERIALS AND METHODS

Ninety two Nelore-Santa Gertrudis-Aberdeen Angus cross bred heifers with copper deficiency symptoms were subdivided at random in three groups which received the following treatments: Group 1, 46 heifers, control: Group II, 23 animals, 50 ml of $CuSO_4$ solution 6 per cent weekly: Group II, 23 animals 5 ml subcutaneously of Cu-complex (INDUVET-VASCUNIN) 100 mg Cu given at the beginning of the experiment and 70 days later.

Every week during the first three months and once a month during the rest of the experimental period the following determinations were made:- blood copper by the caeruloplasmin method (5); haemoglobin by cyanmethemoglobin and haematocrit by a micro-method.

The clinical condition of the animals was monitored daily and the reproductive performance and liveweight gain of animals recorded four times.

The animals were grazing on 150 ha. of Melilotus albus, pasture seeded in 1976 and re-seeded naturally in 1977, the year the experiment was carried out. The content of copper (8); molybdenum (1) and inorganic sulphate (4) was periodically determined in the Melilotus albus.

RESULTS

The mean values of haematocrit, haemoglobin, serum copper, liveweight gain, ovarian activity and percentage of heifers pregnant in the oral and subcutaneous copper administration and untreated groups are reported in Table 1.

The haematocrit and haemoglobin values were statistically significant ($P<0.05$) (Table 1). The effect of oral and subcutaneous copper administration on Cu serum values of hypocupraemic heifers during the experimental period were also statistically significant ($P<0.05$) and the variations are presented (Figure 1).

Differences in ovarian activity and percentage of pregnancies were observed (Table 1) but the differences found were not statistically significant. The difference found on liveweight gain of 0.885 and 0.798 kg/animal/day for the subcutaneous copper treated and control animals, was not statistically significant.

Table 1: Mean values of haematocrit, haemoglobin, serum copper, liveweight gain, ovarian activity and percentage of pregnant heifers.

	Haematocrit %	Haemoglobin g /100ml	Serum Cu µg/ml	Liveweight gain kg/day	Ovarian Activity %	Pregnancy %
Control	38.8	12.0	0.15	0.798	23.9	76.0
Oral Cu	42.1	12.9	0.43	0.774	43.4	82.2
Subcutaneous Cu	41.1	12.8	0.52	0.885	39.1	91.3

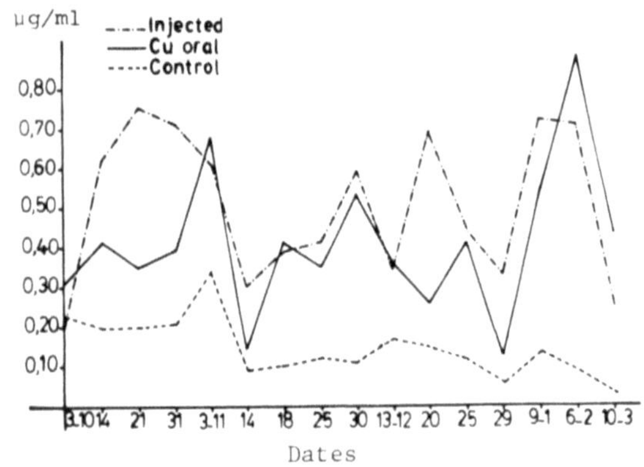

Figure 1: Effect of oral and subcutaneous Cu administration on Cu serum values of hypocupremic heifers.

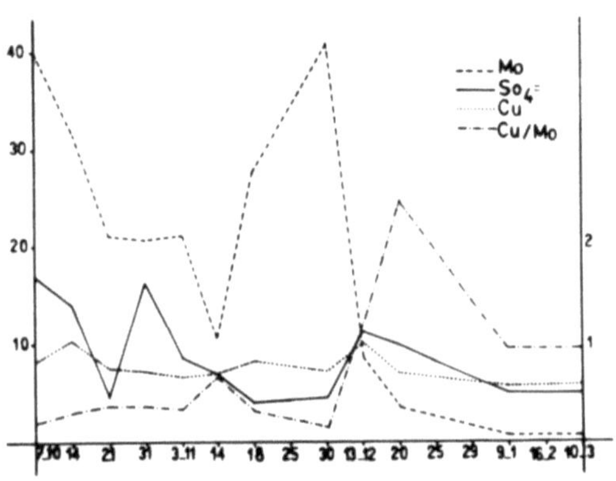

Figure 2: Variation of copper (Cu), molybdenum (Mo), copper/molybdenum ration (Cu/Mo) and inorganic sulphate (SO_4) levels in the Melilotus albus. The concentration of copper was in the range 7 to 10 ppm while the molybdenum concentration presented seasonal variation between 40.5 and 10 ppm, and the organic sulphate concentration was also over 0.5 percent.

DISCUSSION

The copper treatments were not sufficient to maintain the blood copper concentration of the heifers above the desired value of 50 µg/ml (Figure 1). These values confirmed the previous results where it was determined that the treatment utilized was effective only for 60-70 days to increase and maintain the normal concentration of copper in blood of animals in the south-east of Buenos Aires province (6).

The liveweight gains were not statistically significant which confirmed the previous works at the Balcarce Experimental Station, where the observations made, have indicated that although plasma copper concentration was elevated by the use of various slow release compounds, they were not always reflected in a substantial increase in body weight, in spite of the absence of other limiting factors (6).

Generally, there is no clear relationship between copper in blood and fertility (11, 15), in some herds of cattle with low blood copper, fertility has been impaired, while in others conception rate has been high (13). In herds in New Zealand in which blood copper was marginally subnormal, a single administration of 400 mg of copper glycinate improved the conception rate from 53 to 72 per cent for the copper treated cows (10). In Canada an increase in the copper: molybdenum ratio was associated with an improvement of fertility in cows fed silage. The opposite results were obtained in Argentina by the administration of 1000 and 2000 mg/animal/day of molybdenum to the heifers resulting in 19.2% of silent heats and 15.4% anaestrus in the treated animals while the control animals did not present any abnormal symptoms (9).

The experimental animals were grazing on *Melilotus albus* pasture with copper concentration between 7 and 10 ppm, while the molybdenum was between 10 and 40.5 ppm and the inorganic sulphate more than 0.50%. The conception rate found in this experiment was improved from 76% for the control animals to 82.2% and 91.3% for the oral and subcutaneous treated animals, indicating an important increase of fertility, but this was not statistically significant. Copper deficiency is widespread in vast regions of the country and it is generally of the conditioned type associated with an excess of molybdenum with probably other complicating dietary and genetic factors. The approach of determining copper requirement and its relationship with fertility, will not be fully realized until more information on the other factors is available.

REFERENCES

1. Bingley, J.B., 1959, J. Agr. Fd. Chem. 7:269.
2. Bingley, J.B., and B.J. Carrillo, 1964, Est. Exp. Reg. Agrop. Balcarce Publ. Tec. No.15 INTA.
3. Bingley, J.B., and B.J. Carrillo, 1966, Nature 209:834.
4. Bingley, J.B., and A.T. Dick, 1967, J. Agr. Fd. Chem. 15:539.
5. Bingley, J.B., B.E. Ruksan and B.J. Carrillo, 1978, Rev. Med. Vet. 59:3.
6. Carrillo, B.J., J.B. Bingley and B.E. Ruksan, 1978, Prod. Anim. 6:612.
7. Culot, J.B., E. Fernandez Tuñon, A. Bolaño, R.I. Petroni and M.E. Garces, 1967, Est. Exp. Reg. Agrop. Balcarce, Publ. Tec. No.53.
8. Eden, A., and H.H. Green, 1940, Biochem. J. 34:102.
9. Homse, A.C., 1980. III Congreso Argentino de Ciencias Veterinarias. Bs. As. Argentina.
10. Hunter, A.P., 1977, New Zealand Vet. J. 25:305.
11. Littlejohn, A.I., and G. Lewis, 1960, Vet. Rec. 72:1137.
12. Peterson, R.G., and D.E. Walderm, 1977, Can. J. Anim. Sci. 57:395.
13. Rowlands, G.J., W. Little and B.A. Kitchenham, 1977, J. Dairy Res. Sci. 44:1.
14. Ruksan, B.E., B.J. Carrillo and J.B. Bingley, 1978, Prod. Anim. 6:584.
15. Seekles, L., and J. Claessens, 1967, Schweiz. Arch. Tier. 109:76.

DISCUSSION

HOWELL (Australia). What was the total period of the experiment?

RUKSAN (Argentina). It was from October to March.

HOWELL (Australia). How did you assess the conception rates - did you pregnancy test or wait for the calvings?

RUKSAN (Argentina). By rectal palpation.

HOWELL (Australia). And at what time in the experiment please?

RUKSAN (Argentina). The first was about November, at 60-70 days of the experiment and the second by the end of March.

MILLS (Scotland). In Scotland we have tried a big investigation of the effects on fertility. This was carried out on herds that had a low copper status. The difference between our Scottish study and yours was that ours had to be carried out on herds for which there was no evidence of gross clinical deficiency of the type you have indicated. We saw no effects on conception and on calving rate and so on. Do you meet situations where you may have herds where, although they may have a low blood copper they have no clinical signs, and in which you get no response in terms of conception rate?

RUKSAN (Argentina). Yes we have. In one experiment with heifers and cows, we had positive effects on heifers but not on cows, with 92% of pregnancy.

SUTTLE (Scotland). Are these also areas in which the molybdenum levels in the pasture are high?

RUKSAN (Argentina). Generally yes.

SUTTLE (Scotland). In the UK there is no experimental evidence that infertility is commonly caused by copper deficiency certainly not with high molybdenum areas. It is interesting that Dr. Ruksan should have presented this case for us in which molybdenum is involved.

HOWELL (Australia). Perhaps we should take up the suggestion that in the production situation we should talk about copper responsive and copper non-responsive conditions.

BRAY (Australia). I find your inorganic sulphate levels particularly high in this species - what about total sulphur too?

RUKSAN (Argentina). We have not done that. We analysed just inorganic sulphates.

JUDSON (Australia). I was rather surprised at the large variation in the molybdenum levels in your pasture - there were rapid changes within a fortnight. I would like you to comment on this, but before doing so, I want to draw another issue to the fore; we have been rather concerned about the comment in the literature from time to time, about copper molybdenum ratio as an index of inducing a copper responsive condition. We have just recently completed a trace element survey involving trace element laboratories in Australia and New Zealand and the thing that amazed us, incidentally this was using liver copper as a substrate, was the inaccuracy of molybdenum estimation - the precision was okay but the accuracy left much to be desired.

RUKSAN (Argentina). We thought that the first decrease was because we had rainfall of 200 millimetres in the first month of the experiment, that was October and in November and December we had 100 millimetres of rain.

HYPOCUPROSIS IN CATTLE IN CENTRAL QUEENSLAND, AUSTRALIA

G.M. Murphy*, C.K. Dimmock*, T.P. Kennedy+, M.S. O'Bryan*, A.W. Plasto‡, E.E. Powell+, J.O. Twist*, G.S. Wright+, and R.J.W. Gartner*
Queensland Department of Primary Industries,
*ANIMAL RESEARCH INSTITUTE, YEERONGPILLY, 4105
+MILES, 4115
‡TOOWOOMBA, 4350
AUSTRALIA

Previous reports showed that beef cattle grazing *Paspalum dilitatum, Brachiaria multica* and *Sporobolus virginicus* pastures in south eastern (1), northern coastal (2) and central Queensland (3) had low blood and liver Cu levels unassociated with clinical signs. Supplementation with Cu did not effect weight gains in a number of experiments (1) whereas significant responses were obtained in weaner but not in adult cattle (2,3). Analyses indicated that the low Cu status was not due to either a low Cu content of the grasses or to Mo and S interaction. However, the grasses grazed contained from 10-23% crude protein and it has been shown (4) that the capacity of the rumen flora to form sulphide is increased by a relative protein excess in diets of high digestibility. Under these conditions the possibility of formation of poorly utilizable CuS will be increased.

General ill thrift in suckling calves and weaner cattle was recently investigated in central Queensland and attributed to Cu deficiency. As the etiology of the disease differs from the circumstances mentioned above, a summary of findings is given in this paper.

Survey results of serum Cu levels in weaner calves from this area are shown in Table 1.

Table 1. Copper status of weaner cattle and mean (±SE) serum Cu levels within groups.

Cu status	No. properties	Serum Cu (mg/l)	No. bled	Total cattle (x 10^3)	Total area (ha x 10^3)
Deficient	13	0.35 ± 0.110	267	22.2	91.4
Marginal	11	0.63 ± 0.090	206	13.8	56.1
Adequate	6	0.75 ± 0.012	114	6.8	33.2

Clinical signs in Hereford calves at about three months of age were characterised by a yellow coat colour and a stifflegged, humped back appearance. We attribute the humping of the thoracic vertebral region to the limb stiffness and stilted gait. Unthrifty weaner cattle (10-15 months) were reported, but there did not appear to be much difficulty in fattening adult cattle.

A pilot trial was done on a property in the area ("Red Range") to examine the effects of Cu deficiency and the response to Cu treatment. Seventy-six Hereford steers, 18-20 months old were randomly allocated on a liveweight basis to two equal groups one of which received a parenteral Cu injection. A summary of results is shown in Table 2.

Table 2. Mean liveweight, liver, blood and serum analyses of steers at "Red Range".

Sampling period	Liveweight (kg)	Liver Cu (mg/kg)	Blood Cu (mg/l)	Serum Cp (ΔA/60min)	Liver Fe (mg/kg)	Serum Fe (mg/100ml)
			CONTROL			
Initial (Feb.)	239	9	0.28	.089	746	89
12-53 weeks*	269	5	0.33	.076	694	108
67 weeks	377	4	0.38	.097	367	117
			TREATED+			
Initial (Feb.)	233	4	0.24	.036	687	85
12-53 weeks*	278	7	0.48	.175	326	147
67 weeks	390	32	0.83	.394	239	126

*Mean of samplings at 12, 38 and 53 weeks.
+120 mg Cu as Cu-glycinate on weeks 0, 12, 38, 53, 59, 67 and 73.

Although Cu levels were low in the control steers, their rate of gain of 0.68 kg/d from November to the end of May is considered satisfactory for this class of animal and country. There was no liveweight response to treatment between sampling periods. However, at slaughter (92 weeks) the treated group was significantly heavier (455 v. 439 kg; $P<0.01$). Blood and serum Cu analyses were done at each sampling after the first one. Blood levels were higher than serum levels by a mean of 0.16 and 0.20 mg/l for the control and treated groups respectively. Liver Cu reserves were not elevated by injecting 120 mg Cu at 12 weekly intervals. When the frequency was increased to 6 weekly intervals (between 53 and 67 weeks), liver and blood levels were maintained at normal ranges. Treated steers appeared in better condition judged on general activity and coat colour.

At the commencement of the trial 18% of the steers were anaemic (mean Hb 9.6 g/dl). At 53 weeks there was no evidence of anaemia in either group although values for Hb and MCV of the animals originally showing anaemia were higher in those treated with Cu than in the controls. Our results generally support the biochemical processes that Cu deficiency results in low serum ceruloplasmin (Cp) and Fe, reduced Fe mobilization, anaemia, despite adequate Fe storage in the liver (5).

Seven 3 months old calves on the property had a microcytic anaemia (Hb 6.4 - 9.9 g/dl) associated with low blood Cu levels (0.19 - 0.36 mg/l). Response to Cu treatment was a dramatic alleviation of clinical signs. However, due to constraints of experimentation on producers' properties all calves were treated with Cu and could not be used for comparative studies.

Further experiments were done in three successive years on another property in the area ("Flagstone"). The treatments were: control, Cu-EDTA parenteral injection at twice the recommended dose (125 mg Cu/100 kg with a maximum dose of 250 mg Cu), cupric oxide particles (6) in a gelatine capsule given orally (16 g CuO/100 kg with a maximum dose of 24 g CuO = 19 g Cu). Each year a new draft of 24 weaner cattle (8-10 months of age) was randomly assigned to each treatment, the first year's draft was maintained for the three years, the second year's draft for two years. The Cu status of the first draft of steers is shown in Table 3 for the first two years.

Table 3. Serum and liver Cu concentrations of steers at "Flagstone".

Weeks	Serum Cu (mg/l)			Liver Cu (mg/kg)		
	Control	Cu-EDTA*	CuO+	Control	Cu-EDTA*	CuO+
0 (Oct.)	0.51	0.49	0.50	11	14	11
12	0.25	0.76	0.83	5	133	274
24	0.31	0.77	0.78	6	150	124
36	0.37	0.63	0.63	8	83	58
50	0.41	0.50	0.51	9	153	99
64	0.34	0.64	0.86	4	36	136
74	0.38	0.71	0.73	6	145	158
86	0.44	0.60	0.65	7	78	160
98	0.46	0.55	0.52	8	172	114
104	0.34	0.72	0.78	6	70	83

*Injected on weeks 0, 12, 43, 68, 92.
+Dosed on weeks 0, 43, 74.

Intervals between treatments were based on monitoring Cu depletion determined by 6 weekly blood and 12 weekly liver analyses. In general, Cu-EDTA injections maintained hepatic reserves in excess of about 50 mg Cu/kg when given at intervals of about 7 months and the CuO particles when dosed at intervals of about 10 months. Of the two methods of treatment, the injection method is simpler in comparison to oral application.

The analytical findings were similar to those at "Red Range" in that the control animals showed all the indices of a low Cu status. Supplementation with Cu restored values to normal levels but there was no weight response in any of the groups.

Haematological findings were similar to those at "Red Range" in that at the commencement of the experiment there was evidence of ineffective erythropoiesis (microcytic hypochromic red cells with or without anaemia) in approximately 50% of the steers. Hb values increased similarly in both treated and control groups but MCV and MCH values were lower in the control than in both the treated groups.

Grasses from 14 sites in the experimental paddock were sampled at 6 weekly intervals and separated into leaf and stem. Over the first 36 weeks the leaf samples from the predominant species contained the following mean concentrations of components (DM) known to affect Cu metabolism:

	Cu (mg/kg)	Mo (mg/kg)	S (g/kg)	Crude protein (%)
Native grasses	3.3 (1.4 - 6.7)	1.0 (0.4 - 2.3)	0.6 (0.4 - 0.9)	5.8 (2.1 - 9.2)
Cenchrus ciliaris	5.9 (2.7 - 11.7)	1.3 (0.5 - 2.6)	1.5 (0.7 - 2.3)	10.6 (5.9 - 21.3)
Chloris gayana	4.3 (2.2 - 10.4)	1.9 (0.5 - 7.0)	1.8 (1.3 - 2.4)	9.1 (3.9 - 21.6)

Serum Cu in the control group (Table 3) tended to be lowest in summer (end December to mid March) corresponding to a) the period of highest rainfall, maximum growth of pasture and highest liveweight gain, b) the period when *Chloris gayana* contained the highest Mo levels and was selectively grazed, c) when protein and Cu were at their highest levels.

Soil samples taken from four sites at 0-10 cm depth showed a mean Cu concentration of 2 p.p.m. (0.7 - 3.2). These levels are not normally associated with low Cu levels in herbage. However, as the soils were alkaline (mean pH 7.8, range 7.6 - 8.0), the Cu may have been unavailable except when moisture was not limiting in summer.

CONCLUSION

The low Cu status of cattle in central Queensland is the result of low Cu content of herbage. The lowest serum Cu in summer can be attributed to an interaction of Mo with S in limiting the utilization of Cu (7) and possibly there is an additional effect of high protein (4). The growth rate of unsupplemented cattle was normal and basically these experiments report non-clinical hypocuporsis. It has been suggested (8) that hypocupraemia *per se* would have little or no effect on health and productivity unless stress factors are also operating in the herd. We used a light stocking rate in these experiments. If heavy stocking rates result in stress and lowered productivity, then Cu supplementation may result in a productive response.

REFERENCES

1. Harvey, J.M., J.W. Ryley, R.M. Beames, and M.S. O'Bryan, 1961. Qd. J. agric. Sci. 18:85-104.

2. Donaldson, L.E., J.M. Harvey, A.W. Beattie, G.I. Alexander, and M.A. Burns, 1964. Qd. J. agric. Sci. 21:167-179.

3. Alexander, G.I., J.M. Harvey, J.H. Lee, and W.C. Stubbs, 1967. Aust. J. agric. Res. 18:169-181.

4. Hartmans, J., and M.S.M. Bosman, 1970. In "Trace Element Metabolism in Animals" (ed. C.F. Mills)(Livingstone, Edinburgh and London) pp 362-365.

5. Frieden, E., 1978. In "Trace Element Metabolism in Man and Animals - 3" (ed. M. Kirchgessner)(Freising - Weihenstephan) pp 36-39.

6. Dewey, D.W., 1977. Search 8:326-327.

7. Suttle, N.F., 1974. In "Trace Element Metabolism in Man and Animals - 2" (ed. W.G. Hoekstra, J.W. Suttle, H.E. Ganther and W. Mertz)(University Park Press, Baltimore) pp 612-614.

8. Rogers, P.A.M., and D.B.H. Poole, 1978. In "Trace Element Metabolism in Man and Animals - 3" (ed. M. Kirchgessner)(Freising - Weihenstephan) pp 481-485.

DISCUSSION

PAYNTER (Australia). I would like to draw the speakers attention to some differences we have observed between serum and plasma both in terms of copper concentrations, and ceruloplasmin activity. Serum shows a depression of some 20% in ceruloplasmin activity and 15% in copper concentration. In individual animals decreases in serum ceruloplasmin activities can be as high as 50%. I would advise some caution in extrapolating serum values to those obtained with plasma and in trying to compare the two.

GARTNER (Australia). In a number of these experiments we did in fact analyse both blood and serum copper levels and in general the blood levels were higher than serum levels by a mean of 0.16 and 0.2 milligrams per litre for the control and treated animals respectively.

PAYNTER (Australia). The difference I refer to is purely between serum and plasma.

SUTTLE (Scotland). I think we could bracket plasma and serum together as equally unreliable indicators of the functional copper status of the animal which I think is what you are showing here.

LEWIS (England). In Britain we have never really associated copper deficiency in cattle with anaemia although this is often mentioned in the literature as a clinical feature. Do you find that your animals have to be a severely copper deficient for a long time before you find anaemia?

GARTNER (Australia). Anaemia was more prevalent in the younger animals and particularly in the calves. Some of these calves had a microcytic anaemia with haemoglobins between 6.4 - 9.9 grammes per decilitre, so there is some evidence that we do in fact have anaemia in this class of animal.

LEWIS (England). We can produce anaemia in sheep if we have them on a semi-purified diet of low copper concentration for a very long time but normally we do not have anaemia in our sheep associated with low blood copper levels even in areas where swayback is endemic. Under practical conditions the same applied in cattle so it is very interesting to see this as a significant feature in your animals which are so very severely depleted.

GOONERATNE (Australia). Considering the copper/molybdenum/sulphur ratio what would be the optimum levels of these elements you would recommend in the feed?

KUBOTA (USA). I do not see any sense in looking for ratios - we don't find any difference in copper concentration in plants between areas of molybdenosis and areas without - the plants could not care less. I think that what is happening is that the molybdenum enters the plant by mass flow, and the copper enters by a different pathway. So there should not be any relationship between the two elements, and any ratios and their significance seems meaningless.

OXIDISED COPPER WIRE FOR ORAL COPPER THERAPY IN CATTLE

G.J. Judson[1], D.W. Dewey[2], J.D. McFarlane[3] and M.J. Riley[3]

1. Institute of Medical and Veterinary Science, Frome Road, ADELAIDE, S.A. 5000, AUSTRALIA
2. 91 Cross Road, HAWTHORN, S.A. 5062, AUSTRALIA
3. Department of Agriculture, P O Box 618, NARACOORTE, S.A. 5271, AUSTRALIA

The efficacy of an oral dose of oxidised copper wire was compared with that of a subcutaneous injection of copper glycinate ('Cuprate' - Ethnor Vet) as a means of raising the copper status of cattle. The comparison was undertaken on a property near Robe, South Australia, where liveweight responses of calves to injections of Cuprate have been recorded. The oxidised copper wire used in these studies had a specific gravity of about eight and approximately one fifth of the copper was in the oxidised form.

In a preliminary trial, plasma copper concentrations were measured in cows 30 weeks after treatment with 50 g oxidised copper wire per cow or with 120 mg injected copper per cow. At this stage, the plasma copper concentrations in the two copper treated groups were elevated above those in the untreated group but only in the group treated with oxidised copper wire was this elevation significant (Trial 1, Table 1). Later studies showed that 25 g oxidised copper wire per cow was a sufficient dose to sustain elevated plasma copper concentrations, similar to those observed in cows each given 120 mg injected copper, for periods up to 33 weeks post-treatment (Trial 2, Table 1).

Table 1. Responses in plasma copper concentrations (umol/l) of Hereford cows to copper supplementation (Mean±SD).

Weeks Post Treatment	Treatment: Nil	Treatment: Cuprate 120 mg copper	Treatment: Copper Oxide Trial 1=50 g;Trial 2=25 g
	TRIAL 1		
0 (Nov., 1977)	14 ± 2.4 (4)	16 ± 2.2 (4)	12 ± 3.1 (5)
30	7 ± 2.7^{a} (7)	8 ± 2.4 (7)	10 ± 1.5^{a} (9)
	TRIAL 2		
0 (Sept., 1979)	9 ± 1.5 (12)	8 ± 1.5 (6)	8 ± 1.6 (6)
6	7 ± 2.8^{a} (11)	12 ± 1.2^{ab} (6)	8 ± 3.4^{b} (6)
12	5 ± 1.9^{ab} (12)	12 ± 2.8^{a} (6)	9 ± 4.8^{b} (6)
18	4 ± 2.1^{ab} (11)	9 ± 2.1^{a} (6)	10 ± 3.3^{b} (5)
22	4 ± 1.4^{ab} (10)	7 ± 2.6^{a} (5)	9 ± 2.0^{b} (5)
33	4 ± 0.9^{ab} (11)	7 ± 1.6^{a} (6)	7 ± 2.8^{b} (6)

On each sampling occasion, means with similar superscripts differed at P=0.05 (Student's 't' test). In brackets, the number of cows sampled.

Table 2. Responses in plasma, hair and faecal constituents of calves to copper dosing*

Treatment Group	Pre-treatment (July, 1979)	Weeks Post-treatment 6	12	20	29	35	41
Plasma Copper, umol/1 (TCA soluble)							
Nil	10±3.6	6±1.7^{ab}	7±2.4^{ab}	11±2.3	7±2.5	7±1.5^{a}	4±0.9^{ab} (6)
Cuprate	9±3.3	10±0.9^{a}	11±1.1^{a}	12±2.3	7±1.9	8±1.4^{b}	6±1.1^{a}
Cu Oxide	8±4.8	10±2.1^{b}	12±1.7^{b}	12±1.0	10±2.3	11±2.1^{ab}	7±2.1^{b} (7)
Plasma Copper, umol/1 (Total)							
Nil	10±3.8	6±1.7^{ab}	7±2.2^{ab}	12±1.7	8±2.1^{a}	8±1.5^{a}	5±0.9^{a} (6)
Cuprate	9±3.5	11±1.2^{a}	11±1.0^{a}	12±2.3	9±1.7^{b}	7±1.4^{b}	5±1.0^{b}
Cu Oxide	9±4.7	11±2.0^{b}	12±1.4^{b}	12±0.8	12±1.8^{ab}	11±2.0^{ab}	7±1.9^{ab} (7)
Plasma Pepsinogen, U/1							
Nil	4.1±1.2	5.0±1.3	5.7±1.4	4.6±1.3	3.5±1.4	2.8±0.7 (5)	4.4±2.2 (6)
Cuprate	3.9±0.9	4.4±0.7	6.0±1.7	4.0±1.8	3.4±1.5	3.0±1.0	4.3±0.8 (7)
Cu Oxide	3.8±0.5	5.0±1.0	5.9±1.0	5.0±2.0	2.9±0.9	2.6±0.8 (7)	3.7±0.3 (7)
Hair Copper, umol/kg							
Nil		79±20^{a}	63±16^{ab} (6)	71±11^{a}	70±14	50±5^{a} (7)	48±14 (5)
Cuprate		89±15	92±13^{a} (7)	82±17	82±19	55±7^{b}	39±7^{a}
Cu Oxide		102±20^{a}	104±10^{b}	104±18^{a}	85±21	70±16^{ab}	54±16^{a} (7)
Faecal Copper mmol/kg							
Nil	0.27±0.4 (4)	0.17±0.03 (4)	0.38±0.04 (4)	0.32±0.02 (4)	0.15±0.04 (4)	0.25±0.04 (4)	0.09 (2)
Cu Oxide		0.98±1.31 (4)	1.17±0.91 (4)	0.35±0.02 (4)			0.32±0.26 (3)
Pasture, mmol/kg							
Copper		0.12±0.02 (9)	0.11±0.03 (7)	0.08±0.03 (7)	0.05±0.02		0.07±0.02
Molybdenum		0.09±0.05 (9)	0.12±0.06 (7)	0.12±0.05 (7)	0.11±0.05		0.11±0.05
Sulphur		94±11 (9)	90±16 (7)	78±18 (7)	69±9		75±20

* Mean±SD of 8 values except where indicated in brackets. On each sampling occasion and for each constituent, those values with the same superscript differed ($P<0.05$; Student's 't' test).

The efficacy of a 25 g dose of oxidised copper wire as means of raising the copper status of calves three to five months of age was also investigated. The calves, stratified by liveweight within sex were allocated at random to one of three treatment groups each of 21 calves, viz.: no copper, 'Nil'; 25 g oxidised copper wire per calf; or 60 mg copper as Cuprate per calf. Animals were bled and weighed at intervals of six to nine weeks post-treatment for 41 weeks.

In calves given the oxidised copper wire or Cuprate, the initial responses in plasma and hair copper concentrations were similar and were significantly greater than values for Nil calves (Table 2). The results indicate that at 35 to 41 weeks post-treatment the oxidised copper wire was more effective than Cuprate in maintaining a raised copper status of calves. No liveweight responses to copper supplementation were observed over the 41 weeks of the trial during which the mean liveweight of the calves increased from 87 to 256 kg.

The calves were depastured on calcareous sands supporting largely strawberry clover, medic and lucerne. The concentrations of copper, molybdenum and sulphur in the pasture (Table 2) suggest that the hypocupraemia of the Nil calves may have been due to the high molybdenum content reducing the availability of pasture copper although this hypothesis was not supported by the blood data (Table 2). The trichloroacetic acid (TCA) - soluble copper and the total plasma copper were similar suggesting that little blood copper was 'tied-up' as a copper-molybdenum complex - a complex which is apparently insoluble in 10% (w/v) TCA (4).

Faecal copper concentrations were raised in some calves for up to 12 weeks post-dosing of oxidised copper wire (Table 2) confirming the observations of others (1,5) that the particles are retained in the alimentary tract of cattle for some weeks after dosing. Studies with sheep have shown that the copper oxide is retained in the acid environment of the abomasum for intervals of time sufficient to permit the partial solubilization of copper (3). This retention of oxidised copper wire in calves does not appear to have damaged the abomasum as assessed by plasma pepsinogen activity (Table 2) nor reduced the absorption of dietary zinc as assessed by whole blood, plasma and hair zinc concentrations.

Our evidence and that of others (1,2,5) support the view that oral dosing of oxidised copper wire is a useful and safe alternative to a parenteral product for prophylaxis in cattle.

We thank Messrs W and M Quinlan-Watson, the owners of the property, for providing cattle and facilities and North Bulli Colleries Pty. Ltd. for the supply of oxidised copper wire.

1. Costigan, P., and K.J. Ellis, 1980, Proc. Aust. Soc. Anim. Prod. 13:451.

2. Deland, M.P.B., P. Cunningham, M.L. Milne and D.W. Dewey, 1979, Aust. Vet. J. 5:493 - 494.

3. Dewey, D.W., 1977, Search 8:326 - 237.

4. Dick, A.T., D.W. Dewey and J.M. Gawthorne, 1975, J. Agric. Sci., Camb. 85:567 - 568.

5. Suttle, N.F., 1979, Proc. Nutr. Soc. 38:135A.

DISCUSSION

SUTTLE (Scotland). Our work in the UK would endorse these findings entirely. However it would be interesting if you would check the site of retention of the needles in the bovine because our studies suggest that the reticulum can be the major site of retention in that species.

JUDSON (Australia). We have not checked it with the bovine but Mr. Dewey has checked it with sheep and found that you have to be careful when removing the intestinal tract from the animal because if copper oxide is located in the abomasum by the time you remove the gut from the animal some of that copper can be in the fore stomach.

SUTTLE (Scotland). We use the same techniques for sheep as we do for cattle. I think it is to do with the morphology of the reticulum in the two species.

WALKER (Australia). Why don't you use copper 1 oxide pellets rather than oxidised copper 2 wire. Pellets being a better shape and if they result in copper metal as well as copper 2 they will keep on going.

JUDSON (Australia). The wire was used because it was readily available. Cupric oxide was used as the coating as opposed to cuprous oxide, because of the cost of manufacture. Cupric oxide is formed at about 300 degrees whereas Cuprous is about 1000 degrees.

WALKER (Australia). That would be by oxidation. You can produce solutions of copper quite easily to precipitate copper 1 oxide, and then I think the cost of it would be negligible compared to that of the copper glycinate complex.

GRAHAM (Australia). Could you tell an agronomist how this copper is spread around the paddock after it goes through the animal because it is really quite a significant amount of copper?

JUDSON (Australia). We have had a look at this, but with sheep. We are currently undertaking a dose response trial with animals receiving doses of 2½ grammes to 20 grammes per head. Four weeks after dosing we checked the presence of copper oxide in untreated animals and the treated animals. We found no evidence of copper oxide in the untreated animals and grazing intensity was quite high during this period. We have not reached the stage where we can recommend a dose for cattle but perhaps for sheep it might be even less than 5 grammes per head. That is not a lot of copper going on to the paddock although it could be localised.

GRAHAM (Australia). Would about 95% of it end up on the pasture?

JUDSON (Australia). Yes.

TREATMENT OF COBALT DEFICIENCY IN CALVES

G.J. Judson[1], J.D. McFarlane[2], M.J. Riley[2], M.L. Milne[2] and A.C. Horne[1]

1. Institute of Medical and Veterinary Science, FROME ROAD, ADELAIDE, S.A. 5000, AUSTRALIA
2. Department of Agriculture, P O BOX 618, NARACOORTE, S.A. 5271, AUSTRALIA

The effect of subcutaneous injections of vitamin B_{12} or of oral cobalt pellet therapy on liveweight and blood constituents of Hereford calves was examined in three trials conducted on a farm in the south-east of South Australia, on coastal soils known to produce cobalt deficiency in sheep (4).

Trial 1 commenced in April, 1977, Trial 2 in June, 1978 and Trial 3 in October 1979. In each trial, calves one to three months old were stratified by liveweight within sex and allocated at random to treatment groups. Anthelmintic was given to calves at each weighing to minimize infestations of gastro-intestinal parasites and subcutaneous injections of 60 mg copper per calf ('Cuprate' - Ethnor Vet) were given as required to maintain normal serum copper concentrations. Vitamin B_{12} was given subcutaneously as follows: Trial 1, 6 mg cyanocobalamin per 50 kg liveweight; Trial 2, 2 mg hydroxocobalamin ('Cobalex') per 50 kg liveweight and Trial 3, 1 ml 'Ringer 2000' (1.8 mg hydroxocobalamin + 0.2 mg cyanocobalamin) per calf. Calves given regular injections of vitamin B_{12}, 'Regular-B_{12}' treatment group, were given the vitamin B_{12} at weighing, which was every five to eight weeks. In Trials 2 and 3 a 'Single-B_{12}' treatment group was included to assess the efficacy of a single dose of vitamin B_{12} in preventing or correcting cobalt deficiency in calves. During the course of each trial, vitamin B_{12} was also given to calves in the 'Nil' treatment group as an emergency measure to improve liveweight gains. A group of calves each given a single oral dose of two 10 g cobalt pellets, 'Cobalt Pellet' treatment group, was included in Trial 2.

Table 1 summarizes the liveweight gains of the calves.

In Trial 1, calves in the Regular-B_{12} group gained in liveweight at a faster rate than calves in the Nil group for the first 30 weeks of the trial. From week 30, vitamin B_{12} was given to the Nil calves and over the ensuring 13 weeks, a liveweight response to the supplementation was observed.

In Trial 2, liveweight gains of the calves in the Cobalt Pellet group were similar to gains of calves in the Regular-B_{12} group for up to 20 weeks. Thereafter liveweight gains tended to be greater in the Regular-B_{12} group than in the Cobalt Pellet group. The decline in serum vitamin B_{12} concentrations over the 20 to 26 weeks post-cobalt therapy (Table 2), also observed elsewhere with steers (5), is indicative of a limited life of a cobalt pellet in cattle. A 4 mg dose of vitamin B_{12} to the Single-B_{12} group of calves at week 20, when liveweight gain of this group was inferior to that of the Regular-B_{12} group, was found to last at least 12 weeks under the conditions of the experiment.

In Trial 3, the calves in the Single-B_{12} group were each given a 2 mg dose of vitamin B_{12} at week 0 and their liveweight gain was similar to that of the calves in the Regular-B_{12} group for 23 weeks but during the next seven weeks cobalt deficiency became evident. A further dose of 2 mg vitamin B_{12} to the Single-B_{12} calves was then given which corrected the cobalt deficiency, as assessed by liveweight gain, for at least 15 weeks. A 2 mg dose of vitamin B_{12} was also given to each of the Nil calves at week 8 of the trial, at which time marginal vitamin B_{12} deficiency was suspected, and this dose supplied the vitamin B_{12} requirements of the calves for growth for up to 29 weeks. Over the final eight weeks of the trial, weeks 37 to 45, the mean liveweight gain of the Nil calves was negligible indicating that the vitamin B_{12} reserves of these calves were depleted.

Serum vitamin B_{12} concentrations were responsive to vitamin B_{12} injections over the short-term with values in the order of 10 to 16 ug/l at one to two hours post-injection of 2 mg vitamin B_{12} (Trial 3) and about 0.4 ug/l six days after an injection of 16 mg vitamin B_{12} (Trial 1). At six weeks post-injection of vitamin B_{12}, the serum vitamin B_{12} concentrations usually varied from 0.1 to 0.2 ug/l and at 12 weeks were usually less than 0.1 ug/l (Table 2). At 12 weeks the serum vitamin B_{12} values were similar to those values found in Nil calves although it was evident from this work that vitamin B_{12} reserves were

Table 1. Effect of vitamin B_{12} injections and cobalt pellets on mean liveweight gain (kg) of Hereford Calves*

TRIAL 1	Weeks: 0-6	6-12	12-18	18-24	24-30	30-35	35-43	43-50	0-50
Nil	19^{a}	8^{a}	15^{a}	12^{a}	13^{a}	$*39^{a}$	$*37^{a}$	*12	154^{a}
Regular-B_{12}	23^{a}	14^{a}	23^{a}	34^{a}	44^{a}	31^{a}	27^{a}	10	206^{a}

TRIAL 2	Weeks: 0-7	7-14	14-20	20-26	26-32	0-32
Nil	17	31^{ab}	27^{ab}	18^{abc}	$*42^{abc}$	134^{ab}
Regular-B_{12}	19	37^{ac}	44^{ac}	37^{ad}	31^{a}	168^{ac}
Single-B_{12}	15	32^{cd}	30^{cd}	$*43^{be}$	32^{bd}	152^{b}
Cobalt Pellets	15	36^{bd}	40^{bd}	28^{cde}	25^{cd}	145^{c}

TRIAL 3	Weeks: 0-8	8-17	17-23	23-30	30-37	37-45	0-45
Nil	41	$*53^{a}$	33^{a}	14^{a}	6^{a}	1^{ab}	146
Single-B_{12}	*47	43^{a}	24^{a}	3^{ab}	$*18^{a}$	15^{a}	150
Regular-B_{12}	48	49	29	16^{b}	12	20^{b}	174

+ At week 0, mean liveweight of each treatment group (and number of calves/group) was 68 kg (12) Trial 1, 67 kg (15) Trial 2 and 68 kg (11) Trial 3. Regular-B_{12} calves were given vitamin B_{12} at each weighing and '*' indicates when calves in other treatment groups were given vitamin B_{12}. In each Trial and at each weighing, values with the same superscript differed from each other ($P<0.05$; Two-Way Analysis of Variance, Duncan's Multiple Range Test).

adequate for growth in the vitamin B_{12} treated calves. This suggests that serum vitamin B_{12} concentrations may not be a reliable indicator of vitamin B_{12} status of vitamin B_{12} treated calves, a finding also observed with young sheep (1).

A feature of vitamin B_{12} deficiency in these calves was a compensated microcytic normochromic anaemia (3). In severely vitamin B_{12} deficient calves in Trial 1, the anaemia was associated with depressed white cell counts (2).

Table 2. Mean serum vitamin B_{12} concentrations (ug/l) of Hereford calves at different stages of the trial[+]

TRIAL 1	Weeks:								
	0	6	12	18	24	30	35	43	50
Nil	.19	.12[a]	.06[a]	.05[a]	.07[a]	.08[a]*	.08[a]*	.17*	.08[a]
Regular-B_{12}	.19	.17[a]	.12[a]	.13[a]	.13[a]	.15[a]	.15[a]	.27	.13[a]

TRIAL 2	Weeks:					
	0	7	14	20	26	32
Nil	.18	.07[ab]	.05[ab]	.05[abc]	.05[a]*	.14
Regular-B_{12}	.17	.14[ac]	.10[ac]	.09[a]	.12[a]	.15[a]
Single-B_{12}	.22	.08[d]	.06[cd]	.08[b]*	.08	.10[a]
Cobalt Pellets	.24	.28[bdc]	.17[bd]	.17[c]	.11	.15

TRIAL 3	Weeks:						
	0	8	17	23	30	37	45
Nil	.07	.10*	.09	.05	.09	.07[a]	.08
Single-B_{12}	.11*	.15	.09	.06	.09*	.09[a]	.09
Regular-B_{12}	.11	.11	.10	.09	.14	.11	.11

[+] In each Trial and on each sampling, mean values (usually of 5 samples) with the same superscript differed ($P<0.05$; Student's 't' test). Regular-B_{12} calves were given vitamin B_{12} after each sampling and '*' indicates when vitamin B_{12} was given to other treatment groups.

The liveweight responses to vitamin B_{12} supplementation suggest that the calves were born with sufficient vitamin B_{12} reserves for the first few months of life. After this period the vitamin B_{12} of the dam's milk, which was usually less than 1 ug/l milk but ranged from 0.1 to 4.0 ug/l, together with the cobalt from ingested pasture was not sufficient to meet their vitamin B_{12} requirements. In such situations a single injection of 2 mg vitamin B_{12} will probably satisfy the vitamin B_{12} requirements of calves for about 20 weeks.

We thank Messrs W and M Quinlan-Watson, the owners of the property, for providing cattle and facilities. We are grateful to V.R. Laboratories for the 'Cobalex' and to J.D. Luscombe Pty. Ltd. for the 'Ringer 2000'.

1. Hannam, R.J., G.J. Judson, D.J. Reuter, L.D. McLaren and J.D. McFarlane, 1980, Aust. J. Agric. Res. 31:347 - 355.
2. Judson, G.J., 1978. In "Recent Advances in Animal Nutrition 1978" (ed. D.J. Farrell) (University of New England Publishing Unit, Armidale) pp 77-88.
3. Judson, G.J., and K.E. Gifford, 1979, Aust. Vet. J. 55:504 - 505.
4. Lee, H.J., 1951, J. Dep. Agric. S. Aust. 54:475 - 532.
5. MacPherson, A., and J. Dixon, 1980, Anim. Prod. 30:373 - 381.

DISCUSSION

LLOYD DAVIES (Australia). This is a very important paper illustrating practical points on the treatment of young calves for cobalt deficiency. It can be seen that in no way is the cobalt requirement an academic problem it is a particularly important field problem in Australia.

MacPHERSON (Scotland). I would like to congratulate both Dr Judson and Dr Garton on the quality of the pictures showing the clinical symptoms of vitamin B_{12} deficiency and copper deficiency. In the UK we cannot distinguish between the two clinically; I wonder if you would like to comment on whether you can here.

JUDSON (Australia). In the trials we did with cattle poor growth and a very rough coat were two obvious symptoms when the nil versus the B_{12} treatments were compared. Sometimes in the nil group we also appeared to have depigmentation of the hair. The farmers in this area commonly and immediately think of a copper deficiency, but in some cases it might be primarily a B_{12} deficiency and after this is corrected then, perhaps, they might get a copper deficiency or copper responsive condition.

HUTAGALUNG (Malaysia). In Malaysia we have been using cobalt bullets for weaned calves and for cows and we have found that the results were not up to what the manufacturers claimed. One year after the bullet was implanted we found that there was no difference in the weight of the bullet. What is your experience of the effectiveness of cobalt bullets in supplying cobalt for weaned calves and cows.

JUDSON (Australia). We want to investigate it further in South Australia. I intend to start a trial later this year with mature cattle with cobalt supplementation by pellet to see how effective this is not only in raising the B_{12} status of the animal but also in the blood and B_{12} status of the calf. We should be cautious, I suppose I am relying in some instances on serum B_{12} levels and there was a question earlier regarding the value of serum B_{12} levels. From a preliminary survey we have found that when liver B_{12} levels are greater than 0.6 µg/g wet liver they don't seem to bear any relationship to serum B_{12} levels, but this is only on a survey of about 120 animals at this stage.

CONTROLLED RELEASE GLASSES FOR SELENIUM SUPPLEMENTATION

W. M. Allen, P. R. Moore and B. F. Sansom

Agricultural Research Council,
Institute for Research on Animal Diseases,
Compton, Newbury, Berkshire, RG16 ONN, U.K.

INTRODUCTION

Severe selenium (Se) deficiency in sheep causes white muscle disease and less severe deficiency is associated with ill-thrift and infertility. A possible method for overcoming these problems may be to implant sheep with controlled release glasses (CRG's) containing selenium and designed to dissolve at a steady rate over a period of months. These glasses could provide sheep with some or all of their daily requirement for selenium during that period.

Controlled release glasses are phosphate based glasses into which essential elements can be incorporated. By varying the composition of the glasses it is possible to control the rate at which they dissolve and therefore the rate at which the active elements are released. Thus, different glasses can be manufactured to suit different circumstances and they can be made to last the whole of the grazing season.

In order to establish the efficacy of CRG's in alleviating selenium deficiency it is necessary to establish how much of the released selenium is retained by the animal. It was therefore decided to carry out metabolic balance experiments on sheep which had been implanted with selenium CRG. The first of these trials is described here.

MATERIALS AND METHODS

1. The selenium glass. Each glass pellet was cylindrical (3.2 mm x 3.0 mm diameter), weighed approximately 110 mg and contained 43.7% Se w/w. When suspended in water at 30°C each pellet released approximately 10 μg Se/hour.

2. Treatment. Five 1-year old Swaledale sheep were used. Two sheep received 2 pellets ($\triangleq$ 100 mg Se) either subcutaneously (s.c.) behind the ear or intramuscularly (i.m.) into the neck and 2 sheep received 4 pellets ($\triangleq$ 200 mg Se) in the same sites. The fifth animal received only a sham implant. The pellets were implanted with a standard Hoechst implanting gun.

3. Experimental procedure. Four weeks before the experiment began the sheep were placed in metabolism cages so that their urine and faeces could be collected separately. Weighed amounts of food (0.5 kg cake, 0.5 kg hay) of known selenium content were offered daily and all uneaten food was collected and weighed. Water was available *ad libitum*.

For 10 days before and 30 days after implantation each animal's total daily excretion of urine and faeces were collected, measured and subsampled. The subsamples taken during 5 day periods were bulked before being analysed for selenium content.

Jugular blood samples were taken from each animal into heparin before implantation, daily for 1 week after implantation and thereafter weekly for 19 weeks.

Selenium was analysed by the method described by Little, Vagg, Collis, Shaw & Gleed (1979) using 2 ml samples for analysis of blood and urine and 1 g samples for feed and faeces.

After the balance the sheep were removed to pens until slaughter 155 days after implantation. All the pellets implanted intramuscularly were recovered at slaughter but only 1 pellet was recovered from the sheep which received 4 pellets subcutaneously and none from the sheep which received 2 subcutaneously. The weights of selenium lost from the recovered pellets were determined.

RESULTS

1. Balance data. The sheep cake contained 330 ng Se/g DM and the hay contained 66 ng Se/g DM. All the cake offered was eaten but frequently up to 100 g of hay was uneaten. The mean dietary intake of selenium was calculated to be 0.82 mg/5-day period.

The results suggest that in the animal which received 2 pellets intramuscularly (i.m.) the pellets stopped releasing selenium within a few days of implantation. In the 3 other treated animals there was no significant difference between the results in spite of the differences between the doses and routes of administration. In view of this the mean data from the 2 sheep which received 4 pellets and the sheep which received 2 pellets subcutaneously will be presented.

Despite a constant level of selenium in the diet, excretion of selenium in the faeces increased steadily during the balance from 0.53 mg Se/period initially to 0.95 mg Se/period at the end in both the control and treated animals.

Before treatment between 0.1 and 0.2 mg Se/period was excreted in urine and the control sheep continued to excrete selenium at this rate throughout the balance. In the first period after implantation 1.69 mg Se was excreted in the urine of the sheep with 2 pellets i.m. but in subsequent periods the amount decreased to pre-treatment levels. In the remaining 3 sheep a mean of 2.92 mg Se was excreted in the first period after implantation but the rate of excretion decreased rapidly over the subsequent 2 periods and then remained between 0.3-0.4 mg Se/period.

During the experiment the control sheep was in slight negative selenium balance, excreting over the 40 days 1.41 mg more selenium than it consumed (an average of 0.18 mg/period). The results for the treated animals presented graphically in Figures 1 and 2 were corrected for this 'background' effect. All the treated animals were in positive selenium balance before implantation. In the first period after implantation, however, the sheep with 2 pellets i.m. reached a cumulative selenium balance of -0.96 mg which increased to -0.59 mg by the end of the balance. In the remaining 3 sheep a mean cumulative selenium balance of -2.50 mg was calculated by the end of the first period after implantation after which the rate of net loss of selenium decreased, giving a cumulative selenium balance of -3.65 mg by the end of the balance.

FIGURE 1

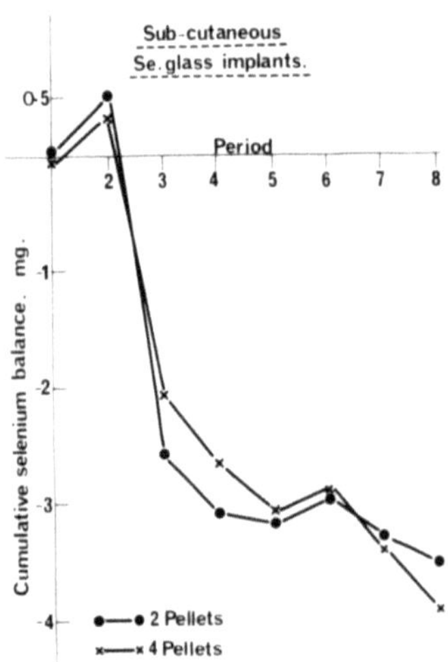

FIGURE 2

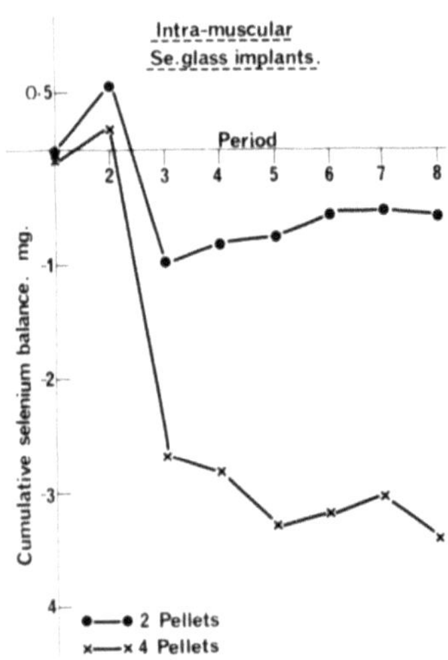

Analysis of the recovered pellets showed a mean weight loss of 18.3 mg/pellet (SD 4.61 mg) and a mean selenium loss of 17.6 mg/pellet (SD 0.76 mg).

2. Blood. Whole blood selenium concentrations in the treated animals increased from pre-dose levels of 170-200 ng/ml to 400 and 475 ng/ml in the 2 sheep with 4 pellets and 400 ng/ml in the 2 sheep with 2 pellets 3 weeks after treatment. Subsequently levels decreased to 315 and 385 ng/ml in the 2 sheep with 4 pellets and 290 and 300 ng/ml in the 2 sheep with 2 pellets at 5 weeks. Over the same period whole blood selenium levels in the control sheep increased from 117 ng/ml to 270 ng/ml. From week 6 onwards the control sheep and the sheep with 2 pellets i.m. maintained levels at approximately 240 ng/ml while the remaining 3 sheep maintained levels at approximately 290 ng/ml.

DISCUSSION

The metabolic balance data show that the 3 treated animals which had a mean balance of -2.50 mg Se in the first post-treatment period continued to show a net loss of selenium at a slower but consistent rate of approximately 0.23 mg Se/period over the remainder of the balance. Assuming that the net selenium loss continued at this rate, over the 155 days up to slaughter, the net cumulative balance would have been -9.4 mg Se.

According to the analysis of the residual glasses, the animals which were given 4 pellets would have received 70.6 mg Se from dissolution of the glass while the animals given 2 pellets would have received 35.2 mg Se. Thus the apparent net retention of selenium by the treated animals was 61.0 mg Se from 4 pellets and 25.6 mg Se from 2 pellets.

The data suggest that there was an initial rapid release of selenium from the glass during the first few days after implantation, which was excreted mainly through the urine, after which the rate of selenium release decreased rapidly before stabilising at a reduced rate. A pink deposit was found on the exterior of the recovered glasses and this may have been due to reprecipitation of the dissolution products of the glass. This deposit was included when the glasses were weighed and analysed for selenium and may explain the fact that 96.7% of the weight lost was selenium. The surface deposit might also explain why the release rate of selenium decreased rapidly to low levels and why there was little variation due to dose rate and route of administration. If the deposit were sufficiently heavy it might seal off the glass completely, thus explaining the anomalous results obtained from the sheep which received 2 pellets i.m.

The apparent large retentions of selenium did not produce large or prolonged increases in the whole blood selenium concentrations. Significant increases were found for only 5 weeks after treatment. This may also have been associated with a change in the release rate of selenium from the glass.

Only selenium excreted in the urine and faeces was measured in this experiment. Another route of excretion has been reported by Imbach & Sternberg (1967) who showed that rats exhaled, as dimethyl selenide, up to 13% of selenium administered by intravenous, intraperitoneal or subcutaneous injections of sodium selenite. Their experiment used a single dose of a readily available source of selenium and no information is available on the amount of selenium which might be excreted through the lungs from a source, such as a CRG, which releases selenium more gradually.

Allowing for the loss of approximately 10-20% of the selenium apparently retained by the sheep by exhalation and assuming a requirement of about 0.1 mg Se/day for a sheep, 2 pellets (100 mg Se in total) would provide an adequate supply of selenium for up to 7 months. New formulations of selenium CRG are being developed to overcome the problem of reprecipitation and give a more linear rate of release of selenium. In this way the released selenium may be more efficiently retained and the dose of CRG reduced.

REFERENCES

1. Imbach, A. & Sternberg, J. 1967. Journal of Applied Radiation and Isotopes. 18: 545-556.

2. Little, W., Vagg, M. J., Collis, K. A., Shaw, S. R. & Gleed, P. T. 1979. Research in Veterinary Science. 26: 193-197.

ACKNOWLEDGEMENTS

This project has been developed in association with I.T.T. Components Group, (C.R.G.), West Road, Harlow, U.K. CM20 2BP.

DISCUSSION

HUDSON (Australia). Could you tell us what the chemical nature of the selenium was in these glass pellets and perhaps give us some opinion as to how the selenium was actually released from the pellet to the animal.

ALLEN (UK). The selenium is incorporated as an integral part of the glass, it is incorporated into the glass as selenium oxide. In this particular experiment there was precipitation of zinc phosphate on the surface, the zinc being one of the constituents which controls the solubility of this particular glass.

BUSSLER (Germany). Why could you have not fed your animals directly with soluble selenium sulphate.

ALLEN (UK). The point was referred to by the last speaker who emphasised that if we incorporate these trace elements in the diet there are competitive difficulties with molybdenum, sulphur and copper. We think that implantation is a more satisfactory way than mixing with the diet. It is a less costly operation to inject these animals with selenium glass than to incorporate selenium into the diet.

EFFECT OF ZN-CONTAINING PELLETS ON SELENIUM PELLET FUNCTION

D.W. Peter[1], A.W. Mann[2] and R.A. Hunter[1]

[1] CSIRO, Division of Animal Production, Private Bag, P.O., Wembley, Western Australia 6014.

[2] CSIRO, Division of Mineralogy, Private Bag, P.O., Wembley, Western Australia, 6014.

INTRODUCTION

Sheep which received an intraruminal zinc pellet (Zn/Fe), composed of 5 g of high purity Zn shot and 5 g of coarse iron (Fe) filings, and a selenium pellet (Se/Fe)* containing 0.5 g of Se and 9.5 g of granular sponge-iron, were previously found to have significantly higher concentrations of Se in blood than those given a Se/Fe pellet alone (1). Studies were conducted with grazing sheep to determine if this "Zn/Fe pellet effect" was 1) due to an increased intake or absorption of dietary Se or 2) resulted from increased absorption or release of Se from the Se/Fe pellet and 3) if the effect was specific to Zn/Fe pellets. An electrochemical mechanism is proposed to explain both the release of Se from Se/Fe pellets and the effect that a second non-Se containing pellet may have on Se release.

MATERIALS AND METHODS

Thirty mature wether sheep were randomly divided into 10 groups of 3 animals and single high density intraruminal pellets administered to animals in each group as follows: 1. No pellet, 2. Se/Fe pellet* (ICI (Aust.) Pty Ltd), 3. Zn/Fe pellet, 4. Zn pellet (pellet cast from 10 g pure Zn), 5. Fe pellet (10 g of Fe filings as used in Zn/Fe pellet), 6. Se/Fe pellet and Zn pellet, 7. Se/Fe pellet and Zn/Fe pellet, 8. Se/Fe pellet and Fe pellet, 9. Se/Fe pellet and Co/Fe pellet (commercial cobalt pellet weighing approximately 10 g and containing approximately 57% cobalt oxide and 38% Fe grit; ICI (Aust.) Pty Ltd.) and 10. Se/Fe pellet and Zn/Ni pellet (5 g of large Nickel shot cast in 5 g pure Zn). All sheep grazed together for 80 days and blood samples were collected to obtain plasma samples for Se analysis (2) at 0, 26, 56 and 80 days.

RESULTS

The results, as shown in Fig. 1, confirm the previously reported findings (1), with significantly higher ($P<0.05$) plasma concentrations of Se at day 26 in sheep given both Zn/Fe and Se/Fe pellets (group 7) than in those given a Se/Fe pellet alone (group 2). Administration of either a pure Zn pellet or a Zn/Ni pellet with a Se/Fe pellet, groups 6 and 10 respectively, produced similar results. Plasma Se concentrations, on day 26, in animals given an Fe pellet and a Se/Fe pellet (group 8) were also significantly higher ($P<0.05$) than in animals given a Se/Fe pellet alone, though lower than those in groups given Zn containing and Se/Fe pellets together. Inclusion of a Co/Fe pellet with a Se/Fe pellet (group 9) resulted in marginally higher (non-significant) plasma Se concentrations on day 26.

There were no significant differences in plasma Se concentration between untreated animals (group 1) and those given non-Se containing pellets alone (groups 3, 4 and 5), though there were small increases in concentration in all these groups between day 0 and day 26 or 56, followed by a steady decline. More pronounced decreases occurred in all Se/Fe pellet treated groups between days 26 and 80, with the most rapid decline in groups which also received Zn containing pellets. By day 80 there were no significant differences between any of the groups given Se/Fe pellets, but plasma Se concentrations were still significantly higher ($P<0.05$) than those found in groups not given Se/Fe pellets.

---

* Se/Fe pellets contained fine-grained Se with an average size of approximately 4μm (range 1-50 μm).

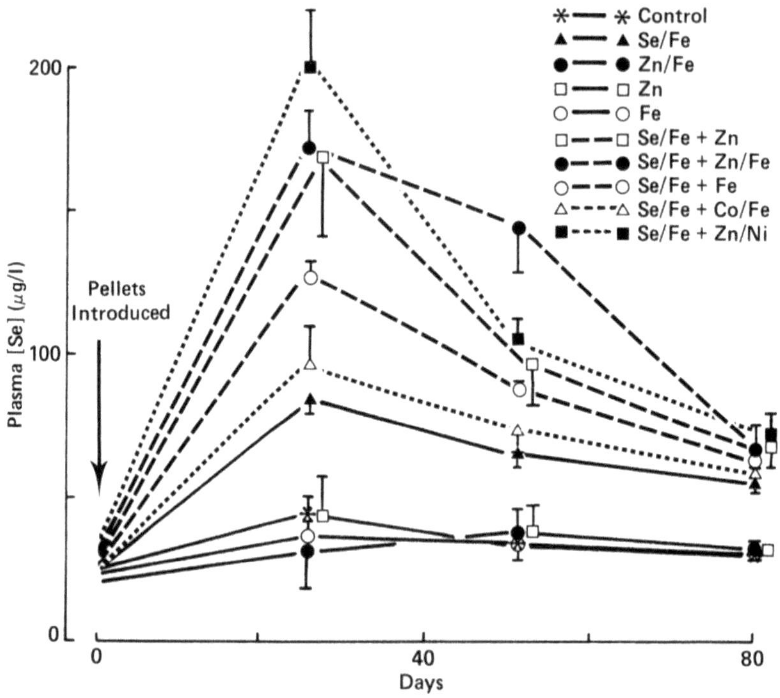

Fig. 1. Changes in plasma Se concentration of sheep given Se/Fe or other high density, intraruminal pellets alone, or together.

DISCUSSION

It is apparent from these results that the previously reported "Zn/Fe pellet effect" is not specific to Zn/Fe pellets alone, though Zn containing pellets had by far the greatest effect of any pellet tested. Since there were no significant differences in plasma Se concentrations between all groups not given Se/Fe pellets, it is unlikely that the "second pellet effect" is caused by alterations in either the intake or absorption of dietary Se. Such an effect must therefore result from either increased absorption of Se released from the Se/Fe pellet and or an increase in the rate of release.

Hudson et al. (3, 4) have proposed that the release of Se from a Se/Fe pellet occurs concomitant with observed alteration of elemental Se to a compound which probably contains "iron selenide" and hydrated forms of Fe. On the basis of published oxidation-reduction potentials for pure Fe and pure Se, values which have been verified 'in vitro' using rumen liquor as an electrolyte solution (unpublished observations), it is postulated that the formation of the "iron selenide" occurs via an electrochemical reaction. In this reaction Fe acts as an electron donor, being oxidised, and elemental Se accepts electrons, being reduced, with some Se being released into solution. A possible composite reaction is as follows:

$$2Fe + 3Se + 2H_2O \rightleftharpoons Fe(OH)_2 + FeSe_2 + H_2Se$$

It can be demonstrated, 'in vitro', that an electrochemical cell is formed between a Zn/Fe pellet and a Se/Fe pellet with Se being released into solution at a faster rate than from a Se/Fe pellet alone (unpublished observations). The latter may be due to the actual rate of the electrochemical reaction and/or the fact that less Se is involved in compound ($FeSe_2$) formation. It is therefore postulated that if a Zn containing pellet comes in contact with a Se/Fe pellet in the rumen of the sheep, an electrochemical cell is formed with Zn as the anode, the release of Se increases and higher plasma Se concentrations subsequently occur. During periods when separate Fe or Fe containing pellets may be in contact with a Se/Fe pellet, the increase in cathodic surface area or the partial separation of anodic and cathodic processes may likewise enhance the rate of reduction and release of Se.

This hypothesis does not take account of a variety of other complex processes which may also be involved, a full discussion of which are beyond the scope of this paper. Such processes, through effects on the rate of alteration and loss of elemental Se in the Se/Fe pellet (see ref. 4), are likely to account for the observed differences in the rates of decline in plasma Se concentration beyond day 26.

ACKNOWLEDGEMENTS

The authors wish to acknowledge the competetent technical assistance of Messrs D. Buscall and P. Young and to thank Dr. D.R. Hudson for time spent in discussion.

REFERENCES

1. Peter, D.W., and R.A. Hunter, 1979, Proc. Nutr. Soc. Aust. 4:145.

2. Watkinson, J.M., 1979, Anal. Chim. Acta. 105:319 - 315.

3. Hudson, D.R., R.A. Hunter, and D.W. Peter, 1981, Aust. J. Agric. Res. (In Press).

4. Hudson, D.R., R.A. Hunter, and D.W. Peter, 1981. Proc. IVth Internat. Conf. on "Trace Element Metabolism in Man and Animals (This issue).

DISCUSSION

LANGLANDS (Australia). We also have had experience with these selenium pellets and again we find that the mechanism by which they are used is very complex. One of the things that we have tried is milling up pellets that have appeared to have finished their useful life, and if this is done, we find that we do in fact get a response in blood selenium. This, I guess, is the same thing that Dr Peter is going to talk about tomorrow. The interesting thing is that if you take a new pellet, a pellet that has never been in the sheep, and do exactly the same thing to it you do not get any response in selenium. Quite clearly there are very complicated reactions going on with these selenium pellets far more complicated I would suspect than was thought when they were first marketed and generally tested.

PETER (Australia). I do not have anything to add to that except that, yes, it is obviously a complex process. As I pointed out elemental selenium in the rumen does not affect plasma selenium concentration in any way whatsoever. We do not know at this stage what effect or what release occurs from the iron selenide that is formed in selenium pellets and I guess I should conclude by saying that until recently everybody looked upon the iron in the selenium pellet as more or less as a dead weight to keep this pellet in the rumen and was probably not involved in anyway in selenium release. But certainly it is involved in selenium release and very deeply involved. I should point out that the iron affect is probably also via an electrochemical process, in other words we have increased the surface area of the anode and thereby improved the rate of reaction with selenium.

MacPHERSON (Scotland). You have found these results working with sheep, have you any comparative results with cattle, because in the UK we have some evidence that selenium pellets are more effective in cattle than they are in sheep.

PETER (Australia). We have no results with cattle.

JUDSON (Australia). We have found that the pellets last for up to a year in cattle but I have not compared the same pellets with sheep.

PAYNTER (Australia). I am interested in some of your work regarding the elemental selenium administered to the rumen and its apparent lack of effects on blood selenium concentration. Isn't this a little in contrast with some of the original work done by Dr Godwin's group in Adelaide using barium sulphate as a binder for pellets containing elemental selenium where they obtained significant increases in blood selenium concentrations using these pellets without iron in them? My second question would be are you sure that these apparent interactions between zinc and iron pellets are not simply due to an abrasion affect on pellet surfaces; can you completely discount this possibility?

PETER (Australia). I am treating the last point first; no I guess we can't absolutely discard it. The facts are that we get such big differences between zinc and iron pellets but the abrasive affects would be similar in both cases. We are getting this different effect, so I don't think that abrasion is very important. Also including grinders i.e. a steel grinder in with the pellet does not have a great deal of effect. Concerning the first question, as far as sulphides go I don't think we have got the answers by any means. There is probably some release of selenium from the iron selenide mixture but I don't think that there necessarily has to be an iron-selenide "mixture" formed for release to occur. In fact we can't exclude the possibility that we are seeing a galvanising effect of the zinc on the iron and this is actually preventing the formation of iron selenide. The selenium release may be occurring during a direct zinc-selenium reaction.

EFFECT OF SELENIUM AND VITAMIN E ON CALF WEIGHT GAINS

Walter H. Johnson, Ben B. Norman* and John R. Dunbar*

Cooperative Extension, University of California
2430 HOSPITAL LANE, ROOM 55, REDDING, CALIFORNIA 96001
*Cooperative Extension, University of California, DAVIS, CALIFORNIA 95616
U.S.A.

Injectable combinations of selenium and vitamin E were the only drugs approved and available for treating selenium deficiency problems in cattle in the U.S.A. until 1979. Earlier work in northern California indicated that vitamin E did not cause any significant increase in weight gains of wintering weaner and yearling beef cattle [Albaugh, Meyer, and Smith, Calif. Agric. 17:12 (1963)] . Cattlemen of the area have been commonly treating new born calves but not weaned calves, with Se + E injections. Field trials were established to determine the effect of various selenium and vitamin E treatments on weaned calves. Treatments without vitamin E were tested because if they were successful they would reduce the cost of treatment. Additionally, it was expected that approval would be given in the future for the addition of selenium to salt-mineral mixtures, and that the use of selenium pellets might prove successful and be approved at some future date. Experience with injections of selenium alone would be helpful in planning for future use of selenium alone either in salt or in pellets.

Fourteen field trials were conducted on two cooperating ranches during the four year period 1974-1977. Test animals consisted of 510 weaned calves of Hereford or Angus breeding. Their weights ranged from 350 to 550 pounds. The number of calves per treatment group varied from 8 to 35 depending on the particular trial. The calves were fed either alfalfa-grass hay or meadow hay grown on the ranch where the trial was conducted. Occasionally limited concentrate supplements were fed, and the supplements came from outside the test area. Both ranches had experienced losses from white muscle disease prior to instituting treatment at birth. All weaned calves in the trials had been treated with two to three mg Se and 100 to 150 mg vitamin E at birth or shortly thereafter. The treatments used were control, selenium alone, vitamin E alone, combinations of selenium and vitamin E, and combinations of selenium, vitamin E, vitamin A, and vitamin D. Treatment details are shown in the accompanying tables. Selenium was provided by sodium selenite, and vitamin E was provided by d-alpha tocopheryl acetate. The materials were administered by subcutaneous injections in the shoulder and neck region. The general objective was to supply approximately 5 mg selenium per 200 pounds body weight, with vitamin E rates varying in different trials. Because of varying animal weights, the selenium treatment was not completely consistent among animals. Weight gain response was the primary criterion for evaluating treatments. Limited analyses of blood and feed samples for selenium were provided by Oregon State University on a cooperative basis.

Eight trials had a control group. Six trials did not have a control group; and compared materials, dosages, or timing of treatment. Every trial without a control group was run concurrently with a trial with control animals in which significant responses to selenium treatment were measured. Therefore, weight gains in the six trials without controls were considered to be a result of selenium deficient calves responding to various Se or Se + E or Se + E + A + D treatments.

A significant weight gain response to treatment with selenium alone and/or selenium-vitamin E combinations and/or selenium-vitamins A, D, and E combinations occurred in all eight trials having a control group. The increase in weight gain ranged from 16% to 59%, with an average of 43% (TABLE 1). Four trials (3, 7, 10, and 13) compared treatment with selenium alone to treatment with selenium-vitamin E combinations and to no treatment; both treatments produced significant increases in weight gains, and there was no significant difference in the weight gains of the two treated groups. Vitamin E alone was used in trials 3, 7, and 10. In trial 7 there was a significant weight gain response to this treatment, but this response was significantly less than that for the Se and Se + E treatments. In the other two trials there was no significant effect from vitamin E alone.

In the six trials without controls there were no significant differences in weight gain except in trial 14 where difference selenium dosages were given (TABLE 2). Four of these trials compared selenium treatment with Se + E combinations, and the addition of vitamin E caused no significant increase in weight gain. The addition of vitamins A, D, and E to selenium in one trial was of no benefit.

Limited observations on the two ranches indicated treatment was effective for 90-130 days. Retreatment after 40 days in trials 1 and 2 produced no benefit in the next 37 days. Calves in trial 3 had a small treatment of Se + E 100 days prior to the start of the trial and had significant weight gain response to Se and Se + E treatments for the 112 day trial period. Retreatment after 91 days in trial 5 increased weight gains, though not significantly, in the following 58 days. In trial 12 there was no significant difference in gain resulting from retreatment after 125 days in a 181 day trial. In trial 14 on the same ranch the following year a significant increase in weight gain resulted from retreatment at 96 days and 117 days.

Blood samples taken from calves in trial 10, 94 days and 123 days after treatment, were analyzed for selenium. All of the control and vitamin E-treated heifers were below .04 ppm 94 days into the trial, and 10 of the 17 selenium-treated heifers were below .04 ppm. After 123 days all trial calves except one treated with Se alone were below .04 ppm.

TABLE 1. Weight Gains in Trials with Control Groups

		Treatment		Gain/head		
Trial no.	No. days	Materials	Day(s) given	Total (lbs)	Lb/day	% Increase
1	77	control		59	.77a	
		12.5 mg Se + 125 mg E	1	94	1.22b****	58
		12.5 mg Se + 125 mg E	1,40	93	1.21b****	57
2	77	control		105	1.37a	
		12.5 mg Se + 125 mg E	1	133	1.73b****	26
		12.5 mg Se + 125 mg E	1,40	129	1.67b****	22
3[1]	112	control		43	.39a	
		10 mg Se	1	65	.58b*	49
		100 mg E	1	40	.36a	-8
		10 mg Se + 500 mg E	1	60	.53b*	36
6	49	control		57	1.17a	
		10 mg Se + 100 mg E	1	78	1.60b****	37
7	112	control		136	1.21a	
		12.5 mg Se	1	201	1.79c*	48
		100 mg E	1	166	1.48b*	22
		12.5 mg Se + 100 mg E	1	197	1.76c*	45
8	112	control		112	1.00a	
		10 mg Se, 100 mg E, 100,000 units A, 10,000 units D	1	154	1.38b***	38
		12.5 mg Se + 125 mg E	68			
10	123	control		91	.74a	
		10 mg Se	1	145	1.18**	59
		100 mg E	1	73	.59a	-20
		10 mg Se + 100 mg E	1	128	1.04b**	41
13	159	control		185	1.16a	
		10 mg Se	1	215	1.35b**	16
		10 mg Se	117			
		5 mg Se + 250 mg E	1			
		5 mg Se + 250 mg E	32	238	1.50b**	29
		10 mg Se + 500 mg E	117			

1 All calves in trial 3 received 4 mg Se + 200 mg E 100 days prior to the trial start.
*P ≤ .05; **P ≤ .01; *** P ≤ .005; ****P ≤ .001
a,b, c Gains with different superscripts in each trial are significantly different as analyzed by Duncan's multiple range test.

Treatment with selenium produced significant increases in weight gain. The addition of vitamin E to Se treatment was of no benefit in increasing weight gain. The addition of a combination of vitamins A, D, and E to Se treatment was of no benefit to weight gain. Treatment with vitamin E alone produced a significant increase in gain in one of three comparisons, but the increase was significantly less than that from Se alone. Although results are variable, it appears that retreatment in 90 days would be beneficial.

Under these conditions treatment with selenium alone was entirely satisfactory, so it might be expected that the use of salt or pellets containing selenium without vitamin E would be satisfactory.

TABLE 2. Weight Gains in Trials without Control Groups

Trial No.	No. days	Treatment: Materials	Treatment: Day(s) given[1]	Gain per head: Total (lbs)	Gain per head: Lbs per day
4[2]	112	10 mg Se	1	119	1.06
		10 mg Se + 500 mg E	1	116	1.03
		10 mg Se + 500 mg E + 100,000 units A + 10,000 units D	1	104	.92
5	58	10 mg Se	1	68	1.17
		10 mg Se	-91,1	88	1.51
		10 mg Se + 100 mg E	-91,1	77	1.33
9	86	10 mg Se + 100 mg E	1	90	1.05
		10 mg Se + 500 mg E	1	100	1.16
11	119	10 mg Se 15 mg Se + 150 mg E	1 +91	213	1.79
		10 mg Se + 100 mg E 15 mg Se + 150 mg E	1 +91	219	1.84
12	181	10 mg Se	1	282	1.56
		10 mg Se 15 mg Se	1 +125	281	1.55
		10 mg Se + 100 mg E 15 mg Se + 150 mg E	1 +125	276	1.52
14	159	10 mg Se	1		1.62a
		10 mg Se 10 mg Se	1 +117	295	1.85b
		5 mg Se + 250 mg E 5 mg Se + 250 mg E	1 +96	254	1.60a

1 Day 1 = trial start; minus = days before trial start; plus = days after trial start.
2 All calves in trial 4 received 4 mg Se + 200 mg E 100 days before trial start.
a, b Gains with different superscripts are significantly different at $P \leq .05$.

TABLE 3. Weight Gains of Single- and Double-Treated Calves - Trial 14

Group	Treatment	Gain per head, lbs: First 117 days*	Next 42 days*	Total 159 days**
1	10 mg Se day 1	201a	57a	258a
2	10 mg Se day 1 and day 117	202a	93b	295b
3	5 mg Se + 250 mg E day 1 and day 96	172a	79c	254a

*Significant at $P \leq .001$; **Significant at $P \leq .05$

DISCUSSION

JANGHORBANI (U.S.A.). Have you tried any forms of selenium other than selenite? If yes, would you have any feeling for what effect you might observe?

JOHNSON (U.S.A.). Injectable selenium came only from sodium selenite, we did not use any others. The approved materials are injectibles with selenium from sodium selenite, so I don't know what would happen with another material. The Food and Drug Administration has within the last two years approved selenium supplementation of cattle rations and salt licks. Addition of selenium to supplements and to total rations is aimed at providing 2 mg selenium per day, and 20 ppm selenium is permitted in salt-mineral mixtures. Our experience with 20 ppm in salt has been unsatisfactory and we find that blood levels of selenium go down when these weaner cattle have free access to 20 ppm selenium-containing salt. On the other hand people can mix it on their own properties and have done so experimentally at amounts ranging from 100 to 200 ppm and have been satisfied with that. The legally approved commercial products have not satisfied our needs.

HILL (U.K.). Were all these trials carried out in the North of your area where the herbage values were stated to be in the 0.03 ppm level, or were some in areas of higher natural selenium concentrations in pastures?

JOHNSON (U.S.A.). There have been trials in several northern counties. The figure of 0.03 ppm is an average and it varies from one farm to another. For the trials that I reported here, I am sorry to say I cannot tell you with any real accuracy what the forage levels are - we had no ability to analyse in our laboratory for selenium levels in the forage. Oregon State University was helpful to us in running some analyses. Where these animals were fed mostly hay, the hay contained somewhere around 0.01 ppm selenium.

ARORA (India). Would you please comment on the vitamin E levels of the diets.

JOHNSON (U.S.A.). We were not able to have any feed analysed for vitamin E and do not know the levels.

ALLEN (U.K.) Have you any information on the blood selenium concentrations? At what blood selenium concentration would you expect that type of large response which you have seen, and how long did the protection last from that single injection?

JOHNSON (U.S.A.). On these two ranches blood levels of weaner calves were in the range of 0.01 µg/ml. As far as the extent of the length of effectiveness from injection, we see a difference we think in 70 days, 100 days, 120 days. It varies from year to year - we don't know exactly why but we suspect that part of it is related to fertilisation practices of the pasture and hay crops.

SELENIUM AND VITAMIN E FOR WEANLING SWINE

D. C. Mahan and A. L. Moxon

Animal Science Department, Ohio Agricultural Research and Development Center
and The Ohio State University, WOOSTER, OHIO 44691 U.S.A.

In determining the dietary Se needs of an animal one must also consider vitamin E since the biological roles of both nutrients are interrelated. Considerable attention has been given to the establishment of the Se requirements of animals since 1957, often without consideration of animal's vitamin E status.

Midwestern grown feeds, especially corn and soybeans, have been shown to be low in their Se content when compared with many areas in the United States. Simplified swine diets composed of these feedstuffs are inadequate in their Se content, especially for weanling pigs. The U.S. Food and Drug Administration thus approved the addition of .1 ppm inorganic Se to swine diets in 1974. In spite of the use of supplemental levels at .1 ppm deficiencies were still encountered in weanling swine. The problem is most apparent during the postweaning period when depletion of both nutrients has been demonstrated (1,2). In an Ohio study deaths related to Se deficiency were not encountered while the pigs were nursing their dams but did occur an average of 6.6 days postweaning (2).

Increased levels of Se in creep and starter diets will increase the serum and tissue Se levels in weanling pigs (2,3,4). It has consequently been suggested that the level of glutathione peroxidase (GSH-Px) in blood and tissues may be useful in determining the Se requirement of animals (5,6).

The objective of this experiment was to determine the efficacy of various dietary Se and vitamin E levels for weanling pigs.

EXPERIMENTAL PROCEDURE

Three replicates of a 3 x 5 factorial experiment in a RCB design were conducted to evaluate supplemental dietary inorganic Se and vitamin E levels over a 5-week postweaning period. A 20% protein corn-soybean meal diet was provided with vitamin E added at 0, 10, or 20 IU/kg and inorganic Se added as sodium selenite at .1, .3, .5, 1.0 or 2.0 ppm.

A total of 360 4-week old pigs, from sows fed diets containing .1 ppm Se and 22 IU vitamin E/kg, were allotted upon weaning by weight, litter, and sex into pens of 12 each. Six animals, selected at random from each pen, were bled by cardiac puncture initially and at weekly intervals and frozen for subsequent analyses. At 35 days two boar pigs from each pen were sacrificed. Heart, liver, kidney, pancreas, testis and *longissimus* muscle tissues were collected from each and frozen for Se analysis. A separate liver sample from each was collected and immediately frozen in liquid nitrogen for GSH-Px activity.

From the second replicate 10 pigs were selected from each Se treatment level and fed to market weight a corn-soybean meal diet supplemented with .1 ppm Se and 10 IU vitamin E/kg to determine the effect of the postweaning treatment on residual Se carry-over in various tissues at approximately 100 kg body weight. Liver, kidney and *longissimus* muscle samples were collected for Se analyses.

Plasma and liver crude homogenate GSH-Px total activities were determined by a modified version of the method by Lawrence and Burk (7). Se independent GSH-Px values were determined by difference. Plasma tocopherol was determined by the colorimetric method of Fabianek *et al.* (8). A composite sample of the basal diet analyzed for α-tocopherol by the method of Bro- Rasmussen *et al.* (9) assayed 6.9 mg/kg. Diets, tissues and plasma Se were determined by the fluorometric method of Olson *et al.* (10). Each diet mixture was analyzed for Se and was found to be approximately .05 to .08 ppm above the supplemental level, attributable to the natural Se in the feedstuff mixture. Tabular data and statistical analyses of the results are reported elsewhere (11).

RESULTS AND DISCUSSION

The effects of dietary Se and vitamin E on postweaning gain and feed consumption indicate that performances responses were similar for all treatment groups. Gain and feed intake were somewhat lower when the 2.0 ppm Se diet was fed but the difference was not significant. It appears that if diets are very low in their natural Se or vitamin E content and the animals are extremely depleted of tissue reservoirs, improved performance may result with supplemental Se and vitamin E; otherwise no response would be anticipated.

Initial plasma Se and tocopherol values were similar in all animals. These values are higher than those in pigs from sows without supplemental dietary Se (3,12) indicating that at the onset of this experiment the pigs were not depleted of either nutrient. After 7 days on trial, plasma Se levels increased linearly ($P<.01$) in all groups except for the group fed .1 ppm Se which showed a decrease. During each succeeding week plasma Se increased cubically ($P<.01$) as dietary Se increased with a distinct rise occurring in the group fed the 2.0 ppm Se diet. Breakpoint regression analyses, omitting the data for the 2.0 ppm group, indicated a plateau at .475, .450, .450 and .440 supplemental Se levels for weeks 2, 3, 4 and 5 respectively. Scott and Thompson (13) suggested that blood Se plateaued when chicks were fed at dietary levels near their nutritional requirement. The plateau of plasma Se and the upward inflection with the 2.0 ppm Se diet group provide evidence that the capacity of body tissues to retain absorbed Se was probably exceeded at the 2.0 ppm dietary level.

Although plasma tocopherol levels were similar at weaning there was a marked decline in all groups ($P<.01$) within 1-week postweaning. Pigs fed diets without added vitamin E had lower ($P<.01$) plasma tocopherol levels at each measurement period than those with 10 or 20 IU/kg vitamin E added. There was a progressive increase in tocopherol levels after the first week postweaning in the vitamin E supplemented groups but none attained the plasma levels at weaning.

The Se content of tissues, in order of decreasing concentrations were the kidney cortex, liver, pancreas, testis, heart and *longissimus* muscle. Liver and pancreas Se concentrations increased in a cubic ($P<.01$) pattern with increasing dietary levels; the testis, heart, kidney cortex and *longissimus* muscle responded quadratically ($P<.01$). Breakpoint regression analyses of these tissues at 5 weeks postweaning were attained at dietary Se levels of .410, .300, .435, .450, .300 and .455 ppm, respectively.

Liver and plasma Se concentrations increased as dietary Se increased almost paralleling one another, particularly above the dietary breakpoint level. It appears that as dietary Se increases, the liver has a greater ability to retain this element while other tissues do not. The data suggest that at 2.0 ppm Se the excretory kidney pathway was over loaded with Se accumulating in the liver, moreso than in other tissues.

As supplemental vitamin E level increased, Se concentrations in kidney, liver, pancreas, testis and heart decreased ($P<.01$), whereas *longissimus* muscle Se values were similar at all levels of vitamin E. These results indicate that Se retention is lower when vitamin E is added to the diet.

Initial plasma GSH-Px activities were similar with dietary Se levels influencing GSH-Px activity each ensuing week. As dietary Se level increased a quadratic response ($P<.05$) resulted in GSH-Px activity for weeks 1 and 5, and a cubic response ($P<.01$) for weeks 2, 3 and 4. Breakpoint regression analyses of GSH-Px/ml plasma occurred at .525, .400, .305, .315 and .285 ppm Se on days 7, 14, 21, 28 and 35, respectively.

Total and Se-dependent liver GSH-Px activities at 35 days postweaning reported on a per mg protein or a per g wet tissue basis resulted in an increased total and Se-dependent GSH-Px activity as dietary levels increased; whereas Se independent liver GSH-Px activity was similar for all treatment groups. Breakpoint regression analyses demonstrated a dietary Se plateau at .405 and .300 ppm for total and .305 and .310 ppm Se-dependent activity on a per mg protein and a per g wet tissue basis, respectively. Lawrence and Burk (7) also reported the presence of both Se-dependent and independent GSH-Px activities in pig livers.

The effects of tissue Se concentrations at market weight were similar suggesting that postweaning diets containing up to 1.0 ppm selenium and/or 20 IU vitamin-E/kg from 8 to 20 kg have no carryover effect.

Use of the breakpoint regression analyses of plasma and tissue GSH-Px activities as indices for establishing the Se requirement of the weanling the results suggest that the Se requirement declines progressively during the postweaning period. Considering the natural Se in the feedstuffs used in this experiment, the total dietary requirement appears to be at least .35 ppm Se by 5-weeks postweaning although a higher supplemental level may be advantageous earlier in the postweaning period. This level of supplemental inorganic Se might be needed only in deficient regions where natural levels are low and may not be warranted in areas where natural Se is higher in feedstuffs. Swine fed on grains higher in natural organic Se may have a higher initial Se status at weaning, negating the necessity for higher fortification levels.

Although postweaning pig mortality was not encountered in this experiment when the .1 ppm Se diet was fed, it has previously been encountered (2,3). It should be noted that when pigs in this experiment were sacrificed for tissue collection, pale muscle was observed in a number of animals, particularly in those fed the lower Se levels.

REFERENCES

1. Mahan, D. C., A. L. Moxon and M. Hubbard. 1977. Efficacy of inorganic selenium supplementation to sow diets on resulting carry-over to their progeny. J. Anim. Sci. 46:738.

2. Mahan, D. C. and A. L. Moxon. 1980. Effect of dietary selenium and injectable vitamin E-selenium for weanling swine. Nutrition Reports International. 21 (N.6) 829.

3. Mahan, D. C. and A. L. Moxon. 1978. Effect of increasing the level of inorganic selenium supplementation in the postweaning diets of swine. J. Anim. Sci. 46:384.

4. Mahan, D. C. and A. L. Moxon. 1978. Effects of adding inorganic or organic selenium sources to the diets of young swine. J. Anim. Sci. 47:456.

5. Hakkarainen, J., P. Lindberg, G. Bengtsson and L. Johnsson. 1978. Serum glutathione peroxidase activity and blood selenium in pigs. Acta Vet. Scand. 19:269.

6. Chaveg, E. R. 1979. Effects of dietary selenium on glutathione peroxidase activity in piglets. Can. J. Anim. Sci. 59:67.

7. Lawrence, R. A. and R. F. Burk. 1976. Glutathione peroxidase activity in selenium-deficient rat liver. Biochem. Biophys. Res. Comm. 71:952.

8. Fabianek, J., J. Defilippi, T. Richards and A. Herp. 1968. Micro method for tocopherol determination in blood serum. Clin. Chem. 14:456.

9. Bro- Rasmussen, Flin and Willy Hjarde. 1957. Determination of α-tocopherol by chromatography on secondary magnesium phosphate. Acta Chem. Scand. 11:34.

10. Olson, O. E., I. S. Palmer and E. E. Cary. 1975. Modification of the official fluorometric method for Se in plants. J. Assoc. Official Anal. Chem. 58:117.

11. Meyer, W. R., D. C. Mahan and A. L. Moxon. 1981. Value of dietary selenium and vitamin E for weanling swine as measured by performance and tissue selenium and glutathione peroxidase activities. J. Anim. Sci. 52:302.

12. Peplowski, M. A., D. C. Mahan, F. A. Murray, A. L. Moxon, A. H. Cantor and K. E. Ekstrom. 1980. Effect of dietary and injectable vitamin E and selenium in weanling swine antigenically challenged with sheep red blood cells. J. Anim. Sci. 51:344.

13. Scott, M. L. and J. N. Thompson. 1971. Selenium content of feedstuffs and effects of dietary selenium upon tissue selenium in chicks and poults. Poul. Sci. 50:1742.

SELENIUM AND TOCOPHEROL TREATMENT OF OVINE WEANER NUTRITIONAL MYOPATHY

P. Steele, D.P. McKenzie, S. Skirrow, R.L. Peet & G. Doncon

Animal Health Laboratory,
Dept. of Agriculture, South Perth, 6151, Western Australia

Nutritional myopathy usually occurs throughout the world in young livestock and in Western Australia lambs may be affected in susceptible areas of high rainfall and low pasture selenium content. The major losses through nutritional myopathy however occur in weaner sheep of 8 to 12 months of age during dry summer-autumn grazing, and this "delayed form" of myopathy has also been reported elsewhere in Australia. In general, the losses through ovine nutritional myopathy have been reduced through the widespread use of selenium pellets and drenches. Outbreaks of ovine weaner nutritional myopathy also occur in the southwest of Western Australia outside the recognised selenium responsive areas (Steele et al 1980). In 1979-81, a form of nutritional myopathy had occurred on several properties in this area and in all cases examined the sheep had an adequate selenium status as judged by liver selenium content and liver and red cell glutathione peroxidase activity. These sheep had low liver α-tocopherol levels compared with other groups of grazing animals and control fed laboratory animals. In this study, the results of limited field therapy trials carried out on the properties where the disease was first found are reported. On three properties the efficacy of oral α -tocopherol treatment was assessed by observing flocks and monitoring plasma creatine kinase activity in control and treated groups before and after α-tocopherol administration. On a fourth property, treatment effects of selenium, or selenium plus α-tocopherol, or α-tocopherol alone were compared in a controlled experiment.

MATERIALS AND METHODS

In trials 1,2, and 3 groups of 50 randomly selected sheep in clinically affected flocks were given a single oral dose of 4000 International units of α-tocopherol acetate as a water dispersable drench. In the fourth trial, Group 1 (n=260) received no treatment; Group 2 (n=80) received a single intramuscular injection of 5 mg selenium as sodium selenite; Group 3 (n=80) received a single intramuscular injection of 2.5 mg selenium as sodium selenite and 750 mg α-tocopherol acetate; and Group 4 (n=80) received a single intramuscular injection of 1650 mg α-tocopherol acetate. In all the trials the sheep were maintained on the paddocks where cardiac and skeletal myopathy had occurred. Blood samples were taken at intervals to monitor the effects of treatment in both clinically affected and subclinically affected sheep. Whole blood was collected by jugular puncture into heparinised tubes which were held chilled until separated. Assay of glutathione peroxidase activity in erythrocytes was carried out by the Andrewartha et al (1979) modification of the method of Paglia and Valentine (1967), with the use of a Pye-Unicam AC-1 automatic chemistry unit. GSHPx activity was expressed in International units as μmoles NADPH oxidised per minute per gram haemoglobin at 25°C. Plasma α-tocopherol was extracted and separated by the reverse phase high performance lipid chromatographic method of De Leenheer et al (1978). Tocol was used as an internal standard and the α-tocopherol detected at 280 nm was quantified using a Waters Model 720 integrating plotter. Creatine kinase activity was assayed using an automated spectrophotometric method with Calbiochem CPK-SVR reagent. Results were expressed as milli International units per millilitre plasma at 25°C.

RESULTS

The concentration of α-tocopherol, selenium, and GSHPx in liver of nutritional myopathy affected weaner sheep in the field outbreaks are shown in Table 1. Both the liver selenium contents ranging from 0.38 to 1.04 μg/g dry weight, and the liver glutathione peroxidase activities from 25.2 to 46.1 nanomol NADPH/min/mg protein are considered to be adequate and are comparable to those of selenium pelleted sheep or laboratory fed sheep receiving a selenium supplemented diet. The liver tocopherol levels assayed ranged from 0.26 to 0.62 μg/g/ wet weight, and these values are extremely low.

TABLE 1

Concentrations of α-tocopherol, Se, and GSHPx activity in liver from nutritional myopathy affected weaner sheep

		n	α-TOCOPHEROL ug/g wet tissue	SELENIUM ug/g dry weight	GSHPx nmol NADPH oxidised/ min/mg protein
OUTBREAK 1	★	2	0·55	1·04	40·0
OUTBREAK 2	◆	2	0·58	0·42	n.a.
OUTBREAK 3	●	4	0·53	0·70	45·7
OUTBREAK 4	▲	2	0·47	0·43	n.a.
OUTBREAK 5	■	1	0·62	0·55	46·1
OUTBREAK 6	◓	1	0·26	0·38	25·2

n.a. not available Values represent means

★ Se-treated affected sheep, York ▲ Untreated affected sheep, Shackleton

◆ Untreated affected sheep, Meckering ■ " " " Beverley

● " " " Greenhills ◓ " " " Southern Brook

The results of orally administered α-tocopherol on plasma CPK, red cell GSHPx and plasma α-tocopherol in trials 1,2 and 3 are shown in Table 2. The significant increase in plasma tocopherol content in treated sheep was accompanied by a significant fall in plasma creatine kinase indicating that the severity of the myopathy was reduced.

TABLE 2

Effect of oral α-tocopherol treatment on plasma CPK, red cell GSH Px and plasma α-tocopherol in weaner sheep with nutritional myopathy.

			α-TOCOPHEROL ug/ml		CPK		
		n	pre-treatment	5 days post treatment	pre-treatment	5 days post treatment	GSH Px IU/g Hb
TRIAL 1	Control	10	0·49	0·36	436	425	153
	Tocopherol treated	10	0·51	1·84	484	157	152
TRIAL 2	Control	10	0·86	0·81	276	280	142
	Tocopherol treated	10	0·73	2·52	253	84 •••	135
TRIAL 3	Control	10	0·85	0·75	577	526	129
	Tocopherol treated	10	0·87	2·50	454	117 •••	138
	Tocopherol treated clinically affected	9	0·59	2·36	613	99 •••	143

Values represent means ••• $p < 0.001$

In the third trial, clinically affected sheep having a mean red cell glutathione peroxidase activity well in excess of the marginal level of 25 IU/g Hb did not respond to selenium administration. On this property, both clinically affected and subclinically affected sheep responded to treatment with α-tocopherol. Of the 9 clinically affected sheep which were treated with 4000 international units of α-tocopherol, 8 responded to therapy but the remaining animal died. Oral treatment was effective in preventing further losses in the flocks. The optimal dose rate and maximum period of protection from a single dose has yet to be defined. Preliminary observations have shown the period is greater than 6 weeks.

In the fourth trial, groups of sheep were given intramuscular injections of selenium or tocopherol and the flock was maintained on an oaten stubble paddock on a site of marginal selenium sufficiency. The results of treatment of the sheep on plasma α-tocopherol and creatine kinase levels are shown in Figure 1 and 2. The sheep showed no clinical evidence of myopathy and the data is consistent with the existence of a subclinical form of nutritonal myopathy in weaner sheep grazing in Western Australia. The sheep responded biochemically to administration of both selenium and α-tocopherol.

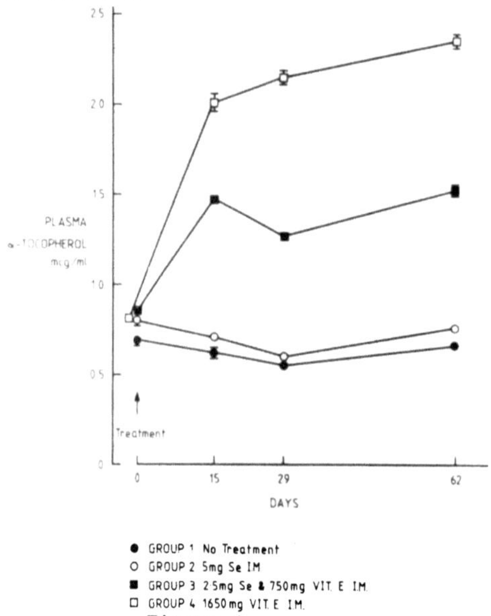

Figure 1

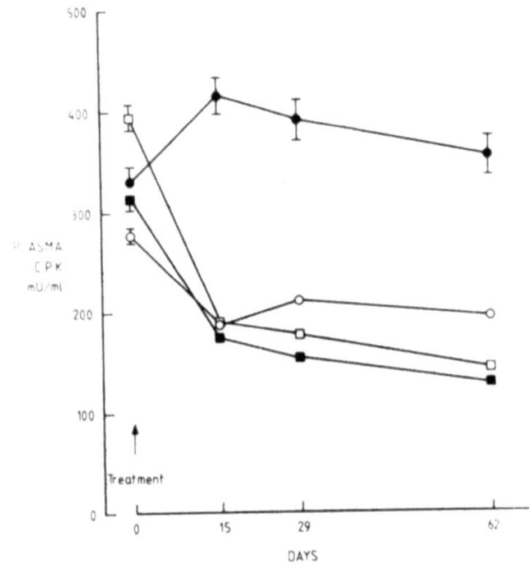

Figure 2

SUMMARY

Nutritional myopathy may occur spontaneously in selenium replete weaner sheep grazing dry cereal stubble of low α-tocopherol content in Western Australia. In four experiments both clinically affected and subclinically affected sheep responded to α-tocopherol acetate treatment when administered orally or intramuscularly.

REFERENCES

Andrewartha, K.A., Edwards, S.J.A., Brain, G.J., and Caple, I.W., 1979, Aust. J. Med. Tech. 10: 87.

De Leenheer, A.P., De Bevere, V.O., Cruyl, A.A., and Claeys, A.E., 1978, Clin. Chem. 24: 585.

Paglia, D.E. and Valentine, W.N. 1967, J.Lab. Clin. Med. 70: 158.

Steele, P., Peet, R.L., Skirrow, S., Hopkinson, W. and Masters, H/G., 1980, Aust. vet. J. 56: 529.

ACKNOWLEDGEMENTS

Supported by Australian Extension Services Grant 491.0637 and Australian Meat Research Committee Grant 492.0022.

DISCUSSION

WHANGER (U.S.A.). Your abstract says that you looked at the different forms of tocopherol in the plasma.

STEELE (Australia). Our assay technique is aimed at measurement of the α-tocopherol isomer. We use a high performance liquid chromatography separation technique which will detect other isomers, but in view of the high biological activity of that isomer we have confined our quantification to the α-tocopherol isomer.

WHITE (Australia). I would like to make a comment on Paul Steele's results. His observation that there is clearly more than one cause of white muscle disease is very important. It appears that the traditional approach of treating white muscle disease in livestock simply as a selenium - responsive disease will have to be altered in the light of this new knowledge. People at Murdoch University are seeing white muscle in sheep that have been given phosphate infusions, and the disease is histologically the same as selenium - responsive white muscle disease. There are also people at the Department of Agriculture in W.A. who are observing white muscle disease in lupinosis-affected sheep that is again histologically the same as selenium-responsible white muscle disease. However, selenium levels in both the phosphate-treated and lupinosis-affected sheep are apparently normal. Besides these two non-selenium responsive forms of white muscle disease, we have in W.A. the traditional selenium-responsive white muscle disease and non-selenium-responsive genetic muscular dystrophy together with this new form reported by Paul Steele. I believe that we are going to have to look a lot further than at CPK values and histological assessments to unravel the complexity of this important problem.

STEELE (Australia). If I may just make one comment on that and add to the list of myopathies we have just heard about. I think we are now confronting this year another myopathy which is associated with adequate vitamin E status, adequate selenium, and not associated with lupinosis.

LEVANDER (U.S.A.). I might just say that Oscar Olson has been in touch with me within the last couple of weeks and indicated they are seeing a type of white muscle disease in South Dakota in the U.S.A., and of course that is not classically a low selenium area.

HUCKER (Australia). In the western district of Victoria we found similar findings to Paul Steele. There was a wide outbreak of white muscle disease in the dry period of 1980 and 1981 in the months February, March and April. The glutathione peroxidase levels were found to be mostly normal, although there were some low ones so we started looking at vitamin E levels of which some were found to be low. This year, we have also found a few sheep showing classical histological lesions of white muscle disease but these sheep have normal glutathione peroxidase levels and vitamin E levels.

J. ALLEN (Australia). Can I just ask Paul to clarify for the people here the dangers of using the term white muscle disease particularly when you start talking about inherited conditions as well.

STEELE (Australia). I think that question might also refer to the use of the words "muscular dystrophy". A dystrophy is indeed a genetic form of myopathy and we would certainly recommend the use of the term "nutritional myopathy" rather than "white muscle disease".

THE EFFECT OF SE GRAIN-SIZE ON THE EFFICACY OF SELENIUM PELLETS

D.R. Hudson[A], R.A. Hunter[B], and D.W. Peter[B]

A. Division of Mineralogy, CSIRO, Wembley, Western Australia, 6014.
B. Division of Animal Production, CSIRO, Wembley, Western Australia, 6014.

INTRODUCTION

The CSIRO prototype intraruminal selenium pellet [1], consisting of about 5% elemental selenium dispersed in an iron grit, provided sheep with a small continuous supplement of selenium for at least 4 years [2]. Recent studies with commercially manufactured pellets showed that they may not be as effective in maintaining an elevated selenium status as the CSIRO pellet [3], and that this could not be attributed to surface coatings on pellets, season of treatment, or age of sheep [4]. The latter experiments indicated a difference in behaviour between two brands of commercially manufactured pellets that could not be explained by any gross differences in their physical properties [5]. This paper reports the results of experiments and microscope observations that show that selenium grain-size has a controlling influence on pellet efficacy. Some suggestions are made on the likely mechanisms of alteration and release of selenium from pellets.

MATERIALS AND METHODS

Experiment 1: Effect of Brand of Pellet on Selenium Status of Sheep.

Twelve sheep, from a flock of thirty-two 5-month-old Merino wethers grazing at the CSIRO Research Station at Bakers Hill, Western Australia, were divided into 3 groups of 4 sheep per group. One group was treated with pellets of brand 1, another with pellets of brand 2, and the third group was untreated. The remaining sheep in the flock had various treatments imposed that are described elsewhere [4]. Samples of blood were obtained by vein puncture at the commencement of the experiment and at intervals thereafter. Plasma obtained by centrifugation, was stored at -15°C. Selenium concentration was determined by the method of Watkinson [6].

Experiment 2: Effect of Grain-size on Reaction of Elemental Selenium in Pellets.

Five nylon bags, each containing one pellet 1 and one pellet 2, were suspended, using the method of Hunter *et al.* [7], in the rumens of Merino wethers which grazed irrigated lucerne (*Medicago sativa*) pasture. One nylon bag was removed from the rumen on each of the following days after the commencement of the experiment: 2, 4, 8, 16, 28. Polished, transverse sections were prepared for microscope and electron microprobe study, using the methods described by Hudson *et al.* [5].

RESULTS

Experiment 1:

The measured plasma selenium concentrations are shown in Figure 1. Both pellets caused a rapid increase in selenium, with maximum values being measured one week after treatment; peak values were presumably reached within the first two weeks. Concentrations then fell irregularly, with the rate of decay being much greater for sheep treated with pellet 1. After 12 weeks, sheep treated with pellet 2 had plasma concentrations about 0.07 mg/l compared to sheep treated with pellet 1 with values about 0.03 mg/l. After one year, sheep treated with pellet 2 had plasma selenium concentrations 0.026 mg/l above sheep treated with pellet 1, and 0.030 mg/l above untreated sheep.

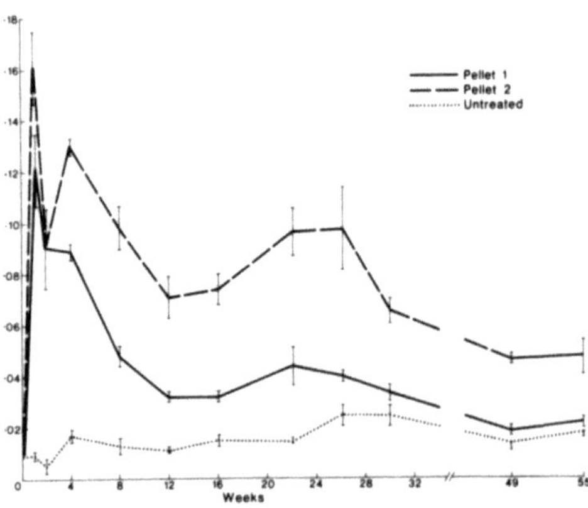

Fig. 1. Plasma Se concentration (±SEM) of sheep given different commercial pellets.

Experiment 2:

Polished sections of unused pellets, viewed under reflected light (Fig. 2), contain roughly equidimensional grains of the grey hexagonal allotrope of selenium dispersed through the interstices of an aggregate of subspherical grains of sponge iron. Selenium in pellet 1 (average grain-size 4μm, range <1 to 50μm) is about one order of magnitude finer than that in pellet 2 (average 40μm, range 5-100μm).

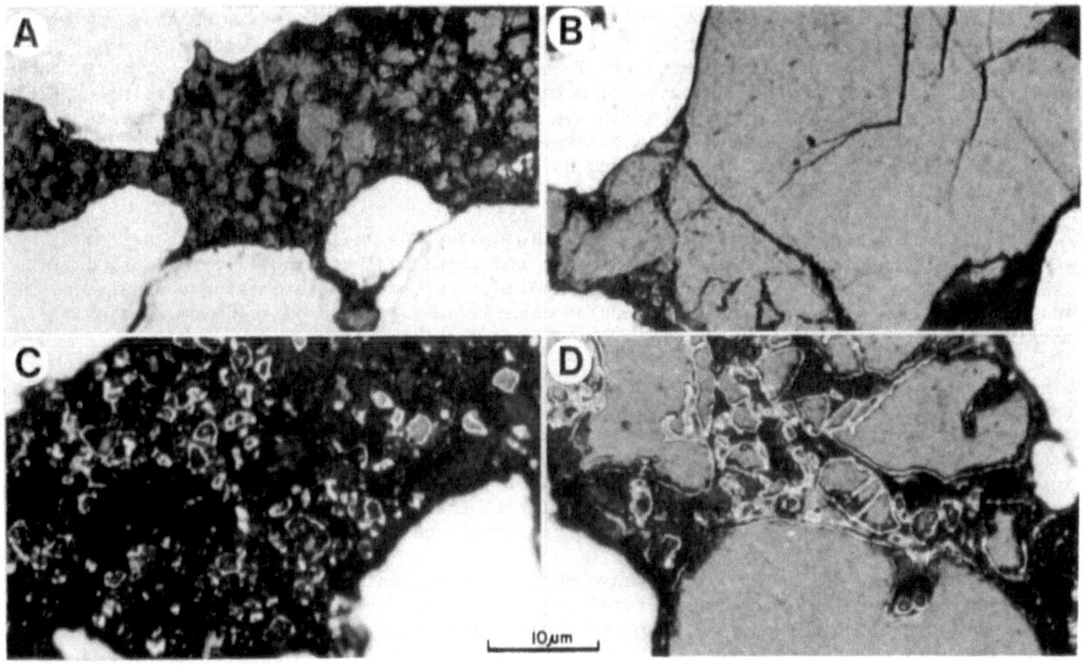

Fig. 2. Photomicrographs of selenium pellet 1 (A new, C used) and pellet 2 (B new, D used)

Used pellets contain selenium that has partially or completely reacted to a cream-coloured phase (Fig. 2), consisting dominantly of iron and selenium. Electron microprobe analyses give an average composition (wt. %) of Fe = 33.7, Se = 51.3, O (by difference) = 15.0. Replacement of selenium commences at grain surfaces and open fractures by development of rims of alteration which vary in thickness from 0.2 μm to 2.0μm after one week in the rumen. For large grains (ca 40 μm) this represents only a small percentage of the original selenium, but for small grains (ca 4 μm), over half of the elemental selenium has reacted (Fig. 3). Concurrent with the alteration of margins of larger grains is an anastomosing to vermiform replacement that pervades the interiors of grains. Some of this replacement may be exploiting incipient fractures or grain boundaries, but other narrow tongues of replacement have no observable control, and may be following submicroscopic features such as defect structures, or boundaries of the sub-grain mosaic. After 28 days, only a small percentage of elemental selenium remains in pellets with fine-grained selenium (Fig. 3), whereas about 50% remains in pellets with coarse-grained selenium. X-ray diffractometer traces of used pellets confirm the progressive disappearance of elemental selenium as a function of time spent in the rumen [5].

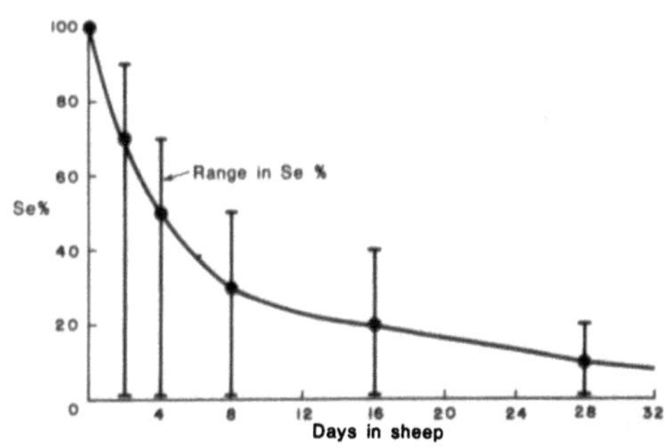

Fig. 3. Percentage of original elemental selenium remaining in used pellets, which contained fine-grained (~4μm) selenium.

DISCUSSION

The only major difference identified between the two commercially manufactured selenium pellets examined in this study was the grain size of the elemental selenium. Differences observed in the long-term effectiveness of the two pellets are believed to be due largely to this difference in grain size. Some tentative conclusions can be drawn by correlating the microscope observations with the measured plasma selenium concentrations in sheep treated with different brands of pellet.

Alteration of selenium begins along the exposed surfaces of selenium grains, and along open fractures. For both brands of pellet this alteration is rapid and correlates with the peak selenium concentrations measured in plasma. The effect of this reaction appears to be controlled by the difference in size, or, more appropriately, surface-area of selenium grains. Reaction is rapid for fine grains in pellet 1 and results in a high percentage loss of elemental selenium within a few weeks. The effect in pellet 2 is not as marked, with only thin surface reaction rims developing on large selenium grains. Rare, large selenium grains in pellet 1 are similarly preserved.

Continued reaction of elemental selenium is slower, as reflected by reduced plasma selenium concentrations after the second week, and is thought to proceed by diffusion of iron through the alteration rim and along a developing system of anastomosing channelways. These channelways enlarge, spread and coalesce until only islands of selenium remain in a matrix of 'iron selenide'.

The relative contribution made to plasma selenium levels by the surface reaction, the anastomosing replacement, or by the breakdown of selenium alteration products is not known. It is clear, however, that those pellets that retain a high percentage of elemental selenium perform better than those that lose most of their elemental selenium within a month. It is important to note that the CSIRO prototype pellet, for which long-term effectiveness was established, has coarse-grained selenium similar to that in pellet 2, although it differs in other respects, such as the nature of the supporting iron matrix.

Experiments by Hudson et al. [5] have shown that elemental selenium, in mixtures of crushed glass and selenium, is not altered or dissolved in the rumen. However, selenium, in mixtures of iron filings and selenium, undergoes alteration similar to that in pellets. These observations are consistent with the experimental data of Pourbaix et al. [8] who showed that Se is stable in the system Se-H_2O between pH 5 and 7, and Eh +0.2 and -0.4 at 25°C. Elemental iron is not stable under these conditions, reacting to form iron hydroxides and ferrous ions in the system Fe-H_2O [8] and iron selenide, $FeSe_2$, in the system Fe-Se-H_2O [9]. Details of the mechanisms of alteration of selenium and release of soluble selenium compounds are not fully understood, and experiments are in progress to test a number of hypotheses. It is clear, however, that diffusion of Fe ions to selenium grain surfaces and along channelways is required to effect alteration of selenium to 'iron selenide'. Reaction of selenium to mixtures of $FeSe_2$ and FeO(OH), involves an increase in volume or a loss of selenium. As there is no evidence for the former it is believed that when alteration occurs there is a concomitant release of selenium to the sheep.

REFERENCES

1. Kuchel, R.E., and R.A. Buckley, 1969, Aust. J. Agric. Res. 20:1099-1107.
2. Anon., 1974, Rural Research 86:19-23.
3. Wilkins, J.F., and B.A. Hamilton, 1980, Aust. Vet. J. 56:87-89.
4. Hunter, R.A., D.W. Peter, D.R. Hudson, and B.S. Chandler, 1981, Aust. J. Agric. Res. (in press).
5. Hudson, D.R., Hunter, R.A., and D.W. Peter, 1981, Aust. J. Agric. Res. (in press).
6. Watkinson, J.M., 1979, Anal. Chim. Acta. 105:319-325.
7. Hunter, R.A., D.B. Purser, and R.J.W. Barron, 1981, Aust. J. Exp. Agric. Anim. Husb. 21:52-54.
8. Pourbaix, M.J.N., J. van Muylden, and N. deZhoubov, 1963, "Atlas d'Equilibres Électrochimiques a 25°C", (Gauthier-Villars, Paris).
9. Bur'yanova, E.Z., 1969, Geokhimiya 12:1451-1465 (Russ.).

DISCUSSION

ULLREY (U.S.A.). Do you use a grub screw in association with these pellets like you do with cobalt pellets?

HUDSON (Australia). As a result of early experiments in which the selenium supplement was found to be much less than the CSIRO prototypes there were suggestions that a phosphate build up was occurring on the outside of the pellet, and as a consequence they were administered with grub screws to try and grind the outside of the selenium pellet. However, we found in a series of experiments where we removed the selenium pellets from sheep and then filed the surface, that there was only a very marginal improvement in the selenium status. It didn't appear that the surface coating was causing the marked decline in selenium values. One thing that I should have mentioned is that our experiments were conducted on pellets that were available about a year ago. As a result of these experiments, all commercial manufacturers are using coarse selenium in their pellets.

ANON. Were the commercial pellets tested on Kangaroo Island, and if so, how did they compare with the CSIRO pellets?

HUDSON (Australia). When the pellets were considered as a commercial proposition there were a number of experiments that were conducted in New Zealand and in Australia; some of these may have been on Kangaroo Island. The results from these experiments, in most cases, were not as good as those first reported for the CSIRO prototype, and at that time it was thought that it was the nature of the supporting iron matrix that was causing this. The original CSIRO pellets were made with iron filings whereas, you would have seen in the slides, the commercial pellets are made with granules of sponge iron.

ANON. Did you observe any variations in pellet performance as a function of diet?

PETER (Australia). There does appear to be variability in the response you get to the selenium pellet with time of year, so presumably there are factors in the diet that do make changes, but these are relatively insignificant overall.

PAYNTER (Australia). In the experiment where glass-selenium and iron-selenium mixtures were suspended in nylon bags in the rumen, you said that degradation of elemental selenium only occurred in the presence of iron. Wouldn't you expect other ions within the rumen from the natural diet that you are feeding to these animals, to have some degradative effect on selenium, or are you proposing that you are getting the effect because of the high concentration of iron in the bag in which you had Fe plus the elemental selenium?

HUDSON (Australia). I think the alteration of selenium is partly due to the very high concentration of ferrous irons, generated by the iron breaking down in the rumen in close proximity to the selenium grains. It is also due to the fact that you have iron actually in contact with selenium in the bag and as the iron breaks down it releases electrons that are conducted across the actual physical contacts with the selenium grains. This will solubilise the selenium as a reduced selenium species. I think that unless you have electron donors that can react with the elemental selenium it is unlikely that you will have any alteration of elemental selenium.

WHITE (Australia). Did you measure the blood selenium concentrations in those sheep that were given glass and elemental selenium into the rumen? Is it possible that the elemental selenium was being dissolved, by bacterial processes for example, but not showing up as any changes under the microscope?

HUDSON (Australia). I think it is always possible that there could have been some dissolution by bacterial processes. There were no measurements of blood selenium values on those sheep in which we suspended the bags with glass and selenium. However, blood selenium values were measured in experiments using elemental selenium in gellatin capsules; no response was found. Similar measurements were made by Godwin in Adelaide, who inserted mixtures of iron and selenium in gellatin capsules, and he also found no response. Dr. Langlands reported yesterday that he had treated sheep with crushed new pellets in gellatin capsules, and again there was no response in the blood and presumably when the capsules dissolved the iron and selenium were separated. I think it is fairly well demonstrated that unless the iron and the selenium are in contact the response is not high.

OPTIMUM Se GRAIN SIZE FOR SELENIUM PELLETS

D.W. Peter[a], R.A. Hunter[a] and D.R. Hudson[b]

[a]CSIRO, Division of Animal Production, Private Bag, P.O. Wembley, Western Australia 6014
[b]CSIRO, Division of Mineralogy, Private Bag, P.O. Wembley, Western Australia 6014

INTRODUCTION

Recent studies with commercially manufactured selenium (Se) pellets indicated that the grain size of elemental Se in the pellets may account for observed differences in the effectiveness of such pellets in maintaining an elevated selenium status of sheep (1, 2). This paper reports the results of microscope examinations of pellets manufactured using elemental Se of different grain sizes and an experiment in which the effectiveness of these pellets in producing and maintaining elevated plasma concentrations of Se was examined. The results are discussed in relation to the alteration and release of Se from pellets and an optimum grain size for the most effective release is suggested.

MATERIALS AND METHODS

A small number of Se pellets were manufactured by ICI (Aust) Pty Ltd using elemental Se with grain sizes < 45μm, 45-75μm, 75-150μm and > 150μm. Pellets containing elemental Se of about 4μm average grain size (range 1-50μm) were used for comparison.

Experiment 1:

Five nylon bags, each containing a single Se pellet manufactured with the grain size designated above, were suspended using the technique of Hunter *et al.* (3), in the rumens of Merino wethers kept indoors and fed a pelleted lucern-chaff/wheaten-chaff ration; the nylon bags were removed after 28 days. Polished transverse sections of each pellet were prepared for microscope examination using the methods described by Hudson *et al.* (1).

Experiment 2:

Twentyfour 12-month-old Merino wethers grazing a mixed subterranean clover/annual grass pasture were divided into groups of 4 sheep per group. Single pellets of one particular Se grain size were administered to individual animals in each of 5 groups with the sixth group remaining untreated. Blood samples were collected to obtain plasma, for Se analysis (4), prior to treatment and at intervals thereafter.

RESULTS

Experiment 1:

The general appearance of polished sections of pellets recovered from nylon bags suspended in the rumen for 28 days was similar to that previously described (2). Fine-to-medium-grained Se (< 100μm) occurred within the interstices of an aggregate of subspherical granules of sponge iron, whereas coarse-grained Se formed an interlocking aggregate with the iron granules (Fig. 1). Although some fracturing and pressure granulation of some large Se grains was observed, no substantial change of Se grain size had apparently occurred as a result of compression during pelletisation. Grain size variation between pellets was retained (Fig. 1); it was observed that the average grain size in pellets with Se grains < 45μm tended towards the upper size limit, and was thus much greater than the 4μm average in the fine-grained pellets used for comparison.

Selenium grains in all pellets were partially altered to a pure cream-coloured phase, previously identified as "iron selenide" (1, 2). Alteration was similar in all the pellets containing Se with grain sizes close to or greater than 45μm, and consisted of an irregular zone or replacement along Se grain boundaries and cracks (Fig. 1). The percentage of elemental selenium that remained was greatest in pellets with Se grains > 45μm, marginally less in pellets with < 45μm, and, as previously described (1, 2) only a very small percentage remained in pellets with an average grain size of 4μm.

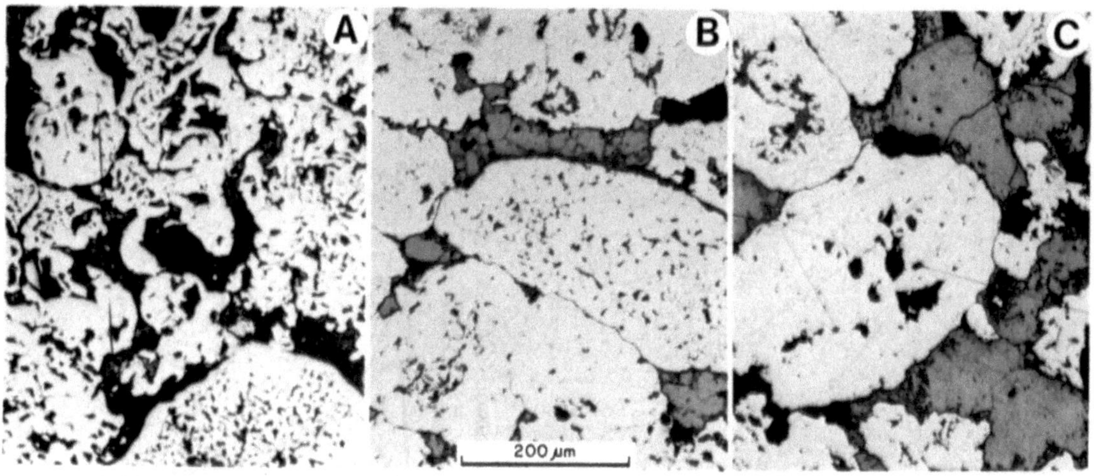

Fig. 1 Photomicrographs of sections of Se pellets with elemental Se of different grain sizes. (A = 4μm average, B = < 45μm, C = 75-150μm; Se, grey; sponge iron, off-white.)

Experiment 2:

Pellets with Se grains > 45μm all produced similar elevated plasma Se concentrations, with measured concentrations at 14 days approximately twice those found with the commercial, fine-grained pellet (Fig. 2). Plasma concentrations in sheep given pellets with Se grains < 45μm were initially lower than those found with the larger grained pellets but again significantly higher ($P < 0.01$) than produced by the commercial, fine-grained pellets. After 3 months, plasma concentrations in all animals, except those with commercial fine-grained pellet and in the untreated group, were similar.

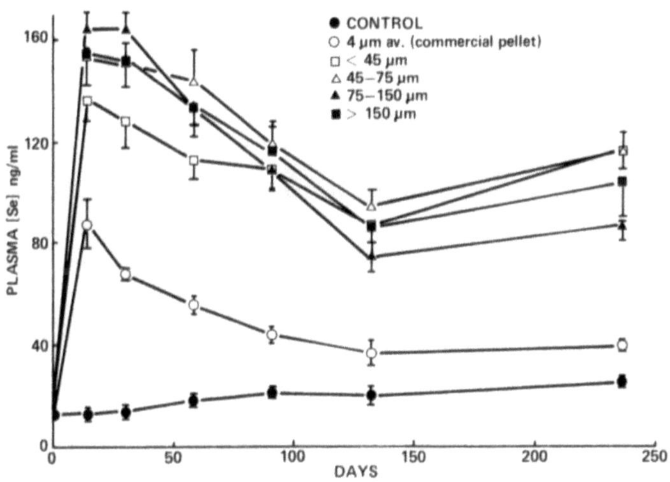

Fig. 2
Plasma Se concentrations (± SEM) of sheep given Se pellets containing elemental Se of different grain sizes.

DISCUSSION

These results confirm the proposal (2) that pellets which retain a high percentage of elemental Se perform better than those that lose most of their elemental Se in a month. The alteration of elemental Se, and presumably the concomitant release of Se to the sheep (1, 2) begins on the exposed surfaces of the Se grains and along open fractures, and therefore the rates of reaction, reduction of amount of elemental Se and release of Se depend on the surface area of the grains. Since the total surface area of elemental Se is initially greatest in pellets containing fine grained elemental Se, the initial loss of elemental Se should be highest in these pellets. However, the formation of a rim of altered Se rapidly reduces both the surface area and amount of the residual elemental Se.

This is readily illustrated by simple calculations based on the assumptions that grains are spherical in shape and selenium replacement occurs only on the grain surface (Fig. 3).

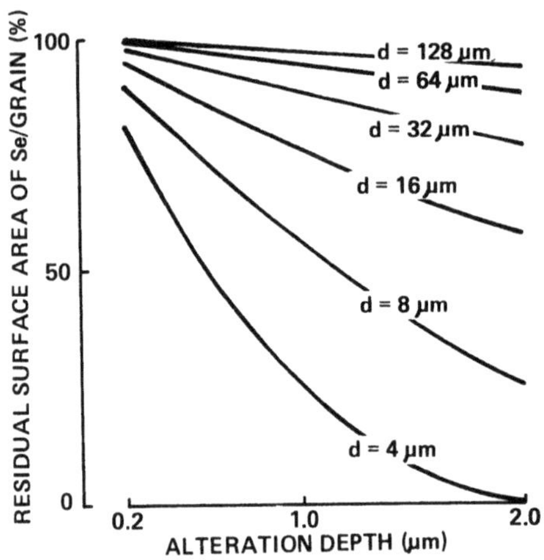

Fig. 3 Calculated residual surface area of elemental Se in spherical grains (% of original area/grain) following surface alteration.

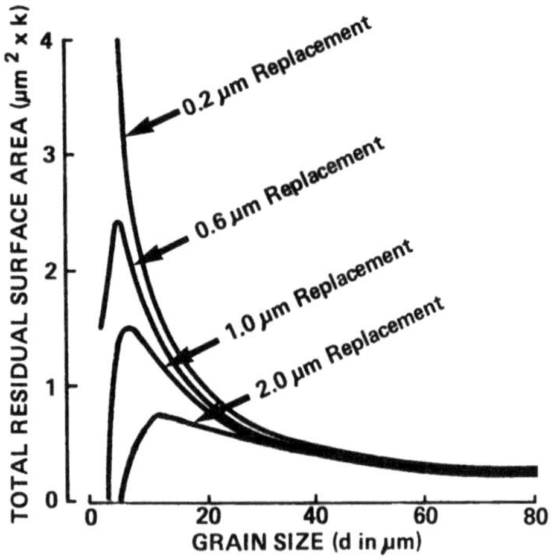

Fig. 4 Calculated total residual surface area of elemental Se in hypothetical Se pellets of different grain size.

If all Se grains in a pellet were identical in size, and the total initial quantity of Se in any pellet were constant, the total surface area of residual elemental selenium would vary as a function of the original grain diameter and the thickness of the alteration product, as shown in Figure 4. These data suggest that if the initial rapid alteration of surface Se occurs to maximum depth of approximately 2μm, then the minimum effective grain size for retaining the largest possible total surface area of unaltered elemental Se would be of the order of 10-12μm. However, the alteration of selenium occurs in an irregular and complex manner, exploiting fractures and open channelways in grains that are themselves irregular in both size and shape. This obviously places some constraints on the ideal model suggested above. Furthermore, it is not known whether the thickness of the alteration product effects the subsequent rate of reaction and release of elemental Se or whether some Se may be released from the alteration product itself. It would also appear that the rate of release can initially exceed the capacity of the biological system to assimilate or utilise the Se.

In order to provide both an immediately usable and continuously adequate release of Se it is therefore suggested that Se grains should not be smaller than 10μm and that the optimum size is probably of the order of 35-45μm. In theory larger grain sizes could lead to a loss in effectiveness because of a loss in surface area, though in practice grain fracturing during pelletisation may prevent this from occurring.

REFERENCES

1. Hudson, D.R., R.A. Hunter and D.W. Peter, 1981, Aust. J. Agric. Sci. (In press).
2. Hudson, D.R., R.A. Hunter and D.W. Peter, 1981, Proc. IVth International Conf. on "Trace element metabolism in Man and Animals" (This issue).
3. Hunter, R.A., D.B. Purser and R.J.W. Barron, 1981, Aust. J. Exptl. Agric. Anim. Husb. 21:52-54.
4. Watkinson, J.M., 1979, Anal. Chim. Acta, 105:319-325.

ACKNOWLEDGEMENTS

The authors wish to thank Messrs D. Buscall and P. Young for their competent technical assistance and ICI (Aust) Pty Ltd for manufacturing the pellets containing elemental Se of different grain sizes.

DISCUSSION

OLDFIELD (U.S.A.). I was just going to ask has the implications of your study been picked up and applied in the production of commercial pellets?

PETER (Australia). Yes as pointed out by Dr. Hudson, commercial pellets produced at this stage all tend to satisfy grain size requirements.

ANON. With regard to these large grain size pellets, do you have any estimates on how long or what's your gut feeling about how long these last, once you put them in the rumen as far as maintaining blood levels and preventing deficiency?

PETER (Australia). Theoretically I don't see any reason why they shouldn't last just about indefinitely, at least for the life of the sheep.

ULLREY (U.S.A.). Are these pellets being used for cattle, and, if so is there a different pellet size or a different length of life?

PETER (Australia). We have done no work whatsoever with selenium pellets in cattle although I believe a selenium cattle pellet is available. Geoff Reed of ICI may be able to answer that question for us.

REED (Australia). There are a number of cattle pellets we have punched out for experimental purposes. Its on sale in the U.K. They are not registered for use in Australia and quantities of the pellets have been sent to some people doing research in the united States. It is a 10% elemental selenium, 30 gram pellet, 2 pellets being required, giving efficacy up to 18 months.

MACPHERSON (U.K.). We have tried out these selenium pellets in cattle in the U.K. and found them to be highly effective using only 1 pellet. We have only used them up to 5 months but they have maintained selenium status or raised selenium status from deficient levels to adequate levels very effectively and they seem to be more effective in the U.K. than the sheep pellett.

JANGHORBANI (U.S.A.). Has anybody done calculations to find out about the absolute efficiency of retention of the selenium that solubilises from the pellets? What percent of it is retained by the animal?

PETER (Australia). I think the answer to that is no. No one has done those measurements. Dr. Langlands has looked at the amount of selenium that has remained in pellets after given times in the animal and this presumably includes reacted selenium, not elemental selenium alone. In other words total selenium, and I think his figures are somewhere around the order of 80-95%. So in other words the level of selenium in the pellet is probably quite adequate if what remains is elemental selenium. We don't know, as I said before, what happens once it gets to the selenide form, what sort of release rates we get or how available that selenium is.

A final point which I also meant to make yesterday in my talk concerns the increase we tend to find in control animals that are running with pellet-treated sheep. In other words what appears to be happening is that we are getting some excretion of selenium and transfer of that selenium to untreated animals; it is more noticeable in some experiments than others.

INTENSIVE GRAZING PRACTICES AND SELENIUM AND VITAMIN B_{12} NUTRITION OF SHEEP

Chris. Halpin, Ivan Caple, Peter Schroder and Ross McKenzie

Department of Agriculture, Attwood Veterinary Research Laboratory, Mickleham Road, WESTMEADOWS, AUSTRALIA 3047

Applications of fertilizers containing sulphur and phosphorus (superphosphate) to improved pastures stocked with sheep at high densities (greater than 12 ha^{-1}) are essential for increasing wool production in the higher rainfall areas of Australia. This treatment of pasture may affect the trace element nutrition of sheep[1]. An increased incidence of selenium-responsive myopathy in sheep has been associated with increased stocking density and use of superphosphate[2]. Increases in wool production of sheep supplemented with selenium in some areas of Australia have been reported[3], and it is likely that these agricultural practices may exacerbate an already marginal selenium deficiency.

In a trial designed to assess the optimal ratio of stocking density and superphosphate application to pasture for wool production in the Western District of Victoria, the selenium and cobalt nutrition of sheep was monitored to determine the magnitude and time course of any effects on the trace element nutrition of sheep.

MATERIALS AND METHODS

Twenty-five plots were prepared in 1975 on a pasture which had received a total of 2.75 tonnes superphosphate ha^{-1} for the preceding 28 years. Each plot received a basal treatment of muriate of potash ranging from 60 kg ha^{-1} in 1975 to 200 kg ha^{-1} in 1980. In 1977, each plot received 105 kg ha^{-1} of potash containing copper (2%) and molybdenum (0.05%).

The design of the trial incorporated 5 stocking densities (12,15,18,21 and 24 sheep ha^{-1}) randomised amongst plots which received 5 superphosphate application rates (0,5,10,15 and 25 kg P. ha^{-1} yr^{-1}). Ten Polwarth wethers grazed each plot from 1975 and were replaced after 3 years. The selenium nutrition of the first group of wethers was monitored on 6 plots during 1977[2], and both the selenium and cobalt nutrition of the second group of wethers were examined on all 25 plots during 1980.

Selenium and cobalt nutrition of the sheep was assessed by erythrocyte glutathione peroxidase (GSHPx) activity[4] and plasma vitamin B_{12} concentration[5] respectively. Pasture and soil samples were collected from all plots in November 1980. Pasture selenium, phosphorus and sulphur concentrations were measured in samples dried at 100°C for 24 hr; pH and available phosphorus were measured in soil.

RESULTS

Effect on Selenium Nutrition of Sheep

There was a marked seasonal variation in mean erythrocyte GSHPx activity of sheep on each plot, and the selenium nutrition of the sheep was lower in December than in July (P<0.001 Figure 1). The mean GSHPx activity (1 unit = 1 µmole NADPH oxidised min^{-1} g Hb^{-1} at 25°C)

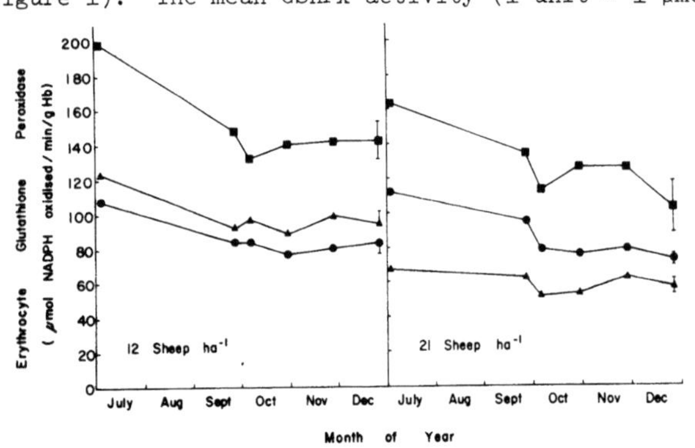

Figure 1: Changes in Selenium Nutrition of Sheep. Mean erythrocyte glutathione peroxidase activity in groups of Polwarth wethers grazing plots which received 0 (■), 10 (●) and 25 (▲) kg.P.$ha^{-1}yr^{-1}$ and were stocked at 12 and 25 sheep ha^{-1}. Bars represent standard error of means.

was lowest in sheep on plots which received the highest applications of superphosphate (Figure 1). Mean GSHPx activities of sheep grazing plots receiving 0 and 25 kg P. ha^{-1} yr^{-1} were 172 and 90 units respectively in July 1980 and 112 and 68 units respectively in December. The regression analysis indicated that erythrocyte GSHPx activity in the sheep was negatively related to the rate of application of superphosphate to pasture ($P<0.01$), but was not significantly affected by the stocking density on each plot.

The effects of superphosphate application on the selenium nutrition of the sheep were evident within 3 years. Comparison of erythrocyte GSHPx activities in sheep in 1977[2] with those in 1980 indicated that no further reductions in selenium nutrition of the sheep grazing the plots receiving the highest applications of superphosphate had occurred after 6 years.

Effect on Cobalt Nutrition of Sheep

Plasma vitamin B_{12} levels in sheep on each plot were significantly higher in December than in July ($P<0.001$, Figure 2). Mean concentrations in sheep grazing plots receiving 0 and 25 kg P. ha^{-1} yr^{-1} were 4.2 and 2.4 µg/l respectively in July and 6.2 and 5.0 µg/l respectively in December. Regression analysis indicated that serum vitamin B_{12} levels were negatively related to the application rate of superphosphate to the plots ($P<0.01$), but there was no significant interaction with stocking density

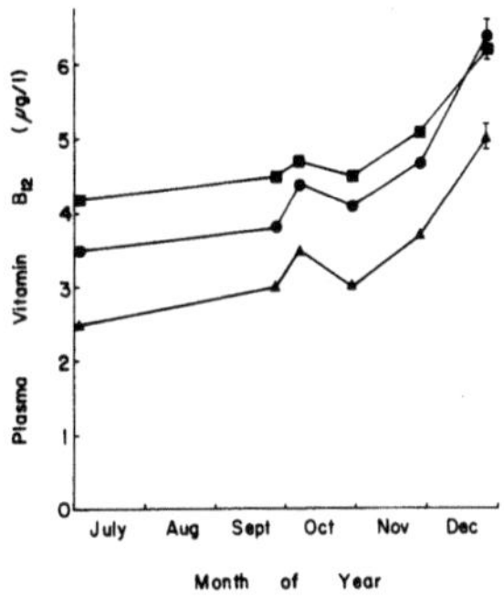

Figure 2: Changes in Cobalt Nutrition of Sheep.
Mean plasma vitamin B_{12} concentration in groups of Polwarth wethers grazing plots which received 0 (■), 10 (●) and 25 (▲) kg P. ha^{-1} yr^{-1} Bars represent standard error of means for all stocking densities.

Effect on Pasture and Soil

Pasture production on the plots increased with increasing applications of superphosphate (Table 1). The selenium concentration in pasture was negatively related to the superphosphate application rate ($P<0.05$), but stocking density had no apparent effect. There was no significant difference between the selenium content of grasses and clovers in each pasture. Grass on plots receiving 0 and 25 kg P. ha^{-1} yr^{-1} contained 0.059 and 0.023 mg/kg respectively when data for all stocking densities were grouped and clover selenium concentrations were 0.048 and 0.019 mg/kg respectively. Pasture sulphur concentrations were variable between plots, and were not related to any treatment (Table 1). Pasture phosphorus concentrations and available soil phosphorus increased with increasing fertiliser application rate (Table 1).

TABLE 1: Mean Soil pH and Available Phosphorus, and Pasture Yield, Composition and Concentrations of Sulphur, Phosphorus and Selenium on Plots Receiving Different Amounts of Superphosphate for 6 years.

Fertilizer [a] Application kg.P.$ha^{-1}yr^{-1}$	Soil pH	Soil Available P [b] mg/kg	Pasture Yield [c] t.D.M.ha^{-1}	Pasture % grass	Pasture S g.kg	Pasture P g.kg	Pasture Se µg/kg
0	5.3	5.7	4.0	52	2.0	2.0	5.7
5	5.3	7.2	5.3	57	1.6	2.3	4.1
10	5.2	7.8	5.8	57	2.0	2.7	4.0
15	5.2	10.3	6.4	62	1.9	2.7	2.7
25	5.2	15.8	6.7	71	1.8	3.2	2.2

a Within each fertilizer treatment plots had been grazed continuously by 10 sheep at stocking densities ranging from 12 to 24 sheep ha^{-1}

b Estimated as Ohlsen-P

c tonnes of dry matter per hectare

DISCUSSION

Allaway (1973) proposed that any soil-plant-animal chain that is operating on acid soils will ultimately become depleted of 'biologically-effective' forms of selenium[6]: the forms of selenium excreted in the faeces and urine of animals may be bound in acid soils and only slowly converted to the available forms. In this trial the application of superphosphate to the pasture significantly decreased erythrocyte GSHPx activities in sheep. This effect was apparent within 3 years. Stocking density had no effect on GSHPx within the 6 year period.

Pasture selenium concentrations were significantly decreased on plots receiving highest applications of superphosphate after 6 years. The decreases in GSHPx in the sheep were related to decreased pasture selenium, and to increased pasture growth and phosphorus concentrations. Decreased pasture selenium may be due to dilution effects associated with the increased plant growth following superphosphate treatment[7]. Pasture selenium has also been reported to be affected by botanical composition[8]. In this study grass content of the pasture was increased following fertilizer treatment, but there were no differences between grass and clover selenium concentrations on plots receiving the same fertilizer treatment. Sulphate has been shown to reduce[9], and phosphate to increase[10], uptake of selenium by legumes. Increased sulphur in the diet may also reduce selenium retention in sheep when sulphur and selenium intakes are low[11]. In this trial the sulphur content of the pastures was not affected by fertilizer treatment and the sheep were receiving adequate dietary sulphur. On the acid soils of these plots increased phosphate application was associated with decreased selenium content of pastures.

Sheep grazing plots receiving the highest application of superphosphate also had decreased plasma vitamin B_{12} levels. Fertilizer treatment may have affected cobalt nutrition by reducing either pasture cobalt availability or vitamin B_{12} synthesis in the rumen. It was postulated that part of the treatment effects on cobalt and selenium nutrition were due to reduced incorporation of these trace elements into their biologically-available forms in the rumen. Supplementation of 5 of the 10 wethers on each plot with intraruminal selenium and cobalt pellets indicated that this mechanism was not involved; the rates of increase and maximum levels of blood GSHPx and vitamin B_{12} following treatment were similar for sheep on all plots.

The interactions of selenium and cobalt in the soil-plant-animal chain are complex and often contradictory, and the effects of intensive agricultural practices are difficult to predict. The most effective way to examine the significance of these interactions is to monitor the selenium and cobalt nutrition of the animals.

ACKNOWLEDGMENT

We are grateful to Mrs. Simpson, 'Corea', Dunkeld, for allowing us to use her property and sheep for this trial. Part of the financial support for this work was provided by the Australian Wool Research Trust Fund and Commonwealth Extension Services Grant.

REFERENCES

1. Gardiner,M.R.,1969. Outlook in Agric. 6:19-28.
2. Caple,I.W.,K.A.Andrewartha,S.J.A.Edwards and C.G.Halpin,1980. Aust.vet.J.56:160-167.
3. McDonald,J.W., 1975. Aust.vet.J. 51:433-435.
4. Andrewartha,K.A.,S.J.A.Edwards,G.J.Brain and I.W.Caple,1979. Aust.J.Med.Technol.10:87-91.
5. Lau,K.S.,C.Gottlieb,L.R.Wasserman and V.Herbert,1965. Blood 26:202-210.
6. Allaway,W.H., 1973. Cornell Vet. 63:151-170.
7. Westermann,D.T. and C.W.Robbins,1974. Agron. J. 66:207.
8. Davies,E.B. and J.H.Watkinson,1966. N.Z.J.Agric. 9:317-327.
9. Pratley,J.E. and J.D.McFarlane,1974. Aust.J.Exp.Agric.Husb. 14:533-538.
10. Carter,D.L.,C.W.Robbins and M.J.Brown,1972. Soil Sci.Soc.Amer.Proc.36:624-628.
11. White,CL and M.Somers, 1977. Aust.J.biol.Sci. 30:47-56.

DISCUSSION

COSTA (Australia). Do you think phosphate reduces selenium availability to the animal directly or do you think the phosphate is having an affect on GSHPx activity independent of selenium? You really do not differentiate that.

HALPIN (Australia). We were hoping to differentiate that by administration of selenium pellets and as I have shown in the last slide, the rates of increase and maximum levels of GSHPx activity in sheep were similar for all plots. Therefore phosphate did not appear to affect selenium within the sheep, the affect was on selenium concentration of pasture.

BRAY (Australia). Were the grazing pressures in your heaviest grazing plots heavy enough to make for significant soil ingestion? I understand selenium tends to accumulate in the organic matter on the surface of the soil.

HALPIN (Australia). Obviously, the heaviest stocked plots grew less grass, and sheep on these plots would have ingested more soil. One of the limitations of our experiment was that the trial was aimed at optimizing the economics of wool production. Small amounts of supplementary feed were required to maintain sheep body weights in winter on some of these plots. These aspects reinforce our conclusion that it is most useful to monitor selenium nutrition of the sheep to assess the actual available intake.

EFFECT OF SUPPLEMENTAL SELENIUM ON THE PERFORMANCE AND ^{75}Se METABOLISM IN CHICKS

Jenny S. Edwardly

National Atomic Energy Agency
Centre for the Application of Isotopes and Radiation
P.O. Box 2 Kebayoran Lama, Jakarta Selatan, INDONESIA

Selenium was recognized as an essential nutrient for animals in 1957, and is needed for the normal functions of biological systems (1,2). A deficiency of selenium in growing chicks was shown to cause exudative diathesis, nutritional muscular dystrophy and encephalomalacia (3,4). There is still limited information on the influence of additional selenium in the diet on the onset of Se deficiency disease in chicks. Several authors have reported on the effect of additional selenium on blood and tissue selenium levels in chicks, swine and laying hens (5,6,7). However there is still a need to study the ^{75}Se metabolism in unsupplemented and repleted chicks. The goal of this work was to study the selenium status of poultry; and to study the effect of supplemental selenium on the performance and ^{75}Se metabolism in chicks.

MATERIALS AND METHODS

In experiment 1, 240 commercial broiler chicks were housed in environmentally controlled cages and with wire floors. All groups were given feed and fresh water ad-libitum. The chicks were alloted to 8 groups. Three replicate lots of 10 chicks were randomly assigned to treatments consisting of a local basal diet and four levels of selenium supplementation (viz: 0, 0.05, 0.10 and 0.15 ppm Se as Na_2SeO_3). The local basal diet contained less than 0.06 ppm Se and was analysed using neutron activation. (see Table 1).

Experiment 1 lasted 4 weeks during which observations were made on mortality, body weight, feed intake and selenium level in blood. After 3 and 4 weeks on the experimental diets, 3.5 μCi $Na_2{}^{75}SeO_3$ (in isotonic saline) was administered orally to each chick. After administration of the isotope, chicks were individually caged in battery cages for 48 hours. Blood samples were obtained by anterior heart puncture at 2, 4, 6, 24 and 48 hours after dosing. Faeces were collected at 24 and 48 hours after dosing. Chicks were then killed and liver and tissue samples were taken for histological examination and counting.

TABLE 1. Composition of basal diet.

Ingredient	% of diet
Corn	50
Rice bran	8
Coconut cake	10
Soya bean cake	30
Mineral (Grit + limestone)	1,75
Vitamin (Premix A)*	0,25
	100

*Pfizer Premix A, provided the following as amount per Kilogram of premix : 2000.000 IU vitamin A, 200.000 IU vitamin D_3, 1.400 mg vitamin K_3 (menadion bisulphite) 200 mg vitamin B_6 (pyridoxin), 2000 mg Niacin (Nicotinic acid), 1100 mg Pantothenic acid (Ca-pant), 2000 mg Choline Cloride, 800 mg vitamin B_{12} (cyanocobalamin), 45,400 mg DL-methionine, 2000 mg Antioxidant-Santoquin, 10,000 mg Mg, 2000 mg Fe, 1000 mg Cu, 3000 mg Mn, 2000 mg Zn, 20 mg I.

In experiment 2, chicks were fed the experimental diets to study the red cell uptake of ^{75}Se. Four weeks after feeding the experimental diet, the blood was taken by cardiac puncture and 4-5 ml of heparinized blood was incubated in air at room temperature with 20 μCi of ^{75}Se as H_2SeO_3. At 30, 60, 120 and 180 minutes, 0.5 ml was removed from the incubation mixture. The red cells were washed twice in 5 ml of isotonic saline solution before counting. The uptake was normalized to a PCV of 40 in all cases (8). Radioactivity was counted using a Packard Gamma Scintillation Spectrometer.

Data were analysed by analysis of variance (9).

RESULTS AND DISCUSSION

In experiment 1, the addition of selenium at 0.05, 0.10 and 0.15 ppm to the basal diet significantly increased weight gain ($p < 0.05$) (Table 2). No exudative diathesis was seen in birds given selenium supplemented feed. The incidence of exudative diathesis in the basal diet group was 20% ($p < 0.05$). Increasing selenium supplements above 0.05 ppm did not effect growth. This indicates that the addition of 0.05 ppm Se is more than adequate for maximum growth.

The mortality of chicks fed the basal diet was 25% ($p < 0.05$). Five percent mortality found in the Se supplemented group was caused by the weakness of the individual chicks. Symptoms of Se deficiency were not found. These results showed that the addition of Se at 0.05 ppm was effective in preventing symptoms of Se deficiency.

TABLE 2. Effect of selenium deficient and selenium supplemented diets on the performance of chicks.

Group	Feeding Treatment (4 week periods)	Weight Gain (g)	Exudative diathesis (%)	Mortality %
I	Basal diet	167.36	20	25
II	Basal diet + 0.05 ppm Se	249.08	-	5
III	Basal diet + 0.10 ppm Se	219.20	-	-
IV	Basal diet + 0.15 ppm Se	221.60	-	-

In experiment 2, the effect of selenium supplementation on ^{75}Se metabolism was observed. The activity of ^{75}Se in plasma is shown in Table 3. The activity of ^{75}Se in the plasma of the unsupplemented group was higher than in the supplemented groups (Table 3). The maximum activity was found at 4 hours after isotope administration. These results could indicate that the absorption of selenium is higher in Se unsupplemented than Se supplemented chicks.

The excretion of ^{75}Se in supplemented chicks was higher ($p < 0.05$) than in unsupplemented birds (Table 4). This effect has been observed by others (10).

TABLE 3. Effect of selenium intake on ^{75}Se activity in plasma of chicks.

Time after oral isotope	Group 1 Basal Diet	Group II Basal Diet + 0.05 ppm Se	Group III Basal Diet + 0.1 ppm Se	Group IV Basal Diet + 0.15 ppm Se
		(In % dose/Ml plasma)		
2 hours	0.066	0.051	0.052	0.050
4 hours	0.098	0.078	0.080	0.077
6 hours	0.057	0.042	0.046	0.043
24 hours	0.041	0.032	0.034	0.036
48 hours	0.032	0.018	0.020	0.021

TABLE 4. Effect of supplemental selenium on ^{75}Se excretion in faeces and urine (in % dose) and on ^{75}Se Activity in tissues (in % dose/g).

Treatment	^{75}Se Excreted 0-48 hour	Tissue ^{75}Se concentration Liver	Heart	Muscle
Basal Diet	19.8	0.69	0.27	0.09
B.D. + 0.05 ppm Se	29.3	0.29	0.10	0.03
B.D. + 0.10 ppm Se	30.6	0.27	0.07	0.03
B.D. + 0.15 ppm Se	33.1	0.28	0.09	0.03

The liver contained the highest level of ^{75}Se activity, followed by the heart and muscle (Table 4). Tissue levels of radioactivity were also highest in the chicks from the basal group compared with those from the three supplemented treatments ($p < 0.05$). The increased tissue concentration shows a response to greater demands for selenium. No differences between the supplemented groups were found. The results above agreed with those reported by Underwood (11) on the work of Jensen and Hopkin (1976). Therefore, plasma activity of ^{75}Se and retention of ^{75}Se could be useful indicators of selenium deficiency in chicks.

The blood level of selenium decreased to 0.04 ppm Se in the basal group after 4 weeks of feeding. Histological changes in liver and muscle were mainly observed in the unsupplemented chicks. Changes in the liver were characterized by a dissociation of hepatic cells. The pectoralis muscle of selenium deficient chicks showed a degeneration of muscle fibre. Fibres also broken in the muscle of chicks given 0.05 ppm Se. No damage was seen in chicks given 0.15 ppm Se. Red cell uptake of ^{75}Se showed a clear difference between the unsupplemented and supplemented chicks (Fig 1). The peak red cell ^{75}Se uptake values of either the unsupplemented chicks, those supplemented with 0.05 ppm Se or supplemented with vitamin E was above 23%, while the peak uptake values of animals supplemented with either 0.10 or 0.15 ppm Se was 15% or less. In unsupplemented animals and animals supplemented with either 0.05 ppm Se or vitamin E showed a peak uptake at 120 minutes. Animals supplemented either with 0.10 or 0.15 ppm Se had a peak at 30 minutes. It may be that 0.15 ppm Se supplementation of the basal diet is not sufficient to provide adequate selenium. Further studies are needed in this area. The results above show that red cell uptake of ^{75}Se can be used as an indicator of selenium deficiency in chicks.

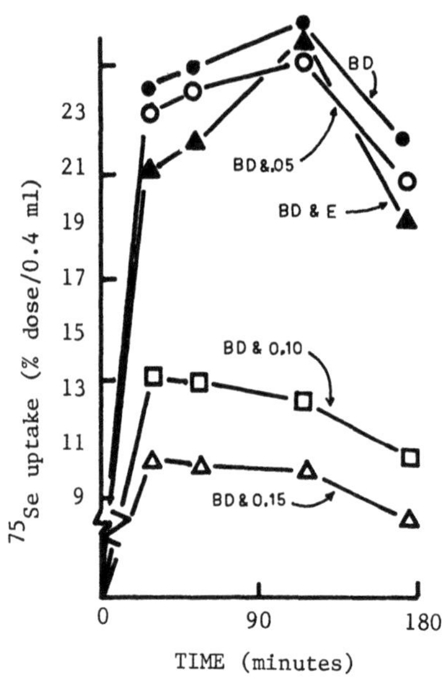

FIGURE 1. Red cell uptake of ^{75}Se in relation to Se supply.

SUMMARY

Experiments were conducted to study the effect of various levels of supplemental selenium on the performance and metabolism of ^{75}Se in commercial broiler chicks. A significant increase in weight gain was observed at the 0.05 ppm level of additional selenium. Percentage mortality was higher in the group given the basal diet. No incidence of exudative diathesis was seen in the supplemented groups. A significant increase of ^{75}Se uptake in red cells and tissues was observed in the selenium deficient chicks compared with supplemented chicks. ^{75}Se in plasma of supplemented animals was also less than unsupplemented birds. The whole body retention of ^{75}Se was significantly higher in deficient chicks. All these results indicate that selenium deficiency in chicks could be detected earlier by studying ^{75}Se metabolism than by waiting for clinical symptoms to appear.

REFERENCES

1. Schwartz, K., and C.M. Foltz, 1957, J. Amer. Chem. Soc. 79 : 3292.
2. Scott, M.L., J.C. Bieri, G.N. Briggs and K. Schwartz, 1957, Poultry Sci. 36 : 1155.
3. Gries, C.L., and M.L. Scott, 1972. J. Nutr. 102 : 2187-2196.
4. Nesheim, M.C. and M.L. Scott, 1961. J. Nutr. 65 : 601-618.
5. Scott, M.L., and A.H. Canter, 1971. Fed. Proc. 30 : 237.
6. Wilkinson, J.E., M.C. Bell., J.A. Bacon., and F.B. Masincupp, 1977. J. Anim. Sci. 44 : 224-228.
7. Arnold, R.L., O.E. Olson., and C.W. Carlson., 1973. Poultry Sci. 52 : 847-854.
8. Wright, P.L., and M.C. Bell., 1963. Proc. Soc. Exp. Biol. and Med. 379-382.
9. Snedecor, G.W., 1956. Statistical Methods. The Iowa State University Press, 5th Ed.
10. Jensen, L.C., E.D. Welter., and J.S. Dunlop., 1963. Proc. Soc. Exp. Biol. Med. 112, 889.
11. Underwood, E.J., 1971. Trace Elements in Humans and Animal Nutrition. 3rd Ed. Acad. Press.

A NEW METHOD FOR PREVENTING IRON DEFICIENCY ANAEMIA IN PIGLETS

P. T. Gleed, B. F. Sansom and W. M. Allen
Agricultural Research Council,
Institute for Research on Animal Diseases,
Compton, Newbury, Berkshire, RG16 ONN, U.K.

INTRODUCTION

A sow's milk provides only about one-tenth of the iron requirements of her piglets. To protect them from anaemia nearly all piglets kept under modern husbandry systems are therefore given an intramuscular injection of an iron dextran compound before they are 4 days old. These injections are effective but they are costly and time consuming, and can be a cause of abscesses, ham staining and sudden deaths. Young piglets have no difficulty in absorbing iron from their gut if it is available, for, under free-range conditions, they obtain their requirement for iron from soil (Venn *et al.*, 1947). Enriching the sow's diet with iron does not increase the concentration of iron in her milk but has been shown to protect the piglets to some extent from anaemia, possibly because they ingest some of her faeces (Veum *et al.*, 1965; Spruill *et al.*, 1971, Brady *et al.*, 1978). However, the quantities of faeces ingested have not been measured and no deliberate attempt has been made to prevent anaemia in piglets by feeding sows iron-enriched diets. The following experiments were designed first to measure the amount of faeces ingested by suckling piglets and secondly to measure the degree of protection against anaemia conferred by supplementing sows' diets with 2000 mg Fe/kg diet.

MATERIALS AND METHODS - 1

Eight sows and their litters were used to measure the intake of faeces by piglets. Sows 1-6 were housed in solid-floored farrowing pens and sows 7 and 8 in farrowing crates with a slatted floor. Each sow received a daily oral dose of approximately 200 µCi ^{198}Au (GCS IP Radiochemical Centre, Amersham), sows 1-3 from 2 days before expected farrowing until their piglets were 10 days old, sows 4-6 from 10-20 days after farrowing and sows 7 and 8 for the 20 days after farrowing. The sows' faeces became highly radioactive, but no radioactivity was transferred into their milk. In the solid-floored farrowing pens 2 kg of sawdust was used daily as bedding and the pens were cleaned daily. The total radioactivity in the faeces and bedding together was measured daily in a whole-body counter (Sansom *et al.*, 1971). No bedding was used in the slatted farrowing crates and faeces were collected daily from within and beneath the crate for measurement of radioactivity. The radioactivity in the piglets was also measured daily in the whole-body counter after they had been thoroughly washed to remove external contamination.

RESULTS

The mean radioactivity in the piglets and in the faeces and bedding, and the calculated mean retentions of faeces in the piglets are shown in Table 1.

Table 1

Litter (no. of piglets)	Mean radioactivity in piglets (counts/sec ± SEM)	Mean radioactivity in faeces and bedding (counts/sec/g)	Mean retention of faeces and bedding* (g)
1 (9)	513 ± 56	20	26
2 (9)	316 ± 23	16	20
3 (10)	164 ± 16	16	10
4 (7)	1207 ± 138	20	60
5 (10)	220 ± 20	22	10
6 (9)	191 ± 13	21	9
7 (8)	527 ± 81	56	9.4
8 (9)	226 ± 30	30	7.5

*Faeces only in litters 7 and 8

DISCUSSION

There were wide variations within and among litters but all the piglets ingested some faeces and there was a qualitative correlation between the "dirtiness" of the sow and the mean radioactivity of her piglets. The piglets on slotted floors ingested smaller weights, on average, than piglets on solid floors but the material that they ingested was not diluted with bedding. At slaughter at either 10 or 20 days of age over 90% of the radioactivity was present in the gastrointestinal tract of the piglets. The mean values for the weights of faeces retained are related to the mean daily intake of faeces by the piglets for, if the average passage time of the radioactive material through the gastrointestinal tract were 24 hours, the calculated values would indicate the amounts of faeces ingested daily. Using this assumption it can be calculated that if the sow's faeces had contained 2 mg of iron/g, every piglet in these experiments would have ingested at least 10 mg of iron daily, sufficient, if it had been absorbed, to protect them from anaemia. Experiments were therefore planned to test whether feeding sows a diet containing 2 g Fe/kg dry matter would prevent their piglets from becoming anaemic.

MATERIALS AND METHODS - 2

Three experiments have been performed to compare the growth-rate and haematology of piglets from conventionally fed sows with those of piglets from sows fed 2000 mg Fe/kg. In the first two experiments the sows were housed in solid-floored farrowing pens, but in the third they were housed in crates with slatted floors.

Experiment 1. 5 sows were fed a diet containing an additional 2 g Fe/kg as $FeSO_4.7H_2O$ from 2 days before farrowing until the piglets were weaned at 3 weeks. These piglets received no iron dextran injections. At the same time equal numbers of piglets from 8 sows receiving a conventional diet (200 mg Fe/kg) were injected intramuscularly after birth with either 100 or 200 mg Fe dextran. The piglets were weighed at birth, 7, 14 and 21 days and blood samples were taken for haematological analysis at 7, 14 and 21 days.

Experiment 2. 20 sows received the iron enriched diet. 16 sows received a conventional diet (200 mg Fe/kg) and piglets from these sows received 200 mg Fe dextran. The piglets were weighed and blood samples were taken at birth and at weaning at 3 weeks.

Experiment 3. 16 sows received the iron enriched diet. 14 sows received a conventional diet containing 408 mg Fe/kg and their piglets received 200 mg Fe dextran. The piglets were weighed and blood samples taken when they were weaned at 5 weeks of age.

RESULTS

Experiment 1. There were no significant differences between the mean body weights of the groups at any stage. The mean haemoglobin concentrations (g/100 ml blood) are shown in Table 2.

Table 2

Age	Piglets from sows fed 200 mg Fe/kg		Piglets from sows
(Days)	100 mg Fe	200 mg Fe	fed 2000 mg Fe/kg
7	9.4	9.5	8.9
14	10.1	11.0	9.7
21	9.7	11.7	10.7

SE for differences between groups = 0.63

Experiment 2. Table 3 shows the mean weights, haemoglobin concentrations and PCV's of the piglets in the two groups at birth and at weaning at 3 weeks. At 3 weeks there were 131 piglets in the control group and 167 piglets from sows fed the high-iron diet.

Table 3

Age	Piglets from sows fed 200 mg Fe/kg and receiving 200 mg Fe dextran			Piglets from sows fed 2000 mg Fe/kg		
	Weight (kg)	Hb (g/100 ml)	PCV (%)	Weight (kg)	Hb (g/100 ml)	PCV (%)
Birth	1.51	10.1	32.3	1.52	10.1	31.9
3 weeks	5.17	12.5	38.5	5.03	10.9	33.8
SEM at 3 weeks	0.10	0.12	0.4	0.12	0.16	0.5

Experiment 3. Table 4 shows the mean weights, haemoglobin concentrations and PCV's of the piglets when they were weaned at 5 weeks. There were 127 piglets in the control group and 163 piglets from sows fed the high-iron diet.

Table 4

	Piglets from 14 sows fed 408 mg Fe/kg diet and receiving 200 mg Fe dextran			Piglets from 16 sows fed 2000 mg Fe /kg diet		
	Weight (kg)	Hb (g/100 ml)	PCV (%)	Weight (kg)	Hb (g/100 ml)	PCV (%)
	7.83	13.2	42.6	7.97	11.6	36.1
SEM	0.16	0.1	0.4	0.15	0.13	0.5

DISCUSSION

The sows accepted the diets containing 2000 mg Fe/kg readily. In Experiment 1 their piglets maintained adequate and increasing haemoglobin concentrations until weaning at 21 days when their haemoglobin concentration lay between those of the groups injected with either 100 mg or 200 mg Fe dextran. In the larger scale experiments there were no differences in body weights at weaning and, although the mean Hb concentration was significantly lower among the piglets from sows fed the high iron diet, none could be considered anaemic. The level of haemoglobin below which there are believed to be adverse effects on the growth rate and health of piglets is disputed, but concentrations above 8 g/100 ml are accepted to be satisfactory (ARC, 1967) and some authorities believe that levels above 6 g/100 ml are also adequate.

Supplementing the sow's diet with iron protected piglets weaned between 3 and 5 weeks of age from anaemia, whether they were reared on solid or slatted floors. The method is cheap, and avoids the work, trauma and other possible hazards associated with iron dextran injections.

ACKNOWLEDGEMENTS

We thank Mrs M. Price for her help with these experiments.

REFERENCES

Brady, P. S., Ku, P. K., Ullrey, D. E. & Miller, E. R. 1978. J. Anim. Sci. 47: 1135-1140.

Sansom, B. F., Taylor, P. J., Wheelock, D. & Vagg, M. J. 1971. In "Mineral Studies with Isotopes in Domestic Animals". International Atomic Energy Agency, Vienna, pp. 125-133

Spruill, D. G., Hays, W. W. & Cromwell, G. L. 1971. J. Anim. Sci. 33: 376-384.

Venn, J. A., McCance, R. A. & Widdowson, E. M. 1947. J. Comp. Path. 57: 314-325.

Veum, T. L., Gallo, J. T., Pond, W. G., Van Vleck, L. D. & Loosli, J. K. 1965. J. Anim. Sci. 24: 1169-1173.

DISCUSSION

HILL (England). Was there any effect on the intake of food by the sows with this high level of iron?

ALLEN (UK). When we supplement we actually add 7 kilograms of ferrous sulphate per ton of food. At 2 thousand parts per million with ferrous sulphate there is no reduction in the dry matter intake by the sow. At 3 thousand and above parts per million of ferrous sulphate there is a reduced intake of ferrous sulphate but our more recent work which is not in our summary shows that there are other iron salts which are even more available to the baby piglet than ferrous sulphate. We have already looked at ferric chloride which is more readily available than the ferrous sulphate, but ferric chloride has an undesirable taste, thus there probably would be a similar inhibition of intake at these high levels. But quite amazingly ferric oxide or the iron ore added to the diet of the sow has several advantages. In initial experiments its availability appears to be even greater than ferrous sulphate and it neither discolours the diet nor the sow in the same way that ferrous sulphate does.

WHEELER (Australia). I am curious to know if you measured the iron concentration of the sows milk after supplementing her diet, would this have been altered in any way?

ALLEN (UK). Yes, we know that the radioactivity was not detectable in the sows milk. Before we measured radioactivity in the baby piglet we washed both the sow and baby piglet and we found that no radioactivity came through into the sow's milk. The second point was we did find isotope around the sow's udder, but you remember from the photograph of that dirty sow which I showed you,that faeces contaminates the sow's udder so that the piglet probably takes in an appreciable amount of faeces when it suckles or noses around the teat, but the intake is via the faeces and not from the milk.

WHEELER (Australia). I wondered if the iron content of milk was increased because in the cow the udder does tend to concentrate some minerals.

ALLEN (UK). No, the intake of iron via the milk represents about 1mg of iron per day and you cannot increase the ingestion of iron via the sows milk even though you are heavily loading the sow with iron.

ANON. Did you adjust any of the other minerals supplementally in the ration. I was thinking particularly of copper, zinc and perhaps manganese.

ALLEN (UK). No, not in these experiments. The only alteration to a normal diet has been the addition of ferrous sulphate as a commercial ferrous sulphate preparation.

The retention and distribution of ^{59}Fe in chicks with fowl typhoid given diets containing added iron in different forms

R. Hill and I.M. Smith

The Royal Veterinary College, (University of London)
Boltons Park, Potters Bar, Hertfordshire, ENGLAND.

INTRODUCTION

Diets containing supplementary iron influenced the survival of chicks infected with Salmonella gallinarum (Smith, Hill and Licence, 1977) and mice infected with Salmonella typhimurium (Smith, Hill and Brothwell, 1980). The effect varied with the iron compound given and the results were different for chicks and mice. With the compounds studied there was a much greater positive effect on survival in chicks than in mice. The addition of 400mg Fe/kg to the diet of chicks as ferric ethylenediaminetetraacetate or ferric nitrilotriacetate gave survival values several fold greater than those from a similar quantity of iron as ferric chloride or ferric fructose (Smith et al, 1977).

In the experiment described here, the retention and distribution of orally administered ^{59}Fe was determined in the tissues of chicks given diets containing the above four iron compounds and infected with S. gallinarum. The observations were made when the infection was developing rapidly but before there were any striking symptoms of disease.

EXPERIMENTAL

Sixty-four day-old broiler cockerels were reared to 12 days of age on a commercial-type starter diet, MPD, described previously (Mohammadi, Hill, Smith and Licence, 1976) and were divided randomly at that stage into four groups. Each group was given the same basal diet to which 400mg Fe/kg had been added in one of the following forms, ferric chloride (FeC), ferric ethylenediaminetetraacetate (FeE), ferric nitrilotriacetate (FeN) and ferric fructose (FeF). These compounds were prepared from ferric chloride as described by Helbock and Saltman (1967).

When the birds were 15 days of age, half the number of each dietary group (eight birds) were infected orally with S. gallinarum (Smith, 1955) and the other half were given sterile broth instead of the culture.

On days three and four after inoculation, all birds were dosed orally with 0.5ml of a solution of $^{59}FeCl_3$ containing about 1 uc radioactivity. On day five after inoculation the birds were killed and samples of blood, liver and spleen were taken for determination of radioactivity. Blood values presented here are estimates for the total volume of blood calculated from body weights when the birds were killed. The remainder of the body was ashed in a muffle furnace and samples of the ash taken for determination of radioactivity. The count for body ash was reduced to take account of the blood estimated to be present. The results of six birds were discarded for reasons unconnected with the experimental treatments.

On day five after inoculation, when this part of the experiment ended, most of the birds were still eating well and showed little clinical evidence of the disease.

The counts of radioactivity were analysed statistically by variance ratio after conversion to their logarithms. The number surviving were analysed by the Chi squared test.

Other groups of birds were given the same diets and were infected to provide corresponding results for survival.

RESULTS

Mean ^{59}Fe counts for control and infected birds given each dietary supplement are shown in Table 1. Blood ^{59}Fe was influenced significantly by the iron supplements, FeC giving the lowest mean and FeN the highest, about twice the FeC value. Infected birds had markedly lower blood counts than control birds though the proportional difference varied among forms of iron in the diet; the proportional difference between infected and control was greatest for FeC and least for FeE. These differences gave an F value of 2.1 for the interaction between iron supplement and infection but this was significant at only P<0.20 (Table 2). Liver ^{59}Fe counts were all much lower than those for blood. The effects of iron supplement and infection on liver counts were similar to those for blood except that the overall difference in infected birds was less than for blood and the proportional difference was more variable among iron supplements. For FeC and FeF liver ^{59}Fe values for infected birds were about 40% of control values but for diets supplemented with FeE and FeN the corresponding proportion was about 70%: the difference gave an F value for the interaction that was significant at P<0.01.

TABLE 1 Mean ^{59}Fe counts, adjusted to a dose of $1x10^6$

	Iron Supplements							
	FeC		FeE		FeN		FeF	
	C	I	C	I	C	I	C	I
No. of birds:	6	6	8	7	8	7	8	8
Blood	13,381	1,665	14,380	6,871	22,579	8,523	15,670	4,014
Liver	1,713	673	2,144	1,631	3,489	2,232	3,954	1,720
Spleen	24	79	38	127	60	249	42	211
Remainder	14,943	24,157	29,377	33,307	30,854	27,990	16,424	20,750
Retention %	3.01	2.66	4.59	4.19	5.70	3.90	3.61	2.67
Supplement mean	2.84		4.39		4.80		3.14	

TABLE 2 Results of Statistical Analyses

	Supplement		Infection		Interaction	
	F	P	F	P	F	P
Blood	7.53	<0.001	73.2	<0.001	2.10	<0.20
Liver	7.14	<0.001	15.2	<0.001	4.45	<0.01
Spleen	5.48	<0.01	89.9	<0.001	<1	NS
Remainder	5.09	<0.01	<1	NS	<1	NS
Retention %	5.43	<0.01	5.3	<0.05	<1	NS

Spleen ^{59}Fe counts were all relatively small and unlike those for blood and liver, values for infected birds were much greater than those for control uninfected birds. The high values from infected birds were associated with much enlarged spleens. The overall effect of iron supplement was significant, the mean count for FeN being greatest and FeC least. In the remainder of the body, there were large ^{59}Fe counts: those for birds given FeE and FeN were significantly greater than those for FeC and FeF, but there was no significant effect of infection nor an interaction between supplement and infection.

The proportion of the dose of radioactive iron retained in the birds varied among treatments between 2.7 and 5.7%. Overall retention was greater for supplements FeE and FeN than for FeF and the infected birds retained significantly less than non-infected.

The survival of birds given diets containing the same four iron supplements and infected with *S. gallinarum* is shown in table 3. The addition of 400mg Fe/kg as FeE and FeN gave greater survival than the same quantity of

TABLE 3 The survival of chicks with fowl typhoid given diets with added iron in different forms.

Fe supplement 400mg Fe/kg diet	Number of birds	Survival %
FeC	105	8
FeE	140	57
FeN	140	63
FeF	105	11

iron as FeC or FeF. The difference was large and highly significant.

DISCUSSION

The values for retention given here are very similar to absorption: there were very low ^{59}Fe counts in the contents of the digestive tract when the birds were killed and there was very little time after dosing with ^{59}Fe in which absorbed iron could be excreted. The retention, or absorption, values recorded here were about half those reported earlier in similar birds (Hill, Smith, Mohammadi and Licence, 1977): the lower values of the present study were probably associated with the large Fe supplement included in each diet. The diet used in the previous work had a high Fe content, about 200mg/kg but no supplement was added. The effect of infection on absorption of Fe was similar in both experiments.

No information was obtained in this study on the location of absorbed ^{59}Fe other than that in blood, liver or spleen. Radioactive iron in the remainder of the body, whatever its location, was not lower in infected than control birds and hence as a proportion of absorbed iron was considerably greater in infected than control birds. Whether this iron was needed in these other tissues in infected birds or could not be mobilised from them is not known.

Survival of infected birds was influenced by the form of iron given as a dietary supplement, FeE and FeN giving about five times the survival level of FeC and FeF. The absorption and retention of ^{59}Fe in birds given FeE and FeN was high whether infected or not and suggests that this was an important characteristic of dietary Fe that promoted survival. The high availability for absorption and retention of dietary FeE and FeN more than compensated for the decrease in absorption of iron that accompanied infection: the amounts of radioactive iron absorbed and retained in infected birds given FeE or FeN were greater than the corresponding amounts for healthy birds given FeC or FeF. Absorbed ^{59}Fe in infected birds given added FeE or FeN was retained largely in tissues other than blood, liver and spleen, but ^{59}Fe counts for blood in these birds were high, about 7,000 and 8,000 respectively compared with 1700 and 4000 for those given FeC and FeF. These high blood counts were for birds of groups in which survival was high suggesting that forms of iron that could be rapidly absorbed and incorporated into blood increased the birds resistance to infection.

SUMMARY

Retention of ^{59}Fe was greater when the diet contained added Fe as FeE or FeN than as FeC or FeF.

Retention was lower in infected than uninfected birds in blood and liver but greater in the spleen. Infection did not significantly affect retention in the remainder of the body but total retention was lower in infected than non-infected birds.

Survival of infected birds was associated with forms of iron that were retained at the highest levels in the whole body and in the blood.

REFERENCES

Helbock, H.J. and P. Saltman, 1967, Biochimica Biophysica Acta, 135: 979-990

Hill, R., Smith, I.M., Mohammadi, H. and S.T. Licence, 1977, Res. Vet. Sci., 22: 371-375

Mohammadi, H., Hill, R., Smith, I.M. and S.T. Licence, 1976, Avian Path., 5: 71-80.

Smith, H.W., 1955, J. Comp. Path., 65: 37-54.

Smith, I.M., Hill, R. and E. Brothwell, 1980, Res. Vet. Sci., 28: 161-167.

Smith, I.M., Hill, R. and S.T. Licence, 1977, Res. Vet. Sci., 23: 263-268

ZINC, COPPER, AND MAGNESIUM IN SERUM OF WOMEN GIVEN A CALCIUM SUPPLEMENT

N.E. Johnson, J.L. Greger, K. Kusserow,
F. Seyedabi and E.L. Smith
Department of Nutritional Sciences
College of Agricultural and Life Sciences
University of Wisconsin
Madison, WI 53706 USA

Interrelationships between mineral elements are a well known phenomena in the field of nutrition. Investigators have demonstrated in animal studies that high calcium intakes affect zinc utilization in several species (1,2) but clear evidence of calcium-zinc antagonism in man has not been demonstrated (3). Similary, calcium-magnesium antagonisms are well known in animals (2) and there are reports (4) that calcium supplementation may contribute to excessive losses of magnesium in man. Copper utilization is also affected by calcium supplementation. Suttle and Mills (5) found that high calcium levels induced a marginal zinc deficiency which favored the development of copper toxicosis by reducing the zinc-copper antagonism.

Serum zinc, copper and magnesium were monitored over a two year period during a clinical study of the effects of calcium supplementation on bone density of women between the ages of 40 and 65 years. One group was supplemented with Os-Cal 500[R] (Marion Laboratories), the other served as a control. The supplement provided a daily total of 1500 mg of elemental calcium in the form of $CaCO_3$; this was in addition to the usual intake of dietary calcium. The Os-Cal tablet was analyzed for trace elements and found to contain the following amounts of mineral: Mg, 4.69 ± 0.04 mg, Zn, 18 ± 1 µg; Cu 16 ± 1 µg; Mn 892 ± 3 µg; Fe 1.96 mg. The placebo tablet contained lactose and negligible amounts of these minerals.

Serum samples were diluted with 0.1% lathinum chloride. Glycerol (5%) was added to zinc and copper standards so that flow rates of the samples were the same. All samples were analyzed by atomic absorption spectrophotometry.

The average serum zinc, copper and magnesium values for supplemented and non-supplemented women are given in Table 1. There were no significant differences between

Table 1. Average serum Zn, Cu, and Mg values for supplemented and non-supplemented groups

Year of Study	Group	Zn µg/100 ml	Cu µg/100 ml	Mg mg/100 ml
Before Study	A (n=65)	(-)[1]	122 ± 20[2]	1.90 ± 0.20
	B (n=59)	(-)[1]	122 ± 20	1.91 ± 0.22
Year 1	A (n=65)	93 ± 14	134 ± 20	1.97 ± 0.15
	B (n=59)	92 ± 13	133 ± 21	1.99 ± 0.17
Year 2	A (n=65)	95 ± 19	126 ± 23	2.12 ± 0.15
	B (n=59)	94 ± 18	124 ± 23	2.13 ± 0.16

[1](-) no values

[2]Mean ± standard deviations

groups for any of the minerals but there was a significant ($P < .001$) effect of time on serum magnesium. The increase in serum magnesium with time occurred in both groups and may be an age-related effect.

About 25% of the women were taking the hypertensive drugs presented in Table 2.

Table 2. Hypertensive drugs used by women

Diazide	Aldomet
Propalonal	Aldorel
Regroton	Hygroton
Hydrochlorothiazide	Hydrocuril
Ismilid	Aldactazide
Endurol	Daicardin
Enduron	Deozide

There was a significant ($P < .001$) and consistent elevation (12%) in serum copper levels each year of the study and in the pre-study period for women taking these drugs. Also, there was a tendency ($P < .077$) for serum magnesium to be lower in women taking hypertensive drugs (Table 3).

Table 3. Effect of hypertensive medication on serum Zn, Cu, and Mg

Year of Study	Group	Zn µg/100 ml	Cu µg/100 ml	Mg mg/100 ml
Before Study	Non-user (n=98)	(-)[1]	118 ± 19[2]	1.92 ± 0.22
	User (n=26)	(-)[1]	132 ± 18	1.86 ± 0.16
Year 1	Non-user (n=98)	92 ± 13	130 ± 21	1.99 ± 0.16
	User (n=26)	92 ± 16	146 ± 15	1.94 ± 0.17
Year 2	Non-user (n=98)	96 ± 19	122 ± 23	2.13 ± 0.16
	User (n=26)	94 ± 19	136 ± 23	2.09 ± 0.14

[1](-) no values

[2]mean ± standard deviation

The reason for the elevation in serum copper of women on hypertensive drugs cannot be explained at this time. Whether the drugs or the disease process cause the elevation is unknown.

Results of this study indicate that calcium supplementation at a level of 1500 mg daily does not significantly affect serum zinc, copper or magnesium. However, serum copper is elevated about 12% in women taking medication for hypertension.

REFERENCES

1. Becker, W.M., and W.G. Hoekstra. 1971. In "Intestinal Absorption of Metal Ions Trace Elements and Radionuclides" (ed. S.C. Skoryna and E.D. Waldron) (Pergamon Press, N.Y.) pp. 229-256.

2. Hoekstra, W.G. 1964. Fed. Proc. 23: 1068-1076.

3. Spencer, H., V. VanKinscott, I. Lewin, and J. Samachson. 1965. J. Nutr. 86: 169-177.

4. Wacker, W.E. and A.F. Parisi. 1968. N. Engl. J. Med. 278: 712-717.

5. Suttle, N.F., and C.F. Mills. 1966. Br. J. Nutr. 20: 149-161.

DISCUSSION

VERSIECK (Belgium). When showing the zinc results in supplemented and non supplemented subjects you mentioned that you have no pre-study values because you supposed that your samples were contaminated. May I ask you to mention briefly what measures you took afterwards to prevent these contaminations. In your opinion, what are the most important precautions to take?

JOHNSON (U.S.A.). I would like to refer this question to my colleague Dr. Gregor who did the analytical work.

GREGER (U.S.A.). When we came into this study the samples had already been drawn and the first year we did not use special vacutainer tubes, so the zinc values were out. In the second year we did have the special blue top tubes that reduce zinc contamination. In terms of other precautions, recoveries were run and internal serum standards were analysed in all three years so that we knew we were consistent in our laboratory.

K. KRISHNAMACHARI (India). I have three points. Firstly, what is the percent absorption of calcium in this study? Secondly, were there any pregnant women during this part of your study, and if so what happened to the level of calcium? Thirdly, what is the affect of calcium supplements on the bone density?

JOHNSON (U.S.A.). We are not studying the absorption of calcium and we are not collecting faeces. To our knowledge there are no pregnant women involved in these studies. The women are 38 and going through the menopausal periods. We do have bone density data but this is a four year longitudinal study and the data are not analysed.

K. KRISHNAMACHARI (India). Abour four years ago we published a study where we gave 600 mg of elemental calcium as a supplement to pregnant women and we monitored the bone density in the mother as well as in the infant. We found an increase in the bone density of the infant born to the mothers belonging to the low socio economic group. Of course they are accustomed to calcium intakes of only 300 mg and we gave them an additional 600 mg. The percent absorption of calcium was 25%.

JOHNSON (U.S.A.). Thank you. We are dealing with a little different population and I suspect in your population the normal intake of dietary calcium may run a couple of hundred mg a day. In the United States our populations would take in probably about 500 mg daily.

ABDULLA (Sweden). Copper levels in serum are elevated in a number of non-specific conditions including cancer and I wonder if you have looked for such diseases during the course of your studies.

JOHNSON (U.S.A.). No. I realise that it is elevated in some diseases but ours was a rather small elevation and I really do not know what it means.

HARLAND (U.S.A.). Could you please elaborate on your observed age-related increase in serum magnesium? Could you also tell me what the protein levels were in the diets of these women.

JOHNSON (U.S.A.). We have not analysed all the dietary data but from other studies we have done we usually find a protein intake of about 70 grams a day. As to the reason for the elevation in magnesium with age; it would seem to me that it might be due to the release of magnesium from osteoporotic bone, since 50% of the magnesium is found in the bone.

TRACE ELEMENTS IN PREGNANCY AND LACTATION

ZINC NUTRITION AND PREGNANCY IN HUMANS

Sten Jameson
Department of Internal Medicine
University Hospital
S-750 14 UPPSALA
SWEDEN

INTRODUCTION

Human zinc requirements seem to be high during periods of rapid growth, such as pregnancy and embryonic life, puberty, and tissue repair. Zinc-responsive growth failure and sexual infantilism have been observed in both sexes. Poor growth, anorexia, and hypogeusia have been reported in children, with improvement following zinc treatment. Acrodermatitis enteropathica (A.E.) is an inherited human zinc-deficiency disorder. Few pregnancies have been reported in women suffering from this disease. Two resulted in malformed infants, one in a baby of low birth weight, and one in abortion. A connection between maternal zinc deficiency and congenital malformations has been postulated in man. Experimental data about zinc deficiency, abnormal gestations, and congenital malformations in animals are abundant and convincing.

ZINC BALANCE

Normal weight gain in human pregnancy amounts to 12 - 13 kilo, which means that additional 375 mg zinc is needed from some source. Endogenous daily zinc losses in young women were 1.6 mg/day (1). Sweat losses in preadolescent girls were 1.4 mg/day (2). Dermal losses ranged from 2.5 - 3.5 mg/day with a mean of 2.8 mg zinc/day in four pregnant teenagers in the third trimester (3). Zinc deficiency developed within 35 days in healthy young women through depletion of accessible body zinc stores when a low zinc diet was given (4).

The recommended daily allowances (R.D.A.) are 20 mg zinc/day in pregnancy and 25 mg in lactation. The calculated daily zinc intake in a random sample of healthy Swedish primigravidae was 9.4 mg/day. Zinc content parallelled protein and energy intake (5). The calculated daily zinc intake in Mexican women ranged between 9.4 mg/day in early pregnancy and 10 mg in the third trimester (6). Estimations of zinc contents of several diets have been summarized (7). More recent studies showed that the analyzed daily zinc intake in Scottish women was 7.6 mg (8). The analyzed mean daily zinc intake in New Zealand women was 8.9 mg (9). American college women had a mean daily dietary intake of 13.8 mg zinc (10).

Zinc balance studies revealed that preadolescent girls needed zinc intakes of 7 - 8 mg/day if zinc availability was 20 - 30 % in order to achieve a zinc retention (2). In nonpregnant university women an intake of 11.5 mg zinc/day equalled fecal and urinary excretions (11). This must indicate a negative balance, as other possible normal zinc losses were not taken into account. An intake of 16 mg zinc/day was sufficient to maintain adult healthy nonpregnant young women in positive zinc balance or at equilibrium, if sweat losses were not great (12). Net absorption of zinc was 25 % of the mean intake of 29 mg zinc/day in four pregnant teenagers studied during the third trimester (3). In order to allow for the normal pregnancy weight gain 1.5 mg zinc would have to be retained per day. Others have calculated lower figures and estimated the daily zinc requirements in pregnancy to 0.75 mg/day (13).

BLOOD ZINC LEVELS

Several studies of serum- or plasma- zinc concentrations in human pregnancy have shown that mean zinc levels decrease with increasing duration of gestation (14). The significance of the findings have remained obscure, however. We found that normotensive, nonanemic pregnant women, who subsequently delivered normally, at normal time and had normal babies, showed stable serum zinc levels throughout pregnancy, within normal range of nonpregnant healty fertile women without any medication (15). Serum zinc levels were significantly lower in poorly nourished pregnant Turkish women than in women of high socioeconomic class. Mean zinc levels in healthy nonpregnant women did not differ from mean zinc levels of each trimester in the well nourished group (16). In Mexican low income pregnant women, however, serum zinc levels were not lower in the group who reported low zinc intakes than in those with higher zinc intakes (6). In another study low plasma zinc concentrations occurred with a significant greater frequency at the puerperium in a low income group women (17). From India it is stated that serum zinc estimations have a very limited use as diagnostic and prognostic aid in obstetrics (18). In any case they report three infants with severe congenital malformations, all with very low maternal zinc levels in pregnancy.

We reported that women with normal pregnancies and normal deliveries with normal babies had significantly higher mean serum zinc in early pregnancy than women who had complications, especially prolonged gestation, inefficient labor, atonic bleeding and who got postmature babies as well as babies with malformations (15). Others could not find altered maternal or fetal blood zinc levels in samples collected at delivery from low birth weight gestations when compared to normals (19). Zinc analyses from parturition may be difficult to interpret as they are collected during a very stressful situation. Thus, an ACTH-infusion reduces serum zinc markedly, but not in those whose serum zinc content initially was already low (20). Maternal plasma zinc and umbilical plasma zinc were lower in pre-eclampsia and significantly lower in severe eclampsia (21). Serum zinc was significantly lower and serum copper higher in severe toxemia during the last weeks of pregnancy (22). These data were not verified in India, however (18).

ZINC AND MALFORMATIONS

Anencephaly is one of two reported malformations in A.E. It is recently reported from Turkey that serum zinc and umbilical blood zinc levels were significantly lower in mothers who gave birth to anencephalic babies (23). We reported one mother whose serum zinc levels dropped markedly during pregnancy. Her infant had a myelomeningocoele and had one of the lowest cord serum zinc values that we have recorded (24). We also reported six of ten malformations with the lowest recorded maternal serum zinc in early pregnancy (15). Others could not find altered serum zinc levels or amniotic fluid zinc levels in women having an increased risk of producing a child with a fetal anomaly (25). Alcoholic mothers had significantly lower plasma zinc levels at delivery than non-alcoholic controls. Fetal cord plasma levels were also lower. Manifestations of fetal dysmorphogenesis appears to relate to zinc status (26). There is accumulating evidence of an interaction between vitamin A and zinc promoting the mobilization of retinol from hepatic stores (27). Vitamin A concentrations in fetal livers from cases with CNS-malformations have been found to be higher than in normal controls (28). These data support the suspicion that human zinc deficiency may act teratogenically, directly or indirectly via free radicals, induced by toxic agents, alcohol, or metals such as Hg, Cd, and Pb.

Zinc absorption is reduced in coeliac disease and in ileal by-pass patients, treated for obesity. Signs and symptoms of zinc deficiency are regularly found in both groups. Three pregnancies reported from by-pass patients resulted in malformed infants, two in fetal death, and 20 % in babies small for gestational age (29). Among seven women with coeliac disease were two who previously had given birth to infants with congenital malformations (15).

AMNIOTIC FLUID ZINC

A positive correlation between zinc levels in the amniotic fluid and fetal weight has been registered (30). Protein content of the amniotic fluid decreased between the 25th to 39th week of gestation, zinc concentrations increased in the course of normal pregnancies (31). A very comprehensive and careful study has shown zinc levels in amniotic fluid to be significantly lower in cases of mild or severe fetal hypotrophy from the 38th week of gestation and onwards. Zinc content of amniotic fluid increased from the 37th week. No difference in protein levels was found. These data open new possibilities for the biochemical control of the fetal condition (32).

Amniotic fluid contains a low molecular weight fraction, with broadspectrum antimicrobial activity. The bactericidal activity of these peptides is zinc dependent (33). Underpriviliged Ethopian women had very low mean amniotic fluid zinc concentrations (0.19 µg/ml) compared to the privileged (0.35 µg/ml). Most of the samples were non-inhibitory to bacterial growth (34). Mean zinc concentrations were significantly higher in bactericidal than in non-inhibitory amniotic fluids. If zinc was added non-inhibitory fluids became inhibitory (35). Amniotic fluid phosphate-to-zinc ratio was accurate in predicting the bacterial growth inhibiting capacity and may be a predictor to determine the fetus at risk of intrauterine bacterial infection (36). Others were not able to confirm the value of the phosphate-to-zinc ratio but reported a trial of zinc-gluconate supplementation in low oral dosage, 30 - 90 µg/day, from the 20th week of gestation. Amniotic zinc levels were higher in supplemented groups but still low. No significant increase of inhibitory activity was found (37).

ZINC AND HAIR

Hair zinc analysis is another possible way of assessing zinc status. Hair changes in pregnancy are common and postpartum alopecia is a handbook complication (38). Hypogeusia, short stature, and low hair zinc in poorly nourished children are well known facts which have recently been confirmed (39). In young healthy well-nourished women hair zinc levels were lower in the last month of pregnancy than in early pregnancy (40). Zinc concentrations in hair were significantly lower in pregnant women of low income groups in Iran than in women of higher socioeconomic status (17). Conflicting data have also been reported. Full term low birth weight infants had not different copper, zinc, manganese, or vanadium levels in hair compared to full term normal infants (41). Hair zinc in controls differed from other published data, however. Mean hair zinc concentrations were significantly higher in mothers of newborn infants with spina bifida, compared to controls. Zinc concentrations increased during pregnancy and were positively correlated to the hair zinc concentrations of the newborn infants (42). These data are comparable with findings in A.E., where hair zinc concentrations were very high when zinc deficiency was extreme and hair growth rate slow. In contrast parents of, and infants with achondroplasia, a malformation also reported in A.E., showed significantly reduced zinc and increased copper concentrations in their hair, compared to controls (43).

FETAL ZINC

Estimates of the daily rate of accumulation of zinc by a human fetus in utero have been calculated (44). Zinc accumulates in the liver of the developing fetus and the liver accounts for about one-quarter of total body zinc at term (60 mg) (45). Zinc concentrations increased sevenfold between the 31st and 35th day of gestation suggesting that during these early stages of development morphogenetic events occur which require the presence of elemental zinc for biochemical and enzymatic reactions. During the second trimester there was more than 50-fold increase in the total hepatic zinc. Zinc concentrations in liver did not change significantly, however, from 85th to 185th day of gestation (46). Zinc concentrations in fetal liver decreased with gestational age in samples collected from abnormal gestations (47). These data might indicate a common situation of deficiency in these pregnancies.

Zinc content of human placenta varies between 58 - 69 ppm/dry weight (48, 49). No data from abnormal gestations seem to have been published.

ZINC THERAPY IN PREGNANCY

All data from investigations on the importance of zinc during pregnancy do not yet allow a definite conclusion about the value of zinc for an undisturbed pregnancy (50). Therapeutic results could be one answer to that question. When we found that women with completely normal gestations had higher mean serum zinc levels in early pregnancy than those who had complications we started a pilot study. Twenty highly selected women with low serum zinc and low urinary zinc excretion took part. Zinc sulphate (Solvezinc) 0.2 mg twice daily was given to seven women during the third trimester and seemed to reduce fetal and maternal complications without side effects (15). In a second study we gave 0.2 g zinc sulphate from the 14th week of gestation to 64 women with initially low serum zinc in an open randomized series. Zinc therapy reduced the number of complications and 63 % of the women delivered normally and got normal babies, compared to 48 % in the low zinc control group. Twenty two percent of all the untreated women achieved spontaneously unchanged or increased serum zinc levels (14th - 36th week), indicating zinc equilibrium or positive balance, 63 % of these women delivered normally and got normal babies. By contrast women with initially low serum zinc and subsequently decreasing levels, indicating disequilibrium or negative balance, delivered normally and got normal babies in only 26 % of the cases (24). According to our data it is favourable to achieve a positive zinc balance in human pregnancy. Zinc sulphate therapy corresponding to 45 mg zinc/day can reduce maternal and fetal complications, when initial serum zinc levels indicate zinc deficiency.

Warnings against indiscriminate use of zinc therapy have been issued. Interactions with other essential trace metals are possible. A therapeutic trial in India was discontinued when four consequent births were premature and one infant was stillborn. Supplements of 0.1 g zinc sulphate were given during the third trimester (51). Russian authors report successful results when using a combination therapy consisting of 5 mg zinc, 4.5 mg manganese, and 6.2 g magnesium daily to anemic pregnant women. Compared to controls the treated group suffered fewer surgical interventions at delivery, fewer asphyctic complications, and had a reduced number of delayed labor. Maternal blood values were better and the number of low birth weights was reduced (52).

SUMMARY

Normal weight gain in pregnancy entails a need for additional 375 mg zinc. Female zinc depots seem to be small. Zinc content in food parallels protein and energy intake. Estimated dietary zinc intakes in pregnancy lie considerably below RDA. Mean blood zinc levels decrease during pregnancy but not in normal pregnancies or in mothers of high socioeconomic class. Low zinc levels are found in amniotic fluid when fetal hypotrophy is present. Antimicrobial activity of amniotic fluid is zinc-dependent and low in poorly nourished mothers. Zinc accumulates in the fetal liver in normal gestations. Low maternal zinc levels, low cord blood zinc, and abnormal hair zinc concentrations correlate to fetal malformations, especially CNS-defects. Low maternal zinc levels correlate to fetal alcohol dysmorphogenesis. Normal gestations and normal deliveries with normal babies are frequent when zinc balance is positive in pregnancy. Zinc sulphate therapy can reduce the frequency of maternal and fetal complications if zinc deficiency is indicated in early pregnancy. Interactions with other essential trace metals are possible. Obviously better dietary habits are necessary in pregnancy.

REFERENCES

2. Ritchey, S.J., and M.K. Korslund, 1979, Am. J. Clin. Nutr. 32:799-803.
3. Schraer, S.K., and D.H. Calloway, 1974, Nutr. Metab. 17:205 - 212.
4. Hess, F.M., and J.C. King, 1977, J. Nutr. 107:2219 - 2227.
5. Jameson, S., and G. Svärdström, 1977, Hygiea. 86 (4), 409 (Abstract).
6. Hunt, I.F., and N.J. Murphy, 1979, Am. J. Clin. Nutr. 32:1511 - 1518.
8. Lyon, T.D.B., and H. Smith, 1979, Br. J. Nutr. 42:413 - 416.
9. Guthrie, B.E., and M.F. Robinson, 1977, Br. J. Nutr. 38:55 - 63.
10. White, H.S., 1976, J. Am. Diet. Assoc. 68:243 - 245.
12. Taper, L.J., and M.L. Hinners, 1980, Am. J. Clin. Nutr. 33:1077 - 1082.
14. Brandes, J.M., and A. Lightman, 1980, Biol. Neonate. 38:66 - 70.
15. Jameson, S., 1976, Acta Med. Scand. Suppl. 593:1 - 89.
16. Cavdar, A.O., and E. Babacan, 1980, Am. J. Clin. Nutr. 33:542 - 544.
18. Prema, K., 1980, Indian J. Med. Res. 71:554 - 560.
19. Bogden, J.D., and I.S. Thind, 1978, J. Lab. Clin. Med. 92:455 - 462.
20. Falchuk, K.H.,1977, N. Engl. J. Med. 296:1129 - 1134.
21. Bassiouni, B.A., and A.I. Foda, 1979, Europ. J. Obstet. Gynec. Reprod. Biol. 75 - 80.
23. Cavdar, A.O., and A. Arcasoy, 1980, Teratology 22:141.
24. Jameson, S. 1980. In "Zinc in the Environment", part 2. (ed. J.O. Nriagu) (Wiley, N.Y.) pp 183-197.
25. Shearer, T.R., and E.W. Lis, 1979, Nutr. Rep. Internat. 19:209 - 213.
26. Flynn, A., and S.I. Miller, 1981, Lancet i:572 - 575.
27. Solomons, N.W., and R.M. Russell, 1980, Am. J. Clin. Nutr. 33:2031 - 2040.
28. Gal, I., and I.M. Sharman, 1972. In "Advances in Teratology", vol. 5. (ed. D. Woollam) (Logos Press, London) pp 143-159.
29. Ingardia, C.J., and J.R. Fischer, 1978, Obstet. Gynecol. 52:215 - 218.
32. Kynast, G., and E. Saling, 1979, J. Perinat. Med. 7:69 - 77.
33. Sachs, B.P., and C.M. Stern, 1979, Br. J. Obstet. Gynecol. 86:81 - 96.
35. Appelbaum, P.C., and G. Shulman, 1980, Am. J. Obstet. Gynecol. 173:579 - 582.
38. Lindelöf, B., 1979, Z. Hautkr. 54:959 - 971.
39. Buzina, R., and M. Jusic, 1980, Am. J. Clin. Nutr. 33:2262 - 2267.
40. Hambidge, K.M.; and W. Droegemueller, 1974, Obstet. Gynecol. 44:666 - 672.
41. Gibson, R.S., and M.S. DeWolfe, 1979, Am. J. Clin. Nutr. 32:1728 - 1733.

42. Bergmann, K.E., and G. Makosch, 1980, Am. J. Clin. Nutr. 33:2145 - 2150.
43. Collipp, P.J., and S.Y. Chen, 1979, J. Pediatr. 94:609 - 610.
44. Shaw, J.C., 1979, Am. J. Dis. Child. 133:1260 - 1268.
46. Chaube, S., and H. Nishimura, 1973, Arch. Environ. Health. 26:237 - 240.
47. Casey, C.E., and M.F. Robinson, 1978, Br. J. Nutr. 39:639 - 647.
49. Riemschneider, R., and A.F. Martins, 1978, Geburtsh. Frauenheilk. 38:971 - 977.
50. Kynast, G., and E. Saling, 1980. J. Perinat. Med. 8:171 - 182.
52. Tymoshenko, L.V., and A.I. Shamyk, 1980, Ukr. Biokhim. Zh. 52:200 - 204.

A full reference list can be obtained from the author.

DISCUSSION

PERLMAN (Israel). You said, Dr Jameson that pregnancy requires an additional 375 mg of zinc but only 60 mg of that goes to the foetus - where does the rest of the zinc go?

JAMESON (Sweden). In our country there is an increase of between 12 and 13 kilos in total weight during pregnancy and if you count 30 mg Zn per kilo that comes to 375 mg.

PERVAN (Australia). Can you comment on the nutritional status of the placenta in your low zinc mothers versus your normal zinc mothers?

JAMESON (Sweden). There were no differences in placental weight between the groups - that is all I can say.

BRATTER (Germany). Can you please give some data on the zinc level of amniotic fluid in the normal case and the diseased case?

JAMESON (Sweden). No I cannot. All the data are from German sources and were published by Kynast.

ABDULLA (Sweden). I wonder if you differentiate between the physiological lowering of zinc during pregnancy and that due to nutritional deficiency. We know that there are certain substances like folic acid which go very low during pregnancy, because the form in which it is complexed and going into the foetus is different and cannot come out. Can it be some mechanism like that accounting for the lower zinc levels that you find in pregnancy.

JAMESON (Sweden). I do not know. However Giroux and his colleagues have been looking at zinc binding capacity during pregnancy.

DREOSTI (Australia). Did the children born from the low zinc status mothers show evidence of behavioural abnormalities as they grew up? There is a lot of evidence that in animals there is a behavioural abnormality as well?

JAMESON (Sweden). We hope to be able to report on the 4 and 5 year control which all Swedish children pass through and in which we can compare the seperate groups. It is now about 4 and 5 years since the study and the children will enter these controls now.

GOLDEN B E (Jamaica). You said that the first group of women that you supplemented with zinc were highly selected. What were you comparing them with, what was your control group?

JAMESON (Sweden). That was the first time that zinc supplementation was allowed. In Sweden we have to report to our FDA and get their acceptance. They scrutinize the work. That first time we had picked out 20 women during the third trimester who had, as far as we could measure, zinc deficiency with very low serum zinc levels, with very low urinery zinc excretion and many of them had a sore mouth and taste disorders. These 20 were randomised. Seven went into the treatment group and the other 13 into the control group.

O'DELL (USA). It is commonly stated in the literature that plasma zinc levels drop during pregnancy and that copper rises. Your data seemed to show very little change in plasma zinc during the course of gestation. Would you like to comment on that? Also did you measure plasma copper levels?

JAMESON (Sweden). Yes we made plasma copper estimations in all these patients. When we treat with zinc we get no difference in serum copper between the control group and the treated group in late pregnancy. We have tried to work with the copper/zinc index and we have not found it useful. The figure that I showed illustrated results for women who had quite normal gestations and when we picked them out they did not have hypertension, nor anaemia and all had delivery within normal time with quite normal babies. They had almost stable zinc levels compared to normal nonpregnant women.

GOLDEN B E (Jamaica). The Indian result of course, is very interesting and extremely worrying. These mothers were supplemented with zinc and one after the other they aborted. Putting that together with the previously reported physiological drop in plasma zinc and the data which shows a relationship between zinc and immune competence - I hypothesise that the physiological drop in zinc may be related to the physiological immune competence of pregnancy and the prevention of foetal rejection on this basis. My question is have you in any way studied the immunology particularly the cell mediated phenenoma in any of your patients?

JAMESON (Sweden). No we have not and I take that point. I think that is is very important not to enter substitution studies unless you can show that dietary intake is low and other measurements indicate zinc deficiency. I will add that in Sweden taste defects are very common during pregnancy. About 25% of first pregnancies have taste defects, which could be just another zinc deficiency sign or symptom.

TRACE ELEMENTS IN MILK FROM VARIOUS SPECIES

B. Lönnerdal, C.L. Keen, and L.S. Hurley

Department of Nutrition
University of California
Davis, CA 95616, USA

The important role of trace elements such as iron, copper and zinc in the development of the neonate has been demonstrated in many species (1). Deficiencies of these nutrients can lead to high mortality, abnormal development, and poor growth. As the only food of the newborn mammal is milk, its capacity to provide sufficient amounts of trace elements to meet the needs of the neonate is crucial. We have analyzed the milk of several species with regard to the concentrations of these trace elements and their changes during lactation (2-4). The iron content of human milk decreases from 0.6-0.8 µg/ml in the early part of lactation (colostrum) to around 0.2 µg/ml in mature breast milk. Similar patterns were found in rat and dog milk, although the total concentration of iron was considerably higher in these species (5 and 6 µg/ml in mature milk, respectively). Copper in human milk also decreased during the lactation period, from 0.8 µg/ml to 0.2 µg/ml. Rat milk exhibited a pattern similar to human milk, although values were approximately ten times higher. Copper in dog milk did not exhibit a pronounced declining pattern; values were about 1 µg/ml throughout lactation. Zinc concentrations decreased significantly during the lactation period for all three species. In this study we have investigated the localization of these elements in different fractions of the milk.

EXPERIMENTAL DESIGN

Colostrum and mature milk from 9 species (human, cow, goat, sheep, horse, pig, rat, dog, and cat) were collected into acid washed plastic vials and stored frozen until analysis. A portion of each sample was analyzed for Fe, Cu and Zn concentration by flame atomic absorption spectrophotometry (Perkin-Elmer 370) following wet digestion (16 N nitric acid) of the samples. A second portion was fractionated into fat, casein and whey by ultracentrifugation at 140,000 g for 1 h at 4°C. The fat layer at the top of the centrifuged sample was transferred with a spatula into an ashing flask, skim milk was decanted into a second flask and the remaining casein pellet was transferred into a third flask. All samples were wet ashed and trace elements determined. Relative distribution among fractions is expressed as % of total content of each sample. Whey samples were subjected to gel filtration chromatography on Sephadex G-50 to separate trace element binding complexes in the whey. The gel was treated with $NaBH_4$ to reduce surface changes originally present on the gel (5). The column was equilibrated in 0.1 M ammonium acetate buffer at pH 6.5 and fractions analyzed for trace element content. Whey samples were also analyzed for citrate concentration using a spectrophotometric assay.

RESULTS AND DISCUSSION

The total content of Fe, Cu and Zn in colostrum and mature milk is shown in Table 1. As can be seen, there is a large variation in Fe content among species. It can be noted that the Fe content of mature milk was lower than that of colostrum for all species. Concentrations of Cu were also found to vary among species, with rat, pig, dog and cat having the highest values. In all species except the dog, Cu concentrations were lower in mature milk than in colostrum. In general, the concentration of Zn in milk was highest among the three elements investigated. In all species, except the cat, Zn content was found to be significantly lower in mature milk than in colostrum.

Iron in milk was found in all three fractions (fat, casein, whey) analyzed, regardless of species (Table 2). For each species there was a large variation in the amount of Fe in each fraction, with the fat fraction seeming to have the lower amount. Fe in human milk fat has been shown to be bound to proteins in the fat globule membrane (6) and it is reasonable that this is true for other species as well. With the exception of human milk, the majority of milk Fe was found in the casein fraction. It has been demonstrated that Fe in cow's milk casein is bound to phosphoserine groups (7). The majority of Cu in milk from all species, except human and pig, was found in the casein fraction. In human

Table 1. Concentrations of Fe, Cu and Zn in Milk

Species	Fe (μg/ml)		Cu (μg/ml)		Zn (μg/ml)	
	CM	MM	CM	MM	CM	MM
Human	0.8	0.2-0.4	0.6	0.2-0.3	5.7	0.7-1.6
Cow	0.2	0.1-0.2	0.3	0.1-0.2	7.2	3-5
Goat	1.4	0.3-0.5	0.4	0.1-0.2	13.1	5-6
Sheep	0.5	0.2-0.5	1.4	0.2-0.3	13.6	1-2
Horse	1.3	0.3-1.0	0.6	0.2-0.4	2.4	1-2
Pig	1.3	1-2	6.3	0.5-1.0	10.7	4-6
Rat	6.6	5-10	6.1	1-2	16.8	10-14
Dog	10.7	2-8	1.6	0.8-1.3	16.2	7-8
Cat	4.1	1-4	0.5	0.6-0.8	6.7	5.6

CM: colostrum milk MM: mature milk

Table 2. Distribution of Fe, Cu and Zn in Milk

Species	Fat (%)			Casein (%)			Whey (%)		
	Fe	Cu	Zn	Fe	Cu	Zn	Fe	Cu	Zn
Human	19-26	20-28	30-41	2-14	1-9	2-10	65-81	45-57	45-58
Cow	13-18	3-11	10-26	61-73	59-80	48-60	9-15	11-35	18-31
Goat	12-20	1-13	2-3	33-49	38-78	60-73	28-51	31-38	11-33
Sheep	20-32	2-8	8-20	37-49	43-50	55-75	25-36	38-44	18-23
Horse	10-43	7-17	5-10	30-80	43-78	49-71	5-33	7-33	21-33
Pig	18-43	38-47	10-19	32-52	30-39	71-86	35-50	20-25	10-15
Rat	3-25	2-9	2-10	34-42	50-64	63-76	17-65	20-50	13-21
Dog	5-19	2-6	2-3	63-79	54-76	65-74	6-26	22-32	13-25
Cat	2-18	1-2	1-5	4-68	53-76	51-84	40-85	20-41	4-37

Table 3. Citrate Content of Whey

Species	Citrate Content (mM)	Low Molecular Weight Zinc
Human	0.5 - 1.4	+++
Cow	1.4 - 2.1	+
Goat	3.4 - 5.7	+
Sheep	3.6 - 5.5	+
Horse	1.9 - 2.6	++
Pig	2.0 - 2.4	++
Rat	0	--
Dog	4.6 - 6.2	+++
Cat	5.1 - 5.9	+++

and pig milk, a considerable amount of Cu was found in the fat fraction. The major proportion of Zn in milk from most species was found in the casein fraction, with the lowest amount being found in the fat fraction. However, human milk had a low amount of Zn in the casein fraction. This may be due to the low casein:whey ratio found in human milk (8).

These data demonstrate that there is wide variation among species in regard to compartmentalization of trace elements. In particular, the distribution in human milk differed considerably from all other species investigated. To answer the question whether variation exists among different species within each fraction, we have examined the localization of Zn in whey as a model.

We have earlier shown that a significant part (30%) of the Zn in human milk is bound to a low molecular weight (LMW) ligand, namely citrate (9). In all species examined, except the rat, a LMW Zn complex was found by gel filtration chromatography (Table 3). In all of these species citrate was present in significant amounts. However, the rat, which does not have LMW Zn, did not have any citrate in the whey. It should be noted that the proportion of Zn bound to citrate is not proportional to the amount of citrate present. Other binding factors, such as casein and other high MW binding compounds and their relative proportions will influence the distribution of the elements in milk.

SUMMARY

This work demonstrates three factors that must be considered when evaluating data on sucklings from different species: (1) the concentration of a trace element can be significantly different from one species to another; (2) the distribution of a trace element among the major milk fractions can vary between species that may have similar total amounts of the element; (3) within a given fraction the molecular localization of an element can be different even if the concentration is similar. Knowledge of these differences among species and elements will allow us to investigate further and to understand trace element metabolism in the neonatal period.

REFERENCES

1. Hurley, L.S. 1980. In "Developmental Nutrition" (ed. L.S. Hurley) (Prentice-Hall, N.J).

2. Lönnerdal, B., M. Clegg, and L.S. Hurley, 1981, Am. J. Clin. Nutr. 34:640.

3. Keen, C.L., B. Lönnerdal, M. Clegg, and L.S. Hurley, 1981, J. Nutr. 111:226 - 230.

4. Lönnerdal, B., C.L. Keen, L.S. Hurley, and G.L. Fisher, 1981, Am. J. Vet. Res. 42: 662 - 666.

5. Lönnerdal, B. 1980. In "Trace Element Analytical Chemistry in Medicine and Biology" (eds. P. Brätter, and P. Schramel) (Walter de Gruyter, Berlin). p439.

6. Fransson, G.-B., and B. Lönnerdal, 1980, Näringsforskning 24:66 - 67.

7. Hegenauer, J., P. Saltman, D. Ludwig, L. Ripley, and A. Ley, 1979, J. Agric. Food Chem. 27:1294 - 1301.

8. Lönnerdal, B., and E. Forsum 1980. In "Nutrition in Europe", Proceedings of the Third European Nutrition Conference (ed. L. Hambraeus) (Almquist & Wiksell , Stockholm) p 139.

9. Lönnerdal, B., A.G. Stanislowski, and L.S. Hurley, 1980, J. Inorg. Biochem. 12:71 - 78.

DISCUSSION

O'DELL (USA). I was intrigued by the distribution of both copper and zinc in the milk fractions particularly in the rather high amount that was associated with the lipid. Do you have any comments about how zinc is bound in lipid and how much protein might be in the lipid fraction.

LÖNNERDAL (USA). The amount of protein in the fraction is very small - in fact if you do a Kjeldahl determination of the whole milk and the skim milk we cannot see any significant difference. However, the trace element concentration is quite significant as you said. We have investigated the milk fat by carefully solublizing the membrane surrounding the triglyceride core of these globules and it is obvious that the outer membrane, the more hydrophilic component of the fat globule membrane, contains these trace elements and we have also evidence that they are bound to proteins and to my knowledge most of them are enzymes. We do not have a very good candidate for copper yet, but for iron and zinc it seems to be xanthine oxidase and alkaline phosphatase respectively. These enzymes have been reported to be present in the fat globule membrane.

GOLDEN B E (Jamaica). When you express your results as percent of total it seems that there is a tremendous amount in the lipid fraction, but in the human milk the total content of the zinc and copper was much lower. Is the absolute amount of zinc in the lipid fraction very much different in human milk than in the other milks?

LÖNNERDAL (USA). The absolute amount in the milk fat is comparable to other species, but relatively it is much higher.

MOMČILOVIĆ (Yugoslavia). I just want to ask you if you have found that the total volume of milk changes during the lactation period so at the late phase of lactation you would have a greater volume of milk produced. Did you correlate your data on concentration of zinc in given volumes with the total milk volume produced at the different stages of lactation?

LÖNNERDAL (USA). We have not tried to correlate it for two of the species, the rat and the dog because we do not have any good data yet on the amount of milk produced. But in the human case we have studied the volume produced and after about 20 or 30 days we do not see a significant increase in the volume produced during 24 hours using weighing before and after feeding for a 24 hour period. In that case it is not a very significant increase in volume produced.

ADBULLA (Sweden). Since there is a lot of confusion regarding zinc binding ligands I wonder if you could comment on the high molecular weight fraction?

LÖNNERDAL (USA). The high molecular weight fraction is the proteins in the milk. As I said casein is one of the factors binding zinc and copper and we have identified another protein which binds zinc and copper, namely serum albumin, which binds these elements in serum, and those are the proteins among the high molecular weight compounds. There has been a confusion about the low molecular weight ligand binding zinc and in support of our results other groups confirm that citrate is in fact the ligand binding zinc in milk. There are other potential ligands that may bind zinc but they don't do it in the milk because their concentrations are so much lower than that of citrate.

ZINC-BINDING LIGANDS IN HUMAN MILK

S.V. Rumball, E.W. Ainscough, A.M. Brodie, J.E. Plowman, H.M. Baker and H.J.L. Brooks

Departments of Chemistry, Biochemistry & Biophysics and Microbiology
Massey University
PALMERSTON NORTH
NEW ZEALAND

Specific zinc-binding ligands (ZBL) present in human milk but not in bovine milk, may be responsible for the higher bioavailability of zinc in human milk and hence its therapeutic value in alleviating the symptoms of acrodermatitis enteropathica (AE), a hereditary disease related to impaired zinc absorption (1). ZBL may also be involved in zinc absorption in infants during the neonatal period (2). In defatted human milk a significant fraction of the zinc is associated with low molecular weight ZBL while zinc in bovine milk is bound almost exclusively to high molecular weight compounds (1). Although macromolecules such as casein bind zinc ions, these macromolecules are generally degraded in the stomach and small intestine after digestion. The high molecular weight fraction is thus regarded as less significant in the transport of zinc ions into the cells of the gastrointestinal tract.

Various attempts have been made to characterise the low molecular weight ZBL of human milk. It has in turn been identified as prostaglandin E_2 (3), picolinic acid (4) and citrate (5). The most recent suggestion is that ZBL of human milk is a heterogeneous group of compounds, each with its own zinc-binding characteristics (6). This is based on evidence that the distribution of zinc between low and high molecular weight fractions of fat-free human milk is influenced by the total zinc content of the milk and possibly by the total amount and type of protein present. It seems that zinc-binding sites on higher molecular weight ligands must first be saturated before the zinc-binding sites on low molecular weight ligands are used (6).

Zinc and protein content vary both between species and with the stage of lactation. The zinc content of human colostrum has been reported as high as 20 μg/ml (7) decreasing to 0.14-3.95 μg/ml in mature milk (8). There is considerable variation between individuals. Zinc content of mature bovine milk is around 2 μg/ml (9). The protein content of colostrum is about twice as high as that of mature human milk (10). Mature human milk has a lower total protein concentration than bovine milk and a lower casein: whey protein ratio. Analysis of the whey proteins shows that the major species are different in human and bovine milks. The prominent human whey proteins are lactoferrin and α-lactalbumin, whereas β-lactoglobulin is the prominent bovine whey protein.

Lactoferrin is a single chain protein (M.W. ∿ 80,000) capable of binding two atoms of ferric iron per molecule of protein (11). In samples of colostrum collected in this area only about 10% of the iron-binding capacity is used. Higher saturation values are often implied. Lactoferrin is present in colostrum at a concentration of around 3.5 mg/ml decreasing to about 1.5 mg/ml in mature milk (10). The concentration in mature bovine milk is in the region 0.02-0.2 mg/ml (12). Lactoferrin has also been found in a variety of other bodily secretions (13) including the luminal surface of the small intestine (14) and in intracellular components (15). Because of its avidity for Fe(III) it has been postulated to have a bacteriostatic function in depriving microorganisms of essential iron required for their growth (16). A nutritional role as an iron carrier has also been suggested (17), particularly in view of its ability to donate iron to human intestinal tissue (18) and its ability to withstand degradation by digestive enzymes (18).

In the course of our studies on lactoferrin we have become interested in the possibility of lactoferrin acting as a carrier for trace elements other than iron. We have therefore investigated the association of zinc with lactoferrin in human milk (19). We have also begun preliminary investigations on the effect of zinc on bacterial growth in human milk. Zinc, like iron, is an essential element for microbial metabolism and competition for zinc may be fundamentally important in host/parasite interactions (20). Since it is not known at present whether microorganisms can produce zinc sequestering agents (analogous to the bacterial siderophores which sequester for iron), the binding of zinc in human milk by zinc-binding ligands might render the element unavailable for microbial metabolism in a manner similar to that of the iron/lactoferrin system.

EXPERIMENTAL

Lactoferrin was isolated from fresh human milk collected during the first week of lactation, using methods previously published (19). Final purification was achieved by chromatography on CM-Sephadex C-50 eluting with phosphate buffer (pH 7.6, 0.01 M) and a NaCl gradient (0 to 0.5 M). Protein concentration in the various fractions was estimated from the 280 nm absorbance (19) and the zinc concentration was estimated by atomic absorption spectrometry. Uptake of iron by lactoferrin in the presence of two moles of zinc per mole of lactoferrin was investigated by following the increase in absorbance at 465 nm as the characteristic salmon pink colour develops (19).

RESULTS

A typical elution curve (Fig. 1) shows the zinc concentration to maximise with the lactoferrin concentration. The tailing of the zinc elution curve after the lactoferrin maximum indicates the zinc is not tightly bound as does the fact that the zinc/lactoferrin ratio drops upon ultrafiltration. The zinc passes through the diaflow membrane while the protein is retained and concentrated. An average value of 0.33 moles of zinc per mole of lactoferrin is observed. The presence of zinc does not impair the iron binding function of lactoferrin but in fact enhances the uptake rate. The time required to half saturate the iron binding site with iron (added as a 1:1 iron cirate complex) is about 5 min at 25 C, whereas in the absence of zinc it is about 12 min. Whereas fresh human colostrum was bacteriostatic to the growth of an enteropathogenic strain of *Escherichia coli*, initial results suggest an enhancement of growth on addition of zinc (as zinc sulphate).

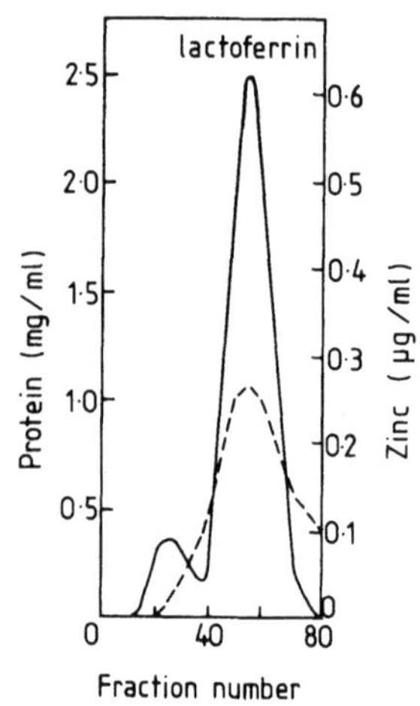

FIG. 1. Elution of zinc (---) and protein (——) from CM-Sephadex C-50.
Eluent: phosphate buffer pH 7.6, 0.01 M, 0 to 0.5 M NaCl gradient.

DISCUSSION

The fact that a proportion of zinc is found loosely bound to lactoferrin suggests that lactoferrin may be a high molecular weight ligand which must be saturated with zinc before the low molecular weight ZBL is used (6). Since lactoferrin is found intact in the intestine, it may therefore play a role in transfer of zinc from human milk to intestinal cells, particularly from colostrum where the lactoferrin concentration is high. Possibly, in view of the presence of lactoferrin in intestinal secretions it may also play a more general role in zinc transport in a manner similar to that of the related protein, transferrin, which although predominantly a serum iron-transport protein, is also known to transport zinc in the portal plasma (21). Binding constants for zinc and lactoferrin are not yet available so that no comparison can be made between the avidity for zinc of lactoferrin and the various low molecular weight ligands proposed. Since the concentration of lactoferrin and possibly the low molecular weight ZBL varies depending on the stage of lactation, the distribution of zinc amongst the various ligands will presumably also alter as lactation proceeds.

The apparent enhancement of bacterial growth in human milk on the addition of free zinc suggests that the amount of free zinc present in human milk may be insufficient for microbial metabolism. However, if the concept that iron is a limiting factor is true, this growth enhancement implies that the addition of zinc also renders iron more available (for bacterial metabolism); zinc and iron are involved in different metabolic pathways and it seems unlikely that zinc could act as a substitute for iron. The binding of zinc by zinc-binding ligands in human milk could limit the availability of free zinc in the same way as binding of iron by lactoferrin reduces the free iron concentration to levels insufficient to sustain microbial metabolism. The binding of zinc to lactoferrin may thus enhance bacteriostasis by lowering the concentration of free zinc (in concert with other zinc-binding ligands), and by enhancing the uptake of iron by lactoferrin.

REFERENCES

1. Eckhert, C.D., M.V. Sloan, J.R. Duncan, and L.S. Hurley, 1977, Science 195:789-790.
2. Duncan, J.R., and L.S. Hurley, 1978, Am. J. Physiol. 235:E556-E559.
3. Evans, G.W., and P.E. Johnson, 1977, Lancet 1:52.
4. Evans, G.W., and P.E. Johnson, 1979, Federation Proc. 38:2501.
5. Lönnerdal, B., A.G. Stanislowski and L.S. Hurley, 1980, J. Inorg. Biochem. 12:71-78.
6. Cousins, R.J., and K.T. Smith, 1980, Am. J. Clin. Nutr. 33:1083-1087.
7. Jelliffe, D.B. and E.F.P. Jelliffe, 1977. In "Human Milk in the Modern World" (O.U.P.) p. 51.
8. Picciano, M.F., and H.A. Guthrie, 1976, Amer. J. Clin. Nutr. 29:242-254.
9. Johnson, P.E., and G.W. Evans, 1978, Amer. J. Clin. Nutr. 31:416-421.
10. Lönnerdal, B., E. Forsum and L. Hambraeus, 1976, Amer. J. Clin. Nutr. 29:1127-1133.
11. Querinjean, P., P.L. Masson, and J.F. Heremans, 1971, Europ. J. Biochem. 20:420-425.
12. Masson, P.L., and J.F. Heremans, 1971, Comp. Biochem. Physiol. 39B:119-129.
13. Masson, P.L., J.F. Heremans and C. Dive, 1966, Clin. Chim. Acta 14:735-739.
14. Masson, P.L., J.F. Heremans, E. Schonne and P.A. Crabbe, 1968. In "Protides Biol. Fluids" (ed. H. Peeters) (Pergamon Press, Oxford) pp 633-638.
15. Masson, P.L., J.F. Heremans, and E. Schonne, 1969, J. Exp. Med. 130:643-658.
16. Bullen, J.J., H.J. Rogers, and E. Griffiths, 1974. In "Microbial Iron Metabolism" (ed. J.B. Nielands) (Academic Press, N.Y.) Chapter 20.
17. McMillan, J.A., S.A. Landaw and F.A. Oski, 1976, Pediatrics 58 (5):686-691.
18. Cox, T.M., J. Mazurier, G. Spik, J. Montreuil, and T.J. Peters, 1979, Biochim. et Biophys. Acta 588:120-128.
19. Ainscough, E.W., A.M. Brodie, and J.E. Plowman, 1980, Amer. J. Clin. Nutr. 33:1314-1315, and references therein.
20. Failla, M.L. 1977. In "Microorganisms and Minerals" (ed. E.D. Weinberg) (Marcel Dekker, N.Y.) Chapter 4.
21. Evans, G.W., and T.W. Winter, 1975, Biochem. and Biophys. Res. Comm. 66(4):1218-1224.

DISCUSSION

LÖNNERDAL (USA). I have a couple of comments to make here. The first one regarding the presence of binding compounds in the milk. It is true that Cousins and Smith reported that there are a lot of compounds that can bind zinc but I do not think that we should give up just because of that fact. The binding constants and the concentration of each compound are going to govern the distribution. We are not going to have a very fluid situation, we are going to have a distribution of zinc to the various binding compounds and I think it is possible for us to determine this distribution. I agree with you that a lot of this confusion emanates from methodological problems. We tried to repeat the study you did with the ion-exchange chromatography of lactoferrin and zinc and we got the same results you did but zinc and lactoferrin are not in fact associated with each other. You said there was a very loose association but if you make an ultrafiltration of zinc and lactoferrin these compounds will part and the association must be very low if any. In fact what I think is happening is that you have an ion-exchanger that is picking up everything that is positively charged, that is the lactoferrin which is a fairly high isoelectric point compound. It would also pick up zinc from the buffer even if you have a supposedly zinc-free buffer, so you will have zinc accumulating to the column. In fact when we repeat the experiment without applying a sample on the column we get a zinc peak when the column is eluted with a high ionic strength buffer - so I think that this is a methodological problem, but not of the sort we were talking about before. In fact, if you use an anion exchanger of a cation-exchanger lactoferring will pass through and the zinc will not be associated with lactoferring at all. We have found as I said before that zinc is associated with serum albumin in human milk which is consistent with the metal-binding properties of this protein.

RUMBALL (New Zealand). Thank you Dr Lönnerdal for sharing your results. The literature has certainly made it clear that there is a great need for caution interpreting results obtained by these techniques.

MAY (Wales). Can I ask you in what form you administered the iron to lactoferrin? I think that with transferrin it has been shown that the complex that you use is very important. Did you administer the iron as ferrous chloride or a citrate or NTA?

RUMBALL (New Zealand). We use iron NTA.

THE ZINC STATUS OF BREAST AND BOTTLE FED INFANTS

L. MacDonald, R.S. Gibson, and D. Shapcott*

Applied Human Nutrition, Department of Family Studies
University of Guelph, Guelph, Ontario, Canada

*Human Nutrition Research Unit, Faculty of Medicine
University of Sherbrooke, Sherbrooke, Quebec, Canada

The most frequently used parameters to monitor zinc status during early infancy, have been serum, plasma and hair zinc levels (1,2). However, few previous investigators have calculated dietary zinc intakes and distinguished between breast and artificial milk formulas. This is unfortunate in view of the recently recognized superior bioavailability of zinc from human milk compared to cows' milk based formulae (3,4). Clearly these differences may have significant effects on the zinc status of breast vs. bottle fed infants.

Therefore, this longitudinal study compared the zinc status of breast and bottle fed infants during the first six months of infancy, using hair zinc levels, and dietary zinc intakes as the principal indices.

SUBJECTS AND METHODS

Thirty-five breast fed infants (20 males and 15 females) and 25 bottle fed infants (14 males and 11 females) from Guelph, Ontario, Canada, were included in the study. All infants were healthy and full term (38-42 weeks gestation) with mean birthweight 3509 ± 278 g. Breast fed infants were those whose only source of milk from birth, was from nursing. Bottle fed infants received only proprietory formula or cows' milk by bottle, from birth.

Dietary zinc intakes were calculated from 3-day food records collected at 30 ± 2, 90 ± 4 and 180 ± 4 days after birth, using company product information, baby food composition data from the newly revised USDA Handbook No 8 (5), and the literature. A small sub-sample of breast fed infants - 25%, were test weighed at each month, by the mothers, at each feeding, for one day. These results were used to assess the mean daily volume of breast milk consumed by the breast fed infants, at one, three, and six months of age. Hair samples, cut close to the occipital region of the scalp, 1-2 cms long, were collected at 30±2, 90±4 and 180±4 days of age, washed, and analysed by instrumental neutron activation procedures (6). Weight (to the nearest 25 g) and supine length of infants (to nearest 0.1 cm) were also recorded at monthly intervals.

RESULTS

Mean daily intakes of zinc and mean intakes per kg/body weight are shown in table 1. Dietary zinc intakes increased only slightly with age. The bottle fed infants received significantly higher zinc intakes at all ages ($p < 0.01$) than the breast fed group. This is not surprising as the majority of the bottle fed infants received milk formula supplemented with zinc (5.6 mg/L) and were introduced to non-milk foods earlier than the breast fed infants.

The breast fed group failed to meet the Canadian Dietary Standard (CDS) and Recommended Dietary Allowance (RDA) for zinc at all ages. In contrast, the bottle fed group met the CDS and RDA at all times. Milk remained the principal dietary source of zinc for breast and bottle fed groups throughout the period of study. By six months of age, non-milk foods supplied 14% and 20% of total dietary zinc for the breast and bottle fed groups, respectively.

Table 2 presents median hair zinc concentrations for the two groups. A decline in median hair zinc concentrations from one to six months was most marked in the male bottle fed infants. Application of the non-parametric sign test (table 3) revealed that only the male bottle fed infants showed a significant decline in hair zinc levels from one to six months ($p=0.004$). No significant changes in hair zinc concentrations were apparent in the female bottle fed infants or the male and female breast fed infants.

Table 1. Zinc Intakes (mean ± S.D.)

Group	Age	n	Zinc mg/day	Zinc mg/kg
Breast Fed	1 month	34	1.9 ± .2	.44 ± .06
	3 months	27	2.1 ± .3	.35 ± .04
	6 months	16	2.7 ± .5	.35 ± .07
Bottle Fed	1 month	25	3.6 ± .6	.82 ± .11
	3 months	25	4.2 ± .7	.69 ± .12
	6 months	23	4.6 ± .7	.58 ± .11

(n=number of infants)

Table 2. Hair Zinc Concentrations

Group	Age	Sex	n	Zinc ug/g Median	Zinc ug/g Quantiles 25th - 75th
Breast Fed	1 month	M	16	147	125 - 175
		F	13	153	127 - 253
	3 months	M	11	153	74 - 221
		F	12	162	109 - 253
	6 months	M	5	136	115 - 163
		F	7	177	153 - 344
Bottle Fed	1 month	M	13	134	112 - 155
		F	10	173	136 - 193
	3 months	M	9	105	82 - 153
		F	10	167	114 - 181
	6 months	M	10	105	64 - 119
		F	9	163	137 - 182

(n=number of infants)

Table 3. Hair Zinc Changes From One to Six Months

	Number of infants showing increasing hair zinc levels.	Number of infants showing decreasing hair zinc levels.	P
Male bottle fed infants	0	9	0.004
Female bottle fed infants	5	4	NS
Male breast fed infants	2	2	NS
Female breast fed infants	4	2	NS

DISCUSSION

The decline in hair zinc levels in the male bottle fed infants was not correlated with growth velocities of these male infants. No comparable systematic decline in hair zinc concentrations was apparent in the female bottle fed infants. This effect of sex on hair zinc concentrations has been noted by others (2,7) and provides further support for the suggestion that the requirement for zinc at this stage of development is sex dependent.

The absence of any significant decline in hair zinc levels for the breast fed infants is attributed to their adequate intake of readily available zinc. Even at six months, these infants received 86% of their total dietary zinc from breast milk. Hence the findings of this study provide further support to the conclusion that the bioavailability of zinc from breast milk is superior to that of cows' milk based formula, presumably due to the presence of a low molecular weight zinc binding ligand in breast milk (4). It is also of interest that the female bottle fed infants received sufficient available zinc from the zinc supplemented formula to prevent a significant decline in hair zinc levels.

The absence of any significant decline in hair zinc levels in both male and female breast fed infants and female bottle fed infants suggests that the fall in hair zinc levels noted during early infancy in previous studies (2,7), may be associated with sub-optimal zinc nutrition and may not represent entirely a physiological decline with age.

REFERENCES

1. Ohtake, M., 1977, Tohuku J. Exp. Med. 123:265 - 270.
2. Gibson, R.S., and DeWolfe, M.S., 1980, Nut. Rep. Internat. 21:341 - 349.
3. Momcilovic, B., Belonge, B., Giroux, A., and Shah, B.G., 1976, J. Nutr. 106:913 - 917.
4. Eckhert, C.D., Sloan, M.V., Duncan, J.R., and Hurley, L.S., 1977, Science 195:789 - 790.
5. Agriculture Handbook No. 8-3 (Rev. 1976), Composition of Foods: Baby Foods, Raw-Processed and Prepared. Agriculture Research Service, Consumer and Food Economics Institute, USDA, 1977.
6. Gibson, R.S., and DeWolfe, M.S., 1979, Pediatr. Res. 13:959 - 962.
7. Walravens, P.A., Hambidge, K.M., 1976, Am. J. Clin. Nutr. 29:1114 - 1120.

DISCUSSION

LÖNNERDAL (USA). How was the zinc concentration on the breast milk determined?

SHAPCOTT (Canada). From the tables in the USDA Handbook for human breast milk.

LÖNNERDAL (USA). That is what I thought, because if you compare the results you had here with the data we have on the concentration of zinc in breast milk it gives your results even more strength because the zinc intake of the formula fed infants would be 5-6 times higher than the breast fed infants rather than 1.7-2. I think those tables are over-estimating the zinc content in breast milk.

KRISHNAMACHARI (India). Your observation that the hair Zn levels are sex dependent is very interesting because when we did the serum profile of breast fed infants 3-6 months of age there was no sex difference at all.

MOMČILOVIĆ (Yugoslavia). I just wonder which type of statistical test you used to observe the significance in hair zinc changes from one to six months because the groups were small.

SHAPCOTT (Canada). This was a rank sum test.

JAMESON (Sweden). Those boys with low zinc intake and low hair zinc - did they behave differently than their comrades?

SHAPCOTT (Canada). These were all perfectly healthy children, there was no obvious difference in any of them. Physical growth was comparable for all of the children. There was no search made for physiological differences.

SUTTLE (Scotland). Is there a sex difference in the rate of hair growth at this age?

SHAPCOTT (Canada). I really could not say because the rate of growth is so variable, some of these children didn't have enough hair to sample. I don't think that there is any, and I have not heard of any information on hair growth at this age.

HURLEY (USA). I would like to make a comment on the remark in the paper concerning the recommended dietary allowance for zinc for infants. You made the statement that none of the breast fed infants reached the dietary recommendations for zinc. I think it is important to remember that these dietary recommendations for infants are really based upon the analysis of breast milk, and then since it isn't known what the availability of all the nutrients would be from other materials that were not breast milk a safety factor was added in the usual manner. Thus there is no way that a breast fed infant can ever reach the dietary intake of the recommended allowances, and I think that is something that we tend to forget.

SHAPCOTT (Canada). Yes, I think that this study re-enforces the old saying it is very true that the "breast is best".

TRACE ELEMENT CONTENT OF BREAST MILK OF INDIAN WOMEN

Rajalakshmi Krishnamachari

National Institute of Nutrition
Indian Council of Medical Research
Jamai-Osmania (P.O.) Hyderabad - 500 007, A.P.
India

INTRODUCTION

There is now growing evidence of the importance of trace elements in human nutrition especially during infancy. Breast milk plays a very important role in the infant nutrition in the developing countries, like India, where it is almost the only source of food available to majority of infants during their first year of life. In view of this fact, it is very essential to know the trace element content of human milk. However, there is paucity of information regarding the trace element content of breast milk of Indian women. Hence a study was undertaken to obtain such data, as well as to study the influence of various factors such as maternal age, parity, income and nutritional status and the duration of lactation on the composition of trace elements in milk. An attempt was also made to study the serum profile of trace elements of some of the breast-fed infants.

MATERIAL AND METHODS

Milk samples were collected from 412 women attending a government hospital in Hyderabad and from 222 women living in villages near Hyderabad, and also from 100 women attending a private nursing home in the city. Serum and milk samples were collected from 152 women of low income group attending the hospital. 41 paired samples of mothers' milk as well as their infants' serum were collected from the hospital.

Milk samples were collected directly from the breast into polyethylene containers after local cleaning of the breast, 2 hrs after the last feed. Blood samples were collected by disposable polyethylene syringes with stainless steel needles. Blood samples from the infants were collected by polyethylene scalpvein needles.

The milk and serum samples were analysed for copper and zinc by Varian Techtron-1000 flame Atomic Absorption Spectrophotometer, after suitable dilution. Molybdenum, cobalt, chromium and manganese levels were determined by lyophilizing and concentrating the samples to 5 times and by using a flameless, Varian Techtron Carbon-Rod atomizer, Model AA6, CRA-90.

RESULTS

Fig 1 and Table 1 show the results of cross-sectional studies. There was a steady decline in the levels of all trace elements from the period of colostrum to that of nearly 12 months. The levels were the highest in colostrum. Copper levels fell from a concentration of 0.46 μg/ml to 0.17 μg/ml by 7-12 months and zinc levels fell from 5.32 μg/ml to 1.16 μg/ml by 12 months. The levels were stabilised after 12 months. The levels of molybdenum, chromium and cobalt showed a similar trend to that of copper and zinc; chromium levels were stabilised by 4-6 months. However, manganese levels were the same in colostrum as well as in mature milk. (Table 2)

There were no differences in the concentrations of copper and zinc in the milk of rural and urban low income group of women. (Table 3)

Significant differences were seen in the concentrations of copper and zinc in the milk of women from low and high income groups. (Figure 1)

Concentrations of copper and zinc in mothers' milk did not correlate with those in serum thereby showing that there is an active transfer of these elements from milk. (Table 4)

The calculated daily intake of copper and zinc by these breast-fed infants was very low compared to the R.D.A. Hence it was decided to determine the serum concentrations of copper and zinc in these infants. 41 paired samples of mothers' milk and blood samples of their infants of 3-6 months of age were analysed. The serum levels of copper and zinc of these infants were within normal limits and there was no correlation between the concentrations in the mothers' milk and their infants' serum. (Table 5)

Analysis done to find out the relationship between the factors such as maternal age, parity and nutritional status to the concentrations of copper and zinc in milk showed that there was no association between these factors and the levels of copper and zinc in milk.

There were no significant day to day or diurnal variations in the concentrations of copper and zinc in milk, thereby showing that one sample can be representative of the contents of these elements in milk.

DISCUSSION

This study shows that duration of lactation plays an important factor influencing the concentrations of trace elements in milk. This finding is similar to that of other studies involving the major constituents of milk (1,2,3).

The absence of correlation between the concentrations in serum and milk clearly point out to a very specific and active type of transfer involved in this context. Though there are some reports available on the zinc binding ligands in milk, (4,5), these reports do not explain the transfer mechanism of these trace elements in milk.

The next important factor to influence the concentrations of these elements in milk seems to be the income group.

Even though the calculated daily intakes of copper and zinc by these infants were lower than the R.D.A. (6), the serum levels seem to be satisfactory and also these infants do not exhibit any clinical deficiencies, This report is similar to the report already available (7). Since there are no data available on the storage of these elements in the body tissues of Indian infants, it is very difficult to comment on this factor. However, as reported earlier (8), the bioavailability of zinc from human milk is better than that from others milks and this may play a role in maintaining the nutritional status of these infants.

REFERENCES

1. Belavady, B. and C. Gopalan, 1959. Chemical composition of milk. Ind. J. Med. Res. 47:234.
2. Gunthur, M. 1973. Breast feeding. In "Nutritional problems in a changing world" (ed. Dorothy Hillingworth and Margaret Russell). (Applied Science Publishers Ltd., London) pp 93-99.
3. Karmarkar, M.G., J. Kapur, A.D. Deodhar and C.V. Ramakrishnan, 1959. Studies on human lactation. Part 1. Ind. J. Med. Res. 47:344.
4. Hurley, L.S., J.R. Duncan, M.V. Sloan and C.D. Eckhert, 1977. Zinc binding ligands in milk and intestine. Proc. Natl. Acad. Sci. USA. 74:3547.
5. Eckhert, C.D., M.V. Sloan, J.R. Duncan and L.S. Hurley, 1977. Zinc binding: a difference between human and bovine milk. Science, 195: 789.
6. Recommended dietary allowances (8th ed.) Washington D.C., Food and Nutrition Board, National Research Council. Natl. Acad. Sci., 1974.
7. K. Michael Hambidge, Philip A. Walravens, Clare E. Casey, Ronald M. Brown and Connie Bender, 1979. Plasma Zinc concentrations of breast-fed infants. Jour. Ped. 94 (4) : 607-608.
8. Johnson, P.E., and Evans, G.W. Relative zinc availability in human breast milk, infant formulas and cow's milk. Am. J. Clin. Nutr. 31:416, 1978.

ACKNOWLEDGEMENT

The author wishes to thank Dr. P.G. Tulpule, Director, National Institute of Nutrition for permission granted for presenting the paper in the TEMA-4-symposium. Thanks are due to Dr. S.G. Srikantia for his guidance and support during the time this study was undertaken. To Dr. Devyani Dangoria for her co-operation in obtaining milk samples from high income group of women, and to Dr. K.A.V.R. Krishnamachari for obtaining milk samples from rural women. Also to Dr. K. Visweshwara Rao and Mr. G. Radhaiah for their help in the statistical analysis. The technical assistance rendered by Mr. M.M. Khan, Mr. K. Somaraju and Mr. Ch.V.B. Swamy is duly appreciated.

Figure 1.

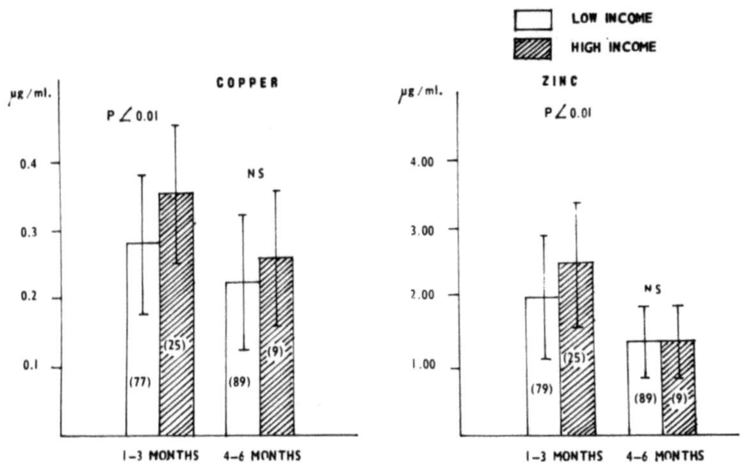

MEAN LEVELS OF COPPER AND ZINC IN MATURE MILK OF WOMEN OF LOW AND HIGH INCOME GROUPS VALUES ARE MEAN ± SD.

Table 1.

S.NO.	DURATION OF LACTATION	NO.	COPPER	NO.	ZINC
1.	COLOSTRUM	76	0.46 ± 0.019	75	5.32 ± 0.312
2.	TRANSITIONAL	31	0.50 ± 0.028	31	4.72 ± 0.274
3.	11 DAYS – 1 MONTH	28	0.45 ± 0.028	28	3.33 ± 0.265
4.	ONE – 6 MONTHS	166	0.25 ± 0.008	168	1.65 ± 0.061
5.	SEVEN – 12 MONTHS	88	0.17 ± 0.090	87	1.12 ± 0.051
6.	≥ THIRTEEN MONTHS	23	0.16 ± 0.014	23	1.16 ± 0.015
	TOTAL	412		412	

VALUES ARE MEAN ± S.E.

COPPER AND ZINC CONTENTS OF HUMAN MILK (µg/ml) CROSS SECTIONAL STUDY

Table 2.

DURATION OF LACTATION IN MONTHS	TRACE ELEMENTS (ng/ml)							
	MOLYBDENUM NO.		MANGANESE NO.		CHROMIUM NO.		COBALT NO.	
C & T	23	34.5 ± 5.09	25	8.2 ± 0.65	24	8.2 ± 1.81	24	95.8 ± 11.62
1 – 3	10	22.5 ± 4.93	9	6.6 ± 1.22	17	4.7 ± 0.88	13	27.9 ± 9.16
3 – 6	13	28.7 ± 4.37	12	7.9 ± 0.90	21	4.6 ± 0.82	16	16.4 ± 1.88
6 – 12	17	19.1 ± 1.79	22	7.0 ± 0.63	25	6.4 ± 1.46	22	11.6 ± 1.31
12 AND ABOVE	7	13.5 ± 2.32	10	4.7 ± 0.68	8	4.7 ± 1.03	8	9.4 ± 1.22
TOTAL...	70		78		95		83	

VALUES ARE MEAN ± SEM.

TRACE ELEMENT CONTENTS OF HUMAN MILK (CROSS SECTIONAL STUDY)

Table 3.

TRACE ELEMENT	DURATION OF LACTATION (mo)	n	URBAN	n	RURAL
COPPER	1 – 3	77	0.29 ± 0.010[a]	22	0.29 ± 0.023
	4 – 6	89	0.21 ± 0.010	41	0.21 ± 0.019
	7 – 12	88	0.17 ± 0.001	73	0.17 ± 0.010
	13 AND ABOVE	23	0.16 ± 0.014	86	0.15 ± 0.013
ZINC	1 – 3	79	2.00 ± 0.101	22	1.88 ± 0.164
	4 – 6	89	1.33 ± 0.055	41	1.32 ± 0.112
	7 – 12	87	1.12 ± 0.052	73	1.03 ± 0.067
	13 AND ABOVE	23	1.16 ± 0.116	86	1.03 ± 0.110

a VALUES ARE MEAN ± SEM.

MEAN LEVELS OF COPPER AND ZINC IN HUMAN MILK FROM URBAN AND RURAL AREAS (μg/ml)

Table 4.

DURATION OF LACTATION	1 – 6 mo.	7 – 12 mo.
n	67	38
TRACE ELEMENT		
COPPER IN MILK	0.29 ± 0.014	0.19 ± 0.016
COPPER IN SERUM	129.31 ± 3.035	127.32 ± 3.555
ZINC IN MILK	1.73 ± 0.123	1.14 ± 0.096
ZINC IN SERUM	118.10 ± 3.990	127.29 ± 5.256

VALUES ARE MEAN ± SEM.

	CORRELATION COEFFICIENTS	
COPPER IN MILK Vs. COPPER IN SERUM	– 0.19	– 0.26
ZINK IN MILK Vs. ZINK IN SERUM	– 0.12	0.13

RELATIONSHIP BETWEEN LEVELS OF COPPER AND ZINC IN MATURE MILK (μg/ml) AND SERUM (μg/100 ml)

Table 5.

MOTHERS' MILK		INFANTS' SERUM	
COPPER (μg/ml)	ZINC (μg/ml)	COPPER (μg/100 ml)	ZINC (μg/100 ml)
0.34 ± 0.029	1.74 ± 0.205	128.9 ± 4.97	109.8 ± 6.55

VALUES ARE MEAN ± S E M

CORRELATION COEFFICIENTS	'r' VALUE	
COPPER IN MOTHERS' MILK Vs COPPER IN INFANTS' SERUM	0.27	NS
ZINC IN MOTHERS' MILK Vs ZINC IN INFANTS' SERUM	0.154	NS

RELATION BETWEEN THE CONCENTRATIONS OF COPPER AND ZINC IN MOTHERS' MILK AND THEIR INFANTS' BLOOD (n = 41 pairs)

SOME ASPECTS OF THE EFFECTS OF ORAL CONTRACEPTIVES ON TRACE ELEMENT METABOLISM

Fiona J. Cumming*, M.H. Briggs* and J.J. Fardy**

* Division of Biological and Health Sciences,
Deakin University,
Geelong, Victoria, 3217,
Australia.

** Chemical Technology Division,
Australian Atomic Energy Commission,
Private Mail Bag, Sutherland, N.S.W.
Australia.

Many studies have demonstrated that oral contraceptive (O.C.) use by women leads to altered trace element status, in particular with respect to iron, copper and zinc (Margen and King, 1975). Recently, Kirksey *et al* (1979) reported that the long-term use of O.C. before pregnancy was associated with significantly lower manganese concentration in colostrum and milk on days 3 and 14 of lactation. However, little is known of the possible implications to trace element metabolism arising from the use of O.C. by lactating women.

The aim of this current investigation was to determine the effect of O.C. on iron, copper, zinc and manganese concentrations in the plasma and milk of lactating women and therefore, the possible influence on the nutritional status of the nursing infant. Unfortunately, due to an unscheduled shut-down of the nuclear reactor used for the trace element determinations, these results have not yet been obtained. In this paper, some of the evidence prompting initiation of the study and the methodology in use will be discussed.

A prospective study was begun on two groups of lactating women. Group I, the control group, did not use O.C. for the duration of the study, while women in Group II commenced using the progestagen-only type O.C. at about six weeks post-partum. From approximately four weeks post-partum, fasting blood and milk samples were obtained from women, along with a record of food ingested the day before the collection of samples. Samples were collected on a weekly basis, for a period of 20 weeks.

In an attempt to avoid possible variations in trace element composition of the milk with the stage of the feed, milk samples were collected 30 seconds after the let-down reflex was noted by the mother. Plasma and milk samples were stored at -20°C until analysis.

Due to the low concentrations of the trace elements, especially manganese, in plasma and mature human milk (Table 1), the most appropriate analytical method was neutron activation analysis (NAA), a highly sensitive technique performed at the Australian Atomic Energy Commission (A.A.E.C.) Research Establishment, Lucas Heights, N.S.W.

TABLE 1: APPROXIMATE TRACE ELEMENT CONCENTRATIONS IN WHOLE BLOOD AND MATURE HUMAN MILK (Iyengar, 1979).

(Units : blood in µmol/L, milk in µmol/kg or L).

	BLOOD	MILK
IRON	6266 - 8953	5.4 - 8.9
COPPER	10 - 20	4.7 - 6.3
ZINC	76 - 138	15 - 46
MANGANESE	0.04 - 0.91	0.36

Multi-element analysis of plasma and milk samples was possible using a radiochemical separation NAA technique (Nakahara *et al*, 1979), following 24-hour irradiation in the X6 facility of the HIFAR reactor at Lucas Heights in a neutron flux of 3×10^{12} $n.cm^{-2}.sec^{-1}$. The chemical separation method adopted was based on the selectivity of an organic chelate reagent, ammonium pyrrolidine dithiocarbamate (APDC) for metals which form soluble sulphides and are extracted by a variety of organic solvents.

Gamma-ray spectrometry measurements will be made using a 100ml coaxial Ge (Li) detector (Optec Model 8011-2019ST, FWHM 1.9 keV at 1.33 MeV) coupled to a 4096 channel pulse-height analyser. The spectra will be processed on an IBM 370 computer, using an A.A.E.C. derived program. The program employs a sensitive and specific search algorithm based on a smooth second differential for locating peaks.

Preliminary work using atomic absorption spectrometry (AAS) (Briggs, unpublished) has indicated that marked changes in plasma trace element concentrations occur with the use of the combined oestrogen-progestagen type O.C., while alterations with the use of the progestagen-only type O.C. (mini-pill) are minimal (Table 2).

TABLE 2: PLASMA TRACE ELEMENTS IN O.C.-USERS AND CONTROLS. (µmol/L)

ELEMENT	UNTREATED*	COMBINED O.C.*	MINI-PILL*
IRON	16.6 ± 4.5	24.1 ± 4.7 (+45%)	17.1 ± 4.8
COPPER	18.9 ± 2.3	34.4 ± 7.3 (+82%)	18.4 ± 3.2
ZINC	15.2 ± 2.1	12.6 ± 2.3 (-17%)	16.8 ± 2.4
MANGANESE	0.15 ± 0.04	0.21 ± 0.05 (+40%)	0.14 ± 0.04

* MEAN $\pm$ S.D. for 50 healthy young women in late cycle.

The raised plasma concentration of iron and copper following treatment with the combined O.C. is probably due to increased hepatic synthesis of the metalloproteins, transferrin and ceruloplasmin. Reduced haptoglobin synthesis also affects plasma iron. Table 3 shows the changes observed in plasma protein concentrations with O.C. use (Briggs & Briggs, 1979).

TABLE 3: CHANGES IN PLASMA PROTEIN CONCENTRATIONS WITH O.C. USE

PLASMA PROTEIN (g/L)	UNTREATED n=56	COMBINED O.C. n=36	MINI-PILL n=20
TRANSFERRIN	2.82 ± 0.56	4.59 ± 0.85	3.02 ± 0.82
CERULOPLASMIN	0.40 ± 0.06	0.81 ± 0.11	0.35 ± 0.09
HAPTOGLOBIN	1.10 ± 0.10	0.87 ± 0.19	1.01 ± 0.11

The results in Table 4 (Briggs, unpublished) indicate that the lower plasma zinc seen in combined O.C. users is due to a reduction in the free zinc rather than albumin-bound zinc. Sephadex G-100 was used to separate protein-bound zinc (fraction I) from unbound zinc (fraction II).

TABLE 4: CHANGES IN PROTEIN-BOUND AND FREE PLASMA ZINC CONCENTRATIONS WITH O.C. USE (μmol/L)

	I	II
UNTREATED	5.8 ± 1.3	11.8 ± 1.7
COMBINED O.C.	5.7 ± 1.5	7.0 ± 1.9
MINI-PILL	5.5 ± 1.3	11.6 ± 2.0

The reason for the increase in plasma manganese concentration with combined O.C. use is unknown, but again probably reflects changes in carrier proteins.

It is presumably the oestrogen component of the combined O.C. which leads to the observed changes in plasma trace elements, as the concentrations are not significantly altered by the progestagen-only O.C. Alterations in other biochemical parameters have also been attributed to the O.C. oestrogen dose (Ram and Bamji, 1979; Prasad *et al*, 1975).

In the area of the present study (Geelong, Victoria) the O.C. commonly used by lactating women is the progestagen-only type, although a minority are prescribed the combined type. For this reason it was considered necessary to study the effects of the progestagen-only O.C. on plasma and milk trace elements using the highly sensitive NAA technique, and later, the effects of the combined O.C. Even small alterations in the milk trace element concentrations could lead to altered trace element status in the nursing infants, requiring corrective measures.

The results arising from this study will be twofold; an extraction technique for use, with NAA in biological samples will be employed for the first time in this country, and nutritional implications of O.C. use during lactation will be elucidated.

1. Margen, S., and J.C. King, 1975, Amer. J. Clin. Nutr. 28:392-402.

2. Kirksey, A., *et al.*, 1979, Amer. J. Clin. Nutr. 32:30-39.

3. Iyengar, G.V., 1979. "Elemental Composition of Human and Animal Milk. A Review", (prepared for the International Atomic Energy Agency, Vienna) pp. 152.

4. Nakahara, H., *et al.*, 1979, J. Radioanal. Chem. 54:183-190.

5. Briggs M.H., and M. Briggs, 1979, J. Steroid Biochem. 11:425-428.

6. Ram, M.M., and M.S. Bamji, 1979, Amer. J. Obstet. Gynecol. 135:470-472.

7. Prasad, A.S., *et al.*, 1975, Amer. J. Clin. Nutr. 28:385-391.

DISCUSSION

WILLIAMS (Wales). Did you record the diet of the women?

CUMMING (Australia). I visited the women once a week, on the same day of each week, for 20 weeks. They kept a written record of everything they had eaten on the day before my visit, including measurements of the quantities of dietary components. I also took a record of their weekly food intake at the beginning and the end of the study.

VERSIECK (Belgium). You found a plasma manganese level of 0.15 μ mol/l in healthy young women in the late phase of their menstrual cycle. We found similar concentrations initially, but after refining our sampling procedure, reducing contamination, we obtained values that were about ten times lower. This has been confirmed by several other investigators in different parts of the world, who find manganese concentrations of 0.01 μ mol/l.

LOW ZINC DIET: EFFECT ON PREGNANT AND LACTATING EWE

J. Apgar, W. A. House, and R. M. Welch
U.S. Department of Agriculture
U.S. Plant, Soil, and Nutrition Laboratory
ITHACA, NEW YORK 14853 U.S.A.

A low zinc (Zn) intake during only the last third of pregnancy in the rat leads to excessive bleeding, prolonged labor, and sometimes death during delivery (1). The severe effects of a low Zn diet on the pregnant rat occur even though the rat has a fair amount of Zn in the body. It is apparently unable to mobilize Zn rapidly enough to meet the demands of pregnancy. The rat, however, produces a large litter in a short period of time. Perhaps animals with a smaller conceptus size and a longer gestation are better able to use existing body stores to meet the demands of pregnancy. We are, therefore, interested in two questions. Do animals other than the rat require Zn in order to deliver normally? If other animals require Zn for normal parturition, are they more able than rats to meet the Zn requirement by utilizing existing body stores? These questions are particularly relevant to human pregnancy since studies of diets consumed by pregnant women indicate that many women, particularly those in lower socioeconomic classes, consume less than half the Recommended Dietary Allowance for Zn for pregnant women.

We used the ewe to study the questions of a need for Zn for normal parturition by animals other than the rat and the ability of other animals to use body stores of Zn to meet such a need. Conceptus size in sheep, approximately 8% of the ewe's weight, is more nearly comparable to the 6-8% in pregnant women; the conceptus size in rats reaches 25-30% of the weight of the female. Gestation time in sheep is also much longer, approximately 5 months compared to 3 weeks in the rat.

Nineteen pregnant ewes were fed a low Zn diet +100 g hay per day for the last third of pregnancy (2). The low Zn diet was not given sooner because rats given a low Zn diet early in pregnancy either fail to implant or resorb the embryos. Nine ewes were given 100 ppm Zn in the drinking water. There was no difference in weight gain during pregnancy between the +Zn and -Zn ewes (Fig. 1). One +Zn and 4 -Zn ewes died. The +Zn ewe died shortly before delivery was expected. One -Zn ewe died well before lambing was expected; 1 died during lambing; and 2 became ill shortly before lambing, survived for a couple weeks with force feeding, but never recovered and finally died. Two -Zn ewes and 1 +Zn ewe required assistance with delivery. There did not appear to be a gross difference in blood loss between the two groups.

Since 4 -Zn ewes died compared to 1 +Zn ewe, it would appear that lack of Zn for the last third of gestation had an adverse effect on the ewe's ability to deliver. We cannot draw a firm conclusion, however, since the sheep on the -Zn treatment had twice as many twins as did the +Zn ewes--7 -Zn and 4 +Zn ewes had twins. Ewes with twins are more subject to pregnancy disease than are ewes with singles so it is difficult to separate the effects of twinning from the effects of low Zn.

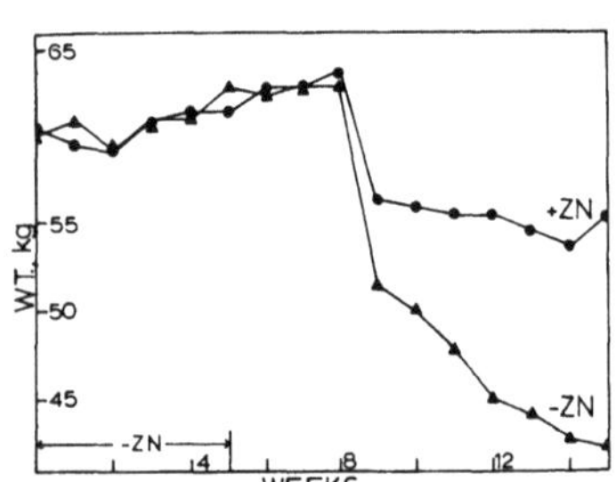

Fig. 1. Weight gain of pregnant ewes given low Zn diet with (●) or without (▲) Zn in drinking water.

The ewes were maintained on the treatments during lactation. The +Zn ewes maintained their weights after parturition whereas the -Zn ewes lost weight steadily (Fig. 1), which suggests that lack of Zn may well have contributed to the deaths in the -Zn group. By the third week of lactation there were clear signs of deficiency in the -Zn ewes--excessive salivation and scabbing and skin lesions on areas that were rubbed on the pen. Birthweight of the lambs was not different, but at 6 weeks the weights of twin lambs from -Zn ewes were significantly less than those of twins from +Zn ewes. No malformations were seen although malformations have been reported in rat pups from dams given a Zn-deficient diet (3).

Plasma Zn was consistently lower in the -Zn ewes than in +Zn ewes (4). Since weight gains were not different during pregnancy, the low plasma Zn at this time was apparently due to the low Zn intake and did not necessarily reflect Zn status. Although plasma Zn was significantly depressed during pregnancy, alkaline phosphatase activity did not differ significantly until just prior to parturition. It was significantly depressed throughout lactation. Since alkaline phosphatase was decreasing at the time that the ewes were losing weight and reached its minimum at the time that definite signs of Zn deficiency were apparent, it is possible that alkaline phosphatase activity reflected the depletion of mobilizable Zn from Zn stores. Food consumption was also declining during this time, however; since we did not have pair-fed controls, we do not know to what extent the reduced food intake contributed to the decrease in alkaline phosphatase activity.

Since we would like to use the ewe in future studies of the role of Zn in pregnancy, we have tried to improve the experimental conditions. We were interested specifically in 1) whether the ewes could be maintained on a semi-purified diet throughout pregnancy, 2) if incorporation of straw into the diet would improve feed consumption, 3) whether Zn deficiency could still be induced if straw were added, and 4) whether X-ray prior to day 100 would distinguish between ewes with single or multiple lambs. For other work in which we are interested, we also wanted to know what effect the incorporation of 1 1/2% phytate in the diet would have on the pregnant ewe.

Estrus was synchronized in approximately 15 Finn ewes. After breeding, 4 ewes were penned individually and fed to appetite a low Zn semi-purified diet (2) mixed with water to which Zn acetate was added. Average daily consumption of Zn from the water was 10 mg. Wheat straw containing 5 ppm Zn was added to the diet: 40 g straw for every 100 g diet. The remaining ewes were housed together and given hay and corn until day 95 of gestation. On day 95 all ewes were X-rayed and 6 additional ewes selected. These were also penned individually and given the same dietary treatment except that 4 ewes had no Zn added to the water. Ewes were weighed and bled weekly. Frequent observations were made at the time parturition was expected. Newborn lambs were removed from the ewes and placed in an artificial rearing system.

1) Three of the 4 ewes fed the semi-purified diet throughout pregnancy delivered normal lambs at the appropriate time. The fourth ewe died approximately 1 week prior to delivery. Chronic pneumonia and lung parasitism were diagnosed at necropsy. Weight gains of the ewes during the first 6 weeks were equivalent to those of the ewes given a hay-corn ration. After that feed consumption became erratic, and body weight at the end of pregnancy was only 10% greater than that at breeding. Neither injection of B-vitamins nor addition of sodium bicarbonate to the diet improved feed consumption.

2) Adding straw to the diet made it possible to feed the diet as a wet mash, which reduced spillage. It did not, however, increase intake of the semi-purified diet above the level that we had obtained previously by feeding the dry diet with a 100 g/day hay supplement.

3) Since Zn concentration of the hay used previously was approximately 15 ppm and that of the straw used in this experiment was 5 ppm, the ewes could consume three times as much straw as hay without increasing their Zn intake above the level at which a Zn deficiency had previously been induced. Feeding the straw in proportion to the diet made the Zn intake more variable, and two of the -Zn ewes' Zn intake from the straw would occasionally have been as much as 20% higher than it would have been from 100 g of hay. Plasma Zn levels in the -Zn ewes in this experiment were not as low as those in the previous experiment and were only significantly different (P<.05) from the +Zn ewes at one interval. Plasma Zn concentrations were generally 0.8±.1 and 0.6±.1 μg/ml in the +Zn and -Zn ewes, respectively. At

Table 1. Effect of a low Zn diet and of Zn and sodium phytate on birthweight of lambs

Treatment	Lamb Weight, kg
-Zn	2.23, 2.01. 3.91 2.50, 2.27, 2.02, 1.54, 2.06 3.10, 3.60 average 2.59
+Zn	2.87, 1.75, 3.04 2.14, 2.41, 2.50 average 2.45
+Zn+Phytate	1.81, 1.89, 1.70 2.13, 1.66, 1.08 2.52 average 1.83

parturition 5 +Zn ewes delivered 13 lambs with only 1 ewe requiring assistance with 1 of 3 lambs. Four -Zn ewes delivered 12 lambs. One ewe delivered twins without assistance; 1 required assistance with 1 twin; 1 delivered 5 lambs without assistance over a 5 hr period; 1 was down for 8 days prior to delivery, and 2 of her lambs had to be pulled. The ewe died 14 h later. Suppurative bronchopneumonia was diagnosed at necropsy.

4) X-ray day 95 correctly predicted the number of fetuses for 6 of 11 ewes; the other 4 were underestimated.

The ewes given 1 1/2% sodium phytate in the diet (3 of the +Zn ewes) did not differ from the other ewes in any of the parameters measured except birthweight of the lambs. Lambs from ewes receiving phytate weighed markedly less than the other lambs. As we found in our previous experiment, birthweights of lambs of -Zn ewes were similar to those of +Zn ewes not receiving phytate. Because of the small number of animals, individual lamb weights are presented in Table 1 without statistical analysis.

In conclusion, it is clear from our earlier work that the lactating ewe is susceptible to Zn deficiency. The rapid deterioration in the -Zn ewes' condition after lambing suggests that the difficulties at lambing were at least partially due to depleted Zn stores at the end of pregnancy. In a subsequent experiment with fewer animals -Zn ewes also had more difficulty at lambing than did +Zn ewes. The ewe apparently requires Zn to deliver normally as does the rat. A low Zn diet that will allow better weight gain in the pregnant ewe is still needed. Incorporation of 1 1/2% phytate in the diet may affect lamb birthweight adversely.

REFERENCES

1. Apgar, J., 1972, J. Nutr. 102:343-348.

2. Apgar, J., and H.F. Travis, 1979, J. Animal. Sci. 48:1234-1238.

3. Hurley, L.S., and H. Swenerton, 1966, Proc. Soc. Exp. Biol. Med. 123:692-696.

4. Apgar, J., 1979, Nutr. Rep. Internat. 19:371-376.

DISCUSSION

HURLEY (USA). I would like to comment that you mention that there were no malformations but one wouldn't expect to get malformations when you gave the zinc deficient diet only during the last third of pregnancy. That would be too late in development to produce any malformations. I would also like to comment on the effect of zinc deficiency on milk production. We showed in rats that even a short period of zinc deficiency at the end of pregnancy could affect the development of the mammary gland and depress milk production.

MASTERS (Australia). You didn't mention how much zinc was in your low zinc diet, and I just wondered what sort of differences in intake you were getting between the two groups when they were on the treatment diets?

WELCH (USA). I can't tell you what the differences were in the intakes but it was a 5 part per million diet.

SUTTLE (Scotland). I don't like the terminology that you use in relation to your criteria of depletion. Zinc is not an element which is readily stored and my interpretation would be that the drop in plasma zinc is evidence that the reserve is inadequate. The changes you see in alkaline phosphatase are an orderly linear decline and I suggest that they simply reflect the slower turnover of an enzyme form of zinc against the rapid turnover of plasma zinc.

WELCH (USA). That is a very good possibility.

SINCLAIR (Australia). I would just like to add a comment to Dr Suttle's. In cattle with ostatagia you can see incredibly rapid changes in alkaline phosphatase. Whilst what Dr Suttle said with respect to Dr Welch's experiment is true, you still can see these very rapid changes.

TRACE ELEMENTS AND THE DEVELOPMENT OF ORGANS AND TISSUES

CHANGES IN BRAIN ENZYMES ACCOMPANYING DEFICIENCIES OF THE TRACE ELEMENTS, COPPER, SELENIUM, OR ZINC

Joseph R. Prohaska
Department of Biochemistry, University of Minnesota, Duluth
Duluth, Minnesota 55812 U.S.A.

Most trace elements accumulate in the central nervous system of mammals during a period when cells are rapidly dividing (hyperplasia) and growing (hypertrophy). This period of rapid brain growth is dependent upon species, but usually occurs during the last trimester of pregnancy or shortly after birth. Therefore, deficiencies of trace elements, like other essential nutrients, are most deleterious to brain when they occur during gestation and lactation. Deficiencies may result from environmental factors such as naturally occurring deficiencies of trace elements in the forage of grazing animals or failure to provide trace mineral supplements to premature infants or those on total parenteral nutrition. In certain cases, genetic factors lead to trace metal deficiencies--Menkes' steely hair disease to copper deficiency and acrodermatitis enteropathica to zinc deficiency. However, most of the information concerning trace element deficiencies in developing brain are due to experiments with laboratory animals fed purified diets.

The functions of trace elements in brain are generally attributed to their cofactor roles in metalloenzymes, since the concentrations of the individual elements, such as Cu, Zn, and Se, are relatively low (10^{-4}-10^{-7} M). When the pool size of an element becomes limited, there is a competition set up and the resultant distribution of the element, and thus metalloenzyme activities, depend on both the affinities of the brain ligands (apoproteins) and their relative concentrations. A major factor in the resistance of brain to trace element deficiencies is the relatively slow turnover of the metalloenzymes.

Deficiencies of trace elements may result in changes in brain enzymes that are not directly dependent on a particular element. These protein alterations usually reflect metabolic consequences of the deficiency--atrophy, for example. In some cases there may be metalloenzyme changes that are due to secondary interactions between two elements, such as copper and iron.

COPPER

Copper is heterogeneously distributed in brain, reflecting cell specificity for certain cuproproteins. Several cuproproteins have been isolated from nervous tissue of many species, including man. Currently, however, only three of these cuproproteins (cytochrome oxidase, dopamine-β-hydroxylase, and superoxide dismutase) are classified as cuproenzymes. Little is known concerning the other cuproproteins' enzymatic potential, or if they might be altered during copper deficiency.

Our knowledge of changes in brain enzymes accompanying copper deficiency comes from the published results of many laboratories investigating both environmental and genetic copper deficiency. Following Hart's classical report describing the essentiality of copper, work continued in Elvehjem's laboratory and established that rat brain oxidase activity and the cytochrome "A component" were both dependent upon dietary copper (1). In the past five decades, numerous studies have examined the effects of dietary copper deficiency on brain enzymes in rats (2-8), sheep (9-13) and mice (14). Many of these studies were patterned after models examining other aspects of copper deficiency and brain development (15-17). Other information regarding copper-dependent enzymes has come from studies on genetic copper deficiency in humans (Menkes' disease) (18) and mice (Brindled) (19, 20). Two other mouse mutants, Quaking and Crinkled, characterized by neurological damage, were reported to result from abnormalities in copper metabolism (21, 22); however, copper levels are normal in brains of both Quaking (23) and Crinkled mice (24).

Cytochrome Oxidase. The first copper-dependent enzyme shown to be reduced in brain tissue of deficient animals was cytochrome oxidase (1). The decrease in activity is much greater in young animals deprived of copper during development (4, 5, 7, 9, 10, 13, 14, 20) than in brain tissue of animals initiated on a purified diet after lactation (1, 2, 8).

Nevertheless, most investigators feel that the atrophic and pathologic effects of brain copper deprivation are due to decreased cytochrome oxidase activity. However, the ability of rat brain mitochondria to oxidize substrates is only moderately altered by copper deficiency (2, 5), and preliminary studies indicated that rat brain ATP levels are not lowered by dietary copper deficiency (5). However, these same studies demonstrated that brain energy metabolism ($NADH/NAD^+$) was altered by dietary copper deficiency (5), but the mechanism may be independent of cytochrome oxidase.

Mammalian cytochrome oxidase consists of 7 polypeptides, 2 moles of heme a and 2 moles of copper found in subunits I and II (25). Brain cytochrome oxidase activity can be increased by chronic elevation of brain copper (23). Conversely, the loss of brain copper is sufficient to explain the loss of enzyme activity; however, it was recognized early (1, 2) that heme a is decreased in copper-deficient brain mitochondria, and this loss of heme a is correlated with the loss of enzyme activity (5, 13). Copper is necessary for both the enzyme activity of cytochrome oxidase and conversion of porphyrin a to heme a (26). Decrease in heme a from brain mitochondria of a subject with Menkes' disease (18) suggests that the neurochemical aspects of genetic copper deficiency in humans are similar to those observed in experimental animals.

Dopamine-β-Hydroxylase. A second copper-dependent enzyme in brain is dopamine-β-hydroxylase, which catalyzes conversion of dopamine to norepinephrine. When the enzyme is inhibited in vivo by a copper chelator (diethyldithiocarbamate), the levels of brain norepinephrine fall, while dopamine synthesis increases (27). Thus, it was not surprising to find that norepinephrine was decreased in brain of copper-deficient rats (4, 5) and lambs (12). However, O'Dell and coworkers found that brain dopamine was also lowered by copper deficiency in lambs (12) and in rats (7, 28) and suggested that depression in tyrosine hydroxylase, normally the rate-limiting enzyme in norepinephrine synthesis, was responsible for this observation (7). However, tyrosine hydroxylase is most likely iron-dependent. In contrast to these reports on dietary copper deficiency, the work of Hunt and coworkers with Brindled mice (19, 20, 29, 30) supports the concept that dopamine-β-hydroxylase activity is rate-limiting in copper-deficient brain. They found lower norepinephrine levels in brain, as expected, but found dopamine to be normal (29) and tyrosine hydroxylase to be elevated (30). Direct measurements of dopamine-β-hydroxylase in vitro suggest that the apoenzyme may be made during copper deficiency, since the in vitro activity was elevated in Brindled mouse brain when exogenous copper was added to the reaction medium (19).

In order to resolve the apparent enigma between dietary copper deficiency in rats (lower dopamine) and genetic copper deficiency in mice (normal dopamine), a study was conducted in a single species (C3H/HeJ) to compare dietary and genetic copper deficiency. The comparisons were made when mice were 11 days of age (Fig. 1). Copper deficiency, genetic

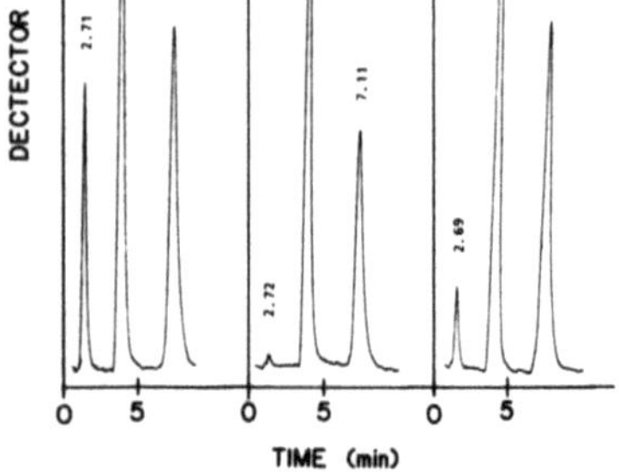

Fig. 1

Mouse brain catecholamine analysis. Male mice (C3H/HeJ) were compared when 11 days old and were derived from either normal dams ($Mo^{+/+}$) fed a low copper diet (-Cu) or dams heterozygous for the brindled allele on the X-chromosome ($Mo^{br/+}$) fed a non-purified diet containing adequate copper. Details of the experimental design are described elsewhere (14). Brains were homogenized in $HClO_4$ containing dihydroxybenzylamine (DHBA) as internal standard. The catecholamines were isolated by adsorption on alumina and subsequently analyzed by reverse-phase ion-pair liquid chromatography with electrochemical detection (49). Norepinephrine (RT = 2.7 min), DHBA (RT = 4.5) and dopamine (RT = 7.1) were quantified by peak integration. Both dietary (-Cu) and genetic ($Mo^{br/y}$) copper deficiency greatly depress brain norepinephrine compared to controls ($Mo^{+/y}$), while dopamine levels remain normal.

or dietary, decreases brain norepinephrine but does not decrease brain dopamine (unpublished), in agreement with the work of Hunt and coworkers. When dietary copper deficiency was extended to 7-week-old mice, the same age as rats previously reported to have low brain dopamine levels (7, 28), we found a significant reduction ($P < 0.01$) in brain norepinephrine; 463±13.7 ng/g (N = 5) in copper-supplemented animals vs 411±18.1 (N = 4) in copper-deficient animals, whereas dopamine levels were not different, 811±87.5 ng/g in copper-supplemented mice vs 908±33.5 ng/g in the copper-deficient mice. These data suggest that copper deficiency in mice selectively affects dopamine-β-hydroxylase (norepinephrine synthesis). We also have been unable to detect depression in brain dopamine levels in copper-deficient rats aged 12, 20, or 34 days (unpublished) but have not extended our studies to 7-week-old rats.

The consequences of chronic norepinephrine depletion by copper deficiency are unknown, but by analogy to studies in which dopamine-β-hydroxylase was chemically inhibited, they are likely serious both to function and development.

Superoxide Dismutase. This cuprozinc-protein, found in the cytoplasm, was first isolated in 1957 by Porter and Folch and named cerebrocuprein (31). The enzyme requires copper and catalyzes the disproportionation (dismutation) of superoxide anions (O_2^-). The loss of activity in brain following dietary copper deficiency (4, 5, 7) or genetic copper deficiency (20) is less pronounced than the loss of cytochrome oxidase activity and is unaffected in brain when dietary copper deficiency is started after the period of rapid brain growth (8). One of the possible consequences of lower superoxide dismutase in brain would be evidence of lipid peroxidation, such as the decreases in unsaturated fatty acids reported in Menkes' disease (18). However, lipid peroxidation did not appear to be evident in brain of copper-deficient rats with decreased superoxide dismutase activity (5).

Cuproproteins. Several other cuproproteins have been isolated from brain, including albocuprein I (MW = 72,000), albocuprein II (MW = 14,000) (32), and neurocuprein (MW = 9,000) (33). Brain may also contain copper-thionein (MW = 6,000). Since the enzymatic functions of these proteins are unknown, the consequences of their depletion remain to be established.

Secondary Effects on Other Metalloenzymes. Deficiency of copper may influence metabolism of other trace elements. Some preliminary data indicated that rat brain iron levels are lowered by dietary copper deficiency (5). This may explain why tyrosine hydroxylase activity was lowered in copper-deficient rat brain (7).

Another iron-dependent enzyme that might possibly be altered in brain tissue is stearoyl coenzyme A desaturase, since the desaturase activity of rat liver is known to be decreased by dietary copper deficiency (34). To examine this possibility, brain myelin was isolated from copper-deficient rat brain and from brain of supplemented rats, and the fatty acid methyl esters were purified and subjected to analysis by gas liquid chromatography (Fig. 2). Stearoyl-CoA desaturase activity was not directly measured. Myelin was chosen because it is a membrane rich in oleic acid, the product of stearoyl-CoA desaturase, and one whose synthesis depends upon an adequate supply of copper (3, 4, 6, 11, 16-18). As can be seen (Fig. 2), the chromatogram obtained from copper-deficient brain myelin displays an extra peak which has not yet been identified. However, the peak does not appear to be the methyl ester of fatty acid--19:0, 18:1 ω7, or 18:2 ω6 (trans). The areas of the unknown peak plus oleic acid are equivalent to the area of oleic acid alone in the chromatogram from the copper-supplemented sample. No other apparent difference in fatty acid composition were evident. The cause and significance of this unidentified peak from myelin samples of copper-deficient rats are presently unknown.

There was a recent report that dietary copper deficiency in rats decreases the activity of the selenoenzyme, glutathione peroxidase, in liver (35). However, we find no effect of dietary (rats and mice) or genetic (mice) copper deficiency on glutathione peroxidase activity in brain, kidney, blood, or liver (36). It is doubtful if lack of copper significantly influences selenium metabolism.

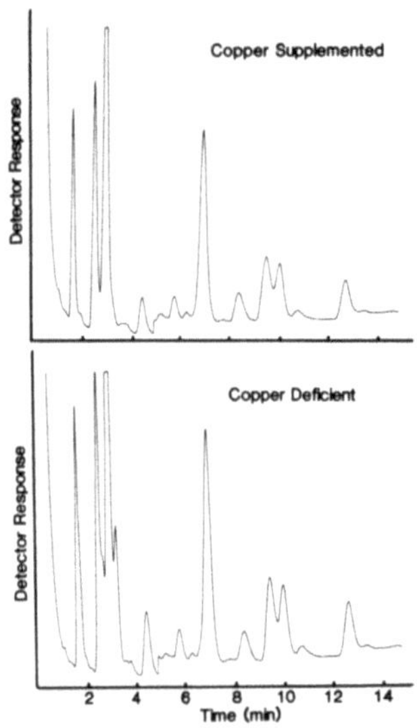

Fig. 2

Gas chromatographic analysis of the fatty acid methyl esters (FAME) isolated from purified myelin of adult female rats. Copper deficiency was produced by feeding a low copper diet during gestation and lactation (7 weeks) to female rats, initial body weight 200 g; copper supplemented dams received $CuSO_4$ in the drinking water (14). The average serum ceruloplasmin activities of supplemented females was 415 units/ℓ compared to 1.05 units/ℓ in the deficient rats. Myelin was purified from the individual brains of several rats, and the lipids were transmethylated with BF_3 and methanol. The FAME were isolated by thin layer chromatography on silica gel and analyzed by gas chromatography after elution from the TLC plates. Methyl esters were separated on a column packed with 10% SP2330 (Supelco), held at 205°C for 4 min and then programmed at 2°/min to 225°C. The FID detector output from the GC was interfaced with a recording integrator for quantitative analysis of the peaks; recorder attenuation was increased after 5 min. Note the extra peak (RT = 3.24 min) in the chromatogram from the copper-deficient myelin sample which follows the three major myelin components: FAME 16:0 (RT = 1.67), 18:0 (RT = 2.51), 18:1 (RT = 2.94).

Secondary Effects on Other Enzymes. Deficit of copper during the perinatal period reduces the growth of the brain. This growth deficit affects both glial and neuronal cells and explains why specific enzymes, not dependent on copper, are reduced in activity in brain tissue. Enzymes associated with the oligodendroglial cell are significantly reduced, such as 2′,3′-cyclic nucleotide 3′-phosphohydrolase (4-6) and UDP-galactose: HFA-ceramide galactosyl transferase (3, 11). A marker enzyme for neuronal cells, which is located on synaptic mitochondria, "latent" hexokinase, is also significantly delayed in appearance as a result of dietary copper deficiency (Fig. 3). These alterations reflect the atrophic response to copper deficiency, hypomyelination and delayed neuronal maturation.

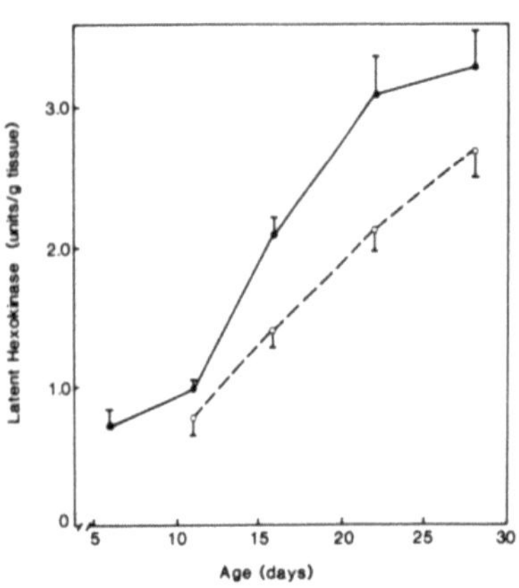

Fig. 3

The postnatal development of cerebellar "latent" hexokinase in control (●) and copper-deficient (○) rats. Copper deficiency was initiated during gestation by feeding a purified diet low in copper, as described previously (4). At the indicated ages, 4 rats of each treatment group were randomly selected and, following decapitation, the cerebella were analyzed for hexokinase activity according to the method of Knull et al. (1973, J. Biol. Chem. 248: 5414-5417). "Latent" hexokinase activity is defined as that amount obtained by subtracting the crude homogenate activity from the total activity obtained following Triton X-100 treatment. At all ages compared, the "latent" hexokinase activities from copper-deficient cerebella are different from control values, $P < 0.05$, suggesting a delay in synaptogenesis.

Over the past five decades, nervous tissue from copper-deficient animals has been assayed for many enzymes whose activities do not change. Some of the enzymes investigated, such as monoamine oxidase (5, 12, 20, 30, 37) and α-glycerol-phosphate acyl transferase (38), were originally thought to be altered by copper deficiency. Others, such as fumarase (5), succinate dehydrogenase (5, 37), malate dehydrogenase (37), and ribonuclease (10), were studied for other reasons.

In general, copper deficiency results in decreases in copper-dependent enzymes and in certain others associated with growth and differentiation. The reductions in cuproenzyme activities may be directly responsible for the neurological alterations accompanying both environmental and genetic copper deficiency.

SELENIUM

In contrast to the plethora of facts dealing with copper deficiency and the central nervous system, little is known concerning selenium and brain. Selenium is rather uniformly distributed throughout the central nervous system in the μM range, with somewhat lower concentrations in myelinated nerve tracts (39, 40). The distribution of selenium between gray and white matter reflects protein concentration, since most of the selenium in brain is bound to protein (40, 41). In rat brain, subcellular studies also indicate that selenium closely followed protein, being slightly enriched in mitochondria and low in the cytoplasm (40). The perinatal accumulation of selenium is very modest and occurs during a brief period; in rat brain, this takes place during the first two postnatal weeks (40). Therefore, selenium depletion studies beginning with weanling rats reveal minimal effects on brain function (40-42). Furthermore, the brain is very resistant to selenium depletion because of the slow selenium turnover (half-life = 45 days) (40).

Glutathione Peroxidase. Although never isolated from brain, the selenoenzyme, glutathione peroxidase, is likely present. It has been demonstrated that following gel filtration, rat brain cytosol contains a peak of glutathione peroxidase activity cochromatographing with ^{75}Se (MW = 94,000) (40), which has kinetic properties similar to those of the enzyme isolated from erythrocytes (42). Following a 13-week depletion period in which rats were fed a low selenium diet beginning at weaning, brain selenium levels are only decreased by 13%, in comparison to rats supplemented with sodium selenite (0.131±0.011 μg/g vs 0.114 ±0.008 μg/g; $P < 0.01$). In other studies there was no significant decrease in brain selenium (40). It is not surprising, therefore, that brain glutathione peroxidase activity does not decrease under these conditions (40). Furthermore, glutathione peroxidase only accounts for about 20% of the total brain selenium, and it is possible that the other selenoproteins in brain might be altered before glutathione peroxidase (40). Further studies are needed to determine the consequences of selenium depletion on early brain development and the other neurochemical functions of selenium.

ZINC

Like copper, there are many studies dealing with the effects of zinc deprivation on brain. Hurley and Swenerton (43) established the essential role of zinc for normal brain development. Zinc is necessary for both protein and nucleic acid synthesis in brain (44). Much of the information concerning the role of zinc in perinatal brain comes from the laboratories of Hurley and Sandstead.

The postnatal accretion of brain zinc is modest and occurs during the suckling period in mice (24) and rats (45). Zinc has a heterogenous distribution in brain (39) which may be reflective of its functions. It is essential for brain growth, and deprivation of zinc during the last trimester of pregnancy (46) or during lactation (45) results in reduced brain size; however, the concentration of zinc (μg/g tissue) remains normal.

Zinc-Dependent Enzymes. Since the level of brain zinc does not change during dietary zinc deficiency, it seems unlikely that activities of zinc-dependent enzymes should change. When compared to pair-fed, but zinc supplemented, rats, the activities of brain superoxide dismutase (45), glutamate dehydrogenase (45, 47), and carbonic anhydrase (47) were normal in zinc-deficient animals. It is difficult to hypothesize that the permanent behavior changes observed in zinc-repleted animals is due to changes in zinc-dependent enzymes. Perhaps zinc has a more direct effect in neuronal function, such as that described by Hesse in the hippocampal mossy fibers (48). The neurochemical role of this important trace metal remains to be established.

ACKNOWLEDGEMENT: I would like to thank the Minnesota Medical Foundation and the Graduate School, University of Minnesota for their financial help.

REFERENCES

1. Cohen, E., and C.A. Elvehjem, 1934, J. Biol. Chem. 107:97-105.
2. Gallagher, C.H., Judah, J.D., and K.R. Rees, 1956, Proc. Roy. Soc. 145(B):134-149.
3. DiPaolo, R.V., Kanfer, J.N., and P.M. Newberne, 1974, J. Neuropath. Exp. Neurol. 33: 226-236.
4. Prohaska, J.R., and W.W. Wells, 1974, J. Neurochem. 23:91-98.
5. Prohaska, J.R., and W.W. Wells, 1975, J. Neurochem. 25:221-228.
6. Zimmerman, A.W., Matthieu, J.M., Quarles, R.H., Brady, R.O., and J.M. Hsu, 1976, Arch. Neurol. 33:111-119.
7. Morgan, R.F., and B.L. O'Dell, 1977, J. Neurochem. 28:207-213.
8. Paynter, D.I., Moir, R.J., and E.J. Underwood, 1979, J. Nutr. 109:1570-1576.
9. Howell, J.McC., and A.N. Davison, 1959, Biochem. J. 72:365-368.
10. Mills, C.F., and R.B. Williams, 1962, Biochem. J. 85:629-632.
11. Patterson, D.S.P., Foulkes, J.A., Sweasey, D., Glancy, E.M., and S. Terlecki, 1974, J. Neurochem. 23:1245-1253.
12. O'Dell, B.L., Smith, R.M., and R.A. King, 1976, J. Neurochem. 26:451-455.
13. Smith, R.M., Osborne-White, W.S., and B.L. O'Dell, 1976, J. Neurochem. 26:1145-1148.
14. Prohaska, J.R., 1981, Nutr. Res. 1: in press.
15. Bennetts, H.W., and F.E. Chapman, 1937, Aust. Vet. J. 13:138-149.
16. Everson, G.J., Tsai, H.C., and T. Wang, 1967, J. Nutr. 93:533-540.
17. Carlton, W.W., and W.A. Kelly, 1969, J. Nutr. 97:42-52.
18. French, J.H., Sherard, E.S., Lubell, H. Brotz, M., and C.L. Moore, 1972, Arch. Neurol. 26:229-244.
19. Hunt, D.M., 1974, Nature 249:852-854.
20. Hunt, D.M., 1977, Comp. Biochem. Physiol. 57C;79-83.
21. Keen, C.L., and L.S. Hurley, 1976, Science, 193:244-246.
22. Hurley, L.S.,and L.T. Bell, 1975, Proc. Soc. Exp. Biol. Med. 149:830-834.
23. Prohaska, J.R., 1980, Life Science 26:731-735.
24. Keen, C.L., and L.S. Hurley, 1979, J. Inorg. Biochem. 11:269-277.
25. Winter, D.B., Bruyninckx, W.J., Foulke, F.G., Grinich, N.P., and H.S. Mason, 1980, J. Biol. Chem. 256:11408-11414.
26. Keyhani E., and J. Keyhani, 1980, Biochim. Biophys. Acta 633:211-227.
27. Musacchio, J.M., Goldstein, M., Anagnoste, B., Poch, G., and I.J. Kopin, 1966, J. Pharmacol. Exp. Therap. 152:56-61.
28. Feller, D.J., and B.L. O'Dell, 1980, J. Neurochem. 34:1259-1263.
29. Hunt, D.M., and D.R. Johnson, 1972, Biochem. Genet. 6:31-40.
30. Hunt, D.M., and D.R. Johnson, 1972, J. Neurochem. 19:2811-2819.
31. Porter, H., and J. Folch, 1957, J. Neurochem. 1:260-271.
32. Fushimi, H., Hamison, C.R., and H.A. Ravin, 1971, J. Biochem. 69:1041-1054.
33. Sharoyan, S.G., Shaljian, A.A., Nalbandyan, R.M., and H. CH. Buniatian, 1977, Biochim. Biophys. Acta 493:478-487.
34. Wahle, K.W.J., and N.T. Davies, 1975, Br. J. Nutr. 34:105-112.
35. Jenkinson, S.G., Williams, D.M., Lawrence, R.A., and R.F. Burk, 1980, Fed. Proc. 39: 1528 abs.
36. Prohaska, J.R. Smith, T.E., and D.E. Gutsch, 1981. In "New Zealand Trace Element Group-3", in press.
37. Kelly, W.A., Kesterson, J.W., and W.W. Carlton, 1974, Exp. Mol. Path. 20:40-56.
38. DiPaolo, R.V., and P.M. Newberne, 1972. In "Trace Substances in Environmental Health" (ed. D.D. Hemphill) (Univ. Missouri Press, Columbia) pp. 177-196.
39. Höck, A., Demmel, U., Schicha, U., Kasperek, K., and L.E. Feinendegen, 1975, Brain 98: 49-64.
40. Prohaska, J.R., and H.E. Ganther, 1976, J. Neurochem. 27:1379-1387.
41. Trapp, G.A., and J. Millam, 1975, J. Neurochem. 24:593-595.
42. Ganther, H.E., Hafeman, D.G., Lawrence, R.A. Serfass, R.E., and W.G. Hoekstra, 1976. In "Trace Elements in Human Health and Disease II" (ed. A. Prasad) (Academic Press, NY) pp. 165-234.
43. Hurley, L.S., and H. Swenerton, 1966, Proc. Soc. Exp. Biol. Med. 123:692-696.
44. Sandstead, H.H., Gillespie, D.D., and R.N. Brady, 1972, Pediat. Res. 6:119-125.
45. Prohaska, J.R., Luecke, R.W., and R. Jasinski, 1974, J. Nutr. 104:1525-1531.
46. McKenzie, J.M., Fosmire, G., and H.H. Sandstead, 1975, J. Nutr. 105:1466-1475.
47. Huber, A.M., and S.N. Gershoff, 1973, J. Nutr. 103:1175-1181.
48. Hesse, G.W., 1979, Science 205:1005-1007.
49. Felice, L.J., Felice J.D., and P.T. Kissinger, 1978. J. Neurochem. 31:1461-1465.

DISCUSSION

SMITH (Australia). I have been very interested in your figures for the ratio of lactate to pyruvate in the brains of your copper deficient animals. They show very big differences. We did not measure those, we measured ATP:ADP ratios and reduced and oxidised pyridine nucleotides in the brains of copper deficient foetal and neonatal lambs and we were not able to show very much effect. There is evidence of vascular stasis in some of these copper deficient states, it has been reported in rats as well as sheep. I think that the sort of measurements which you report would be very critically dependent on freeze-clamping times,so that,if after killing the animal the oxygen supply should fail faster in the deficient animal than the control then that, perhaps, might be an explanation of the changes which you see.

PROHASKA (USA). You point out a very interesting problem that we have in the Se studies and that is to rapidly inactivate the tissue. In fact in the case of the copper deficient animal you have to run two kinds of controls: one is a body weight control and oen is an age control. If you take a copper deficient rat and a control rat of the same chronological age, there is a difference in body weight. The way which we freeze the brain is to plunge the animals into liquid nitrogen. It takes longer to freeze the brain of the control animal because he is bigger than a copper deficient one. In order to compare the metabolites, therefore, we have actually compared two different types of animals. The levels of brain lactate were approximately 1 μmole per gram which are very close to the levels of brain lactate obtained by using the brain blower which is thought to be the most rapid way of freezing the nervous tissue. So, at least for the lactate values, I have a great deal of confidence in the fact that they really were elevated in copper deficiency.

DREOSTI (Australia). Your zinc enzyme work on the brain was mainly on the cerebellum. I think I have read papers of yours where you have looked at enzymes in the hippocampus. With the higher level of zinc in the hippocampus would that not be a good area to look?

PROHASKA (USA). I have not done that. You point to the interesting fact that the distribution of zinc in the brain is very heterogenous and that the hippocampus is very rich in zinc. There is some electrophysiological data from Hesse at Harvard (Science, 1979) to suggest that the zinc deficient brain is electrophysiologically different in the hipp-campus. Here there may be a direct effect of zinc rather than an effect working through some enzyme.

REEVES (USA). With regard to copper how do you account for the effect of food intake on the different parameters that you are measuring in these small rats?

PROHASKA (USA). As far as we can tell, the dams that are copper deficient consume the same amount of food. We don't see much of a difference in food intake once they are put into their cages, but I have not done a pair feeding study.

REEVES (USA). The copper deficient rats are smaller are they not?

PROHASKA (USA). Yes, I feel copper is necessary for growth, but I don't think food intake explains the growth failure.

O'DELL (USA). You raised the question relative to the basis for the differences in our observations on brain dopamine levels in copper deficient rats. This is an important question. Were your measurements made on cerebellum or on the total brain other than cerebellum?

PROHASKA (USA). The catecholamine analyses in the rat are on a piece of tissue that does not have the cerebellum nor the olfactory bulbs, and in the mice it is the whole brain.

O'DELL (USA). That is the same tissue that we analysed and I want to make the point that not all parts of the brain are the same. It is made up of many compartments and it is difficult to analyse the individual components. Nevertheless, we have analysed not only the portion which you describe, but also some discrete areas, including the corpus striatum, which as the highest concentration of dopamine, and the hypothalamus, which has the highest concentration of norepinephrine. We find the greatest difference in dopamine in the striatum and no difference in norepinephrine in the hypothalamus. There is a difference in norepinephrine concentrations in areas other than hypothalamus. My explanation, for the difference in dopamine is that the norepinephrine level is decreased by copper deficiency and you can restore that rather quickly by supplementary copper. Dopamine levels are not restored by copper supplementation. This leads me to believe that dopamine is down because of a morphological defect that arises during development. In order to get a dopamine effect you must induce a significant copper deficiency in the dam during the gestation period. Thus the copper effect is not on an enzyme system for example such as on dopamine-β-hydroxylase in the case of norepinephrine.

PAYNTER (Australia). Have you ever attempted to characterize the third ^{75}Se peak that you are picking up in your chromatograms and do you know if it has any similarities to the peak observed by Hoekstra's group in the rat erythrocyte?

PROHASKA (USA). Are you talking about the smaller molecular weight peak? I have not done that. There is a well identified ^{75}Se polypeptide in testes that has approximately the same molecular weight and I am not certain whether this protein is related to the brain ^{75}Se peak.

LINDER (USA). You are probably aware of the work of Dallman on iron deficiency in brain and one of the points he makes is that it is irreversible. That is, even if you replete the animal in the early part of the suckling period the iron levels in the brain cannot be restored to normal. Have you looked at repletion of copper and its effects on the various brain enzymes you have mentioned?

PROHASKA (USA). I have done a little of that and so has Dr O'Dell. When you give copper back to the animal the norepinephrine levels actually overshoot. The cytochrome oxidase comes back rather slowly. I think even after 4 weeks there is still a significant decrease in brain copper.

SUTTLE (Scotland). The dry matter content of the brain increases - certainly in the ovine in the early stages of life - does this happen in the rat, and if it does so, could it account for those smaller changes that you see in elements such as zinc?

PROHASKA (USA). I don't really know to what extent the dry weight does change, nor if it would make any changes to small effects we see with zinc.

TRACE ELEMENTS, GENETIC FACTORS, AND DRUGS IN DEVELOPMENTAL ABNORMALITIES

L.S. Hurley, C.L. Keen, B. Lönnerdal, P. Mark-Savage, and R. Hackman

Department of Nutrition
University of California
Davis, California 95616, USA

Previous work from this laboratory has demonstrated that certain chemicals or drugs may interact with essential trace elements in influencing prenatal development. Thus, the teratogenic effects in rats produced by a large amount of dietary EDTA were remarkably similar to those produced by severe dietary zinc deficiency, and they could be prevented by supplementation of the diet with zinc (1). Another example of such an interaction is the influence of dietary zinc concentration on the teratogenic effects of the drug 6-mercaptopurine (2). When pregnant rats were fed a diet marginally deficient in zinc, the frequencies of fetal death and malformation in drug-treated rats were higher than in normally fed drug treated rats. Conversely, when the diet contained supplementary zinc, fetal death and malformation were lower than in normally fed drug treated rats. We have now studied the effect of two additional drugs, salicylate and D-penicillamine, on teratogenesis and trace element status. We compared salicylate-induced teratogenesis in two different strains of rats fed varying amounts of zinc. Salicylate, the main metabolite of aspirin, produces teratogenic effects in a number of species similar to those resulting from zinc deficiency.

Female rats of the Sprague-Dawley or Wistar strain were bred and fed throughout gestation a purified diet containing various amounts of zinc. The animals were given a single dose of salicylate by gavage on day 9 and pregnancy was terminated on day 21. Fetuses were removed and scored for malformations or resorptions. The purified diet contained 0, 4.5, 9, 100, or 1000 ppm zinc. Salicylate was given at levels of 250, 500, or 750 mg/kg body weight. A variety of malformations was observed.

The data were analyzed using a factorial model. Each litter was coded by strain and by levels of zinc and salicylate. Since the responses were expressed as proportions, a logistic transformation was used to reduce the variance within litters. The response was written as a function of the unadjusted mean, plus the influence contributed by the single, two-way, and three-way interactions. Chi-square analysis was used to determine whether omitting any term in the model affected the response.

The frequency of resorptions was influenced by the strain, the zinc and salicylate levels, and the strain-salicylate, zinc-salicylate, and strain-zinc-salicylate interactions. The frequency of malformations was influenced by dietary zinc, salicylate level, and the strain-zinc, and zinc-salicylate interactions. The proportion of implantation sites affected was influenced by the same factors as the resorption rate, namely the three single factors as well as the strain-salicylate, zinc-salicylate, and strain-zinc-salicylate interactions.

Additional studies indicate that another teratogenic drug, acetazolamide, also interacts with zinc in prenatal development in a similar way.

D-penicillamine (dimethylcysteine) is a drug that can chelate divalent cations such as copper. It was introduced in 1956 by Walshe (3) for clinical use in the treatment of Wilson's disease, a genetic disorder characterized by high tissue concentrations of copper. Penicillamine is also prescribed for the treatment of several other disorders, including cystinuria and rheumatoid arthritis. Two reports (4,5) have described congenital abnormalities in the infants of women treated with D-penicillamine during pregnancy for these disorders. Connective tissue defects were a salient feature. Because of the copper chelating properties of this drug we formulated the hypothesis that penicillamine may be teratogenic due to its chelation of copper, resulting in low copper concentration in maternal plasma.

Sprague-Dawley rats were fed throughout pregnancy a complete purified diet containing casein as the protein source. Some rats were fed D-penicillamine in the diet at 0.17%, 0.83%, or 1.66%. These levels of penicillamine are similar to the dosages given to humans as reported in the literature. On day 21 of gestation, fetuses were removed by caesarian section and implantation sites were counted. The fetuses were examined for grossly visible external and internal malformations. Copper, zinc, iron, and manganese were measured by atomic absorption spectrophotometry in maternal liver and plasma, whole fetus, and fetal liver.

Food intake was not reduced by feeding penicillamine. Weight gain was also similar in all groups except the one receiving the high level of drug. The weight gain of these rats was slightly lower (10%) than in the other groups.

The frequency of resorptions and malformations increased with increasing levels of the drug. The resorption rate, or the percentage of implantation sites which had dead or resorbed fetuses, was highest in the animals fed the highest level of the drug, 1.66% penicillamine (10%), while the controls had 1.4% resorptions. The rats given the low level of drug, 0.17% penicillamine, had a resorption rate of 6.7%, also higher than the control value.

The rate of malformations also increased with increasing levels of drug. Few malformations were found in fetuses from dams fed 0.17% penicillamine. However, in dams fed 0.83% penicillamine, 30% of the fetuses were malformed; in those fed 1.66% penicillamine, 65% of the fetuses were malformed. The types of malformations included cutis laxis (or loose skin), abdominal herniations, spina bifida occulta, cleft palate, and bone abnormalities.

The copper concentration of both maternal plasma and liver was significantly lower in drug treated females than in controls. Thus, plasma copper concentrations were 35% lower in the 0.17% dams, 65% lower in the 0.83% dams, and 74% lower in the 1.66% dams than in the controls. A similar pattern was seen for liver copper values with the higher values found in controls, and the lowest in rats receiving 1.66%. It is of interest that maternal tissue copper concentration was depressed by penicillamine at the lowest level of the drug fed, even though the fetuses of this group were not visibly malformed.

The copper concentration of the whole fetus, and of fetal liver, was also significantly lower in all the drug treated groups than in controls. As with maternal tissues, the higher the level of drug fed, the lower was the tissue copper concentration. Thus liver copper concentration was 28% lower in the 0.17% fetuses, 43% lower in the 0.83% fetuses, and 61% lower in the 1.66% fetuses. A similar pattern was seen for whole body copper content. Therefore, like maternal tissues, fetal tissue copper concentration was affected by penicillamine at the lowest level of drug fed, although these fetuses were not grossly malformed.

The concentration of zinc in maternal plasma was not different between control females and those receiving 0.17% penicillamine or 0.83% penicillamine. However, rats receiving 1.66% penicillamine had significantly lower plasma zinc levels than controls. Similarly, liver zinc concentration in females fed the highest level of penicillamine was significantly lower than in controls.

The concentration of zinc in the whole fetus and in fetal liver was slightly higher than normal with higher doses of penicillamine.

Maternal plasma iron concentration was not affected by the drug regardless of level fed. However, liver iron concentration was significantly higher in drug fed females than in controls, and increased with increasing levels of drug. In contrast to the dams, fetuses did not show any consistent effect of D-penicillamine on iron concentration.

Liver manganese concentration was not appreciably affected by penicillamine, either in maternal or fetal tissues.

Thus, at all levels of D-penicillamine tested, copper concentration of both maternal and fetal tissues was significantly lower than in controls. The greater the amount of penicillamine fed, the more the concentration of tissue copper was affected. D-penicillamine was teratogenic at both 0.83% and 1.66% of the diet, and was most severe at the highest dose. The concentration of zinc in some maternal and fetal tissues was lower than normal in rats given 1.66% dietary D-penicillamine. The concentration of iron in maternal tissues was higher than normal in rats given 1.66% dietary D-penicillamine.

These findings support, but do not prove, the hypothesis that the teratogenicity of D-penicillamine may be due to its copper chelating ability, resulting in a lower than normal concentration of maternal plasma copper. These results also suggest that the copper and zinc status of women receiving D-penicillamine, as well as that of their infants, should be evaluated. However, women with Wilson's disease would probably be unlikely to have problems, as their liver copper concentration is abnormally high.

The mechanisms by which the drugs that I have discussed interact with the elements concerned are unknown, but it is probable that the chelating properties of these drugs are involved. The site or sites of chelation, are many, however. It is possible that a drug may chelate copper or zinc in the intestinal lumen, and (depending on the amounts given) either depress or enhance absorption. Chelation at the tissue level is also a possibility, altering normal rates of transport across membranes. Increasing the rate of urinary excretion of these trace elements is another way in which chelators could influence body zinc or copper status.

Although we do not as yet understand the mechanisms involved, we can conclude that drugs may interact with trace elements (or other essential nutrients) in affecting prenatal development and that these responses are influenced by genetic factors as well.

In addition to these single factors it is important to consider the interactions among genetic factors, various dietary factors, and the type and amount of drugs used when assessing risk to the fetus. We postulate that some developmental abnormalities in humans may result from multifactorial interactions which may involve disturbances in trace element metabolism.

REFERENCES

1. Swenerton, H., and L.S. Hurley, 1971, Science 173:62 - 64.

2. Hirsch, K.S., and L.S. Hurley, 1978, Teratology 17:303 - 314.

3. Walshe, J.M., 1956, Amer. J. Med. 21:487 - 495.

4. Mjølnerød, D.K., S.A. Dommerud, K. Rasmussen, and S.T. Gjeruldse, 1971, Lancet i: 673 - 675.

5. Solomon, L.G., G. Abrams, D. Dinner, and L. Berman, 1977, N. Engl. J. Med. 296:54 - 55.

DISCUSSION

WALKER (Australia). Dr Hurley did you notice that the urine of these rats was coloured perhaps violet. Freeman in Sydney has worked out the structure of the copper penicillamine complex and it is a mixed oxidation state of copper 1 and copper 2. I would have thought that the urine might have been coloured.

HURLEY (USA). We did not record the colour of the urine.

ELMES (Wales). Were the two human cases of malformed infants born of mothers given penicillamine therapy for Wilsons Disease, rheumatoid arthritis or cystinuria?

HURLEY (USA). One was a case of cystinuria and one of rheumatoid arthritis. They were not Wilsons Disease. In fact, some surveys that have been made of pregnancies in Wilsons Disease patients indicate that they do not have problems of teratogenicity.

HETZEL (Australia). Could I ask about the salicylate. Were the dosage and the plasma level comparable to pharmacological levels?

HURLEY (USA). One single dose was given by gavage. The amounts were related to those that have been used in studies of teratogenicity in rats. I am sorry I cannot relate this to levels in the blood - this is something that we're working on now.

LINDER (USA). Did zinc counteract the effect of salicylate? Zinc deficiency exacerbated the toxicity. Did increased zinc in the presence of increased salicylate help?

HURLEY (USA). In general, with the higher amount of zinc in the diet there were less effects of the salicylate and conversely, with the lower amount of zinc in the diet there was a higher effect of the drug.

MAY (Wales). I wonder under what conditions the copper 2 penicillamate complex forms that you showed on the slide. Whenever we tried to form this, we found that the copper is reduced.

HURLEY (USA). We did not form this copper complex - it is something that is taken from the literature to illustrate the composition of penicillamine and the possible way in which copper is bound to it.

MAY (Wales). This is a longstanding error in the literature. The point is that copper is not bound to penicillamine in this way.

DEVELOPMENTAL CHANGES IN ZINC AND COPPER IN MOUSE AND RAT TISSUES

C.L. Keen, B. Lönnerdal, and L.S. Hurley

Department of Nutrition
University of California
Davis, CA 95616, USA

The essentiality of the trace elements zinc and copper has been well established (1). Characterization of developmental changes in tissue trace element concentration and localization may help to identify critical time periods for alterations in trace element metabolism. Such underlying metabolic changes can then be investigated. Liver zinc concentration in the mouse does not change dramatically with age, while liver copper concentration decreases during the first weeks of life (2). In rat liver, zinc concentration decreases with age, while liver copper concentration increases during the two weeks of life and then decreases until day 25 (3). These differences in developmental patterns suggest that the metabolism of these two elements may be different in the neonates of these species. We have started to investigate the molecular localization of zinc and copper in liver and intestine from mouse and rat from birth to day 60.

EXPERIMENTAL DESIGN

C57Bl/6J mice and Sprague-Dawley rats were used. Animals were killed and liver and intestine were removed. The intestine was perfused with cold saline saturated with nitrogen and then frozen in liquid nitrogen. Intestines were homogenized (25% w/v) in 0.1 M ammonium acetate buffer and centrifuged at 10,000 g for 10 min. The supernatants were subjected to gel filtration chromatography on Sephadex G-50 in 0.1 M ammonium acetate buffer, pH 6.5 at 4°C. The gel had been treated with sodium borohydride in order to reduce surface charges originally present on the gel (4). Eluted fractions were analyzed for Zn and Cu by flame atomic absorption spectrophotometry (Perkin-Elmer 370). For liver the same procedure was followed except that this tissue was not perfused.

RESULTS AND DISCUSSION

In mouse liver, Zn was found in two peaks at all ages studied (Fig. 1A). The first peak had a molecular weight (MW) of $\geq$30,000 while the second peak had a MW of 6-8,000, a MW similar to that reported for metallothionein (5). The size of the high MW peak did not change appreciably with age. At all times this was the predominant peak. The 6-8,000 peak decreased in size from day 2 to day 10. After day 10 the 6-8,000 peak was either very small or missing (day 60). The observation that the overall peak area is not changing appreciably between day 1 and 60 is consistent with the observation that liver Zn concentration does not change during this time period.

The distribution of Zn in rat liver (Fig. 1B) is different from that found for the mouse. On day 1 Zn was found in both the high MW peak and the 6-8,000 MW peak. The 6-8,000 MW peak was considerably greater than the high MW peak. The amount of Zn in the high MW peak is similar to that found in the corresponding peak in mouse liver. Thus, the difference in liver Zn content between these species at birth (rat 100 µg/g; mouse 25 µg/g) is primarily due to a difference in Zn content of the 6-8,000 MW fraction. This peak decreased in rat liver resulting in a pattern similar to that of the mouse by day 30. At this time period the two species have similar liver Zn concentrations (25 µg/g).

The distribution of Cu in mouse liver (Fig. 2A) is similar to that of Zn with Cu present in two peaks with the same MWs as above. The 6-8,000 MW peak decreased with age. The $\geq$30,000 MW peak increased between days 2 to 10 and then decreased by day 20. Cu in rat liver was also distributed between two peaks (Fig. 2B) similar to those found in the mouse. The amount of Cu in both peaks increased from day 1 to day 10 and then decreased. This is consistent with the observation that Cu concentration in rat liver increases postnatally (20 µg/g day 1; 50 µg/g day 14). The distribution pattern for liver Cu in the rat at day 60 was similar to that of the mouse at this age. These species both have liver Cu concentrations of around 5 µg/g at this time.

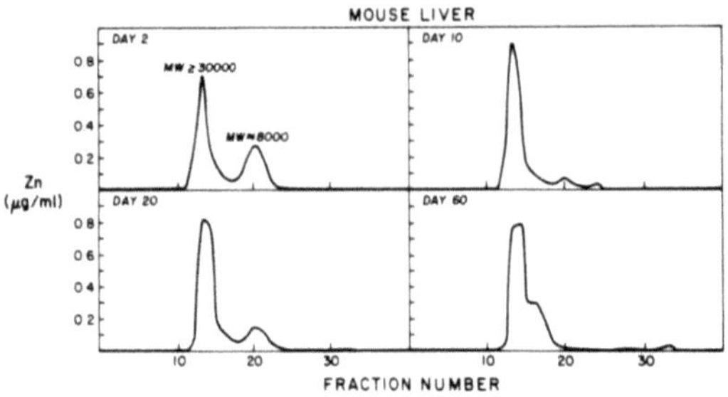

Fig. 1A

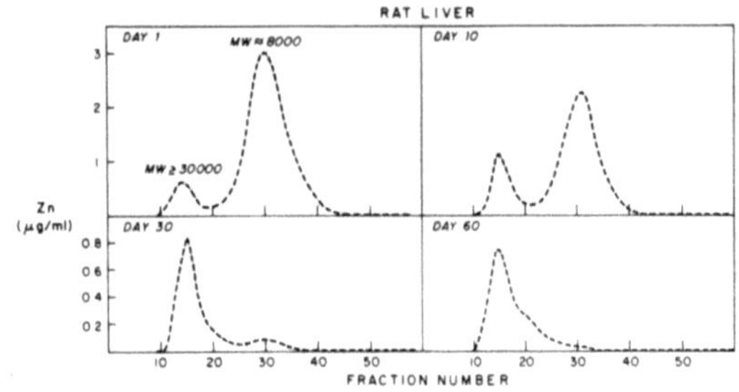

Fig. 1B

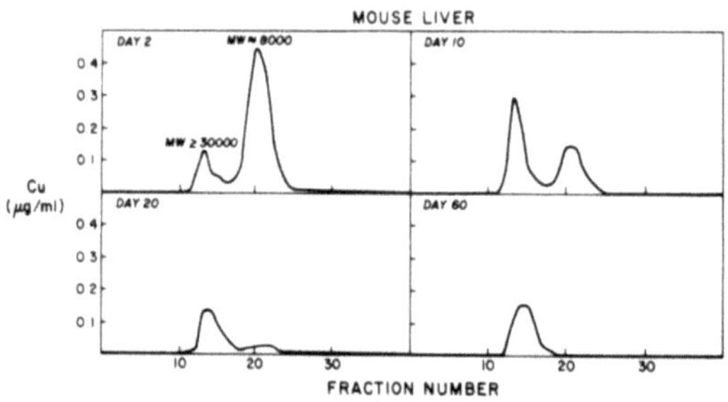

Fig. 2A

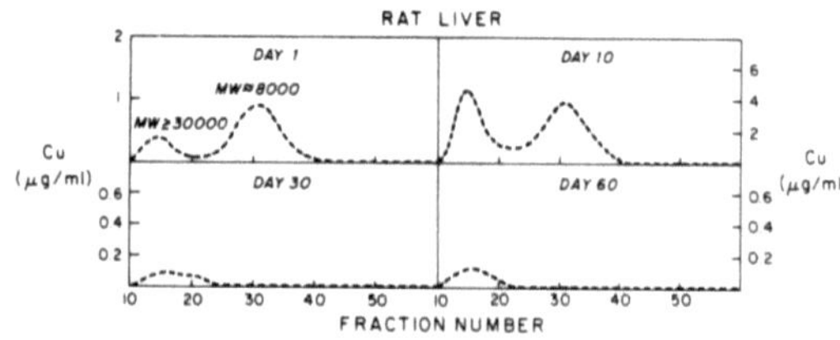

Fig. 2B

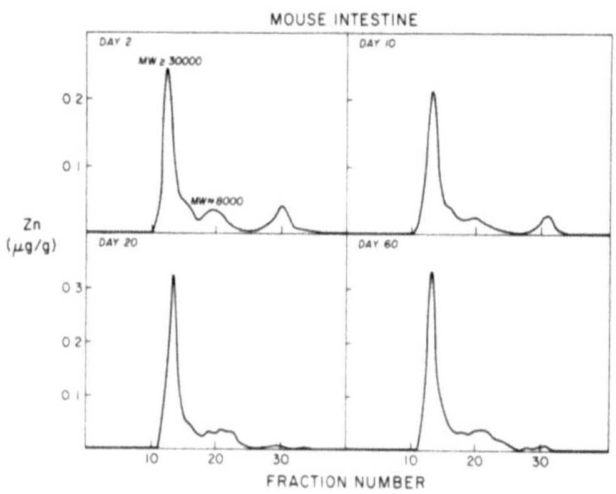

Fig. 3A

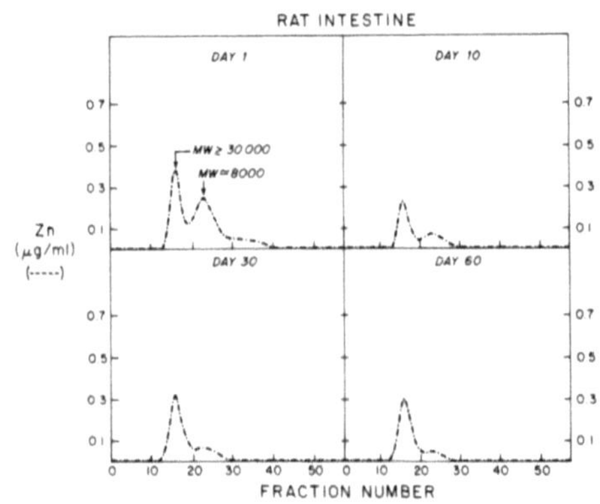

Fig. 3B

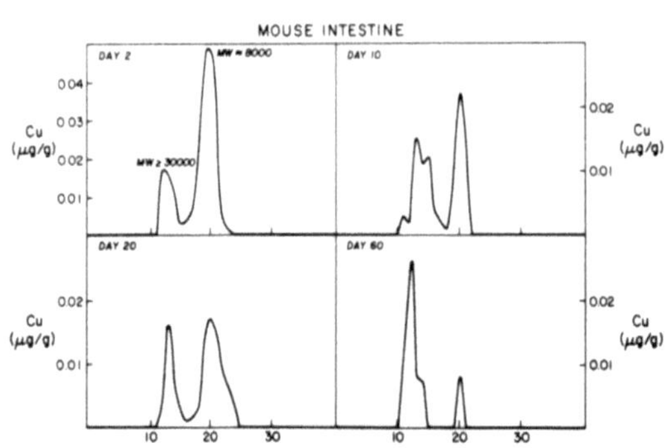

Fig. 4A

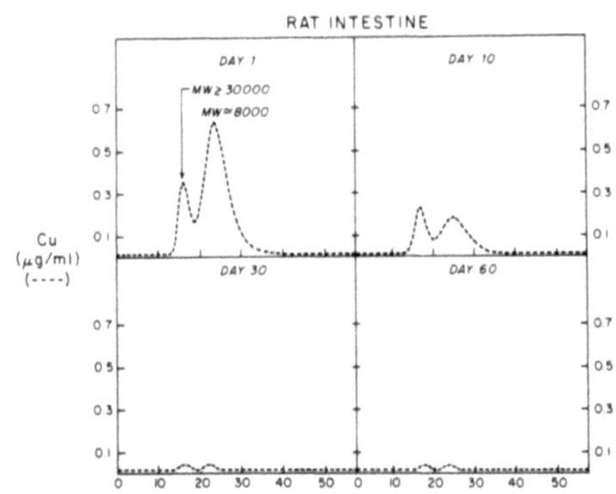

Fig. 4B

Zn in mouse intestine (Fig. 3A) eluted in a ≥30,000 MW peak and a 6-8,000 MW peak. The amount of Zn in the high MW peak increased slightly with age while there was little change in the 6-8,000 MW peak. At the age of 2 to 10 days, a low MW Zn peak (≤1,500) was observed, while in older animals no such peak occurred.

A high MW Zn peak was also found in rat intestine (Fig. 3B) at all ages. The size of this peak was similar to that of the mouse. Like the mouse, a 6-8,000 MW peak was observed in days 1 and 10 rat intestine. However, the amount of Zn in this peak was much higher than that of mouse intestine. Day 30 and 60 rat intestine had a pattern similar to that of the mouse.

The amount of Cu in mouse intestine (Fig. 4A) was considerably lower than that of Zn. The high MW peak of day 2 and day 10 animals was smaller than the 6-8,000 peak. No low MW peak was observed. By day 20 the 6-8,000 MW peak was similar in size to the high MW peak. The amount of Cu found in rat intestine at days 1 and 10 (Fig. 4B) was 10 times higher than in mouse intestine, although the distribution between the two peaks was similar. This is consistent with the observation that rat liver Cu concentration increases during the early postnatal period, while mouse liver Cu concentration does not. After day 10 the amount of Cu in the peaks was similar.

SUMMARY

In summary, these data have demonstrated that: 1) a salient feature of zinc and copper metabolism in the neonate is a series of sizeable alterations that occur in the concentrations of zinc and copper in various body tissues during the course of development; 2) the patterns for these changes can be different for different body tissues; 3) the patterns within a tissue can be different for zinc and copper; 4) the patterns for these changes regarding the whole tissue level and the molecular level can be considerably different among species. Knowledge of these changes and the time they occur enables us to choose optimal time periods to study developmental changes in the metabolism of trace elements and their binding ligands.

REFERENCES

1. Hurley, L.S. 1980. In "Developmental Nutrition" (ed. L.S. Hurley) (Prentice-Hall, N.J.).

2. Keen, C.L., and L.S. Hurley, 1980, Mech. Ageing and Develop. 13:161 - 176.

3. Dungan, D.D., C.L. Keen, B. Lönnerdal, and L.S. Hurley, 1980, Fed. Proc. 39:903.

4. Lönnerdal, B. 1980. In "Trace Element Analytical Chemistry in Medicine and Biology" (eds. P. Brätter, and P. Schramel) (Walter de Gruyter, Berlin).

5. Kojima, Y., C. Berger, B.L. Vallee, and J.H.R. Kägi, 1976, Proc. Natl. Acad. Sci. USA 73:3413 - 3417.

DISCUSSION

WIENER (Scotland). I wonder whether you have had an opportunity to look at these species differences at equal stages of maturity, instead of equal ages.

KEEN (USA). With regard to changes in liver maturity, I am not sure whether the mouse at day one is very different from the rat at day one. I don't think you will see liver concentrations in the mouse or molecular localization patterns similar to the rat until the weaning period of each species has been passed.

LINDER (USA). In your abstract you mentioned something about a 2000 Dalton copper component in intestine at some stage of development. Is that something you are still finding?

KEEN (USA). In a couple of cases especially between days 5 and 15 we seem to have a low MW complex, but we don't find it consistently.

SUTTLE (Scotland). Do you find differences between strains of mice and strains of rat, and how many strains must you look at before you can talk about species differences?

KEEN (USA). We haven't concentrated on strain differences either in the rat or the mouse. We have chromatographed C57 black and C3H mice and they appear to have similar changes in patterns. However the absolute change can be different. The observation between the difference of the mouse and the rat, I don't think is going to be changed by looking at one strain or another strain unless it is a mutant.

SUTTLE (Scotland). The reason for asking the question is that in our attempts to deplete mice at Moredun we find that the Swiss white strain to be much more susceptible to deficiency than the other strains that we looked at.

NEWBERNE (USA). That is a very good point, there are strains of mice, as you know, that are much more susceptible to either infectious disease or to carcinogenesis and different levels of trace elements are found in the liver.

LINDER (USA). What is the difference in the copper and zinc content of the milk between mouse and rat? Is there anything there that might give you a clue as to why the liver stores should differ in these two species at birth.

KEEN (USA). We only have some very preliminary data on mouse milk. The amount of zinc in mouse milk appears to be somewhat similar to that found in rat milk, though there does appear to be some distribution changes. There is a small low MW fraction containing zinc found in mouse milk which we do not observe in rat milk. With regard to the copper concentration that is quite different, in that while in the rat there is a very high amount of copper in the colostrum, the samples that we have looked at would suggest that this is not true for the mouse. The amount of copper in colostrum in the mouse is relatively low compared to other species. This might explain why we don't see a postnatal increase in liver copper concentration in the mouse.

PREVENTION OF DELAYED SWAYBACK IN LAMBS

Gwyneth Lewis, S. Terlecki, B.N.J. Parker and Pamela L. Don

Ministry of Agriculture, Fisheries and Food, Central Veterinary Laboratory,
NEW HAW, WEYBRIDGE, SURREY, KT15 3NB,
ENGLAND

Early work in Australia and New Zealand on enzootic ataxia indicated that the onset and course of the disease in susceptible lambs could be prevented or arrested by repeated oral dosing with dilute copper sulphate solutions from 3 days to 3-4 months of age (1,2). Cunningham (2) stated that "the effectiveness of post-natal administration of copper would suggest, that in the New Zealand disease, the final stages of the pathological processes which give rise to ataxia occur after birth".

In the delayed form of swayback, which is essentially similar to enzootic ataxia, lambs appear clinically normal at birth but develop signs of the disease from about 1 week to 6 months of age. The essential lesion consists of degeneration and necrosis of the motor neurones of the brain stem and spinal cord and degeneration of the nerve fibres of the spinal cord. There are no significant lesions in the cerebrum. It has been shown that the lesion of delayed swayback is not present at birth but develops post-natally (3).

This raises the possibility of prevention of the disease, particularly on farms where the ewes have not been given copper during pregnancy and where the first lambs born have either the congenital form of the disease or develop delayed swayback at an early age. To be fully effective and practical under farm conditions, any recommended preventive measure should be a single, safe dose of copper which could be given parenterally to lambs from a few days of age.

The young lamb is extremely susceptible to copper poisoning and the safety margin between a therapeutic and a toxic dose is very small. Deaths occurred in lambs 2 days to 7 weeks of age given 10-50 mg copper subcutaneously as copper calcium EDTA (4).

The objects of this work were to determine (a) whether the onset of delayed swayback, in a highly susceptible population, could be prevented by parenteral administration of copper to the young lamb and (b) an effective dose rate which would give a margin of safety.

Lambs were obtained from ewes of assorted breeds, mainly Swaledale, Derbyshire Gritstone or Swaledale crosses, which had been maintained for at least two years on the semi-purified diet containing about 2 mg copper/kg dry matter previously described (3). The lambs were reared on this diet.

Two proprietary preparations suitable for parenteral administration were tested in successive years.

In the first year, histological examination of 38 foetuses, still-born lambs or lambs dying shortly after birth showed that 22 (58%) were cases of congenital swayback. Fifteen lambs from this highly susceptible population were reared and weaned onto the semi-purified diet. Copper calcium EDTA (Coprin. Glaxo Ltd.), given subcutaneously (s/c) was used as the source of copper.

In 4 sets of twins, one lamb was given 5 mg copper between 7 and 10 days of age, the other was left untreated as control. Four single lambs were each given 10 mg copper when 7-10 days old, 2 undosed single lambs were reared as controls. Lambs were killed by intravenous injection of pentobarbitone sodium when showing clinical signs of delayed swayback or at 6-8 months of age. Material for pathological studies was removed as soon as possible after death: brain and spinal cord were fixed in 10% neutral buffered formalin. Histological procedures and the pathological criteria for the diagnosis of swayback were as previously described (3).

None of the 4 twin lambs receiving 5 mg copper or the 4 single lambs receiving 10 mg copper showed clinical or histological signs of swayback. However 3 of the 4 undosed twins developed delayed swayback between 62 and 203 days of age and the fourth was recorded as "doubtful" histologically. One of the two single lambs left untreated developed clinical swayback at 114 days, which was confirmed histologically.

In our experience, and that of other workers, in the case of twins where one is affected and if no treatment is given the other lamb will invariably develop the disease although there may be considerable difference between them in the time of onset.

The results indicated that 5 mg copper as copper calcium EDTA given subcutaneously at 7-10 days of age safely and effectively prevented the onset of delayed swayback in a highly susceptible population of lambs maintained on a low level of dietary copper. Although 10 mg copper produced no toxic effects in these young copper deficient lambs toxicity has been reported in commercial flocks when this dose was given to lambs 4 days - 4 weeks of age. This preparation is presented as a cartridge containing 100 mg copper in an effective length of about 4.5 cm. It is not easy to mark the tube into small divisions representing 5 mg copper and it would be difficult to give accurate, small doses to young lambs under field conditions.

In the second trial a methionine copper complex (Copavet. C-vet Ltd.) was tested. This preparation is in the form of a suspension and the instructions for use suggest intramuscular (i/m) injections for cattle or sheep or subcutaneous injection into the brisket for sheep. From 14 sets of twins one lamb was injected between 4 and 10 days of age: as follows

5 mg Cu	subcutaneously	6 lambs	10 mg Cu	subcutaneously	3 lambs
5 mg Cu	intramuscularly	3 lambs	10 mg Cu	intramuscularly	2 lambs

The other lamb of the pair had either died at birth or was left untreated as control.

From 2 sets of triplets in which all three lambs were reared, one of each set was given 5 mg Cu s/c, one 5 mg i/m and the third left as control. In a third set of triplets one died at birth, one was reared as control and the third received 5 mg Cu i/m.

Of six single lambs, one received 10 mg Cu i/m, one 10 mg Cu s/c and the other 4 served as controls. As before, lambs were killed when showing clinical signs or at 7-10 months of age.

There were 22 untreated control lambs, of these 3 which died at birth, showed no visible lesions but eleven of the 19 reared (58%) developed delayed swayback in 17-254 days. Injection of 5 mg copper either s/c or i/m did not prevent development of swayback in every lamb whose control twin or triplet had earlier succumbed to the disease. However it appeared to ameliorate the condition and delay the onset of clinical signs, since 4/14 lambs (28.6%) given 5 mg Cu developed the disease in 149-263 days. Intra-muscular injection of the preparation seemed to be more effective than the subcutaneous route, although we consider the latter to be more desirable under farm conditions if young lambs are treated by unskilled hands.

Injection of 10 mg copper either s/c or i/m appeared to prevent the onset of disease, there were mild lesions in only one lamb, killed at nine months of age. No toxic effects were observed.

It was easier to measure and inject an accurate small dose of this preparation (5 mg Cu in 0.25 ml). Although under the experimental conditions of severe copper deficiency the 5 mg dose did not completely prevent the onset of delayed swayback, in practical farm situations where supplementary feed is frequently given, 5 mg copper would probably be effective with the advantage of a safety margin if the dose was accidentally exceeded.

The method of choice for the prevention of swayback is still by treatment of the copper deficient pregnant ewe. However if this has not been done and in a year with a high prevalence of the disease, the first lambs born have either the congenital or early delayed form of the disease, then although it is probably too late to prevent further cases of congenital swayback, the delayed disease may be prevented by injection of the young lamb.

Some controlled field trials have shown that this approach can be of practical value under farm conditions where ewes have not received copper during pregnancy (5).

SUMMARY

It has been demonstrated that the onset of delayed swayback can be prevented or retarded or the condition ameliorated in a highly susceptible population of copper deficient lambs by parenteral injection of small doses of copper complexes.

It is suggested that the subcutaneous route of injection is to be preferred to the intramuscular route for young lambs.

It is suggested that the dose should be limited to 5 mg copper for lambs of 3-10 days of age. No toxic effects were observed.

Neither of the preparations tested were particularly suitable for giving accurate small doses to young lambs under farm conditions.

Since young lambs are particularly susceptible to copper toxicity, copper therapy should only be given to animals known to be copper deficient and not receiving supplementary copper from any other source.

REFERENCES

1. Bennetts, H.W., and A.B. Beck, 1942, Commonwealth of Australia Council for Scientific and Industrial Research, Bulletin No 147

2. Cunningham, I.J. 1949, N.Z.J. Sci. Tech. A31:42

3. Lewis, G., S. Terlecki, and B.N.J. Parker, 1974, Vet. Rec. 95: 313-316

4. Allcroft, R., D. Buntain, and W.T. Rowlands, 1965, Vet. Rec. 22:634

5. Buntain, D., G. Lewis, and S. Terlecki. in preparation.

DISCUSSION

SUTTLE (Scotland). The observation that methionate gives variable and inconsistent protection is consistent with our own findings on the ability of these different complexes to alleviate hypocupremia in the adult ewe. In the adult you get quite considerable local reactions to methionate, do you see reactions in the young lambs given small doses? In the neonates of most species including lambs, the absorption of copper is very efficient, near to 100%. Have you considered the oral route of copper supplementation in preventing delayed swayback?

LEWIS (England). We gave methionate subcutaneously behind the shoulder and in small amounts - we did not notice any local reaction. There was no reaction to the intramuscular infection. One worries about this in the farm situation because it is possible to get sterile abscesses and other problems, but we didn't see this in our lambs. At this stage we haven't tried oral copper but particularly with the work we are hearing now on the use of copper oxide needles particularly, I think oral copper supplementation will be very interesting. We have tried giving copper sulphate orally and that works effectively but you usually have to give two doses.

SMITH (Australia). For how many days after birth of the lamb is copper effective in preventing the development of delayed swayback.

LEWIS (England). I don't know the ultimate limit. In some field trials where they were anxious to avoid any possibility of copper toxicity, lambs were treated at 3 to 4 weeks of age with copper methionate, and there was a statistically significant reduction in cases of swayback. Administration before 20 days is probably the most effective as this is when the second peak of myelination occurs in the spinal cord. In our area the lambs are moved to the hill at an early age, so that we feel, under practical conditons, they should be treated early while they are close at hand.

COPPER AND THE PATHOGENESIS OF ENZOOTIC ATAXIA

R.M. Smith, R.A. King, W.S. Osborne-White and F.J. Fraser

CSIRO, Division of Human Nutrition
Kintore Avenue, ADELAIDE, SOUTH AUSTRALIA 5000

Enzootic ataxia or swayback in lambs is a well recognized disease characterised at the clinical level by motor inco-ordination of the hind limbs and recognizable at the microscopic level by several histopathological changes in the central nervous system. The disease is associated with low levels of copper in the brain and liver of affected lambs and may be prevented by treatment of the ewe with copper during the latter half of pregnancy (1). The essential pathology includes specific changes in certain regions of spinal cord white matter which are perceptible by the Marchi method at an early stage as degenerative changes in myelin and only later appear as areas of frank demyelination (3). Experiments carried out on areas of affected land in South Australia (2,3) have shown that this degradative process may be seen in the fetal spinal cord by 100 days gestational age, very shortly after myelination has begun. These experiments also revealed that a second feature of the classical pathology of the disease, that of chromatolysis and necrosis of large multipolar neurons in the spinal cord, could not be detected until several weeks after birth. This suggested that the neuronal pathology may have resulted from prior damage to the nerve fibres and not the converse, as had previously been supposed (4). The occurrence of early fetal lesions was also difficult to reconcile with the view that the primary biochemical event was a depletion of mitochondrial cytochrome oxidase activity since at that stage there is little activity of that enzyme in the brain or spinal cord (5). The present experiments set out to examine the time course of development of the spinal cord lesions in relation to the synthesis of myelin in the spinal cord and the activity of cytochrome oxidase in the brain. The results in general support the view that development of the spinal cord lesions cannot readily be attributed to depleted cytochrome oxidase activity. Occurrence of the classical lesion in two neonatal lambs with normal copper status leads to a reassessment of the role of copper in preventing enzootic ataxia.

EXPERIMENTAL

A group of 50 Australian Merino ewes, 2 years of age, was transferred to copper-deficient land at Robe in South Australia, and grazed there with the treatments described below for 1 year before mating with normal rams (2). All of the ewes received a cobalt pellet and grinder and 28 of them received a weekly drench of 70 mg copper as $CuSO_4$ in water. Fetal lambs of known conceptual age from 60 d to 140 d and neonatal lambs up to 20 d were recovered and killed as previously described (2,3). Samples of cerebal cortex (motor area), spinal cord (cervical enlargement) and liver were obtained immediately after killing for estimation of cytochrome oxidase activity (6) and copper content (atomic absorption spectophotometry) and, where appropriate, for myelin content by use of the marker enzyme 2',3' - cyclic nucleotide 3' - phosphohydrolase (CNP) (7). Histological techniques were as previously described (2,3). Spinal cord lesions were assigned an intensity rating between 1 (mild) and 4 (very severe).

RESULTS

The ewes and their fetal or neonatal lambs were killed after the former had grazed the copper-deficient pasture for periods ranging from 13 to 16 months. At termination maternal blood copper levels were 0.22±0.02 mg/l (20) for deficient ewes and 0.93±0.03 mg/l (26) for treated ewes. Mean copper concentrations in the brains of the deficient lambs were: fetal, 4.3±0.7 μg/g dry (14), neonatal, 3.4±0.3 μg/g dry (5) and for lambs of copper-treated ewes: fetal, 10.4±0.6 μg/g dry (18), neonatal, 11.6±0.7 μg/g dry (6) respectively. Table 1 shows the incidence and intensity of Marchi-positive lesions in the characteristic regions of the spinal cord and the brain Cu levels of those lambs that showed a positive Marchi lesion. All of the postnatal lambs in the deficient group showed the spinal cord lesion but the lesion was also seen in two control lambs which were shown to have a normal copper status. The validity of this histopathological observation was confirmed in longitudinal sections of the spinal cord (2).

TABLE 1. INCIDENCE AND INTENSITY OF MARCHI POSITIVE LESIONS BELOW THE DORSAL HORNS IN THE CERVICAL SPINAL CORD OF FETAL AND NEONATAL LAMBS

Age, range (F, fetal; N, neonatal days)	Treatment					
	Copper deficient			Copper treated		
	Incidence	Intensity[a]	Brain Cu (μg/g dry)	Incidence	Intensity[a]	Brain Cu (μg/g dry)
60-89 F	0/3	-	-	0/1	-	-
90-119 F	1/6	1	4.3	0/8	-	-
120-149 F	2/4	2,3	2.7±0	0/6	-	-
1-20 N	5/5	1,2,3,3,4	3.4±0.8	2/6	2,2	13.1±0.9

[a] 1=mild, 2=moderate, 3=severe, 4=very severe

Activity of cytochrome oxidase in the brain of the fetal and neonatal lambs is shown in Fig. 1. At ages below 100 d activity was very low and there were no effects of treatment. After 100 days cytochrome oxidase activity increased but significant effects of treatment did not emerge until after birth. Both of the deficient fetuses examined at 129 days showed gross cavitation of cerebral white matter as well as moderate to severe Marchi lesions in the spinal cord, but in only one of these was the cytochrome oxidase activity below the normal range. Postnatally one case of clinical ataxia was found. The brain cytochrome oxidase activity of the postnatal deficient lambs was significantly below that of the controls ($P<0.001$). Measurement of CNP activity showed rapid development from 60 days fetal age to 20 d postnatal in both deficient and control spinal cord with total activity in the cord increasing by 4 orders of magnitude (Fig. 2). The results show no measurable effect of Cu deficiency to depress or delay this development.

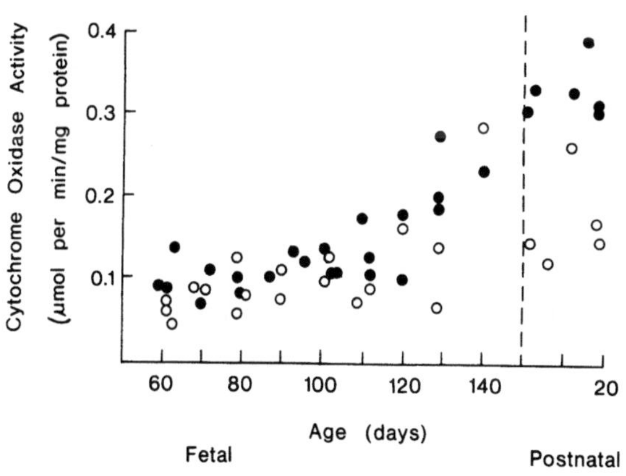

Fig. 1. Cytochrome oxidase activity in homogenates of cerebral cortex of fetal and neonatal lambs. Open symbols, copper deficient ewes; closed symbols, copper treated ewes.

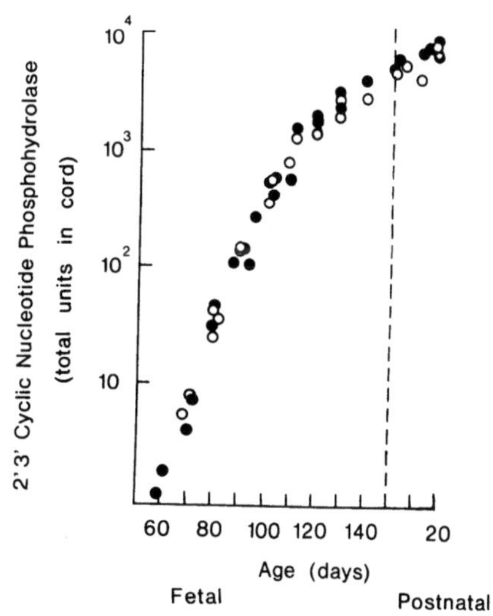

Fig. 2. Myelination in the fetal and neonatal spinal cord as measured by total activity of the myelin marker enzyme 2', 3'-cyclic nucleotide 3' - phosphohydrolase. Open symbols, copper deficient ewes; closed symbols, copper treated ewes.

DISCUSSION

Barlow (8) had previously noted that the spinal cord lesion was found occasionally in normal neonatal lambs less than 10 d of age but the copper status of these was not reported. The finding of distinctive Marchi lesions in the characteristic regions of the spinal cord of neonatal lambs with established brain Cu status renders untenable the view that the degradative process results from the failure of some normal process of development that requires copper in order to occur and this is borne out both by the normal development of myelin in the deficient cord indicated by CNp activity and by the modes reduction in cytochrome oxidase activity. The view must now be taken that the function of copper in these circumstances is to protect against damage to the spinal cord, possibly by some environmental agent and that such protection, at least in the neonatal lamb, is imperfect. It is one of the purposes of this communication to suggest that the agent responsible may be molecular oxygen.

First the common feature of all known biological functions of copper is an involvement with molecular oxygen. The intrinsic toxicity of oxygen led Frieden (9) to propose that appearance of the copper-enzyme superoxide dismutase (SOD) may have been a crucial event in the evolution of aerobic organisms. The superoxide radical O_2^- arises as a byproduct of several biological oxidative processes or its formation may be catalysed by ferrous (although not by ferric) iron. The activity of SOD in fetal tissues is low but after birth increases rapidly, especially in the brain (10).

In the extracellular environment a second copper protein, caeruloplasmin, is effective as scavenger of the O_2^- radical and, like SOD, activity develops postnatally (11). Thus when it first encounters gaseous oxygen the newborn animal may not be well equipped to counter its potentially toxic effects and a deficit of available copper may exacerbate this hazard. A decreased activity of SOD has been reported in the cerebellum of rats that were copper-deficient during fetal and neonatal life (12) and reduced caeroplasmin activity is a general feature of copper-deficiency.

Another function of caeruloplasmin in the circulation is to catalyse the oxidation of ferrous to ferric iron. Thus the copper-deficient neonate may be at additional risk from an enhanced potential to generate O_2^- due to an elevated Fe^{++}/Fe^{+++} ratio.

Finally the degenerative lesions in the spinal cord occur peripherally in the ventrolateral columns of white matter just below the dorsal horns of grey matter and in the vental cleft. These sites are adjacent to the three spinal arteries and it is possible that a local abundance of oxygen accounts for the susceptibility of these regions to degradative change, especially in conditions of copper deficiency.

We thank Mr V Dawson of Robe, South Australia, for care of the animals and Mr P F Rogers for the copper analyses.

REFERENCES

1. Allcroft, R., F.G. Clegg and O. Uvarov, 1959, Vet. Rec. 71:884-889.
2. Smith, R.M., F.J. Fraser, G.R. Russell and J.S. Robertson, 1977, J. Comp. Path. 87:119-128.
3. Smith, R.M., F.J. Fraser and J.S. Robertson, 1978, J. Comp. Path. 88:401-408.
4. Cancilla, D.A., and R.M. Barlow, 1968, Acta Neuropath. 11:294-300.
5. Barlow, R.M. 1969, J. Comp. Neurol. 135:249-262.
6. Smith, L. 1955, Meth. Biochem. Anal. 2:427-434.
7. Kurihara, T., and Y. Takahashi, 1973, J. Neurochem. 20:719-727.
8. Barlow, R.M., 1958, Proc. Roy. Soc. (Lond.) 51:748-752.
9. Frieden, E. 1974, Adv. Exp. Med. Biol. 48:1-31.
10. Mavelli, I., B. Mondovi, R. Federico and G. Rotilio, 1978, J. Neurochem. 31:363-364.
11. Chang, I.C., D.C. Milholland and G. Matrone, 1976, J. Nutr. 106:1343-1350.
12. Prohaska, J.R., and W.W. Wells, 1974, J. Neurochem. 23:91-98.

DISCUSSION

O'DELL (USA). Did you measure the activity of superoxide dismutase?

SMITH (Australia). No. We haven't yet done that. However it is apparent that the biochemistry of copper and oxygen are irrevocably intermingled; that superoxide dismutase does prevent damage to brain by oxygen. It can be demonstrated that brain lipids are particularly susceptible to oxygen damage and that this can be prevented by superoxide dismutase, by caeruloplasmin (which is a superoxide scavenger) and by chelators of ferrous iron. Ferrous iron is a catalyst for the formation of the superoxide anion. It is noticeable that all of these things, superoxide dismutase and caeruloplasmin and the capacity to oxidase ferrous iron to ferric iron by caeruloplasmin emerge postnatally in those animal species that have been examined and that the risk of damage therefore in a copper deficient animal is greatly enhanced over that of a normal animal. The normal animal itself must be at some risk of oxygen toxicity but the copper deficient animal is at much greater risk.

PROHASKA (USA). I would like to comment on your suggestion, and I think it is a reasonable one, because there would appear to be species difference as to how the tissues change in regard to, say for example, superoxide dismutases. Yesterday we heard a talk (Paynter and Allen) that brought this point out in terms of superoxide dismutases. We know in Menkes' disease that there appears to be a decrease in the polyunsaturated fatty acids both in brain and in red blood cells which suggests some sort of lipid peroxidation. But I caution you with the following, in the case of the copper deficient rat-brain where we have low brain copper, low cytochrome oxidase and low superoxide dismutase, we really don't see any changes in lipid peroxidation. I still think the hypothesis is reasonable but obviously I think you should measure the superoxide dismutase and lipid hydroperoxides in the tissue.

SUTTLE (Scotland). These cases I think, are natural cases from an area where cobalt deficiency is also endemic. What steps were taken to ensure that cobalt was not a limiting element too, and had produced neurological effects?

SMITH (Australia). All the animals received a cobalt pellet plus a grinder throughout the experiment.

SWAYBACK LESIONS AND VULNERABLE PERIODS OF DEVELOPMENT

J. McC. Howell, D. A. Pass

Division of Veterinary Biology,
School of Veterinary Studies, Murdoch University,
MURDOCH, WESTERN AUSTRALIA, 6150

and

S. Terlecki
Ministry of Agriculture, Fisheries and Food
Central Veterinary Laboratory, NEW HAW, WEYBRIDGE, SURREY, ENGLAND

Swayback or enzootic ataxia of lambs occurs as two clinical syndromes known as congenital and delayed swayback. In the congenital form the symptoms are present at birth and a variable number of the lambs have cavities or areas of gelatinous softening in the cerebral white matter (1). Neuronal changes are present in the brain stem and grey matter of the spinal cord, their predilection sites being the red and vestibular nuclei, reticular formation and the ventral horns of the spinal cord. However "whilst damaged cells in these areas were not uncommon in newborn lambs, there was a progressive increase in animals up to 8 weeks of age" (2). Recent evidence (3) was thought to indicate that the neuronal changes in the spinal cord did not appear until some weeks after birth.

Cases of delayed swayback are clinically normal for several weeks before the development of ataxia and all have lesions in large neurones of the brain stem and spinal cord and in the white matter of the spinal cord. There is good evidence (4) to suggest that these lesions develop after birth. Barlow (2) recorded that in the youngest of his swayback lambs changes in Purkinje cells and neurones of the central nuclei of the cerebellum were occasionally seen. All other workers have failed to report changes in the cerebellum of swayback lambs.

Congenital swayback has been described (5, 6) in the goat and one of these cases (5) might have had cavitation of the cerebral white matter. Delayed swayback has also been reported in goats (5, 6, 7, 8). The last two papers described well developed histological lesions in the cerebellum. These lesions were present in eleven of the sixteen cases described in one report and "patchy cerebellar cortical hypoplasia involving both granule and Purkinje cells" was described (8).

This paper describes lesions in the cerebrum of goats and in the cerebellum in sheep and goats with swayback. A hypothesis is given for the development of the lesions of swayback which could encompass the major changes seen in the central nervous system (CNS) of congenital and delayed cases.

Two outbreaks of swayback were investigated in goats in Western Australia. Eighty Angora cross goats were present on the first property and swayback was diagnosed in two kids. Hypoplasia of the cerebellar granular layer was seen in one of these. On the second property 475 goats were at risk and swayback was diagnosed in 6 kids. Three of the animals had the congenital form of the disease and two of them had cavitation and gelatinous lesions in the cerebral white matter and small abnormal cerebella. The cerebellar lesions consisted of patchy loss of cells from the external and internal granular layers and dysplasia of Purkinje cells which were sometimes necrotic or vacuolated. The spinal cord was available from two of these 1 day old kids and in both of them chromatolysis was present in the neurones and changes were present in the white matter. In these outbreaks of swayback in goats the presence of lesions in the cerebral white matter was unequivocally demonstrated for the first time and the importance of cerebellar lesions was confirmed. Cerebellar lesions have been found by one of us (S.T.) in 8 experimental and 4 field cases of swayback in lambs. These lambs also had cerebral cavitation. The cerebellar lesions were almost entirely restricted to the vermis where lobules 4, 5, 6 and 7 were consistently affected. In these areas the internal granular layer contained few cells and Purkinje cell changes were seen. The overlying external granular layer was well developed in most areas and was not uncommonly more than six cells deep. This is the first time that a developmental abnormality has been described in the cerebellum of swayback lambs.

Any hypothesis seeking to explain the lesions of swayback must be able to account for the congenital and delayed form of the disease and the presence of lesions in cerebral white matter, the Purkinje and granular cells of the cerebellum, the neurones of the brain stem and spinal cord and the white matter of the spinal cord. It is suggested that the lesions or group of lesions that appear in swayback occur when the deficiency of available copper is severe enough to produce a failure of biochemical processes or cell injury in the foetus at the time when developmental changes are at their vulnerable periods.

The CNS develops in a number of stages. Initially there is organogenesis and neuronal multiplication and the shape of the brain develops. Thereafter cells mature and axons and dendrites grow. At this time glial multiplication and myelination occur. The final stage is a period of growth in size. These events occur at different times for various parts of the CNS. The parts grow in a series of growth spurts, the processes involved are more vulnerable during a growth spurt and this has led to the concept of "vulnerable periods of development".

Foetal brain development in the sheep has been studied (9, 10, 11). The major growth phase of the cerebrum was found to be in utero but occurred after birth for the spinal cord. The spinal cord of the lamb more than doubled its size from 5 weeks of age to maturity by the deposition of lipid and without an increase in cell numbers. Two periods of peak activity for myelination were found, one occurring 20 days before and the other approximately 20 days after birth. In the second phase the spinal cord had a much greater activity than other parts of the CNS (10). The most rapid increase in DNA occurred in the cerebral hemispheres from 40-90 days in utero and confirmed a biphasic pattern of brain growth for lambs, neuroblast multiplication occurred from 0-90 days and neuroglial multiplication and myelination occurred from 90 days onwards. The cerebellum was the most cellular component of the brain from 80 days onward (11).

If the critical period of copper deficiency occurred in early or mid pregnancy the cerebral hemispheres would be most vulnerable. Work with attenuated bluetongue virus vaccine (12) has confirmed this concept for if the vaccine was given to pregnant ewes between 40-80 days of gestation severe cerebral damage was induced. Cerebral lesions have been found in copper deficient lambs at 99 days of foetal age (13). Thus the lesion may be produced during early foetal development.

If the critical period of copper deficiency occurred in late pregnancy myelination might be affected in brain and spinal cord and cellular multiplication and division may be affected, particularly in the cerebellum, producing a lesion in myelinated structures in cerebrum and cord and changes in granule and Purkinje cells in the cerebellum.

If the critical period of copper deficiency occurred after birth the spinal cord would be most severely affected. This is comparable to the situation in the delayed case.

The process of myelination in foetal sheep is preceeded by the development of maximum cytochrome oxidase activity in the interfascicular glia (9). It has been suggested (14, 15, 16) that reduced myelin formation in swayback is related to impaired lipid biosynthesis attributable to copper and cytochrome oxidase deficiencies. It has been shown (15, 17) that the spinal cord of swayback lambs is deficient in myelin lipids and that the myelin formed was deficient in copper and in one of its basic proteins (17). Such myelin would be expected to be unstable and might even be Marchi positive.

Neurones of the brain stem and spinal cord migrate into their positions early in development but their grwoth in size accompanies the overall growth of tissue. They are sensitive cells and it is of relevance that when pregnant ewes were fed molybdenum and sulphate to induce copper deficiency, the CNS of their offspring was deficient in copper and cytochrome oxidase but the only lesion observed in the CNS was chromatolysis of neurones in the red nucleus and ventral horns of the spinal cord (18). It has been considered (3) that the primary event in the spinal cord white matter lesion may be in the axon rather than in the neurone cell body. However if the primary event is an impairment of the metabolism of the neurone cell body and it cannot maintain the integrity of the axon then the initial morphological change will be seen within the axon.

It is suggested that the various lesions seen in swayback animals are all due to copper deficiency and that the variation depends on the existence of critical copper deficiency at the time when a particular structure is at its vulnerable stage of development.

REFERENCES

1. Howell, J.McC., 1968, Proc. Nutr. Soc. 27:85 - 88.

2. Barlow, R.M., 1963, J. Comp. Path. 73:51 - 60.

3. Smith, R.M., F.J. Fraser and J.S. Robertson, 1978, J. Comp. Path. 88:401 - 408.

4. Lewis, G., S. Terlecki,and B.N.J. Parker, 1974, Vet. Rec. 95:313 - 316.

5. Hedger, R.S., D.A. Howard, and M.L. Burdin, 1964, Vet. Rec. 76:493 - 497.

6. Owen, E.C., R. Proudfoot, J.M. Robertson, R.M. Barlow, E.J. Butler and B.S.W. Smith, 1965, J. Comp. Path. 75:241 - 251.

7. O'Sullivan, B.M., 1977, Aust. vet. J. 53:455 - 456.

8. Cordy, D.R., and H.D. Knight, 1978, Vet. Pathol. 15:179 - 185.

9. Barlow, R.M., 1969, J. Comp. Neurol. 135:249 - 262.

10. Patterson, D.S.P., D. Sweasey, and C.N. Herbert, 1971, J. Neurochem. 18:2027 - 2040.

11. McIntosh, G.H., K.I. Baghurst, B.J. Potter, and B.S. Hetzel, 1979, Neuropath. appl. Neurobiol. 5:103 - 114.

12. Osburn, B.I., A.M. Silverstein, R.A. Prendergast, R.D. Johnson, and C.J. Parshall, 1971. Lab. Invest. 25:197-210.

13. Smith, R.M., F.J. Fraser, G.R. Russell, and J.S. Robertson, 1977, J. Comp. Path. 87:119 - 128.

14. Howell, J.McC., A.N. Davidson, and J. Oxberry, 1964, Res. vet. Sci. 5:376 - 384.

15. Howell, J.McC., A.N. Davidson, and J. Oxberry, 1969, Acta neuropath. 12:33 - 41.

16. Gallagher, C.H., and V.E. Reeve, 1971, Aust. J. exp. Biol. med. Sci. 49:21 - 31.

17. Patterson, D.S.P., J.A. Foulkes, D. Sweasey, E.M. Glancy, and S. Terlecki, 1974, J. Neurochem. 23:1245 - 1253.

18. Fell, B.F., C.F. Mills, and R. Boyne, 1965, Res. vet. Sci. 6:170 - 177.

DISCUSSION

NEWBERNE (USA). I think one of your slides pointed out an interesting aspect, namely that growth of the brain stem parallels that of the cerebellum.

HOWELL (Australia). Yes. The lesions in the brain stem involve chromatolysis of large neurones and occasionally you will see the changes in the spinal cord proceeding through the medulla and up the cerebellar peduncles. However the most common lesion is chromatolysis of large neurons, and I have always thought that it was significant and extremely interesting that when Fell, Mills and Boyne (1965) attempted to produce swayback in sheep by feeding the pregnant ewes rations containing molybdenum and sulphate they didn't produce clinical signs of swayback, but the lambs had chromatolysis and swelling of large neurones in the brain stem. These workers also demonstrated a significant reduction in cytochrome oxidase activity in those neurones.

SMITH (Australia). I would like to point our that there may very well be a difference between the way in which the cerebral lesions and those in spinal cord arise. This difference emerges in the fact that one can't demonstrate the degradative process with the Marchi stain in or around the cerebral lesions. There is a school of thought for which one of the proponents of is Spais in Greece that poor vascularisation of the brain, in the deeper parts of the cerebral cortex, is responsible for the failure to myelinate in that part of the copper deficient brain. Secondly that there hasn't, as far as I know, previously been any demonstration of demyelination in the cerebellum. Could you give us some idea of how the demyelinating process of the cerebellum might relate to a failure of vascularisation - where does the vascular supply come from?

HOWELL (Australia). Yes, the idea of a vascular malfunction in the pathogenesis of swayback has previously been considered (Behrens, H. and Schulz, L-CL, 1959, Dtsch. tierarztl. Wschr. 66:502; Schulz, L-CL and Behrens, H. 1960, Beitr. path. Anat. 122:282). As Dr Lewis has said earlier, a considerable amount of work has been done in Weybridge where they produced swayback experimentally in animals with known conception dates and Dr Terlecki is currently working up this material so that he will have, using H and E techniques and myelin techniques, a demonstration of the changes in relation to feotal age. I am sure, that Dr Terlecki will look at the blood vessels with great care. I know of no evidence that suggests a primary defect in the blood vessel walls in these animals. With regard to the condition of the cerebellar myelin, we did not carry out biochemical assays but histochemical stains such as Luxol fast blue did not demonstrate changes in our animals. The lesion in the cerebellum was a reduction in the number of cells in the internal granular layer. A similar picture is seen in kittens for example when a Parvo virus, which requires actively dividing cells for replication, attacks the cells of the external granular layer in the neonate. The cells die and are no longer available for migration to form the cells of the internal granular layer. Hence the kitten develops cerebella hypoplasia. Copper deficiency may in some way interfere with the cells in the cerebellum of the lamb when they are actively dividing - when they are at a vulnerable stage of development. This is part of the hypothesis that we are putting forward.

MILLS (Scotland). Could I just make two points. I think Dr Terlecki has already been quizzed on the question of changes in vascularity at a meeting of the Ciba Foundation about 9 months ago, and his view then was that there was no evidence in the swayback brain for a decrease in vascularity relevant to the lesion. Could I just raise one point in relation to the discussion. Nobody has touched upon the fact that the gross clinical signs of swayback can come on so very abruptly. Now we are not talking about weeks - we are talking about hours, or sometimes quarters of an hour. There has been very little direct experimental work done upon this, but in the course of studies that have been done in Aberdeen, on at least five occasions we have noticed, and I think it has been reported from the field too, a situation where a sudden increase in oxygen demand has precipitated gross clinical signs. I find this very difficult to understand in terms of lipid degeneration. It seems to me that we must be considering the abrupt failure and the permanent failure of a process which is reflecting mitochondrial activity. I think, certainly susceptibility must change, as you pointed out with developmental age, but I really come back to the point that whatever it is, can occur extremely quickly - the population of cells can be on the brink and an increase of oxygen demand, can cause permanent damage. Now what I mean by an increase in oxygen demand, is either physical exercise or emotional stress, including animals being moved to an area where they have never been before.

HOWELL (Australia). Certainly I would agree that it is surprising that the animal can be clinically normal one day and the next day has the signs of swayback - that has been seen many times in the delayed form. I agree with your observations but it needs to be looked at critically for if you make a lamb that looks clinically normal, but which has swayback, run quickly and turn, it will show the signs, it will then stand still and it won't show the signs. However, Dr Lewis has shown, in some of her studies with twins that if you kill an animal that looks clinically normal it may have the lesions of swayback. If its twin is allowed to survive it will subsequently develop the clinical signs of swayback. This is an interesting aspect of the problem; there is still much to do.

COPPER DEFICIENCY AND TIMING OF HYPOMYELINATION

M. Perlman, B. Buchanan and B. Sarkar

Departments of Pediatrics and Biochemistry
The Hospital for Sick Children and University of Toronto
555 University Avenue, TORONTO ONT.M5G 1X8
CANADA

Nutritional copper deficiency (CuD) occurs in human infants at an age which coincides with the "critical period of myelination" (2). Owing to the association of CuD with hypomyelination (3-4), maldevelopment of the central nervous system (CNS) may be implicated in human CuD. This study deals with the timing of reduced myelin synthesis in the sequence of pathological events arising from CuD. In this rat model, conditioned CuD was synchronized with the period of active myelination.

MATERIAL AND METHODS

Pregnant Sprague-Dawley rats received a commercially available CuD diet (ICN, Cu 0.7 ppm) and double-deioned water (CUD group), or the same diet with added Cu (15 ppm) and tap water (control group) from day 2 of gestation. Rats were housed in individual stainless steel cages with grated floors until 15 days of gestation, when stainless steel plate floors were placed in the cages, without bedding. On day 2 after birth CuD litters were reduced to 8 and control litters to 10-12.

Animals were selected for study of CNS uptake of isotopes by matching body weights of 2-3 CuD littermates with body weights of 2-3 control littermates. (^{35}S)-sulphate (0.3 uCi/g body weight) was given i.p. on days 6, 12, 18 and 24. (^{32}P)-orthophosphate (0.15 uCi/g body weight) was injected i.p. on days 6, 12, 14, 16, 18 and 24. Preliminary experiments had established that brain and spinal cord uptakes of both isotopes were proportionate to dosage. Cu-refeeding experiments (200 ppm in drinking water) were performed in 19-day rats and in weaned 23-day rats. Animals were killed by decapitation under halothane anaesthesia 24 hours after injection of isotopes. The CNS was rapidly excised, divided into parts indicated (figures 1,2), dried by blotting,and weighed in preweighed homogenization tubes kept on ice. Cerebellum, brain stem and spinal cord samples from littermates were pooled.

After homogenization CNS lipids were extracted by the method of Folch (5). Upper phase contained less than 10 percent of radioactivity. Aliquots of lower phase were evapourated to dryness by nitrogen in scintillation vials and counted. Counts were corrected for decay and expressed as cpm per gram wet weight of tissue. Thin layer chromatography (TLC) (6) was performed on aliquots of selected samples and identified phospholipid fractions scraped into scintillation vials and counted. Serum ceruloplasmin (Cp) values were determined (7).

RESULTS

Serum Cp and hematocrit values (Table) were significantly reduced at 7 days of age.

Table: Serial hematocrit and Cp values in CuD and control rats. (± standard deviation)

		7d	13d	15d	17d	19d	25d	29d
Hemato-crit %	N	37.6±2.1	23.4±1.4			29.7±1.0	36.1±2.3	37.7±1.0
	CuD	30.3±1.8*	23.4±3.3			21.3±3.1	22.6±1.3	30.8±1.0
Cerulo-plasmin mg/dl	N	15.2±2.0	10.9±2.6	12.7±0.9	12.5±1.9	14.8±1.7	21.0±2.2	23.0±2.7
	CuD	9.7±0.8*	5.1±1.6	2.6±0.5	1.9±0.6	1.5±0.7	1.1±0.6	1.3±0.6

* $p < 0.005$. p value of all other differences was <0.0005.

The (^{35}S) uptake in CuD rats was significantly reduced in brain at 13 days and in spinal cord at 13 and 19 days (fig.1). Uptakes after Cu-refeeding from 19 days were similar to uptakes seen in controls between 19 and 21.5 days of age.

The (^{32}P) uptakes in CuD were reduced in all CNS parts from 15 days, differences were maximal at 17 days, and a trend to recovery was noted at 19 and 25 days (fig 2). Cu-refed weanlings (from 23 to 25 days) had 35 per cent greater cerebral uptakes of ^{32}P than weight-matched littermates maintained CuD ($p<0.05$). Counts recovered from lipid extracts of CNS of rats aged between 13 and 19 days and applied to TLC were 74-110 per cent in phospholipids; no differences in the distribution of the individual phospholipids was discerned in CuD compared with controls.

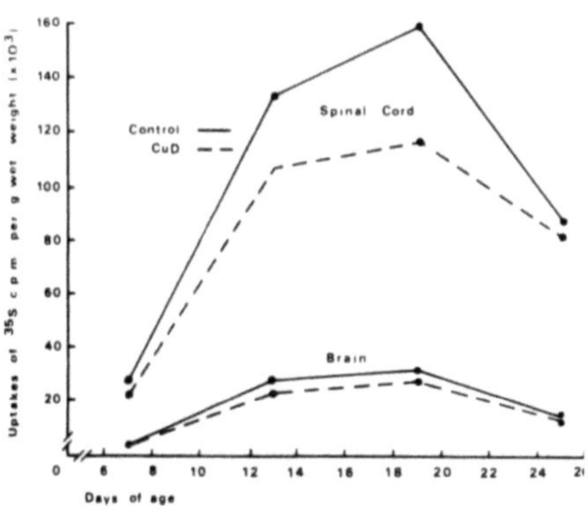

Fig. 1 -Rat CNS uptake of (^{35}S) in CuD and controls of 7-25 days of age.

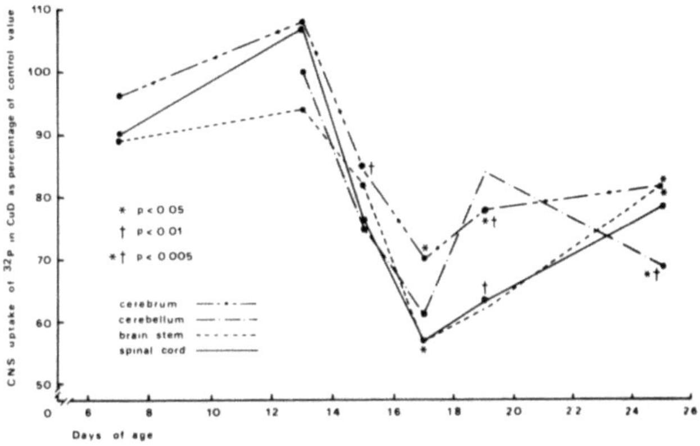

Fig. 2 -Rat CNS uptakes of (^{32}P) in CuD at 7-25 days of age as percentage of control uptakes.

DISCUSSION

In previously published studies of myelination in CuD (3,4) animals were severely CuD prior to as well as during the period of active myelination. As active myelination is preceded by a period of proliferation of oligodendroglia, observed "hypomyelination" may have been secondary to a CuD lesion of oligodendroglia. The present study was intended as a model of CuD as is seen in human infants in that the CuD was focused on the period of active myelination. In previous studies CuD animals were severely growth retarded compared to control animals. The present study controlled for body weight and brain weights as well as body weights proved to be similar in CuD and controls at all ages.

Uptake of ^{35}S has been used as a marker of the rate of synthesis of sulphatide which is a relatively "myelin-specific" lipid (8). Uptake of ^{32}P by the lipid fraction of CNS is a marker of phospholipid synthesis, a quantitatively important component of myelin (9). Owing to the predominance of myelin synthesis as a metabolic activity during peak myelination, ^{32}P uptake probably largely reflects myelin synthesis between 13-25 days of age.

Significant reduction in haematocrit and Cp values were already well established by 7 days of age whereas initial reductions of CNS uptakes of ^{35}S and ^{32}P were observed at hypomyelination seems to be a relatively late event.

Of interest was the observed trend for the uptakes of both ^{35}S and ^{32}P to normalize with increasing age (figs. 1-2), despite evidence of a persistent CuD state (Table), and the need for Cu accretion owing to continued growth. A similar phenomenon is evident in the haematocrit time courses. One can speculate that Cu is less essential at later ages, or, that an "escape phenomenon" exists which indicates the development of alternative pathways to circumvent the effects of CuD in a number of different systems.

REFERENCES

1. Shaw, J.C.L., 1980, Amer. J. Dis. Child. 134:74-81.

2. Davison, A.N., and J. Dobbing, 1966, Brit. Med. Bull, 22:40-44.

3. Zimmerman, A.W., J.-M. Matthieu, R.H. Quarles, R.O. Brady and J.M. Hsu, 1976, Arch. Neurol. 33:111-119.

4. Dipaolo, R.V., J.N. Kanfer, P.M. Newberne, 1974, J. Neuropathol, Exp. Neurol. 33:226-336.

5. Folch, J., M. Lees and G. Sloane-Stanley, 1957, J. Biol. Chem. 226:497-508.

6. Phillips, F., and O.S. Privett, 1979, Lipids 14:590-595.

7. Holmberg, C.G. and C.B. Laurell, 1951, Scand. J. Clin. Lab. Invest. 3:103-107.

8. McKhann, G.M. and W. Ho, 1967, J. Neurochem., 14:717-724.

9. Abdel-Latif, A.A. and L.G. Abood, 1965, J. Neurochem. 2:157-166.

DISCUSSION

NEWBERNE (USA). I would like to ask Professor Howell if these observations correlate with his observations in the goat?

HOWELL (Australia). I think the involvement of vulnerable periods fits in very well in that the stress was applied during the period of active myelination. What also fits quite well is the sequence of neuronal proliferation, neuroglial proliferation and active myelination, which occur in sequence, no matter which part of the brain, but at different times in relation to birth. Barlow has shown that in fetal lambs the neuroglial proliferation precedes myelination activity and has shown histochemically that the neuroglial proliferation is accompanied by the appearance of cytochrome oxidase activity in those cells.

PERLMAN (Israel). I would like to mention that in rodents myelination is a monophasic process rather than the biphasic process seen in ungulates.

PROHASKA (USA). I think we see very similar results in terms of this self-healing process. After a period of time the copper-deficient animals are able to recover. We see exactly the same kind of trend with respect to hematocrit and even with regard to the norepinephrine levels. The question: Were you able to demonstrate hypomyelination in terms of the yield or by actually measuring the concentration of myelin components?

PERLMAN (Israel). We did not quantitate myelin synthesis by other means.

FAILURE TO CONFIRM COPPER DEFICIENCY IN CRINKLED MICE

J.R. Mann*, J. Camakaris†, J.M. Gillespie§, B. Koellreuterϕ, J.M. Matthieuϕ, P.M. Royce* and D.M. Danks*
*Genetics Research Unit, Royal Children's Hospital,
†Department of Genetics, University of Melbourne,
§Division of Protein Chemistry, C.S.I.R.O.,
PARKVILLE, 3052, VICTORIA, AUSTRALIA and
ϕService de Pediatrie, Centre Hopitalier Universitaire,
VAUDOIS, CH1011, LAUSANNE, SWITZERLAND.

INTRODUCTION

Crinkled (*cr*) is an autosomal recessive mutant gene in *Mus musculus*. Various phenotypic abnormalities in homozygous mice, (*cr/cr*), which include abnormal hair growth, delayed pigmentation and thin skin (1), have been claimed to resemble those observed in copper deficient animals and in the inherited copper deficiency disease in humans, Menkes' syndrome (2). A moderate reduction in liver copper and the activity of the copper-dependent enzyme superoxide-dismutase has been reported in neonatal *cr/cr* mice (3,4), as well as alterations in the concentrations of some chemical components of brain myelin (5). Many of the phenotypic abnormalities have been reported to be ameliorated by copper supplementation (2,4,6).

We could not find any evidence in *cr/cr* mice for the presence of any abnormalities known to exist in nutritionally copper deficient animals (7), and in the inherited diseases of copper deficiency in mottled mice and human Menkes' syndrome (8).

MATERIALS AND METHODS

Crinkled mice were obtained from the Jackson Laboratory, Bar Harbour, Maine, U.S.A. The *cr* gene was maintained on a hybrid background (C3H X C57B1/6J). All the abnormalities in our *cr/cr* mice corresponded to those previously described (1,2). Phenotypically normal littermates (+/?) were used as controls.

Copper concentrations in livers and cultured cells were determined by the method of Stevens (9), and caeruloplasmin oxidase activity assayed by the method of Camakaris (10). The concentration of hair sulphydryl groups was measured by the methods of Leach (11) and Inglis *et al* (12), skin lysyl oxidase activity was assayed by an adaptation of the method of Royce *et al* (13), and chemical analyses of brain myelin were carried out according to Zimmerman *et al* (14). Skin was sectioned and stained with haemotoxylin and eosin for histological observation.

In copper supplementation experiments, *cr*/+ females were mated with *cr*/+ males and fed during pregnancy and lactation, either 1) normal diet: food cubes containing 11.8 ± 4.7 μg of copper/g and water containing 0.27 ± 0.02 μg of copper/ml, or 2) high copper diet: the same diet as 1), except that the drinking water contained 220 μg of copper/ml as copper chloride. Litters were observed each day up to 10 days of age after birth to monitor the effects. In another experiment, three *cr/cr* mice were injected with 25 μg of copper as cupric chloride in 0.9% sodium chloride (aq) at 3 days of age.

RESULTS AND DISCUSSION

No differences in the copper concentrations in the livers of 4-5 day old +/? and *cr/cr* mice were found, whether comparing between or within litters (Table 1). This result differs from the report of liver copper deficiency in *cr/cr* mice (3). However, this was only observed when comparing within litters, and the liver copper was only reduced to 73% of +/? mice, prior to weaning age. An accompanying reduced activity of the copper-dependent enzyme superoxide-dismutase was also reported (4), however, this observation should be compared with non-copper-dependent enzymes to eliminate a possible non-specific effect of *cr*. Caeruloplasmin oxidase activity was also found to be normal in 6-9 day old *cr/cr* mice. [A_{530}: +/?, 0.326 (3); *cr/cr*, 0.345 (3). Mean (n)].

Table 1. Copper Concentrations in Livers
µg of copper/g dry weight of individual whole livers.

	+/?	*cr/cr*
Litter 1	143,142	125,151
2	202	163
3	130,171	202,197
4	193	230
5	240	239
Mean ± S.D.	174 ± 39	187 ± 42

The copper concentrations in *cr/cr* cultured cells 2 days after incubation were normal in both normal and copper-loaded medium (Table 2). This differs from the high concentrations of copper observed in cultured cells of mottled mice and patients with Menkes' syndrome (15). Furthermore, over 24 hours, *cr/cr* cultured cells accumulated normal amounts of ^{64}Cu added to the medium.

Table 2. Copper Concentrations in Cultured Cells
µg of copper/10^9 cells ± S.D. (n). FbMM : Fibroblast Maintenance Medium.

		FbMM 0.6 µg/ml Cu	FbMM + 1.5 g/ml Cu	FbMM + 3.0 g/ml Cu
lung	+/?	8 ± 4(8)	18 ± 8(7)	32 ± 10(3)
fibroblasts	*cr/cr*	11 ± 8(6)	14 ± 9(7)	21 ± 9(3)
kidney epithelial	+/?	11 ± 8(9)	19 ± 8(7)	57 ± 48(3)
cells	*cr/cr*	17 ± 5(5)	19 ± 16(4)	47 ± 14(2)

Copper supplementation did not reduce the high rate of early mortality in *cr/cr* mice. The survival up to 20 days of age on normal and high copper diets respectively was 100% and 91% in +/? mice, and 29% and 8% in *cr/cr* mice. The age of death of *cr/cr* mice on normal diet was 7.2 ± 1.9 (21) days and on high copper diet 5.1 ± 2.5(12) days [Mean ± S.D. (n)]. It was also apparent that the phenotypic abnormalities did not improve. Hurley and Bell (2) have reported an increased survival rate and an amlioration of phenotypic abnormalities in *cr/cr* mice by copper supplementation, but the effects were not marked. Copper-nitriloacetate added to the drinking water has been reported to be more effective (6). However, in these experiments (2,6), the low rate of survival even in +/? mice was also improved by the copper supplementation, suggesting the strain of mice used by this group had a higher requirement for copper than was provided in the normal diet.

It has been regarded that the presence of only straight hair in *cr/cr* mice is similar to the loss of crimp that occurs in the wool of copper deficient sheep (2). In sheep, the effect is on growing hair. However, in *cr/cr* mice, the abnormality is an absence of three normal types of hair, two of which are normally bent, leaving only the normally straight awls (1). Furthermore, in mottled mice and Menkes' syndrome, the effect of copper deficiency on hair is to cause twisting and kinking. The hair abnormalities in copper deficiency diseases are associated with a defective formation of keratin disulphides, leaving a high concentration of sulphydryl groups. The concentration of hair sulphydryl groups in *cr/cr* mice is normal, in contrast to the great increase seen in mottled mice (Table 3), Menkes' syndrome (8) and copper deficient sheep (7).

Table 3. Concentration of Hair Sulphydryl Groups
µmole - SH/g of hair. Means ± S.D. (n). Mice were 50-65 days of age.

crinkled *(cr/cr)*	: 63 ± 16(3)	normal (+/?)	: 56 ± 2(3)
mottled (*Mo*/Y)	: 275 ± 5(2)	normal (+/Y)	: 66 ± 12(4)

It has also been claimed that *cr/cr* mice are similar to copper deficient animals because of delayed skin pigmentation during the first few days after birth (2). However, this is only a manifestation of the abnormality of a slow growth of hair, as the pigment seen in the skin of neonatal mice, which are born without fur, is actually contained in the hair yet to emerge. The hair that *cr/cr* mice do possess is normally pigmented (1).

Thus it was not surprising that this delayed pigmentation was not improved in the three *cr/cr* mice that received a subcutaneous injection of copper. This method markedly enhances hair pigmentation in mottled mice (16).

It was commented that the thin skin of *cr/cr* mice is similar to the situation in copper deficient rats (2). However, it is not apparent in the literature that copper deficient rats have thin skin. Skin abnormalities in copper deficient animals, such as decreased tensile strength and elasticity, are associated with a decreased activity of the copper-dependent enzyme, lysyl oxidase. This enzyme initiates cross-link formation in collagen and elastin of the epidermis and dermis (17). Histological examination revealed that the thin skin of *cr/cr* mice was manifest as a reduction in the layer of subcutaneous fat. The thickness of the epidermis and dermis was normal. Furthermore, lysyl oxidase activities in extracts of skin from 6 day old *cr/cr* mice were found to be normal. [Activity (c.p.m./mg protein): +/?, 1614 ± 429 (6); *cr/cr*, 2120 ± 660 (6). Means ± S.D.(n)].

The analyses of brain myelin showed no differences between +/? and *cr/cr* mice with regard to myelin, phosphorous, galactolipid, sulphatide, cholesterol, cholesterol esters, glycoprotein and 2',3'-cyclic nucleotide 3'-phosphohydrolase concentrations. Analysis of 5% and 15% polyacrylamide gels failed to reveal any difference in the myelin protein fractions or in the major myelin glycoprotein. In particular, the 21 day old animals did not show the increase in myelin sulphatide content described by Theriault *et al* (5). None of the animals were comparable to the 17 month old mice in which these authors described increased brain cerebrosides.

It is possible that our failure to confirm a copper deficiency in *cr/cr* mice could be explained by the different genetic backgrounds used to maintain the *cr* gene. C3H X C57Bl/6J by ourselves and CBA X C57Bl (2-6). However the phenotypic abnormalities and the pattern of mortality were the same on both genetic backgrounds. This suggests that any condition of copper deficiency that might be associated with some stocks of crinkled mice is secondary and unimportant.

REFERENCES

1. Falconer, D.S., A.S. Sobey and J.W.B. King, 1952, J.Genetics, 50: 324-346.
2. Hurley, L.S., and L.T. Bell, 1975, Proc.Soc.Exp.Biol.Med. 149: 830-834.
3. Keen, C.L. and Hurley, 1979, J.Inorg. Biochem. 11: 269-277.
4. Keen, C.L. and L.S. Hurley, 1979, Proc. Soc. Exp. Biol. Med. 162: 152-156.
5. Theriault, L.L., D.D. Dungan, S. Simons, C.L. Keen, and L.S. Hurley, 1977, Proc. Soc. Exp. Biol. Med. 155: 549-553.
6. Keen, C.L., P. Saltman, and L.S. Hurley, 1980, Am. J. Clin. Nutr. 33: 1789-1800.
7. Underwood, E.J. 1977. In "Trace Elements in Human and Animal Nutrition" (Academic Press, N.Y., S.F. Lond.) pp 77-80.
8. Danks, D.M., 1977, Inorg. Perspect. Biol. Med. 1: 73-100.
9. Stevens, B.J., 1972, Clin. Chem. 18: 1374-1384.
10. Camakaris, J., J.R. Mann, and D.M. Danks, 1979, Biochem. J. 180: 597-604.
11. Leach, S.J., 1960, Aust. J. Chem. 13: 547-566.
12. Inglis, A.S., D.T.W. McMahon, C.M. Roxburgh and H. Takayanagi, 1976, Anal. Biochem. 72: 86-94.
13. Royce, P.M., J. Camakaris, and D.M. Danks, 1980, Biochem.J. 192: 579-586.
14. Zimmerman, A.W., J.M. Matthieu, R.M. Quarles, R.O. Brady, and J.M. Hsu, 1976, Arch. Neurol. 33: 111-119.
15. Camakaris, J., D.M. Danks, L. Ackland, E. Cartwright, P. Borger, and R.G.H. Cotton, 1980, Biochem. Gen. 18: 117-131.
16. Mann, J.R., J. Camakaris, and D.M. Danks, 1979, Biochem.J. 180: 605-612.
17. Siegel, R.C., 1979, Int. Rev. Connect. Tissue Res. 8: 73-118.

DISCUSSION

PROHASKA (USA). Is this an unusual strain of mouse? It appears as though your liver copper values are higher than those reported by Keen (Tema-4). Although, I confess that I have seen liver coppers in normal mice as high 50 μg/g (wet wt.) in our mice (C3H/HeJ) (Nutr. Res. in press). Was the genetic background of the crinkled mice the same as the mottled mice, in this study?

MANN (Australia). The strain maintaining *cr* is a hybrid C3H x C57BC/6J. These liver copper concentrations (dry wt.) are similar to those we obtain in mice of similar age in our strains maintaining Mo^{br} and Mo^{blo}. If these results are higher than other observations, this could be due to diet or the methods of analysis. Our strains maintaining Mo^{br} and Mo^{blo} are unknown but are nearest to CBA by isosyme analysis. Altering the genetic background of Mo^{blo} and Mo^{br} does not alter their defective copper metabolism.

FETAL IODINE DEFICIENCY

Basil S. Hetzel

CSIRO Division of Human Nutrition
KINTORE AVENUE, ADELAIDE, SOUTH AUSTRALIA 5000, AUSTRALIA

Fetal Iodine Deficiency (F.I.D.) has now been clearly defined by epidemiological study with confirmation in an experimental model. The condition occurs in man in association with endemic goitre of high incidence with severe dietary iodine deficiency usually in mountainous areas (Himalayas, the European Alps, and in the Andes). Severe endemias have also been described in Indonesia, Papua New Guinea and in Zaire.

ENDEMIC CRETINISM - CLINICAL FEATURES

Clinicially the condition occurs as "endemic cretinism" (1). The term "cretinism" was used in the latter half of the eighteenth century by European authors - and notably in Diderot's Encyclopedie (1754) to describe "an imbecile who is deaf, dumb, with a goiter hanging down to the waist", known to be widely prevalent in Switzerland, Southern France and Northern Italy - among people living in the valleys of the Alps and Pyrenees. Typical examples of the Alpine cretins are shown in illustrations to the Report of the Sardinian Royal Commission published in 1848.

The constant association with endemic goitre led to use of the term "endemic cretinism". In fact, some did not have a palpable thyroid, though usually a goitre was present.

The condition known as "sporadic cretinism" dates from the recognition by the English clinicians Curling and Fagge of a condition of thyroid deficiency associated with retarded physical and mental development which is corrected by use of thyroid preparations.

In recent years, Costa and his colleagues have studied cretins from the Alpine valleys of Northern Italy and have shown clearly that they differ from sporadic cretins. In general the Alpine cretins were clinically euthyroid, unlike the sporadic cretin. The alpine cretins were mentally retarded deaf-mutes with gross neurological deformities similar to those originally described by the Sardinian Commission from the same area.

Since 1960, there has been a spate of studies made outside Europe - notably in the Congo, the Himalayas, New Guinea and South America.

In a series of 254 patients (2) from the Huon Peninsula in Papua New Guinea deaf mutism was observed in 70%, a vacant facies suggesting mental deficiency (64%), and signs of upper motor neurone damage (brisk reflexes 61%), flexural deformities (28%), muscular incordination (26%) and dwarfism (26%). The visible goitre rate was 26%.

There was no clinical evidence of hypothyroidism and Xray of limbs showed no retardation of epiphysial fusion (exception 1 case) and a normal electrocardiagram in all except 2 cases which showed changes of doubtful significance.

Studies of thyroid function revealed elevated I^{131} uptake (mean 63%) lowered serum PBI (mean 2.6μg%) and a normal response to administration of perchlorate (failure of discharge of I^{131} indicating no defect in organic binding).

These findings are similar to those observed in South America (Brazil, Ecuador, Peru), and in the Himalayas and in Indonesia though there is more evidence of hypothyroidism (10-20%) in the latter two areas. They differ from the predominant picture seen in Zaire which is usually one of severe hypothyroidism with growth retardation; the neurological features occur in only 10% of cases in association with normal thyroid status clinically.

RELATION TO IODINE DEFICIENCY OF THE FETUS

Review of the global epidemiology of endemic cretinism indicates that the condition is constantly associated with high goitre rates and low urinary excretion of iodine (usually less than 20μg per day - normal 80-120μg per day) suggesting severe iodine deficiency (3).

Furthermore in Switzerland the use of iodised salt in some cantons after 1926 was followed by a fall in the incidence of deaf mutism. However such a change also occurred in cantons where iodised salt was not used (4). Between 1920 and 1960 endemic cretinism disappeared spontaneously from various parts of Southern Europe suggesting no specific relationship to iodine deficiency.

The considerations led to the setting up of a controlled trial on the prevention of endemic cretinism in the Western Highlands of Papua New Guinea at the time of the first census in 1966. Previous studies had shown that a single injection of iodised oil provided a satisfactory correction of severe iodine deficiency for 3-5 years (5). Alternate families were therefore injected with iodised oil and saline in a total population of 8,000. Subsequent follow up (6) over 4 years revealed 26 endemic cretins out of a total of 534 children born to mothers who had not received iodised oil; the mothers of 5 of these cretins were pregnant at the start of the trial. In comparison, 7 cases of endemic cretinism occurred among 498 children born to mothers who had been treated with iodised oil; in 6 of these 7 cases, the mother was pregnant when the trial commenced. It was concluded that intramuscular iodised oil was effective in the prevention of endemic cretinism and that, for it to be effective, it should be given prior to conception. This suggested that severe iodine deficiency in the mother produces neurological damage during fetal development. Mass injection programs have followed in PNG and other countries.

TABLE I: CHILDREN BORN IN JIMI RIVER SUBDISTRICT, WESTERN HIGHLANDS OF PAPUA NEW GUINEA CLASSIFIED ACCORDING TO TREATMENT RECEIVED BY MOTHER (6)

Treatment received by mother	Total No. of new births	No. of children examined	No. of deaths recorded	No. of endemic cretins
Iodised oil	498	412	66	7*
Untreated	534	406	97	26+

*6 already pregnant when injected with oil
+5 already pregnant when injected with saline solution

More recently studies (7) of motor development in apparently normal children aged 6-12 years from the same population in the Western Highlands of Papua New Guinea revealed defects in coordination in those (79 children) born to iodine deficient mothers compared to those (115 children) born to mothers whose iodine deficiency had been corrected by injection of iodised oil. These defects were demonstrable after correction for age in tasks requiring manual dexterity such as threading beads ($p<0.01$) and inserting pegs on a peg board ($p<0.05$).

These findings indicated subclinical defects in the whole population exposed to severe iodine deficiency during fetal development.

There is therefore a continuum of defects arising from iodine deficiency of the fetus varying from the gross neurologic defects of mental deficiency, deaf mutism, and spastic diplegia to subclinical coordination defects. These findings further indicate that the whole continuum of defects can be prevented by correction of the fetal iodine deficiency before pregnancy in the iodine deficient mother. The experience in Papua New Guinea has been confirmed by studies in South America and Indonesia.

MECHANISM OF THE EFFECT OF FETAL IODINE DEFICIENCY ON BRAIN DEVELOPMENT

Severe iodine deficiency is associated with lowered levels of maternal and fetal thyroid hormones. Maternal hypothyroidism due to various causes has not been associated with children exhibiting the clinical features of endemic cretinism (6). Fetal hypothyroidism occurs most severely in Zaire but as already mentioned this endemic is not typically associated with the neurological form of endemic cretinism. These observations have led to the suggestion that iodine deficiency itself may affect early fetal brain development independent of the effect on maternal and fetal thyroid function (6).

Further investigation clearly required the development of an animal model. Initial studies in the mouse, rat and guinea pig have had only limited success. However the sheep has now provided a model showing the effect of severe iodine deficiency in the pregnant ewe on fetal development (8). A condition of severe fetal hypothyroidism has been produced characterised by absence of wool growth, skeletal retardation and retarded brain development.

A similar but less severe condition has been produced by thyroidectomy of the lamb fetus in utero at 50-60 days gestation (9). Maternal hypothyroidism produced by thyroidectomy has been associated with a transient retarding effect on fetal brain development at mid gestation.

Brain retardation at 140 days gestation (normal period of gestation in sheep 150 days) indicated by reduced brain weight, brain DNA and brain protein occurred in both cerebral hemispheres and cerebellum. Histological studies indicate increased cell density in the hippocampus. In the cerebellum there is thickening of the external granular layer due to a slowing of the normal internal migration of cells, a retardation of dendrite arborisation of the Purkinje cells, and a reduced thickness of the molecular layer. (Table II)

TABLE II: MORPHOLOGICAL DIFFERENCES IN IODINE DEFICIENT FETAL SHEEP BRAIN+* (per cent) (140 days gestation)

Cerebellar cortex		Cerebral cortex		Hippocampus	
molecular layer/total area	-11*	Motor area cells per mm^2	+16*	CA_1 cells per mm^2	+30*
medulla layer/total area	+24*	Visual area cells per mm^2	+34*	CA_4 cells per mm^2	+19*
Purkinje cells per mm	+13*				

+See Cha Hua et al. (10)
*$p<0.05$
Compared with control (iodine suplemented) sheep

These histological findings could explain the defects in coordination that have been observed in apparently normal children as well as the more severe mental and coordination defects characteristic of endemic cretinism.

However no effects on cranial nerves IV and VI have been demonstrated. It has not been possible so far to observe the iodine deficient fetal lambs following delivery due to the high risk of loss during labour.

However the observation of gross hypothyroidism in these fetal lambs indicates a startling difference from the usual form of human fetal iodine deficiency. However the condition observed in Zaire does closely resemble that of the iodine deficient fetal lamb.

It is concluded that the clinical and epidemiological entity of fetal iodine deficiency has been demonstrated in human populations subjected to severe dietary iodine deficiency. Experimental studies in animals have established a model in the sheep which resembles one form of human fetal iodine deficiency. Further studies are indicated.

REFERENCES

1. Hetzel, B.S., 1972, In Human Development and the Thyroid Gland, (ed. J.B. Stanbury and R.L. Kroc) (Plenum, N.Y.) p. 119-134.
2. Buttfield, I.H. and Hetzel, B.S., 1969, Aust. Ann. Med. 18:217-221.
3. Querido, A., 1971, In Endemic Cretinism Monograph Series No. 2 (eds. Hetzel, B.S. and Pharoah, P.O.D.) (Institute of Human Biology, Papua New Guinea. pp. 132-133).
4. Wespi, H.J., 1945, Schweiz, Med. Wochenschr. 75:625-629.
5. Buttfield, I.H. and Hetzel, B.S., 1967, Bulletin WHO 36:243-262.

6. Pharoah, P.O.D., Buttfield, I.H. and Hetzel, B.S., 1971, Lancet, 1:308-310.

7. Connolly, K.J., Pharoah, P.O.D. and Hetzel, B.S., 1979, Lancet, 2:1149-1151.

8. Potter, B.J., McIntosh, B.H., Mano, M.T., Belling, G.B., Marshall, J., Wellby, M.L., Marshall, J. and Hetzel, B.S., 1980, Thyroid VIII p. 12-15.

9. McIntosh, G.H., Baghurst, K.I., Potter, B.J. and Hetzel, B.S., 1979, Neuropath. and App. Neurobiology, 5:363-376.

10. Cha Hua, Cragg, B., Potter, B.J., McIntosh, G.H. and Hetzel, B.S., 1980, Thyroid VIII p. 16 - 19.

DISCUSSION

KRISHNAMACHARI (India). We are very much aware of Dr Hetzel's work, but in India where, as you know, goitre is an endemic problem, the cretinism is not as much seen as it is in the endemic areas of Papua New Guinea, Indonesia and Zaire. Professor Howell presented his findings in the brains of copper deficient sheep and goats where changes in the cerebellum appear to be similar to those seen in iodine deficiency. Therefore probably in the present context we will have to discuss the possibility of interaction between iodine deficiency and deficiencies of other trace element salts, which will explain why in some endemic areas cretinism is more common although iodine deficiency is almost of the same magnitude.

HETZEL (Australia). It is a very interesting suggestion. I think the immediate point of course is the severity of the deficiency of iodine. We would say that you do not see cretinism unless the iodine intake is less than 20% that is 20 gamma a day, you get goitre below 50 gamma a day I would say. Cretinism certainly only occurs with severe iodine deficiency, and is characteristic of mountainous areas, and villages well up the slopes so to speak, goitre is a much wider problem. Cretinism is of more limited occurrence and is easily prevented if you can get above the 20 gamma intake. This accounts for its spontaneous disappearance.

HURLEY (USA). I wonder if you think that there was a genetic factor involved in these geographical differences with the people in these isolated mountain valleys having a greater rate of inter-marriage? Secondly, if a decline in the close inter-marriages might account for the decline in the frequency of cretinism even before the iodisation of salt?

HETZEL (Australia). Well I think those are very reasonable suggestions. The genetic component certainly was postulated earlier and I think in fact the view that you have just put forward was the current view about cretinism I would say up until 1965. I believe that the geographic differences can be largely accounted for by differences in iodine intake. I think that diet is diversified for these people when they became less remote and a very small increase in iodine would be sufficient to prevent the condition. So my hypothesis is still along the ideas of the environment and the iodine although one could not of course exclude some genetic factor, but I think it is a minor one.

REVERSAL OF BRAIN RETARDATION IN IODINE-DEFICIENT FETAL SHEEP

B.J. Potter, M.T. Mano, G.B. Belling, P.F. Rogers, D.M. Martin and B.S. Hetzel

CSIRO Division of Human Nutrition,
KINTORE AVENUE, ADELAIDE, SOUTH AUSTRALIA 5000
AUSTRALIA

Epidemiological studies in Papua New Guinea have established a relationship between iodine deficiency and neurological endemic cretinism in man (1,2). The fact that cretinism is often unaccompanied by hypothyroidism has led to the suggestion that iodine may have an essential role in fetal brain development independent of any action it may have via the thyroid gland. This is supported by reports of controlled human studies in which severe iodine deficiency was corrected by a single intramuscular injection of iodized oil (2.15 g iodine) and that endemic cretinism could be prevented provided iodine deficiency was corrected before pregnancy. The claim that the deficiency cannot be corrected by iodized oil after conception suggests that iodine is necessary for normal brain development in the first trimester (1,3).

The role of iodine in fetal brain development and the effect of iodized oil administration during pregnancy on brain growth have now been studied in iodine-deficient sheep. The sheep was chosen as the experimental animal model for these investigations because of its close resemblance to man in thyroid hormone metabolism particularly in relation to maternal-fetal function (4).

MATERIALS AND METHODS

A group of Merino ewes aged three years was fed a low-iodine diet of crushed maize and pelleted pea pollard (5), which contained only 5-8 µg iodine per day, and another group (controls) on the same diet received a single injection of iodized oil at commencement. It was established in a previous trial that one injection of iodized oil containing 400 mg iodine was an effective supplement in mature severely iodine-deficient sheep for at least two years.

After 5 months, when iodine deficiency was established, the ewes from both the deficient and control groups were mated with normal fertile rams. At 56, 70, 98 and 140 days gestation, fetuses were delivered by hysterotomy while the ewes were under pethidine hydrochloride/acepromazine sedation and epidural anaesthesia. In the iodine-deficient group, some of the ewes received an intramuscular injection of iodized oil at the end of the second trimester (100 days gestation) and pregnancy was terminated 40 days later when the fetuses were weighed, bled and killed. The fetal brain was dissected into cerebral hemispheres, cerebellum and brain stem for deoxyribonucleic acid (DNA) and protein (5) estimations. The thyroid gland of each fetus was also removed and weighed and its iodine content determined (5). Plasma thyroxine (T4) and thyroid stimulating hormone (TSH) concentrations were estimated by radioimmunoassay (5).

RESULTS

Thyroid histology revealed hyperplasia from 56 days gestation in the iodine-deficient fetuses and, from 70 days onwards, the follicles were replaced by narrow spaces with barely detectable amounts of colloidal material and the epithelial cells lining the follicular spaces by columnar rather than the normal cuboidal cells. Values for fetal thyroid weight, thyroid iodine, plasma T4 and TSH are shown in Table 1.

A significant reduction in body weight, brain weight, brain DNA and brain protein occurred in the deficient fetuses and as shown in Table 2 this first became obvious at 70 days gestation. The main changes were found to be in the cerebral hemispheres and cerebellum. Brain weight in the control fetuses followed an identical growth pattern to that observed in fetuses from grazing ewes: the changes therefore were related only to lack of iodine in the diet.

At 140 days the iodine-deficient fetuses showed absence of wool coats, goitre, varying

TABLE 1: EFFECT OF SEVERE DIETARY IODINE DEFICIENCY ON FETAL THYROID FUNCTION IN SHEEP

Gestational age		56 days	70 days	98 days	140 days
Thyroid weight (g)	I-defic.	(5)0.01	(7)0.33	(5)4.30	(7)12.69
	Control	(3)0.05	(4)0.08	(5)0.28	(3) 0.99
Thyroid iodine (μg)	I-defic.	0.06	1.17	8.7	29
	Control	0.24	8.79	146.9	2636
Plasma thyroxine (nmol/l)	I-defic.	<4	4	6	7
	Control	†10	25	125	190
Plasma TSH (ng/ml)	I-defic.	56	>160	>160	>160
	Control	†12	13	13	12

Number of observations shown in parentheses. † Number of observations (2).

TABLE 2: EFFECT OF SEVERE DIETARY IODINE DEFICIENCY ON FETAL BRAIN DEVELOPMENT IN SHEEP

Gestational age		56 days	70 days	98 days	140 days
Brain weight (g)	I-defic.	(5)1.79	(7)4.20*	(5)19.0	(7)46.4**
	Control	(3)1.68	(4)5.01	(5)22.1	(6)53.8
Brain DNA (mg)	I-defic.	8.86	14.2*	27.8**	62.6**
	Control	8.37	16.2	32.5	74.6
Brain Protein (mg)	I-defic.	80.4	176*	702**	2522***
	Control	75.3	207	863	3277
Body weight (g)	I-defic.	31.7	101*	662	2930**
	Control	32.2	129	753	3820

Number of observations shown in parentheses. *P<0.05 **P<0.01 ***P<0.001

degrees of subluxation of the foot joints, frequent deformation of the skull shown by a dome-like appearance of the cranium and retrognathia. There also was radiological evidence of delayed bone maturation from 98 days.

The effect of iodized oil at 100 days gestation on brain development in iodine-deficient fetuses is shown in Table 3.

TABLE 3. EFFECT OF IODIZED OIL INJECTED AT 100 DAYS ON BRAIN DEVELOPMENT IN IODINE-DEFICIENT FETUSES (140 DAYS GESTATIONAL AGE)

		Body weight(kg)	Total Brain Weight(g)	Total Brain DNA(mg)	Total Brain Protein(mg)	T4 (nmol/l) Fetal	T4 (nmol/l) Maternal
Control	(6)	3.82	53.79	74.55	3277	211	127
I-deficient	(7)	2.93**	46.36**	62.58**	2522***	<7***	19***
+ I at 100 days	(5)	3.38	51.46++	76.20++	3529+++	177+++	169+++

Number of observations shown in parentheses.

	P<0.01	P<0.001
I-Deficient vs. control	**	***
I-Deficient vs. + I at 100 days	++	+++

Iodine given at 100 days produced in the next 40 days rapid body growth which was considerably greater than that of the iodine-deficient group and only 11.5% less than that of the controls. This rapid growth response was even more prominent for the fetal brain, in which the increase was significantly more than that of the iodine-deficient fetuses and only 4.3% less than that of the controls. Similar rapid increases also occurred in DNA and protein content of the brains of the fetuses that received the oil and maternal plasma T4 concentrations rose from 19 to 169 nmol/l and fetal T4 values from <7 to 177 nmol/l (Table 3).

DISCUSSION

The sheep model demonstrates experimentally the necessity of iodine for normal develoment of the fetal brain. This is clearly relevant to the human problem of endemic cretinism. In iodine deficient fetuses the lowered brain weight and brain DNA at 70 days gestation indicates reduced cell numbers probably due to slower neuroblast multiplication which normally occurs from 40-80 days in the sheep (6). Although brain protein also was reduced the protein:DNA ratio remained unchanged, thereby indicating no significant change in size of the brain cells. After 80 days, the development of the neuroglia appears to be slowed also in iodine deficiency, as both DNA and protein in 98 and 140 day fetal brains were significantly less than in controls. Evidence of fetal hypothyroidism was provided by low fetal thyroid iodine and plasma T4 values, significant reduction in body weight coinciding with the brain retardation and absence of wool growth and skeletal retardation near term.

Similar findings to those observed here for iodine deficiency have been reported in new-born lambs which were thyroidectomized as 60 day fetuses, but the retardation in brain growth in these lambs was less marked and significant only at term (7). The fetal thyroid, therefore, from the time it becomes active late in the second trimester (8), could play a prominent role in the normal development of the fetal brain. This suggestion is supported by the present findings that iodized oil injected at 100 days restored brain growth to normal within 40 days despite the retardation in the fetal brain seen at 98 days. Maternal and fetal plasma T4 levels also were rapidly restored to normal by the iodized oil. It is possible, in view of the fetal plasma T4 levels, that the iodized oil restored thyroid function to normal soon enough to prevent the effects of iodine deficiency seen at 140 days.

The restoration of fetal brain structure in response to iodized oil indicates a degree of plasticity of the brain unexpected in the light of current concepts (9). The findings may indicate that catch-up neuroblast multiplication can occur at a later stage in pregnancy in the sheep, which is a prenatal developer, than in a perinatal (human) or postnatal developer (rat). This may partly explain also the effect of iodized oil in restoring brain growth in the fetal sheep but apparently not in the human during pregnancy.

REFERENCES

1. Pharoah, P.O.D., Buttfield, I.H. and B.S. Hetzel, 1971, Lancet 1: 308-310.
2. Stanbury, J.B. and B.S. Hetzel (eds.), 1980, In "Endemic Goitre and Endemic Cretinism" (Wiley Biomed, NY).
3. Hetzel, B.S., Thilly, C.H., Fierro-Benitez, R., Pretell, E.A., Buttfield, I.H. and J.B. Stanbury, 1980, In "Endemic Goitre and Endemic Cretinism" (J.B. Stanbury and B.S. Hetzel eds.) (Wiley Biomed, NY), pp. 513-532.
4. Fisher, D.A., Dussault, J.H., Erenberg, A. and R.W. Lam, 1972, Pediat. Res. 12: 894-899.
5. Potter, B.J., Jones, G.B., Buckley, R.A., Belling, G.B., McIntosh, G.H. and B.S. Hetzel, 1980, Aust. J. Biol. Sci. 33: 561.
6. McIntosh, G.H., Baghurst, K.I., Potter, B.J. and B.S. Hetzel, 1979, Neuropath. Appl. Neurobiol. 5: 103-114.
7. McIntosh, G.H., Baghurst, K.I., Potter, B.J. and B.S. Hetzel, 1979, Neuropath. Appl. Neurobiol. 5: 363-376.
8. Fisher, D.A., Dussault, J.H., Sack, J. and I.J. Chopra, 1977, Recent Progr. Hormone Res. 33: 59-116.
9. Dobbing, J., 1974, In "Scientific Foundations of Paediatrics" (J.A. Davis and J. Dobbing, eds.) (Heinemann Med. Books, London, UK) pp. 565-577.

DISCUSSION

BRAY (Australia). Was the brain development retarded more than the overall body development?

POTTER (Australia). No there was a brain-sparing effect. This was quite interesting the changes were greater in the body, retardation in body weight being greater than that in brain weight.

BRAY (Australia). The second point is that with our work, although we could show histologically that all thyroids were affected in iodine-deficient lambs, they were not always greater in size. Were yours?

POTTER (Australia). Ours were always greater in size.

O'DELL (USA). I would like to ask how one could separate the effect on thyroid from a direct effect on the brain? How could you prove it experimentally?

POTTER (Australia). You mean thyroid effect as against iodine? Well, there are several reasons for this, I suppose I should say that in some other experiments we have done we have actually thyroidectomised mothers and then mated them and had a look at the fetal thyroids and fetal brains at various days of gestation. We found an interesting situation arose in those animals. In those fetuses in which the mothers were thyroidectomised before conception, if we looked at the brain weights from 70 days up to 140, we found that there was a weight depression at 84 days which was not there at 98 or 125 days, so in other words the brain growth caught up. Now we put this down to the fact that the fetal thyroid gland itself does not function until after 70 days, probably around about 84-90 days. So what we think happened was that, when there were no maternal thyroid hormones available to the developing fetus before its own thyroid took over, a depression in brain weight occurred, once the fetal thyroid took over, the brain caught up again. We checked this by doing thyroidectomies on the fetuses of normal mothers at 98 days gestation and we found that these fetuses were very similar to those which we obtained with iodine-deficiency, but the retardation was not nearly as drastic. We believe that we have a situation where the fetal brain relies on maternal thyroid hormones initially, fetal thyroid hormones subsequently and maybe retardation is what happens when you deplete both of them of iodine.

WHEELER (Australia). Firstly, how did you measure iodine in such a small quantity as 1.17 μg, secondly how much iodine was administered in the iodised oil?

POTTER (Australia). I will answer the second first. We gave about 400 milligrammes in one injection. Now measuring the iodine, we published this in a paper last year, Potter, B.J., Jones, G.B., Buckley, R.A., Belling, G.B., McIntosh, G.H. and Hetzel, B.S. Australian Journal of Biological Science 33: 53-61, 1980. The whole method is completely written up and we have been able to reproduce our results over and over again.

TRACE ELEMENT DEFICIENCIES

METABOLIC FUNCTIONS OF ZINC - A NEW LOOK

B.L. O'Dell

Department of Biochemistry
University of Missouri
Columbia, MO 65211 USA

I. INTRODUCTION

Trace element research encompasses a wide spectrum of interests but a major one is concerned with the metabolic functions of these important nutrients. It is generally conceded that knowledge of their mechanisms of action contributes significantly to the diagnosis and treatment of specific deficiency signs that arise from any of several physiological insults. One of the goals of nutritional biochemists is to identify enzyme or metabolic defects with specific pathology of deficiency. For some elements, e.g. copper, this has been achieved to a large extent. In the case of zinc many years of effort have failed to produce satisfying relationships in animals between biochemical function and pathology. This failure is particularly evident in the case of the incipient signs of zinc deficiency.

If one considers all species, approximately 100 zinc dependent enzymes have been described, but the number of confirmed zinc metalloenzymes found in vertebrate animals is much more limited. As indicated in Table 1, the activities of only a few zinc metalloenzymes are significantly depressed by dietary zinc deficiency.

TABLE I. Established Vertebrate Zinc Metalloenzymes.[a] Effect of Zinc Deficiency on Their Activities in Soft Tissues

Enzyme	Activity Decreased[b]
Alkaline Phosphatase	+
Carboxypeptidase A and B	+
Carbonic Anhydrase	0
Alcohol Dehydrogenase	0
Leucine Aminopeptidase	0
Superoxide Dismutase	0
Dipeptidase	?

a. This tabulation includes observations made in two or more laboratories or species (1).
b. A value judgment based primarily upon the summary of Kirchgessner et al. (2).

Other zinc metalloenzymes have been found in lower organisms and the activities of some of their animal counterparts are decreased by dietary zinc deficiency (3). The activities of a few non-metalloenzymes in animals are affected by zinc deficiency; thymidine kinase is decreased and ribonuclease is increased. Many of the latter group of enzymes, listed in Table III, are concerned with nucleic acid metabolism. Several investigators (3) have proposed that impairment of nucleic acid biosynthesis accounts for the slow growth rate and wound healing associated with zinc deficiency. However, these enzymes do not afford a reasonable explanation for the early signs of zinc deficiency.

TABLE II. Non-Vertebrate Zinc Metalloenzymes and Others Affected by Zinc Deficiency in Animals

Enzyme	Activity Decreased
DNA Polymerase[a]	+
RNA Polymerase[a]	+
Collagenase[b]	+
Thymidine Kinase	+
δ-Aminolevulinate Dehydratase	+
Ribonuclease	-

a. Zinc metalloenzymes in lower forms (3).
b. A mammalian metalloenzyme (4) whose activity is decreased in chick bone (5).

Rats fed an otherwise adequate diet but deficient in zinc experience a rapid drop in plasma zinc (6) and this is associated within a few days with depressed and cyclical food intake (7). Within a few days after zinc deprivation rats exhibit a bleeding tendency which is rapidly reversed by oral zinc (8). In the chick the packed cell volume increases after a few days and this is rapidly reversible (17). During the short period required to produce these pathological signs there is little or no change in zinc concentrations of most soft tissues. Even platelets, which have a short life span, exhibit normal zinc concentration after function is impaired (8).

These observations have led us to postulate that extracellular zinc is of key metabolic importance. This compartment of zinc may be in equilibrium with a small pool of cellular zinc, such as, that in the plasma membrane. In any case a relatively small pool of zinc is critical to cell function. The purpose here is to present data which support the concept that zinc plays a role in the maintenance of membrane integrity. Such a role may involve membrane-bound enzymes, either zinc dependent or not. The discussion will consider zinc deficiency pathology that may relate to: (a) arachidonic acid and prostaglandin metabolism, (b) water and cation balance and (c) membrane peroxidation.

II. PATHOLOGY POSSIBLY RELATED TO THE EICOSANOIDS

1. Skin lesions in mammals. Essentially all animals subjected to zinc deficiency develop dermal lesions of varying intensity. This is exemplified most dramatically by parakeratosis in the pig and acrodermatitis enteropathica in the human infant. The latter disease is of genetic origin and the pathology results from zinc deficiency. The skin lesions associated with deficiencies of zinc are grossly similar to those of essential fatty acid (EFA) deficiency. Both lesions are characterized by a scaly appearance with varying degrees of hyperkeratosis and acanthosis. The dermis has dilated capillaries and the epidermis fails to keratinize. Parakeratosis in swine responds promptly to zinc and, to a lesser extent, to soybean oil (9). The dermal lesions of EFA deficiency respond to percutaneous treatment with prostaglandin (PG) E_2 (10). The effect of PGE_2 in the zinc deficient rat has not been determined, but low dietary zinc accentuates the signs of EFA deficiency (11). It is worthy of note that skin of the zinc deficient rats in the latter study contained a higher proportion of arachidonic acid that of ad libitum fed controls. Although these experiments involving skin pathology suggest a relationship between zinc and EFA metabolism, they are not definitive in this respect.

2. Impairment of reproduction in the female rat; relation to aspirin toxicity. Several reproductive processes are impaired in the zinc-deficient pregnant rat (12). The pathology includes an extended gestation period, difficult and prolonged parturition, excessive blood loss and failure to retrieve and care for the pups. The signs of the resulting post-parturient illness resemble those of physiological shock and include low body temperature, low blood pressure and hyperkalemia (13). Highly analogous pathology is observed in female rats given toxic doses of aspirin near term (13). See Table III.

TABLE III. Analogous Pathology of Zinc Deficiency and Aspirin Toxicity in Reproducing Female Rats

General Prostaglandin Effects

1. Extended Gestation Period
   a. $PGF_{2\alpha}$ is luteolytic
2. Prolonged and Difficult Parturition
   a. PGE_2 and $PGF_{2\alpha}$ induce labor
3. Depressed Blood Pressure
   a. PGI_2 and PGE_2 are vasodilators; $PGF_{2\alpha}$ is a vasopressor
4. Low Body Temperature
5. Blood Loss at Parturition
   a. PGI_2 depresses and TxA_2 promotes platelet aggregation

While analogous pathology does not prove a common cause, the results suggest that zinc deficiency impairs PG production or function in as much as aspirin is a well recognized inhibitor of PG synthetase. In at least some species $PGF_{2\alpha}$ exerts a luteolytic effect and terminates pregnancy. Both E_2 and $F_{2\alpha}$ induce labor and E_2 is a vasodilator while $F_{2\alpha}$ is a vasopressor. PGI_2 or prostacyclin is a vasodilator and an inhibitor of platelet aggregation. An eicosanoid relative of the PGs, thromboxane(TX) A_2, promotes platelet aggregation.

3. Bleeding tendency. Not only is there excessive blood loss in the zinc deficient rat at parturition, but blood loss from a tail wound is also increased in parous female (13) and growing male rats (8). Bleeding time is significantly increased within 4 days after zinc

deprivation and is restored to normal within 4 hours after oral zinc therapy. This hemostatic defect has been traced to an impairment of platelet aggregation (14). Platelet rich plasma from zinc deficient rats is less sensitive than normal to such platelet aggregating agents as ADP, collagen and arachidonic acid but zinc deficient platelets respond normally to the divalent cation ionophore, A23187. Recently, similar observations have been made in human subjects who consumed an otherwise adequate formula diet low in zinc (15). These results could be explained by failure of TxA_2 production and/or function, or by excess production of prostacyclin. Data presented below indicate that it is not due to lack of prostacyclin production.

4. Prostaglandin metabolites. If zinc is directly involved in prostaglandin and thromboxane biosynthesis, the plasma levels of these compounds might be expected to decrease in zinc deficiency. Unfortunately, the primary prostaglandins, E_2 and $F_{2\alpha}$, are rapidly destroyed in the lungs and the plasma concentrations show wide diurnal variations. Hence, the concentrations of their respective metabolites, 13,14-dihydro-15-keto-PGE_2, ($DHKE_2$) and 13,14-dihydro-15-keto-$PGF_{2\alpha}$ ($DHKF_{2\alpha}$), and the prostacyclin (PGI_2) metabolite, 6-keto-$PGF_{1\alpha}$ ($KF_{1\alpha}$), in plasma reflect production more reliably than do the primary PGs. The relationships of these metabolites to arachidonic acid, the primary prostanoids, and the endoperoxide intermediates are shown in Figure 1.

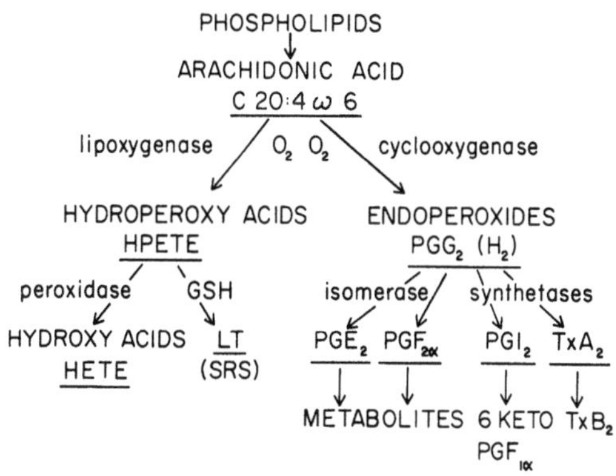

Figure 1. Arachidonic acid metabolites of physiological significance

To assess the effect of zinc deficiency on PG biosynthesis in the growing rat we have measured the concentrations of the plasma metabolites by radioimmunoassays (16). The results obtained for male rats fed for a 3-week period are shown in Table IV.

TABLE IV. Effect of Zinc Status on Plasma Levels of Prostaglandin Metabolites[1]

Dietary Zn ppm	$DHKE_2$ pg/ml	$DHKF_{2\alpha}$ pg/ml	6-$KF_{1\alpha}$ pg/ml
0.5-AL (0.45)[2]	327±23[a]	178±18[a]	325±22[a]
100-AL (1.32)	246±19[b]	148± 8[a]	539±38[b]
100-PF (1.34)	182±15[c]	95±12[b]	405±35[a]

1. Weanling male rats fed a soybean protein based diet for a period of 3 weeks. The PF rats were fed the amount consumed ad libitum by individuals fed the low Zn (0.5 ppm) diet. Values are means ± S.E.M.; different superscripts indicate statistical significance, p = 0.02 or less.
2. Plasma zinc in ppm.

The concentrations of the PGE_2 and $F_{2\alpha}$ metabolites are significantly higher in plasma of deficient than in that of pair-fed control rats. The concentration of the prostacyclin metabolite in deficient rats is not different from pair-fed controls but is lower in the deficients than in ad libitum controls. While the metabolite concentrations would be affected by the rates of both production and catabolism of the primary PGs, they probably offer the best available estimate of in vivo PG biosynthesis. Contrary to earlier prediction, the results show no impairment of PG production. At present no data are available relative to the functional capacity of the prostanoids in zinc deficiency. Defective prostanoid receptors in plasma membranes would explain the pathology.

III. PATHOLOGY POSSIBLY RELATED TO WATER AND CATION BALANCE

1. Water balance. One of the early recognized and typical signs of zinc deficiency in most species is a dehydrated appearance. Associated with the apparent dehydration are elevated hematocrits and diarrhea. Bettger et al. (17) have determined body water and its distribution in zinc-deficient chicks. They found no change in total tissue water but a marked shift of water from extra- to intra-cellular compartments. Extra-cellular water in zinc deficiency changed from 29.4 to 19.6% of body weight and plasma volume from 6.0 to 3.4%. The decrease in plasma volume largely explains the high hematocrits observed in chicks within 5 days after deprivation of dietary zinc. Oral treatment with zinc and saline solutions restore the hematocrit to normal within 4 hours while zinc alone or saline alone is ineffective. These results suggest an impairment of electrolyte balance and may explain the apparent preference of zinc deficient rats for salt solutions over that for water (18).

2. Cation balance. Studies with reproducing female rats (13) and with malnourished children (19) show a derangement of electrolyte balance associated with zinc deficiency. Plasma potassium is elevated in the zinc deficient rat (13). Zinc supplementation of children who have low plasma zinc decreases their plasma K and increases the rate of Na transport by their leucocytes (19). The Na/K ratios in liver, muscle and skin of zinc deficient chicks are significantly elevated, although that in plasma is unchanged (17). Such a shift of Na into tissues may explain the higher concentration of water in the cells of major tissues. In the zinc deficient chicks, the driving force for water movement is unknown but it is not the plasma electrolyte concentration. The deranged Na/K concentrations in tissues suggest a change in membrane permeability, "leaky membranes", or a defective sodium pump. The ready reversibility of the pathology points up the critical role of extracellular zinc, or of a labile cellular pool, which helps to maintain membrane integrity.

IV. PATHOLOGY POSSIBLY RELATED TO PEROXIDATION OF MEMBRANES

1. Skin and hock lesions in chicks; polyunsaturated fatty acids (PUFA). Chicks fed a low zinc diet develop skin lesions, particularly on the feet and shanks, which differ somewhat from those seen in mammals. Hyperkeratosis predominates, accompanied by development of deep fissures in the foot pad. Avian species also develop abnormal hocks and an arthritic gait which suggests pathology of the joints. The latter is alleviated in part by histamine and non-steroidal anti-inflammatory agents, such as aspirin (20). The beneficial effect of PG inhibitors suggests the involvement of eicosanoids, but in our experience, percutaneous treatment with PGE_2 or $F_{2\alpha}$ had no effect on the dermal lesions (21). If failure of PG synthesis were the basis of the dermatitis, fat-free diets or diet based on saturated fat should accentuate the lesion. In fact, feeding either hydrogenated coconut oil (HCO), a fatty acid mixture simulating HCO, or a fat-free diet, largely eliminates the severe dermal lesions observed when corn oil, soybean oil or the fatty acids simulating these vegetable oils are fed (21). The mechanism by which the PUFA accentuate skin lesions, or the saturated fatty acids protect, is unknown, but there is no effect on zinc status. As expected the skin of chicks fed PUFA contains higher proportions of PUFA than those fed HCO, and skin from those fed the low zinc diet had a higher proportion of arachidonic acid than ad libitum controls. The latter difference may relate to decreased food intake and body development. If the dermal lesions are the result of oxidative damage, as suggested here, consumption of PUFA would tend to aggravate the pathology compared to that observed in chicks fed saturated-fat or low-fat diets. It should be pointed out that chickens are highly resistant to essential fatty acid deficiency and possibly utilize PUFA in a different manner than mammals. Regardless, a possible explanation for the skin pathology in zinc deficient chicks is that *in vivo* zinc exerts an antioxidant effect.

2. Protective effects of antioxidants. The plausibility of an antioxidant effect of zinc is substantiated by the observation that vitamin E and, to a lesser extent, other fat-soluble antioxidants protect zinc-deficient chicks against leg and skin disorders (22). The ameliorative effect is proportional to the dietary vitamin E level up to 500 IU/kg. Furthermore, the rate of *in vitro* peroxidation, as measured by malondialdehyde production, is higher in skin slices from zinc deficient chicks than from controls. Under the conditions used, dietary vitamin E decreased the rate of malondialdehyde production in zinc deficient skin but had no effect in control skin.

Additional support for an interaction of vitamin E and zinc comes from work with the female producing rat. As mentioned above the zinc-deficient rat undergoes "physiological shock and depression" at or near parturition. The most objective parameters of the syndrome are low blood pressure and body temperature. As shown in Figure 2, graded dietary levels of vitamin E improve the physiological status of zinc-deficient dams without affecting the controls (23). The basis for the beneficial effect of excess vitamin E is not clear and the problem requires additional research. Explanations include the possibility that vitamin E improves prostaglandin metabolism and/or protects membranes prone to peroxidation as a result of zinc deficiency.

Red blood cells from zinc-deficient rats are more fragile to osmotic stress than controls while those from copper-deficient rats are more stable than normal (24). Cupric ion accentuates hemolysis and peroxidation of cells in vitro whereas zinc ions either have no effect or are protective. The two ions may compete for the same or similar binding sites on plasma membranes. The incorporation of zinc into sickle cells improves their filterability at intermediate oxygen tensions and antagonizes calcium binding by red cell membranes (25). Feeding rat excessive levels of zinc (1000 ppm) protects their erythrocytes against hemolysis in hypotonic solution (26). On a protein basis, rat red cell plasma membranes contain more zinc than the cytoplasm (27). Zinc is a component of the red cell ghost, being linked to both lipid and protein (28).

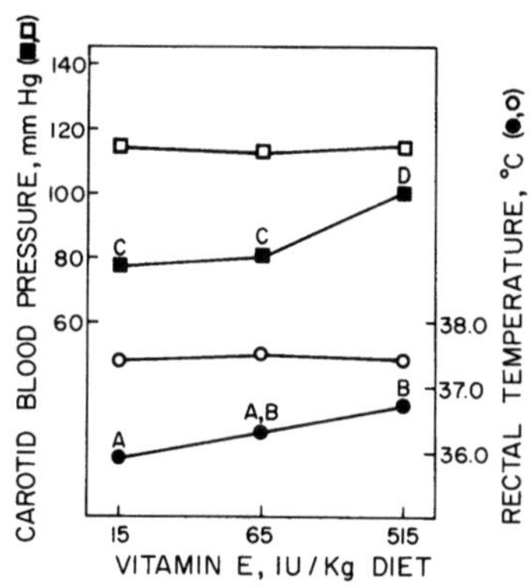

Figure 2. Interrelationship of dietary vitamin E and zinc.
Closed squares represent blood pressure and closed circles rectal temperature in zinc deficient rats; open symbols, controls.

V. FUNCTION OF ZINC IN MEMBRANES

Chvapil et al. (20) first proposed that zinc stabilizes membranes either by inhibiting lipid peroxidation or thiol oxidation. Using supra-physiological levels, both in vivo and in vitro, zinc has been shown to stabilize liver lysosomes (30). However, high concentrations of zinc inactivate or immobilize many cell types, such as mast cells, platelets, macrophages, leucocytes and spermatozoa (31). From these observations, Chvapil proposed a model which depicts how zinc may function at the membrane level. He proposed that zinc binds to the membrane surface and thereby modifies its fluidity and increases its stability. Zinc may modulate various membrane enzymes such as ATPase. According to the model zinc may also modulate the function of various membrane receptors (31).

Considering the pathological effects of zinc deficiency summarized in this paper, one is led to speculate that zinc exerts a key function in cell membranes. The data suggest an antioxidant effect and this might be visualized primarily as protection of sulfhydryl groups in membranes as depicted in Figure 3. In this model zinc is reversibly complexed with vicinal sulfhydryls and protects them against oxidation to the disulfide. When plasma zinc decreases, zinc dissociates from the membrane allowing association of cations of variable valence. The latter promote oxidation to the disulfide, a form which does not bind zinc readily and is not reduced. This concept is based primarily on the properties of a zinc dependent enzyme, δ-aminolevulinate dehydratase. The activity of this enzyme is maintained by zinc when it is exposed to air, but zinc is not essential for activity under anaerobic conditions (32,33). Two SH groups at the active site are essential for activity, and both catalytic activity and zinc are lost upon oxidation to the disulfide bond. Sulfhydryl groups, but not zinc, are necessary for activity; zinc is required to prevent auto-oxidation.

The membrane protective effect of zinc may involve the lipid component of the membrane as well as sulfhydryl groups. In any case, the speculation that zinc is involved in the maintenance of membrane function and integrity explains many of the signs of short-term zinc deficiency. According to the model, plasma zinc level would be critical to membrane structure and thus to the function of hormone receptors, absorption sites and membrane bound enzymes.

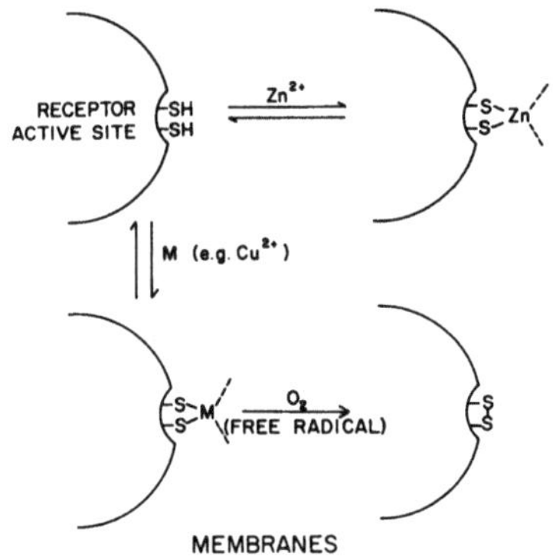

Figure 3. Tentative model of zinc function in membranes.

References

1. Vallee, B.L., and W.E.C. Wacker, 1976 in Handbook of Biochemistry and Molecular Biology. (ed. G.D. Fasman) 3rd Ed. Proteins, Vol. II, p. 276 (CRC Press, Inc.).
2. Kirchgessner, M., H.P. Roth, and E. Weigand, 1976, in Trace Elements in Human Health and Disease (ed. A.S. Prasad and D. Oberleas) Vol. 1 (ed. A.S. Prasad and D. Oberleas) Vol. 1 (Academic Press, NY) pp. 189-225.
3. Roth, H.P. and M. Kirchgessner, 1980 World Rev. Nutr. Diet. 34:144-160.
4. Seltzer, J.L., J.J. Jeffrey, and A.Z. Eisen, 1977 Biochem. Biophys. Acta 485:179-187.
5. Starcher, B.C., C.H. Hill, and J.G. Madaras, 1980 J. Nutr. 110:2095-2102.
6. Hurley, L.S., P.R. Gordon, C.L. Keen, and L. Merkhofer, 1980 Fed. Proc. 39:431.
7. Chesters, J.K.,and M. Will, 1973 Br. J. Nutr. 30:555-566.
8. Gordon, P.R.,and B.L. O'Dell, 1980 Proc. Soc. Exp. Biol. Med. 163:240-244.
9. Hanson, L.J., D.K. Sorensen,and H.C.H. Kernkamp, 1958 Am. J. Vet. Res. 19:921-930.
10. Ziboh, U.A., and S.L. Hsia, 1972 J. Lipid Res. 13:458-467.
11. Bettger, W.J., P.G. Reeves, E.A. Moscatelli, G. Reynolds, and B.L. O'Dell, 1979 J. Nutr. 109:480-488.
12. Apgar. J. 1968 Am. J. Physiol. 215:160-163.
13. O'Dell, B.L., G. Reynolds, and P.G. Reeves, 1977 J. Nutr. 107:1222-1228.
14. Gordon, P.R., and B.L. O'Dell, 1980 J. Nutr. 110:2125-2129.
15. Gordon, P.R., C.W. Woodruff, H.L. Anderson, and B.L. O'Dell, 1981 Fed. Proc. 40:839.
16. Unpublished.
17. Bettger, W.J., J.E. Savage, and B.L. O'Dell, 1981 J. Nutr. In Press.
18. McConnell, S.D., and R.I. Henkin, 1974 J. Nutr. 104:1108-1114.
19. Patrick, J., B.E. Golden, and M.H.N. Golden, 1980 Am. J. Clin. Nutr. 33:617-620.
20. Nielsen, F.H., M.L. Sunde, and W.G. Hoekstra, 1968 J. Nutr. 94:527-533.
21. Bettger, W.J., P.G. Reeves, E.A. Moscatelli, J.E. Savage, and B.L. O'Dell, 1980 J. Nutr. 110:50-58.
22. Bettger, W.J., P.G. Reeves, J.E. Savage, and B.L. O'Dell, 1980 Proc. Soc. Exp. Biol. Med. 163:432-436.
23. Unpublished.
24. Bettger, W.J., T.J. Fish, and B.L. O'Dell, 1978 Proc. Soc. Exp. Biol. Med. 158:279-282.
25. Brewer, G.J., and F.J. Oelschlegel, 1974 Biochem. Biophys. Res. Commun. 58:854-861.
26. Chvapil, M., Y.M. Peng, A.L. Aronson, and C. Zukoski, 1974 J. Nutr. 104:434-443.
27. Bettger, W.J., M.F. Fernandez, and B.L. O'Dell, 1980 Fed. Proc. 29:896.
28. Chvapil, M., D. Montgomery, J.C. Ludwig, and C.F. Zukoski, 1979 Proc. Soc. Exp. Biol. Med. 162:480-484.
29. Chvapil, M., J.N. Ryan, and Z. Brada, 1972 Biochem. Pharm. 21:1097-1105.
30. Ludwig, J.C., and M. Chvapil, 1980 J. Nutr. 110:945-953.
31. Chvapil, M. 1976 Medical Clinics of North America 60:799-812.
32. Tsukamoto, I., T. Yoshinaga, and S. Sano, 1979 Biochim. Biophys. Acta 570:167-178.
33. Bevan, D.R., P. Bodlaender, and D. Shemin, 1980 J. Biol. Chem. 255:2030-2035.

A contribution from the Missouri Agricultural Experiment Station. Supported in part by a NIH Grant, No. HL11614, of the U.S. Public Health Service.

DISCUSSION

GARNYS (Australia). ALAD is of course the important enzyme in lead toxicity and zinc has been noted to de-toxify lead to a certain extent. Would you care to speculate on the mechanism?

O'DELL (U.S.A.). One would expect lead to be tenaciously associated with sulphydryl groups probably more tightly than zinc. You must remember that it is not a static system as shown in a diagram. There is an equilibrium with many possible cations competing for the same sites for complexation. I suspect that lead in high concentrations would compete very well for sulphydryls and would eliminate zinc from such sites. I suspect that zinc is not bound to these sites as strongly as lead or even copper, but high concentrations of zinc in the extracellular medium would protect such sites from oxidation or lead inactivation.

LINDER (U.S.A.). Have you looked at plasma protein concentrations in zinc deficiency? When you see large changes in intra- and extracellular fluid volumes there may be some marked changes in protein concentration.

O'DELL (U.S.A.). Yes that is a possibility. We have not looked at plasma proteins. Others have and in some animals there is a tendency for albumen to decrease in animals that are maintained on zinc-deficient diets for several weeks. I doubt that this applies in the short term deficiency we are seeing.

SABBIONI (Italy). In our laboratory we have isolated a calf thymus deoxynucleotidyltransferase enzyme which is a DNA polymerase, and we have demonstrated that this is a zinc enzyme using neutron activation coupled with biochemical techniques. We have also isolated the DNA from calf thymus and we have found significant amounts of zinc in the nucleic acid. Would you please give a short comment on the metabolic function of zinc in DNA metabolism in mammals so we can put our results in the right perspective.

O'DELL (U.S.A.). I did not know of your work showing that calf thymus DNA polymerase is a zinc metalloenzyme. Given that it is, there is no doubt about its importance in metabolism. However in the context of the short-term deficiencies described here, I think this could not be explained by such enzymes as DNA polymerase, RNA polymerase and thymidine kinase. There must be a more rapid change due to extracellular zinc. I am not saying that the zinc metalloenzymes are not important but that they do not seem to explain the short term effects.

ABDULLA (Sweden). There seems to be a lot of confusion regarding the use of plasma as an indicator of the zinc status and I wonder if you have looked at the zinc levels in any other cells like lymphocytes or granulocytes.

O'DELL (U.S.A.). We have not looked at leucocytes. Platelets, which we have used in these studies, were analysed for zinc. Their zinc content was normal.

HURLEY (U.S.A.). Your talk illustrates very clearly that we cannot have a simplistic approach to the functions of zinc. There certainly are different functions of zinc in these rapid changes versus more long term effects. This is especially so in the case of the developing embryo. For example, we have shown changes as early as day 3 of pregnancy, and in this case there are changes in thymidine kinase and DNA polymerase. Perhaps these enzyme changes may be secondary to some other more rapid changes, but as you have indicated, I think it's important to remember that there are different kinds of functions at different stages of development and at different periods of the deficiency.

O'DELL (U.S.A.). Thanks very much for your comments. I think you put it in the right perspective. One wonders whether or not, in the developing egg blastocysts you examined, the membranes might also be damaged perhaps by peroxidation.

WHITE (Australia). On the basis of your results with the prostaglandins am I right in assuming that you have given away the hypothesis that zinc may be involved in prostaglandin synthesis?

O'DELL (U.S.A.). Yes, we have given up the idea that zinc deficiency decreases prostaglandin production. Our data suggests either no effect on production or an actual increase.

MILLS (Scotland). Two quick questions. Firstly, with regard to the controversial prostaglandin hypothesis. What prospects do you see for precursor-product studies in this field? You have shown an increase in metabolites, but this may be due to an increased production rate or a decreased degradation rate. My second question relates to the food intake data that you showed. The treatment with arachadonic acid and prostaglandins seems to be partially effective in reducing the skin lesions of zinc deficiency, but how does this relate to the rapid restoration of food intake when zinc is given? We are talking about 60 to 120 minute responses in these animals.

O'DELL (U.S.A.). This is a complex situation and we have looked at a very broad spectrum. I don't think that it involves polyunsaturated fatty acids in the case of food intake. Food intake could be affected by receptor function. If you'll tell me what the signal is to regulate food intake I'll give you a more specific answer. I hope some people here can shed some light on that question as we proceed through this meeting. If receptors in plasma membranes are defective then that signal might not get through. Obviously the circulating levels of prostaglandins will depend on both production and degradation. All we have measured is circulating or equilibrium levels and cannot say whether there was truly increased production or decreased degradation. All of the pathology suggested that circulating levels should have been down, but in most cases they were high, contrary to the hypothesis.

ELMES (Wales). Just a very brief suggestion. In future anti-prostaglandin drug work why not use the newer prostaglandin inhibitors that are much more specific in their action. For instance indomethacin and ibuprofen which might be more valuable than using aspirin which is rather a blunder-buss type drug.

B. GOLDEN (Jamaica). On a different aspect. Your sodium-potassium results fit in with inhibition of the sodium pump. Have you considered that; could you speculate on its role?

O'DELL (U.S.A.). Yes, I am familiar with your work in Jamaica. It fits in very well with our observations. If the membranes are leaky, then the sodium-potassium ATPase would have to work harder to maintain balance. Its activity would increase and one might predict that the sodium pump or Na-K-ATPase activity would be increased. We do not have data on this point.

BINNERTS (The Netherlands). What about physiology Dr. O'Dell? You are always speaking from the biochemical stand point which I appreciate very much of course, but when cell volume gets larger, you must expect a very large signal from the brain where the normal cell volume is regulated. Wouldn't it be worthwhile to look for the hormones and the enzymes involved?

O'DELL (U.S.A.). It's a good suggestion but we haven't had an opportunity to look at this.

PROBLEMS AND POSSIBILITIES IN DIAGNOSING ZINC DEFICIENCY

M. Kirchgeßner and H.-P. Roth

Institut für Ernährungsphysiologie
Technische Universität München
D-8050 FREISING-WEIHENSTEPHAN
FED. REP. GERMANY

The animal body possesses a high capacity for regulating the metabolism of trace elements. This homeostatic adaptation makes it possible to maintain the concentration of zinc in tissues at a relatively constant level, because both absorption and endogenous excretion homeostatically change over a wide range of different zinc intake. For example, the zinc concentration in the bone pool shows little dependence upon the dietary zinc content in the case of normal supply. Long-lasting severe undernutrition with zinc, however, eventually brings about a pronounced reduction of the zinc concentration in bones (2). Such extreme and distinct zinc deficiency states as could be induced in experimental studies are, however, unlikely to occur under practical conditions. More often, the latent forms of zinc deficiency that are difficult to diagnose because they do not manifest themselves in clinical deficiency symptoms, such as growth cessation and skin lesions, are encountered in the field. They, however, also impair health, performance and reproduction, and the bodily development of man and animals. Therefore, some sensitive parameters are needed to diagnose these deficiency states. As newer studies show, these states of marginal deficiency are more widespread than thus far believed.

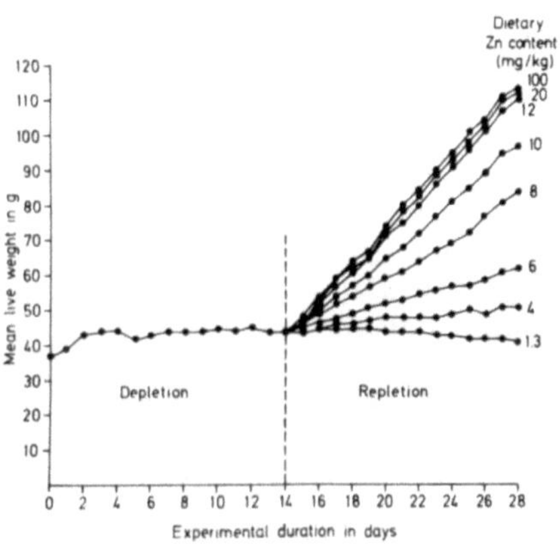

Growth response of rats following zinc supplementation at graded levels

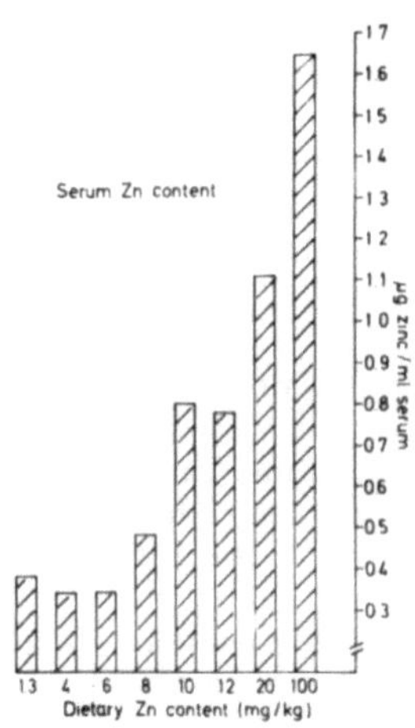

Lately, the serum or plasma zinc content has been proposed repeatedly as an indicator of zinc status. Under practical conditions these values greatly vary and are influenced by many other factors. The greatest disadvantage of using the serum zinc content for assessing the supply status is, however, the fact that serum zinc steadily rises with increasing dietary zinc supply (Fig. 1), and that it does not assume a plateau when zinc supply is adequate in contrast to growth that does no longer differ between 12 and 100 ppm Zn in the diet.

Not even the in vitro ^{65}Zn uptake by erythrocytes, as proposed by Chesters and Will (1) at the TEMA-3 symposium, showed any dependence upon the zinc status in our studies (3) with rats given diets with different zinc contents (Table 1). When these values are corrected to a uniform hematocrit of 40 %, because the hematocrit can in part be greatly increased during zinc deficiency, it is evident that there is an increased ^{65}Zn uptake in the case of the deficient to suboptimal supply ranging from 6 to 12 ppm dietary zinc as compared to animals given the higher supplies of 20 or 100 ppm dietary zinc. The low ^{65}Zn uptake in the extreme zinc deficiency range between 1.3 and 4 ppm may be attributed to the damage that has already occurred to the erythrocyte membrane. Obviously, this in vitro radioisotope method is of no major diagnostic value either.

Table 1. In vitro ^{65}Zn uptake by erythrocytes of rats with varied dietary Zn supply

Dietary Zn content (ppm)	^{65}Zn uptake by erythrocytes in % absolute	corrected to 40 % hematocrit	hematocrit
1.3	26.2 ± 9.3	23.5 ± 8.4	44.8 ± 4.9
4	23.7 ± 4.4	23.5 ± 4.0	40.1 ± 2.0
6	32.9 ± 4.8	35.2 ± 3.7	36.1 ± 3.0
8	25.7 ± 3.9	36.3 ± 7.2	28.4 ± 1.2
10	21.7 ± 2.5	30.7 ± 3.3	28.3 ± 1.9
12	23.5 ± 2.9	34.9 ± 5.0	27.1 ± 3.3
20	18.0 ± 2.6	24.8 ± 2.1	28.9 ± 1.9
100	17.4 ± 1.7	22.0 ± 2.9	31.3 ± 3.1

In our model studies (4) we, therefore, gave preferrence to the activity of the serum alkaline phosphatase (Table 2), a zinc metalloenzyme that quickly responds to dietary zinc depletion with a marked loss of activity. Under practical conditions, however, the activity of the alkaline phosphatase, similarly to the serum zinc content, not only greatly varies but is also affected by many other factors, so that it is not possible to state a normal activity that would indicate adequacy of supply. This great problem of designating norm values was to be avoided by measuring the activity before and three days after zinc injection. Rats were brought to definite supply states by giving them diets with different dietary zinc contents to investigate whether suboptimum zinc supply can be diagnosed by this response technique. There was a close correspondence between the enzyme's activity and the zinc supply in these rats (Table 2). The range between minimum and optimum supply covered the factor of 10. Even in the range of extreme deficiency from 1.3 to 6 ppm zinc in the diet, the alkaline phosphatase showed a corresponding grading of its activity and it increased manifold following zinc injection. Similar to the response of the serum zinc, the alkaline phosphatase did not yet reach an optimal activity when the dietary zinc content ranged between 12 and 20 ppm. In the case of 20 ppm zinc, the activity still rose by about 40 % following zinc injection. The optimal level of the activity of the alkaline phosphatase in serum most likely amounts to about 250 mU/ml in these studies. Accordingly, zinc injection did not bring about an increased activity in the animals given the diet with 100 ppm zinc. By contrast, the serum zinc content does not assume this plateau.

Although the activity of the alkaline phosphatase decreases with advancing age, especially toward the end of the growing phase, a reliable evaluation of the zinc supply status was also possible for rats at three different ages (5). The activity of the

TABLE 2. Activity of the alkaline phosphatase in serum of rats before and three days after zinc injection (0.8 mg zinc per animal)

Dietary Zn content (ppm)	Alkaline Phosphatase (mU/ml) before Zn injection	after Zn injection	Difference (ΔmU/ml)	Relative activity increase (%)
1.3	23 ± 8	152 ± 36	129	560
4	39 ± 10	162 ± 35	123	315
6	88 ± 41	195 ± 61	107	122
8	116 ± 19	185 ± 72	69	59
10	139 ± 29	193 ± 33	54	39
12	183 ± 44	245 ± 62	62	34
20	175 ± 39	249 ± 66	74	42
100	239 ± 27	246 ± 50	7	3

alkaline phosphatase increased in rats initially weighing 50 g as well as in those weighing 100 g in response to zinc injection when the diet supplied between 1.3 and 9 ppm zinc. In the case of the animals given the adequate supply of 60 ppm zinc, the activity did not increase in either age group, while the serum zinc content still continued to rise. Even in the group with 300 g heavy rats, the response was basically the same. Zinc injection into those animals given the diets with 1.3, 2, 3 and 5 ppm zinc resulted in a higher activity of the alkaline phosphatase three days later, whereas in the animals with 60 ppm zinc in their diet, the activity was unaffected. Ergo, if zinc injection causes an increase in activity, a more or less pronounced zinc deficiency must have been present depending upon the extent of the response, since additional zinc does not bring about an increased activity in the case of optimum supply.

Table 3. Percent zinc-binding capacity of serum from rats before and three days after zinc injection (0.8 mg zinc per animal)

Dietary Zn content (ppm)	Zinc-binding capacity (%) before Zn injection	after Zn injection
1.3	87.1 ± 2.7	70.4 ± 2.0
4	89.1 ± 3.5	76.6 ± 2.3
6	84.3 ± 2.1	71.4 ± 3.8
8	83.8 ± 1.1	74.0 ± 3.5
10	74.2 ± 4.8	67.4 ± 3.1
12	74.8 ± 3.1	59.8 ± 2.4
20	69.2 ± 3.7	70.0 ± 3.3
100	58.9 ± 5.3	61.2 ± 3.8

As a rule, however, it is not satisfactory to diagnose such a complex criterion as the zinc supply status by just one parameter. Therefore we proposed another method based on the percent zinc-binding capacity of serum (6). As is shown in Table 3, the percent zinc-binding capacity of rats was nearly 90 in the range of extreme zinc deficiency with diets containing between 1.3 and 4 ppm Zn and it decreased to somewhat less than 60 as the dietary zinc content increased. Hence, the percent zinc-binding capacity of serum exhibited an inversely proportional response to the dietary zinc content. Within three days after zinc injection, the zinc-binding capacity had significantly decreased to values near 70 % in all the animals of the groups given the diets with 1.3 - 12 ppm zinc. In the rats on the diets with 20 and 100 ppm zinc, the percent zinc-binding capacity of serum was not altered by zinc injection. Because of the constancy of the percent zinc-binding capacity of serum, as it became evident from the studies with different ages, it is not necessary to inject zinc into the test animals. Thus, the zinc supply status of animals or of humans can be assessed by an in vitro test with serum. This means that the percent zinc-binding capacity should reach a value between 60 and 70 % when zinc supply is optimal. Basically, however, it must yet be investigated to what extent other factors that, for example, influence serum zinc concentration also might affect the percent zinc-binding capacity.

In summary, it may be stated that, under the experimental conditions shown, the percent zinc-binding capacity of serum and the activity of the alkaline phosphatase reflect rather realistically the alimentary zinc status. We, therefore, consider both methods as suitable for diagnosing the zinc supply status, specifically in the marginal range.

REFERENCES

1. Chesters, J.K., and M. Will, 1978. In "Trace Element Metabolism in Man and Animals-3" (ed. M. Kirchgeßner) (ATW-Freising-Weihenstephan) pp 211 - 214
2. Pallauf, J., and M. Kirchgeßner, 1972, Z. Tierphysiol., Tierernährg. u. Futtermittelkde. 30: 193 - 202
3. Roth, H.-P., and M. Kirchgeßner, 1979, Z. Tierphysiol., Tierernährg. u. Futtermittelkde. 42: 95 - 101
4. Roth, H.-P., and M. Kirchgeßner, 1979, Res. Exp. Med. 174: 283 - 300
5. Roth, H.-P., and M. Kirchgeßner, 1980, Zbl. Vet. Med. A, 27: 290 - 297
6. Roth, H.-P., and M. Kirchgeßner, 1980, Res. Exp. Med. 177: 213 - 219.

DISCUSSION

M. GOLDEN (Jamaica). Three days in clinical practice is a very long time. Profound metabolic changes can occur in the patients we want to diagnose for zinc deficiency within that time. What is the time course of the repletion rise in alkaline phosphatase? It seems to me that it would need to be much more rapid than three days to provide a useful test of zinc deficiency.

KIRCHGESSNER (Germany). We have not yet tried to assess how much time would be necessary for the alkaline phosphatase to rise after Zn injection. We have found, however, that a good response occurred after half a day. I cannot say whether or not this is sufficient time for the full increase in the enzyme's activity, so this was another reason why we have proposed an alternative method that does not require a zinc injection. This zinc-binding capacity method, can be assessed within less than an hour.

PRASAD (U.S.A.). I would like to make a few comments about human experimental zinc deficiency and also naturally occurring zinc deficiency diseases such as sickle cell disease and chronic uremia. We find that the determination of zinc and alkaline phosphatase in the neutrophil is a very good and sensitive indicator. We also find plasma zinc to be a useful indicator, despite comments made by others concerning its limitations. If you can rule out some acute distress like myocardial infarction or perhaps acute infection or complete starvation then in my experience plasma zinc has not been a bad indicator for zinc deficiency.

HURLEY (U.S.A.). Could you please tell us how you measured the serum zinc binding capacity?

KIRCHGESSNER (Germany). This method was first proposed by Kincaid and Cronrath (1979). At first, the serum zinc content is measured. Another aliquot of serum is mixed with an excess of $ZnCl_2$ and the unbound zinc is precipitated with Mg carbonate. Zinc concentration is then measured in the supernatant after centrifugation.

PALLAUF (Germany). Would you please comment on the possibilities of diagnosing zinc deficiency by analysing the dietary zinc content and the hair zinc content.

KIRCHGESSNER (Germany). As to the first question, we have repeatedly tried to diagnose zinc deficiency by analysing the hair zinc content because the literature states that this is a good method for zinc. While we obtained an excellent response in growth and other Zn deficiency symptoms to varying Zn supply in rats, there was no definite relationship in the case of hair Zn content. This is also true for other species. In dairy cows, for instance, we noted a decrease in hair zinc content of only about 10% after a 6-week severe dietary Zn depletion, whereas other criteria decreased by approximately 50%. Regarding the second question, it is important to realize that the dietary Zn content alone does not account for differences in the absorptive efficiency or in the subsequent use of the absorbed amount in intermediary metabolism. We know that a number of factors can affect these efficiencies. In this context, the dietary zinc concentration is of some help as well as the serum zinc concentration as Dr. Prasad has pointed out, but they do not really allow a reliable diagnosis of Zn deficiency.

LOW ZINC AND REPRODUCTION IN THE EWE

D.G. Masters

Department of Animal Science and Production
University of Western Australia
Nedlands, WESTERN AUSTRALIA 6009

The effects of severe zinc deficiency (1) including those on reproduction (2,3) have been described for many animal species. However the influence of a marginal zinc deficiency on reproduction is not well defined. Work from our laboratory has shown that the zinc status of ewes in Western Australia is low during pregnancy in autumn and winter (4). Furthermore, ewes grazing pastures with between 15 and 30 μg Zn/g have responded to supplemental zinc by producing more and heavier lambs (5). This response occurred in the absence of any clinical signs of zinc deficiency and indicated that the requirement for zinc during pregnancy may be higher than at other times. This suggestion is supported by the results of Williams *et al.* (6), who reported that the developing foetus may accumulate a significant proportion of the ewe's zinc intake.

In the experiment described here, this possibly increased requirement for zinc was investigated by feeding a marginal zinc diet to pregnant ewes.

METHODS

Thirty mature age Merino ewes were allocated according to body weight into groups of 15. One group was fed a basal diet containing 9 μg Zn/g (LZ diet) and the other the basal diet supplemented with zinc sulphate and containing approximately 50 μg Zn/g (HZ diet). The basal diet consisted of hammer-milled oaten hay plus a mineral and trace element mix. For the last 60 days of pregnancy the basal diet also included 19% wheat starch and 3.8% casein. Throughout the experiment the sheep were fed *ad libitum* up to 0.95 kg per day.

After two weeks on the diets, the ewes were artificially inseminated from pooled semen but were maintained in two groups in pens lined with pressed hardboard. Pregnancy was diagnosed 35 days after mating by endoscopic examination of the uterus. Ten pregnant ewes from each group were transferred to individual metabolism crates for at least the last 75 days of gestation. Plasma zinc concentrations and live weight were measured every two weeks.

Six of the high zinc and four of the low zinc ewes were slaughtered after parturition together with their lambs before they had sucked. Samples of liver and brain were homogenized in 5 ml 0.25 N Tris HCl buffer (pH 8.0) and centrifuged for 15 minutes at 700 g. The supernatant was then assayed for thymidine kinase by a modified procedure described by Duncan and Hurley (7) and for alkaline phosphatase using the colorimetric method with sodium p-nitrophenylphosphate as the substrate.

Samples of liver, kidney, brain and pancreas were lyophilized and then digested in HNO_3, H_2SO_4 and $HClO_4$ (10:0.5:0.5 ml). Colostrum was similarly digested without lyophilization. After dilution of the digested material zinc concentration was estimated using atomic absorption spectrophotometry.

Results for zinc concentration in lamb liver and pancreas were statistically analysed using Wilcoxon's two sample rank test. Other comparisons were made using Students' t-test.

RESULTS

All ewes maintained pregnancy to term; four of the low zinc and one of the high zinc ewes required assistance at parturition. Zinc concentration in the liver, kidney and pancreas of the lambs from high zinc mothers were higher than those from the other group ($P < 0.02$) (see Table 1).

Table 1. Zinc concentration in lamb tissues (μg Zn/g dry tissue ± SE)

	Liver	Kidney	Brain	Pancreas
High Zn lambs	235.5 ± 47.7	80.7 ± 2.8	67.6 ± 4.3	178.6 ± 21.4
Low Zn lambs	64.0 ± 9.7	70.0 ± 2.2	73.3 ± 5.5	109.3 ± 7.9
Significance	$P < 0.01$	$P < 0.02$	n.s.	$P < 0.01$

No differences in thymidine kinase or alkaline phosphatase activities were observed (see Table 2). Plasma zinc fell during the 21 weeks (HZ 0.74 ± 0.04 to 0.59 ± 0.03 and LZ 0.64 ± 0.03 to 0.47 ± 0.03 μg/ml) but the differences between groups remained constant. Birth weight of lambs, live weight of the ewes and their wool and tissue zinc concentrations did not differ when the two groups were compared. Zinc in colostrum (HZ 47.5 ± 5.7, LZ 39.7 ± 6.8 μg/ml) was 5 to 10 times higher than in milk after one week of lactation.

Table 2. Alkaline phosphatase and thymidine kinase activities in lamb tissues (± SE)

	Alkaline phosphatase[a]		Thymidine kinase[b]	
	Liver	Brain	Liver	Brain
High Zn lambs	26.8 ± 3.82	41.3 ± 4.13	19.2 ± 2.7	10.2 ± 2.6
Low Zn lambs	30.0 ± 6.52	45.0 ± 5.33	21.1 ± 5.0	12.6 ± 4.2

a mU/mg protein where one unit (U) is the enzyme activity that causes 1 μmole of substrate to react in 1 minute at 25°C.

b p mol [^{3}H] thymidine incorporated into thymidine monophosphate/mg protein/hour.

DISCUSSION

There were no significant differences in the zinc status of the two groups of ewes at any stage of pregnancy, as indicated by the variables measured. Nevertheless, some of the additional zinc fed to the ewes was preferentially transported to the tissues of their lambs. The concentration of zinc in the liver, kidney and pancreas of these lambs was higher than in the lambs from the LZ group. In the rat foetal brain the activity of thymidine kinase and alkaline phosphatase is depressed by maternal zinc deprivation (8). However, despite the lower concentration of zinc in the liver of the LZ lambs, the activity of neither enzyme declined in the brain or liver. The amount of zinc present in these tissues in both groups was enough to allow optimal enzyme activity.

Further research is required to determine whether the low zinc content of the diet was involved in the number of difficult births. The numbers of animals used in this experiment were inadequate to draw any conclusions.

Under the conditions of this experiment a diet containing as little as 9 μg Zn/g diet was adequate to maintain ewes throughout pregnancy. As neither ewes nor lambs exhibited any clinical signs of zinc deficiency, absorption of zinc from the diet must have exceeded the 13% reported by Hansard and Mohammed (9) or the lamb must have accrued less zinc over the last third of foetal development than found by Williams *et al.* (6) (i.e. < 1.0 mg/day) or a combination of both occurred.

REFERENCES

1. Underwood, E.J., 1977, "Trace Elements in Human and Animal Nutrition", 4th Ed. (Academic Press, N.Y.)

2. Hurley, L.S. and H. Swenerton, 1966, Proc. Soc. Exp. Biol. Med. 123: 692-696.

3. Apgar, J., 1970, J. Nutr. 100: 470-476.

4. Masters, D.G. and M. Somers, 1980, Aust. J. Exp. Agric. Anim. Husb. 20: 20-24.

5. Masters, D.G. and H.E. Fels, 1980, Biological Trace Element Research 2: 281-290.

6. Williams, R.G., I. McDonald and I. Bremner, 1978, Brit. J. Nutr. 40: 377-386.

7. Duncan, J.R. and L.S. Hurley, 1978, Proc. Soc. Exp. Biol. Med. 159: 39-43.

8. Dreosti, I.E., I.R. Record and S.J. Manuel, 1980, Biological Trace Element Research 2: 21-29.

9. Hansard, S.L. and A.S. Mohammed, 1968, J. Anim. Sci. 27: 807-813.

DISCUSSION

WIENER (Scotland). You said at the beginning of your talk that zinc supplementation led to increased numbers of lambs, 14% I think you said. What is the source of loss in the case of the zinc deficiency? Is it lower ovulation rate or where does the loss occur?

MASTERS (Australia). We don't know exactly where the loss occurs. We laparoscoped the ewes early in one experiment and there were no differences in ovulation rate. We postulated that losses occur sometime in the first half of pregnancy, but we could not pinpoint it any closer.

ARORA (India). What is your opinion about the plasma zinc index in relation to the results you have obtained in the tissues?

MASTERS (Australia). I don't know. Before I started the experiment I expected there would be fairly large differences in plasma zinc values. We did not get them in this experiment although we got them in earlier field experiments where we also saw responses to zinc supplementation. So where we have seen growth responses to zinc we have also seen responses in plasma zinc levels. In the current experiment I presume that the sheep were absorbing enough zinc to maintain their plasma zinc levels and that perhaps they were not deficient.

PERLMAN (Israel). Do you have any idea what the optimal concentration of zinc is in the neonatal liver?

MASTERS (Australia). No. As I mentioned these animals were slaughtered before they had a chance to take any milk in. I am not aware of any reported values for the amount of zinc in tissues of lambs at this stage. Some of our preliminary results indicate there are large changes in the amount of zinc in these tissues after the lambs have consumed colostrum, but otherwise I do not know exactly what levels to expect.

MILLS (Scotland). Two points that I think are extremely interesting with this study. Work with really high zinc intakes has indicated the failure of the placenta to protect the foetus against zinc accumulation. Your experiments indicate that this is not just simply an overload situation that developed in these high zinc studies but was a general inability to discriminate. I think it is very relevant to aspects of zinc therapy or zinc prophylaxis during pregnancy in almost any species. There have been reports already of abortion caused by excessive uses of zinc during the terminal stages of human pregnancy. The second point I think is very interesting is that your results may well support what has been found with the rat; that the efficiency of absorption of zinc does in fact increase very markedly during late stages of pregnancy and also during lactation.

ABDULLA (Sweden). You mentioned there was a decreased level of zinc in the pancreas of the deficient animals. I wonder if you have looked at the insulin levels in the blood of these animals?

MASTERS (Australia). No, we have not.

NUTRITIONAL ZN DEFICIENCY AND METABOLISM OF INSULIN AND GROWTH HORMONE

H.-P. Roth and M. Kirchgeßner

Institut für Ernährungsphysiologie
Technische Universität München
D-8050 FREISING-WEIHENSTEPHAN
FED. REP. GERMANY

The pancreas is one of the most sensitive tissues of zinc metabolism, and its ß-cells of the islets of Langerhans have a high capacity for accumulating, turn-over and storing the retained zinc. This tissue responds to dietary zinc deprivation with a quick and severe zinc loss (4) and of the activity of digestive enzymes synthesized there, specifically the carboxypeptidase A and B (3). Since Scott (1934) discovered that crystalline insulin contains appreciable amounts of zinc, numerous studies have indicated a close functional and morphological relationship between zinc and insulin. Although the etiology of disorders of this endocrine system has been investigated extensively, only few studies are available regarding the influence of dietary factors on insulin metabolism, for example, the influence of the dietary zinc content. Changes in zinc status of the body could therefore affect the synthesis, storage, secretion and hormonal potency of insulin. Since alimentary zinc deficiency leads to growth depression in both young humans and animals, it was also the objective to determine the concentration of growth hormone.

The biosynthesis of insulin proceeds via proinsulin in the pancreas, the precursor of the insulin molecule, consisting of a single peptide chain. The pancreatic enzymes identified to be able to convert proinsulin into insulin are trypsin and carboxypeptidase B. However, the carboxypeptidase B of the pancreas is a zinc-metalloenzyme that loses about 50 % of its activity in zinc-deficient rats (3). It is thus possible that in zinc-deficient rats, the capacity of the pancreatic islets to convert proinsulin to insulin may be diminished or that a predominant secretion of proinsulin occurs directly from the ß-cells by a route which evades the granules where the conversion to insulin normally takes place. By contrast, the trypsin-like activity in the duodenal fluid of rats is almost doubled during zinc deficiency (6). The fact that proinsulin possesses just a slight biological activity has led to the speculation that the carbohydrate intolerance as found in earlier studies (5) is possibly the consequence of an augmented proinsulin secretion or of a reduced secretion or biological potency of insulin.

Table 1 shows the radioimmunologically estimated insulin concentrations in the serum of zinc-deficient and pair-fed control rats after an intraperitoneal glucose injection of 200 mg/100 g metabolic body size. In the pair-fed control animals, serum insulin levels showed a typical biphasic response. This characteristic insulin release pattern is usually induced by glucose stimulation alone as has been often reported. Twenty minutes after glucose injection, the serum insulin concentration of the pair-fed rats had almost returned to the initial level. In the case of the zinc-deficient animals, however, there was no additional insulin secretion in response to glucose injection.

TABLE 1. Serum insulin concentration of zinc-deficient and pair-fed control rats after i.p. glucose injection (200 mg/100 g metabolic body size

Time	(μU Insulin/ml Serum)	
(Minutes)	Depleted animals	Pair-fed controls
0	24.0 ± 11.4	15.0 ± 4.2
5	16.9 ± 6.8	44.5 ± 28.8
10	24.9 ± 17.3	14.3 ± 4.5
15	20.2 ± 12.0	40.7 ± 23.9
20	19.9 ± 7.7	19.8 ± 14.7
25	16.6 ± 5.9	17.0 ± 10.1
30	20.1 ± 5.0	19.3 ± 5.8

The proinsulin contents in serum (see Table 2) did, however, not show any differ-

TABLE 2. Serum proinsulin concentration of zinc-deficient and pair-fed control rats after i.p. glucose injection

Time (Minutes)	(ng Proinsulin/ml Serum) Depleted animals	Pair-fed controls
0	2.4 ± 1.4	1.1 ± 0.5
5	1.9 ± 0.7	2.3 ± 0.7
10	2.5 ± 0.8	2.5 ± 0.5
15	2.1 ± 0.6	2.5 ± 0.9
20	3.5 ± 1.3	3.5 ± 2.0
25	2.4 ± 1.2	2.0 ± 1.1
30	2.6 ± 1.5	2.8 ± 1.8

ence in response to zinc deficiency during the course of the glucose tolerance test with depleted and control rats. In either group, there was a slight increase after 20 minutes. Hence, the conversion of proinsulin to insulin is largely unaffected by dietary zinc content and by glucose stimulation. If insulin should require zinc to maintain its physiological activity, changes or shifts in the zinc concentrations of the tissues are likely to occur along with the secretion of substantial amounts of insulin following glucose stimulation. In the relevant studies, the serum zinc concentration of the depleted animals was reduced by about half in comparison with that of the pair-fed control animals (see Table 3). In the zinc-depleted animals, there was no major change in the serum zinc content nor in the insulin levels within 30 minutes following glucose injection. In the case of the pair-fed control animals, however, the serum zinc level markedly increased above its initial value at the start of the experiment within 10 to 15 minutes after glucose injection. It was also exactly during this time span of the high zinc concentration that the highest blood glucose and insulin levels were recorded. After 20 minutes the serum zinc content had returned again to normal. This suggests that the secretion of insulin leads to elevated serum zinc contents. The extent to which insulin would require zinc for maintaining its physiological potency is difficult to elevate.

TABLE 3. Zinc concentration in serum of zinc deficient and pair-fed control rats after i.p. glucose injection

Time (Minutes)	(μg Zinc/ml Serum) Depleted animals	Pair-fed controls
0	0.77 ± 0.17	1.49 ± 0.17
5	0.70 ± 0.11	1.56 ± 0.21
10	0.73 ± 0.20	1.83 ± 0.08
15	0.86 ± 0.06	1.90 ± 0.13
20	0.81 ± 0.08	1.42 ± 0.29
25	0.68 ± 0.03	1.61 ± 0.14
30	0.80 ± 0.05	1.61 ± 0.17

Since the dietary chromium content may also affect glucose tolerance, the chromium concentration in serum was determined as well during glucose stimulation (see Table 4). The chromium concentration in serum of the pair-fed control rats showed a similar response curve following glucose stimulation as did the zinc level. It had reached its maximum within 10 minutes. In contrast to zinc, chromium concentration also significantly increased in serum of deficient animals, especially toward the end of the glucose tolerance test. Overall, the serum chromium content, which is 1000 times lower than the zinc content, and hence lies in the ppb range, was somewhat lower in the zinc-deficient rats than in the control animals, although their diets both supplied chromium at the same level of 2.2 ppm. In view of the scarce information available, it is not possible to speculate on a specific role of chromium in insulin secretion nor on its mode of action.

TABLE 4. Chromium concentration in serum of zinc-deficient and pair-fed control rats after i.p. glucose injection

Time (Minutes)	(ng Chromium/ml Serum) Depleted animals	Pair-fed controls
0	0.77 ± 0.23	1.03 ± 0.03
5	0.95 ± 0.12	1.18 ± 0.10
10	0.88 ± 0.14	1.28 ± 0.09
15	0.79 ± 0.14	1.14 ± 0.09
20	1.00 ± 0.08	1.10 ± 0.14
25	1.00 ± 0.06	1.08 ± 0.09
30	1.16 ± 0.07	1.04 ± 0.05

Since the radioimmunological insulin determination does not measure the actual biological potency of insulin, the total insulinlike activity (TILA) in serum was determined in relation to dietary zinc supply by using isolated cells of epididymal adipose tissue (see Table 5). A significantly higher TILA was found in zinc-deficient animals than in pair-fed controls between 10 and 15 minutes after glucose stimulation. Evidently, the bodies of zinc-deficient animals may attempt to partially compensate for the lower insulin levels by secreting and activating other insulinlike substances during a glucose load.

TABLE 5. Total insulinlike activity in serum of zinc-deficient and pair-fed control rats after i. p. glucose injection

Time	(c p m)	
(Minutes)	Depleted animals	Pair-fed controls
0	363 ± 116	443 ± 153
5	247 ± 84	292 ± 89
10	416 ± 100	252 ± 121
15	314 ± 95	216 ± 47
20	214 ± 74	223 ± 61
25	311 ± 126	228 ± 71
30	261 ± 59	319 ± 151

Hence growth depression is known as an early and characteristic deficiency symptom when young animals are subjected to alimentary zinc depletion, the concentration of growth hormone in serum was determined radioimmunologically in two experimental series (see Table 6). The growth hormone concentration of the zinc-deficient rats was greatly reduced as compared to the ad libitum-fed control animals. The same reduction, however, was also noted in the case of the pair-fed control animals. Accordingly, the reduction of the growth hormone concentration must be regarded as the result of the diminished feed intake and not of zinc deficiency per se.

In summary, it may be concluded on the basis of these experimental results (7, 8, 9) that zinc has an important role in insulin metabolism, since we also observed reduced insulin contents in the case of zinc-deficient cows (2), which response could be reversed by zinc supplementation. By contrast, the lowered concentrations of growth hormone in serum of zinc-deficient rats must be regarded as a secondary effect following the restricted feed intake.

TABLE 6. Growth hormone (GH) concentration in serum of zinc-deficient, pair-fed and ad libitum control rats

		(ng GH/ml Serum)	
	Depleted animals (1.3 mg Zn/kg DM)	Pair-fed controls (100 mg Zn/kg DM)	ad libitum controls (100 mg Zn/kg DM)
a)	32.7 ± 18.5	30.0 ± 6.7	93.5 ± 82.5
b)	59.8 ± 16.1	55.3 ± 9.5	93.6 ± 47.7

REFERENCES

1. Scott, D.A., 1934, Biochem. J. 28: 1592 - 1602
2. Kirchgeßner, M., Roth, H.-P., and W.A. Schwarz, 1976, Z. Tierphysiol., Tierernährg. u. Futtermittelkde. 36: 175 - 179
3. Roth, H.-P., and M. Kirchgeßner, 1974, Z. Tierphysiol., Tierernährg. u. Futtermittelkde. 33: 62 - 67
4. Roth, H.-P., and M. Kirchgeßner, 1975, Z. Vit. Ern. Forsch. 45: 201 - 208
5. Roth, H.-P., Schneider, U., and M. Kirchgeßner, 1975, Arch. Tierernährg. 25: 545 - 549
6. Roth, H.-P., and M. Kirchgeßner, 1977, Internat. Z. Vit. Ern. Forsch. 47: 277 - 283
7. Roth, H.-P., and M. Kirchgeßner, 1979, Z. Tierphysiol., Tierernährg. u. Futtermittelkde. 42: 277 - 286
8. Roth, H.-P., and M. Kirchgeßner, 1979, Z. Tierphysiol., Tierernährg. u. Futtermittelkde. 42: 287 - 299
9. Roth, H.-P., and M. Kirchgeßner, 1981, Biological Trace Element Research, in press.

DISCUSSION

DREOSTI (Australia). Would you care to comment on the idea that zinc is involved with the binding of insulin to the insulin receptors. Do you think that might be a metabolic mechanism?

ROTH (Germany). It is possible that zinc deficiency affects the receptor for insulin binding but we have no results at this stage.

ABDULLA (Sweden). In our Institute we have been trying to find the response in healthy human subjects to oral zinc supplements and we found that they have a transient increase in blood glucose whereas the glucose tolerance test was normal. I wonder if you have a comment?

ROTH (Germany). I have no explanation for the response you have mentioned. It is known from the literature that Zn-deficient animals have a reduced glucose tolerance. We have confirmed this in several studies which also showed that the amount of glucose injected is important. If, for example, the glucose dose is too high, the homeostatic regulation maintaining the blood glucose is overloaded in both the Zn-depleted and the control animals, while after the injection of too low a glucose dose, both groups can readily cope with this load, and hence no overt difference occurs.

ELMES (Wales). May I point out that when you are talking about zinc and endocrine cells you must remember that insulin represents only one of the systems that in humans we call the gastro-enteric-pancreatic-endocrine system, and it has been known for a long time that the islet cells that produce insulin and glucagon are high in zinc. We have some preliminary results from microprobe analysis which indicate that some of the endocrine cells in the intestinal mucosa (which are possibly the ones producing CCK PZ) are also exceedingly high in zinc. I think that the zinc story goes back very early to the actual granule formation within the endocrine cell.

AGGETT (Scotland). Just supplementary to Dr. Elmes point, there are species such as the guinea pig and the coypu where zinc is not associated with the presecretory granule. Perhaps we should be using these animals as models for studies on the metabolism of insulin in zinc deficiency.

ROTH (Germany). If you inject zinc-free insulin into an animal, you see that the insulin is active, but this does not necessarily mean that zinc is not needed for insulin activity because in the body you have enough zinc to activate zinc-free insulin. I think zinc is also very important for storing insulin: pro-insulin binds 5 moles of zinc and is soluble. Insulin binds only 1 mole of zinc and is insoluble. This Zn-containing insulin is present in crystalline form in the β-cells. When the insulin is secreted, the zinc content of β cells decreases very fast. However, in the α-cells, where glucagon is synthesised, the zinc content does not decrease.

REGULATION OF PROTEIN INTAKE IN ZINC DEFICIENT RATS

P.G. Reeves and B.L. O'Dell

Department of Biochemistry
University of Missouri
COLUMBIA, MISSOURI 65201
UNITED STATES

Food intake depression is a common phenomenon in rats fed zinc deficient diets. Food intake becomes cyclic in nature with 2 to 3 day periods between high and low intakes. The level of dietary protein is related to this cyclic feeding pattern in that rats fed 20% protein diets cycle but those fed 5% protein do not (1). There is also a relationship between the level of plasma tyrosine (TYR) and the high and low points of the feeding cycle in the deficient rat (2).

Protein and energy intakes can be regulated by rats when they are allowed to self-select between 2 isocaloric diets fed simultaneously and containing different ratios of protein and carbohydrate (3). Recently it was found that rats allowed to self-select zinc deficient diets composed of different ratios of protein and carbohydrate choose more of the low protein diet when compared to rats receiving diets with adequate zinc (4). The ratio of plasma tryptophan (TRP) to the sum of the neutral amino acids valine, leucine, isoleucine, tyrosine and phenylalanine (TRP/ΣNAA) is inversely related to the amount of protein selected. Because these 6 amino acids compete with each other for entry into the brain, the concentration in the brain of both TYR and TRP as well as their metabolites, norepinephrine (NE), dopamine (DA) and serotonin (5HT) are influenced by the relative quantities of these amino acids in the plasma (5). It has been proposed that the regulation of protein as well as total food intake in self-selecting rats is mediated through a neurotransmitter mechanism (5). Whether this is also the mechanism that regulates food intake and nutrient preference in the zinc deficient rat is the subject of this report.

MATERIALS AND METHODS

General. Wistar derived rats from the Biochemistry Department (UMC) colony were used throughout these studies. Rats were preconditioned for 4 days to a semipurified diet nutritionally adequate in zinc (4). They were housed in stainless steel cages with wire mesh bottoms and given water ad libitum in plastic bottles fitted with silicone stoppers and stainless steel sipper tubes. Zinc in the diet and plasma was determined by atomic absorption spectrophotometry, amino acid in plasma and brain by micro-column high pressure chromatography and neurotransmitters in brain by high performance liquid chromatography with electrochemical detection.

Experiment 1. This experiment is presented to show the effects of short-term zinc deficiency on the self-selection of protein and carbohydrate. Twenty male rats were divided into groups of 10 rats each. Individual rats in one group were offered, simultaneously, 2 isocaloric zinc-deficient diets. One diet contained 10% and the other 50% of protein. The source of protein was a soybean isolate treated with EDTA to remove excess zinc (6). Both diets contained 5% corn oil. The control group of rats received similar diet pairs which contained 100 ppm of zinc. Dietary intake and body weight were recorded daily for nine days whereupon plasma and whole brain were collected. Plasma was analyzed for zinc and free amino acid contents and brain tissue for NE, 5HT and free amino acids.

Experiment 2. Some work has shown a strong positive correlation between whole brain 5HT levels and self-selection of protein but no correlation with respect to catecholamines and protein intake (7). By contrast, other studies find food intake decreased when rats are fed a non-self-selecting regimen and treated in ways that increase tryptophan and/or 5HT levels in brain (8). An effect on food intake of rats can be demonstrated when drugs that disrupt catecholamine metabolism are injected into discrete areas of the hypothalamus. There is some agreement that this region of the brain is the control center of food intake regulation in the rat. Therefore, the second experiment in this series was to determine the effects of short-term zinc deficiency on the concentration of 5HT, NE and DA and also the rate of synthesis of 5HT in the hypothalamus. To avoid the possible effect of diet, self-selection was not allowed. To study the rate of synthesis of 5HT, a group of rats, fed a zinc adequate diet, was pair-fed to a group receiving a zinc-deficient diet. After 6 days of feeding, all rats were fasted overnight and then allowed access to food for 3 hours. Each rat was injected intraperitoneally with pargyline, a monoamine oxidase inhibition (100 mg per kg of body weight). The accumulation rate of 5HT and the depletion rate of 5 hydroxy-indolacetic acid (5HIAA), a metabolite of 5HT, was measured in the hypothalamus over a period of 1 hour. To assess the relationship between zinc deficiency and catecholamine metabolism, rats were fed zinc deficient diets ad libitum for 6 days, fasted for 24 hrs and refed the diets for 36 hrs. After a fast during the day light hours the rats were killed and the catecholamine concentration measured in the hypothalamus. The hypothalamus was divided into the anterior and posterior regions by taking 2 mm cross sections through the hypothalamic region of the brain (9).

RESULTS

Rats in all experiments, whether self-selecting protein and carbohydrate or not, showed a decline in both food intake and body weight after consuming the zinc deficient diet for 3 days. Figure 1 shows that short-term zinc deficiency dramatically reduced the percentage of dietary protein but increased the percentage of carbohydrate, voluntarily selected by the rat. Table 1 shows that the ratios of both TRP/ΣNAA and TYR/ΣNAA were elevated in zinc deficient rats compared to rats fed adequate zinc. Contrary to prediction, TRP and 5HT levels in brain were not elevated. However, brain TYR was increased as a consequence of elevated TYR/ΣNAA; there was no difference in NE levels. Further study revealed that zinc deficiency had no effect on the rate of synthesis of 5HT. The rate of synthesis of 5HT in zinc deficient rats was 0.6±0.5 μg/g/hr while the control rate was 0.5±0.03 μg/g/hr. The rate of disappearance of 5HIAA from the hypothalamus was identical in both groups (0.6 μg/g/hr). There was a small but significant elevation of NE in the anterior but not in the posterior hypothalamus of zinc deficient rats not allowed to self-select their diet (Table 2). At the same time, DA levels were significantly lower in the posterior hypothalamus of zinc deficient rats but were unchanged in the anterior hypothalamus when compared to controls. Plasma zinc levels were lower in zinc deficient rats, as expected. Food intake, per unit of body weight, during the previous 36 hours was not different between groups (Data not shown).

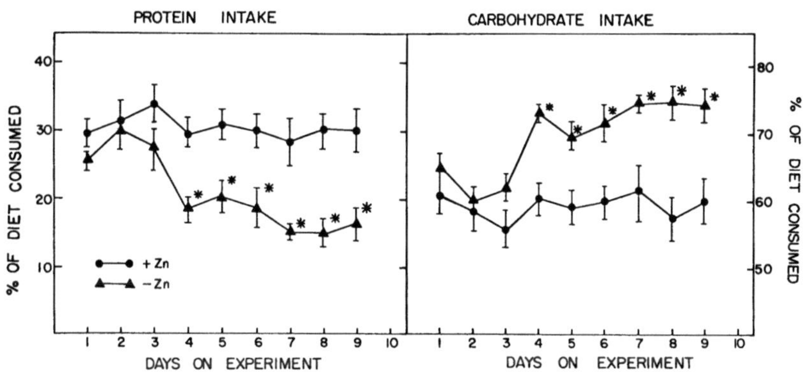

Fig. 1. Effect of short-term zinc deficiency on the percentage of protein and carbohydrate in the diet selected by the rat. *Mean ± S.E.M., 10 reps per period. Groups at periods with asterisks are significantly different from each other with P values of 0.05 or less.

DISCUSSION

These experiments show that rats, deprived of dietary zinc, and allowed to choose between 2 diets varying in the ratio of protein to carbohydrate, select a diet with half as much protein as rats fed adequate zinc. It was postulated that the mechanism for the difference in protein selection was caused by a change in the metabolism of neurotransmitters in the hypothalamus. An evaluation of the TYR/and TRP/ΣNAA ratios in plasma of zinc deprived rats predicts that tryptophan and tyrosine levels in the brain should be elevated. These in turn should elevate 5HT and/or NE and DA. The results showed, however, that the levels and rate of synthesis of 5HT in the hypothalamus were not affected by dietary zinc deprivation and therefore are probably not related to protein selection in the zinc deprived rat. On the other hand, NE and DA levels were affected by zinc deficiency, but inversely in different parts of the hypothalamus. Zinc deprived rats had elevated NE levels in the anterior but not in the posterior portion. DA levels, by contrast, were depressed in the posterior but not in the anterior hypothalamus. These results provide no direct evidence of a functional role of 5HT or the catecholamine in the control of feeding behavior in the zinc

Table 1. Effect of Dietary Zinc on Plasma Amino Acid Ratios, Plasma Zinc and Brain TRP, TYR, 5HT and NE in Self-Selecting Rats.

Dietary Zinc (ppm)	<1	100
Plasma AA Ratios		
TRP/ΣNAA	0.14±0.01[a*]	0.10±0.01[b]
TYR/ΣNAA	0.19±0.01[a]	0.13±0.01[b]
Plasma Zinc (ppm)	0.45±0.02[a]	1.55±0.08[b]
Brain: (μg/g)		
TRP	4.45±0.59	4.33±0.47
TYR	13.03±1.09[a]	10.66±0.60[b]
5HT	0.47±0.02	0.45±0.03
NE	1.20±0.09	1.20±0.12

*Mean ± S.E.M., 10 replications each. Means between diet groups with different superscripts are significantly different with a P value of 0.05 or less.

Table 2. Effect of Short-term Feeding of Low Zinc Diets on NE and DA Concentrations in the Rat Hypothalamus.*

		Anterior		Posterior	
Dietary Zn ppm	Plasma Zn ppm	NE μg/g	DA	NE μg/g	DA
<1	0.45±0.04[a]	2.29±0.08[a]	0.37±0.02	2.46±0.09	0.51±0.04[a]
50	1.00±0.03[b]	2.04±0.04[b]	0.35±0.02	2.26±0.07	0.66±0.05[b]

*See footnote Table 1. 8 replications in each of these groups.

deficient rat. Indeed, the higher levels of NE and DA in the hypothalamus of these rats may only reflect the effects of under-nutrition in general. It was reported recently, however, that whole brain NE and DA levels were higher in zinc deficient rats when compared to controls either pair-fed, ad libitum-fed or fasted for 12 hours (10). Whether the changes in the present study are related to food intake and protein selection in zinc deprived rat remains to be determined.

REFERENCES

1. Chesters, J.K., and M. Will, 1973, Br. J. Nutr. 30:555-566.
2. Griffith, P.R., and J.C. Alexander, 1972, Nutr. Rep. Int. 6:9-20.
3. Ashley, D.V.M., and G.H. Anderson, 1975, J. Nutr. 105:1412-1421.
4. Reeves, P.G., and B.L. O'Dell, 1981, J. Nutr. 111:375-383.
5. Fernstrom, J.D., and R.J. Wurtman, 1971, Science 173:149-152.
6. O'Dell, B.L., C.E. Burpo, and J.E. Savage, 1972, J. Nutr. 102:653-660.
7. Anderson, G.H., and D.V.M. Ashley, 1977, Life Sci. 21:1227-1234.
8. Goudie, A.J., E.W. Thornton, and T.J. Wheeler, 1976, J. Pharm. Pharmoc. 28:318-320.
9. Heffner, T.G., J.A. Hartmen, and L.S. Seiden, 1980, Pharm. Biochem. and Behav. 13:453-456.
10. Wallwork, J., and H. Sandstead, 1981, Fed. Proc. 40:(Part II) 939.

DISCUSSION

HURLEY (U.S.A). Did you follow the plasma zinc level, that is, the decline in plasma zinc in the two groups and were there any differences that related to the differences in protein concentrations?

REEVES (U.S.A.). No I did not follow the plasma zinc in this experiment. The slide showing the final level of plasma zinc is for the time the animals were killed, and the levels were significantly lower in the zinc deficient rats. I didn't follow it specifically anytime during the experiments, or at any time just prior to killing the animals.

HURLEY (U.S.A.). I wondered if the difference in the selection of the low protein diet might be related to the type of phenomonen that we have observed with protein deficient diets and with either starvation or with low calcium diets. That is if the protein deficiency was severe enough to cause a greater rate of catabolism of the soft tissues, this would allow for the release of zinc from the tissues and would make the rat "feel better".

REEVES (U.S.A.). That is highly possible. However, in these experiments where rats were given a choice of both high and low protein diets, the diet chosen by the zinc deprived rats contained approximately 15% protein. Under normal circumstances, this diet would not be deficient in protein. It has been our experience that these animals do not cycle as readily as those on a no-choice regimen. The plasma zinc under these circumstances probably would not fluctuate.

An additional comment. The differences between zinc deficient and control rats in norepinephrine levels in the hypothalamus may seem too small to possibly affect a change in feeding behaviour. We have, however, in other similar experiments observed as much as 34% of an increase in zinc deprived rats over zinc adequate controls.

A STUDY ON EXPERIMENTAL ACRODERMATITIS ENTEROPATHICA

SHIGERU YAGI AND TAKAFUMI MORISHIMA
DEPARTMENT OF DERMATOLOGY, NIHON UNIVERSITY SCHOOL OF MEDICINE, TOKYO, JAPAN.

Acrodermatitis enteropathica (AE) is a genetically determined disorder characterized by psoriasiform or impetiginous skin lesions on the acral areas and around the natural orifices, alopecia and diarrhea. Recently AE could be classified into two types, hereditary and acquired (1). The acquired form can develop from intravenous hyperalimentation (2) and longterm lactose-free milk alimentation (1). Concerning the pathogenesis of AE, the prevailing view is that AE results in a zinc deficiency in humans. Some investigators believe, however, that it occurs due to the metabolic abnormality of fatty acids.

In order to determine the etiology of AE, we tried to produce AE-like lesions experimentally in pigs fed on a low zinc diet.

MATERIALS AND METHODS

A low zinc diet was devised by removing as much zinc as possible from the basal swine diet (Zn content, 320 ppm) without any changes in the basic composition. The zinc content in the low zinc diet was 46 ppm. It's value was similar to that in the experiment described by Tucker and Salmon (3).

Experiment 1. Four SPF miniatured pigs which had been weaned 1 month earlier were used for this experiment. Two of them were fed on the basal swine diet as the control group. As the experimental group, the other two were fed on the low zinc diet.

Experiment 2. Two newly-weaned Hampshire X Landrace pigs were fed on the low zinc diet, containing 2% $CaCO_3$ and 2% phytic acid, in order to inhibit the absorption of zinc (4).

All tests were conducted on concrete-floored pens without bedding. The water given to the pigs was city water and it's zinc content was under 1 ppm. The zinc levels in the serum and the skin were measured by atomic absorption spectrometry.

RESULTS AND COMMENTS

Experiment 1. This experiment was continued for 6 months. The pigs in the experimental group grew favourably and did not develop any AE-like lesions. The results indicated that the zinc deficiency state could not occurred easily.

Experiment 2. In this experiment (Fig 1), AE-like lesions appeared for the first time. About 20 days after the beginning of the experiment the pigs showed anorexia and diarrhea. About 1 week after that impetiginous or psoriasiform skin lesions developed on the acral areas (Fig 2a, b), around the anus and on the abdomen (Fig 2c), which were susceptive to mechanical stimulation. The zinc level in the serum during the dermatitis stage significantly decreased to 28 μg/dl. The body weight decreased from 8 kg at the beginning of the experiment to 7.5 kg 1 month later. At this time, the body weight of the pigs in the control group was about 30 kg. Histological findings of skin lesions were similar to those of psoriasis. These clinical and histological findings were consistent with those of AE in humans.

Seven weeks after the beginning of the experiment, one pig with skin lesions was fed on swine milk containing zinc at a concentration of 220 ppm instead of the low zinc diet containing calcium and phytic acid, without the addition of zinc. Appetite increased, diarrhea stopped 2-3 days after the change of the diet, and the skin lesions rapidly disappeared 1 week later (Fig 2d). The body weight also increased to 13 kg 10 days after and to 20 kg 25 days after the change of the diet.

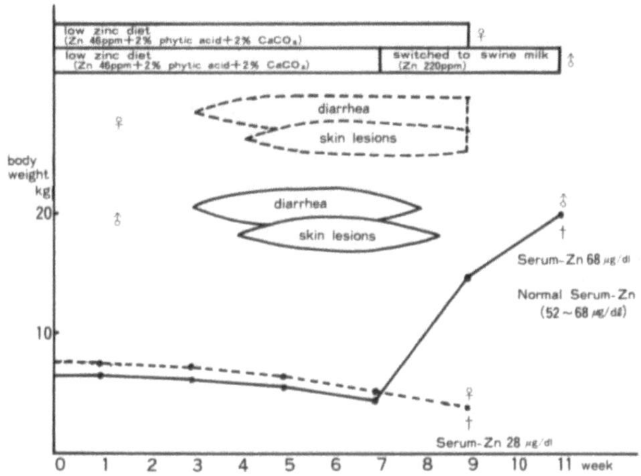

Fig.1(Experiment 2) Clinical course of AE-like lesions in the pigs fed on the low zinc diet. (Hampshire x Landrace)

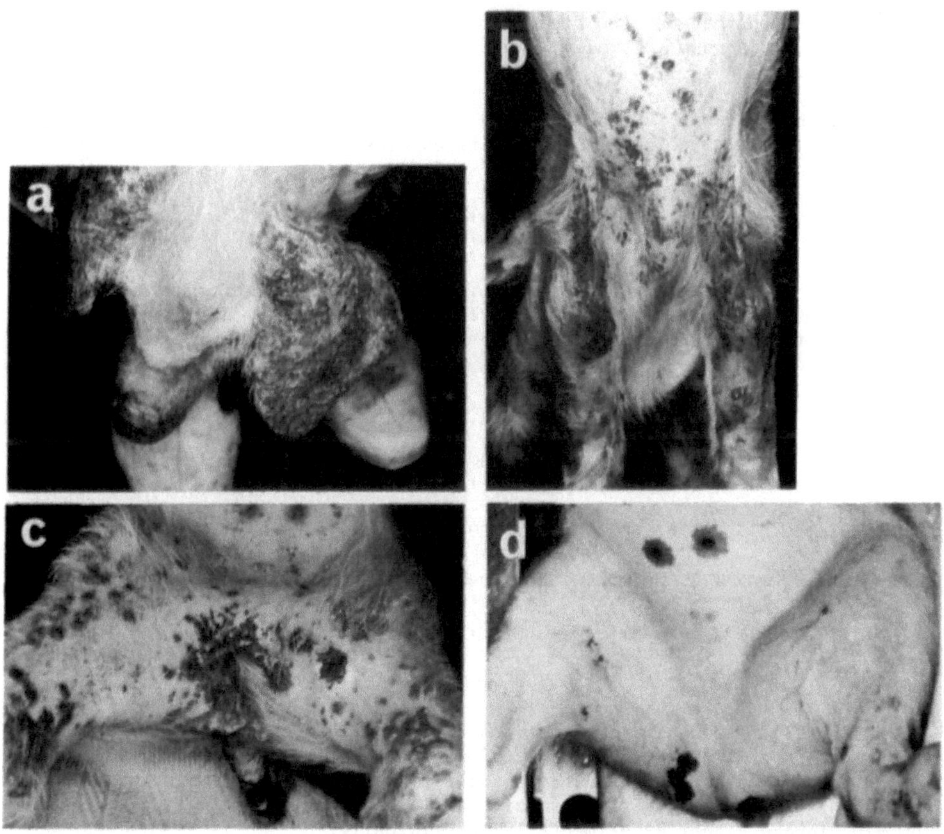

Fig.2 Dermatological findings: During the zinc deficiency stage AE-like lesions developed on the areas susceptive to mechanical stimulation(a,b,c). The skin lesions rapidly disappeared just by swtiching to the usual swine diet without the addition of zinc(d).

The zinc level in the serum during the recovery stage was 68 μg/dl. The zinc levels in the sera of the control pigs ranged from 52 to 68 μg/dl.

The zinc levels in the skin during the dermatitis stage (affected sites; 13.3 μg/g dry weight : non-affected sites; 14.4 μg/g dry weight) were significantly lower than those during the recovery stage (34.4 μg/g dry weight), but there were no differences in the zinc levels in the skin both in the lesions and the non-affected sites. From these results, it is presumed that the skin lesions resulted not only from decreased levels of zinc in the skin but also from mechanical stimulation. On the basis of the above-mentioned results of the experiments, the pathogenesis of AE is considered to result from zinc deficiency.

Some investigators believe that the metabolic abnormality of fatty acids is related to the development of AE. Therefore, fatty acid fractions in the serum and the skin were determined during the dermatitis stage and during the recovery stage. There was no significant difference between these stages regarding the rate of linoleic acid. The rate of arachidonic acid in the experimental group was higher than that in the control group during both the dermatitis stage and the recovery stage. That is, a decrease in arachidonic acid could not be demonstrated. Judged from these results, it is considered that the AE-like lesions in the pigs occurred due to zinc deficiency rather than to the metabolic abnormality of fatty acids.

SUMMARY

The results obtained from the present study were summarized as follows:

(1) The pigs fed on the low zinc diet containing calcium and phytic acid showed anorexia and diarrhea initially. Subsequently AE-like skin lesions appeared.

(2) During the dermatitis stage, the zinc level in the serum was low and the zinc level in the skin was low, both in the lesions and the non-affected sites. Mechanical stimulation was considered to be closely related to the development of eruptions.

(3) The pigs showed remarkable retardation of growth during the zinc deficiency stage.

(4) By switching from the low zinc diet to the usual swine diet, without the addition of zinc, the above-mentioned skin lesions rapidly disappeared, the body weight increased, and the zinc levels in the serum and the skin also returned to normal.

REFERENCES

1. Morishima, T. Yagi, S. Kuwabara, K. Endo, M. and T. Takemura, 1980, J. Dermatol., 7 : 121-125

2. Okada, T. et al., 1976, Surg., 80 : 629-635

3. Tucker, H.F. and W.O. Salmon, 1955, Proc. Soc. Exp. Biol. Med., 88 : 613-616

4. O'Dell, B.L., 1969, Am. J. Clin. Nutr., 22: 1315-1322

DISCUSSION

HURLEY (USA). You have confirmed that zinc deficiency produces some severe skin lesions that are similar to those of acrodermatitis enteropathica. However, it would be a good idea if we agreed to restrict the term acrodermatitis enteropathica to the genetic disorder and reserve the term zinc deficiency for conditions that are produced by low zinc intake either experimentally or clinically.

INFLUENCE OF ZINC ON SYNTHESIS AND ACCUMULATION OF COLLAGEN IN EARLY GRANULATION TISSUE

I. Tengrup, J. Ahonen and B. Zederfeldt

Departments of Surgery, Experimental Research and Nuclear Medicine,
Malmö General Hospital, University of Lund,
214 01 MALMÖ, SWEDEN

Chronic zinc deficiency causes retardation of growth and delay of reparative processes. The rate of repair is normalized by zinc administration. Acute lowering of serum zinc follows trauma. The importance of acute lowering of serum zinc for reparative processes is not clear. Intramuscular zinc administration starting before trauma results in increased collagen accumulation in early granulation tissue (1). The *collagen content* is the net result of a co-existant collagen synthesis and collagenolysis (Fig. 1).

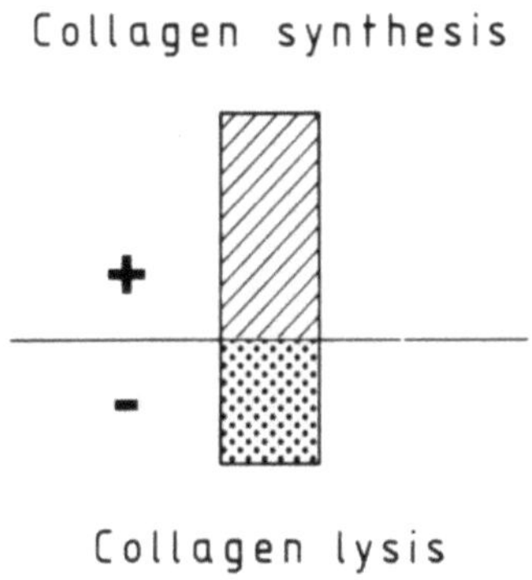

Fig. 1. Collagen accumulation.

Thus, higher collagen content can be explained either by increased collagen synthesis or decreased collagenolysis. The aim of the present study is to elucidate the mechanisms by which zinc administration causes increased collagen accumulation in early granulation tissue. *Collagen synthesis* has been studied in vivo and in vitro by determination of the incorporation of ^{14}C-L-proline into collagen.

Fifty-four Sprague Dawley rats were used. To obtain granulation tissue 6 polyvinyl sponges were implanted subcutaneously on the backs of the rats. Fifty μg zinc were given daily intramuscularly with the first injection half an hour before the operation. In the in vivo study 2.78 MBq ^{14}C-proline was injected intravenously into the tails of the rats 24 hours before removal of the sponges. The sponges were removed 3, 4 and 5 days after implantation. For the in vitro study 2 sponges from each animal, removed 4 days after implantation were cut into slices and incubated with 0.37 MBq ^{14}C-proline in a Krebs-Ringer solution for 2½ hours. The sponges were analyzed for *hydroxyproline content* and the *radioactivity of ^{14}C-hydroxyproline* was measured. Hydroxyproline content was used for calculating *specific activity of ^{14}C-hydroxyproline* expressed as DPM ^{14}C-hydroxyproline/μmol hydroxyproline (2).

We found that the *amount of hydroxyproline* was very low at day 3 but higher at day 4 and 5 (Fig. 2). At day 4 there was a higher amount of hydroxyproline in the zinc treated group than in the control group. The *radioactivity of hydroxyproline* per sponge is low at day 3, somewhat higher at day 4 and very much higher at day 5 (Fig. 3). There are no significant differences between control and zinc treated animals.

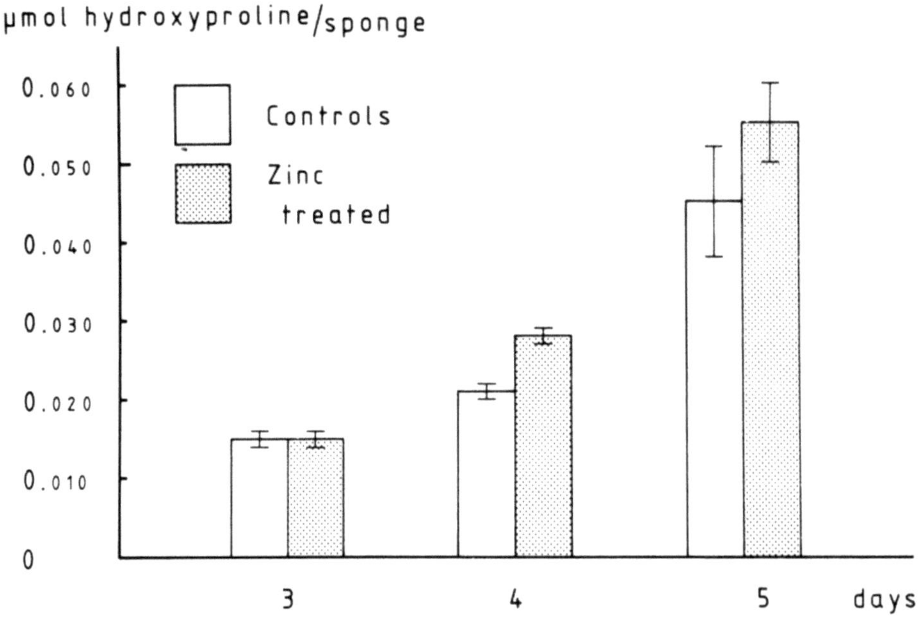

Fig. 2. Hydroxyproline content per granuloma.

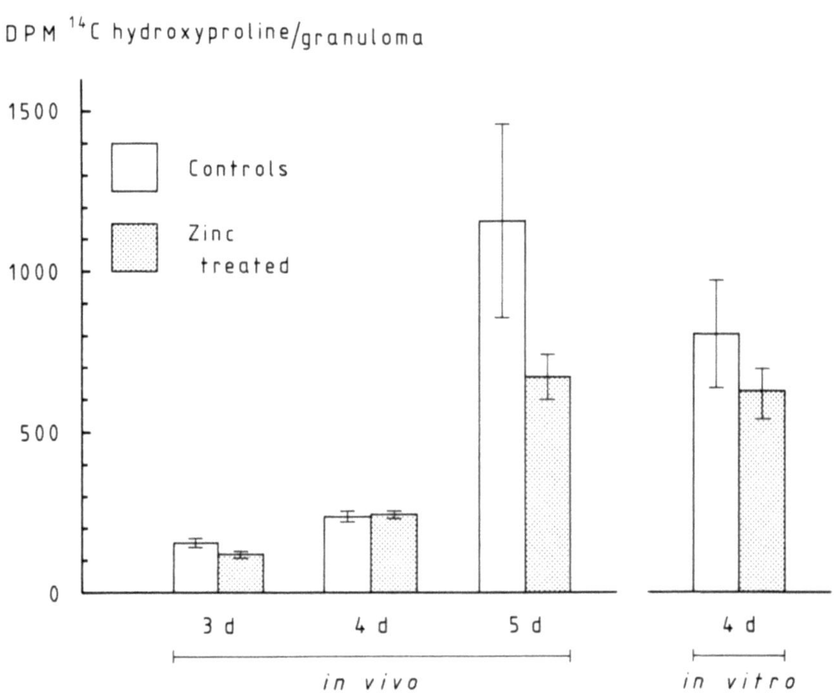

Fig. 3. Radioactivity of hydroxyproline per granuloma.

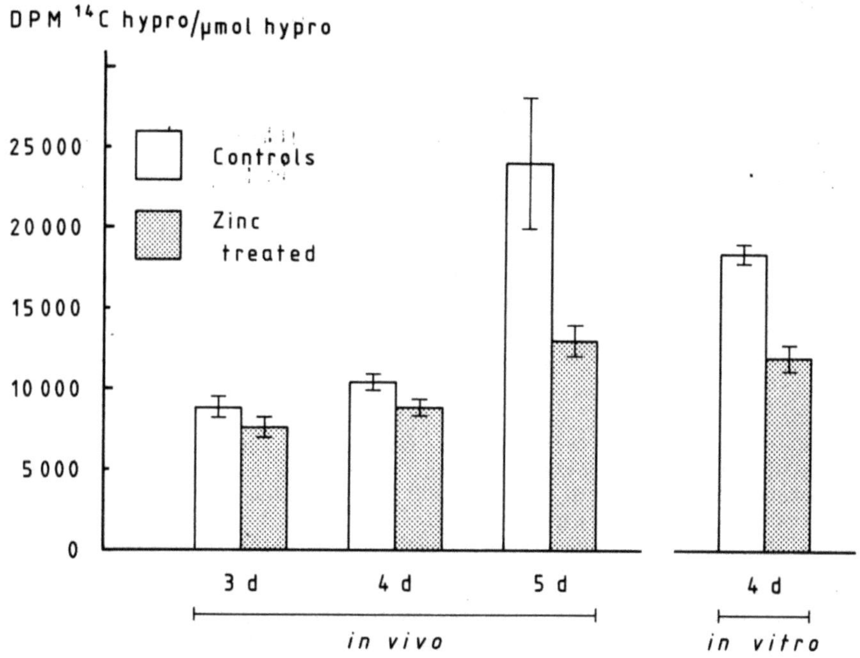

Fig. 4. Specific activity of hydroxyproline.

The *specific activity of hydroxyproline* is lower in the zinc treated animals at day 4 and 5 (Fig. 4).

No difference was found regarding the rate of collagen synthesis in granulation tissue for control and zinc treated animals. The increased amount of collagen found at day 4 thus cannot be explained by increased collagen synthesis. The specific activity of collagen is a measure of newly synthesized collagen in relation to the total amount of collagen in the specimen. Specific activity was found to increase from day 3 to 5 which is a consequence of increase in the rate of collagen synthesis during this period. It was lower in the zinc treated group than in control groups at day 4 and 5. This indicates that after zinc administration there is more of earlier synthesized not labelled collagen present in the granulation tissue. A possible explanation for this finding is that in granulation tissue there is a concomitant synthesis and breakdown of collagen and that the rate of breakdown is decreased by zinc administration.

Polymorphonuclear leucocytes disappear more rapidly from granulation tissue after zinc administration (3). An earlier disappearance of this source of collagenolytic enzymes could explain the decrease in the rate of breakdown of collagen.

REFERENCES

1. Tengrup, I., J. Ahonen, and B. Zederfeldt, 1980, Acta Chir. Scand. 146:1-44.

2. Juva, K., and D. Prockop, 1966, Anal. Biochem. 15:77-83.

3. Tengrup, I., J. Ahonen, F. Rank, and B. Zederfeldt, 1980, Acta Chir. Scand. 146:243-246.

SELENIUM STUDIES - TREPIDATIONS, TISSUE CALCIFICATION AND SELENOAMINO ACIDS

P.D. Whanger and M.A. Bielstein

Department of Agricultural Chemistry, Oregon State University,
(CORVALLIS, OREGON 97330, U.S.A.)

One purpose of this paper is to point out some of the pitfalls in studying selenium metabolism, especially as related to column chromatography and selenoamino acid analysis. Some forms of selenium are very reactive and hence erroneous results can be obtained if special precautions are not taken.

TREPIDATIONS

This term is used to alert researchers of some of the potential problems associated with the study of selenium metabolism. Ganther (1) has reviewed many of the difficulties involved in metabolic studies with selenium. This discussion includes precautions concerning checks on purity of radioactive selenium compounds; exposure of selenium to metallic utensils; storage of selenium compounds; some chromatographic problems with selenium compounds; the affinity of some volatile selenium compounds for various types of containers; precautions for trapping volatile selenium; and some precautions for identification of selenium compound. One reason for the present discussion is to indicate some of the unpublished difficulties we have encountered in our laboratory which were not discussed in this review. As indicated in this review it is critical to check the purity of radioactive selenium compounds. Even though the compounds may be pure when obtained from the supplier, we have found changes to occur upon storage. Fortunately there are some fairly straight-forward procedures to purify various ^{75}Se containing compounds as discussed by Ganther (1). The pH of the buffer used to homogenize the tissue is crucial. Minimum losses of selenium were found when tissues from animals injected with ^{75}Se-selenite were homogenized in phosphate buffer of pH 6.3, but increased losses occurred with increases of pH up to 8.3.

Reducing agents have been found to influence the chromatographic patterns of tissue cytosols subjected to gel filtration. When mercaptoethanol was included in the buffers, it resulted in an increase of selenium eluted in a low molecular weight (MW) area ($\approx$ 1000) and a decrease in a peak corresponding to about 30,000 MW as compared to buffers without this compound. This is in contrast to reduced glutathione or dithiothreitol which do not influence the elution patterns. The inclusion of these reducing agents in solutions of ^{75}Se-proteins, however, is necessary in order to prevent losses when concentrated with Diaflo UM-2 membranes. Up to 40% of the ^{75}Se can be lost during this concentration step when these reducing agents are not present. In contrast, mercaptoethanol does not prevent this loss on these membranes and often increases the losses. Thus, it is important to keep the ^{75}Se-proteins reduced with appropriate reducing agents.

Storage conditions of plasma from animals injected with ^{75}Se have also been shown to influence the elution patterns on gel filtration. The elution patterns of plasma which had been frozen at -70°C or fresh plasma were very similar. Most of the radioactivity eluted with the globulins with the remainder in the albumin area of the chromatogram. In contrast, chromatography of plasma which had been stored at 4°C for 24 hours resulted in the same amount with the globulins, but a decrease in the albumin area of about 50% with a creation of a new peak at about 30,000 MW. Other problems will be mentioned in connection with studies reported below.

TISSUE CLASSIFICATION

One of the pathological conditions in white muscle disease (WMD) is the accumulation of calcium in skeletal tissues (2). In order to investigate this disorder, the ability of sarcotubular reticulum (SR) membranes from muscle of normal and WMD lambs to sequester calcium (3) was studied. As shown in Table 1, the membranes from normal lambs exhibited a 12-fold greater uptake rate than those from the WMD lambs, but very little difference was found in the ATPase activity. Thus, we conclude that the calcium pump is not significantly affected by WMD but it is the binding of calcium by the membranes which is affected. These results offer a possible reason for soft tissue calcification in WMD.

When the crude SR membranes were placed on a linear sucrose gradient of 25 to 45% and centrifuged (4), two peaks of calcium uptake and ATPase activity were obtained. The major one banded at about 27% and the minor one at 36% sucrose. When ATP, however, was included in the homogenization media, all of the activity from normal muscle banded at 36% sucrose, but this activity from WMD muscle banded at both 27 and 36% sucrose with the most active one at the former sucrose level. When membranes from normal lamb muscle that was homogenized without ATP were applied to sucrose gradients, additions of some of the soluble material remaining on top of the gradient to that banding at 36% sucrose stimulated the calcium uptake activity 4-fold whereas this material had no effect on the uptake ability of that banding at 27% sucrose. Further purification of the peak banding at 27% sucrose (from tissues without ATP in the homogenization media) was obtained by applying the most active fractions to a second linear sucrose gradient of 23 to 36%. The peak again sedimented at 27% sucrose.

Table 1. Calcium uptake and ATPase activity of SR membranes from normal and WMD lamb muscle.

Activity	Normal*	WMD*
Ca-uptake (ηmoles/min/mg protein)	253 ± 252	21 ± 14
ATPase (ηmoles PO_4 released/ min/mg protein)	203 ± 164	190 ± 121

* Means of 9 animals ± standard deviation.

When this fraction was subjected to polyacrylamide gel electrophoresis (5), a difference in the intensity of staining of these membranes from normal and WMD muscle was obtained when stained with Amido black, but not when stained with Coomassie blue. The number of proteins which were stained, however, was directly proportional to the severity of WMD -- 6 bands in membranes from normal lambs, 8 bands from muscle of lambs with moderate WMD, and 13 bands from muscle of lambs with severe WMD. This suggests that there may be proteolytic degradation of the membranes in WMD lambs.

SELENOAMINO ACIDS

Selenocysteine has been indicated to be the form of selenium in glutathione peroxidase (6). Thus, we were interested in determining whether selenium was also found as selenocysteine in other selenium containing proteins and some tissue organelles of animals. According to published reports (7, 8) acid hydrolysis will destroy selenocystine, but we have found this amino acid to be very stable to such conditions when oxygen is rigorously excluded from the hydrolysis vessel (9). We routinely recovered up to 90% of selenocystine when subjected to acid hydrolysis in an inert atmosphere. In agreement with earlier reports (7, 8) we found this selenoamino acid was destroyed when traces of oxygen were present during acid hydrolysis.

When a semi-purified preparation of the cardiac low MW seleno protein from a lamb injected with ^{75}Se-selenite (10) was acid hydrolyzed and chromatographed on a column of Dionex DC6A ion exchange resin with citrate buffers, the only selenoamino acid present was selenocysteine (11). It is important to omit thiodiglycol from the loading buffer because exposure of selenite to this compound results in a complex which elutes at the identical position as selenocystine. Selenite in the thiodiglycol-free loading buffer, however, is not retained by this resin. Consequently, if samples contain both selenite and selenocystine, thiodiglycol must be omitted from the loading buffer in order to separate them on this column. Evidence for selenocysteine in this protein was also obtained by two other methods. This partially purified protein was subjected to enzymatic hydrolysis with papain, carboxypeptidase-A and prolidase, and chromatography of this digest revealed the presence of ^{75}Se as selenocysteine. In the other procedure, a carboxymethyl derivative of the selenium residue in this partially purified protein was made by using chloroacetic acid. Chromatography of this acid hydrolyzed protein resulted in a ^{75}Se containing entity which co-eluted with authentic carboxymethyl-selenocysteine (9).

In order to investigate the form of selenium in tissue organelles, the crude nuclear fraction, mitochondria and microsomes were isolated from liver and heart of a lamb injected with ^{75}Se-selenite. These were solubilized in 0.15 M Tris buffer containing a detergent, and centrifuged to remove any insoluble material. Less than 10% of the total ^{75}Se was centrifuged down. The samples were chromatographed on Sephacryl columns with 0.1 M phosphate buffer (pH 7.2) containing a detergent. When the major ^{75}Se-containing peaks were acid hydrolyzed and chromatographed on the column of ion exchange resin as indicated above, selenium was shown to be present as selenocysteine (9). Thus, selenium appears to be present in all mammalian tissue organelles as selenocysteine when inorganic selenium is given.

SUMMARY

Some problems associated with purity of ^{75}Se compounds, chromatography of tissue cytosols from animals injected with ^{75}Se-selenite, storage of plasma samples containing ^{75}Se, tissue calcification, and selenoamino acid analysis are discussed. It is important to keep ^{75}Se-proteins reduced in order to prevent losses of radioactivity upon gel filtration. Sarcotubular membranes from muscle of normal lambs exhibited a 12-fold greater activity to sequester calcium than those from selenium deficient lambs. In contrast to literature reports, selenoamino acids were not found to be destroyed during acid hydrolysis in an inert atomsphere. Selenium was shown to be present as selenocysteine in a low molecular weight protein from heart cytosol and in crude nuclear, mitochondria, and microsomes of liver and heart of lambs injected with selenite.

ACKNOWLEDGEMENT

This work was supported by N.I.H. Grant Number NS 07413.

REFERENCES

1. Ganther, H.E., 1979, In "Advances in Nutritional Research", (ed. H.H. Draper) (Plenum Press, New York & London) pp 107-128.

2. Schubert, J.R., O.H. Muth, J.E. Oldfield and L.F. Remmert, 1961, Fed. Proc. 20: 689-694.

3. Meissner, G., G.E. Connor and S. Fleischer, 1973, Biochem. Biophys. Acta 298: 246-269.

4. Martonosi, A., R. Boland and R.A. Halpin, 1972, Cold Spring Harbor Symp. Quant. Biol. 37: 455-468.

5. Laemmli, U.K., 1970, Nature 227: 680-685.

6. Forstrom, J.W., J.J. Zakowski and A.L. Tappel, 1978, Biochemistry 17: 2639-2644.

7. Huber, R.E. and R.S. Criddle, 1967, Arch. Biochem. Biophys. 122: 164-173.

8. Walter, R., D.H. Schlessinger and I.L. Schwartz, 1969, Analyt. Biochem. 27: 231-243.

9. Beilstein, M.A., M.J. Tripp and P.D. Whanger, 1981, J. Inorganic Biochem. (submitted).

10. Black, R.S., M.J. Tripp, P.D. Whanger and P.H. Weswig, 1978, Bioinorgan. Chem. 8: 161-172.

11. Beilstein, M.A., M.J. Tripp and P.D. Whanger, 1980, Fed. Proc. 29: 338 (abs.).

DISCUSSION

O'DELL (U.S.A.). Can you tell us how the selenium gets into the amino acid?

WHANGER (U.S.A.). There are two possible ways selenium could be incorporated into amino acids. One could be by post-translational modification. Alternatively, there could be a specific t-RNA for the incorporation of selenium into protein. From our data, we cannot say which is the case.

LINDER (U.S.A.). Are you suggesting that most or all of the selenium containing proteins have only selenocysteine and not selenomethionine in them?

WHANGER (U.S.A.). Yes, I am suggesting that this is true if inorganic selenium is given. All the evidence indicates that one of the major forms of selenium in plant material is selenomethionine, and that could quite possible be incorporated as such into protein. However, it looks as though when you have given inorganic selenium, the major form of deposition in tissue is selenocysteine.

BRAY (Australia). Is the calcification with selenium deficiency different to the calcification you get with normal tissue injuries, say with bruising?

WHANGER (U.S.A.). I think calcification is a result of selenium deficiency; that is a secondary affect. It is certainly not a primary affect. Calcification is not a specific indication of selenium deficiency. I think calcification is the result of selenium deficiency and does not cause selenium deficiency.

SODA (Japan). Recently we have developed a procedure to determine L-selenocysteine specifically with an enzyme, selenocysteine reductase coupled with bacterial L-alanine dehydrogenase. L-selenocysteine is degraded to hydrogen selenide and L-alanine by the reductase, than alanine is determined with the L-alanine dehydrogenase by measuring NAD produced from NADH. Alternatively, L-selenocysteine is determined by measuring hydrogen selenide, produced enzymatically, with lead acetate.

WHANGER (U.S.A.). That would certainly be an interesting approach to follow up.

SELENIUM AND REPRODUCTIVE EFFICIENCY IN CATTLE

H. R. Conrad, J. H. Harrison, T. A. Reinhardt, and A. L. Moxon

Department of Dairy Science
Ohio Agricultural Research and Development Center
WOOSTER, OHIO 44691
UNITED STATES

Selenium deficiency has been found in several animal species including cattle, sheep, rats, poultry, and primates (NRC, 1971). Sheep, rats and chickens have been the primary target of laboratory experiments. Because dairy cattle have the greatest daily feed intake with respect to metabolic size of all domestic livestock and there was only limited information on the nutritional importance of selenium in lactating cows, we began experiments in 1973 in Ohio, a selenium deficient area (Moxon and Olson, 1974).

Muscular dystrophy in young ruminants and reproductive failure in adults are often consequences of selenium deficiency (Muth et al., 1958; Julien et al., 1976a). Hartley et al. (1960) first reported beneficial effects of selenium in correcting several reproductive problems in ewes. Buchanan-Smith et al. (1969) found in ewes fed purified diets that both selenium and vitamin E were required to obtain satisfactory reproductive performance in selenium-deficient groups. In dairy cows on low-selenium diets, Trinder first observed a beneficial effect of selenium with vitamin E on the incidence of retained placenta. These workers (Trinder et al., 1969) showed also that blood selenium levels were lower in herds with a higher incidence of the disorder, and that the problem was controlled with a vitamin E-potassium selenate injection three weeks prior to calving.

In Ohio selenium deficiency resulted in a higher incidence of retained placenta (Julien et al., 1976a). Retained placenta occurred in ten of 26 deficient cows consuming 0.24 mg of selenium daily but was entirely absent in comparable cows fed the same diet but raised to 1 mg of selenium intake daily. Injection of 50 mg of selenium with 680 IU of alpha-tocopherol 3 weeks before calving was also effective in eliminating retained placenta. Protein supplementation as soybean meal reduced retention rate from 50% to 20% in unsupplemented cows. This increased protein was associated with an increase in selenium in the ration from .021 ppm to .047 ppm.

Because the numbers of cattle were small in the experiments described above, Julien et al. (1976b) conducted a field trial involving four commercial dairy herds in Ohio with a high incidence of retained placenta in which about 60% were injected with selenium and 40% were left as controls. One-hundred thirteen cows received 50 mg selenium and 680 IU of vitamin E and there were 10 retentions. However, among the 80 cows left as controls there were 41 retentions.

In a recent experiment, Reinhardt et al. (1978) found that prepartal feeding of alfalfa silage reduced the effectiveness of the selenium-vitamin E injections for controlling rate of retained placenta. We have continued research with silage rations with the objective of determining the relative importance of vitamin E and selenium in the etiology of retained placenta.

GENERAL PROCEDURES

Sixty-eight cows were used in four treatment groups. One-half of the cows were grazing on permanent pasture grasses (orchardgrass and bluegrass). The remaining cows were fed alfalfa silage. At 21 days prepartum group 1 was bled via the tail vein and injected with 50 mg of selenium plus 680 IU of vitamin E. Group 2 remained on pasture with no supplement except iodized salt. Group 3 was fed alfalfa-silage containing about 25% orchardgrass by weight and injected as above with selenium and vitamin E. Group 4 received alfalfa silage without supplements. Forty other cows also fed grass-alfalfa silage containing not more than 25% alfalfa was used to compare the effect of vitamin E. One-half of the cows were injected with selenium at the rate of .1 mg/kg of body weight.

Analyses included plasma selenium (Olson, 1975), glutathione peroxidase (Lawrence and Burk, 1976), and vitamin E (Bieri et al., 1979) at calving. Retained placenta was defined as cows failing to expel the placental membranes within 12 hours. Postpartum examinations were made periodically to 12 weeks of the uterus and ovaries for detectable abnormalities.

RESULTS

A comparison of the incidence data indicates that the addition of alfalfa silage during the dry period may have increased the need for Se at parturition to protect against retained placenta (Table 1). In further investigations we found that plasma selenium was declining rapidly (2% per day) during the three weeks prepartum. Glutathione peroxidase in the plasma increased rapidly after selenium injections and sustained a level twice that of control cows through parturition. In cows fed vitamin E, glutathione peroxidase increased most rapidly. Cows fed 1 gram of vitamin E per day had less retained placentas (5%) than cows fed the same diets without vitamin E (25%).

During the course of these experiments we noted that cows injected with selenium 21 days prepartum had very few cystic ovaries (7%) compared with a herd average of 46%. During the preceding lactation most of the cows would have been fed 8 to 11 kg of concentrate daily with some added fat. Thus, they come into the dry period after long periods of oxidative stress. Other factors involved in the carryover of prepartal selenium feeding to postpartal reproductive health are not known at this time.

Table 1. Plasma selenium at parturition and incidence of retained placenta varies in injected cows with the diet during the dry period.

		Se[1]		Retained		Plasma Se
Group	Roughage	Mg	No.	No.	%	μg/ml[2]
1	Pasture[3]	50	17	0[4]	0	.049
2	Pasture	0	17	4	23.5	.020
3	Alfalfa-grass silage	50	17	2	11.8	.040
4	Alfalfa-grass silage	0	17	5	29.4	.023
1 + 3	Combined	50	34	2	5.8	.045
2 + 4	Combined		34	9	26.4	.022

[1]Injected with 50 mg of selenium and 680 IU of vitamin E twenty-one days prepartum.

[2]Standard error of mean was less than 0.003.

[3]The pasture was a mixture of orchardgrass, Dactylis glomerata, and bluegrass, Poa pratensis.

[4]A difference of three observations between groups is statistically significant by Chi-Square test.

SUMMARY AND CONCLUSIONS

Experiments involving dietary deficiencies of selenium demonstrated its essentiality for the prevention of post-parturient reproductive problems. Several experiments were done in which one among paired groups of pregnant cows was treated with selenium three weeks prepartum. It was discovered that a shortage of dietary selenium was an important factor in herds having a high incidence of retained placentas. Incidence of retention of placenta was decreased to none in the experimental group of cows treated with 50 mg of selenium and 680 IU of alpha-tocopherol three weeks postpartum. When prepartum diets were raised from 0.24 mg to 1.0 mg of selenium per day retention of placenta was eliminated. When legume-grass silage was fed during the dry period, injected selenium was less effective. In a herd in which cystic ovaries were experienced by 46% of the cows, prepartal injection of 100 μg of selenium as sodium selenite per kg of body weight and 1 g of alpha-tocopherol daily per head in the diet prevented the occurrence of ovarian cysts during a twelve-week period after calving. These results indicate that eliminating selenium deficiency in the presence of adequate vitamin E improves the post-parturient health of high producing dairy cows.

REFERENCES

1. Bieri, J. G., T. J. Tolliver, and G. L. Catignani, 1979, Amer. J. Clin. Nutr. 32: 2143 - 2149.

2. Buchanan-Smith, J. G., E. C. Nelson, B. I. Osburn, M. E. Wells, and A. D. Tillman, 1969, J. Animal Sci. 29:808 - 815.

3. Hartley, W. J., and A. B. Grant, 1961, Fed. Proc. 20:679 - 688.

4. Julien, W. E., H. R. Conrad, J. E. Jones, and A. L. Moxon, 1976a, J. Dairy Sci. 59: 1954 - 1959.

5. Julien, W. E., H. R. Conrad, and A. L. Moxon, 1976b, J. Dairy Sci. 59:1960 - 1962.

6. Lawrence, R. A., and R. F. Burk, 1976, Biochem. Biophys. Res. Comm. 71:952 - 958.

7. Moxon, A. L., and O. E. Olson. 1974. In "Selenium" (eds. R. A. Zingaro and W. C. Cooper) (Van Nostrand Reinhold) Cincinnati, pp. 675 - 707.

8. Muth, O. H., J. E. Oldfield, L. E. Remmert, and J. R. Schubert, 1958, Science 128: 1090.

9. National Research Council Subcommittee on Selenium, 1971, National Academy of Sciences, Washington, D.C.

10. Olson, O. E., Ivan Palmer, and Erle Cary, 1975, JAOAC 58:117.

11. Reinhardt, T. A., H. R. Conrad, W. E. Julien, and A. L. Moxon, 1978, J. Dairy Sci. 61:185 (Abstr.).

12. Trinder, N., C. D. Woodhouse, and C. P. Rentan, 1969, Vet. Rec. 85:550.

DISCUSSION

JUDSON (Australia). One thing that surprises me about your results is the high plasma selenium level. If you consider whole blood, the bulk of the selenium is in the red cell and I would expect values around 0.04 that you were quoting as being indicative of selenium inadequacy in cows on a whole blood basis but not on a plasma basis. I was wondering if you would like to comment on that please?

MOXON (U.S.A). We would like to have the levels up to 0.1 ppm in the blood for good health in high-producing dairy cattle - they are much different from beef cattle. We need higher levels in dairy cattle because after producing milk for a long period they go into a dry period in a very depleted state and in order to bring them back up it takes a good injection of selenium. If we can get the levels up over 0.06 ppm at least in the blood plasma, we feel quite happy about it, but it is hard to get them up there. If we are offering feed that is high in organic selenium, i.e. naturally occurring selenium in the feed, then it is not much of a problem. However, if you are adding inorganic selenium you know a lot of it is not absorbed because of the action of the rumen bacteria on it - putting it into the elemental state probably where it goes on through the animal and is not absorbed.

ALLEN (U.K.) I would like to support the last speaker. At the current moment in England we have cows swimming in selenium because of some reports of these types of responses but in our own herds where we have some 400 dairy cattle, our plasma selenium levels are below those quoted by Dr. Moxon and in spite of looking back carefully through the records for these cows, back to 1942, the instance of retained placenta in our herds is about 3% with plasma selenium levels of about 20 to 25 nanogram per ml. Injection of these cows with inorganic selenium does raise the plasma selenium concentration but this still does not lower the incidence of retained placenta below 3%, and the increase in plasma Se lasts in our cows throughout their lactation.

MOXON (U.S.A). The problem has increased in Ohio since they have been feeding higher amounts of corn silage and taken out pasture and hay - they have gone to mechanised feeding with corn silage. Concentrate and corn silage is very low in selenium content and I think the vitamin E in silage is also reduced greatly because of the heat generated during the ensiling process, we find that especially true in the grass silages. I think the problem is different in Ohio than many places because of the type of feeding we have in dairy cattle and the very high production level.

OLDFIELD (U.S.A.). Dr. Moxon, do you have any reason to believe that there may be some factors present in these crude feedstuffs including the alfalfa silage that may interfere with selenium metabolism?

MOXON (U.S.A.). We feel that, but we do not know what all the factors are - there is some work going on into that angle of it but we have no results yet.

MYOPATHY IN SELENIUM DEFICIENT CATTLE

J. R. Arthur

Rowett Research Institute,
Greenburn Road, Bucksburn,
ABERDEEN, AB2 9SB, U.K.

INTRODUCTION

Concurrent selenium and tocopherol deficiencies in ruminants can result in numerous disorders (1,2). In young calves selenium/tocopherol deficiency may result in damage to skeletal muscle and the myocardium, the latter, often resulting in death. In older cattle (6 month-2 years of age) Se/tocopherol deficiency may result in delayed myopathy also known as paralytic myoglobinuria. Almost invariably, plasma activity of the muscle enzyme creatine phosphokinase CPK (E.C.2.7.3.2) is greatly elevated in affected animals. In older animals, myopathy is rarely fatal, the heart infrequently being involved; mildly affected animals may recover spontaneously. This paper describes some experiments designed to investigate factors influencing the onset of myopathy in Se deficient cattle over 6 months of age.

RESULTS AND DISCUSSION

Eight (initially two months of age) Friesian calves were maintained on semi-synthetic diets based on Torula yeast, oat husks and maize starch. The basal diet containing <0.01 mg Se/kg DM was offered to 4 calves while the other 4 received the basal diet supplemented with 0.1 mg Se/kg DM as Na_2SeO_3. Animals were housed in wooden pens with slatted floors. At slaughter, after 45 weeks of experiment, tissues from both groups of calves were analysed for selenium concentration and glutathione peroxidase activity (Table 1).

TABLE 1 The Se concentration and glutathione peroxidase (GSHpx) activity in tissues from calves maintained on low Se and Se supplemented diets

Tissue	Basal Diet (low Se)		Se Supplemented Diet	
	Se mg/kg wet wt.	GSHpx activity units/mg protein	Se mg/kg wet wt.	GSHpx units/mg protein
Liver	0.052±0.004	0.006±0.003	0.218±0.014***	0.078±0.013**
Kidney	0.66 ±0.03	0.045±0.008	1.33 ±0.092**	0.460±0.054**
Heart	0.036±0.004	0.015±0.006	0.181±0.012***	0.203±0.031+
Muscle (Triceps)	0.013±0.001	ND	0.106±0.003***	0.027±0.011*
Muscle (Semitendinosis)	0.02 ±0.002	ND	0.096±0.018+	0.022±0.007**

Selenium was determined in the tissues indicated by a fluorimetric method (4). GSHpx activity was measured using a coupled enzyme technique (5) on supernatants from homogenates of tissues spun at 40,000xg for 30 min (results mean ± SEM: four animals/group). ND Not Detected. Significance of difference from basal group +P<0.02, *P<0.01, **P<0.005, ***P<0.001.

Selenium contents of tissues from the low Se group were in the range of published values for calves with nutritional myopathy (6,7). Plasma tocopherol concentrations of the calves in this experiment (<10 μg/100 ml) were below concentrations found in calves with nutritional myopathy (25-180 μg/100 ml) (8,9). Despite the low tocopherol concentrations the Se deficient calves showed no clinical signs of myopathy and exhibited only small increases in plasma CPK activity above those of Se supplemented animals (10). Plasma CPK activity was only slightly increased by 30 minute periods of exercise, no significant differences in this response were evident between low Se and Se supplemented animals.

As clinical myopathy did not occur in this experiment, further investigations were directed at discovering nutritional or physical stresses which could cause the onset of myopathy in Se/tocopherol deficient cattle. Cu deficiency could exacerbate the effects of Se deficiency as both elements are involved in antioxidant systems in the cell; Se in glutathione peroxide and Cu in superoxide dismutase. Four groups of 4 Friesian calves (2 months of age) were maintained for 26 weeks on the basal Torula yeast diet with the following adjustments of Se and Cu concentrations (mg Se and mg Cu/kg DM diet respectively): group A 0.01, 1.5; group B 0.01, 12; group C 0.1, 1.5; group D 0.1, 12. Copper deficiency simultaneously induced with Se/tocopherol deficiency (group A) did not cause subclinical or clinical signs of myopathy to occur in calves. However the Cu deficiency induced in the calves of groups A and C was not severe enough to reduce the superoxide dismutase activity in liver and muscle in comparison to the activity in the tissues from the calves in the Cu supplemented groups B and D. Therefore effects of a more severe Cu deficiency on Se deficient calves are not excluded. Clinical myopathy and/or large rises in plasma C.P.K. were induced in the low Se calves (groups A and B) when they were "turned out" from indoor housing to low Se pasture. Se supplementation of the diet protected calves against these changes (groups C and D) there being no effects of the Cu treatments.(results not shown). Myopathy at turnout has been postulated to be due to unaccustomed exercise, inclement weather (11) or the high concentration of polyunsaturated fatty acid in grass providing the substrate for peroxidation (9,12). Turnout and the change of diet from semi-synthetic to fresh grass caused a tenfold rise in the percentage of linolenic acid in plasma; previous treatment did not influence this response (Figure 1). These data indicate that some linolenic acid from fresh grass had therefore passed through the rumen unhydrogenated to be absorbed from the gut into the plasma.

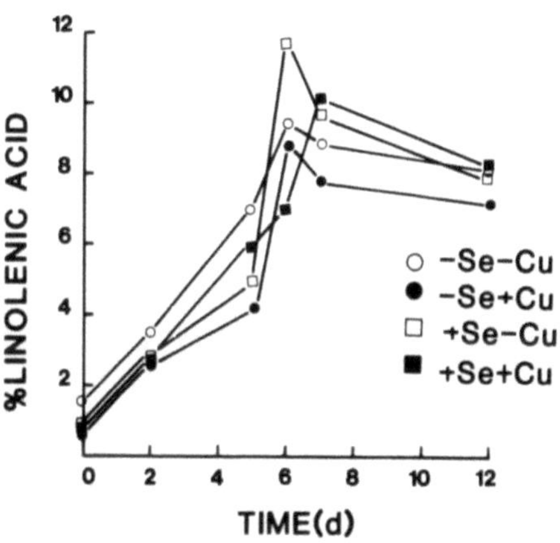

FIGURE 1 Changes in the percentage of linolenic acid in plasma lipids of calves after turnout to pasture.

FIGURE 2 Changes in the percentage of linolenic in plasma lipids of calves (a) consuming fresh grass indoors and (b) after turnout to pasture.

The effects of linolenic acid in low Se grass on plasma fatty acids and CPK activity in calves were investigated in a further experiment. The Se reserves of 6 Friesian calves (2 months of age initially) were depleted by maintaining the animals on Torula yeast-based semisynthetic diet for 5 months. Six calves received the same diet supplemented with 0.1 mg Se/kg DM as Na_2SeO_3. Three calves from each group were turned out to Se deficient pasture, while the remaining 3 animals in each group remained indoors and were offered cut grass. Samples of the grass used in this experiment contained 0.015–0.018 mg Se/kg DM and 40–70% linolenic acid in total lipids. As shown in Figure 2 virtually identical changes in plasma linolenic acid occurred in the calves housed indoors and those "turned out". Despite the rises in plasma linolenic acid there was no rise in plasma CPK activity in the low Se calves housed indoors. Only the low Se calves which were "turned out" had marked prolonged rises in plasma CPK activity (Figure 3a and b) when the low Se calves receiving the cut grass indoors were eventually 'turned out' there were no prolonged rises in plasma CPK activity. However, during the period of grass feeding the calves had been consuming adequate supplies of tocopherols which presumably protected against subsequent muscle damage.

These results suggest that while an increased dietary load of polyunsaturated fatty acid may be relevant to the induction of myopathy in Se/tocopherol deficient calves other unidentified variables are also involved.

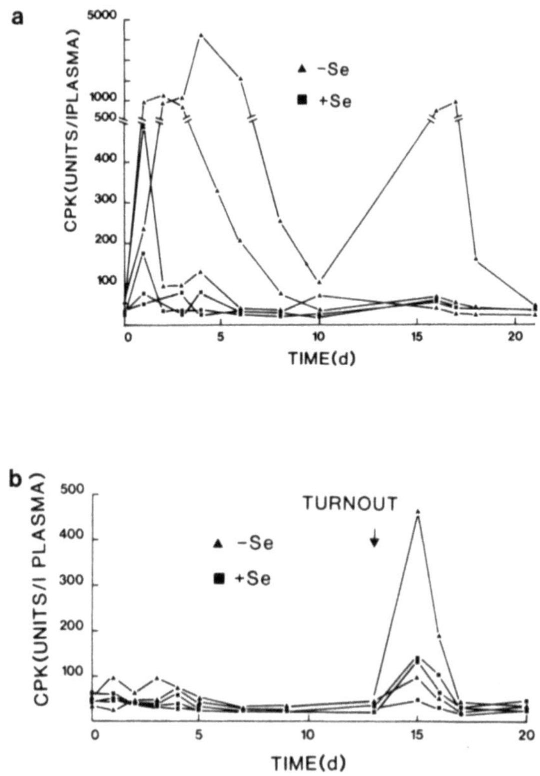

FIGURE 3 CPK activity in plasma of low Se and Se supplemented calves (a) after turnout to pasture and (b) during consumption of fresh grass indoors.

REFERENCES

1. Lannek, N., P. Lindberg, 1975, Adv. Vet. Sci. Comp. Med., 19:127-164.
2. Underwood,,E.J., 1977, Trace Elements in Human and Animal Nutrition 4th Edition (Academic Press, N.Y.).
3. Anderson, P.H., S. Berrett and D.S.P. Patterson, 1976, Vet. Rec., 99:316-318.
4. Olsen, O.E., I.S. Palmer and E.E. Cary, 1975, J. Ass. O.A.C., 58:117-121.
5. Paglia, D.E. and W.N. Valentine, 1967, J. Lab. Clin. Med., 70:158-169.
6. Hidiroglou, M., K. Jenkins, R.B. Carson and G.A. Brossard, 1967, Can. J. Physiol. Pharmacol., 45:568-569.
7. Van Vleet, J.F., 1975, Am. J. Vet. Res., 36:1335-1346.
8. Allen, W.M., 1977, Vet. Sci. Comm., 1:243-250.
9. McMurray, C.H. and McEldowney, 1977, Brit. Vet. J., 133:535-542.
10. Arthur, J.R., 1979, Proc. Nutr. Soc., 38:13A.
11. Allen, W.M., R. Bradley, S. Berrett, W.H. Parr, K. Swannack, C.R.Q. Barton and A. Macphee, 1975, Brit. Vet. J., 131:292-308.
12. McMurray, C.H., W.J. Blanchflower and D.A. Rice, 1980, Proc. Nutr. Soc., 39:65A.

DISCUSSION

WHANGER (U.S.A.). I would have expected more changes in the CPK based on work with sheep. Have you done any with sheep?

ARTHUR (Scotland). No we have never tried measuring CPK activity in sheep plasma. We use the Sigma test kit UV45. Results are usually expressed in international units of enzyme activity/litre. In some clinical cases plasma CPK activities can be as high as 100,000 IU/litre.

EFFECTS IN PIGS FED THE CROPS GROWN IN KESHAN DISEASE AFFECTED PROVINCE OF CHINA

L.Z. Zhu *et al*

Institute of Health, Chinese Academy of Medical Science
Z.H. Lu *et al*
Institute of Animal Science, Chinese Academy of Agricultrual Science
BEIJING, THE PEOPLE'S REPUBLIC OF CHINA

Previous studies have clearly defined that most Keshan Disease affected areas are located in low-selenium regions of China. The selenium status, as indicated by blood selenium, whole blood glutathione peroxidase (GSHpx) and hair selenium, of residents of this area is very low. Our study has implicated a deficiency of selenium as a primary etiologic factor in Keshan Disease. This disease is a cardiomyopathy of unknown cause. The main pathological feature is multiple focal myocardial necrosis scattered throughout the heart muscle with differing degrees of inflammatory cell infiltration and various stages of fibrosis. Ultrastructural observations revealed extensive and conspicuous changes of mitochondria. Histochemical examination showed that acid phosphatase activities sharply increased in myocardiocytes around necrotic foci of subacute Keshan Disease patients. These changes might be the consequence of damage of organelle membranes.

MATERIALS AND METHODS

A series of experiments on mice and rats fed crops grown in Keshan Disease-affected areas have been performed. Damage to mitochondria of these animals was observed. Recently, one experiment was conducted using young pigs. The pigs and rats were divided into four groups and each had a certain diet : (1) Low selenium-containing yeast diet (yeast group) (2) Diet composed of staple cereals grown in a Keshan Disease affected province (Keshan group) (3) Diet composed of staple cereals grown in non-affected area (control group) (4) Na_2SeO_3-Se-supplemented Keshan group (Se-supplemented Keshan group). All animals were fed the diets for 3 months then killed by bleeding.

RESULTS AND DISCUSSION

1. Pathological observations

Pathological changes were observed in the heart, liver and striated muscle of animals in yeast or Keshan groups. No changes were seen in the control group, and Selenium supplementation of the Keshan diet virtually eliminated the lesions. The incidence of each is listed in Table 1. Separated lobulous necrosis characterized the liver damage, it was often more severe at the central portion of the lobuli than at the peripheral. Small necrotic foci were found in myocardium with varying inflammatory cell infiltration; and they somewhat resembled the lesion of Keshan Disease. The striated muscle showed mainly coagulative necrosis of individual fibres. Mitochondrial damage in myocardiocytes also exists as partial wasting of the crista and matrix in some specimens.

Table 1. The incidence of pathological changes of liver, heart and striated muscles of pigs

Group	Heart	Liver	Straited muscle
Yeast group	4/5	3/5	2/5
Keshan group	2/13	3/12	2/10
Control group	0/10	0/10	0/9
Se-supplemented Keshan group	0/10	2/9	0/9

Numerator, Number of pigs affected
Denominator, Number of pigs examined

2. Histochemical observations

Histochemical observations of acid phosphatase, succinate dehydrogenase, adenosine triphosphatase, and monoamine oxidase in heart, liver, and striated muscle was performed.

Acid phosphatase activity could not be demonstrated with Holt's method in normal heart and striated muscle. It was found around the necrotic foci of myocardium and striated muscle in yeast and Keshan groups. The reaction product was a brown particle located in the cytoplasm. The results are shown in Table 2.

Table 2. Histochemical changes of Acid Phosphatase in pigs

Group	Heart	Striated muscle
Yeast group	2/4	2/4
Keshan group	1/11	3/11
Control group	0/10	0/10
Se-supplemented Keshan group	0/10	0/10

Numerator stands for number of positive responses.
Denominator stands for numbers of animals examined.

In addition, the activity of acid phosphatase was greatly reduced or completely disappeared in damaged liver cells.

Succinate dehydrogenase activity, appearing as fine violet granules which line the myofibrils, is distinguishable by Nachlas' method in normal myocardiocytes of whole heart. In normal striated muscles it could only be demonstrated in certain muscles. This reaction was greatly reduced or completely disappeared in damaged myocardiocytes and myocytes.

Acid phosphatase and succinate dehydrogenase reactions observed in heart tissues of the yeast group (which was selenium and vitamin E deficient) were comparable to that of Keshan Disease.

3. Biochemical Behaviour of Mitochondria

Se-GSH-px activities, TBA values and swelling of mitochondria of these four groups were investigated.

a) Se-GSH-px activities of the mitochondria

Se-GSH-px activities of the mitochondria from liver and heart appeared to be consistent with the selenium content of diet. It was significantly lower in the yeast and Keshan groups than in control or Se-supplemented groups. It was similar to the results obtained from rats, although enzyme activity of rat heart mitochondria was much higher than that of pigs (Table 3).

Table 3. Se-GSH-px Activities of Mitochondria of Pigs in Various Groups ($\bar{X} \pm$ S.E.)

Animal	Group	Se content of diet (PPM)	Liver		Heart	
			Se level (μg/g prot.)	SeGSH-px activities (unit)**	Se level (μg/g prot.)	SeGSH-px activities (unit)
	Yeast group	0.02		N.D.(3)		0.14(2)
	Keshan group	0.01		0.79 ± 0.47(8)		0.92 ± 0.48(9)
Pig	Control group	0.15		5.78 ± 1.58(8)		9.88 ± 3.22(8)
	Se-supplemented Keshan group	0.26		7.12 ± 1.45(8)		5.20 ± 1.13(7)

Table 3 (Cont.) SeGSH-px Activities of Mitochondria of Rats in Various Groups ($\bar{X}$ ± S.E.)

Animal	Group	Se Content of diet (PPM)	Liver		Heart	
			Se level (μg/g prot.)	SeGSH-px activities (unit)**	Se level (μg/g prot.)	SeGSH-px activities (unit)
Rat	Keshan group	0.007	0.21 ± 0.095(4)*	N.D.(5)	0.45 ± 0.10(4)	10.39 ± 2.52(5)
	Control group	0.030	0.88 ± 0.045 (6)	2.92 ± 0.66 (5)	1.05 ± 0.06 (4)	35.57 ± 4.37 (5)
	Se-supplemented Keshan group	0.200	3.43 ± 0.25 (5)	7.08 ±0.95 (4)	1.18 ±0.125 (4)	41.17 ±2.74 (4)

* Number in bracket represents the number of determinations

** An enzyme unit of activity was defined as a decrease in the log (GSH) per mg protein per minute after the decrease in log (GSH) per minute of the non-enzymatic reaction was subtracted.

*** N.D. -- not detectable.

b) TBA value of homogenates and mitochondria

The lipid hydroperoxides in tissue homogenates and mitochondria were examined by the TBA method and expressed as TBA value. There was no significant difference between treatment groups, either in liver or in heart. (Table 4).

Table 4. TBA Value** of Homogenates and Mitochondria of Liver and Heart ($\bar{X}$ ± S.E.)

Animal	Group	Liver		Heart	
		Homogenate	Mitochondria	Homogenate	Mitochondria
Pig	Yeast group	2.53(2)*	1.23 ± 0.15(5)	0.290(2)	1.12 ± 0.17(5)
	Keshan group	2.10 ± 0.28(9)	1.59 ± 0.31(10)	0.269 ± 0.040(10)	1.85 ± 0.14(11)
	Control group	1.51 ± 0.28(8)	1.56 ± 0.325(8)	0.213 ± 0.039(8)	1.91 ± 0.74(9)
	Se-supplemented Keshan group	1.90 ± 0.43(7)	1.95 ± 0.223(9)	0.248 ± 0.029(8)	2.49 ± 0.38(9)
Rat	Keshan group	1.70 ± 0.44(14)	0.566 ± 0.046(9)	0.368 ± 0.122(16)	0.950 ± 0.117(7)
	Control	1.87 ± 0.38(16)	0.616 ± 0.063(9)	0.438 ± 0.812(16)	0.814 ± 0.058(9)
	Se-supplemented Keshan group	1.64 ± 0.25(16)	0.673 ± 0.085(9)	0.365 ± 0.023(16)	0.750 ± 0.077(9)

* Number in bracket represents the number of determinations
** TBA value - nmoles MAD/mg protein

c) Swelling of mitochondria

The spontaneous swelling and GSH-induced swelling of mitochondria were investigated in vitro. It was observed that both spontaneous and GSH induced swelling of mitochondria from pigs in the Keshan group were slower than that of the control or the Se-supplemented Keshan group (Fig. 1). This is also true for the liver mitochondria of rats. The swelling rate of mitochondria from the yeast group was faster than from the Se-supplemented Keshan group, except in the case of the spontaneous swelling of heart mitochondria (Fig. 1). This difference may be due to the low vitamin E content of the yeast diet, because the mitochondrial swelling of another yeast group supplemented with vitamin E was slower than that of the unsupplemented group.

Vitamin E deficiency has also obviously enhanced the severity of histological and histochemical abnormalities of these pigs, hence studies of vitamin E status of populations living in Keshan Disease affected areas is warranted.

Ultrastructural observation and histochemical studies revealed similar pathological changes of organelle membranes in both patients with Keshan Disease and in animals fed diets from the Keshan region. It has been presumed that these changes might be a result of the accumulation of lipid hydroperoxides in tissues of low selenium status, but the biochemical findings of this experiment failed to support this idea. However, there are differences of mitochondrial swelling between the Keshan group and Se supplemented groups. This difference might be related to other bio-functions of selenium which remain to be investigated.

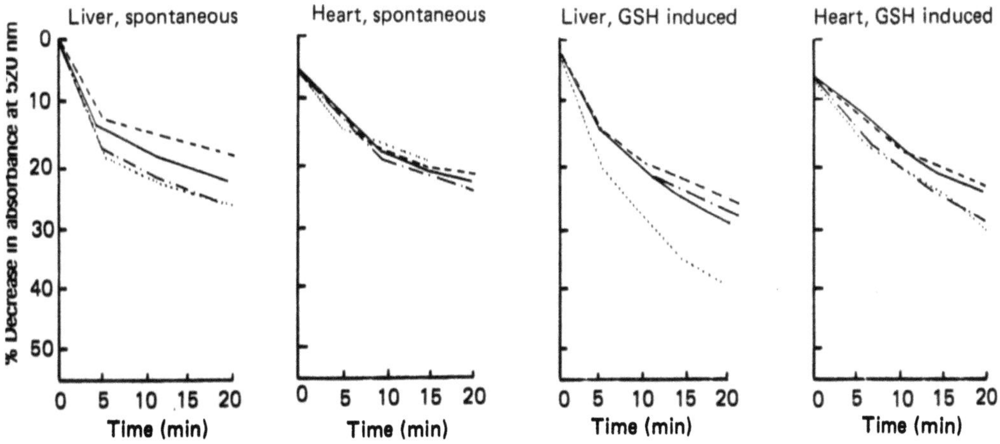

Fig. 1 Mitochondrial swelling of pigs.
.... Yeast group ---- Keshan group —·— Control group —— Se supplemented group

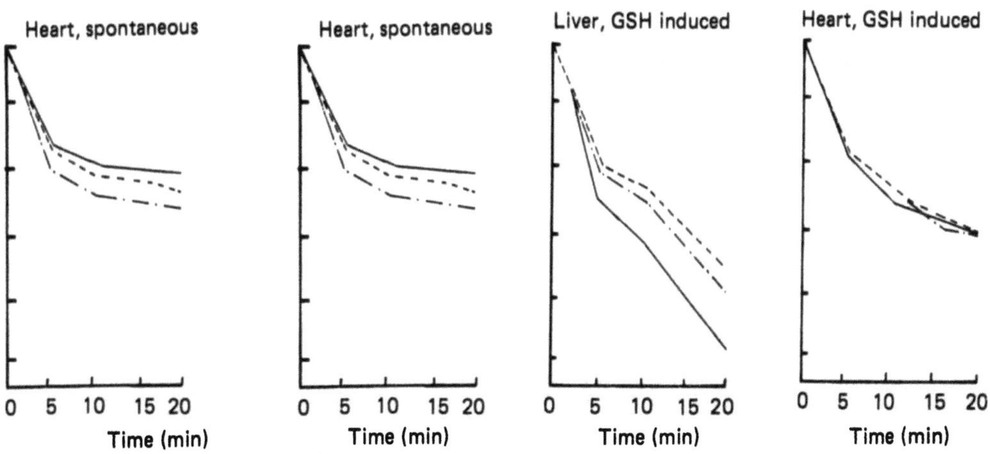

Fig. 2 Mitochondrial swelling of rats.
---- Keshan group —·—. Control group —— Se supplemented group

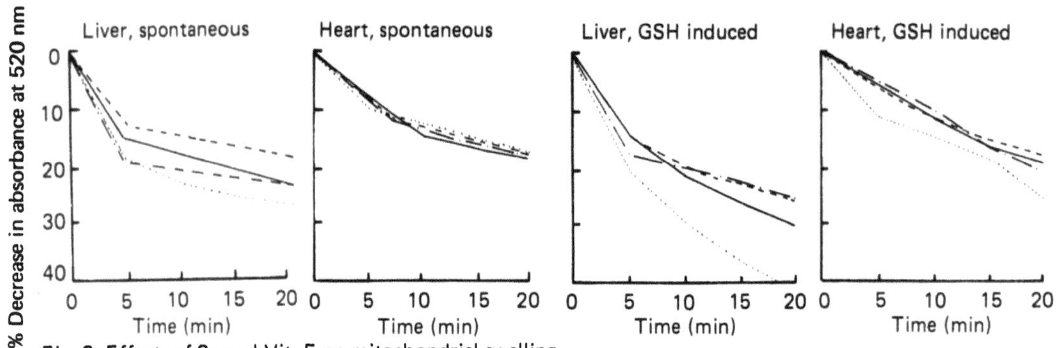

Fig. 3 Effects of Se and Vit. E on mitochondrial swelling.

.... Yeast group ----- Keshan group —·— Vit. E supplemented group —— Se supplemented Keshan group

COMPARATIVE EFFECTS OF SELENIUM DEFICIENCY ON MALE REPRODUCTION IN SEVERAL MAMMALIAN SPECIES

F. R. Pond, M. J. Tripp, A. S. H. Wu, P. D. Whanger and J. E. Oldfield

Oregon State University, Corvallis, Oregon, U. S. A.

One of the early manifestations of selenium deficiency was the demonstration by Cousins and Cairney (1961) and Hartley (1963) of impaired reproduction in sheep. Although estrus, ovulation and fertilization proceeded normally, development of the fetuses was interrupted, usually early in gestation. This frequently resulted in fetal abortion or resorption, giving the impression that the females were unbred. For some years, the Se-responsive reproduction problem in the female was tacitly accepted as the underlying cause of reproductive failure in areas of selenium deficiency and possible effects of selenium in male reproduction went largely uninvestigated. In 1969, however, McCoy and Weswig suggested that sterility in male rats could result from a lack of dietary selenium and they went on to show that male rats had a higher selenium requirement than females. About the same time, Wu et al., (1969, 1979) also working with laboratory rats, showed that selenium deficiency could be identified with certain specific defects of the sperm, and these observations have been confirmed by others (Brown and Burk, 1973). This study was undertaken to provide further information on the nature of selenium's involvement in the male reproductive process in various species, including both laboratory and domestic animals.

Method: Subject animals were cattle and sheep; rats and voles. The effects of selenium deficiency in terms of various parameters were studied in male progeny of several species born to females fed Se-deficient diets during pregnancy. Comparable males were maintained in selenium adequacy, as positive controls. Depending on the species under study, either purified or commercial diets were used, the criterion of Se-deficiency being a total diet dry matter content of < 0.02 ppm. Sperm for study were obtained from the larger species by means of artificial vagina or electroejaculation techniques; in the smaller species they were taken from the epididymis immediately after killing. Sperm samples were examined under both light and electron microscopes. For scanning electron microscopy (SEM), sperm cells were fixed in buffered glutaraldehyde followed by osmium tetroxide and critical point drying with trichlorotrifluoroethane. For tansmission electron microscopy (TEM), sperm preparations were embedded in EPON-812. Glutathione peroxidase ($GSHP_x$) activity was determined in seminal plasma, sperm and cytosols from various accessory glands, using the coupled enzyme procedure of Paglia and Valentine (1967) modified by Whanger, et al., (1977). ^{75}Se incorporation studies were done on two, 9-month old rams, one each Se-deficient and adequate.

Intramuscular injections with 500 μCi ^{75}Se as $Na_2{}^{75}SeO_3$ were made, followed by semen collections over a 9-week period. Radioactivity of the collections was measured with a Beckman 8000 gamma counter. Either SEPHADEX G-150 or SEPHACRYL S-200 packing was used in columns for gel filtration studies. Protein was determined by the method of Lowry, et al., (1951) using bovine serum albumin as the standard. Sperm cells from the testis and epididymis were fixed in buffered glutaraldehyde followed by osmium tetroxide, dehydration and embedding in EPON-812. Autoradiographs were prepared, after Salpeter and Bachmann (1964) using Kodak NTE emulsion. The developed, ultra-thin sections were examined using a Phillips model 300 TEM.

Results and Discussion: In most instances the various species studied reacted similarly to Se-deficiency. Total body, testicular and epididymal weights were depressed. Histology of the testis in sheep, rats and voles showed that active spermatogenesis could occur but that sperm motility was severely impaired in the rat and vole, due apparently to sperm membrane damage, especially in the region of the mid-piece of the tail. $GSHP_x$ could protect against such damage by destroying lipid peroxides, thus preventing oxidative destruction. Dietary supplementation of rats with vitamin E or other antioxidants did not protect against the sperm damage, which appeared to be Se-specific. Rat and vole sperm exhibited the characteristic midpiece damage but ram sperm did not under conditions of this study. Se-dependent $GSHP_x$ was present in both bull and ram seminal plasma but not in the sperm cells, per se. Among the accessory glands the prostate made the largest contribution of $GSHP_x$ to bull and ram seminal plasma, while Cowpers glands apparently did not add significant amounts. The seminal vesicles showed high $GSHP_x$ activity, and are also known to possess high prostaglandin concentrations. The implication of $GSHP_x$ in prostaglandin synthesis (Lands, et al., 1971) suggests that $GSHP_x$ in seminal vesicles may function in prostaglandin synthesis rather than, or perhaps in addition to being secreted in the seminal plasma.

Labelled ^{75}Se was found in both seminal plasma and sperm cells of the bull and ram. At 21 days post-injection of the label, some 93% of the ^{75}Se counts in the ram ejaculate were associated with the cells, suggesting that selenium may be incorporated into sperm during various stages of spermatogenesis, however peak radioactivity at 30-49 days post-injection indicates greatest incorporation during or prior to the spermatid stage. Radioactivity in ram sperm cells at peak levels was nearly tenfold that in the seminal plasma; in the bull, others (Smith et al., 1979) found the counts in the seminal plasma were nearly twice those in the sperm cells. Our studies did not establish whether these differences were real or may have involved leakages of seleno-protein from the sperm cells of the two species. Four distinct seleno-proteins were identified by gel filtration of seminal plasma from the ram and two from the bull. In each case, one of these was identifiable with $GSHP_x$

Autoradiography showed generally similar localization of ^{75}Se across all species studied. The expected concentration of radioselenium over cellular and intracellular membranes of the sperm midpiece was observed but additionally areas of heavy ^{75}Se exposure were found in nuclei. This occurrence needs confirmation, due to the observation of Pallini and Bacci (1979) that presence of Se in nuclear fractions of bovine sperm was due to mitochondrial contamination.

References:

Brown, D. G. and R. F. Burk. 1973. Selenium retention in tissues and sperm of rats fed a torula yeast diet. J. Nutr. 103: 102-108.

Cousins, F. B. and I. M. Cairney. 1961. Some aspects of selenium metabolism in sheep. Aust. J. Agr. Res. 12: 927-943.

Hartley, W. J. 1963. Selenium and ewe fertility. Proc. N.Z. Soc. Animal Prod. 23: 20-27.

Lands, W., R. Lee and W. Smith. 1971. Factors regulating the biosynthesis of various postaglandins. Ann. N.Y. Acad. Sci., 180: 107-122.

Lowry, o. H., N. G. Rosebrough, A. L. Farr and R. J. Randall. 1951. Protein measurement with the Folin phenol reagent. J. Biol. Chem. 193: 265-275.

McCoy, K. E. M. and P. H. Weswig. 1969. Some selenium responses in the rat not related to vitamin E. J. Nutr. 98: 383-389.

Paglia, D. E. and W. N. Valentine. 1967. Studies on the quantitative and qualitative characterization of erythrocyte glutathione peroxidase. J. Lab. Clinical Med. 70: 158-169.

Pallini, V. and E. Bacci. 1979. Bull sperm selenium is bound to a structural protein of mitochondria. J. Submicr. Cytol. 11: 165-170.

Salpeter, M. M. and L. Bachmann. 1964. Autoradiograph with the electron microscope: a procedure for improving resolution, sensitivity and contrast. J. Cell Biol. 22: 469.

Simth, D. G., P. L. Senger, J. F. McCutchan and C. A. Landa. 1979. Selenium and glutathione peroxidase distribution in bovine semen and selenium-75 retention by the tissues of the reproductive tract in the bull. Biol. Repr. 20: 377-383.

Whanger, P. D., P. H. Weswig, J. A. Schmitz and J. E. Oldfield. 1977. Effect of selenium and vitamin E on blood selenium level, tissue glutathione peroxidase activity and white muscle disease fed on purified or hay diets. J. Nutr. 107: 1298-1306.

Wu, A. S. H., J. E. Oldfield, O. H. Muth, P. D. Whanger and P. H. Weswig. 1969. Effect of selenium on reproduction. Proc. West. Sec. Am. Soc. Animal Sci. 20: 85-89.

Wu, A. S. H., J. E. Oldfield, L. R. Shull and P. R. Cheeke. 1979. Specific effect of selenium deficiency on rat sperm. Biol. Repr. 20: 793-798.

NEUTROPHILS FROM Se AND Cu DEFICIENT CATTLE

J.R. Arthur, R. Boyne, M.J. Okolow-Zubkowska* and H.A.O. Hill*

Rowett Research Institute, Bucksburn, Aberdeen AB2 9SB, U.K. and
*Inorganic Chemistry Laboratory, South Parks Road, Oxford OX1 3QR, U.K.

INTRODUCTION

Neutrophils (polymorphonuclear leucocytes) have an important role in the defence of the body against infection. These phagocytic cells ingest and may kill potentially injurious microorganisms in the bloodstream. Neutrophils from selenium deficient rats (1) and selenium (2) and copper-deficient cattle (3) retain their ability to ingest cells of the yeast C. albicans. However, the ability of the phagocytic cells to kill the ingested yeast cells is impaired in comparison to neutrophils from rats or cattle consuming diets supplemented with Se or Cu. In stimulated neutrophils from low Se and low Cu cattle there is indirect histochemical evidence for impaired production of superoxide (O_2^-). Production of O_2^- and other oxygen derived radicals is thought to be essential for the microbicidal activity of the neutrophils (4). The selenoenzyme glutathione peroxidase and the cuproenzyme superoxide dismutase may help protect the neutrophil from the damaging effects of the toxic oxygen species which are produced during microbicidal activity (5). We now report the effects of Se and Cu deficiency on the activities of these enzymes and other biochemical variables in neutrophils from bovine blood.

METHODS, RESULTS AND DISCUSSION

All the investigations reported here were carried out on neutrophils prepared from bovine venous blood by the method of Carlson and Kaneko (6). Cell preparations were always >95% pure and viability assessed by trypan blue exclusion was greater than 98%. The Se status and the Cu status of the cattle was altered by using semisynthetic diets based on torula yeast and oat husks (2,3). Unsupplemented diets contained <0.01 mg Se/kg DM and 1.5 mg Cu/kg DM; where appropriate Se was added as Na_2SeO_3 to give a final concentration of 0.1 mg/kg and Cu was added as $CuSO_4.5H_2O$ to a concentration of 12 mg/kg. Copper depletion in the cattle was monitored by measurement of plasma and liver Cu concentrations. Selenium depletion was monitored by measurement of whole blood and liver Se concentrations and glutathione peroxidase activities.

FIGURE 1

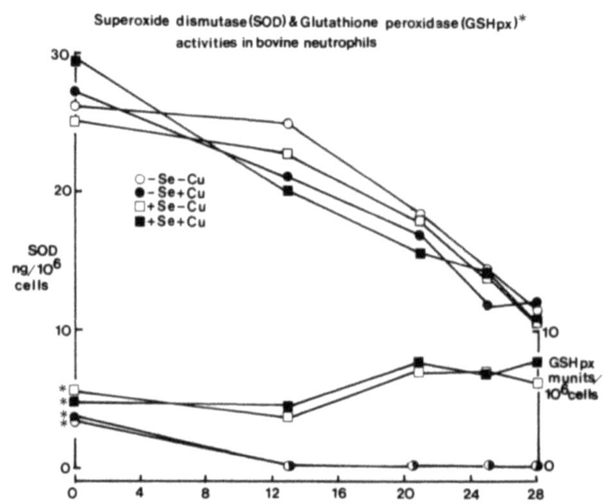

Glutathione peroxidase and superoxide dismutase activities were assayed (7,8) in the supernatants obtained by centrifugation (1200xg, 10 mins at 4°C) of neutrophils disrupted by freezing and thawing (x3).

Figure 1 shows glutathione peroxidase and superoxide dismutase activities measured in cytosolic fractions prepared from neutrophils of cattle being made Se or Cu or Se and Cu deficient over a 28 week period. The loss of microbicidal activity (3) in Se deficient neutrophils is therefore accompanied by a loss of glutathione peroxidase activity. Superoxide dismutase activity declined to a similar extent in both Cu supplemented and Cu deficient groups and thus was not apparently related to the loss of microbicidal activity (3). Selenium deficiency significantly decreased $^{14}CO_2$ production from [1-^{14}C] glucose by zymosan stimulated neutrophils; Cu deficiency had no significant effect (Table 1). Glucose is oxidised via the hexose-monophosphate shunt in stimulated neutrophils to provide NADPH used during O_2^- production (4).

TABLE 1 Release of $^{14}CO_2$ from [1-^{14}C] glucose by bovine neutrophils

Group	Week of Experiment 17*	25**
-Se -Cu	8220	5714
-Se +Cu	7458	5506
+Se -Cu	11293	7167
+Se +Cu	11158	6416
SE of differences	1636	465

Results are counts/5 min released by $2x10^6$ neutrophils in 30 minute incubations. Conditions of incubation as described by Bass et al (9). Significance of Se effects $*P<0.02$, $**P<0.005$.

Thus selenium deficiency impairs an antioxidant system, namely glutathione peroxidase activity and at least one metabolic pathway of the neutrophil. In contrast the degree of Cu deficiency induced in this experiment neither affected the superoxide dismutase antioxidant system nor hexose monophosphate shunt activity. This suggests that the two deficiencies may possibly exert their effects on microbicidal activity (2,3) by quite different mechanisms.

To examine the effects of selenium deficiency on neutrophil function in more detail, neutrophils were prepared from the blood of calves maintained on Se deficient and Se supplemented diets for at least 20 weeks. Blood Se concentrations in these animals were <0.01 and 0.08 – 0.1 mg Se/litre for deficient and supplemented animals respectively (6 animals/group). Reduction of cytochrome c and 2-(p-iodophenol)-3-(p-nitrophenyl)-5-phenyl tetrazolium chloride (INT) were used as measures of superoxide production by stimulated neutrophils. Reduction of both compounds was inhibited over 95% by addition of superoxide dismutase to incubations. Oxygen uptake of neutrophils was also measured using a Clark electrode attached to a Gilson Oxygraph. Results are shown in Table 2.

Table 2 Oxygen consumption, reduction of cytochrome c and of INT by stimulated neutrophils from Se supplemented and Se deficient cattle

	Se deficient	Se supplemented
O_2 consumption nmoles/10^6 cells/min	1.78 ± 0.14(5)	2.83 ± 0.27(6)*
Cytochrome c reduction nmoles/10^6 cells/min	0.266 ± 0.048(5)	0.519 ± 0.035(6)**
INT Reduction nmoles/10^6 cells/15 min	1.89 ± 0.24(5)	3.31 ± 0.26(6)**

Oxygen uptake was measured using a Clark oxygen electrode (10). Incubations were carried out using $2x10^6$ neutrophils/1 ml in Hanks balanced salt solution with 1 mg cytochrome c or 15 μg INT/ml. The neutrophils were stimulated by addition of zymosan (0.5 mg/ml final conc) and cytochrome c reduction monitored at 550 nm using a recording spectrophotometer. Reduced INT was determined after 15 minutes incubation by the absorbance at 500 nm in the supernatant obtained after the neutrophils had been sedimented by centrifugation at 1500 x g for 5 minutes. Results are mean ± SEM number of determinations are in parenthesis. Significance of differences between groups $*P<0.01$, $**P<0.005$ Student's t test.

FIGURE 2 E.p.r. spectra obtained after incubation at 37°C of bovine blood neutrophils with DMPO and opsonised zymosan

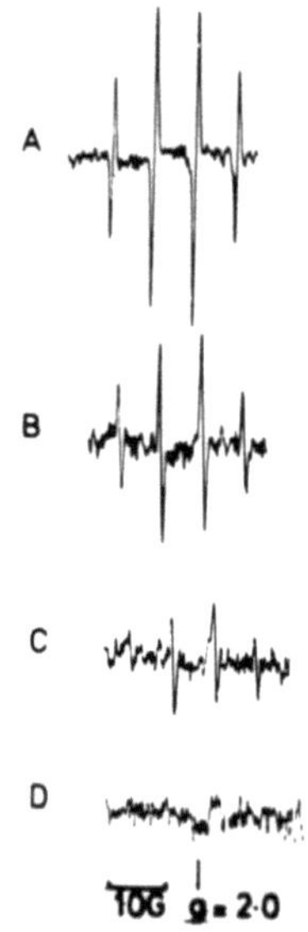

(A) Normal cells + zymosan.

(B) Se deficient cells + zymosan.

(C) Se deficient cells + zymosan + mannitol (10 mM).

(D) Se deficient cells + zymosan + superoxide dismutase (100 μg/ml).

Neutrophil oxygen uptake was inhibited by selenium deficiency as was reduction of cytochrome c and INT. Since reduction of both cytochrome c and INT was almost completely inhibited by superoxide dismutase, these changes suggested that the neutrophils were less able to produce superoxide.

Production of toxic oxygen species was furthex investigated in the neutrophils using spin trapping techniques (10). The results of these studies are shown in figure 2. When neutrophils were stimulated with zymosan they produced 25% less hydroxyl radical adduct of 5'5' dimethyl pyrroline-N-oxide (DMPO) than cells from Se supplemented animals. These signals were completely abolished by inclusion of 100 μg/ml superoxide dismutase in the incubations. Mannitol a hydroxyl radical scavanger at concentrations (30 mM) which do not interfere with the respiratory burst reduced the amount of DMPO hydroxyl adduct formed.

The experiments reported here suggest that neutrophils from Se deficient cattle have a defect in O_2^- production which may impair their microbicidal activity. This could result from low activity of the selenoenzyme glutathione peroxidase allowing toxic peroxides produced in the neutrophil to damage vital enzyme systems. Production of NADPH via the hexose monophosphate shunt may be one such system affected. Cu deficiency decreases neutrophil microbicidal activity and there is histochemical evidence for impaired O_2^- production similar to Se deficiency (3). Preliminary results indicate, however, that some of the biochemical changes induced in the Cu deficient neutrophil may differ to those of Se deficiency.

REFERENCES

1. Serfass, R.E. and H.E. Ganther, 1976, Nature, 255:640-641.
2. Boyne, R. and J.R. Arthur, 1979, J. Comp. Path., 89:151-158.
3. Boyne, R. and J.R. Arthur, 1981, J. Comp. Path., 91:in press.
4. Roos, D., 1977, Trends in Biochem. Sci., 2:61-64.
5. Serfass, R.E. and H.E. Ganther, 1976, Life Sci., 19:1139-1144.
6. Carlson, G.P. and J.J. Kaneko, 1973, Proc. Soc. Exp. Biol. and Med., 142:853-860.
7. Paglia, D.E. and W.N. Valentine, 1967, J. Lab. Clin. Med., 70:158-169.
8. Beauchamp, C. and I. Fridovich, 1971, Anal. Biochem., 44:276-287.
9. Bass, D.A., L.R. De Chatelet, R.F. Burk, P. Shirley and P. Szejda, 1977, Infect. and Immun., 18:78-84.
10. Green, M.R., H.A.O. Hill, M.J. Okolow-Zubkowska and A.W. Segal, 1979, FEBS Lett., 100: 23-26.

BIOCHEMICAL LESIONS IN TISSUES OF Cu-DEFICIENT CATTLE

N. T. Davies, C. B. Lawrence, Lorna Miller, A. C. Dalgarno & C. F. Mills

Rowett Research Institute, Bucksburn, Aberdeen AB2 9SB, U.K.

Overt clinical signs of Cu-deficiency in cattle include reduced weight gain and food conversion efficiency, diarrhoea and abnormalities of gait. Many pathological changes have been observed at post-mortem examination including long bone rarefaction and widespread osteoporosis, aortic thinning and extensive disorganisation certainly of elastin and possibly of collagen. It is also evident that such lesions may develop in individual animals exhibiting no overt signs of Cu deficiency and that existing indices of Cu status provide little indication of the extent of their development or imminence. To investigate the nature of biochemical lesions possibly responsible for such defects, 12 Friesian steers were maintained for 370 d from weaning on a semi-synthetic ration supplying either <1 mg Cu/kg (Cu-deficient) or the same diet supplemented with $CuSO_4.5H_2O$ to provide 12 mg Cu/kg (control).

Determination of plasma benzylamine oxidase activity and benzylamine and lysyl oxidase activities of aorta and ligamentum nuchae

Since the connective tissue lesions described above are consistent with defective cross-linking of elastin and collagen, soluble and membrane-bound benzylamine oxidase activities and lysyl oxidase activity of aorta and ligamentum nuchae were assayed post mortem. Benzylamine oxidase activities were assayed by the methods described by Harris and O'Dell (1). Lysyl oxidase activity was measured by the method of Melet et al (2) on homogenates of aorta or ligamentum nuchae extracted with 6 M urea. A collagen substrate was used, prepared by incubating 17-day-old chick embryo calvaria parietal bones in a tissue culture medium containing L-[4-5^{3}H] lysine (Radiochemical Centre, Amersham, Bucks.) and the lysyl oxidase inhibitor β-amino-isopropionitrile (BAPN) (3). Assays were carried out for 4 h in triplicate and the difference in tritiated water released in the absence of BAPN compared with BAPN inhibited controls was taken as a measure of lysyl oxidase activity.

Plasma monoamine oxidase was monitored throughout the experiment and was significantly lower in the Cu-deficient animals compared with controls after 180 d ($P<0.01$) and continued to decline throughout the duration of the experiment. However neither total, soluble nor membrane-bound benzylamine oxidase activities of either aorta or ligamentum nuchae were affected by the dietary treatments.

The lysyl oxidase activities of aorta and ligamentum nuchae are shown in Table 1.

TABLE 1 Lysyl oxidase activity in the aorta and ligamentum nuchae of copper deficient and control steers. Results cpm ^3H_2O released/g fresh wt/4 hrs

Aorta	Expt. 1	Expt. 2
Copper deficient	14,233 (2)	10,652 ± 483
Control	13,663 (3)	10,040 ± 1,029
Ligamentum nuchae		
Copper deficient	6,308 ± 2,619 (3)	5,570 ± 1,547 (5)***
Control	11,091 ± 649 (3)	13,713 ± 1,291 (6)

Means ± SEM. Nos. of animals in parentheses. ***, $P<0.001$.

Whereas aortic lysyl oxidase was unaffected, lysyl oxidase activity of the ligamentum nuchae from Cu-deficient animals was 60% lower ($P<0.001$) than that of controls. These findings are consistent with previous observations with cattle that the effects of Cu-deficiency on the histological integrity of the ligamentum nuchae is more severe than on the aorta (4).

Beyond the observation that both plasma benzylamine oxidase activity and ligamentum nuchae lysyl oxidase activity were reduced in Cu-deficient cattle no clear relationship between the activity of these two Cu-dependent enzymes was evident.

The effect of Cu-deficiency on the nor-adrenaline concentration of duodenal and colonic tissues

Histopathological and biochemical lesions that have been observed in the intestinal mucosa of Cu-deficient cattle include decreased cytochrome oxidase activity, mitochondrial fragility and total or partial villus atrophy. However the relationship of these to the pathogenesis of diarrhoea has yet to be established. Another possible but hitherto unconsidered cause of the diarrhoea could be changes in tone of the intestinal musculature. Accordingly, a study was made of the noradrenaline concentrations of both the muscle and mucosa of duodenum and colon of Cu-deficient and control steers at slaughter.

Samples of tissue were removed, immediately frozen in liquid nitrogen and stored at -20°C until analyses could be completed. Before thawing, the mucosa was scraped from the muscle and all samples were homogenised in ice cold perchloric acid, centrifuged and the supernatant analysed for noradrenaline using a radio-enzymatic assay (5) based on the conversion of noradrenaline by the action of phenylethanolamine-N-methyl transferase (PNMT, E.C.2.1.1.28) and ^{3}H methyl S-adenosyl methionine to tritiated adrenaline. Results are presented in Table 2. Whereas Cu-deficiency produced no change in the noradrenaline concentration of the mucosa from either intestinal site it resulted in a 41% ($P<0.05$) fall in noradrenaline concentration in duodenal muscle. A similar 59% decrease ($P<0.05$) was observed in colonic muscle.

TABLE 2 Noradrenaline concentrations in the intestine of copper deficient and control steers. Results ng/g wet tissue

	Duodenal Muscle	Colonic Muscle	Duodenal Muscle	Colonic Muscle
Copper Deficient (4)	88 ± 21.7*	65 ± 10.4*	28 ± 1.7	32 ± 4.6
Control (5)	150 ± 12.7	158 ± 40.3	38 ± 7.5	43 ± 9.8

Means ± SEM. Nos. of animals in parentheses. *, $P<0.05$.

Considerable evidence indicates that nor-adrenaline markedly affects intestinal motility. However, the significance of our observations in relation to the diarrhoea observed in Cu-deficient cattle has yet to be established.

Concurrent studies on the cattle used in this experiment have shown that Cu-deficiency induced a similar decline in the concentration of nor-adrenaline in the adrenal medulla (6). Similarly, Cu-deficiency has been shown to decrease nor-adrenaline concentrations in rat brain (7) and in the anterior brain stem of ataxic lambs (8) indicating that a defect influencing the synthesis or metabolism of nor-adrenaline is widely distributed in tissues of Cu-deficient animals. However, the biochemical nature of the defect has not been elucidated. Clearly, further studies are warranted to establish the biological significance, if any, of these findings to the pathological defects induced by Cu-deficiency.

REFERENCES

1. Harris, E.D. & B.L. O'Dell, 1973, Adv. exp. Med. Biol. 48:267-284.
2. Melet, J., G.D. Vianden & B.N. Bachra, 1977, 77:141-146.
3. Siegel, R.C., 1974, Proc. Natl. Acad. Sci., 71:4826-4830.
4. Mills, C.F., A.C. Dalgarno & G. Wenham, 1976, Br. J. Nutr. 35:309-331.
5. Henry, D.P., J. Barbra, D.G. Johnson & R.H. Williams, 1975. Life Sci. 16:375-384.
6. Hesketh, J.E., 1980, Biochem. Soc. Trans. 8:342.
7. Morgan, R.F. & B.L. O'Dell, 1977, J. Neurochem. 28:207-213.
8. O'Dell, B.L., R.M. Smith & R.A. King, 1974, Fed. Proc. 33:668.

DISCUSSION

O'DELL (USA). Have you seen any gross pathology in the ligamentum nuchae that would correlate to the head droop that you see.

MILLS (Scotland). We invariably see head droop, but we do not see gross external changes except that the ligamentum is often longer and thinner.

O'DELL (USA). You use calvaria as a source of substrate, that is primarily a collagen source. I wonder if you had tried elastin sources such as aortic tissues as a source of substrate.

MILLS (Scotland). No we are very much aware of the work you are doing on different substrates in relation to activity but calvaria have been the only substrate we have used.

ELMES (Wales). With regard to the mechanisms of diarrhoea, this is primarily determined by colonic reception of the contents of the distal small intestine, so whereas your colonic muscle results are of great value in understanding actual diarrhoea - I really think that to look at the duodenum is a bit far off from the point of action. I would have thought that you would have been far better off to have looked at the ileum or the lower jejunum.

SUTTLE (Scotland). In the copper deficient lamb which is showing diarrhoea, we found histological signs of damage in the colon in the form of oedema and at no other site.

ALLEN (UK). With regard to the first question Gallagher J.A. in England (Personal Communication) has demonstrated lesions in the epiphyseal end plate in copper deficient animals, I wonder has Dr. Mills had an opportunity to look at lysyl oxidase activity in the epiphyseal plate.

MILLS (Scotland). No, I have not. We are doing a lot of conformational studies on bone at the moment. Changes in leg bone conformation seems to be a remarkably early indicator but we are not doing any biochemistry yet.

COPPER-SUPEROXIDE DISMUTASE AND COPPER DEFICIENCY IN RUMINANTS

David I. Paynter[1] and John D. Allen[2]

1. 'Attwood' Veterinary Research Laboratory, Department of Agriculture, WESTMEADOWS, VICTORIA, AUSTRALIA, 3047
2. School of Veterinary Studies, Murdoch University, PERTH, WESTERN AUSTRALIA, 6150

Blood copper superoxide dismutase (CuSOD) and ceruloplasmin activities may provide useful long and short term indicators of copper nutrition in animals. The dependence of CuSOD activity in erythrocytes and other tissues on dietary copper availability has been established in several species (Williams *et al* 1975; Paynter *et al* 1979; Andrewartha and Caple 1980). In copper depleted and repleted rats, changes in erythrocyte CuSOD are considerably slower than those observed for plasma ceruloplasmin activity; differential rates of depletion are also observed for CuSOD and cytochrome oxidase activities in a number of tissues of this species (Paynter *et al* 1979). In cattle, changes in cytochrome oxidase activity have been implicated in several of the pathological tissue changes associated with copper deficiency, including heart hypertrophy (Leigh 1975) and a copper responsive diarrhoea (Fell *et al* 1975). The possible involvement of CuSOD changes in the pathogenesis of these conditions has not generally been considered. In the present experiments, the effects of copper nutrition on the relative activity changes of ceruloplasmin and CuSOD in blood and on cytochrome oxidase and CuSOD activities in heart, liver and intestinal mucosa were investigated using groups of supplemented (monthly or 3-monthly injections of 120mg of Cu as Cu-glycinate, starting from approximately 1 month of age) and non-supplemented calves, 7 months of age, suckling cows grazing pastures low in copper. Copper/molybdenum concentration ratios in these pastures were generally greater than 3, with approximately 0.25% w/w sulphur.

Activity of CuSOD was measured by a modification of the method of Misra and Fridovich (1977). Whole blood, packed cells or tissue homogenates (20% w/v in 0.2% v/v triton x-100) were treated with 0.25 volumes of ethanol and 0.15 volumes of chloroform, followed by centrifugation. Samples (<50μl) of the clear supernatant were added to a reaction mixture composed of 0.95 ml of 50 mM K_2HPO_4 (pH 7.8), 0.1 mM EDTA, 13μM riboflavin and 0.2 mM o-dianisidine di-HCl. After an illumination period of 8 minutes the reaction was terminated by addition of 2ml 9M H_2SO_4 and absorbance of the final solution read at 540 nm. One activity unit is defined as that activity equivalent to 1μg of purified bovine CuSOD. Cytochrome oxidase activities in tissue and mucosal homogenates (20% w/v in 0.2% v/v triton x-100) and ceruloplasmin activities were determined as described previously (Paynter *et al* 1979; Paynter 1980a). Copper was measured by atomic absorption spectrometry.

At the start of the experimental period, plasma ceruloplasmin activities were low in non-supplemented calves and several had activities of <1 mU/ml, compared with mean activities of >70 mU/ml in supplemented calves (Table 1). At this time serum copper concentrations in non-supplemented calves, which had been monitored for several months previously, also reached their lowest values.

Table 1. Erythrocyte CuSOD activities, plasma ceruloplasmin activities and weight gains of cattle during the experimental period.

		No Cu	3-monthly Cu	1-monthly Cu
CuSOD activity (U/g Hb)	Week 0	466 ± 21.1[a]	533 ± 16.4[b]	556 ± 11.7[b]
	4	459 ± 23.8[a]	570 ± 18.2[b]	604 ± 14.0[b]
	8	484 ± 16.5[a]	604 ± 20.3[b]	632 ± 9.1[b]
Ceruloplasmin activity (mU/ml)	Week 0	9 ± 2.5[a]	78 ± 3.1[b]	72 ± 3.7[b]
	4	18 ± 2.9[a]	74 ± 3.0[b]	66 ± 3.3[b]
	8	32 ± 4.0[a]	85 ± 4.3[b]	84 ± 3.8[b]
Weight gain (kg)	Weeks -4 to 0	33 ± 1.2[a]	31 ± 1.5[a]	34 ± 1.4[a]
	0 to 4	31 ± 1.5[a]	30 ± 2.1[a]	30 ± 1.5[a]

Mean values (± SEM) of 14 calves for each treatment. Means not followed by a common superscript are significantly different (P<0.05).

Erythrocyte CuSOD activities were lower in non-supplemented calves, but not to the extent observed with ceruloplasmin (Table 1). The comparatively high activity of CuSOD to that of ceruloplasmin in the blood of untreated calves, indicated that these animals had not been fully depleted of copper. The lack of effect of copper supplementation on growth confirmed this finding. It would appear that for growth of cattle to be affected by a simple copper deficiency, plasma ceruloplasmin activity must be near zero and erythrocyte CuSOD should also be decreased (<75% of normal activity). In this context, erythrocyte CuSOD appears to be a better indicator of tissue copper depletion than is ceruloplasmin.

In erythrocytes, CuSOD activity was correlated with copper concentration ($r = 0.83$, $P<0.001$, $n = 39$), and accounted for 77% of the cellular copper. This distribution was independent of copper treatment. For several other tissues examined, the amount of tissue copper accounted for by CuSOD was dependent on copper treatment. This was most pronounced in liver where CuSOD activity accounted for 45% of tissue copper in non-supplemented animals, but less than 4% in supplemented animals (Table 2).

Table 2. Effect of copper treatment on tissue enzyme activities and copper concentrations.

		Cu treatment		
		No Cu[1]	3-monthly Cu[2]	1-monthly Cu[2]
Liver	Cu[3]	2.3 ± 0.22^a	44.3 ± 6.42^b	76.1 ± 5.10^c
	Cyt. Ox.[4]	48.6 ± 3.59^a	64.9 ± 4.56^b	63.8 ± 2.60^b
	CuSOD[5]	267 ± 13.4 ^{a}	501 ± 15.4 ^{b}	389 ± 24.2 ^{b}
	% Cu as CuSOD[6]	45.1	4.3	1.9
Heart	Cu	3.9 ± 0.05^a	4.5 ± 0.14^b	4.6 ± 0.14^b
	Cyt. Ox.	149 ± 7.8 ^{a}	138 ± 9.3 ^{a}	159 ± 13.5 ^{a}
	CuSOD	123 ± 7.0 ^{a}	115 ± 8.7 ^{a}	124 ± 5.8 ^{a}
	% Cu as CuSOD	11.9	9.8	10.2
Duodenum	Cu	1.1 ± 0.04^a	1.4 ± 0.14^a	3.2 ± 1.50^b
	Cyt. Ox.	41.1 ± 1.25^a	43.0 ± 1.67^a	41.0 ± 3.37^a
	CuSOD	57.9 ± 4.50^a	73.0 ± 5.41^a	66.8 ± 10.22^a
	% Cu as CuSOD	20.2	20.5	8.0
Jejunum	Cu	0.8 ± 0.04^a	1.1 ± 0.06^a	1.6 ± 0.42^b
	Cyt. Ox.	29.9 ± 1.21^a	42.4 ± 2.67^b	35.9 ± 2.06^b
	CuSOD	88.3 ± 4.47^a	105.2 ± 3.72^b	89.8 ± 1.93^a
	% Cu as CuSOD	41.8	35.9	22.4
Ileum	Cu	0.7 ± 0.04^a	1.0 ± 0.05^b	1.0 ± 0.11^b
	Cyt. Ox.	26.5 ± 1.05^a	30.3 ± 2.38^a	28.1 ± 2.51^a
	CuSOD	107 ± 5.0 ^{a}	128 ± 6.7 ^{b}	113 ± 4.6 ^{a}
	% Cu as CuSOD	58.9	47.7	42.5
Colon	Cu	1.4 ± 0.34^a	1.3 ± 0.09^a	1.2 ± 0.09^a
	Cyt. Ox.	27.5 ± 1.75^a	30.9 ± 1.41^a	31.4 ± 2.63^a
	CuSOD	113 ± 3.6 ^{a}	104 ± 9.0 ^{a}	107 ± 8.3 ^{a}
	% Cu as CuSOD	30.9	30.9	33.3

1. Values are means (± SEM) of 14 calves. 2. Values are means (± SEM) of 5 animals. 3. µg Cu/g wet weight. 4. µmoles cytochrome c oxidized/min/g wet weight. 5. U/g wet weight. 6. Calculated from 0.38% Cu in purified bovine CuSOD. Means not followed by a common superscript are significantly different ($P<0.05$).

Cytochrome oxidase and CuSOD activities were both reduced in the livers of non-supplemented animals, with CuSOD showing the greatest reduction. No changes in either enzyme were evident in heart tissue. In the intestinal mucosa, activity of CuSOD increased along the small intestine and through to the colon; the reverse was observed with cytochrome oxidase activities. Only minor changes in the activities of these enzymes in the intestinal mucosa were evident with copper deficiency (Table 2).

Changes in activities of cytochrome oxidase have been proposed as the primary biochemical lesion in copper-induced heart hypertrophy in cattle (Leigh 1975). This was not supported by the results of the present experiment in which copper deficiency reduced cytochrome oxidase in several tissues but not in heart, even though copper concentrations were reduced in this tissue. This indicates, as has been proposed in other species (Paynter 1980b), that changes in factors other than heart cytochrome oxidase or CuSOD activities with copper deficiency may be implicated in the hypertrophy induced in this tissue. The finding that a concurrent dietary manganese deficiency prevents copper-induced heart hypertrophy in rats (Paynter 1980b), further supports this concept, and strongly suggests that biochemical pathways requiring not only copper but also manganese may be involved in this hypertrophy.

Fell *et al* (1975) associated copper-responsive diarrhoea in cattle with large decreases in small intestinal epithelium cytochrome oxidase activities. The results of the present experiment, in which diarrhoea was not observed in any of the calves and decreases in intestinal mucosa cytochrome oxidase activity were relatively small, are in agreement with these initial findings. However, it is also apparent that CuSOD activity accounts for a large proportion of the cellular copper at several of the sites sampled along the intestine and that this activity is also responsive to dietary copper. This indicates that changes in this enzyme should also be considered in relation to the mechanisms involved in enterocyte mitochondrial abnormalities and villus atrophy changes associated with copper-responsive diarrhoea in cattle.

1. Andrewartha, K.A., and I.W. Caple, 1980, Res. Vet. Sci. 28:101-104
2. Fell, B.F., D. Dinsdale and C.F. Mills, 1975, Res. Vet. Sci. 18:274-281
3. Leigh, L.C., 1975, Res. Vet. Sci. 18:282-287
4. Misra, H.P., and I. Fridovich, 1977, Arch. Biochem. Biophys. 181:308-312
5. Paynter, D.I., R.J. Moir and E.J. Underwood, 1979, J. Nutr. 109:1570-1576
6. Paynter, D.I., 1980a, Proc. Nutr. Soc. Aust. 5:183
7. Paynter, D.I., 1980b, Biol. Trace Element Res. 2:121-135
8. Williams, D.M., R.E. Lynch, G.R. Lee and G.E. Cartwright, 1975, Proc. Soc. Exp. Biol. Med. 149:534-536

DISCUSSION

MILLS (Scotland). One of the interesting results of your paper is the difference between the behaviour of your cattle heart and that of the rat heart. In Paynter's study cytochrome oxidase in the heart was particularly sensitive to depletion.

ALLEN (Australia). Yes the greatest reduction in CuSOD occurred in the liver of the copper depleted rats with only minor changes in cytochrome oxidase, whereas in heart cytochrome oxidase was considerably depleted. Using o-dianisidine as substrate with ceruloplasmin assay there were several calves with activities of less than one unit per ml. These animals were still quite healthy. There was no correlation between total haemoglobin or packed cell volume in the animals with ceruloplasmin values of this level. Presumably such animals can live for several weeks with very low ceruloplasmin levels without an effect on iron metabolism.

DANKS (Australia). I am a little confused by your suggestion that CuSOD is a much better indicator than ceruloplasmin. Are you suggesting that ceruloplasmin over indicates copper deficiency? If so I don't know quite what your ultimate criterion of copper deficiency is. It seems to me one can either take the attitude of saying that it would be good to have all of your copper enzymes at normal levels and that any fall indicates less than ideal nutrition for copper, or else you can say that "if you stay alive, that's fine and I am not too interested in what chemical differences there are".

ALLEN (Australia). Certainly, by biochemical estimates we have a copper deficiency in these calves. However, by productivity they were not deficient. The non-supplemented animals grew just as well and generally had a similar appearance to those of the supplemented animals. As for the question dealing with CuSOD the point I was trying to make is that ceruloplasmin activities are a measure of the liver copper reserves in the animal whereas blood CuSOD is more a measure of tissue copper enzyme depletion.

CATECHOLAMINE SYNTHESIS IN COPPER-DEFICIENCY

John E. Hesketh

Rowett Research Institute, BUCKSBURN, ABERDEEN AB2 9SB, SCOTLAND, U.K.

Copper is an essential nutrient and its deficiency leads to lesions in a variety of tissues (1) including the central nervous system; in young lambs copper deficiency is responsible for a neonatal ataxia (swayback) associated with neuronal necrosis and in the rat and guinea-pig copper deficiency has been found to lead to abnormal myelination and other neurological lesions. The biochemical role of copper in brain metabolism, however, is not well understood and the neurological features of copper deficiency cannot as yet be explained in terms of known cuproproteins. One cuproprotein specific to nervous tissue is dopamine-β-hydroxylase (3, 4 dihydroxyphenyl-ethylamine, ascorbate:O_2 oxidoreductase (β-hydroxylating) E.C.1.14.17.1), an enzyme involved in the synthesis of the catecholamines adrenaline and noradrenaline (2). Although control of catecholamine synthesis is regarded as primarily involving tyrosine hydroxylase, administration of copper-chelating substances in vivo has been found to lead to a reduction in noradrenaline synthesis and the activity of dopamine-β-hydroxylase becomes rate-limiting (3). Dietary deprivation of copper has been shown to reduce brain noradrenaline (4,5,6) but little attention has been given to study of dopamine-β-hydroxylase under these circumstances. This report presents evidence that in copper deficient cattle and rats there is a reduction in adrenal noradrenaline content. In cattle this can only partly be explained by a reduction in activity of dopamine-β-hydroxylase; furthermore hydroxylase activity is not rate-limiting but increased in the copper-deficient rat.

15 Friesian steers were fed a skimmed milk, oat and maize starch diet for 355 days. The average copper content of the diet was 11.7 µg/g dry matter (copper-supplemented) and 1.32 µg/g (copper-deficient). Rats were weanling Hooded Listers, 40-50 g in weight. They were housed on a group basis in stainless-steel/polypropylene cages and fed the basal diet of Davies and Reid (7) with and without copper supplements. The copper content of the deficient diet was less than 0.5 µg/g and that of the supplemented diet 5 µg/g.

Rats were anaesthetised using thiopentone sodium, bled from the heart and their adrenal glands were immediately removed and dropped into liquid N_2. Glands were subsequently stored at -20°C. Cattle were killed using a captive bolt pistol and bled from the main blood vessels of the neck. The adrenals were removed and immediately chilled. Medulla was dissected from cortex, minced and frozen at -20°C. For catecholamine estimations tissue was homogenised in 0.4 M perchloric acid (50 ml/g tissue) and the extract diluted 100-fold with bidistilled water before fluorimetric analysis (8). Dopamine-β-hydroxylase was measured by the method of Nagatsu and Udenfriend (9) in the presence of 2.5×10^{-2}M N-ethyl-maleimide; tissue was homogenised in 20 volumes 5 mM Tris-HCl, pH 6.5, 0.1% Triton and diluted 20-fold in distilled water.

Adrenal medulla from copper-deficient cattle contained 40% less noradrenaline than medullae from copper-supplemented cattle (10). There was no statistically significant difference between dopamine-β-hydroxylase activity in adrenal homogenates between copper-deficient and control cattle (10) although there was a tendency for the copper-deficient animals to show reduced enzyme activity. There was a correlation between adrenal medulla noradrenaline content and homogenate specific dopamine-β-hydroxylase activity in copper-deficient cattle ($0.05 < p < 0.1$) but no such correlation occurred in the controls. However for a given hydroxylase activity the noradrenaline content appeared higher in the control animals. Thus although the decrease in noradrenaline in the copper-deficient cattle can be partially explained in terms of a reduction in dopamine-β-hydroxylase activity the results suggest that some other factor is also involved in leading to the reduced noradrenaline content. Adrenals from rats nutritionally deprived of copper for periods of 2-12 weeks showed a marked reduction (37%) in their mean noradrenaline content but there was no difference in the adrenaline content (Figure 1). There was no difference in the adrenal weights of the two groups of animals (mean weight was 26 mg for each group). Statistical analysis showed an overall significant reduction ($p < 0.01$) of noradrenaline content in the copper-deficient animals. The differential effect on noradrenaline and adrenaline is similar to that observed in cattle (10). The lack of effect on adrenaline suggests that the reduction in noradrenaline is a real effect of copper deficiency and not a non-specific effect of stress.

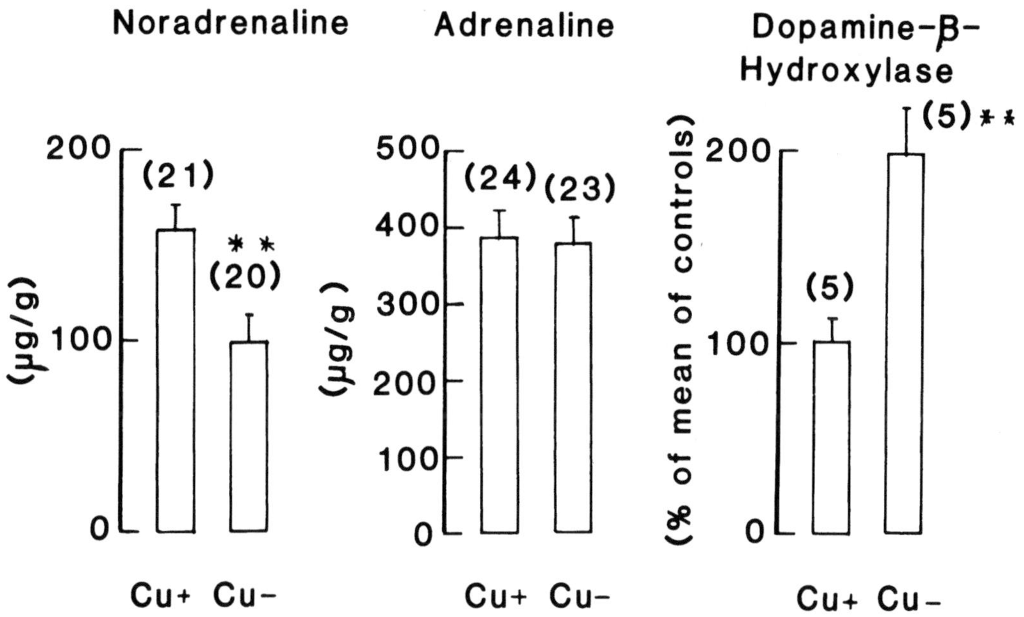

Figure 1. Rat adrenal catecholamine content and dopamine-β-hydroxylase specific activity. Histograms represent mean values from animals fed the copper-deficient or copper-supplemented diet for 2-12 weeks (noradrenaline, adrenaline) or 12 weeks (DβH). Bars represent s.e.m. **, $p < 0.01$ compared to copper-supplemented controls. Number of animals is given in parentheses.

Figure 1 shows dopamine-β-hydroxylase activities in adrenal homogenates from copper-deficient and control rats. The results are quite distinct from those found in cattle, there being a marked increase in the hydroxylase activity in rats fed a copper-deficient diet. Evidently in the copper-deficient rat the observed decrease in noradrenaline cannot be explained in terms of reduced dopamine-β-hydroxylase activity; again some other factor seems to be involved in reducing the noradrenaline content of the gland. The increase in dopamine-β-hydroxylase is similar to that described in the brain and adrenal glands of mutant mottled mice which are unable to absorb copper (11). Hunt included copper in the incubation media when assaying dopamine-β-hydroxylase and this could have activated apo-enzyme present in the homogenate; such a possibility is less likely in the present experiments since N-ethyl-maleimide and not copper was added to allow full exhibition of enzyme activity.

The increased dopamine-β-hydroxylase activity and the reduction of noradrenaline concentration in the copper-deficient rat medulla, suggest that depletion of noradrenaline exerts feedback control to increase the activity of the synthesizing enzyme dopamine-β-hydroxylase. Kinetic studies showed that the K_m of dopamine-β-hydroxylase for tyramine was decreased in adrenal homogenates from copper-deficient rats (4.7 mM) compared with control preparations (9 mM). This decrease in K_m is unlikely to be due to a loss of endogenous inhibitors since measurement of the enzyme activity at sub-optimal N-ethyl-maleimide concentrations showed no evidence for a change in the concentration of such inhibitors in the copper-deficient preparations. The change in K_m suggests therefore that there is an increased affinity of dopamine-β-hydroxylase for its substrate tyramine in the copper-deficient rat. This suggests an activation of the enzyme and provides evidence for some regulation of the enzyme activity.

Adrenaline and noradrenaline are synthesized in different cells of the adrenal medulla, the two cell types showing a common biosynthetic pathway as far as noradrenaline, and the adrenaline cells in addition containing the enzyme phenylethanolamine-N-methyltransferase which converts noradrenaline to adrenaline. The observation that copper deficiency reduces adrenal noradrenaline but not adrenaline shows that the sequence of events initiated by copper deprivation is different in the two cell types.

In conclusion, the results have shown that in both rats and cattle, copper deficiency results in a reduction in adrenal noradrenaline although the mechanism by which copper deprivation causes such a decrease in noradrenaline is not clear; it appears not to be accounted for only by changes in dopamine-β-hydroxylase activity but might involve a defect in storage of noradrenaline, an increase in catabolism by monoamine oxidase (12) or a reduction in synthesis due to reduced tyrosine hydroxylase activity (13). Such changes in noradrenaline metabolism may be related to lesions of the peripheral nervous system (Hesketh and Fell, in preparation) and the brain found in copper-deficient animals. It is interesting to speculate that lesions of catecholamine metabolism in the peripheral nervous system may be important factors in the aetiology of several symptoms of Cu-deficiency; cardiac enlargement, "falling disease" of cattle associated with sudden heart failure and diarrhoea (1) could be related to lesions of noradrenergic innervation.

REFERENCES

1. Underwood, E. J., 1977, Trace Elements in Human and Animal Nutrition. 4th Edit. Academic Press, New York.
2. Blaschko, H., 1973, Br. Med. Bull. 29:105-109.
3. Goldstein, M., Anagouste, B., Lauber, E., and M. R. McKereghan, 1964, Life Sci. 3:763-767.
4. Prohaska, J. R., and W. W. Wells, 1974, J. Neurochem. 23:91-98.
5. O'Dell, B. L., Smith, R. M., and R. A. King, 1976, J. Neurochem. 26:451-455.
6. Feller, D. J., and B. L. O'Dell, 1980, J. Neurochem. 34:1259-1263.
7. Davies, N. T., and H. Reid, 1979, Br. J. Nutr. 41:579-589.
8. Renzini, V., Brunori, C. A., and C. Valori, 1970, Clin. Chim. Acta 30:587-594.
9. Nagatsu, T., and S. Udenfriend, 1972, Clin. Chem. 18:980-983.
10. Hesketh, J. E., 1980, Biochem. Soc. Trans. 8:342.
11. Hunt, D. M., 1980. In "Biological Roles of Copper" (Ciba Foundation Symposium No. 79) pp 247-266.
12. Kelly, W. A., Kesterton, J. W., and W. W. Carlton, 1974, Exp. Mol. Path. 20:40-56.
13. Morgan, R. F., and B. L. O'Dell, 1977, J. Neurochem. 28:207-213.

DISCUSSION

HOWELL (Australia). Your comment about lesions in peripheral nerve is an interesting one. The peripheral nerves have not been given the study that they deserve. Wallerian degeneration of peripheral nerves is a significant feature in many cases of swayback.

HESKETH (Scotland). The two lesions that I should mention have been seen in the preliminary work on the cattle. Firstly there were enlarged mitochondria (mega mitochondria) similar to those discussed earlier this morning, and also there was the occurence of multivesicular bodies. The large multivescular bodies in the axons could be involved in degeneration of storage vesicles. There were also lesions of the Golgi apparatus suggestive of the reduction of storage vesicles.

PROHASKA (USA). I would like to make one comment concerning monamine oxidase in copper deficiency. I don't know about the cofactors for monamine oxidase in the adrenal medulla, but in the central nervous system, both Hunt, O'Dell and myself have shown that monamine oxidase activity is not increased in copper deficiency. So that explanation for reduced norepinephrine probably, is not a viable one.

HESKETH (Scotland). Kelly's work I believe is on the heart. I haven't looked at it in the adrenals but the materials I still in the freezer.

MILLS (Scotland). Do either Dr Hesketh or Professors Prohaska or O'Dell consider that the origin of this change in dopamine β hydroxylase and its relationship to noradrenalin synthesis may reflect the iron-dependent step at the beginning of the sequence (i.e. in tyrosine 3-hydroxylase). I think that in other tissues of the animals that you were working with, there was a change in tissue iron deposition as a secondary effect of Cu deficiency. The typical situation in the ruminant is an increased deposition of liver iron. This must reflect a defect of iron metabolism. Does anyone know of any similar effects of Fe deficiency upon noradrenalin. I know that Sourkes showed that tyrosine hydroxylase increased in iron deficiency even though it was an iron dependent enzyme.

PROHASKA (USA). I think it has been shown in experimental iron deficiency that brain catacholamines do not change. In the rat studies we did, the concentration of brain iron changed and norepinephrine was decreased but in the mice studies, brain iron was normal and norepinephrine was low so I don't think there is any relationship between brain iron concentration and low norepinephrine.

ASPECTS OF COBALT DEFICIENCY IN EWES AND THEIR LAMBS

G. A. Garton, W. R. H. Duncan and B. F. Fell

Rowett Research Institute,
BUCKSBURN, ABERDEEN AB2 9SB
UNITED KINGDOM

INTRODUCTION

Earlier investigations at this Institute (1,2) showed that, when sheep were fed on diets containing a high proportion of readily-fermentable starch, their tissue lipids contained unusually high proportions of odd-numbered fatty acids (ONFA) and branched-chain fatty acids (BCFA). The amount of propionate arising from ruminal fermentation is apparently in excess of that which can be rapidly metabolised in the liver and so propionyl-CoA becomes available as the primer unit for long-chain fatty acid biosynthesis. The primary metabolite of propionyl-CoA is methylmalonyl-CoA and this can be incorporated (in place of malonyl-CoA) into newly-synthesised fatty acids, thereby giving rise to BCFA containing one or more methyl groups in the acyl chain (3,4). Thus it was concluded that the further metabolism of methylmalonyl-CoA to succinyl-CoA, which involves a vitamin B_{12}-dependent enzyme (methylmalonyl-CoA mutase), was probably the rate-limiting step in the entry of propionate-carbon into the tricarboxylic acid cycle. The question thus arose, is long-chain fatty acid synthesis similarly affected in cobalt deficiency when the activity of methylmalonyl-CoA mutase is known to be reduced (5)? Accordingly, it was decided to investigate the effects of cobalt deficiency in pregnant ewes and their lambs with particular reference to the tissue lipids of lambs born to ewes which, in late pregnancy, were excreting methylmalonic acid (MMA) in their urine. The lipids of lambs were selected for analysis because it was considered that any abnormalities in fatty acid synthesis which might result would be more readily apparent in lambs since all their long-chain fatty acids arise by synthesis *de novo* and so any ONFA and BCFA would not be 'diluted' by pre-existing fatty acids.

EXPERIMENTAL AND RESULTS

Two groups of ewes were fed to appetite throughout pregnancy on a cobalt-deficient diet (0.04 mg Co/kg) consisting of grass hay, maize meal, salts and vitamins; the control animals were given cobalt (as $CoSO_4$) in their drinking water. One group (Group A) was given the diet from the beginning of pregnancy, with the other (Group B) receiving the diet from four months before mating. Cobalt depletion in the ewes was monitored at intervals by determining serum vitamin B_{12} concentration and by estimating urinary excretion of MMA. The state of depletion of the ewes is indicated in Figs. 1 and 2 which show that, three weeks before parturition, serum B_{12} values of the ewes of both groups were low and that one animal in Group A (ewe A7) and all those in Group B were excreting abnormal amounts of MMA. The ewes in Group A did not lose appetite and they lambed at normal term. Twins were born to three of the cobalt-depleted ewes (nos. A4, A5 and A7 in Fig. 1), though those born to ewe A7 were underweight and died at birth; all the other lambs appeared to be normal. With Group B the picture was very different. The ewes began to lose appetite in early pregnancy and, though they too lambed at normal term, their offspring were stillborn, or died at birth, or were killed within a day of birth because they were very weakly and were not being suckled. Liver and adipose tissue was taken for analysis from five or these lambs. Total liver lipids were separated into neutral lipids and phospholipids and triglycerides were extracted from adipose tissue. The fatty acid composition of each lipid fraction, determined by gas chromatography, showed that, compared to similar preparations from control animals, notably high proportions of ONFA (mostly 15:0, 17:0 and 17:1) were present in the tissue lipids of the lambs born to the vitamin B_{12}-depleted ewes which, before parturition, were excreting enhanced amounts of urinary MMA. These findings are illustrated in Figs. 3 and 4 by reference to liver phospholipids and adipose tissue triglycerides respectively. The data given in Figs. 3 and 4 also show that, although no such substantial increment to the BCFA content of the tissue lipids was observed, the proportions of these acids were consistently greater than were present in the corresponding lipids of lambs born to control ewes.

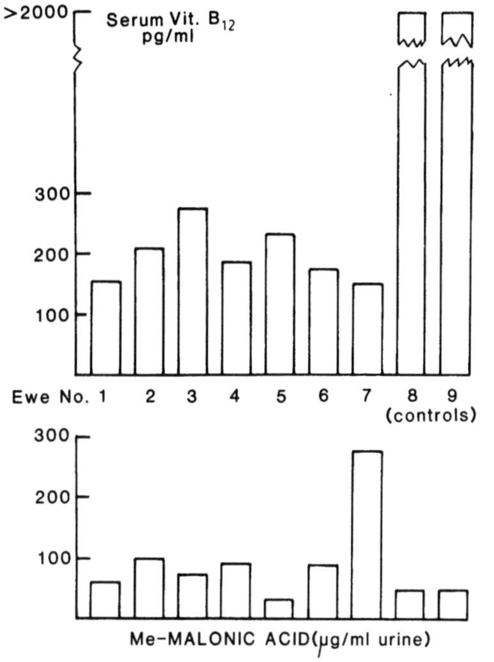

Fig. 1. Serum vitamin B_{12} concentration and urinary content of methylmalonic acid of Group A ewes three weeks before parturition.

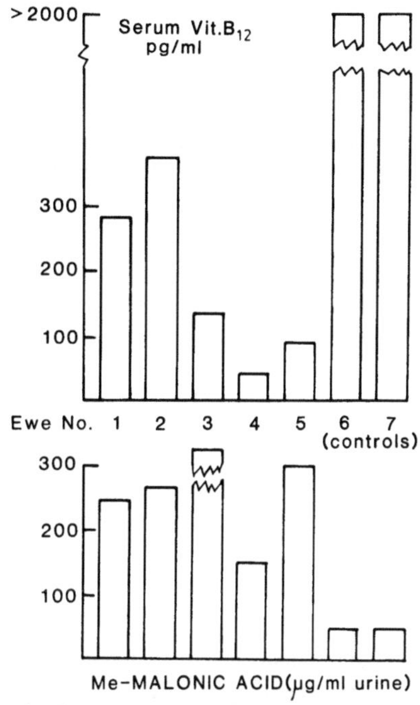

Fig. 2. Serum vitamin B_{12} concentration and urinary content of methylmalonic acid of Group B ewes three weeks before parturition.

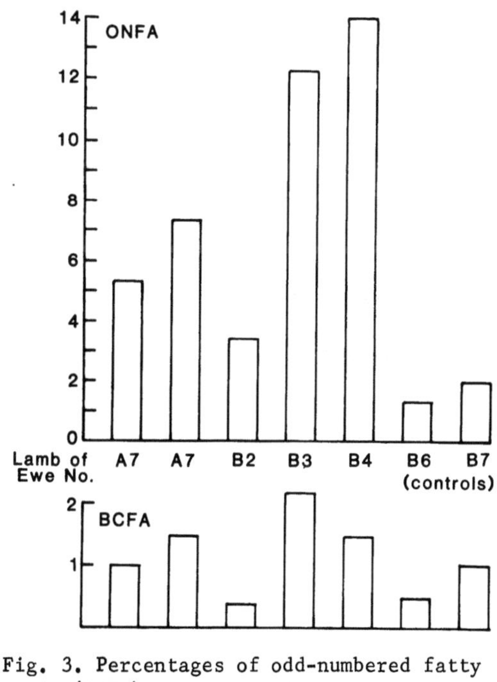

Fig. 3. Percentages of odd-numbered fatty acids (ONFA) and branched-chain fatty acids (BCFA) in liver phospholipids of lambs born to vitamin B_{12}-depleted ewes.

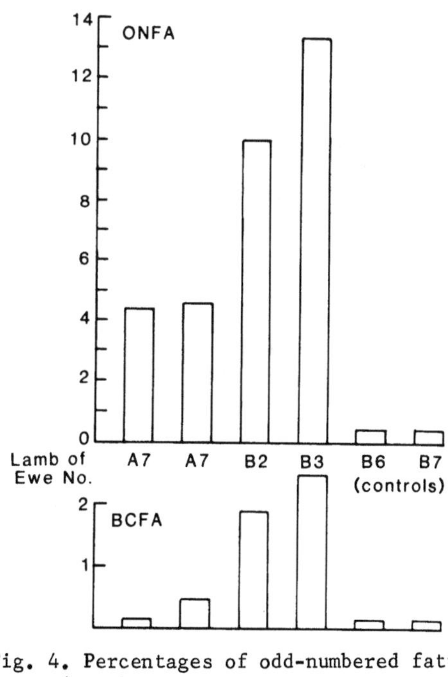

Fig. 4. Percentages of odd-numbered fatty acids (ONFA) and branched-chain fatty acids (BCFA) in adipose tissue triglycerides of lambs born to vitamin B_{12}-depleted ewes.

CONCLUSIONS AND RELATED OBSERVATIONS

The present studies have shown that perinatal mortality and morbidity can be high amongst lambs born to cobalt-deficient ewes when they have received an inadequate intake of cobalt for some time before, and during pregnancy. It may be more than coincidental that enhanced production of MMA (as evidenced by urinary output) is apparently associated with failure of the ewes to produce viable lambs. Enhanced production of ONFA and, to a limited extent, of BCFA is consistent with depressed metabolism of propionate resulting from deficiency of methylmalonyl-CoA mutase. Similar observations with respect to ONFA and BCFA in liver lipids have been reported in B_{12}-deficient baboons (6) and in the nervous tissue of a child with methylmalonic aciduria (7); enhanced synthesis of ONFA has been observed in tissues of B_{12}-depleted rats (8) and nerves of patients with pernicious anaemia (9).

Further investigations at this Institute have shown that, as in B_{12}-deprived rats (10), the conversion of phosphatidylethanolamine (PE) to phosphatidylcholine (PC) is impaired in B_{12}-depleted ewes; the ratio PE:PC in liver lipids was about 0.8 compared with about 0.5 for control ewes (A. K. Lough, unpublished). Decreased production of PC can be ascribed to reduced availability of the methyl donor S-adenosylmethionine, the liver concentration of which is decreased in B_{12}-deficient sheep (11).

Examination post-mortem of the tissues of cobalt-deficient ewes and of lambs born to such ewes is being made. In ewes, lesions have been found in liver and central nervous system and atrophy of skeletal muscle and myocardium is associated with the presence of morphologically abnormal mitochondria. The liver, which appears to be the organ primarily affected, shows fatty change, accumulation of lipofuscin and bile duct proliferation. In nervous tissue atrophy and degeneration of neurones in the cerebral cortex and brain stem were found. In affected lambs, liver damage manifested itself as swelling and hydropic changes of hepatocytes; brain abnormalities included reduced numbers and broadening of cerebral gyri.

REFERENCES

1. Garton, G. A., Hovell, F. D. DeB., and W. R. H. Duncan, 1972, Br. J. Nutr. 28:409-416.
2. Duncan, W. R. H., Lough, A. K., Garton, G. A., and P. Brooks, 1974, Lipids 9:669-673.
3. Scaife, J. R., Wahle, K. W. J., and G. A. Garton, 1978, Biochem. J. 176:799-804.
4. Smith, A., Calder, A. G., Lough, A. K., and W. R. H. Duncan, 1979, Lipids 14:953-960.
5. Marston, H. R., Allen, S. H. and R. M. Smith, 1961, Nature, Lond. 190:1085-1091.
6. Garton, G. A., Scaife, J. R., Smith, A., and R. C. Siddons, 1975, Lipids 10:855-857.
7. Kishimoto, Y., Williams, M., Moser, H. W., Hignite, C., and K. Biemann, 1973, J. Lipid Res. 14:69-77.
8. Fehling, C., Jägerstad, M., Akesson, B., Axelsson, J., and A. Brun, 1978, Br. J. Nutr. 39:501-513.
9. Frenkel, E. P. 1973, J. clin. Invest. 52:1237-1245.
10. Åkesson, B., Fehling, C., and M. Jägerstad, 1978, Br. J. Nutr. 40:521-527.
11. Gawthorne, J. M., and R. M. Smith, 1974, Biochem. J. 142:119-126.

DISCUSSION

SINCLAIR (Australia). Dr Garton, you have presented a number of different indicators which might be useful in the diagnosis of B_{12} deficiency. These were plasma B_{12}, urinary methylmalonic acid, and the possibility of changes to the levels of branched-chain and odd-chain fatty acids. Presumably these could occur in the plasma. I wonder also about the change in the PE to PC ratio. Would you care to comment on which would be the more reliable of these indicators for cobalt deficiency.

GARTON (Scotland). The answer, in short, is I don't know. I think you would perhaps agree with me that serum B_{12} is a prime index which one cannot ignore. As Judson indicates in the paper which follows, serum methylmalonic acid may be a better indicator of B_{12} deficiency than methylmalonic acid in urine. When it comes to PC/PE ratios, this will clearly depend on taking liver biopsy samples and this may not be a practicable thing to do.

SUTTLE (Scotland). Did you observe any loss of appetite in your cobalt deficient ewes?

GARTON (Scotland). Yes we did. We observed loss of appetite in group A ewes only after parturition. Group B, which as you recall, had been fed on the diet for 16 weeks before they became pregnant began to lose their appetite a few weeks before parturition and were consuming only about a kilo of the diet each per day.

SUTTLE (Scotland). I think it is important to realise the possible dangers of studying lipid metabolism in such animals if you haven't got pair-feed controls.

GARTON (Scotland). I appreciate your point and some work is going on now with pair-fed animals. On the other hand, one has got to remember that the B_{12} deficiency 'syndrome' in sheep is essentially bound up with loss of appetite.

HAYWOOD (England). You showed quite extensive pathological changes in the brain of these lambs. I am interested to know how these correlated with the functional disturbance. I am sure there must have been some neurological clinical disturbance manifested in these lambs which should have diagnostic implications. Can you comment on this?

GARTON (Scotland). Yes, the lambs which were born alive to group B ewes were very weakly and they could not stand up at all readily.

HAYWOOD (England). Is the demyelination as extensive as one sees it in swayback?

GARTON (Scotland). Muscular incoordination in adult B_{12}-deficient sheep is not a feature which we have observed, except in extreme conditions of prolonged deficiency.

LEWIS (England). Did the animals in group B that were off their appetite in late pregnancy show ketosis at all?

GARTON (Scotland). No overt signs were apparent.

COBALT DEFICIENCY IN SHEEP

G.J. Judson[1], R.J. Hannam[2], T.H. Benson[1] and D.J. Reuter[2]

1. Institute of Medical and Veterinary Science, FROME ROAD, ADELAIDE. S.A. 5000, AUSTRALIA
2. Department of Agriculture, BOX 1671 GPO, ADELAIDE. S.A. 5001, AUSTRALIA.

Responses in liveweight and fleece growth to cobalt pellet therapy were examined in weaner sheep depastured in the south-east of South Australia on soils known to produce cobalt deficiency in sheep (5). Three trials were conducted over a period of two years. Each trial commenced in spring with weaner sheep off-shears and was terminated at shearing the following spring. In these and other trials the usefulness of serum vitamin B_{12} and methylmalonate in serum and urine as indicators of cobalt deficiency in weaner sheep was also studied.

Increases in liveweight to cobalt therapy (10 g cobalt pellet/animal) were recorded in all trials and the responses were apparent in the spring-early summer period and again in the winter-spring period of the following year (see Table 1). The increase in liveweight gain to cobalt therapy usually coincided with increased fleece growth rates as measured by dye-banding (2), except in the winter-spring period when responses in fleece growth rates preceeded liveweight responses by at least six weeks. The increased fleece growth rate in the cobalt-treated sheep resulted in a significant increase in total fleece weight at shearing (Table 1). A compensatory liveweight response in the untreated sheep following cobalt inadequacy was detected over late summer in both years on Farm 2 (Table 1) but this response was not associated with an increased response in fleece growth rate. The compensatory effect was associated with raised serum vitamin B_{12} concentrations (Table 1) suggesting increased cobalt intake, probably from soil contaminated pasture (1).

In most instances gains in liveweight and fleece growth rates were depressed in untreated sheep when serum vitamin B_{12} concentrations were less than about 0.6 ug/l. Concentrations of vitamin B_{12} less than 0.5 ug/l are usually indicative of cobalt inadequacy in sheep (3). However, in detecting marginal vitamin B_{12} deficiency it is preferable to measure a substrate such as methylmalonate which is dependent on vitamin B_{12} adequacy for its utilization by the tissues of the animal (4).

In this study serum and urinary methylmalonate was measured and preliminary findings showed that raised serum methylmalonate concentrations were only detected in sheep with serum vitamin B_{12} concentrations of less than 0.3 ug/l (Figure 1). Although urinary methylmalonate concentrations were consistently elevated in sheep with serum vitamin B_{12} values of less than 0.3 ug/l, elevated values were also recorded in sheep with serum vitamin B_{12} values up to 1 ug/l. In our view serum methylmalonate is a more reliable indicator of cobalt deficiency in sheep than urinary methylmalonate and is probably more reliable than serum vitamin B_{12} as an indicator of marginal vitamin B_{12} deficiency.

This work was supported by a grant from the Wool Research Trust Fund on the recommendation of the Australian Wool Corporation. We are grateful to Messrs C Emery and V. Gepp, owners of the properties, for providing sheep and facilities. J.D. McFarlane, L.D. McLaren and M. Jordan provided competent technical assistance.

1. Andrews, E.D., B.J. Stephenson, J.P. Anderson and W.C. Faithful, 1958, N.Z. J. Agric. Res. 1: 125 - 139.

2. Chapman, R.E. and J.L. Wheeler, 1963, Aust. J. Sci. 26: 53 -54.

3. Gardiner, M.R., 1977, Dep. Agric. West. Aust., Tech. Bull. No. 36

4. Gawthorne, J.M., 1968, Aust. J. Biol. Sci. 21: 789 - 794.

5. Lee, H.J., 1951, J. Dep. Agric. S. Aust. 54: 475 - 532.

Table 1. Effect of cobalt therapy on mean fleece growth rate, liveweight change and serum vitamin B_{12} concentrations of Merino weaners+

	FARM 1 (1974-75)								Fleece, kg/sheep	
	Weeks: 0-5	5-15	15-22	22-29	29-37	37-44	44-54++	0-49	Greasy	Clean
Clean Fleece, g/d/sheep										
NIL (20)		8.8	11.9	14.5	10.5	8.8	5.1		6.1	4.5
COBALT (20)		9.3	12.6[a]	15.6[b]	11.9[a]	10.6[c]	7.6[b]		6.8[b]	4.9[a]
Liveweight Change, kg										
NIL (20)	1.9	7.0	3.3	2.2	4.5	0.9	1.3	21.2		
COBALT (20)	2.1	7.5	2.9	1.9	5.4	3.7[b]	5.3[c]	28.7[c]		
Serum B_{12}, ug/l										
NIL	2.0	0.5	1.1	0.4	0.3					
COBALT	7.2[c]	1.9[c]	1.9[c]	1.4[c]	0.5[c]					
	FARM 2 (1974-75)									
	Weeks: 0-7	7-17	17-24	24-31	31-39	39-46	46-51	0-51		
Clean Fleece, g/d/sheep										
NIL (20)		8.2	8.3	9.0	8.9	9.2	6.2		4.4	3.3
COBALT (20)		9.0[b]	9.0	9.7	10.1[a]	10.7[b]	7.8[a]		5.1[a]	3.7[a]
Liveweight Change, kg										
NIL (20)	2.3	2.6	1.4	0.0	1.4	1.8	-0.8	8.9		
COBALT (20)	3.6[a]	4.1[a]	0.3[a]	0.2	1.0	4.3[c]	2.5[c]	16.0[c]		
Serum B_{12}, ug/l										
NIL	0.4	2.0	1.3		0.4	0.2	0.2			
COBALT	4.7[c]	4.2[c]	3.0[c]		3.5[c]	3.5[c]	3.3[c]			
	FARM 2 (1975-76)									
	Weeks: 0-9	9-16	16-22	22-31	31-40	40-48	48-57	0-57		
Clean Fleece, g/d/sheep										
NIL (18)			8.5	8.8	11.4	10.9	8.3		5.3	3.8
COBALT (18)			10.0[a]	10.1[b]	12.5	12.2[a]	9.4[a]		6.2[a]	4.5[b]
Liveweight Change, kg										
NIL (18)	2.5	1.9	0.8	2.4	5.8	4.2	0.2	17.8		
COBALT (18)	5.5[c]	4.0[c]	-0.2[b]	2.0	5.1	4.2	3.6[c]	24.1[c]		
Serum B_{12}, ug/l										
NIL			1.4	0.6		0.4	0.6			
COBALT			2.7[c]	1.7[c]		2.1[c]	5.2[c]			

\+ Treatments commenced in October 1974 and in September 1975. Initial mean liveweights (and serum vitamin B_{12}, ug/l) were 25.9 kg (5.1) for Farm 1, 26.5 kg (0.7) for Farm 2, 1974 and 26.2 kg(0.4) for Farm 2, 1975. Mean values of fleece weight and of serum vitamin B_{12} were compared by Student's 't' test and mean values of liveweight change were compared by Paired 't' test. Letters indicate where means differ at P=0.05(a) at P=0.01(b) and at P=0.001(c). In brackets the no. weaners per group. Serum vitamin B_{12} was usually measured in all sheep in each treatment at the end of each interval.

++ Liveweight change measured from 44 to 49 weeks.

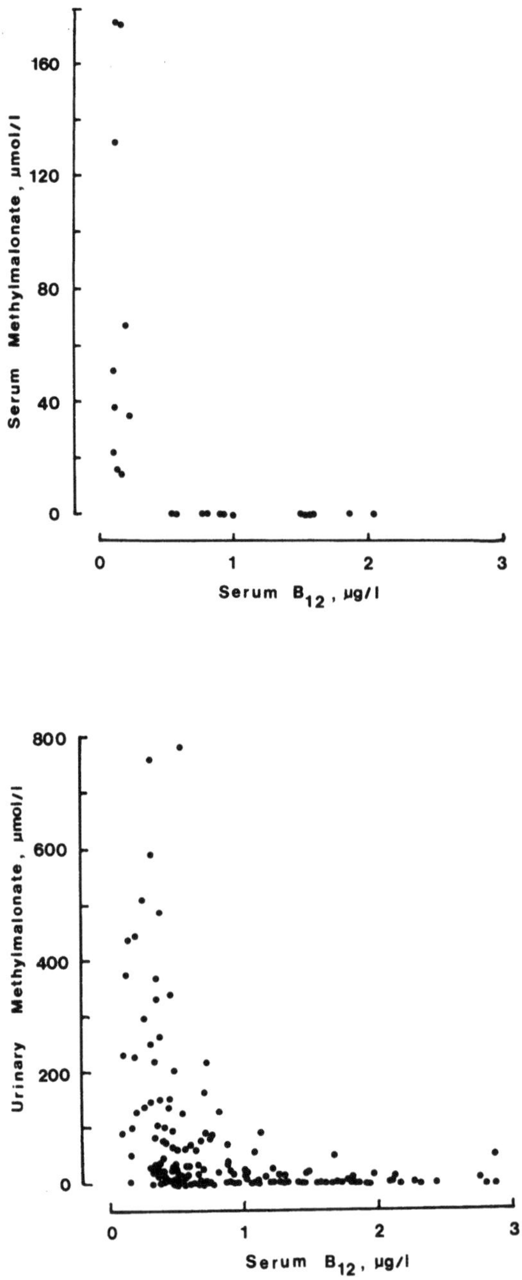

Figure 1. Relationship between serum vitamin B_{12} and methylmalonate in serum and urine of sheep.

DISCUSSION

PETERSON (Australia). Would you comment on the relative ease of doing a serum methylmalonate compared to doing a vitamin B_{12} estimation.

JUDSON (Australia). The collection of the sample is a lot easier; there have been quite amusing stories of people out in the field trying to collect a urine sample from stock. As far as doing the assay, the initial preparation is very similar for urine as opposed to serum. The technique we have been using is an acetone precipitation of protein from the serum, then derivatising the MMA with boron trifluoride in butanol and then applying that to the gas liquid chromatography procedure. Concerning serum B_{12} levels as an indicator of cobalt inadequacy in ruminants, one should be careful with using radio assay procedures as there has been a lot of discussion in the literature about vitamin B_{12} analogues in serum which might give false high values.

GARTON (Scotland). Did you observe decarboxylation of the MMA derivative on the GLC column?

JUDSON (Australia). The person who did the assays is not with us, but he was quite happy with the technique.

GORDON (Australia). Dr Garton mentioned in his paper, that some of his sheep had shed some of their fleece. Have you observed the same thing, because B_{12} deficiency was suggested as a cause of some lambs shedding wool in our institute at Prospect, NSW.

REUTER (Australia). I don't think this was seen in our animals.

I wonder would some aspect of the sulphur pathway be a valuable index for B_{12}; would Dr Garton or one of his colleagues, like to comment?

HESKETH (Scotland). Australian work has shown that urinary formiminoglutamic acid might well be a useful indicator of B_{12} deficiency. We have done a little work on it at the Rowett and it does not appear to be anymore useful than MMA at the moment, although our results are only preliminary.

ANON. Relating the MMA in urine to creatinine excreted might increase the sensitivity of the assay.

ALLEN (UK). Perhaps one of the two previous speakers could explain why if the metabolic blockage is between methylmalonyl CoA and Succinyl CoA why the equilibrium appears to move to the odd-numbered fatty acids rather than the branched chain fatty acids; perhaps you have a simple kinetic explanation.

GARTON (Scotland). No, I haven't. It would appear that whereas propionate can be converted to its CoA ester (and so become available for utilization as a primer unit for long-chain fatty acid synthesis), the carboxylation of propionyl CoA to methylmalonyl CoA does not readily take place in the liver of foetal lambs of cobalt-deficient ewes. We hope to investigate this aspect of foetal metabolism in future studies.

LEAD DEFICIENCY AND ITS EFFECTS ON GROWTH AND METABOLISM

M. Kirchgeßner and Anna M. Reichlmayr-Lais

Institut für Ernährungsphysiologie
Technische Universität München
D-8050 FREISING-WEIHENSTEPHAN
FED. REP. GERMANY

In model studies with rats (Sprague Dawley) the essentiality of the element lead for the organism was established. In these experiments purified, semisynthetic basic diets were used with lead concentrations of 45 ppb or 18 ppb. The control diet was supplemented with 1 ppm lead as lead acetate (suprapure). For a sufficient depletion it was necessary to conduct the experiments over more generations. The rats were kept in metalfree cages. The animal room was maintained at constant temperature (23°C) and relative humidity (55 %). Purification and composition of the diets, experimental designs, and treatment and breeding of the rats were described earlier (4, 5).

First physiological changes resulting from lead depletion could be ascertained in young rats of the f_1-generation, whose mothers were provided with a depletion diet having a concentration of 45 ppb lead (4). On the day of weaning these young rats of lead-depleted mothers showed reduced values of hematocrit (-15 %), hemoglobin (- 11%) and mean corpuscular cell volume (- 9 %) in comparison with control mothers.

As the result of a far lower lead concentration in the diet at the level of 18 ppb, more severe hematological changes occurred (6). In this experiment lead-depleted mothers already showed a trend to decreased values for hematocrit and mean corpuscular cell volume in comparison with control rats. Massive hematological changes occurred in rats of lead-depleted mothers from the f_1-generation. For illustration the changed values in rats from the f_1-generation, three days after weaning, are presented in table 1. After weaning, continued depletion till the 38th day of life made the anemia even more severe.

TABLE 1. Average blood values of 24-day old rats (f_1-generation), experiment with 18 ppb lead in the diet

Group	Erythrocytes ($10^6/\mu l$)	Hematocrit (%)	Hb (g/100 ml)	MCV (μ^3)	MCH (pg)	MCHC (%)
Depletion	5.1[a] ±0.3	27.8[a] ± 1.5	7.0[a] ±0.7	54[a] ± 2	14[a] ± 1	25[a] ± 1
Control	4.9[a] ±0.5	35.5[b] ± 4.0	9.3[b] ±0.9	70[b] ± 4	19[b] ± 1	26[a] ± 1

In addition to these symptoms of anemia occurring in this experiment over generations with 18 ppb lead in the depletion diet, growth depression could be induced for the first time (5). In comparison with young rats of normally lead-supplied mothers, young rats of lead-depleted mothers showed decreased growth during the whole time of nursing. On the day of weaning, the average weight of the young rats of depleted mothers was 39 g and that of young rats of control mothers 43 g. Therefore, the weight difference between these groups was about 11 % ($p < 0.025$). This weight difference could be enlarged by continued depletion after weaning till the 38th day of life. In the period from the day of weaning to the 38th day of life, the average daily growth was 3.9 g in the depletion group and 5.0 g in the control group. This means a depression of growth by about 22 % ($p < 0.001$). On the 38th day of life the average weight of lead-depleted rats was 102 g and that of control rats 126 g.

The hematological changes in lead deficiency, which could be characterized as a microcytic hypochromic anemia in their distinct form indicated disturbances in iron metabolism as the result of lead depletion. Therefore, iron concentrations in serum, liver, and spleen and the total iron-binding capacity in serum (1, 7) were determined, as well as the absorbability and metabolic efficiency of iron in lead deficiency (2). For illustration, the results of 38-day old rats of the f_1-generation from the experiment with 18 ppb lead in the diet are presented in table 2. In comparison with control rats the iron-binding capacity in serum of lead-depleted rats was increased. The causes for these changes and for the anemia are probably a reduced iron absorption in lead deficiency (Table 2).

TABLE 2. Criteria of Fe metabolism determined in 38-day old rats of the f_1-generation (experiment with 18 ppb lead in the diet)

Criteria of Fe metabolism		Depletion	Control	p <
Serum Fe	(μg/ml)	0.6 ± 0.3	3.1 ± 0.6	0.001
IBC	(μg/ml)	7.4 ± 0.5	5.0 ± 0.6	0.001
Fe in liver	(μg/g FW)	30 ± 3	177 ± 28	0.001
Fe in spleen	(μg/g FW)	103 ± 15	147 ± 22	0.001
Absorbability	(%)	24	53	0.001
Metabolic efficiency	(%)	93	78	0.001

In further experiments, in which growth depressions, hematological changes and disturbances in iron metabolism could be repeatedly reproduced, it was found that repletion of young rats from lead-depleted mothers either by having them nursed by a control mother or by weaning them with a lead-supplemented diet led to a normalization of growth (5), hematological abnormalities (6) and of the changes in iron status (1). For example, in table 3 the normalization of blood values in rats repleted with lead starting from the day of birth is demonstrated.

TABLE 3. Average blood values of 30-day old depleted and repleted rats in comparison with control rats

Group	Erythrocytes ($10^6/\mu l$)	Hematocrit (%)	Hb (g/100 ml)	MCV (μ^3)	MCH (pg)	MCHC (%)
Depletion	5.5^a ±0.2	36.6^a ± 2.2	10.4^a ± 0.6	67^a ± 2	18^a ± 1	28^a ± 1
Repletion	5.9^a ±0.5	41.3^b ± 1.7	11.7^b ± 0.4	71^b ± 3	20^b ± 1	28^a ± 0
Control	5.6^a ±0.2	42.9^c ± 1.2	12.3^c ± 0.3	77^c ± 2	22^b ± 1	29^a ± 0

Further physiological changes in lead deficiency were indicated by altered activities of different enzymes (8, 9) and concentrations of diverse metabolites (3). Table 4 demonstrates these changes in the case of 38-day old rats from the f_1-generation of the experiment with 18 ppb lead in the diet.

Overall, the results showed that the lead depletion induced reproduceably growth depressions and further physiological abnormalities. Addition of lead abolished or prevented these abnormalities. Therefore, it could be derived that lead is an essential trace element.

TABLE 4. Enzyme activities and metabolite concentrations in selected organs of lead-depleted 38-day old rats in comparison with control rats (FW = fresh weight)

Enzymes or Metabolites		Organ	Depletion	Control	p <
Alkaline phosphatase	(U/g protein)	Liver	2.9 ± 0.6	1.5 ± 1.5	0.001
		Femur	271.5 ± 71.7	344.5 ± 49.5	0.025
GOT	(U/g protein)	Liver	106.0 ± 33.3	160.9 ± 34.1	0.005
GPT	(U/g protein)	Liver	29.7 ± 6.8	60.2 ± 8.4	0.001
G-6-P-DH	(U/g protein)	Liver	90.4 ± 5.1	101.0 ± 28.4	-
LDH	(U/g protein)	Liver	1756.6 ± 306.8	1806.0 ± 389.6	-
MDH	(U/g protein)	Liver	3618.6 ± 419.8	3446.0 ± 750.6	-
Coeruloplasmin	(E/min ml)	Serum	0.123 ± 0.023	0.097 ± 0.026	0.005
Catalase	(E/ml)	Blood	394.0 ± 72.0	618.0 ± 60.0	0.001
Protein	(g/100 ml)	Serum	5.7 ± 0.2	5.8 ± 0.2	-
	(mg/g FW)	Liver	162.6 ± 8.9	159.4 ± 9.8	-
	(mg/g FW)	Femur	51.1 ± 5.0	44.3 ± 5.5	0.01
Glucose	(mg/g protein)	Liver	86.8 ± 29.0	149.6 ± 15.0	0.001
Triglycerides	(mg/g protein)	Liver	65.7 ± 10.4	135.0 ± 16.8	0.001
Cholesterol	(mg/g protein)	Liver	30.6 ± 3.5	25.3 ± 2.0	0.001
LDL	(mg/g protein)	Liver	25.2 ± 4.6	35.2 ± 10.6	0.025
Phospholipids	(mg/g protein)	Liver	31.0 ± 8.2	85.3 ± 11.8	0.001

ABSTRACT

In depletion experiments over generations, the essentiality of lead for growth and metabolism could be established. Lead deficiency caused growth depressions, microcytic hypochromic anemia, and disturbances in iron metabolism, activities of enzymes and concentrations of metabolites. These physiological abnormalities could be prevented or abolished by addition of lead.

REFERENCES (in press)

1. Kirchgeßner, M., and Anna M. Reichlmayr-Lais, 1981, Changes of iron concentration and iron-binding capacity in serum resulting from alimentary lead deficiency. Biol. Trace Elem. Res.
2. Kirchgeßner, M., and Anna M. Reichlmayr-Lais, 1981, Absorbierbarkeit und intermediäre Verfügbarkeit des Eisens im alimentären Blei-Mangel. Internat. Z. Vit. Ern. Forsch.
3. Kirchgeßner, M., and Anna M. Reichlmayr-Lais, 1981, Konzentrationen verschiedener Stoffwechsel-Metabolite im experimentellen Blei-Mangel. Nutr. Metab.
4. Reichlmayr-Lais, Anna M., and M. Kirchgeßner, 1981, Depletionsstudien zur Essentialität von Blei an wachsenden Ratten. Arch. Tierernährg.
5. Reichlmayr-Lais, Anna M., and M. Kirchgeßner, 1981, Zur Essentialität von Blei für Wachstum und Stoffwechsel. Z. Tierphysiol., Tierernährg. u. Futtermittelkde.
6. Reichlmayr-Lais, Anna M., and M. Kirchgeßner, 1981, Hämatologische Veränderungen im alimentären Blei-Mangel. Nutr. Metab.
7. Reichlmayr-Lais, Anna M., and M. Kirchgeßner, 1981, Eisen-, Kupfer- und Zinkgehalte in Neugeborenen sowie in Leber und Milz wachsender Ratten bei alimentärem Blei-Mangel. Z. Tierphysiol., Tierernährg. u. Futtermittelkde.
8. Reichlmayr-Lais, Anna M., and M. Kirchgeßner, 1981, Katalase- und Coeruloplasmin-Aktivität im Blut bzw. Serum von Ratten bei Blei-Mangel. Zbl. Vet. Med. A.
9. Reichlmayr-Lais, Anna M., and M. Kirchgeßner, 1981, Aktivitäts-Veränderungen verschiedener Enzyme im alimentären Blei-Mangel. Z. Tierphysiol., Tierernährg. u. Futtermittelkde.

DISCUSSION

SABBIONI (Italy). What was the iron concentration in the diets?

REICHLMAYR-LAIS (Germany). It was 50ppm.

MACK (Australia). The anaemia and some other signs are similar to those in copper deficiency. Is there any interference with copper metabolism during lead deficiency?

KIRCHGESSNER (Germany). The changes in iron metabolism induced by copper deficiency are different from those induced by lead deficiency. In copper deficiency the iron concentration in liver increases and the ceruloplasmin oxidase activity decreases. In lead deficiency the opposite occurs.

SALTMAN (USA). I would like to comment on the defect in iron metabolism in copper deficient animals. We published data that shows that iron is absorbed normally in copper deficient mice and rats, but it is taken up by the liver and is not further mobilised for incorporation into red blood cells.

THE METABOLIC LESION IN IRON DEFICIENT MUSCLE

P. Saltman, Y. Ohira, J. Hegenauer, C. S. Chen, L. Strause
Department of Biology, University of California San Diego
H. Beinert
Institute for Enzyme Research, University of Wisconsin, Madison

There is a well documented relationship between iron deficiency and the ability to perform muscular work. This evidence has been accumulated not only in iron-deficient human populations (1), but also in experimental animals (2,3). It has been assumed that the cause of fatigue and cardiac stress during work performance could be directly related to the lower concentration of the oxygen binding proteins, hemoglobin and myoglobin, in blood and muscle respectively. Recently this assumption has been called into question by Edgerton, Finch, and others (2,4-10). Finch et al. (7) were able to demonstrate that iron-deficient animals when transfused with normal blood could still not work at the anticipated level. This and other investigations have led to the proposal that non-hemoglobin related effects are important in work performance.

In our experiments, iron deficiency was produced in weanling, Sprague-Dawley rats by feeding a nutritionally complete liquid milk diet containing less than 0.3μM Fe. The control diet contained 100mM as $FeCl_2$. After five weeks there was significant reduction of hemoglobin concentration and poor work performance on a treadmill test where animals were run to exhaustion (Table 1).

Table 1. Body weight, hemoglobin, and treadmill performance of rats 4 days after iron supplementation.

Iron Status	Weight (gm)	Hemoglobin (g/dl)	Run Time (min)
Normal	192.4 ± 12.2	13.7 ± 0.4	21.8 ± 4.7
Anemic	192.4 ± 17.3	8.9 ± 1.1	17.3 ± 2.6
Anemic + Fe	188.3 ± 16.0	12.2 ± 0.9	21.3 ± 3.9

Iron in various muscle tissues and organs was assayed by electrothermal atomic absorption. Hemoglobin was quantified by the cyanmethemoglobin method. Electron paramagnetic resonance (EPR) of whole tissue was measured and quantitated by signal amplitude. Succinic dehydrogenase (SDH) and cytochrome oxidase were measured (11). There was a significant depression not only of hemoglobin in the iron deficient animals, but also in the concentrations of Fe in various organs (liver, spleen, and kidney) and in some muscles including plantaris and vastus lateralis. Heart and red muscle retained Fe at control levels. Red muscle from iron-deficient animals showed a reduced SDH activity (Table 2). No change in cytochrome oxidase activity could be observed in any

Table 2. Levels of succinic dehydrogenase (SDH) and cytochrome oxidase (cyto ox) in various muscles.

		Heart	Gastrocnemius	Red Vastus Lateralis
Normal	SDH	3031 ± 222	2456 ± 589	2088 ± 59
	cyto ox	14.9 ± 7.4	6.0 ± 3.2	13.6 ± 2.6
Anemic	SDH	1669 ± 299	1263 ± 114	1726 ± 674
	cyto ox	15.4 ± 1.5	5.7 ± 2.4	13.4 ± 1.4
Anemic + Fe	SDH	1917 ± 373	1370 ± 467	1472 ± 153
	cyto ox	16.9 ± 2.8	6.0 ± 1.8	10.9 ± 3.8

tissue but heart. In iron-deficient tissues, there was a decrease in the signal from the EPR spectrum at g=1.92, characteristic of the non-heme-iron electron transport proteins NADH dehydrogenase and SDH. This decrease was observed only in muscles and heart but not in other tissues and organs (Figure 1, Table 3).

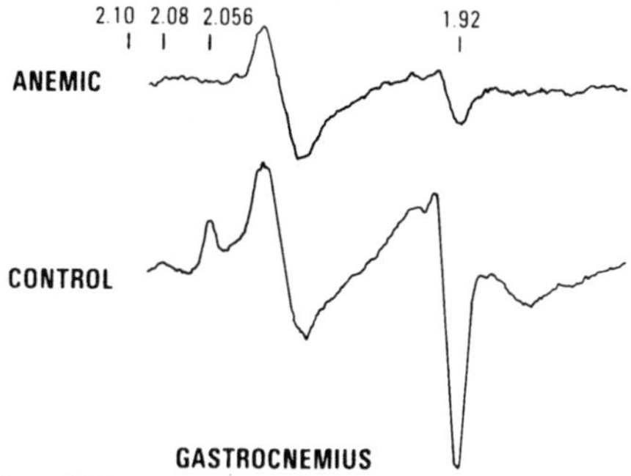

Figure 1. EPR spectrum for anemic and normal rats.

Table 3. EPR signal at g = 1.92 from iron-deficient animals. Values are reported as percent of normal.

Soleus	34
Gastrocnemius	52
Heart	78
All other tissue	100

After five weeks on the deficient diet, half of the anemic animals were fed the complete milk diet containing 100mM $FeCl_2$. A few animals were injected interparitoneally (I.P.) with iron-dextran to compare injected versus oral iron administration. At 4, 10, and 14 days following supplementation, animals were tested for work performance, and several other parameters related to their iron status were measured. Within 4 days following either oral or I.P. supplementation, run times of supplemented animals were comparable to controls. Hemoglobin had significantly increased (Table 1). Neither EPR signal nor SDH values were restored. The EPR method did reveal shifts in the various forms of catalase and increased manganese (superoxide dismutase?) in liver of deficient animals.

These experiments demonstrate unequivocally that iron deficiency can lead directly to the loss of a crucial enzyme system involved in the production of ATP in muscle tissue. Thus, there is not only an impairment of the amount of oxygen available to tissues in iron-deficiency anemia, but also a metabolic lesion induced by the loss of the essential non-heme-iron oxidative enzymes in muscle mitochondria. However, the slow rate of recovery of the EPR signal and SDH activity does not correlate with the rapid return of work performance. A much closer correlation is seen with the rapid rise of hemoglobin when supplemented iron is administered. These results suggest that SDH and NADH dehydrogenase (EPR signal) activity in muscle is not closely correlated with work performance in rats. These enzymes do not appear to be likely candidates for explaining the non-hemoglobin related effects in work performance (7).

REFERENCES

1. Ohira, Y., V. R. Edgerton, G. W. Gardner, B. Senewiratne, B. J. Barnard, & D. R. Simpson (1979). Work Capacity, Heart Rate and Blood Lactate responses to iron treatment. Brit. J. Haematol. 41: 365-372.

2. Edgerton, V. R., S. L. Bryant, C. A. Gillespie, & G. W. Gardner (1972). Iron deficiency anemia and physical performance and activity of rats. J. Nutr. 102: 381-399.

3. Ohira, Y., B. J. Koziol, V. R. Edgerton, & G. A. Brooks (1981). Oxygen consumption and work capacity in iron-deficient anemic rats. J. Nutr. 111: 17-25.

4. Cusack, R. P. & H. C. Schwartz (1965). Iron deficiency in rats: Changes in body and organ weights, plasma proteins, hemoglobins, myoglobins, and catalase. J. Nutr. 86: 383-393.

5. Dallman, P. R. & H. C. Schwartz (1965). Distribution of cytochrome c and myoglobin in rats with dietary iron deficiency. Pediatrics 35: 677-686.

6. Dallman, P. R. & H. C. Schwartz (1965). Myoglobin and cytochrome response during repair of iron deficiency in the rat. J. Clin. Invest. 44: 1631-1638.

7. Finch, C. A., L. R. Miller, A. R. Inamdar, R. Person, K. Deiler, & B. Mackler (1976). Iron deficiency in the rat. Physiological and biochemical studies of muscle dysfunction. J. Clin. Invest. 58: 447-453.

8. Dallman, P. R. (1969). Iron restriction in the nursing rat: Early effects upon tissue heme proteins, hemoglobin, and liver iron. J. Nutr. 97: 475-480.

9. Finch, C. A., P. D. Gollnick, M. P. Hlastala, L. R. Miller, E. Dillman, & B. Mackler (1979). Lactic acidosis as a result of iron deficiency. J. Clin. Invest. 64: 129-137.

10. Koziol, B. J., Y. Ohira, D. R. Simpson, & V. R. Edgerton (1978). Biochemical skeletal Muscle and hematological profiles of moderate and severely iron deficient and anemic adult rats. J. Nutr. 108: 1306-1314.

11. Hatefi, Y. (1978). Resolution of complex II and isolation of succinic dehydrogenase. In Methods of Enzymology, Vol. LIII (S. Fleisher & L. Packer, eds.) Academic Press, N. Y., 27-35.

(Supported by NIH grant AM-12386 and by grants from the National Dairy Council and the Nutrition Foundation).

DISCUSSION

SUNDE (Scotland). Have you tried giving the iron deficient rats additional red blood cells to see if in fact they will run for longer times and have you tried haemolytic drug administration to see if that will cause impaired performance on the treadmill?

SALTMAN (USA). We have not used the haemolytic drugs. We have tried to do transfusions repeating some of Finch's work. We were unable to do so. We find that it is very traumatic for the mouse or rat. I am impressed how quickly you can bring them back with oral iron without changing their mitochondrial respiratory systems. The haemolytic experiment should be done, but we have not done it yet.

MILLS (Scotland). Are there any observations which confirm previous indications that at mild degrees of anaemia it is voluntary work performance rather than imposed work performance that suffers?

SALTMAN (USA). I know of Dr Ohira and Dr Edgerton's work. We have not repeated their experiments but we currently have 45 women in a detailed work performance study that involves different levels of dietary iron.

HAIR ANALYSIS FOR THE DETECTION OF CHROMIUM DEFICIENCY

D. Shapcott, J.S. Vobecky, P.P. Demers, J. Vobecky and M. Côté
Départements de Pédiatrie, de Santé Communautaire et de Médecine,
Faculté de médecine, Université de Sherbrooke,
Sherbrooke, Québec, J1H 5N4

Although chromium is now accepted to be an essential trace element for human metabolism (1) there is as yet no satisfactory method to define chromium deficiency except indirectly by the improvement in a physiological parameter following supplementation with the metal,for example the improvement for glucose tolerance in malnourished infants and in adults with impaired glucose tolerance. Serum analysis has been employed but is fraught with difficulty due to the extremely lowlevels which are at about the detection limit for most techniques,and so measurement is restricted to a few laboratories with the necessary equipment and expertise. In the past decade there has been a decrease of two orders of magnitude in the reported normal serum Cr levels (2). Furthermore, at least part of the circulating chromium appears to be present in a non-metabolically active form (3) and the stringent precautions necessary to avoid contamination pose further difficulties. The same difficulties occur with urine analysis, and in general, although recently some few investigations have arrived at concordant normal values using different techniques there is far from general agreement as to the normal serum or urine chromium levels, nor as to their variation with disease or deficiency.

Hair has been used as an analytical medium historically for toxicological purposes but more recently in the evaluation of trace metal deficiency (4). Hair Cr level (HCL) offers the following advantages.

(a) Collection is not an invasive procedure, a factor of importance on studies with children. (b) The hair chromium content is 2-3 orders of magnitude greater than the serum level making analysis much less difficult. (c)It is not as influenced by short term changes in diet or serum glucose levels as is serum chromium. However, hair analysis has several disadvantages: (a) It is not suitable for the diagnosis of acute changes in chromium status. (b) It is evidently subject to environmental contamination. (c) Hair growth is influenced by other factors such as zinc nutrition.

In practice, environmental contamination of hair has not proven to be a problem in our laboratory except in subjects exposed to chromium in industry.

In contradistinction to serum or urine analysis there is a remarkable degree of agreement between individual investigators as to the HCL in normal adults, particularly in the U.S. and Canada.

HAIR CHROMIUM LEVELS IN HEALTHY ADULTS

INVESTIGATOR	MEAN HAIR CR (NG/G)		MEAN HAIR CR (NG/G)
HAMBIDGE (U.S.)	750 (F)	ASHURBEKOV (USSR)	1030 (M)
HAMBIDGE (U.S.)	540 (M)		800 (F)
OBRUSNIK (U.S.)	480 (F)	MAHALKO (THAI)	309 (F)
BURT (U.S.)	570 (F)	GURSON (TURKEY)	240 (M)
RABINOWITZ (U.S.)	770 (M)		221 (F)
VOBECKY (CANADA)	499 (F)	ROSSON (UK)	260 (F)
SHAPCOTT (CANADA)	567 (F)		250 (M)
SHAPCOTT (CANADA)	490 (M)		
SHAPCOTT (CANADA)	458 (M)		
COTE (CANADA)	635 (M)		

The significance of the differences found between samples from men and women is unclear and more studies are needed to clarify this point which could be important if there is a consistant difference between males and females. The low levels from Turkey are of interest in view of the report that tissue chromium levels are higher in subjects from the Mid-East (5). It remains to be seen if these are purely due to dietary intake or if they represent a different distribution of the metal in the various body pools.

Most investigators have found significant differences in HCL between control subjects and those with diabetes, the diabetic subjects having lower levels. This is also true for pregnancy, the reported HCL being about fifty per cent of those in non-pregnant women. There is less information concerning infants and children but it appears that the level is highest at birth and decreases to a plateau during early childhood in North American children. Several studies on Turkish subjects both children and adults have shown lower HCL in normal subjects (6).

It is obvious that comparisons cannot be made directly between the values reported from different countries or even different investigators and test subjects must always be compared with their appropriate references values.

Those reporting HCL usually give a wide range of values both in normal and deficient populations with the highest level being ten or more times the lowest level. This makes it difficult to assess single samples and makes statistical analysis difficult. For this reason it is preferable to use a simple descriptive proceedure such as percentile since this may show significant differences.

Although clear cut differences have been demonstrated between controls and various test populations it is often difficult to place an individual in either category on the basis of a HCL. We have recently completed a longitudinal study where hair chromium was measured in a group of infants from 1 to 3 years of age.

HAIR CHROMIUM - VARIATION WITH AGE

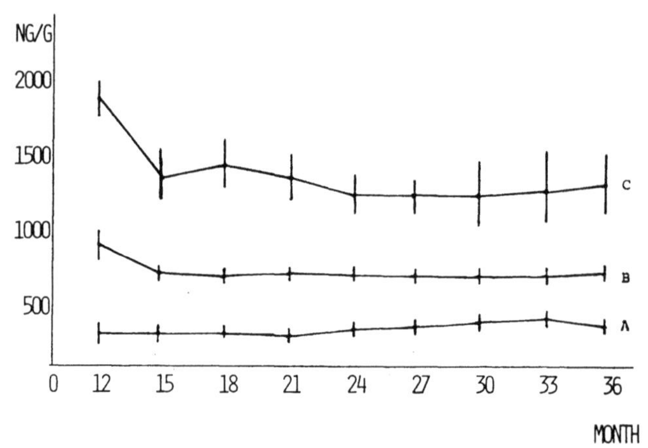

FIGURE

(A) Infants in the 10th percentile at 12 months.

(B) All infants.

(C) Infants in 90th percentile at 12 months.

As can be seen, there was a marked tendency for those infants whose first sample was in the 10 percentile to have subsequent values also in the 10th percentile, in fact a 66% possibility. There was no obvious difference in dietary intake to account for this and we assume that either these children were deficient at birth and unable to gain normal levels or that their metabolism of chromium was defective.

We conclude that given an adequate reference population hair chromium analysis can be used as an indicator of relative chromium deficiency.

REFERENCES:

(1) Mertz W. 1979, in "Chromium in Nutrition and Metabolism" (Eds D. Shapcott and J. Hubert) (Elsevier N.Y.) pp. 1-14.

(2) Hubert J. 1979, in "Chromium in Nutrition and Metabolism" (Eds D. Shapcott and J. Hubert) (Elsevier N.Y.) pp 15-30.

(3) Shapcott D. et al, 1977, Clin. Biochem. 10: 178-180.

(4) Hopps H.C., 1976, Sci. Total Environ. 7: 71-89.

(5) Schroeder J.A. et al 1970, J. Chron. Dis. 23: 123-142.

(6) Saner G., 1980, in "Chromium in Nutrition and Disease" (ed. G. Saner),(A.R. Liss. N.Y.), pp. 54-61.

DISCUSSION

GARNYS (Australia). Have you noticed that chromium varies over the length of the hair and if it does, have you considered using various proximal or distal parts of the hair to try to get better correlation between individuals and within individuals.

SHAPCOTT (Canada). Chromium does vary along the length of the hair with the dietary intake and it does not seem to be due to an uptake of chromium from the atmosphere since washing has no effect.

MINERAL IMBALANCE AND BORON DEFICIENCY

Rex E. Newnham
Bor-rex Pty. Ltd.
PO BOX 61, THORNBURY VICTORIA 3071
AUSTRALIA

SUMMARY

After having taught biology, chemistry, geology, soil science and botany for many years circumstances led me to associate a boron deficient diet with my own arthritis. A dietary supplement containing boron soon corrected the arthritis. Over the last three years over 25,000 people have taken 70,000 bottles of boron and magnesium tablets and about 90% of these have lost all or most of their arthritic troubles. Dogs, cattle and horses have also got better and have stopped limping. An initial analysis has shown that arthritic bones contain less boron than normal bone. Toxicity problems are discussed.

INTRODUCTION

REVIEW OF LITERATURE

Boron is known to be essential for growth and cell division in plants. Its exact role is still uncertain but the latest theory is that it is involved in RNA/DNA metabolism. If correct, this would also apply to men and animals.(7)

Most of the boron research has been to test its toxicity which is described as only mild.(1) Carl C. Pfeiffer claims that we must guard against the elimination of boron in the human diet.(6) Mommsen claims that the variable concentration of boron in different bodily organs points to the fact that boron exercises specific functions in each organ.(4) He also shows that its lack results in skin allergies and that boron plays a role in glycogen formation. Russian workers have shown a relationship between boron and vitamins, enzymes and adrenalin and the thyroid gland. He classes boron among the vital trace elements for humans and animals.(4) Except in very specialized instances boron has not been used recently in treating any pathology.(9) In the past boric acid and borates have been used for burn treatment, as a mild disinfectant, as an emmenagogue and as a slimming agent. As an infant pacifier it was sometimes used too freely and infant deaths have occurred. Its internal use has now almost ceased.(9)

It is not possible to remain alive on a boron free diet as it is found in all sustaining foods. Our intake of boron varies from 0.3mg to 41mg per day.(5) Frenchmen get about 36mg and Americans about 8 mg per day.(10) It is known and accepted that increasing use of superphosphate and potassic fertilizers inhibits the uptake of boron.(7)

THE BENEFITS OF BORON

Seventeen years ago I was eating some food from a boron deficient soil and then developed osteo arthritis. Medication failed to help and I sought a possible cause, and it seemed as if there could be a boron deficiency. I took a daily supplement of sodium borate and in three weeks all the symptoms had gone. A year and a half later a further maintenance dose was needed. Others were told and they improved.

Recently tablets containing magnesium carbonate and sodium borate were made up and the usage rate has been:

2 years	1,000	bottles	Average	weekly	usage	10
13 months	2,000	"	"	"	"	35
6 "	7,000	"	"	"	"	270
8 "	20,000	"	"	"	"	590
6 "	20,000	"	"	"	"	770
4 "	20,000	"	"	"	"	1180

As a food supplement no advertising claims can be made, so it is just that satisfied customers tell others. There have been no undesirable side effects but some people have reported beneficial side effects, the sort that one would expect when a limiting factor has been removed from our diet.

The magnesium carbonate was used as a filler because it was known that arthritic sheep often improved when dolomite was added to their pasture. It seems that magnesium and boron are synergistic.

Carnarvon, 1,000 km north of Perth has more boron in the soil and water than any other inhabited part of Australia. The water supply has not more than .2ppm boron, but some soils and more water in the area have up to 7ppm boron. A recent survey has shown that the incidence of musculo-skeletal diseases is less than half what it is in other parts of the country. Some people go there to enjoy the good climate so as to get rid of their arthritis but the curative factor is the water and not the climate. In 1980 a survey was conducted in Carnarvon to find out how the incidence of arthritis there compared with that in other parts of the country. One per cent of the population was known to the hospital and the local rheumatism association, but a questionnaire taken to people showed that 20% had some arthritic trouble. This agreed with a survey reported in 1975,(2) so the same questionnaire that was used in Carnarvon was used in parts of Melbourne and over 50% of adults had an arthritic complaint. Any conclusions are made difficult by the fact that some people used only packaged food and drink, not locally produced. More thorough research would be beneficial using trained interviewers. Those used were unemployed youths.

Some mineral springs have much more boron. The Ngawha Springs, north of Auckland are well recognised as having a beneficial effect on arthritis, when people soak in its muddy pools. The interesting factor is that the water has about 2000ppm of boron.

Samples of human femur heads were analysed and arthritic femur heads have about half the boron content of a healthy femur head.

5 samples (each 33, 33, 23, 20, 39) mean 29.6ppm boron in Arthritic bone.
1 sample 56ppm boron in healthy bone.

It is very difficult to get healthy human femur heads, there are no donors.

Some people have given the same supplement to their animals and dogs, cattle and horses have all recovered their health.

In some places boron compounds classed as therapeutic substances are S4 poisons (Prescription only) but as a food supplement they are not toxic provided the boron concentration is below 1 or 2%. An adult could safely take 200 of these tablets each day for a month or more and would show no signs of toxicity.(1) A tablet has to be a safe unit dose and these just taste chalky, so are not attractive to children.

People ingest from 0.3 to 41mg of boron per day.(5) Frenchmen take in about 36mg and Americans 8mg.(7) This is because of the different agricultural methods used. Many French farmers used time honoured methods of soil husbandry while American farmers pour on the superphosphate and potash fertilizers and both these inhibit the uptake of boron.(7) In Australia and New Zealand we are nearer the American than the French figures.

A possible explanation of the biochemical role of boron comes from animal research. It has been noticed that when animals are given boron as borate, in their drinking water the amount of urinary phosphate excretion is increased.(3)

It becomes obvious that arthritis and rheumatism is a disease of modern living brought on by a universal exploitation of the soil. A rich organic soil has much microflora activity and these micro organisms tend to release boron from tightly bound combinations so that crop plants can utilize it.

In order to have a healthy population we must have a foundation in a healthy soil. Meanwhile people can often regain their lost health by using the relevant food supplement, generally in the form of a tablet. Until our soil is properly rejuvenated we must get our minerals, and this is not a medication or something added to our intake, it is just replacing what has been removed by exploiting the soil, instead of good soil husbandry.

Tests have shown that both superphosphate and potash fertilizers when applied to the soil in more than minimal concentrations inhibit the plant uptake of boron, copper and zinc. Superphosphate has been used in Australia for 90 years now but since 1940 the rate of application has been greatly increased as has the use of potash fertilizers. Some farmers are now using borated superphosphate. It is possible in many parts of Australia and New Zealand to drive through hundreds of miles of countryside and to continually see boron deficient plants by the roadside, especially in the spring. But this varies from season to season. After a wet summer there are fewer boron deficiency symptoms than after a dry summer. This is because a moist warm soil has much more micro-flora activity than a dry soil, and active micro-flora do release more boron for the growing plants. This is tied up with mycorrhiza activity. Heavy applications of soluble chemical fertilizer inhibits this activity.

Farmers are using increasing amounts of borated superphosphate, but only because their crops will not grow without it. There is normally no thought of increasing the trace element content of crops so that the ultimate consumer may be healthier. Bio-dynamic farmers use compost and bacteria so as to increase the micro-flora activity of a soil, and this does release trace elements in a balanced proportion.

CONCLUSION

The trace mineral balance of our food has been upset by modern agricultural methods and this is a major contributing factor towards the increasing incidence of arthritis and rheumatism. A daily supplement of a few mg of boron in a properly balanced tablet corrects and prevents these crippling diseases.

BIBLIOGRAPHY

1. Adams, R., 1964. In "Boron, Metallo-Boron Compounds and Boranes". (Interscience) pp 700-731.

2. Bridges-Webb, C. 1973. Medical Journal of Australia 108(b) pp 719-721.

3. Green, G.H. and Weeth, H.J. 1977. Journal of Animal Science 45:4 Univ. of Nevada.

4. Mommsen, H. The Physiology and Biochm. of Boron. Tabellarium of Inst. of Biol. of Vital Substances and Nutrition. Univ. of Bremen. p 4 (trans).

5. Schlettwein-Gsell, D. and Mommsen-Straub, S. 1973. Inter. Z. Vit.-Ern.- Forschung 43. pp 93-109.

6. Pfeiffer, C.C. and Jenny, E.G. 1950. Bull. of Nat. Formulary Comm. 18:3-4. pp 57-74

7. Schutte, C. 1964. The Biology of Trace Elements. pp 58 and 186.

8. Sholnik, M. 1965. Inter. Symp. on Min. Nut. in Plants and Animals. Jena.

9. Solis-Cohen. 1928. Pharmacotherapeutics. p 583.

10. Weir, R.J. and Fisher, R.S. 1972. Toxicology and Applied Pharmacology. 23. pp 351-364.

DISCUSSION

McDOWELL (USA). What foods are highest in boron? Are animal products higher in boron than plants?

NEWNHAM (Australia). The green leaf vegetables are highest in boron, especially when grown without chemical fertilizers. Bee pollen and some honeys are also good sources. Animal products are low in boron, but colostrum contains twice as much boron as later milk.

HISTOLOGIC DEVELOPMENT OF BORON DEFICIENCY AND NEOPLASMS

W. BUSSLER

Fachgebiet Pflanzenernährung Technische Universität
Lentzeallee 55-57, 1000 Berlin 33, West Germany

To improve the diagnosis of boron deficiency in plants we examined many species supplied with different amounts of boron. We found that the cause of non-uniform symptoms that appeared later was an increased rate of cell division in meristematic tissues and a loss of differentiation.

This boron-deficient tissue can be described as neoplasmic. The cells have an uncontrolled autonomous growth with unlimited division. They were not under the arranging influence of surrounding tissues. The cells have larger and malformed nuclei, no higher organization and an increased energy for divisions, far from the basic laws for the formation of organs. The cells were starved for oxygen and mineral nutrients because they were not connected with veins. The physiological regulation of growth was completely lost.

These symptoms indicate that one function of boron in metabolism could involve the inhibition of cell division enzymes. Under boron deficiency these enzymes continue to work in the process of cell division. Also catalase and peroxidase change their activities under boron deficiency or in neoplasmic tissues in the same manner.

Mineral nutrients are not excluded as factors involved in the transition from embryonic cells to permanently specialized tissue in animals. It cannot be excluded that boron or another mineral nutrient deficiency is involved in neoplasmic growth. If boron is the minimum factor, cell division in meristematic tissues of plants is not limited but differentiation is completely stopped, as in neoplasms.

It seems improbable that many different carcinogenic substances induce the same irregularities in cell division and differentiation. It is more likely that they disturb a regulating system inside the tissue.

Tissue proliferation under nutrient stress in plants could be homologous to the formation of animal neoplasms.

TRACE ELEMENT ENVIRONMENTAL CONTAMINATION AND TOXICITY

HAIR CADMIUM AND LEAD IN LEARNING DISABLED CHILDREN

Barbara F. Harland, James P. Harwood and Robert W. Thatcher

Food and Drug Administration, Washington DC
University of Maryland, Eastern Shore, Md., USA

ABSTRACT

Several recent studies have attempted to relate nutritional status with human performance, and certain trace elements have been implicated in a number of learning disabilities. A population of school children from the Eastern Shore of Maryland has been examined who exhibited large variations in scholastic performance. Tests have been applied by the Neurometrics Unit of the University of Maryland to assess student performance and we have correlated the results with trace element levels as determined in hair samples. The 99 children, ages 5-15, black and Caucasian, male and female, were divided into 5 groups on the basis of IQ and the Wide Range Achievement Test (WRAT). Socio-economic status, assessed by the Hollingshead procedure, was accounted for in the analyses. The groups were characterized as follows:

(1) Gifted, 11 students, IQ's > 130, normal WRAT school grades > C

(11) Normal, 31 students, IQ's 90-129, normal WRAT school grades > C

(111) Borderline, 23 students, IQ's 85-89, normal WRAT

(1V) Mildly Learning Disabled, 27 students, IQ's > 84, below normal WRAT

(V) Severely Learning Disabled, 7 students, IQ's < 84, below normal WRAT

For mineral analysis, a sample of hair was cut from the nape of the neck and wet-ashed with a $HNO_3:H_2O_2$ mixture. The mineral content was determined by inductively-coupled argon plasma spectrometry. Analysis of hair zinc content revealed no significant differences between the groups. The severely learning disabled children however, had a high hair content of cadmium and lead, both of which were about 4 times the concentration of these elements found in the normal group (11). The existence of these correlative findings is of interest, but caution should be exercized in assuming a direct casual relationship between these minerals and the learning disability. Lead has been implicated in impaired mental development in children. Cadmium is well-recognized as a toxic element in the adult human but has not been closely studied in children, or in relation to mental development. Further investigations will be directed to analysis of other trace elements in hair, and more precise correlations between elevated toxic mineral levels and the learning disabilities. In addition, it is necessary to locate the dietary or other source of these toxic elements.

DISCUSSION

ANON. Do you think any of the vegetables contain any contaminant lead or cadmium.

HARLAND (USA). I don't have another explanation but it is part of the study to investigate this. What you say is a possibility. These three counties are on the eastern shore of Maryland on the Chesapeake Bay, which is primarily a fishing area. We know of the forces of contamination of cadmium. At some point, certain areas of the Chesapeake Bay have had to be closed because of the high level of cadmium, so I think as you suggested, the leafy greens would be a source, perhaps the fish from those waters, and in addition, it's an old, early settled community dependent in a large part on well water and we do know that waters will reflect cadmium levels. We also know that in old communities you will have pipes in which cadmium can be brought into the food supply. But more importantly, as a nutritionist I am very interested in looking at their nutritional intake and this is part of the study. We will closely examine the nutritional intake of the children.

READING (Australia). I measure trace elements, vitamin levels and food allergies in children with learning difficulties, behavioural disorder, autistic children and hyperactive children. I was very interested in your findings. I think you have shown a magnificant correlation between severe learning difficulties and high cadmium levels. In hyperactive and autistic children and children with severe learning disorders, I have noted lead levels either high or towards the high side and aluminium certainly towards the high side and most of them are towards the high side in copper, whereas they tend to be low or on the low side with manganese, zinc, magnesium, iron and calcium, and I think in patients that I see, which come from all over New South Wales and other parts of Australia, the majority of these abnormalities are purely secondary to food allergies. The more severe the deficiencies or the metabolic disturbance the more severe the food allergies. Somehow they preferentially concentrate lead, copper, aluminium and apparently cadmium, I can't answer why, in preference to zinc, iron, calcium etc., and I think it is related to the food allergies I have looked at, which are particularly lgM, lgA and lgG mediated, as seen with cytotoxic testing. They especially involve milk/grains/yeast. I think you might find a strong correlation in severe learning disordered children with food allergies - which somehow preferentially concentrate these heavy metals into the child.

HARLAND (USA). Thank you. I appreciate and agree with everything you have said. In one of the slides, I showed that we have looked at 28 minerals but as you know we didn't show results. Of the rest of the minerals I would particularly look at aluminium - one of those which I would not have suspected, but I would look at those.

SEGAL (Australia). Getting back to the question of lead and cadmium among the farm population, do you use lead arsenate in the United States as an insecticide, also you mentioned before that the lead and cadmium may be contaminating the green leafy vegetables. Now are lead and cadmium a significant component of the washed green leafy vegetable or are they a surface contaminant?

HARLAND (USA). You have presented a number of questions - let's take them step by step. My first answer to every one of those is I don't know for certain - I don't know about the pesticide. I don't think of lead in the leafy greens I only think of cadmium in the leafy greens.

SEGAL (Australia). I think cadmium is actually concentrated mainly in the roots of vegetables rather than the upper parts. When analyses are done of root or leafy vegetables and so forth is there a significant lead or cadmium content in the root or in the fruit or in the vegetable itself?

HARLAND (USA). We have not measured that - that will be something in the future.

MacPHERSON (Scotland). Your results seem to show that there is a high proportion of disabled children on the farms. Your high instance of low IQ children came from farms, but you have children from towns as well. What was the distribution between the farms and between the different centre of population that you looked at?

HARLAND (USA). I cannot answer that yet.

GARNYS (Australia). I find your results very interesting and in fact they fit in with the pattern seen in the survey that my colleagues and myself ran in Sydney among Sydney school children. We studied about 1200 school children and analysed lead. The lead in hair didn't appear to correlate with the health effects, but the significant correlations were with the classroom performance and behavioural abnormalities and so I suggest that you are fitting into that pattern which is nice to see.

WILLIAMS (Wales). It is sad that the most important decision in a child's life is the selection of its parents. Have you Dr Harland noticed any genetic trends in your preliminary studies - do for example, blacks tend to suffer more than caucasians, social class for social class and so on, or perhaps you don't have enough subjects for this.

HARLAND (USA). We are not quite ready to answer that. When we perform additional regression analyses during which we regress out these factors, we can show a greater number of correlation. I should tell you a little more about the study quickly - this was a sub population of 99 - these data represented 99 children. Subsequently we have to look at 150 children who were excluded from the first studies because they could not present a clear history without some neurological potential for damage - for instance, they might have been traumatised at birth or their mothers might have had bleeding during the third trimaster. Again, these are preliminary studies, eventually we will look at 2000 children in this area, then we can answer more securely.

ARSENIC EFFECTS ON HUMAN NERVE CONDUCTION

J.L. Valentine, D.S. Campion, M.D. Schluchter, F.J. Massey

University of California
Los Angeles, Ca. USA 90024

INTRODUCTION

Nerve conduction measurements increasingly are used to detect subclinical health effects after heavy metal exposure (especially lead) via ambient air (1), indoor paint (2,3), drinking water (4) and occupation (5-13). The desire is to have indication of adverse health effects additional to the traditional enzyme and metal body burden deviation from normal.

In clinical situations arsenic has been related to peripheral neuropathy. Only two studies were located which evaluated arsenic effects on nerve conduction. Landau et al. (9) found reduced nerve conduction velocities after occupational exposure to arsenic compared to a control group. Hidmarsh et al. (4) observed nerve conduction velocities in 33 persons with varying exposure to arsenic in drinking water.

Evaluated are nerve conduction velocities of the peroneal, ulnar and sural nerves in four populations exposed to arsenic via drinking water and in two control populations.

METHOD

Persons were taken from 4 communities with arsenic concentrations in the drinking water ranging from 51 to 393 µg/l (Table I). Virginia Foothills with 51 µg/l arsenic at time of survey had values of 100 µg/l for several years prior to installation of a water treatment plant 4 months prior to our testing. The two control communities had levels below the U.S. EPA drinking water standard of 50 µg/l. Participants used the local tap water and had resided in their community for at least one year prior to data collection.

Arsenic determinations on samples of blood, hair, and urine from each participant were determined by the procedure described in (14) and (15). Briefly, generated arsine gas from sodium borohydride reduction was determined on an atomic absorption spectrophotometer.

Motor latency and conduction velocity were measured in ulnar and peroneal nerve and sensory latencies measured in the ulnar and sural nerve. Surface electrodes were used for both stimulation and recording. All measurements were made using a "Neuro-diagnostic" portable machine, with latencies being measured directly from the stored response, at the take off for the motor response and at the peak for the sensory response. The extremities of which nerve measurements were made were always at or above 32°C.

Analyses of covariance with age as covariate and multiple comparisons of adjusted means were performed (16) on all conduction measurements.

RESULTS

Arsenic concentrations in tap water as well as in whole blood, hair and urine are presented in Table I. Arsenic exposed populations had markedly higher concentrations of arsenic in body samples when compared to the controls. The three arsenic exposed communities having similar water concentrations had similar arsenic concentrations for body samples and Edison, having the highest water arsenic concentration, was found to have higher mean concentrations for body samples compared to the other three arsenic exposed populations.

Results of nerve conduction measurements are in Table II. Means and most observations are within the range of normal as previously reported by others (2,9,10,17). Generally, the adjusted means were in the direction of faster velocities or smaller latencies in the control communities. Two exceptions were for Sun Valley where mean velocities for PCV in males and for UCV in females were the slowest of the six communities. Also Edison, with the highest arsenic exposure, was 9 times out of 12 the best of the exposed communities.

Of the 180 city comparisons of adjusted mean conductions 96 were between control and exposed communities and, of these 96, 21 were highly significantly different ($P \leq .01$), all in the direction of higher velocities or smaller latencies in the control cities. Also 4 of the 12 comparisons of Casper and Sun Valley were significant which points to the variability of "control" communities. However, only 4 of the 72 comparisons of adjusted means between exposed cities were highly significant and all four involved Edison.

TABLE I. ARSENIC CONCENTRATIONS IN TAP WATER, BLOOD, URINE AND HAIR OF EXPOSED AND UNEXPOSED GROUP

Community	N	Tap Water (μg/1)	N	Blood (μg/100 ml)	N	Hair (μg/g)	N	Urine (μg/1)
Edison	14	387 ± 23 (328 – 427)	33	1.33 ± 1.18 (0.18 – 5.50)	34	1.16 ± 0.81 (0.36 – 4.31)	34	168.0 ± 116.9 (4.7 – 480)
Fallon	21	98 ± 2 (94 – 101)	33	0.28 ± 0.16 (0.03 – 0.85)	40	0.53 ± 0.36 (0.08 – 1.68)	40	50.03 ± 19.6 (18.8 – 103.7)
Hidden Valley	11	122 ± 16 (107 – 163)	21	0.42 ± 0.17 (0.08 – 0.73)	21	0.50 ± 0.37 (<.01 – 1.39)	22	48.5 ± 21.3 (7.6 – 117.8)
Virginia Foothills	14	50 ± 13 (33 – 76)	27	0.52 ± 0.66 (.01 – 3.5)	24	0.45 ± 0.42 (.07 – 1.74)	27	42.6 ± 43.8 (6.1 – 234)
Sun Valley	24	2.3 ± 0.8 (1.0 – 4.0)	43	< 0.1 (<0.1 – .2)	42	0.03 ± 0.05 (<.01 – 0.14)	42	3.6 ± 5.8 (0.1 – 34.7)
Casper	29	0.6 ± 0.4 (0.3 – 2.9)	46	< 0.1 (–)		Not determined	51	3.1 ± 3.2 (0.6 – 23.1)

TABLE II. NERVE CONDUCTION MEASUREMENTS OF EXPOSED AND UNEXPOSED GROUPS, BY SEX.

	Edison	Fallon	Hidden Valley	Virginia Foothills	Sun Valley	Casper
Females	(N=25)	(N=25)	(N=14)	(N=14)	(N=25)	(N=23)
Ulnar M.L. (msec)	2.72 ± 0.27 (2.74)	3.12 ± 0.55 (3.11)	3.04 ± 0.27 (3.04)	2.88 ± 0.34 (2.87)	2.73 ± 0.37 (2.73)	2.27 ± 0.27 (2.26)
Ulnar S.L. (msec)	2.85 ± 0.23 (2.89)	3.27 ± 0.41 (3.25)	3.38 ± 0.33 (3.37)	3.26 ± 0.43 (3.22)	2.94 ± 0.38 (2.96)	2.66 ± 0.25 (2.65)
Ulnar C.V. (M/sec)	60.5 ± 5.8 (60.9)	60.2 ± 6.3 (60.3)	62.3 ± 4.3 (62.3)	60.6 ± 3.5 (60.9)	58.8 ± 4.8 (58.7)	64.0 ± 5.0 (64.1)
Peroneal M.L. (msec)	4.45 ± 0.51 (4.49)	4.68 ± 0.75 (4.66)	5.01 ± 0.95 (5.00)	4.64 ± 0.75 (4.60)	4.36 ± 0.94 (4.37)	4.37 ± 0.73 (4.36)
Peroneal C.V. (M/sec)	53.0 ± 4.8 (52.2)	50.8 ± 4.5 (51.1)	49.9 ± 3.6 (50.0)	51.1 ± 4.8 (51.8)	52.2 ± 4.9 (52.0)	52.6 ± 4.2 (52.7)
Sural S.L. (msec)	3.83 ± 0.46 (3.92)	4.14 ± 0.30 (4.11)	4.24 ± 0.15 (4.23)	4.04 ± 0.26 (3.96)	3.85 ± 0.56 (3.87)	3.83 ± 0.56 (3.81)
Age	28 8-53	42 14-76	38 9-62	46 26-63	35 8-73	40 9-75
Males	(N=14)	(N=17)	(N=8)	(N=13)	(N=19)	(N=28)
Ulnar M.L. (msec)	2.90 ± 0.24 (2.98)	3.19 ± 0.54 (3.18)	3.03 ± 0.48 (2.99)	3.00 ± 0.41 (2.98)	2.98 ± 0.40 (2.95)	2.51 ± 0.40 (2.52)
Ulnar S.L. (msec)	3.10 ± 0.27 (3.22)	3.66 ± 0.90 (3.65)	3.45 ± 0.48 (3.39)	3.44 ± 0.34 (3.41)	3.21 ± 0.46 (3.16)	2.99 ± 0.36 (3.00)
Ulnar C.V. (M/sec)	59.0 ± 4.7 (58.1)	53.4 ± 7.0 (53.5)	55.6 ± 3.7 (56.1)	55.6 ± 3.7 (55.8)	58.4 ± 6.1 (58.8)	58.8 ± 5.8 (58.8)
Peroneal M.L. (msec)	4.76 ± 0.69 (4.95)	5.02 ± 0.83 (5.01)	4.89 ± 0.69 (4.79)	5.49 ± 0.82 (5.45)	5.00 ± 0.59 (4.92)	4.81 ± 0.84 (4.83)
Peroneal C.V. (M/sec)	52.9 ± 6.7 (51.0)	48.4 ± 4.6 (48.6)	48.9 ± 4.3 (49.8)	49.4 ± 3.9 (49.8)	47.2 ± 4.9 (48.0)	50.4 ± 7.7 (50.3)
Sural S.L. (msec)	3.99 ± 0.51 (4.18)	4.27 ± 0.44 (4.26)	4.18 ± 0.20 (4.07)	4.11 ± 0.34 (4.06)	4.01 ± 0.50 (3.94)	3.89 ± 0.57 (3.90)
Age	23 9-30	38 10-57	44 14-62	40 14-65	43 14-70	36 8-81

TABLE III. PERCENT OF PARTICIPANTS SHOWING MEASUREMENTS OUTSIDE THE UCLA CENTER FOR HEALTH SERVICES NORMAL RANGE

		Males Exposed		Males Controls			Females Exposed		Females Controls		
Ulnar M.L.	> 4.0	0%	0/52	0%	0/47	N.S.	0%	0/73	0%	0/48	N.S.
Ulnar C.V.	< 47	3.8%	2/52	2.1%	1/47	N.S.	0%	0/73	0%	0/48	N.S.
Ulnar S.L.	> 3.5	26.9%	14/52	10.6%	5/47	p < .05	19.4%	14/72	2.1%	1/48	p < .001
Peroneal M.L.	> 7.0	0%	0/51	0%	0/47	N.S.	1.4%	1/72	2.1%	1/48	N.S.
Peroneal C.V.	< 41	1.9%	1/52	10.6%	5/47	N.S.	0%	0/72	0%	0/47	N.S.
Sural S.L.	> 4.2	29.4%	15/51	28.8%	13/47	N.S.	29.2%	21/72	14.6%	7/48	N.S.

Most of the highly significant differences (15/29) were between Casper and the exposed communities and 8 of the 15 were for Ulnar Motor Latency. The UML and UCV values arranged by size are

	MALES						FEMALES					
	CA	SV	ED	VF	HV	FA	CA	SV	ED	VF	HV	FA
UML	2.52	2.95	2.98	2.98	2.99	3.18	2.26	2.73	2.74	2.87	3.04	3.11
	FA	VF	HV	ED	CA	SV	SV	ED	FA	VF	HV	CA
UCV	53.5	55.8	56.1	58.1	58.8	58.8	58.7	60.2	60.3	60.9	62.3	64.1

Here Casper is significantly different from all the other communities (with P values mostly less than .001) but Sun Valley only differs significantly from Fallon for Females. For USL, Casper and Sun Valley are significantly from Fallon only for males while for females Casper is significantly different from all but Edison and Sun Valley is significantly from Hidden Valley and Fallon. For UCV, Casper and Sun Valley are both significantly different from Fallon.

The above includes 18 of the significant differences between control and exposed communities. In addition for SSL Casper is significantly different from Fallon (males) and from Hidden Valley (females) and Sun Valley differs from Hidden Valley (females).

However, some individual nerve measurements were found to be outside the UCLA EMG laboratory range of normal (Table III). The percentages of abnormal ulnar and sural sensory latencies showed significant differences ($p < 0.05$) between the combined exposed and the combined control groups. Fewer, if any, abnormals for the ulnar and peroneal motor latencies and conduction velocities were noted, and were not statistically significant ($p > 0.05$).

DISCUSSION AND CONCLUSION

Our studies seem to indicate that low level arsenic exposure may have an adverse affect on nerve conduction. Comparisons of percentages outside the UCLA normal ranges (Table III) for arsenic exposed vs. nonexposed groups showed significant differences for the ulnar sensory nerve. This finding might be compared with Goldstein _et al._ (18) where it is thought that lead is more apt to produce a motor neuropathy, whereas other metals produce a symmetrical sensory or mixed sensorimotor polyneuropathy that affects the lower more than the upper limbs.

Most findings were in the normal range. However, within this range arsenic exposed communities had nerve conduction measurements that were worse than those determined on control populations. These differences were most prevalent in measurements of ulnar motor and sensory latencies. However, differences found did not mirror arsenic body burden concentrations when Edison was included.

REFERENCES

1. Landrigan, P.J., Baker, E.L., _et al._, 1976, J. Pediatrics 89: 904-910.
2. Feldman, R.G., Haddow, J., _et al._, 1973, In "New Developments in Electromyography and Clinical Neurophysiology," Vol. 2 (ed. J.E. Desmedt) pp. 313-317, (Karger, Basel).
3. Feldman, R.G., Haddow, T., _et al._, 1973, Am. J. Dis. Child. 125: 39-41.
4. Hindmarsh, J.T., McLetchie, O.R., _et al._, 1977, J. Anal. Toxicology 1: 270-276.
5. Verbeck, M.M., 1976, Int. Arch. Occup. Environ. Health 38: 141-143.
6. Araki, S. and Honma, T., 1976, Scand. J. Work Environ. & Health 4: 225-231.
7. Seppalaninen, A.M. and Hernberg, S., 1972, Br. J. Ind. Med. 29: 443-449.
8. Lilis, R., Blumber, W.E., _et al._, 1977, Arch. Environ. Health 32: 256-266.
9. Landau, E., Thompson, D.J., _et al._, 1977, U.S. Environmental Protection Agency, Pub. No. (EPA) 560/6-77-018, Washington, D.C.
10. Baloh, R.W., Spivey, G.H., _et al._, 1979, J. Occup. Med. 21: 490-496.
11. Spivey, G.H., Baloh, R.W., _et al._, 1980, J. Occup. Med. 22: 607-612.
12. Vasilesca, C., 1972, Rev. Roum. Neurol. 9: 345-350.
13. Vasilesca, C., 1973, Rev. Roum. Neurol. 10: 221-226.
14. Kang, H.K. and Valentine, J.L., 1977, Anal. Chem. 49: 1829-1832.
15. Valentine, J.L., Kang, H.K., _et al._, 1979, Environ. Res. 20: 24-32.
16. Dunn, O.T. and Clark, V.A., 1974, "Applied Statistics": John Wiley & Sons, New York.
17. McQuillen, M.P. and Gorin, F.J., 1969, J. Neurol. Neurosurg. Psychiat. 32: 144-148.
18. Goldstein, N.P., McCall, J.T., _et al._, 1975, In "Peripheral Neuropathy," Vol. II (eds. P.J. Dyck, P.K. Thomas and E.H. Lambert)(W.B. Saunders, Philadelphia) pp.
19. Goodgold, J. and Eberstein, A., 1978, Electrodiagnosis of Neuromuscular Diseases. 2nd Ed. Williams and Wilkins Company.

DISCUSSION

GARNYS (Australia). How did you account for possible differences in lead intake because lead has been implicated in the conduction velocity?

VALENTINE (U.S.A.). We gave questionnaires to everyone, so we knew if they worked in a chemical plant or something like that. We also analysed the water for other trace elements and we did not find lead. The only element that was really high was arsenic, and we made sure we checked them.

WESTERMARCK (Finland). It seems that selenium may be an antidote, was there determination of this?

VALENTINE (U.S.A.). We determined selenium in the water supply and it was less than 0.01 mg/kg.

IODINE CONTAMINATION OF MILK BY IODOPHOR DAIRY SANITIZERS

Sally M. Wheeler, G.H. Fleet* and R.J. Ashley

NSW Department of Agriculture,
Dairy Research Centre
PO BOX 217 RICHMOND NSW 2753
AUSTRALIA

*University of New South Wales, Sydney, NSW, Australia

1. INTRODUCTION

Since the early 1960's, iodophors have been found to be effective sanitizers and have been widely used within the Australian dairy industry.

In 1971, an outbreak of thyrotoxicosis in Tasmania prompted an evaluation of the levels of dietary iodine in many foods, including milk. Further investigations correlated elevated iodine concentrations in milk with iodophor use at the farm level, and, in 1975, a report was submitted to the National Health and Medical Research Council on this topic (1). Subsequently, a recommended level for iodine concentration in milk was established at 500 micrograms of iodine per litre of milk.

Surveys conducted in New South Wales suggested that iodophor sanitation of milking machines was a primary source of iodine residues in milk (2). Inadequate drainage of iodophor from these machines (3) and possible retention and desorption of iodophor by milking machine materials (4,5) seemed likely sources of iodine in milk. A major objective of this study was to assess the iodophor sanitation of milking machines as a source of iodine contamination of milk.

Examination of the literature revealed little understanding of the functional chemistry of the components of iodophors, and this was felt to be fundamental to an understanding of the behaviour of iodine in sanitation. Therefore, a second aim of this project was to examine some aspects of the chemistry of iodophors.

Milk contains physiological iodine secreted by the cow, and it has been noted that this concentration may change as the milk is processed (6,7). The fate of iodophor-derived iodine in milk during the various processing operations has not been examined and became a further objective of this study.

2. IODOPHOR CHEMISTRY

The chemistry and uses of iodophors have been reviewed (8,9) to show that formulation and method of preparation affect the properties of the final product. In this study, a laboratory-prepared iodophor of known, constant composition, and checked to be comparable with commercial products, was made by dissolving elemental iodine in hydriodic acid, and adding nonionic surfactant, water and phosphoric acid, at ambient temperature.

At a normal working dilution, this iodophor contained 27.3 mg available iodine (10) per litre (mg avI/L), which was 72% of the formula iodine. Total iodide was measured (11) and found to be 37.6 mgI/L, a calculated recovery of 98.8% of added formula iodine, indicating that no surfactant-iodine binding had taken place.

The role of the surfactant in the iodophor was determined by examining iodophors prepared with surfactant concentrations both greater and less than that of the control formulation. Those containing less surfactant were heterogeneous mixtures, in which the iodine did not undergo complete solubilisation and were not studied further. Increased surfactant concentration did not increase the amount of iodine present in the available form, and did not initiate surfactant-iodine bonding. When the surfactant was omitted completely, 47% of the formula iodine precipitated out, indicating that the surfactant was responsible for the relatively high iodine concentration of these sanitizers, possibly by physically complexing the iodine (9). Omission of the hydriodic acid from the control formulation did not cause any iodine precipitation, indicating that the surfactant was able to solubilise/complex all the formula iodine. The proportion of iodine in the available form was similar to that of the control formulation. However, about 5% of the iodine was bound to the surfactant. The function of the phosphoric acid was studied, and it was considered that its role was only that of pH control for maximum bactericidal efficiency.

3. RETENTION AND RELEASE OF IODOPHOR-IODINE FROM MILKING MACHINE MATERIALS

To measure the potential for various materials to retain iodophor-derived iodine and later release it into milk, model experiments were conducted using small tubes of glass, steel, rubber, plastic and silicone rubber, which were first treated with ^{131}I-labelled laboratory-prepared iodophor (27.3 mg avI/L) and later rinsed with milk. The effects of iodophor contact time (one hour or 18 hours) and water-rinsing were examined, and it was found that, with all materials except rubber, increasing contact time increased the iodine residue in milk. With one hour's contact, rubber gave the greatest residue of iodine to milk (0.099 $\mu gI/cm^2$), compared with silicone rubber (0.038 $\mu gI/cm^2$), plastic (0.013 $\mu gI/cm^2$), glass (0.008 $\mu gI/cm^2$) and steel (0.003 $\mu gI/cm^2$). Water-rinsing reduced the residue for all materials except silicone rubber.

In order to emulate the dairy situation, the materials were repeatedly contacted with iodophor, rinsed with milk, and then washed and dried. Under these conditions, rubber, plastic and silicone rubber absorbed iodine in a linear fashion, while virtually none remained associated with glass or steel.

Using data from these experiments, and by measuring both residual fluid in a milking machine (3), and the relative surface area of each material, it was found that desorption from surfaces was in the same range as the residue contributed by residual fluid iodophor in the machine.

In considering the relative importance of residues from the two sources, it is not possible to extrapolate too far, because the effects of the milking machine vacuum and resulting turbulence within the machine are not known. Nevertheless, it is clear that choice of materials may be as important as milking machine design, when considering residues from iodophors.

3. THE EFFECT OF PROCESSING UPON IODINE IN MILK

a) Effect of Boiling Upon Concentration of Added Iodophor-iodine in Milk

^{131}I-labelled iodophor was added to milk, and the mixture boiled in glass distillation apparatus. By measuring ^{131}I counts in both distillate and residue, it was found that only 0.02% of the label was lost from milk on boiling.

b) Effect of Acid Precipitation of Casein Upon Distribution of Added Iodophor-iodine in Milk

^{131}I-labelled iodophor was added to milk, and casein precipitated by dilute hydrochloric acid, under standard conditions. By counting the ^{131}I label in both whey and casein, it was found that about 3% of the added iodine was associated with the casein fraction and the remaining 97% with the whey.

c) Effect of Milk Processing Upon the Concentration of Iodine in Milk

Whole milk, unadulterated with iodophor, and the same whole milk deliberately adulterated with iodophor (addition of 500 $\mu gI/L$ milk), were subjected to cream separation, pasteurisation (78°C) and spray-drying. Iodine concentration of milk at each step was measured by both a chemical method (12) and an iodide ion-specific electrode method (13).

Statistical analysis showed that processing did not affect the iodine concentration of either adulterated or unadulterated milk ($p<0.05$). However, the results of the two analytical methods were significantly different ($p<0.01$). It is clear that processing cannot be relied upon to lower the iodine concentration of milk.

4. CONCLUSIONS

In iodophors of the type used in these experiments, iodine is in the form of available iodine or iodide only, and, as such, is in an active chemical state. It is clear that iodophor-iodine may be retained by milk-contact surfaces to an extent which is dependent on the type of material, and this iodine may subsequently be released into milk. Milking machine rubber is a significant contributor of iodine residues to milk, even after water-rinsing, and is particularly important because of the large area of rubberware in the milking machine.

In milk, the bulk of iodophor-iodine is present in the whey fraction, and high recoveries of added iodophor-iodine in milk using the iodide specific-ion electrode, would suggest that, in milk, available iodine is converted to iodide. Iodine concentration was unaffected by processing, although the results of these experiments highlighted the problem of measuring trace quantities of iodine in milk.

REFERENCES

1. Dunsmore, D.G., and A.M. Luckhurst, 1975. Iodophor disinfectants in the dairy industry - a report to the National Health and Medical Research Council, N.S.W. Dept. Agriculture.
2. Dunsmore, D.G., and S.M. Wheeler, 1977, Aust. J. Dairy Technol. 32: 166-171.
3. Wheeler, S.M., R.J. Ashley, and T.B. Brock, 1978, Milchwissenschaft. 33: 145-147.
4. Dunsmore, D.G., S.M. Wheeler, and R.N. Barnes, 1977, Aust. J. Dairy Technol. 32: 119-123.
5. Palmer, J., and J. O'Shea, 1973, Ir. J. Agric. Res. 12: 175-185.
6. Berchenko, N.V., and I.Y. Ostrovskii, 1961, Erevan 1960: 67-71.
7. Raymond, A.E., and G.W. Williams, 1964, Radiol. Hlth. Data. 5: 70-72.
8. Twomey, A., 1968, Aust. J. Dairy Technol. 23: 162-165.
9. Schmidt, W., and M. Winicov, 1967, Soap Chem. Spec. 43: 61,66,170,172.
10. Australian Standard AS 1398:1972, Iodophors for Use in the Dairying Industry. Sydney: Standards Association of Australia.
11. Wheeler, S.M., and R.J. Ashley, 1978, N.Z. J. Dairy Sci. Technol. 13: 106-110.
12. Leonards, J.R., and P.R. Davoren, 1962, Ann. N.Y. Acad. Sci. 102: 83-84.
13. Wheeler, S.M., L.R. Fell, G.H. Fleet, and R.J. Ashley, 1980, Aust. J. Dairy Technol. 35: 26-29.

DISCUSSION

BINNERTS (The Netherlands). How did you label the iodophor?

WHEELER (Australia). Na ^{131}I (<0.01 mg I/ml), from the Australian Atomic Energy Commission, was used. 0.01 ml was added to 100 ml of dilute laboratory-prepared iodophor (approximately 25 mg av I/L). The mixture was allowed to equilibrate, in the dark overnight. Partitioning of the iodophor with n-heptane was carried out to check whether the activity was present in both the available iodine and iodide forms of iodine in the iodophor; ^{131}I was found to be distributed in both phases, and the labelling proceedure therefore considered to be suitable for these experiments.

BINNERTS (The Netherlands). What is your chemical method?

WHEELER (Australia). The chemical method was that of Leonards and Davoren (Annal N.Y. Acad. Sci., <u>102</u>, 83, 1962), which uses wet-ashing with chloric acid, followed by As-Se catalysis.

BINNERTS (The Netherlands). Have you an explanation for the large difference with the electrode method in "unadultered milk"? Is the electrode really specific for iodine?

WHEELER (Australia). It is reported that sulphydryl groups in milk may interfere with electrode response (Bruhn, J.C., Franke, A.A., J. Dairy Sci 61,1557). However, our method recovered a mean of 95.5% of added iodophor-iodine.

BINNERTS (The Netherlands). Are you sure that "unadultered milk" has not been adultered by the farmers? Your results are ± 50 x higher than the supposedly normal level.

WHEELER (Australia). The 'unadulterated' milk for these experiments came from two cows - they were not teat-dipped with iodophor, nor was any of the milk-collecting equipment in contact with iodophor.

BINNERTS (The Netherlands). I think your conclusion that iodophor-iodine in free-reacting is very important. It would be interesting to look for the metabolic effects of this iodine, and also for the metabolism of iodophor-iodine used in teat-dipping and treatment of claws. i.e. will it be absorbed through the milk gland or the skin, and will the resulting milk "contamination" be partly caused by iodine that has passed the general metabolism. Do you intend to perform physiological studies with radio-I to check this?

WHEELER (Australia). There is evidence that some of the iodine residues from teat-dipping cows with iodophor becomes absorbed by the cow, and therefore I feel that some of the contamination from this source will be due to iodine which has been through the general metabolism of the cow. We will not be carrying out any physiological studies of this kind.

AIRBORNE PARTICULATE LEAD CONCENTRATIONS - TEMPORAL AND SPATIAL VARIATIONS IN THE METROPOLITAN AREA OF PERTH, WESTERN AUSTRALIA

B. H. O'Connor and W. J. Chang

School of Physics and Geosciences
Western Australian Institute of Technology
Bentley, Western Australia 6102

INTRODUCTION

Airborne particulate Lead (Pb) levels are being studied in Perth, Western Australia, in order to obtain representative data on exposure levels for the general population. Preliminary work has shown that levels vary to such an extent with location and with meteorological conditions that it would be necessary to operate an extensive monitoring network for the data to be acquired by direct measurement. As the costs associated with such an exercise would be prohibitive, we are investigating a more efficient strategy whereby levels will be modelled mathematically by computing the dispersal of atmospheric Pb from the major emission sources as a function of meteorological data.

It has been established previously[1], on the basis of constant bromine/lead (Br/Pb) ratios in Perth air particulate bulk samples, that motor vehicles are in effect entirely responsible for airborne particulate Pb. Therefore it is convenient to employ road traffic density data as source data for modelling purposes, and to incorporate wind and atmospheric inversion parameters for computing Pb concentrations at off-roadway locations. Model validation will be performed with airborne Pb data from a limited number of sampling stations.

This paper summarises preliminary air particulate Pb data for three monitoring stations in metropolitan Perth which give an indication of -
(i) the extent of temporal variations on a weekly and seasonal basis,
and (ii) spatial variations in terms of contrasts between inner-city and suburban locations.

DATA ACQUISITION

Particulate matter has been collected on either fibreglass or cellulose filters by high-volume air filtration at an inner-city sampling station (Perth) and two suburban stations (Shenton Park and Bentley) - see References 2 and 3. The Perth data which have been authorised for publication are those spanning the years 1974-76; the Shenton Park data cover 1971-76; and the Bentley data 1978-79. More recent data for these stations will be published in due course. Samples have also been collected at two stations adjacent to the Kwinana Industrial Area some 50 km south of Perth.[4] The locations of the stations are shown in Fig. 1.

The Pb concentrations for the Perth air filters were determined by atomic absorption spectrometry of solutions prepared by digesting the filter paper with a mixture of nitric and perchloric acid, and hydrogen peroxide, and then made up to volume with dilute nitric acid. Calibration was performed with standard solutions. Filters from the other stations were analysed non-destructively for Br and Pb using x-ray fluorescence spectrometry, with calibration against standard filter papers impregnated with known amounts of a Pb compound.[5]

RESULTS

Typical examples of the weekly average data for Perth, Shenton Park and Bentley are given in Fig. 2 and the corresponding seasonal data in Fig. 3. The Shenton Park and Bentley concentrations both sharply peaked in winter. While the behaviour of the Perth data from Autumn-Spring is similar, the Perth summer values do not show the marked decline observed for Shenton Park and Bentley. This apparent anomaly may be associated with systematic errors in the summer data for Perth as discussed in Reference 3.

The substantial short-term variations ($\leq$ weekly) in the measured Pb concentrations tend to coincide with changes in the general meteorological conditions for Perth, notably with the ventilation parameters-wind speed and mean inversion height. However, it is not possible to account for all variations in this way which we consider may be due to micro-meteorological influences. On this basis, it appears that weekly average data may be unsuitable for simple modelling.

An important aspect of the pronounced variations in weekly average data is the need to collect data over long periods (>>weekly) if the data are to be used in estimating average human exposure levels. Given the reproducibility of the seasonal data from station to station, it may be acceptable to collect data for some sampling stations over a three-month interval and to then infer the concentrations for the other seasons on the basis of continuous measurements made at only one station.

Details of the data for the Kwinana Industrial Area are considered in detail elsewhere.[4] In summary, it has been found that the Pb concentrations are due to the passing traffic and not to plant emissions. Levels are lower than those in the metropolitan area due to the lower traffic density in the vicinity of Kwinana.

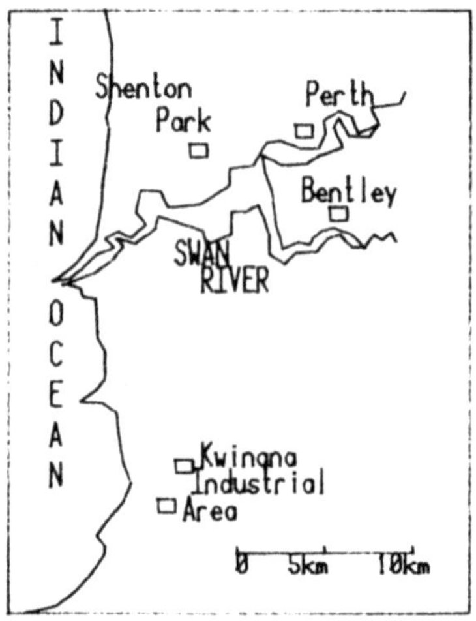

FIG. 1. AIR MONITORING STATIONS IN METROPOLITAN PERTH AND KWINANA INDUSTRIAL AREAS

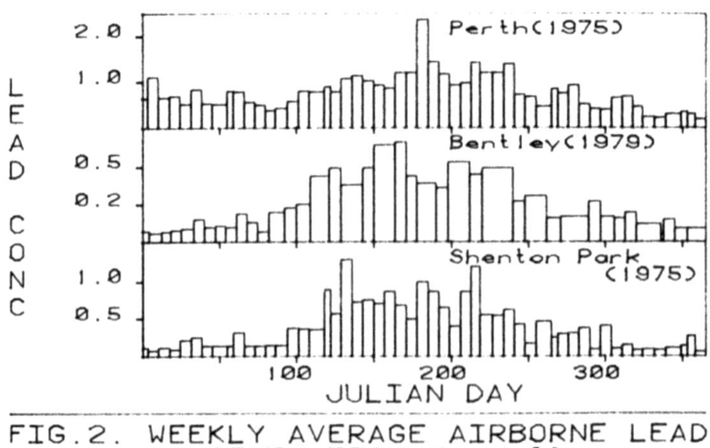

FIG. 2. WEEKLY AVERAGE AIRBORNE LEAD CONCENTRATION (ug/m³)

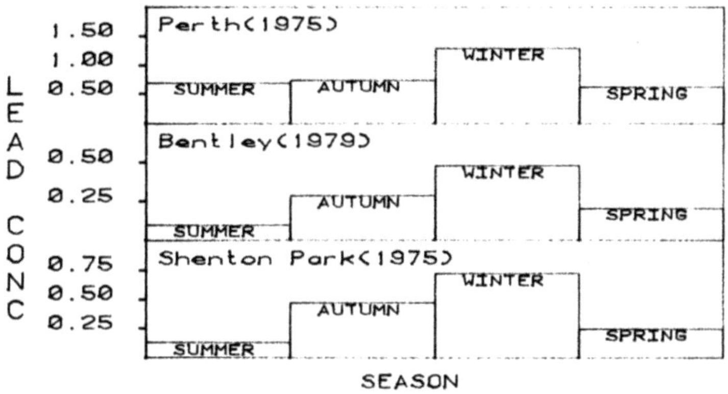

FIG.3. SEASONAL AVERAGE AIRBORNE LEAD CONCENTRATION (ug/m³)

CONCLUSIONS

1. Variations between the weekly average Pb concentrations indicate that micrometeorological factors have an important influence on local Pb values.

2. There is considerable evidence that the relative variations in seasonal concentrations are similar for the stations considered.

3. Conclusions 1 and 2 suggest that it may be necessary to use seasonal data for modelling calculations in order to avoid complications caused by short-term fluctuations in the micrometeorological data.

ACKNOWLEDGEMENTS

The investigation described in this paper has been supported by the Australian Research Grants Committee and the Western Australian Institute of Technology. The Pb concentrations for the Perth Station were provided by Western Australian Department of Health and Medical Services; the air filters for the Shenton Park station by the Western Australian State X-ray Laboratories; and the Kwinana air filters by the Western Australian Department of Conservation and Environment.

REFERENCES

1. O'Connor, B.H., Kerrigan, G.C., Thomas, W.W. and Pearce, A.T., 1977. Atmospheric Environment 11:635-638.

2. O'Connor, B.H., Kerrigan, G.C. and Nouwland, C.R., 1978, Atmospheric Environment 12: 1907-1916.

3. O'Connor, B.H. and Allen, G.H., 1980. Clean Air 14:3-7.

4. O'Connor, B.H., Chang, W.J. and Martin, D. 1981. In "Proceedings of 7th International Clean Air Conference". In print.

5. O'Connor, B.H., Kerrigan, G.C., Thomas, W.W. and Gasseng, R., 1975. X-ray Spectrometry. 4: 190-195.

THE INFLUENCE OF DIET ON LEAD TOXICITY

J.T. Spickett and R.R. Bell

School of Community Health
Western Australian Institute of Technology
Hayman Road BENTLEY 6102
WESTERN AUSTRALIA

The extent of the physiological and biochemical changes as a result of exposure to lead depends on the amount ingested and the proportion of lead that is absorbed and retained in the body tissues. Several factors including species and age of the animal, its nutritional status, the form of the lead and the composition of the diet can influence the absorption and toxicity of lead.

The main source of environmental lead exposure is through the diet and therefore any dietary component which alters the absorption or excretion of lead is of significance. Furthermore dietary factors which influence the susceptibility of animals to lead toxicity are poorly understood (Committee on Toxicology, 1976).

Most of the work in this area to date has dealt with the enhancement of lead absorption as a consequence of increasing or decreasing the amounts of dietary components in animals given either a single dose of radioactive lead or relatively high doses (500-2000μg/g) of dietary lead.

The initial aim of the present work was to investigate the tissue lead levels and biochemical indicators of lead toxicity in animals exposed to low levels of dietary lead. Three groups of male weanling rats (10 in each group) were fed semi purified diets providing protein, vitamins, minerals and fat to meet the levels currently recommended and supplemented with lead (0, 10, 20μg/g as lead acetate). All diets contained equal amounts of Ca, P, Fe and vitamin D. The food intake was monitored and after a 4 week period biochemical tests for lead toxicity were performed. Blood was removed by heart puncture and then the animals were killed and the bone, brain and kidney were removed for lead analysis. Urinary δ-aminolaevulinic acid (ALA-U), free erythrocyte protoporphyrin (FEP), δ-aminolaevulinic acid dehydratase (ALA-D) and tissue lead analyses were performed as described by Bell & Spickett (1981) and results are presented in Table 1.

TABLE 1: Lead Concentrations in Tissues and Biochemical Indicators of Lead Toxicity in Rats Fed Diets Supplemented with 0, 10 or 20μg/g Lead

	Supplemented Dietary Leadμg/g			
	0	10	20	ANOVA
Tissue Lead				
Blood (μg/100ml)	1.85 ± 0.75	6.09 ± 0.76	9.82 ± 2.09	P<0.001
Brain (μg/g)	0.002 ± 0.003	0.021 ± 0.008	0.036 ± 0.012	P<0.001
Liver (μg/g)	0.052 ± 0.018	0.137 ± 0.025	1.266 ± 0.067	P<0.001
Kidney (μg/g)	0.097 ± 0.049	0.728 ± 0.105	1.019 ± 0.145	P<0.001
Femur (μg/g)	0.433 ± 0.287	3.797 ± 1.144	7.401 ± 1.942	P<0.001
Biochemical Indicators				
Urinary ALA (μg.24hr)	26.0 ± 8.2	31.9 ± 8.8	22.7 ± 8.0	NS
FEP (μg/100ml)	34.2 ± 15.1	32.0 ± 2.6	34.9 ± 5.3	NS
ALA Dehydratase (U)*	8.43 ± 1.01	6.76 ± 0.47	5.53 ± 0.77	P<0.001

* U - μmole PBG/min/1RBC

Lead supplementation (10 and 20μg/g) of the diet resulted in a significant increase in blood, bone, brain, kidney and liver lead concentrations.

The increases in lead levels in the bone and kidney were substantially greater than the increase in blood, supporting previous observations that blood lead level is a poor indicator of the lead concentration in other body tissues. Furthermore the blood lead level of 9.8μg/100 ml is substantially less than 30μg/100 ml, the level regarded as "cause for concern" by the NH & MRC (1979) for children.

Our data indicates that the absorption and retention of lead is dose dependent at relatively low concentrations in the diet. Of the three biochemical indicators of lead toxicity only ALA-D showed any significant difference between the three groups.

Recent studies (Bell & Spickett, 1981) have shown that milk consumption increases tissue lead concentrations in rats fed low concentrations of lead and milk simultaneously. In this work three groups of animals were fed diets based on soy protein, whole milk or lactose hydrolysed milk and half the animals in each group were given 20μg/g lead as lead acetate in the diet. All rats given the lead supplemented diets had higher concentrations of lead in the tissues analysed however the increase was significantly greater in animals fed the lead supplemented milk diet. For example the femur lead concentrations in animals given the lead supplemented soy, milk and lactose hydrolysed milk diets were 32.7 ± 4.1, 44.8 ± 5.8 and 28.9 ± 3.9μg/g, respectively. Animals fed a lead supplemented milk diet from which the lactose had been removed by hydrolysis had tissue lead concentrations similar to those of rats fed the lead supplemented soy diet. The results from this work suggest that lactose is the factor in milk responsible for increasing lead absorption and tissue lead concentrations.

Although it has been known for some time that lactose increases the absorption of calcium from the small intestine of many mammals, the mechanism by which this occurs has not been fully explained. Armbrecht & Wasserman (1976) monitored the uptake of ^{45}Ca from isolated everted gut sac preparations and concluded that lactose does not bind with calcium itself but rather that it interacts with the absorptive cells of the small intestine and increases their permeability to calcium. Of the other common sugars investigated only sucrose increased the uptake of ^{45}Ca to the same degree as lactose.

It is possible that lactose may increase the permeability of the mucosal cells to lead as well as calcium and further, that other disaccharides may exhibit a similar effect.

In order to investiage this possibility four groups of male weanling rats were fed diets containing 20g soy protein, 8g corn oil, 2.5g cellulose, 5g mineral mix, 1g vitamin mix, 0.5g DL-Methionine and starch to make 100g. Three of the groups received diets containing 30% maltose, lactose or sucrose in place of an equal amount of starch. Each diet was supplemented with 20μg/g lead as lead acetate. All diets contained equal amounts of Ca, P, Fe and vitamin D. The food intake was monitored and after a 2 week period the animals were killed and the bone, brain, liver and kidney were removed for lead analysis. The results of these analyses are presented in Table 2.

TABLE 2: Tissue Lead Concentrations in Rats Fed Diets Containing Different Sources of Carbohydrate supplemented with 20μg/g Lead

	Diet With 20μg/g Lead			
Tissue	Starch	Maltose	Lactose	Sucrose
Blood (μg/100ml)	8.3 ± 2.7	7.8 ± 2.5	7.7 ± 2.3	7.5 ± 2.1
Brain (ng/g)	22.6 ± 3.4	24.1 ± 3.9	25.5 ± 4.0	24.1 ± 3.8
Liver (μg/g)	0.227 ± 0.069	0.171 ± 0.116	0.173 ± 0.057	0.158 ± 0.082
Kidney (μg/g)	1.14 ± 0.22	0.90 ± 0.21	1.40 ± 0.74	0.933 ± 0.065
Femur (μg/g)	3.95 ± 1.41	3.33 ± 1.14	3.82 ± 1.70	3.98 ± 1.20

No significant difference in lead concentration between treatments for a given tissue, by analysis of variance.

Previous work has indicated that lactose in milk enhanced the absorption and retention of dietary lead and it was considered that if maltose or sucrose were to exhibit the same effect then evidence of lead absorption and retention would be observed after a period of 2 weeks. However as can be seen from Table 1 there was no significant difference in any of the tissue lead concentrations in the four groups of animals.

The results of the experiments described here seem to indicate that lactose in the absence of the other components of milk does not exhibit the same effect of enhancing the absorption and retention of lead under the experimental conditions described. The reason may be due to the shorter period of exposure and that during this time the animals on the lactose diet suffered mild diarrhoea resulting in a smaller weight gain for this group as compared to the starch, maltose and sucrose groups. A further possible explanation is that lactose must be in the milk matrix in the gut in order that the enhancement of lead absorption and retention can take place.

References:

Armbrecht, H.J., and Wasserman, R.H., (1976). J. Nutr. 21, 541.

Bell, R.R., and Spickett, J.T., (1981). Fd. Cosmet. Tox. (In Press).

Committee on Toxicology (1976). Recommendations for Prevention of Lead Poisoning in Children. Assembly of Life Sciences, National Academy of Sciences/National Research Council, Washington, DC, USA.

National Health and Medical Research Council Public Health Advisory Committee. Eighty eighth session, October 1979. Australian Government Publishing Service, Canberra.

DISCUSSION

STEVENS (Australia). What methodology did you use to measure lead in the soft tissues and bone?

SPICKETT (Australia). For blood lead we used the method of Kubasik, Volosin and Murray. Blood is haemolysed with Triton X-100 and analysed for lead in a Varian 575 AAS fitted with a CRA 90 automatic sampler.

Tissue samples were digested in lead-low nitric acid for about 30 minutes and after suitable dilution were analysed for lead with the same instrument. Standard additions were made to calibrate the analysis.

LINDER (USA). How did you produce the lactose-depleted milk.

SPICKETT (Australia). We used milk in which the lactose had been hydrolysed to glucose and galactose enzymatically. It was commercially available.

NIELSEN (USA). Have you ever looked at the effects of feeding the lactose - hydrolysed diet, supplemented with lactose?

SPICKETT (Australia). This is our next experiment.

SALTMAN (USA). I would like to comment that lactose is an effective chelating agent for lead, calcium and other divalent metal ions - it is better than sucrose and some other disaccharides.

RETENTION BY HORSES OF FLUORINE FROM CALCIUM AND PHOSPHORUS SUPPLEMENTS

B.J. Blaney, R.J.W. Gartner and T.A. Head
Queensland Department of Primary Industries,
ANIMAL RESEARCH INSTITUTE, YEERONGPILLY, 4105
AUSTRALIA

In some areas of Queensland, pasture grasses from Africa and the Americas have been introduced into beef producing regions. Many of these grasses contain more than 0.5% oxalates, and this induces calcium (Ca) and phosphorus (P) deficiency in horses used for mustering cattle in the regions (1,2). The deficiency results in Nutritional Secondary Hyperparathyroidism, characterised by lameness, jaw swellings and ill thrift (3). One kg of rock phosphate mixed with 1.5 kg of molasses and offered to each horse once per week was shown to overcome daily Ca and P losses of approximately 20 mg/kg live weight/day and 10 mg/kg live weight/day respectively (4). Since defluorinated rock phosphate was difficult to obtain, we changed our recommendation to 1 kg of a 2:1 (W/W) mixture of dicalcium phosphate plus limestone which provided a similar Ca and P intake to 1 kg of rock phosphate but a lower total fluorine (F) intake.

In attempting to assess the fluorosis hazard to horses given these supplements, we were hampered by the need to relate recommended maximum daily intakes of F based on NaF, to a once weekly dose of mineral supplement containing less soluble forms of F. The only report available to us (5) suggested that horses may be less susceptible to fluorosis than other species. We subsequently determined retention of F and Ca from both supplements in mineral balance experiments, using horses fed oxalate-containing grass hays, and in negative Ca balance.

EXPERIMENTAL

Mineral balance experiments were conducted using six Standardbred geldings in Expt. 1, and five in Expts. 2 and 3. These weighed between 385 and 509 kg. The horses were fed hammer milled hays of various grass species, all containing 1.3% total oxalates. The hays were *Pennisetum clandestinum* (kikuyu) in Expt. 1; *Setaria sphacelata* var. *sericea* c.v. Kazungula (setaria) in Expt. 2; and *Cenchrus ciliaris* (buffel) in Expt. 3. After an equilibration period of 21 days in yards, the horses were moved to metabolism crates for separate urine and faecal collection (6). After an initial collection period of 7 days (period 1), mixtures of mineral supplements and molasses were offered on day 8. Collection was then continued until and including day 14 (period 2). The supplement was consumed by day 10, and fully metabolised by days 13 and 14 (4). The supplements were 1.0 kg rock phosphate plus 1.5 kg molasses in Expt. 1. In Expts. 2 and 3, 1.0 kg of a 2:1 (W/W) mixture of dicalcium phosphate and limestone was substituted for rock phosphate.

Methods of analysis of Ca in feed, faeces, urine and supplements have been described elsewhere (1). F was determined in hydrochloric acid extracts of the samples using a F specific ion electrode, after adjustment of total ionic strength and pH with buffers. Citrate overcame interference by ferric ions in sample extracts containing rock phosphate. The rock phosphate used was Christmas Island rock phosphate precipitator dust and contained 1.50%F, while the dicalcium phosphate contained 0.4%F. The molasses and limestone contained negligable amounts of F.

RESULTS AND DISCUSSION

In all three experiments (Table 1) the supplement in period 2 effectively corrected a large negative Ca balance in period 1 arising from the effect of oxalates in the grass hays. Retentions of Ca and F from the supplements were calculated by dividing the changes in Ca and F balances from period 1 to period 2 by the amounts of Ca and F provided in the supplements, neglecting slight differences in Ca provided by the feed between periods.

The data show that a negligable amount of the F in rock phosphate was retained by the horses, while about half of that in the dicalcium phosphate plus limestone mixture was retained. Since Ca was absorbed from the rock phosphate, but not F, it is logical to assume that upon dissociation of the rock phosphate in the weakly acidic conditions of the

Table 1. Mean Ca and F intakes, outputs and balances (±SE of means) mg/kg live weight/day, and retentions of Ca and F from the supplements (%).

	Experiment 1		Experiment 2		Experiment 3	
	Period 1	Period 2	Period 1	Period 2	Period 1	Period 2
Fluorine						
Intake - feed	0.05 ± 0.001	0.04 ± 0.001	+	+	+	+
Intake - supplement		4.89 ± 0.126		0.90 ± 0.015		0.92 ± 0.038
Output - faeces	0.04 ± 0.003	4.91 ± 0.210	+	0.46 ± 0.023	+	0.33 ± 0.012
Output - urine	0.14 ± 0.010	0.07 ± 0.004	0.07 ± 0.007	0.06 ± 0.004	0.06 ± 0.014	0.05 ± 0.011
Balance	-0.13 ± 0.011	-0.05 ± 0.110	-0.07*± 0.007	0.40*± 0.030	-0.06*± 0.014	0.56*± 0.030
% Retention of supplement F		1.6 ± 1.27		51.3 ± 2.91		67.9 ± 1.29
Calcium						
Intake - feed	46.3 ± 1.24	52.6 ± 1.35	22.2 ± 0.58	25.1 ± 1.03	54.9 ± 2.81	53.5 ± 2.32
Intake - supplement		90.7 ± 2.34		101.9 ± 1.70		104.0 ± 4.23
Output - faeces	52.8 ± 2.73	115.9 ± 4.74	49.5 ± 2.05	120.8 ± 5.62	73.2 ± 4.28	152.2 ± 5.63
Output - urine	11.6 ± 2.16	13.7 ± 0.84	4.0 ± 0.42	9.3 ± 1.39	2.8 ± 0.80	6.7 ± 1.32
Balance	-18.2 ± 2.65	13.6 ± 2.69	-31.3 ± 2.00	-3.1 ± 3.01	-21.1 ± 2.36	-1.4 ± 5.06
% Retention of supplement Ca		35.2 ± 2.56		27.8 ± 1.43		19.0 ± 4.66

\+ Not detected (< 0.04)

* Assuming value where not detected to be ½ maximum detection limit (0.02)

equine stomach, both soluble Ca ions and insoluble CaF_2 are formed, an assumption supported by the stereochemical conformation of F and Ca ions in fluoroapatite crystals (7). The preparation of dicalcium phosphate first involves formation of phosphoric acid by the action of concentrated sulphuric acid on rock phosphate. During this process much of the F is removed by precipitation as CaF_2, but a proportion is converted into soluble F salts, silico fluorides and fluoroaluminates (8). When the crude phosphoric acid is reacted with hydrated lime in the second stage of the process, some of the F then reacts to form insoluble CaF_2 but the remainder, bound as soluble silicon and aluminium complexes does not, and this may be the portion of F absorbed from the dicalcium phosphate.

The total amount of F retained from the dicalcium phosphate, if related to a daily dry matter intake of 10 kg, would represent approximately 25 mg F/kg in the ration. This level should be tolerated by horses in long term feeding (5). Therefore neither this supplement nor rock phosphate should present a hazard of fluorosis when given as recommended.

SUMMARY

Horses grazing grasses containing more than 0.5% oxalates are in negative calcium (Ca) and phosphorus (P) balance, which leads to the production of Nutritional Secondary Hyperparathyroidism. A supplement of 1 kg of either rock phosphate or a 2:1 mixture of dicalcium phosphate plus limestone when given once per week mixed with 1.5 kg molasses, corrects this mineral imbalance. Because of the long term hazard of fluorosis due to the fluorine (F) content of these supplements, the retention of Ca and F from both supplements was measured in mineral balance experiments on horses fed oxalate containing grass hays. None of the F in commercially obtained rock phosphate (1.5%F) was retained while about half of that of dicalcium phosphate (0.4%F) was retained by the horses. The results suggest that neither supplement should cause chronic fluorosis in horses when given as recommended.

REFERENCES

1. McKenzie, R.A., B.J. Blaney and R.J.W. Gartner, 1981. J. Agric. Sci., Camb., 97: in press.

2. Blaney, B.J., R.J.W. Gartner and R.A. McKenzie, 1981. J. Agric. Sci., Camb., 97: in press.

3. Walthall, J.C. and R.A. McKenzie, 1976. Aust. Vet. J. 52: 11-16.

4. Gartner, R.J.W., B.J. Blaney and R.A. McKenzie, 1981. J. Agric. Sci., Camb., 97: in press.

5. Shupe, J.L. and A.E. Olson, 1971. J.A.V.M.A. 158: 167-174.

6. McKenzie, R.A., B.J. Blaney, R.J.W. Gartner, R.D. Dillon and N.F. Standfast, 1979. Equine Vet. J. 11: 232-234.

7. Posner, A.S., 1969. Physiol. Rev. 49: 760-792.

8. Frazier, A.W., J.R. Lehr and V. Dillard, 1977. Bulletin Y113 National Fertilizer Dev. Centre, pp 1-19, R 12.

EFFECTS OF FLUORIDE INGESTION ON DAIRY CATTLE

J.W. Suttie
Department of Biochemistry
University of Wisconsin-Madison
420 Henry Mall
Madison, Wisconsin 53706
USA

The effects of the ingestion of elevated levels of fluoride by dairy cattle have been studied by a number of investigators, and reviews of these studies are available (1,2). Ingestion of sufficient fluoride can cause fluorotic lesions in developing teeth, skeletal lesions, intermittent lameness, and an impairment in food intake. The current tolerance level established for the continual ingestion of a soluble fluoride by young dairy cattle is 40 ppm F in the total ration (1). Although it is possible to observe some fluoride-induced lesions at ingestion levels below this, effects on the productive performance of cattle have not been observed below this level. Much of the current interest in the toxic effects of fluorides results from the fact that fluorides are air pollutants associated with effluents from the aluminum, phosphate, steel, and ceramic industries. Air pollution standards based on the concentration of fluoride in forage consumed by grazing animals have been developed and are used to protect animals in areas surrounding these industries. The adverse effects of fluoride ingestion range from an adverse effect on production at high levels of ingestion, through severe dental fluorosis, to mild non-damaging dental lesions which are seen at very low levels of fluoride intake. Air pollution standards based on forage fluoride concentrations have attempted to protect animals not only from any loss of productive ability but also to prevent any significant incidence of severe dental fluorosis. The fluorotic lesions of bovine incisor teeth can be assessed by an objective scoring system where no. 1 is defined as a normal tooth, no. 2 as a tooth with very mild lesions best seen when the tooth is illuminated from the rear, no. 3 as a completely mottled tooth, and no. 4 as a tooth that shows definite enamel hypoplasia and will wear more rapidly than normal. The developing dentition of the bovine is very sensitive to short-term, high-level exposures to fluoride (3), and these standards have attempted to protect animals from short-term, high-level exposures as well as continuously elevated exposures. Although standards adopted by different states have varied somewhat, many of them fit this general statement: The yearly average forage fluoride shall not exceed 40 ppm, and it shall not exceed 60 ppm for more than two months or 80 ppm for more than one month.

This standard has been developed by the extrapolation of data from a number of experiments, but has not previously been directly tested. An experiment is currently in progress where 4-month-old calves are being fed a number of different patterns of fluoride intake. All of the ingestion patterns fall within the suggested standard, but push it to the maximum. Most of the cattle are fed an average of 40 ppm F in the forage which would be the maximum allowed, and many receive over 100 ppm F for a 1-month period. Changes in the plasma, urine, and skeletal fluoride concentrations are being monitored, and tooth development is followed by periodic dental x-rays. The major emphasis in the experiment is to determine the effect of fluoride on the developing incisors, to assess the incidence of increased molar wear, and to determine the extent of skeletal fluorosis. This experiment is in an intermediate stage at the present time. Sixteen animals, each started at 4 months of age, were maintained on the experimental diet until they were 68 months old and have now been slaughtered. Most of the incisor teeth of these animals had fluorosis scores of no. 2 or 3. There were 6 animals that had a fourth pair of incisors which were scored no. 4. This incisor is formed after the molars are calcified (4); and, in keeping with past experience, the presence of a no. 4 incisor in only the fourth pair of incisors was not associated with increased molar wear. Analyses of the available data indicate that there was no single factor, such as one month at a very high fluoride level, that was responsible for the dental changes. It was apparent that in some individual animals very high fluoride intakes for as long as a month did not cause a severe effect on any of the incisors, while the same level adversely affected another animal. The levels of fluoride fed in this experiment did not influence voluntary hay or grain intake, and the degree of osteofluorosis observed was mild. Plasma fluoride levels were closely monitored, and the variations seen were similar to those observed in previous studies. These studies are being continued with additional animals in an attempt to more clearly define an adequate standard.

A recent field report (5) has suggested changes in thyroid status, a decrease in red cells, and an increase in eosinophiles due to fluoride intake. With the exception of the anemic response, which has been observed in some experimental studies, these conditions have not previously been directly related to fluoride ingestion. Animals from an existing experiment that compares the effects of feeding high amounts of fluoride to young calves (aged 4 to 6 months) for 28 months were utilized to study these responses. Ten calves were fed 100 ppm F as NaF for 4 months and 40 ppm F for the next 8 months of each year to achieve an average intake of 60 ppm F. Blood samples were taken during the 24th month (40 ppm F intake) and during the 28th month (100 ppm F intake), and hemoglobin and thyroxin concentrations and differential white cell counts were determined.

				% Of white cells as			
Cow group	F intake ppm	Hemoglobin gm/dl	Thyroxine T_4 µg/dl	Lympho-cytes	Neutro-cytes	Mono-cytes	Eosin-ophiles
F	40	11.3	4.49	63	28	2.7	5.3
Control	0	12.5	3.40	65	26	1.8	6.1
F	100	12.1	3.57	66	26	1.6	6.2
Control	0	13.1	5.14	60	28	4.0	7.5

The only consistent result was a reduced hemoglobin concentration in calves receiving NaF. The values were, however, in the normal range for both control and fluoride-fed animals. There was no statistically significant difference in the thyroxin concentrations in the control and fluoride-fed calves, and the percentage of white cells as eosinophiles were somewhat elevated in all groups but were not affected by fluoride intake. These data suggest that alterations in white cell distribution observed in the field were related to some unknown variable rather than to fluoride ingestion.

Although much of the data that have been utilized in determining fluoride tolerance for cattle have been made available from controlled studies, it has been possible in some cases to check the expected responses in field situations. The performance of a commercial purebred Ayrshire dairy herd located in close proximity to a primary aluminum smelter has been followed (6). Fluoride emissions from this source increasd the fluoride content of the forage consumed by these animals to above 40 ppm for most of the ten-year period studied. The intake was variable and was often above 100 ppm F in the fall of the year. The supplemental grain mix fed to these animals also contained about 40 ppm F due to the fluoride present in the phosphate added. The ingestion of this amount of fluoride resulted in increased skeletal fluoride retention to the extent that the vertebral ash of 33-month-old cattle contained 5000-6000 ppm F during most of the period studied. Dental fluorosis and osteofluorosis developed to a degree that would have been predicted from the results of experimental studies where similar amounts of fluoride have been fed. There was no indication that the incidence of lameness, breeding performance of the herd, or size of the animals was influenced by the level of fluoride ingested. Because there is no adequate control group in this type of study, it is not possible to make definitive statements about the effects of fluoride on the productivity of the animals studied. However, the average milk production of the herd increased over the period studied, as it did in most Canadian dairy herds during that period. There was no indication that variations in fluoride exposure between different animals, as indicated by the degree of dental fluorosis, had any influence on their productive ability. The data demonstrate that the responses expected from controlled experimental studies are observed in field situations. The data also demonstrate that although it is not advisable, it is possible to maintain a productive dairy herd in a region whre the fluoride intake is higher than would be advised and where it exceeds that allowable by most regulatory agencies.

REFERENCES

1. National Research Council, 1974. "Effects of Fluorides in Animals" (National Academy of Sciences, Washington, D.C.) 70 pp.
2. Suttie, J. W., 1977, J. Occup. Med. 19:40-48.
3. Suttie, J. W., and E. C. Faltin, 1971, Am. J. Vet. Res. 32:217-222.
4. Brown, W. A., P. V. Christofferson, M. Massler, and M. B. Weiss, 1960, Am. J. Vet. Res. 2:7-34.
5. Hillman, D., D. L. Bolenbaugh, and E. M. Convey, 1979, J. Dairy Sci. 62:416-423.
6. Suttie, J. W., 1981. In "Light Metals 1981" (ed. G. M. Bell) (The Metallurgical Society of AIME, Chicago) pp 887-895.

DISCUSSION

SCHENKEL (Germany). I have two questions, firstly is there any additional supplementation of calcium or aluminium salt in your diets on the Mostook Farm and secondly, you have shown the variation of fluoride concentration in the diet during the year, we found a similar pattern for fluoride concentration in forages with peaks in spring and autumn. What is your opinion about such variations because recommended values are average values? Is it possible to give an upper limit related to fixed time? For example might we expect detrimental effects if we applicated diets with about 150 ppm for several weeks in spring and autumn keeping 40 ppm an average for the year.

SUTTIE (USA). There was no excess calcium or aluminium salts fed at this Farm. We used ordinary dairy rations, the ones we normally would use. In terms of protecting animals my personal feeling is that if you talk about an air pollution standard you cannot use a standard that is based simply on average level of fluoride for the year because as you say there are peaks which cannot be tolerated. I personally have advocated a standard based on forage levels, keeping in mind that usually the dry matter intake from the grain ration will dilute this out, but the forage itself should not be allowed to exceed a 40 parts per million annual average, should not be allowed to exceed 60 parts per million for more than two months in a row or 80 parts per million for more than one month. Now for some of the years at Mostook Farm they would have held to under 40 parts per million but they would have exceeded the short term limitations and a couple of months at 130 parts per million is going to give you at least one pair of severely damaged teeth. That should be protected against. I personally do not feel that you need to set an absolute number as a limit to be exceeded, because if you start to play around with numbers you see that it goes very high for one month and you will never meet the other averages anyway. A lot of people feel more comfortable with an absolute number, and I understand that, and perhaps that is the thing to do.

LOW LEVEL DETERMINATIONS OF CADMIUM IN THE ENVIRONMENT

J.R. De Laeter, K.J.R. Rosman and A. Chegwidden
Department of Physics
Western Australian Institute of Technology
South Bentley 6102
WESTERN AUSTRALIA

INTRODUCTION

Cadmium was recognised many years ago as a highly toxic element. However, it was not until recently that concern was expressed over the possible effects to human health by exposure to chronic low level concentrations of cadmium. This situation has developed because of its increasing technological use. Chronic exposure to cadmium contaminated food and water has caused a bone disease in members of a small community in Japan, who live on a river heavily polluted by mining wastes (Singal *et al.* 1975).

A variety of cadmium minerals exist in nature and these are soluble in excess aqueous salt solutions; many cadmium compounds are also soluble in fresh water. Williams and David (1973) have shown that the cadmium content of phosphatic fertilisers and superphosphate used in Australian pastures ranges from 27 to 90 mg kg^{-1} (ppm). Thus cadmium can be transferred to natural waterways and reservoirs and hence find its way to estuaries and the oceans. Ingestion from food or water, and inhalation from the atmosphere are the major routes by which cadmium enters the body. Man, who is at the top of the food chain, con sumes food and water representative of his environment. The long biological halflife of cadmium and its toxicity to a wide variety of body tissues and metabolic pathways leaves little doubt that cadmium can induce morphological as well as biochemical changes in organs involved with reproduction, excretion and general body metabolism (Singhal *et al.* 1975).

THE ISOTOPE DILUTION TECHNIQUE

Using the mass spectrometric isotope dilution technique (MSID), Patterson (1974) has shown that almost all the analytical data on lead in marine waters is inaccurate because of gross contamination from faulty sampling and analytical procedures. It is likely that much of the published work on cadmium in the environment is inaccurate, particularly where the content is at the sub-parts per billion level.

MSID is a well-established method for analysing elements present in trace amounts with excellent sensitivity and high accuracy and precision. This technique measures the concentration of cadmium in a matrix by the addition of a known quantity of the same element, the isotopic composition of which has been artificially altered (the 'spike'). After equilibration of the spike with the sample, the cadmium mixture is chemically extracted by ion exchange in a form suitable for mass spectrometric analysis. The measurement of the abundance ratio of two isotopes of the mixture permits the concentration of cadmium in the sample to be determined (De Laeter and Rosman, 1977). Since only the abundance ratio of two isotopes for the element is required, quantitative separation of cadmium is unnecessary. Interferences which are a source of uncertainty in many other analytical techniques are not a problem since isobaric interferences are avoided.

Recent developments in automated mass spectrometry and ionisation enhancing media have enabled the analysis of cadmium by MSID to be undertaken for a wide range of materials (e.g. Rosman and De Laeter 1974, Chow and Snyder 1977, Hosie *et al.* 1978).

BASELINE CADMIUM LEVELS

In order to make an adequate assessment of the effects of cadmium contamination in the environment, it is necessary to distinguish between natural and anthropogenic cadmium levels. This is not a particularly easy task, but Western Australia, which is an isolated region with little industrial development, has some advantages in establishing baseline values. Rosman and De Laeter (1976) surveyed two river systems in Western Australia and showed that cadmium levels were about 0.1 μg kg^{-1}, being approximately one hundredth of the drinking water limit set by the World Health Organisation. The data indicate that the cadmium content tends to decrease upstream from the mouth of the rivers and that in the protected reservoir catchment areas, the cadmium content is as low as 0.02 μg kg^{-1}. These levels are substantially lower than waterways in many other parts of the world and could therefore serve as a benchmark for cadmium concentrations in river systems in remote areas.

More recently, the authors have determined the level of cadmium in seawater sampled at a depth of 30 m, some 3.3 km west of Rottnest Island, Western Australia in the Indian Ocean. It was necessary to construct a contamination-free sampling pump and develop a high purity chemical extraction technique so that accurate determinations could be made at the sub-ppb level. A summary of the Cd data is given in Table 1. The cadmium levels of approximately 2 ng kg^{-1} are comparable to values from North Pacific seawater samples recently reported by Bruland (1980).

CADMIUM POLLUTION IN COCKBURN SOUND

In the past, it has often been tacitly assumed that effluents containing heavy metals are rapidly dispersed in the open ocean, thereby reducing the metal concentrations to negligible values. Heavy metals are discharged into Cockburn Sound from a number of industries situated on the eastern boundary. It is known that the major contributor of cadmium is a fertiliser plant situated on the southern end of the Sound, which discharges approximately 4.5 kg of cadmium per day into the Sound, present in waste gypsum.

Surface seawater samples were collected from Cockburn Sound, a sheltered body of water to the south of the Port of Fremantle in Western Australia in November and December 1978 and May 1979, Rosman *et al.*(1980). Cockburn Sound is a marine basin with a maximum depth of 22 m. It is separated from the Indian Ocean in the west by the Garden Island Ridge and very shallow banks in the north and the south (Fig. 1). There is restricted water flow through the Sound.

The samples collected were from a depth of 1 m below the surface from 46 positions in Cockburn Sound on a 2 km grid, together with sample stations to the west of Garden Island. The cadmium isopleths for the December 1978 collection are plotted in Figure 1. It is apparent that south-west sea-breeze conditions influenced a strong long-shore movement of cadmium to the northern end of the Sound, with the highest values being found along the eastern margin. There is also evidence of some movement of cadmium to the south of the discharge point. A significant enrichment in cadmium can also be seen in the Woodman Point region. There appears to be minimal leakage of cadmium around the north end of Garden Island and there is an intrusion of oceanic water into the southern end of the Sound. The cadmium concentration isopleths suggest a stable marine environment for the Sound during the summer period. A computer model for the circulation of water in Cockburn Sound was developed using direct measurements of water currents and meteorological data (Steedman and Craig, 1979). The computer model indicated that a number of eddies or gyres develop in the Sound. The major gyre moves anticlockwise during the summer sea-breeze cycle, and reverses to a clockwise direction during low pressure storms. The effect of the gyres is to hold the water within the Sound so that there is only limited exchange of water with the ocean. Even during storm conditions it is unlikely that the rate of exchange is sufficient to flush the Sound's water completely, so that its ability to cope with cadmium which has a long residence time, is limited.

It might be thought that since the levels of cadmium in Cockburn Sound are $<$ 1 ppb (except in the immediate vicinity of the fertiliser plant, point A in Figure 1), that this is an insignificant health hazard to man or animals. However work in Cockburn Sound by Talbot and Chegwidden (1981), has revealed raised cadmium concentrations in mussels and oysters that often exceed the local, as well as the National Health and Medical Research Council recommended health standards. Other high values have been recorded in tissues from the hepatopancreas of crabs, the flesh of polychaete worms and the leaves of the green sea lettuce. A significant amount of cadmium is also incorporated in underlying sediments.

WATER QUALITY CRITERIA

The results presented in this paper show that the baseline cadmium levels for seawater sampled west of Rottnest Island is approximately 2 ng kg^{-1}. Levels in Cockburn Sound were found to be between 20 and 200 ng kg^{-1} except in the mixing zone of the gypsum discharge where values as high as 3.47 µg kg^{-1} have been measured.

A set of Water Quality Criteria is at present being considered by the Environmental Protection Authority for beneficial uses of the receiving waters of estuaries and marine situations in Western Australia. Unfortunately, because little scientific work in this field has been carried out in Australia, it is likely that overseas experience, principally from the United States, will have to be used in setting up Water Quality Criteria for

Western Australia. If, for example, the Water quality Control Plan for Ocean Waters of California (Klapow and Schueller, 1977), is used to derive criteria for acceptable cadmium levels in Cockburn Sound waters, then the maximum allowable 6 month median is less than 3 μg kg^{-1} with no single reading exceeding 8 μg kg^{-1} cadmium. Marine waters conforming to these criteria should then be able to support a number of beneficial uses including harvesting fish and molluscs for food, as well as being suitable for all forms of aquaculture with regard to cadmium.

Our work in Cockburn Sound waters has demonstrated that cadmium concentrations as much as two orders of magnitude lower than the 6 month median figure of 3 μg kg^{-1} has had a significant effect on cadmium levels of the marine biota of the Sound. It is also likely that the single reading level of 8 μg kg^{-1} cadmium would never be exceeded in Cockburn Sound except, perhaps, at a gypsum discharge point, even though the lower levels encountered presently have had a significant pollution impact.

Three factors need to be considered when deriving criteria for cadmium in Cockburn Sound waters. The first consideration should be that the Sound is a unique water body with limited flow through characteristics and needs to be assessed as an individual entity. Secondly, when Californian authorities established criteria they used acute and chronic toxicity data for their biota. Such data should not be extrapolated automatically to Western Australian species which may have differing characteristics. Finally Klapow and Schueller (1977) have used a value of 0.1 μg kg^{-1} cadmium for a baseline figure when establishing their criteria whereas our work has shown that the baseline value is approximately 50 times lower for marine waters west of Rottnest Island.

ACKNOWLEDGEMENTS

This work was supported by the Australian Research Grants Committee and the Department of Conservation and Environment of Western Australia.

Table 1: The *Concentration of Cadmium in Seawater Samples

Collection Period	Sample Bottle	Approx. Sample Size(g)	Concentration ng kg^{-1}
August, 1980	1	450	2.5, 2.1
	2	550	2.4
	Mean		2.3 ± 0.9
December, 1980	1	1200	1.9, 1.8
	2	1200	2.1, 1.9
	Mean		1.9 ± 0.4

*Where replicate analyses of the same bottle were made, these are individually listed.

The analytical blank for these analyses was approximately 0.2 ng kg^{-1}

Figure 1: Contoured representation of cadmium concentrations in surface water from Cockburn Sound, in December 1978 (ppb by weight).

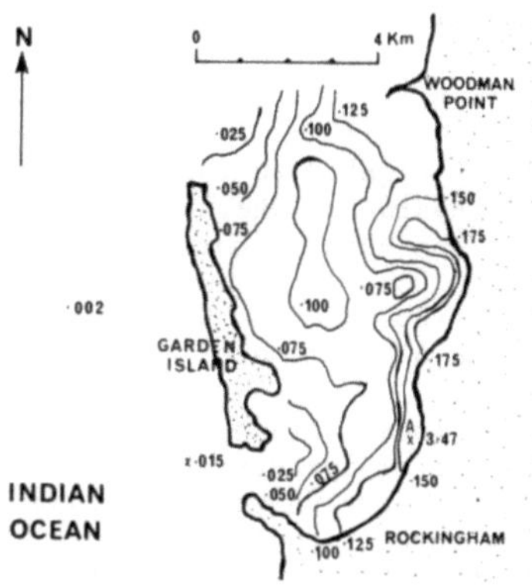

REFERENCES

1. Bruland, K.W., 1980, Earth Planet. Sci. Lett. 47 : 176-198.
2. Chow, T.J., and C.B. Snyder, 1977, Anal. Chim. Acta, 94 : 155-159.
3. De Laeter, J.R., and K.J.R. Rosman, 1977, Geostandards Newslett. 1 : 35-38.
4. Hosie, D.J., A. Bogioas, J.R. De Laeter and K.J.R. Rosman, 1978, Search 9 : 47-49.
5. Klapow, L., and H. Schueller, 1977, State Water Resources Control Board, California.
6. Patterson, C., 1974, Science, 183 : 553-554.
7. Rosman, K.J.R., and J.R. De Laeter, 1974, Geochim. Cosmochim. Acta, 38, 1665-1677.
8. Rosman, K.J.R., and J.R. De Laeter, 1976, Nature, 261 : 685-686.
9. Rosman, K.J.R., J.R. De Laeter and A. Chegwidden, 1980, Sci. Total Envir. 16 : 117-130.
10. Singhal, R.L., Z. Merali, J.A. Thomas and G.R. Knotts, 1975, Contem. Ed. 46 : 100-103.
11. Steedman, R.K., and P.D. Craig, 1979, Dept. Conserv. Envir. Report, Western Australia.
12. Talbot, V., and A. Chegwidden, 1981, Aust. J. of Mar. and Freshwater Res. *(in press)*.
13. Williams, C.H., and D.J. David, 1973 Aust. J. Soil Res. 11 : 43-56.

DISCUSSION

ANON. Dr. de Laeter you mentioned that there was a substantial amount of soluble Cd salt in fertilisers and therefore was likely to be run off into streams and estuaries. I have forgotton the amount of phosphatic fertilizers we actually use here, but it is a fairly considerable quantity. Does this have a noticeable effect on Cd in the water catchment system?

DE LAETER (Australia). Not so much in the water catchment area, but certainly in the Swan River, and more particularly in the Peel Estuary, which is in the State's irrigation area. There the cadmium levels in the water are quite high and the estuary is certainly contaminated. What happens, in effect, is that the cadmium is gradually taken up by the sediments until you get to the stage where an equilibrium situation is established and as much material is released from the sediments back into the water as is being adsorbed. Now we haven't reached that level yet, but sooner or later we certainly will.

NEWNHAM (Australia). I was interested in that you took samples off Rottnest. Some 35 years ago there were 39 train loads of excess ammunition dumped in the 200 fathom mark west of Rottnest - that included hundreds of tons of brass cartridge cases containing zinc and apparently cadmium too. Did you notice any more out there or did you just take surface samples, or did you go to the bottom?

DE LAETER (Australia). We actually sampled at the 30 metre mark. We were trying to obtain a base-line value for cadmium. The base line value we have at the moment is 0.002 parts per billion. I imagine if one analysed ice cores from Antarctica corresponding to a period of time way back beyond the industrial revolution (which has been done for studies in lead), you would probably push that figure down, but certainly the figure of 0.002 ppb is still about 50 times less than people normally use as the base-line value for cadmium. I guess that's the point we are trying to make. So certainly, the .002 might still be too high - I would suspect it would be, but it is a more accurate value than scientists have used up till now as a baseline value.

BRATTER (Germany). How many samples per day can you analyse using your method?

DE LAETER (Australia). The isotope dilution technique is certainly not a technique which is generally used by analytical chemists because most people don't have a mass spectrometer. However you can analyse about 20-30 samples in an 8 hour day. The actual measurement itself is very simple. Once you have mounted the sample in the mass spectrometer it only takes 10 or 15 minutes to collect the data, to give an absolute accuracy of better than 10% right down there at the 0.002 ppb level. The real problem of course is in the chemical separation where you have to take a tremendous amount of care to avoid contamination and this is certainly time consuming.

WOOLLIAMS (Scotland). Did you look at any other trace metals and did you find anything comparable with other metals?

DE LAETER (Australia). Yes we have looked at most of the heavy metals. If you take crabs or mussels or oysters from Cockburn Sound, particularly those that are living near the fairly warm water where the gypsum discharge takes place, you can observe that their skeletal structure is changed, and it is obvious that they have been affected by the contaminants they have absorbed. It has been calculated that if a person eats 6 mussels a week, harvested from that area, that person would exceed the World Health Organisation body burden per week per person.

K. KRISHNAMACHARI (India). Do you have any information about the possible danger to human health, particularly in relation to the kidneys of the people who are living in these areas? A renal type of rickets has been described in Japan, and its thought to be due to the consumption of fish and crab that have been contaminated by cadmium. I ask this because a West Australian nutritionist told me that osteomalacia is observed here.

DE LAETER (Australia). Yes, we are concerned about the amount of contamination going into the Sound because it is used as a recreational and a fishing area. I don't think we have had crabs on the menu at TEMA-4 but certainly one has to be careful, and the Public Health Department in Western Australia routinely analyse for cadmium and a number of other elements. However we are worried about the long term affect on people who traditionally have fished or crabbed that area and have eaten their catch.

CRITICAL EVALUATION OF THE SIGNIFICANCE OF INTERNAL VERSUS EXTERNAL EXPOSURE ON Cd CONTENT OF THE HAIR OF THE LABORATORY RAT.

W.E. Kollmer

Abteilung für Nuklearbiologie
Gesellschaft für Strahlen- und Umweltforschung m.b.H.
D-8042 MÜNCHEN - NEUHERBERG
WEST GERMANY

For various toxic metals hair has been frequently proposed as an indicator of exposure [1,2,3].If the method is practised for the evaluation of the risk to the individual, it is necessary to know 1.) if the rate of internal deposition is sufficient to distinguish different levels of intake by analyzing Cd in hair and 2.) if the deposition is sufficiently distinct, even at low levels of intake, in comparison to the external deposition of the metal. We have previously reported [4,5] on the deposition of Cd in hair after a single intravenous injection of radioactive Cd. Now our findings in animals during chronic exposure are presented. These experiments are based on the combined use of radioactive Cd and stable Cd.

MATERIALS AND METHODS

Male adult rats of 340 g average body weight of a Wistar-derived strain were used. They were subjected to an artificial light-dark cycle of 12 to 12 hours. Food and water were supplied ad libitum. The drinking water was demineralized and supplied with 3, 30 or 300 mg/l Cd as $CdCl_2$ in the respective groups. Radioactive ^{109}Cd was given as a chloride by stomach tube after an overnight fasting. The dose was 50 uCi ^{115m}Cd and 120 uCi ^{109}Cd in 0.5 ml saline, at a pH of 5. The content of stable Cd was adjusted to the concentration of Cd in the drinking water of the respective group. The interference of the radiation of ^{65}Zn was excluded by analysis of the γ-spectrum water consumption was checked.

The main features of the experiment were: The hair was in the growth phase (anaphase) during the time of exposure. This was achieved by inducing a new hair cycle by plucking a square of approximately 3x4 cm in the middorsal region of the animals. This was done under ether anesthesia. On the 11th day after plucking, the new hair erupts through the skin. Radioactivity was given at this time. Ten days later the animals were killed and hair samples were taken by plucking. Total body radioactivity was measured with a whole-body counter, radioactivity of the hair in a well-type γ-counter. Stable Cd in hair was determined by flameless atomic absorption spectrometry after wet ashing under pressure at 150°C. The AAS analysis was checked by an NBS standard.

RESULTS

After the administration of radioactive Cd by stomach tube the excretion of radioactivity came practically to an end within 7 days. The remaining radioactivity was between 0.3 and 1.6 % of the administered dose irrespective of the level of stable Cd supply. Since the urinary excretion of Cd is very small, this 7-day retention was taken to represent the intestinal resorption. The deposition of radioactivity in the hair of 15 animals was 0.050 ± 0.005% of the intestinal resorption of the individual animals. There was no difference between the groups given different levels of stable Cd. The relation between Cd ingested and internal Cd deposition in hair was calculated on the basis of these findings. To avoid underestimation of the intestinal resorption, this was assumed to be 2%. This relation is shown in Fig. 1. It may be seen that 1 mg Cd ingested leads to an internal deposition of Cd in hair of only 0.01 µg/g.

In unexposed animals kept in conventional cages the stable Cd found by atomic absorption spectrometry was roughly 0.6 ± 0.1 ug/g hair. Very similar levels were also found in the control animals of other experiments. We therefore consider this to be the natural background level under the circumstances prevailing in our laboratory. - For comparison: In groups of "unexposed" humans, values averaging between 0.24 and 2.7 ug Cd/g hair have been reported [6]. - We have calculated the total Cd deposition in hair on the basis of these findings, assuming that there was no additional external deposition over and above the "background" level of the unexposed group (Fig. 2). In the same graph we have plotted the Cd concentration actually analysed in the newly grown hair of the animals kept at different levels of exposure. Quite obviously, the data obtained by analysis are considerably higher

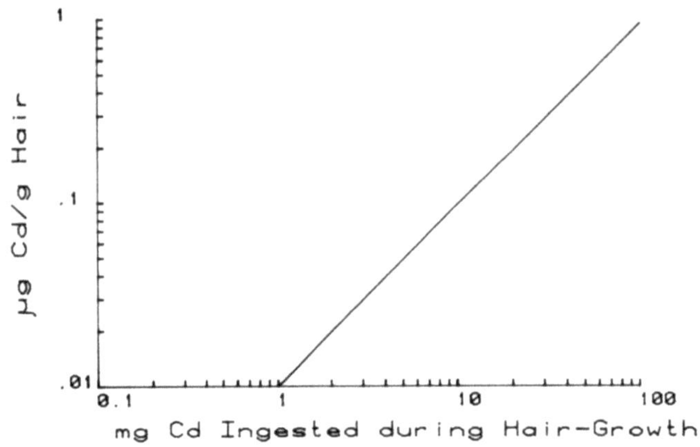

Fig. 1:

Relation between Cd ingestion and internal Cd deposition in hair based on radioactive data
Assumptions:
1. The intestinal resorption is 2% of the ingested amount of Cd.
2. The internal deposition is 0.05% of the resorbed amount of Cd per g hair.

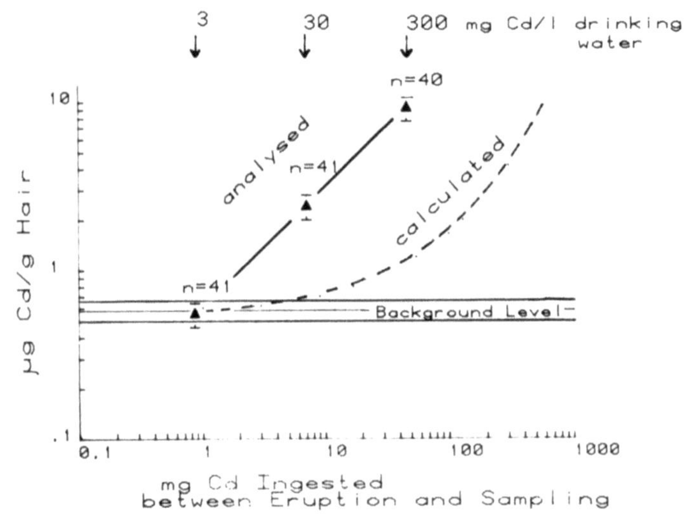

Fig. 2:

Comparison of calculated internal Cd deposition above the background level, and actually analysed total Cd in hair at different levels of exposure to Cd in the drinking water.

than the deposition as calculated on the basis of radioactivity. This controversy may be explained either by a substantial prevalence of external deposition or by the participation of a deep compartment. To check the first mentioned possibility, groups of animals with different degrees of exlusion of external deposition but at the same level of Cd intake were compared. They were kept

1.) in conventional cages containing wood shavings, exposure 2 - 12 weeks
2.) on a stainless steel grid, exposure 2 weeks
3.) in metabolism cages, exposure 2 weeks

All of them except the controls were given drinking water containing 30mg Cd/l.and it seems to be noteworthy that 15mg Cd/l have been reported to cause vomiting in man (3) . A control group was kept in metabolic cages and given demineralized water. It may be assumed that more than 98% of the Cd taken up in the drinking water was excreted with the faeces. Hence all animals given water supplied with Cd excreted most of this Cd in the faeces, much less in the urine. Since the water consumption and the Cd content of the water were approximately equal within the exposure groups, the Cd excreted per animal may also be assumed to have been equal within these groups over the entire time of exposure. In the cages with stainless steel grids the excretions fell through the grid and remained in the wood shavings just below the grid. In both of these the wood shavings were changed twice a week. In the metabolic cages the animals were kept singly, in the other cages there were 5 animals per cage. There was no significant variation of Cd in the hair with time of exposure in the animals kept conventionally. Therefore these data were treated as one group.

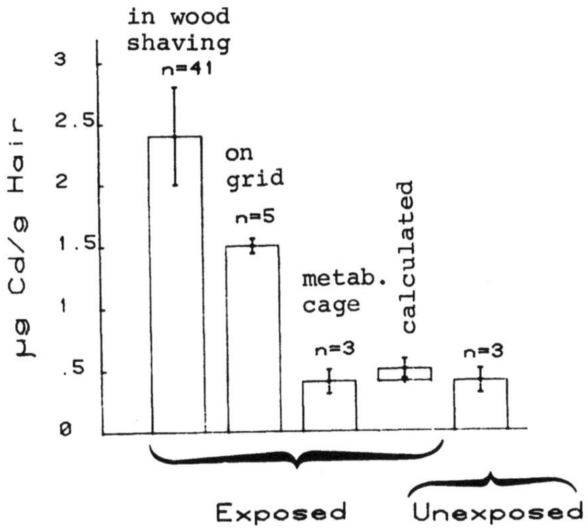

Fig. 3:

Influence of cageing on the total deposition of Cd in growing hair of the rat. All of the exposed animals received 30 mg Cd/l drinking water. The "calculated" deposition is the internal deposition above the background as calculated from radioactivity data.

Fig. 3 demonstrates the stable Cd analysed in the newly grown hair of the animals. It was clearly related to external Cd exposure. Keeping the animals in metabolic cages prohibited (external) Cd deposition in hair. In these animals the amount of stable Cd per g hair could not be distinguished from that of the unexposed controls. This finding is consistent with the calculation based on the radioactivity deposition, which also indicates that the internal deposition at this level of exposure should lie within the standard error of that found in the unexposed group. An internal deposition of Cd in hair significantly exceeding the level of "unexposed" animals can therefore be expected only at an exposure higher than 30 mg Cd/l drinking water. Furthermore, such an internal deposition of stable Cd can only be distinguished in animals kept in metabolic cages.

SUMMARY AND CONCLUSIONS

It has been demonstrated that the internal Cd deposition in hair is very low. In the rat, even a level of intake above that causing vomiting in man is not sufficient to bring about a deposition significantly exceeding that of unexposed animals. It turns out that the radioactivity deposited after a single dose reflects internal deposition very well. The findings corroborate our suspicion of a highly predominant portion of external deposition of Cd in animals kept in conventional cages. Therefore Cd analyses in hair of conventionally kept rats are not representative for internal Cd deposition.

ACKNOWLEDGEMENTS

This work has been done under a Research Agreement with the International Atomic Energy Agency, Vienna. For technical assistance I am grateful to Mrs. G. Treutler and Mr. H. Ruhstorfer. For help with the γ-spectrometry I am indebted to Dr. D. Berg.

REFERENCES

1. Anke, M., and Risch, M., 1979, "Haaranalyse und Spurenelementstatus" (VEB Gustav Fischer Verlag, Jena), pp. 212-227.

2. Valkovic, V., 1977, "Trace Elements in Human Hair" (Garland STPM Press, New York - London), pp. 91-95.

3. Friberg, L., Kjellström, T., Nordberg, G.F., and Piscator, M, 1979, in "Handbook on the Toxicology of Metals" (eds. L. Friberg, G.F. Nordberg and V.B. Vouk) (Elsevier/North Holland, Amsterdam - New York - Oxford), p. 72 and 367.

4. Kollmer, W.E., 1979, J. Radioanalyt. Chem. 52: 189-197.

5. Kollmer, W.E., 1980, J. Radioanalyt. Chem. 57: 535-541.

6. Iyengar, G.V., Kollmer, W.E., and Bowen, H.J.M., 1978, "The Elemental Composition of Human Tissues and Body Fluids" (Verlag Chemie, Weinheim - New York), p. 51.

DISCUSSION

SEGAL (Australia). Did you make a study of the distribution of cadmium over various parts of the body of the animal. The reason I ask this is because a study of trace elements in human hair showed quite a variation in distribution depending on whether it was occipital hair or side hair or from other parts of the scalp (McKenzie, J.M. American Journal of Clinical Nutrition, 31, 470-476, 1978).

KOLLMER (Germany). Yes, we ourselves have published a paper on that subject 2 years ago describing different rates of deposition in the dorsal and the ventral hair of the rat. We do not know if this is due to the different rate of growth or to morphologic peculiarities or others. In this study we have only taken hair from the same dorsal region to assure comparability.

SCHENKEL (Germany). Do you wash the hair before analysis.

KOLLMER (Germany). We made some trials with distilled water and organic soluents. These were unsuccessful, which is in accordance with the literature. Therefore we only apply cleaning by brush.

HARLAND (USA). The nutritionists speaks again. You were giving the pellet diet. Our experience with standard chow in that form is that it is laced with a vast number of high levels of many minerals as a safety factor so that you will not complain to the manufacturers about the lousy diet. So much so that you are not able to detect sensitivity in the experiment. I say this not as a criticism but I want all of us to be aware of this. My second comment is that as a result of your caging differences coprophagy might have been practised. In our experience animals who are very well nourished do not engage in this, but you really did not have a very good way of monitoring it especially with shavings.

KOLLMER (Germany). The diet was a standard diet of high protein and we have also done a cadmium analysis of the diet. It was found that there was cadmium, the cadmium content was between 0.1 and 0.2 ppm. The other constitutents in these standard diets are kept to a certain standard but we do not know about minor changes. If you want to do very precise things then you should mix your own diet of course, which is a very difficult thing to do for laboratory experiments. Coprophagy is usually not observed under our conditions but it would have been of course excluded by the experiments with the animals on the grid and in metabolism cages.

GARNYS (Australia). It strikes me that your half part per million basal level in hair is high for a controlled animal. Can you give me details of how you presented the hair sample to the electrothermal atomiser in other words the dissolution technique and would you also give some detail on what steps you took to make sure of the accuracy of the analysis.

KOLLMER (Germany). I am not sure if I understood you, you want to know something about an analytical matter. Dilution flame atomic absorption.

GARNYS (Australia). What acid did you use to dissolve the cadmium in?

KOLLMER (Germany). We used nitric acid.

^{109}Cd KINETICS AND ACCUMULATION OF Cd IN THE ADRENALS AND THE PANCREAS OF THE RAT DURING CHRONIC EXPOSURE TO Cd.

W.E. Kollmer

Abteilung für Nuklearbiologie
Gesellschaft für Strahlen- und Umweltforschung m.b.H.
D-8042 MÜNCHEN - NEUHERBERG
WEST GERMANY

Functional deviations have been described after Cd administration in various endocrine glands [2-9]. In many cases it is not clear if these effects are caused directly by Cd in these tissues or are secondary to damage in other organs. We have previously reported a relatively small difference between the accumulation of ^{109}Cd in the adrenals and the kidneys of the rat following a single intravenous administration. In this study we have applied single doses of radioactive Cd during long term chronic exposure to stable Cd at different levels.

MATERIALS AND METHODS

Stable $CdCl_2$ was supplied in the drinking water. ^{109}Cd plus ^{115m}Cd as chlorides were administered by stomach tube. The ^{115m}Cd served mainly for whole body measurements to determine the amount resorbed. This was assumed to be equal to the total body retention at 7 days after the administration when the faecal excretion had come to an end. Demineralized water was given as drinking water to the controls. In two other experimental groups the drinking water was supplied with 30 and 300 mg stable Cd/l. The animals were fed commercial food pellets. Samples taken randomly from one lot of this diet contained between 0.1 and 0.2 µg Cd/g. The water consumption was measured twice a week. The radioactivity was administered as a single dose 40 days after the onset of Cd supply in the drinking water. In each group 5 animals were killed at intervals of 30 days thereafter. Stable Cd was analysed in tissues from all groups. Radioactivity was measured in all except the 90 and 150 day groups. The kidneys, the adrenals, and the pancreas were dissected. To avoid cross contamination the animals receiving the lowest amount of Cd were always dissected before those under higher exposure. Scissors, blades and forceps were carefully cleaned after dissection of each tissue and special instruments were taken for the pituitary and the adrenals. With kidneys and adrenals the peritoneal sheath was removed. In the case of the kidneys both organs from one animal were measured together. The adrenals had to be measured as a pooled sample of each group. Radioactivity was measured in a well-type γ-counter. The interference from ^{65}Zn radiation was excluded by analysis of the γ-spectrum. Radioactivities were not measured in the 90 days group. Stable Cd was analysed by flameless atomic absorption spectrometry, after wet ashing under pressure at a temperature of 150°C. The AAS was checked by an NBS standard. Due to their small volume, the pituitaries could not be analyzed by AAS. A median coronal slice of 100-150 mg was taken for analysis from the right kidney. From the pancreas, a subsample of 100-150 mg was analysed.

RADIOACTIVITY (Fig. 1)

In an earlier experiment [1], where ^{109}Cd had been administered by i.v. injection, the ratio in the accumulation of radioactivity between adrenals and kidneys had been found to be comparatively small (factor 5 at 150 days after injection). Indeed, it was much smaller than the ratios 27, 64, and 36 which had been calculated from data of other investigators for unexposed human subjects [10,11,12]. In our experiment, where radioactive Cd was administered by stomach tube, the maximum retention in the tissues may have disappeared already before the first samples were taken at 30 days post administration (Fig. 1). The retention of radioactivity in the adrenals was always practically at the same level as in the pancreas. In both glands it declined only slightly within 215 days. Due to the very long residence time of Cd in the kidneys a constant level was observed in the controls and a slight increase towards the end of the experiment at the lower exposure. At the higher exposure a significant decline of radioactivity occured in the kidneys reflecting a damage of the kidney function. The ratio of the concentration in the kidneys to that in the adrenals was between 14 and 34 in the unexposed animals and between 11 and 36 in the group exposed to 30 mg Cd/l. In both cases it increased gradually with time. In the animals at the high exposure it decreased due to the decline in the kidneys caused by the damage to this organ.

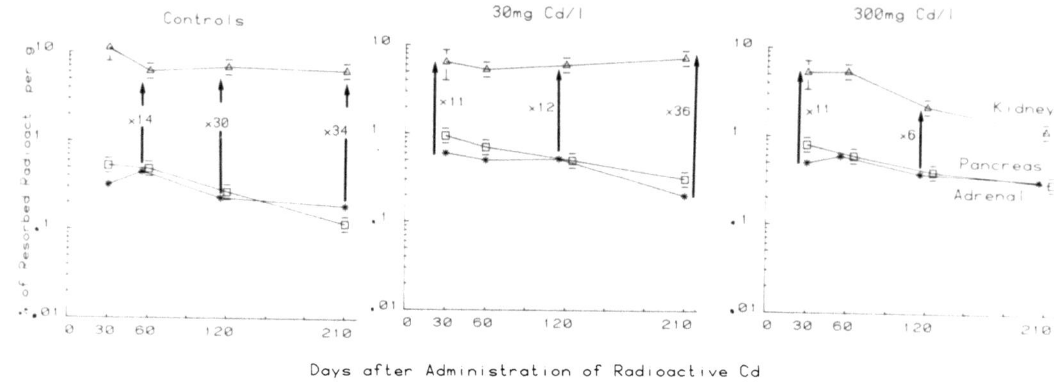

Fig. 1: Accumulation of radioactivity after a single dose ^{109}Cd by stomach tube (n=5)

STABLE CADMIUM (Fig. 2)

Within the initial period of exposure the accumulation of Cd proceeded very quickly. Of course the increase of Cd in the kidneys was most pronounced at the highest exposure. As the kidney increases its potential for Cd storage up to the toxic level, the kidneys cannot be assumed to exhibit a steady state in either of the experimental groups. In the pancreas, approximate steady state conditions prevailed in the controls and at the 30 mg/l level. Due to the analysis of pooled samples the statistical variation of Cd in the adrenals between the individual animals is unknown. In the animals exposed to 300 ppm, the concentration of Cd generally increased with time in both the adrenals and the kidneys. In the animals at 30 ppm exposure, the Cd concentration in the adrenals paralleled that in the pancreas - i.e., at equilibrium - with the exception of the 155 day group which for unknown reasons may be considered to behave not quite typically. In the controls a wide variation of Cd in the adrenals was observed. This may be due to a high sensitivity of this organ to the variation of the Cd intake possibly because of an uneven distribution in the commercial diet.

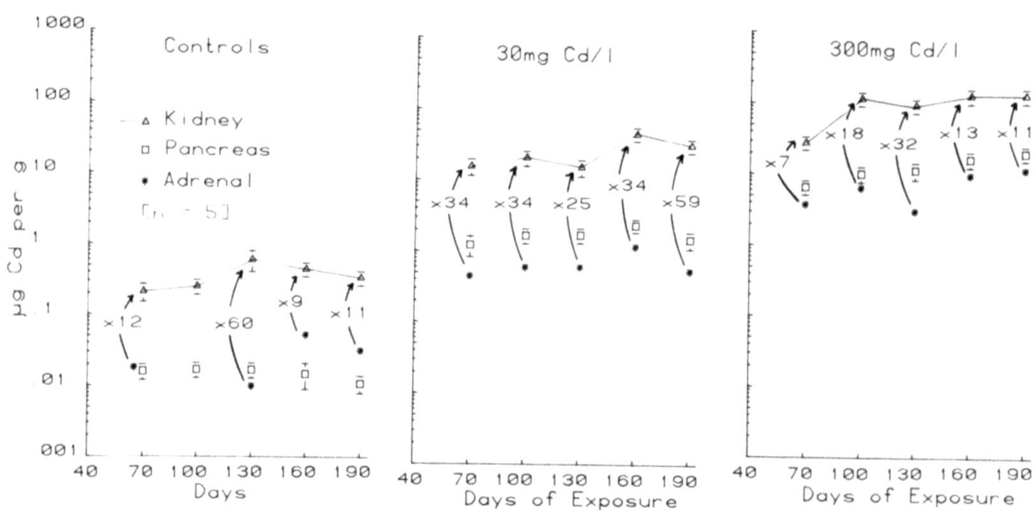

Fig. 2: Accumulation of stable Cd during chronic exposure

For stable Cd the ratio of the concentration in the kidneys to that of the adrenals ranged between 7 and 60. Most frequently it was around 30 especially at the exposure to 30 mg/l. Because at the one hand the Cd in the tissues of the controls was apparently influenced by the variation of the element in the food pellets and at the other hand it was influenced by kidney damage at the 300 mg/l exposure the data recorded for the 30 mg/l group may be indeed the most representative for animals on a steady supply of Cd at a subtoxic or pretoxic level. Therefore it is noteworthy that the Cd concentrations in the tissues of the 30 mg/l group were generally of the same magnitude as those reported for unexposed human beings.

DISCUSSION

After a single dose of radioactive Cd a large part of it is transitorily stored in the liver. There it is "diluted" with stable Cd so that during the following redistribution the radioactivity is much less significant, i.e., the specific activity is lower and depends on the Cd level in the liver. The radioactive data reflect therefore, in the first place, the initial uptake and the release by the individual tissues of the specific amount entering the blood stream within a short time. The present experiment shows that the initial distribution of radioactivity delivered via intestinal resorption to the tissues is different from that of an intravenous dose. The ratio of radioactive Cd concentration in the kidneys to that in the adrenals which was 5 after intravenous administration [1] was between 11 and 36 after administration by stomach tube. Ratios around 30 were dominating for very long periods in those groups without kidney damage. For stable Cd which was supplied in the drinking water for the whole experimental period the concentrations accumulated in the kidneys and the adrenals at different intervals also exhibited ratios between 30 in those animals which were on a steady supply of Cd and which did not accumulate toxic amounts in the kidney. Most strikingly, this corresponds well to the ratio obtained [1] by using the data of Tipton et al. [10] for man, which have also been recorded on the basis of fresh weight.

We have previously reasoned, on the basis of radioactivity distribution, in favour of a direct effect of Cd on the function of the adrenal. Although we have found a higher ratio for the actual accumulation of Cd in the adrenals during chronic exposure, the concentrations were not much below those in the pancreas. If the microdistribution demonstrated by Berlin and Ullberg [13] is taken into consideration a higher accumulation than indicated by our analyses may be expected in the cortex. Therefore a direct effect of Cd on the function of the adrenal cortex may be suspected after chronic exposure.

SUMMARY

The distribution of radioactive Cd administered as a single dose was different if the dose was delivered intravenously or by stomach tube. The ratios for the concentrations of radioactive Cd in the kidneys to that in the adrenals found over a period of 200 days after administration of a single dose by stomach tube were in accordance with those found for stable Cd accumulated during chronical exposure to Cd in the drinking water as well as with the ratio which was calculated from Cd concentrations in unexposed humans. They were appreciably higher than those found for radioactive Cd after intravenous injection. The radioactivity concentration in the pancreas was in the same range as in the adrenals. However, the accumulation of chronically supplied stable Cd was slightly higher in the pancreas than in adrenals with the exception of the controls.

ACKNOWLEDGEMENTS

I am grateful to Mrs. G. Treutler and Mr. H. Ruhstorfer for technical assistance. For help with γ-spectrometry I am grateful to Dr. D. Berg.

REFERENCES

1. Kollmer, W.E., 1980, Int. J. Appl. Radiat. Isotop. 31: 607-609.
2. Anbar, M., and Inbar, M., 1964, Acta Endocrinol. 46: 643-652.
3. Balkrishna, J., 1962, J. sci. ind. Res. India 21: 187-189.
4. Kar, A.B., Dasgupta, P.R., and Das, R.P., 1960, J. sci. ind. Res. India 19: 225-230.
5. Chiappino, G., and Baroni, M., 1969, Med. Lavoro 60: 297-305.
6. Madhyastha, N.A., and Dutt, N.H.G., 1970, Indian J. exp. Biol. 17: 637-639.
7. Parizek, J., and Zahor, Z., 1956, Nature 177: 1036.
8. Richardson, M.E., Fox, M.R.S., and Fry, B.E. Jr., 1974, J. Nutr. 104: 323-338.
9. Singhal, R.L., Merali, Z., and Hrdina, P.D., 1976, Fed. Proc. 35: 75-78.
10. Tipton. I.H., Steiner, R.L., Foland, W.D., Mueller, J., and Stanley, M., 1954, USAEC-ORNL-Report CF-54-12-66.
11. Tipton, I.H., and Cook, M.I., 1963, Health Phys. 9: 103-145.
12. Forssen, A., 1972, Ann. med. exp. Biol. Fenn. 50: 99-162.
13. Berlin, M., and Ullberg, S., 1963, Arch. environ. Health 7: 686-693.

DISCUSSION

SPICKETT (Australia). Did you notice any change in body weight of the animals fed the 300 milligrams per litre cadmium?

KOLLMER (Germany). We observed a lower increase of body weight at increasing levels of exposure. This is also the common experience described in the literature.

SPICKETT (Australia). Have you taken the weights of the kidneys? Do you know the increase of Cd in the whole kidney?

KOLLMER (Germany). Yes, we have taken the weights. We have also calculated the ^{109}Cd in the whole kidneys. Stable Cd for whole kidneys was not calculated due to the procedure of analysing slices. Because we were mainly interested in the concentrations which bear a direct relation to the risk of toxicity. I do not have the data on total kidneys with me. I can send them to you if you are interested.

SCHENKEL (Germany). I would just make a comment. This was a very interesting study. We have done a similar experiment with pigs with dietary concentrations of 0.04ppm and 0.6 ppm cadmium and we killed the animals at a body weight of 30kg and a body weight of 100 kilograms. We got also very different distributions in the body depending on the dose and on the age of the animals and only when the cadmium was given orally. You also showed that there were different distributions when the cadmium was given via the diet or intravenously.

KOLLMER (Germany). Now in our experiments we could see that when we went to a dose of chronic exposure that was lower than 3ppm we did not get a distinct effect when compared to the controls. The standard food, as I told you, has a concentration of 0.1 - 0.2ppm so if you come into that region or even a factor of 10 above that you do not get a distinct deviation from your controls. That is all I can say. In the controls, as we have shown here, there may be fluctuations which you do not control, if you do not control the content of the standard food. On the other hand influences of the dose can be seen again if there is any toxic effect maybe on the liver or maybe on the kidney. It is with the highest dose of course, where kidney damage is well known. Even at the dose of 30ppm you may get kidney damage on long exposure.

STUDIES OF A WORKFORCE EXPOSED TO CADMIUM

R. Scott,* C. Cunningham,* G.S. Fell,* J. Ottoway+

* Department of Urology and Biochemistry, Royal Infirmary, Glasgow
+ University of Strathclyde, Glasgow, Scotland

During 1975 from a survey of a random population it was possible to determine that the prevalence of stone disease in the local community was 3.5%[1]. In addition a variety of biochemical parameters including serum inorganic phosphate were measured and normal ranges established[2]. In parallel with this study, a group of coppersmiths who had been chronically over-exposed to cadmium fume were found to have a prevalence of upper urinary tract stone disease of 18.5%. One of the most significant findings in the coppersmiths was that the serum inorganic phosphate levels were significantly different from control subjects[3]. The process which had caused the chronic cadmium poisoning was of a continuous production line type where brazing rods containing 20% cadmium were used to solder small copper tubes onto a compressor unit. The T L V's for cadmium fume had been regularly exceeded. It was decided that since the brazing process had previously been undertaken at several points in the factory that the general workforce should be investigated.

METHOD

All members of the factory were invited to participate by completing a simple questionnaire and were then asked to submit to a simple blood and urine examination. Blood cadmium, copper, iron and zinc with serum inorganic phosphate were checked in the routine laboratory and urine was checked for protein by the albustix and salycylsulphonic acid tests. 403 subjects (371 male, 32 female) were examined and 29 other subjects completed forms but did not attend for examination.

RESULTS

Careful scrutiny of the factory personnel revealed they were twice as likely to have suffered from renal stone disease as the general population.

TABLE I - FACTOR PERSONNEL v POPULATION SURVEY - RENAL STONE DISEASE

SYMPTOM	FACTORY (403)	RANDOM POPULATION (3378)
Colic	3.2%	3.1%
Stone disease	3.4%	1.1%
Passage of stone/gravel	2.7%	1.4%

PROTEINURIA

The same test applied to both groups revealed that albustix positive results were obtained in 10.66% of the factory scan as compared with 3.6% of the general population.

SERUM INORGANIC PHOSPHATE

In view of the significant difference of this particular biochemical estimation in the coppersmith group particular attention was paid to this measurement in the general factor scan. In 30 of the latter group where the serum inorganic phosphate was less than 1 mmol/litre the prevalence of stone disease was found to be 13.1% (4 of 30 subjects) and in 2 of these individuals the disease was bilateral.

BLOOD CADMIUM RESULTS

When consideration was taken of the smoking habits the mean blood cadmium levels in all of the factory personnel was at least three times greater than those in a non exposed, age matched group of individuals. In any normal population smokers would be expected to have higher blood cadmium levels than non smokers[4] and this pattern was found in those employed for less than 11 years after which time the effect of the pollution in the atmosphere results in this normal pattern being changed and this is demonstrated when male smokers and non smokers are compared (Fig. 1).

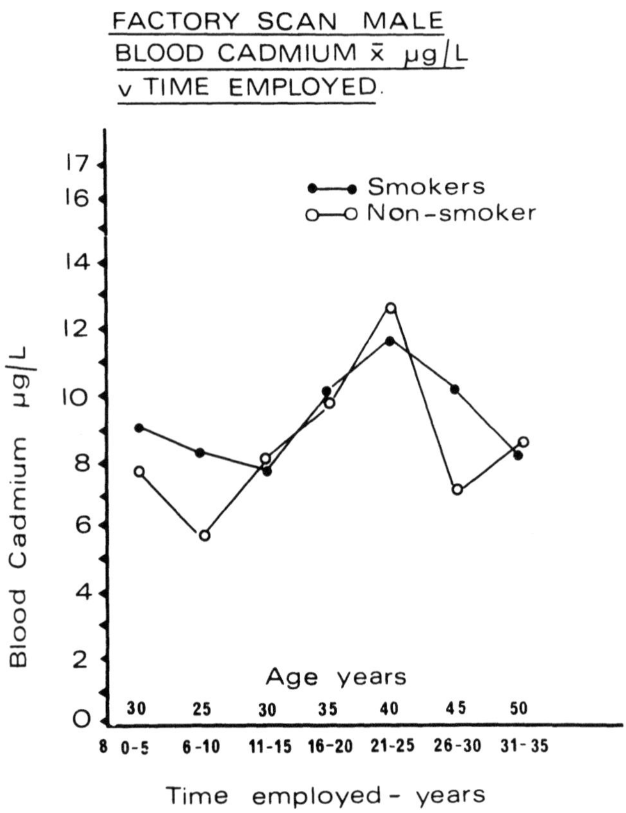

DISCUSSION

When cadmium contaminates a factory atmosphere it is more than likely that the element will affect the total workforce and cause major changes in blood levels of cadmium which can be seen to be significantly different from the normal population. It has been demonstrated that cadmium poisoning occurring as a result of industrial exposure can be studied epidemiologically by measuring the $Beta_2$ microglobulin in the urine[5]. Follow-up studies in such situations often demonstrate the persistence and increased quantities of low molecular weight protein excretion in severely affected individuals[6]. The present study has shown that even a simple dip-stick type of analyses of urine will results in a greater number of albumin positive results being obtained. Low molecular weight proteinuria in chronic cadmium poisoning is accepted as an indication of renal tubular damage but there is a growing awareness that renal stones are a major clinical problem in cadmium exposed workers.

The finding that in the general factory scan there was a greater than normal prevalence of renal stones suggests that in the investigation of an individual thought to be likely to be suffering from chronic cadmium poisoning then a plain abdominal film is essential to exclude possible renal calculi. Where there is evidence that a stone might be present then full radiological investigation of the urinary tract is necessary.

The present studies have shown both within a specialized group such as the coppersmiths and in the general factory scan that a high index of clinical suspicion should be raised when the serum inorganic phosphate is found to be low in an individual subject[7]. Very little attention has been paid to this biochemical measurement in the past but the present studies show that it may be the best indication that the individual has or is likely to develop renal stone disease.

REFERENCES

1 Scott, R. et al., 1979, Stone Symposium, Perth, Australia.

2 Gardner, M. and Scott, R., 1980, J. Clin. Path., 33, 380-385.

3 Scott, R. et al., 1976, Lancet 2, 7882, 396.

4 Lauwerys, R.R., 1978, Trace metals. Commission of the European Committee, pp 43-64.

5 Piscator, M., 1979, Heavy metals in the Environment, London.

6 Kazantsis, G., 1979, Environment Health Perspect. 28, 155-59.

7 Scott, R., et al., 1981, (in press). Renal Stone Symposium, Williamsburg, USA.

DISCUSSION

ABDULLA (Sweden). Do you have any theory of what led to this transformation apart from the low phosphate levels.

SCOTT (Scotland). Yes, a large number of these individuals have got hypercalciuric urine. This has been recorded in the past. We did a particular study in coppersmiths where we measured total body calcium using an in vivo neutron activation analyses and we could show, when we corrected for lean body mass, weight, sex, smoking habits etc., they have got a significant drop in total body calcium, so obviously they are pouring out calcium. They are also pouring out phosphates in their urine.

WILLIAMS (Wales). There are two sides to any trace element dependent condition; there is the "accident prevention" side and there is the "treatment of the injured". I think we have had an excellent paper from you, Gordon Fell and collaborators on the "accident prevention" side. What about the treatment of the patients? Have you tried any ligands for detoxifying these. We particularly commend "Unithiol" (2,3 dimercaptopropanesulphonic acid) this new agent that Dr Planas-Bohne from Karlsruhe has developed?

SCOTT (Scotland). We would be extremely interested to do this. At the present time what I have been doing is giving thiozide diuretics to dimish the urinary calcium excretion, hoping to prevent stone formation in these individuals. I would be very interested to speak to you later because I understand you have a method of actually mobilising and getting rid of cadmium. We have been very wary of this because of the heavy level of contamination in these individuals. Coincidentally the coppersmiths still, after six years, have a mean blood cadmium which is over 8 micrograms per litre, which will give you an idea of the total body burden. We were very wary of trying to mobilise that cadmium in case we produced acute toxicity.

GARNYS (Australia). You mentioned in passing atomic fluorescence analysis - I guess this is an extension of John Ottoway's work; was this by flame or furnace.

SCOTT (Scotland). It is a flame method,it is written up by Gordon Fell who is the biochemist in this team.

ELMES (Wales). I am sorry I missed the first minute of your talk so I did not know how long the coppersmiths had been exposed to cadmium but I noticed on your years of exposure to cadmium there was a peak in the middle of the graph of blood cadmiums and a fall off at the right-hand end. Why is this?

SCOTT (Scotland). I think possibly this could well be the related to the known phenonomen that, as in kidney cadmium which we are showing in the next paper, one gets a rise to 50 years then it falls off for some odd reason. I do not really understand why the blood cadmium should behave as it did. The average working time in the factory for the coppersmiths was 19 years.

BRATTER (Germany). I think that often renal disorders are connected with bone disorders. Do you also consider variation of bone metabolism in connection with the cadmium exposure?

SCOTT (Scotland). Yes we have looked at one particular aspect of calcium metabolism looking for bone disorders. We have undertaken for instance, densiometry studies of bone and have shown no major obvious differences in the male subjects in our group. Now thinking about the ITAI-ITAI story we did whole body in vivo neutron activation analysis measuring total body calcium. Then we did find significant differences related to length of time of exposure and I suspect that if we look at them again probably we may find by now that we are beginning to see bone changes, hoping of course, that having stopped them from further exposure that these changes may not fully develop.

THE CADMIUM CONTENT IN HUMAN KIDNEYS

R. Scott,[+] E. Aughey,* M. Reilly,* C. Cunningham,[+] A. McLelland,[+] G.S. Fell[+]

[+] Departments of Urology and Biochemistry, Royal Infirmary, Glasgow
* Department of Histology and Vetinary Medicine, University of Glasgow

The association between cadmium and human cancer has been of interest since it was noted that cadmium workers had a significantly greater risk of developing prostatic cancer [1,2]. Some workers have disputed this association[3]. Recent work however, has confirmed an increased risk factor for several cancers including a significant increase in nasopharyngeal cancers and a greater than normal risk factor for lung, prostate and colorectal cancer in subjects who have been exposed to cadmium in a work situation[4].

PRESENT STUDY

Kidney samples were obtained from routine post mortems. The lower pole of the left kidney was removed and preserved by storing at -20°C until analyses could be undertaken. After separating the cortex and medulla of the segment of kidney, samples were analysed in the trace element laboratory for cadmium, zinc and calcium content. All three elements were measured in µg/g and expressed as wet and dry weights. Histological specimens were prepared using Haemotoxylin and Eosin, and Masson staining techniques.

DISEASE/AGE/SEX/SMOKING HABITS

All post mortem reports were carefully scrutinised, along with details of the clinical notes for each individual patient. The cause of death was considered to be the major condition which had been present antecedent to death. For example, if someone died following an operation in the course of a prolonged illness then the latter was considered to be the main cause of death and obviously more likely to be affected by previous trace element exposure. Smoking habits were ascertained in 189 individuals from the case sheet information and where possible from next-of-kin. Enquiries were also made into the individuals' occupation but the latter data proved to be very difficult to obtain.

RESULTS

Of the first 300 kidneys analysed it was possible to have paired analyses for all three elements in 290 kidneys. Of the other 10 samples 5 were controls and 5 were either incomplete or missing. Analyses confirmed that there is a highly significant difference ($p < 0.001$) between cortex and medulla for all three elements both when expressed as dry and wet weights.

DISEASE CLASSIFICATION - The cause of death (Table 1) was arbitrarily divided into 7 main categories.

TABLE I - CADMIUM CONTENT v DISEASE

	MALE NO.	$\bar{x}$	FEMALE NO.	$\bar{x}$
Pulmonary	22	19.22	12	21.56
Cancer	56	16.73	52	19.27
Ischaemic Heart Disease	30	17.13	15	14.06
Hepatic Disease	12	25.70	11	17.12
Renal Disease	1	11.00	5	16.14
Cerebro vascular accident	17	14.89	13	25.00
Miscellaneous	17	19.03	27	14.18

The highest mean values of cadmium levels were found in men dying of predominately hepatic causes and in females dying of cerebro vascular accidents (Table I). Statistical analyses however confirmed a non parametric distribution of results. Therefore a Mann

Whitney test was applied to the figures and this revealed that when median values are analysed then medullary cadmium in subjects (male and female) dying from pulmonary or neoplastic causes was significantly higher than those dying from miscellaneous causes. There was no detectable significant difference between those dying of cancer and those dying of ischaemic heart disease, hepatic, renal or cerebro vascular disease.

AGE v CORTICAL CADMIUM (WET WEIGHT)

Comparisons were made of the cortical cadmium in 10 year periods. In subjects of more than 80 years the median renal cortical cadmium values were significantly lower (Mann Whitney test) than in those aged between 50 and 79 years of age.

SMOKING HABITS

Good information was available in 189 subjects. Comparison of the non smokers with smokers (Gp I < 10, Gp II 10-25, Gp III 26-45 cigarettes/day) revealed that the non smokers had significantly less renal cortical cadmium than those who smoked 10-45 cigarettes per day (Table II).

TABLE II - NON SMOKERS v SMOKERS - SIGNIFICANT DIFFERENCES

	DAILY CONSUMPTION OF CIGARETTES		
	GROUP I <10	GROUP II 10-25	GROUP III 26-45
Cadmium cortex (wet weight)	-	***	*
Cadmium medulla (wet weight)	-	***	**
Zinc medulla (wet weight)	*	-	-

* = significant, ** = highly significant, *** = very highly significant

Significant differences were found between heavy cigarette smokers (Gp II and III; Table II) and cigar and pipe smokers.

Medullary zinc was found to be lower in non smokers and this attained significance when compared with all forms of cigarette smoking, being highly significant in the heavy smoking groups. Calcium showed no significant differences in any group. When the smoking habits were applied to the different disease categories then it was noted that when the four largest groups were compared (Table III) there was an obvious difference in the mean value in those suffering from ischaemic heart disease. Because of the non parametric distribution already mentioned the results were analysed using the median values and were found to be significantly different ($p < 0.05$ Mann Whitney test) between smokers and non smokers dying of ischaemic heart disease (cortical cadmium).

TABLE III - CADMIUM - MEAN FOR A VARIETY OF DISEASE - SMOKERS v NON SMOKERS

Mean all disease = 17.95 µg/g wet weight

	TOTAL	MISSING	SMOKERS		NON SMOKERS	
	NO.	NO.	NO.	$\bar{x}$	NO.	$\bar{x}$
Pulmonary	34	7	17	19.80	10	18.20
Ishcaemic heart disease	45	15	19	19.80	11	12.60
C.a. Lung	30	8	16	17.60	6	15.60
C.a. gastro intestinal tract	42	12	13	20.70	17	19.50

DISCUSSION

It has been recognised for some time that there is a greater concentration of cadmium in the cortex of the kidney as compared to the medulla[5]. The present study has confirmed a highly significant difference for cadmium, zinc and calcium when cortical and medullary concentrations are compared. When the overall results in the present study are compared with other sources then there is no obvious difference between the results in Scotland as compared with those from other areas (Table IV).

TABLE IV - CADMIUM CONTENT OF KIDNEYS - REPORTED SERIES μg/g WET WEIGHT

AUTHOR	YEAR	NO. OF SAMPLES	$\bar{x}$ CADMIUM
Elinder et al[6]	1976	292	22.00 μg/g
Milton et al[7]	1976	91	16.80 μg/g
Tsuchiya et al[8]	1976	160	125.30 μg/g
Scott et al	1981	155 (males)	17.87 μg/g
		135 (females)	18.14 μg/g

When compared with the results from Japan the results are markedly lower and suggest that differences in diet play a major part in determining total cadmium intake.

The highest cadmium value obtained in the present survey was 84.8 μg/g (wet weight) which is well below the critical level thought to be likely to produce renal toxic effects. What is of interest however is that the present study has been undertaken in Scotland, a country where cancer of the lung is a common cause of death. Hypertensive disease is also very prevalent in the community especially in the West Central belt where this study was conducted. Hypertensive disease is considered by some authors to be associated with cadmium exposure[9].

When the cadmium content of kidneys is related to the major conditions causing death then the present study has failed to show any significant difference between subjects dying of ischaemic heart disease and other causes of death. What is of interest however is that in subjects dying of cancer or pulmonary disease, there are significant differences in the median values of cadmium expressed as wet weight in the renal medulla. In the early work on the effects of cadmium at a clinical level it was emphasised that the pulmonary effects were paramount[10]. The finding of a significant difference in cancer subjects in an area known for its high cancer morbidity clearly merits further investigation.

Analyses of the cadmium content related to smoking habits again shows that in any interpretation of cadmium levels it is important to realise that smoking markedly effects the content of cadmium in kidneys. The present study undertaken in an area where ischaemic heart disease causes considerable morbidity and mortality reveals significant differences in cortical cadmium values between smokers and non smokers. If cadmium is shown to be a factor in hypertensive disease then clearly the present study adds to the clinical arguments advocating the reduction of smoking in hypertensive subjects.

REFERENCES

1 Potts, C, 1963, Ann. Occup. Hyg. 8, 55.

2 Kipling, M.D., Waterhouse, J.A., 1967, Lancet 1, 730.

3 Malcolm, D., 1972, Ann. Occup. Hyg., 15, 33-36.

4 Kjellstrom, T., Friberg, L., Rahnster, B., 1979, Environ. Health 28, 199-204.

5 Livingstone, H.D., 1972, Clin. Chem. 18, 67.

6 Elinder, C.G., Kjellstrom, T., Friberg, L., 1976, Arch. Environ. Health, 31, 292-300.

7 Miller, G.J., Wyllie, M.J., McKeown, D., 1976, Med. J. Austral. 1, 20-23.

8 Tsychiya, K., Seki, Y., Sugita, M., 1976, Keio J. Med., 25, 83-90.

9 Schroeder, H.A., Brattleboro, V.T., 1964, JAMA, 187, 358.

10 Bonnel, J.A., 1955, Brit. J. Indust. Hyg., 12, 181.

Acknowledgements: This work was supported by the Department of the Environment, London and the Department of Pathology, Royal Infirmary, Glasgow.

DISCUSSION

ANON. May I point out one thing - there has been some suggestions concerning the cadmium contents in the kidney and hypertension, and I wonder if you have particularly looked at this aspect in your study.

SCOTT (Scotland). Yes. I should have emphasised that in the talk, it is very interesting but, we have got big problems with ischaemic heart disease in our particular area but we do not have enough data at the present time to talk about hypertension beforehand. We are continuing with this particular study and we hope eventually to have 1000 kidneys for analyses, which should give us a good idea of the hypertension picture. It is very obvious, for example, that when someone is admitted to hospital as an acute emergency and they die fairly quickly, then you don't have any concept of what the blood pressure has been running at over a long period of time, so we therefore do not have enough data of that type at the moment but we are pursuing this particular aspect.

SCHENKEL (Germany). May I ask if in your material there are also some industrial workers which you have shown in your earlier paper, and may I ask you for the cadmium concentration in the kidney?

SCOTT (Scotland). Again finding details of occupation is extremely difficult. We have been attempting to get the information by writing to relatives. Now this is an extremely emotive time in which to write to relatives and ask about details of jobs. But we are accumulating data and again we hope to eventually be able to report that particular aspect.

GOONERATNE (Australia). Have you got any information on kidney function with the increasing cadmium in the kidney?

SCOTT (Scotland). Well we do know from our own clinical studies that certainly kidney function diminishes - there is no question of that. I don't know the particular relationship to the post mortem specimens. Again we have got that data available and eventually will have it analysed.

GOONERATNE (Australia). By how much does the function diminish?

SCOTT (Scotland). I think this depends on the exposure individuals have had to cadmium. I think there is quite a variable degree of renal function in these individuals. The interesting thing is if one thinks back to clinical signs, then the coppersmiths are showing the classical picture of an increasing proteinuria with diminishing creatinine clearance.

GOONERATNE (Australia). Is this glomerular or tubular dysfunction.

SCOTT (Scotland). I happen to belong to a school of thought which is of the opinion the protein in urine which you get in cadmium poisoning is both tubular and glomerular and we have got very good experimental proof from the animals that you get tremendous glomerular damage and if one looks at the beta globulinuria for example, it is quite obvious that there are a large number of diseases including glomerulonephrites which cause increased $beta_2$ microglobulin levels and I don't think $beta_2$ microglobulin can be specific for cadmium.

SEGAL (Australia). A few years ago the high incidence of heart disease in Scotland was attributed to the hardness of the drinking water. Would you care to comment on that?

SCOTT (Scotland). It is certainly not due to the hardness of water - it is due to the lack of hardness. Our water is extremely soft and acidic. It is generally thought that because of this low degree of hardness with the high acidity, if there is cadmium in the pipes then it is possible for it to leach into the water and therefore be consumed by the general population. This again has not been properly studied and we are hoping that by showing this association of high cadmium content and ischaemic heart disease, this will help solve this problem.

ANON. We are embarking on an organ sampling programme and I would like details of how you take the samples, because we are finding that all of the hospital scalpels that we have used to sample kidney are just loaded with cadmium. What technique do you use to sample kidney in a contamination-free way?

AUGHEY (Scotland). Due to the goodness of the Department of Environment we have in fact a technician who undertakes all this under instruction from Dr Fowell and we have scalpels which are specifically contaminent-free and when the kidneys are taken out of the deep freeze then the superficial area is dissected off. We then dissect off the cortex from the medulla and it is all handled in plastic with minimal cadmium contamination all the way through.

LEWIS (England). Presumably in smokers the cadmium is getting into the body by inhalation. In other parts of the factory, previously mentioned, do you think some of the cadmium comes through the oral route and have you any means of distinguishing between these routes in the tissues you examine?

SCOTT (Scotland). There is no way of determining where cadmium comes from once it is into the tissue, but there is no doubt about it that you can get cadmium in through the gastro intestinal tract from dust.

LEWIS (England). Yes I realise that - I wondered about the levels you were finding in different parts of the factory, particularly in the non-smokers, do you think that these people were getting it by an oral route rather than by inhalation?

SCOTT (Scotland). No - this is going back a paper. I think in fact these chaps were getting it from fumes. We did a few experiments around the factory and discovered that this braizing process had in fact been undertaken in different parts of the factory and that with certain prevailing winds etc., you could see the dust coming down on top of them.

EMBRYONIC UPTAKE OF CADMIUM FOLLOWING MATERNAL EXPOSURE IN MICE

W.S. Webster and J. Christley

Department of Anatomy,
University of Sydney,
SYDNEY, N.S.W. 2006,
AUSTRALIA.

Employees in at least 30 occupations in Australia are in direct risk of industrial exposure to cadmium (Cd). Because of the increasing participation of fertile women in the workforce the risks associated with this exposure to Cd include its effect during pregnancy.

It is well known that single, large doses of Cd salts (2.4mgCd/kg), are teratogenic in rodents, producing a wide range of defects such as exencephaly and limb reduction deformities; but neither the mechanism by which Cd damages the embryo, nor the lowest safe level of maternal exposure, is known. There is even controversy as to whether Cd reaches any part of the embryo other than the gut (Dencker, 1975).

The aims of this investigation were (1) to determine, by gamma scintillation counting, whether maternal Cd exposure in pregnant mice, during the organogenetic period, resulted in embryonic exposure, and, if so, was the relationship dose dependent, and (2) to determine by autoradiography whether Cd reached those tissues of the embryo that subsequently develop abnormally following a teratogenic dose of Cd.

For the gamma scintillation study 48 pregnant mice (C57BL/6J), on day 9 of gestation, received an intraperitoneal injection of $CdCl_2$ laced with 20µCi $Cd^{109}Cl_2$. The amount of non-radioactive $CdCl_2$ varied so that the final dose was either 0.66µgCd/kg, 40µgCd/kg or 2400µgCd/kg, the latter dose was teratogenic (Webster and Messerle, 1980). The pregnant mice were sacrificed at 1, 12, 24 or 240 hours after the Cd exposure and the radioactivity in the embryos was measured using a Packard Autogamma Counter. The results were expressed as percentage of injected dose per embryo (Table 1) or percentage of injected dose per gram dry weight (Table 2).

TABLE 1

Embryonic uptake of Cd, at 1, 12, 24 or 240 hours after maternal exposure, expressed as percentage of injected dose x 10^{-3} per embryo ± S.D.

	Time after injection (hours)			
Dose µg/kg	1	12	24	240
0.66	0.12 ± 0.02	0.04 ± 0.02	0.15 ± 0.10	12.33 ± 6.88
40	0.13 ± 0.04	0.07 ± 0.03	0.08 ± 0.02	6.69 ± 2.47
2400	0.23 ± 0.05	0.06 ± 0.03	0.10 ± 0.03	1.43 ± 0.79

When the Cd content of each embryo was expressed as a percentage of the dose (Table 1) it was evident that there were marked and significant changes in the Cd content of the embryos with time, but, with the exception of the 240 hour figures, differences between doses were not significant. The figures indicate that Cd rapidly entered the embryos following maternal exposure but levels then decreased during the next 11 hours only to rise again by 24 hours and continued to rise for the next nine days. The Cd content of the foetuses after 240 hours was so much higher than at any time measured that it indicates Cd was available for incorporation into the foetuses for the entire 10 day period following maternal exposure. This pattern of incorporation was probably related to the blood Cd levels which have been shown to peak within 1 hour of a single injection then to fall rapidly during the following 12 hours then to rise again between 12 and 96 hours (Nordberg, 1972). Although the embryos continued to accumulate significant amounts of Cd after the initial 24 hour period, this accumulation is unlikely to be associated with teratogenesis because the concentration of Cd in the embryos progressively declined as the embryos increased in weight (Table 2). During the time period studied the mean embryonic dry weight increased from 0.13mg to 100mg.

TABLE 2

Embryonic uptake of Cd, at 1, 12, 24 or 240 hours after maternal exposure, expressed as percentage of injected dose per gram dry weight ± S.D.

	Time after injection (hours)			
Dose μg/kg	1	12	24	240
0.66	0.95 ± 0.46	0.21 ± 0.15	0.23 ± 0.20	0.08 ± 0.04
40	0.84 ± 0.13	0.19 ± 0.08	0.16 ± 0.09	0.05 ± 0.02
2400	1.66 ± 0.67	0.31 ± 0.20	0.28 ± 0.13	0.01 ± 0.01

The Cd concentration in the embryo, was at its highest after 1 hour and then decreased rapidly in the next 11 hours and continued to decrease for the rest of the pregnancy. Differences between Cd concentrations as a percentage of the injected dose were not significant.

For the autoradiographic studies, pregnant mice, on days 7, 8 or 9 of gestation, received an intraperitoneal injection containing 200μCi of $Cd^{109}Cl_2$. Some mice received only this dose which corresponded to 6.6μgCd/kg, others also received 2400μgCd/kg as non-radioactive $CdCl_2$. The animals were sacrificed at 1 or 12 hours and the embryos were prepared for autoradiography after fixation in 2.5% glutaraldehyde and 0.5% tannic acid in 0.1M Millonig's phosphate buffer.

The autoradiography revealed that Cd was present in all tissues of the embryos at both 1 and 12 hours after maternal injection of the teratogenic dose and was particularly concentrated in the cells of those tissues which subsequently became abnormal such as the neural tube or limb buds. In the embryos exposed to the lower dose, the highest Cd accumulation was seen in the embryonic gut, but all tissues showed a low level of labelling.

In conclusion it has been shown that during pregnancy in mice, maternally administered Cd rapidly entered the developing embryos during the organogenetic period. The highest concentration of Cd occurred 1 hour after maternal exposure. The amount of Cd present in the embryos was related to the time after maternal exposure and the dose. Over a wide range of doses a constant percentage entered the embryos. Cadmium was found in all tissues of the embryos but at high doses was particularly concentrated in those tissues that would subsequently develop abnormally.

1. Dencker, L., 1975, J. Reprod. Fert. 44:461-471.

2. Webster, W.S., and K. Messerle, 1980, Teratology 21:79-88.

3. Nordberg, G.F., 1972, Environ. Physiol. Biochem. 2:7-36.

DISCUSSION

BRATTER (Germany). We are also interested in the transfer of cadmium to the foetus in connection with smoking habits and we are presently investigating the cadmium content of amniotic fluid. The aim of your research is also the transfer of maternal cadmium burden to the foetus, but your experiments are carried out using animals. How can you use the results you have obtained with respect to pregnant women. Please give your comment on this.

CHRISTLEY (Australia). We were using the mouse as a model pregnancy, we accept that there are some placental differences between mice and humans but this study gives some indication of the behaviour of cadmium in a reproductive system.

AGGETT (Aberdeen). I would just like to ask did you in fact have a look at the foetuses to see if there were any teratogenic affects.

CHRISTLEY (Australia). Abnormalities were only seen with the highest dose of Cd. At one hour after exposure there was no obvious cellular deformity in the foetuses. However if you looked at twelve hours there were abnormalities in the neural tube and the limb buds, with large amounts of cytoplasmic inclusions and dead cells slaughed off the neural tube. At this time, of course, the embryo did contain less cadmium, but these effects were probably due to the amount of cadmium that was in the neural tube at one hour.

AGGETT (Aberdeen). Did all the embryos, that you anticipated would survive the 240 hours, actually survive.

CHRISTLEY (Australia). There was an increased resorption rate of the embryos. But a significant proportion of the embryos did survive, and did show the abnormalities at birth.

HARLAND (USA). Tell us something about the mouse diet employed, as you know, cadmium in some instances is not taken up to a great extent if certain nutrients are present in the diet.

CHRISTLEY (Australia). It was a standard commercial diet. The cadmium was injected intra peritoneally into the mice where it seemed to be absorbed fairly well and rapidly.

RAT BLOOD CADMIUM LEVELS AND EARLY RENAL LESIONS

[+]E. Aughey, *R. Scott, G. S. Fell, C. Cunningham

[+]University of Glasgow Veterinary School, Glasgow, G61 1QH, SCOTLAND
*Royal Infirmary, Glasgow, G4 OSF, SCOTLAND

Introduction

Cadmium nephrotoxicity has been described in the rat following parenteral injection (Bonnell Ross & King, 1960) and after oral dosage (Itawaka, Abe & Tanaka 1973). Loeser and Leucke (1977) used prolonged oral dosage with graded amounts of Cd up to 30 ppm and although the metal accumulated in both kidney and liver, no obvious pathological changes were found. Human exposure to low levels of Cd can occur from industrial operations or from cigarette smoking but the clinical significance is uncertain (Fell, 1980). We have therefore examined low level exposure in the rat using sensitive histological and biochemical techniques to monitor dose related uptake of cadium and early renal pathology.

Materials and Methods

Three groups of 15 adult male Wistar rats were given, 0, 10 or 50 ppm Cd in drinking water. Records were kept of food and water intake, body weight gain and behavioural pattern. Three rats from each group were killed at 1, 2, 4, 8 and 12 weeks. Blood (2-4 ml) was obtained via the abdominal aorta, and the kidney and liver removed. Light and electron microscopy studies were carried out on kidney sections. Cd determinations were made on tissue samples and in whole blood, red cells and blood plasma using atomic absorption spectrophotometry.

Results

The food and water intakes of all 3 groups of rats were similar, but the 50 ppm Cd group showed a slight reduction in growth rate towards the end of the 12 week study. A dose related accumulation of Cd in kidney and liver was demonstrated (Table 1).

TABLE 1

	KIDNEY µg Cd/g wet weight			LIVER µg Cd/g wet weight		
Week	0	10 ppm Cd	50 ppm Cd	0	10 ppm Cd	50 ppm Cd
1	0.06	0.11	0.38	0.05	0.08	0.39
	0.07	0.13	0.55	0.04	0.13	0.30
	0.06	0.24	0.47	0.02	0.10	0.75
2	0.09	1.29	3.14	0.03	0.23	1.33
	0.15	0.67	1.67	0.05	0.21	0.75
	0.04	0.62	2.60	0.01	0.33	1.14
4	0.07	2.33	6.33	0.06	0.41	4.27
	0.08	1.69	8.05	0.01	0.55	4.08
	0.11	1.88	6.06	0.01	0.64	2.50
8	0.08	3.81	9.63	0.11	1.00	12.65
	0.09	4.18	18.61	0.09	1.28	13.96
	0.09	4.61	17.47	0.09	2.21	12.02
12	0.09	2.46	29.21	0.05	3.46	23.49
	0.06	6.31	25.72	0.08	2.21	11.76
	0.12	6.94	28.20	0.06	2.68	31.36

The concentration of Cd in whole blood, and its distribution between red cells and plasma were measured after 4 weeks (Table 2).

TABLE 2	Whole Blood Cd μg/l			Red Cell Cd μg/l			Plasma Cd μg/l		
Week	0	10	50 ppm	0	10	50 ppm	0	10	50 ppm
4	2.3	8.2	15.7	3.7	15.8	27.8	0.7	2.0	3.2
	1.1	4.7	20.2	1.0	8.6	32.5	0.7	1.2	3.2
	1.1	9.6	15.0	0.8	10.7	51.1	1.0	1.0	2.8
8	0.8	5.7	38.9	1.3	8.8	75.9	0.6	2.6	10.0
	0.8	9.6	36.9	1.0	13.8	43.8	0.6	4.3	31.0
	0.6	9.2	43.0	0.7	9.6	85.8	0.6	8.6	8.0
12	0.6	11.6	57.9	0.4	16.5	80.4	2.4	2.2	7.7
	0.6	9.1	56.5	0.4	11.9	136.5	0.4	2.1	6.5
	0.8	9.2	55.5	0.4	15.0	157.5	0.4	1.0	8.3

Levels at 1 and 3 weeks were too low for reliable measurement by the technique then available. There was a linear increase in Cd concentration with time and with extent of dosage in kidney liver and red cells. However, the level of Cd in blood plasma was low during the first two weeks of dosage, but at week 4 a detectable increase was observed which peaked at week 8 and then fell by week 12 (Table 2). In the 50 ppm animals no histological evidence of cellular damage could be demonstrated after 1 and 2 weeks but after 4 weeks defects were seen in the renal capillary epithelium of both glomerular and interstitial blood vessels. The damage allowed local seepage from these vessels with resultant oedema, platelets were present. After 8 weeks Cd dosage there was fusion of glomerular foot processes with associated endothelial damage. Increased collagen deposition in interstitial connective tissue was noted and the presence of fibroblasts. The basal lamina of the proximal convoluted tubule was thickened and areas of granular deposits seen, but only after 12 weeks exposure. This last group of rats also had protein casts in the proximal convoluted tubules and collecting ducts.

The degenerative changes in the 10 ppm Cd dosage rats developed more slowly but at 12 weeks endothelial lesions were seen, corresponding in severity to those found at 4 weeks in the 50 ppm Cd group.

Discussion

The accumulation of Cd by the liver and kidney is demonstrated here as a function of dose and time and agrees with previously reported work (Friberg, et al 1974). The blood cadium is partitioned between the red cells and the plasma, in the latter carried as Cd metallothionein. This small protein is synthesised in the liver and transported in the blood plasma (Nordberg 1972). Selective toxicity towards the kidney is exhibited by this protein (Nordberg et al 1975). In this experiment renal vascular lesions coincided with the plasma Cd increase at 4 weeks in the high dose rat, and 12 weeks in the low dose rat. These clear pathological changes could be seen at kidney tissue levels well below the quoted critical threshold of 200 μg Cd/g. kidney tissue. It is possible that the observed renal dosage could represent an increased synthesis and release of Cd metallothionein which was then being transported to the kidney in the blood plasma. The selective uptake of the Cd by the kidney would account for the renal toxicity and the observed drop in Cd blood plasma levels. It has been suggested that the toxic effect is caused by a specific antigenic response to the Cd metallothionein (Joshi et al 1981) but immune complex was not seen in this experiment. The direct effect of Cd on endothelium has been reported (Gabbianni. et al 1974) and was observed here. In addition the increase in collagen leading to interstitial fibrosis would have a direct effect on the thickness of blood vessels (Thind, 1974) and account for Schroeder's hypothesis that dieting Cd and hypertension are linked in the rat (1964, 1970b). The fusion of the epithelial foot processes causes filtration defects and the known toxic response by the proximal convoluted tubule to Cd will further aggravate this defect leading to the appearance of tubular casts, the standard signal for well developed Cd toxicity (Bonnell et al 1960).

References

1. Bonnell, J.A., Ross, J.H. and King, E. (1960) Br. J. Ind. Med. 17: 69-80.

2. Fell, G.S., (1980) Chemistry in Britain 16: 323-326.

3. Friberg, L., Piscator, M. Nordberg, G.F. Kellstrom, T. (1974) In "Cadmium in the Environment 2nd Edition CRC Press pp 35-36.

4. Gabbiani, G., Badonnel, M.C. Mathewson, S.M. and Ryan G.B. Lab Invest. (1974) 30: 686-695.

5. Itakawa, Y., Abe, T. and Tanaka S. (1973) Arch. Environ. Health 26: 241-246.

6. Joshi, B.C., Divivendi, C. Powell, A. and Holscher, M. J. Comp. Path. (1981) 91: 11-15.

7. Loeser, E., and Leuke, D. (1977) Toxicology 7: 215-224.

8. Nordberg, G.F., (1972) Environ. Physiol. Biochem. 2: 7-36.

9. Nordberg, G.F., Goyer, R.A. and Nordberg, M. (1975) Arch. Pathol. 99: 192-197.

10. Schroeder, H.A., (1964) An. J. Physiol. 207: 62-66.

11. Schroeder, H.A., Whanger, P.D. and Weswig, P.H. (1970b) Fed. Proc. 29: 301

12. Thind, G.S., (1974) Angiology. 25: 752-763.

DISCUSSION

HAYWOOD (England). I have two questions - first of all, have you any evidence of renal excretion? I am interested because in my studies of copper poisoning in the rat and this aspect interests me. And secondly, as regards the behavioural abnormalities in rats, have you looked at the brain at all because I see this sort of abnormality in my long-term dosing with copper?

AUGHEY (Scotland). No we have not looked at the urine, it is not easy to collect from the rat, we are doing these in metabolism cages and this was a preliminary experiment. We would like to do some studies on renal excretion and look for signs of renal damage in the urine. The behavioural aspect was introduced because we had observed it ourselves so regularly. I have not looked at brains, but I think it is an interesting pattern and it is even more interesting that you are in fact observing it again in the form of metal toxicity that you are studying.

BUSSLER (Germany). Please do you have an idea about the chemical background of the lesions in the endothelium?

AUGHEY (Scotland). No not really. I am not too convinced that it is basically a cadmiumthionein damage to possibly cell membranes that is causing endothelial damage. I wonder if in fact it is an ionic imbalance. This is a hypothesis with no background but we have heard so much in this trace element conference about the effect of imbalance of ions and obviously endothelium is an area of very rapid exchange. But it would not be too difficult to assume that a heavy metal such as cadmium, suddenly being thrown in at fairly high levels, would be sufficient to upset the delicate balance and this would be enough to cause the sort of breakdown which we are looking at. The standard response of tissue to any form of breakdown is to tidy up and treat it as a scar, you get fibroblasts moving in and collagen deposition. So I think one thing is possibly leading to another.

GOONERATNE (Australia). Do you think the plasma cadmium has to reach certain levels to trigger off the mechanism for formation of cadmiumthionein.

AUGHEY (Scotland). I think with Dr Bremner here I would not dare answer that question. I really do not know. Obviously there is a period of induction where the liver recognises the cadmium and the programme is set in motion for the production of the thionein which then collects the cadmium which presumably goes back into the plasma and it is this increase in plasma cadmiumthionein that we are looking at. This is then being leached out by the renal tubules. It is a very small protein and it will go through the gromerulus to be taken into proximal tubules, what happens to it thereafter I am not absolutely sure.

O'DELL (USA). Have you looked at the copper status of these animals? Copper deficiency in rats will cause some abnormal neurological behaviour - I wonder if with the large metallothionein induction it might actually induce a copper deficient state.

AUGHEY (Scotland). No we have not, but it is a thought and it is simple enough to put into the programme of analysis.

SCHENKEL (Germany). Do you expect the same lesions if you give lower concentrations for a longer period and you told that you found a lesion at low cadmium concentrations in the kidney. Do you think that allowable cadmium intake which was given by WHO should be revised therefore?

AUGHEY (Scotland). Yes I do. With Dr Fell, who is a Biochemist doing this work with us, we discussed this in some detail and relating it to the kidney levels that Mr Scott is going to report I think it is quite possible that the WHO figure should be revised, not so much from the renal damage point of view as the possibility of vascular damage. The levels that we are reporting here are the equivalent to the levels found in the smoking population and I think with cadmium you are dealing with a slow prolonged accumulation of a toxic metal over very long periods of time. I do not know what the clinical significance of it is, I do not know if there is any other than acute toxicity but I think it is a fact that we should bear in mind.

BLOOD COPPER, SORBITOL DEHYDROGENASE AND ACID PHOSPHATASE IN COPPER POISONING

KUMARATILAKE, J.S., HOWELL, J.McC. AND GOONERATNE, S.R.

Division of Veterinary Biology, School of Veterinary Studies,
Murdoch University, Western Australia, 6150

Elevated blood copper (Cu) levels and liver necrosis are associated with the haemolytic crisis of chronic copper poisoning in sheep (1,2). The precise relationship of these events to haemolysis is still in doubt and this experiment was carried out to study changes occurring in blood and liver prior to, during and after haemolysis.

MATERIALS AND METHODS

Ten Merino ewes, seven test and three control were used. Test sheep were dosed repeatedly with a 0.2% solution of copper sulphate in order to induce chronic copper poisoning. The following parameters were estimated twice weekly for the first five weeks and thereafter daily:- plasma and whole blood (W.B.) Cu by atomic absorption spectroscopy as described by Gooneratne et al (3), packed cell volume (PCV) by a microhaematocrit method, sorbitol dehydrogenase (SD) activity in plasma as described by Ford (4), and acid phosphatase (AP) activity in serum as described by Walter & Schutt (5) after adjusting the pH of the substrate solution to 4.5 (6).

RESULTS AND DISCUSSION

It was considered that plasma SD activity was elevated when the value obtained was greater than the mean plus two standard deviations i.e. 80.61 I.U. For all other parameters elevations were considered to be values greater than the mean plus three standard deviations i.e. AP 1.06 μ mole/min/l; plasma Cu 1.79 μg/ml; W.B. Cu 1.21 μg/ml and RBC Cu 1.37 μg/ml.

Plasma SD activity:- The first elevations were observed from 6-50 days prior to haemolysis. Thereafter elevations of varying magnitude were observed at different times and the frequency of elevations increased as copper dosing continued (Fig. 1). The highest elevations of SD activity were observed from 1-41 days prior to haemolysis (Table 1).

Serum AP activity:- The first elevations were observed from 2-49 days prior to haemolysis. Thereafter elevations of AP activity were observed at different times and were of varying degrees, but the majority of them were observed towards the latter part of the pre-haemolytic period (Fig. 1). The highest elevations were observed from 1-3 days prior to haemolysis. However, in sheep 320 the highest elevation was observed 12 days prior to haemolysis (Table 1) but on the next day the activity had fallen to normal and remained so until a day prior to haemolysis. During haemolysis measurements were not made as the colour interfered with the assay. After haemolysis the AP activity was still elevated, but gradually decreased.

Plasma Cu:- The first elevations in plasma Cu were observed from 2-19 days prior to haemolysis. Thereafter rises of varying levels were observed at different times and the frequency of rises increased towards the latter part of the pre-haemolytic period (Fig.1). In two sheep marked elevations were observed for 1 day, in two sheep for 2 days, in two sheep for 3 days and in one sheep for 6 days prior to haemolysis. The highest rises of plasma Cu were observed from 1-2 days prior to haemolysis or on the first day of haemolysis. However in sheep 319 and 320 the highest rises were observed 12 days prior to haemolysis but on the next day the levels had fallen and the second highest rise was observed one day prior to haemolysis (Table 1). Marked elevations of plasma Cu were observed during haemolysis and thereafter gradually decreased.

W.B. Cu:- The first elevations were observed from 2-26 days prior to haemolysis. The elevation of W.B. Cu followed the same pattern as that of plasma Cu (Fig. 1). In all but one sheep the highest elevations of W.B. Cu were observed one day prior or on the first day of haemolysis (Table 1). In sheep 319 the highest elevation occurred 12 days prior to haemolysis but on the next day the level had fallen to normal and the second highest

TABLE 1

HIGHEST ELEVATION IN RELATION TO HAEMOLYSIS

SHEEP NO.		SD ACTIVITY I.U.	day	AP ACTIVITY µmole/min/l	day	PLASMA COPPER µg/ml	day	W.B. COPPER µg/ml	day	RBC COPPER µg/ml	day
316		845.2	-5	4.52	-3	4.68	-1	6.60	-1	11.63	-1
318	First Hae.	772.8	-8	2.94	-1	2.69	-1	1.79	-1	NE	
	Second Hae.	413.0	-2	6.42	-2	8.28	-2	8.08	-1	14.77	1
319		272.9	-1	1.76	-2	3.37	-12	2.33	-12	4.41	-3
320		449.2	-41	1.79	-12	5.06	-12	4.56	1	13.65	1
324		784.9	-23	4.29	-2	13.50	-1	8.82	1	13.87	1
325		212.5	-5	1.67	-1	5.40	1	8.34	1	17.96	1
326		483.0	-6	4.26	-2	6.38	-2	6.22	-1	6.25	-1

Figures given on the right hand side of each column relate to the time of haemolysis, eg -2 = 2 days prior to; 1 = first day of haemolysis. AP - serum acid phosphatase; SD - plasma sorbitol dehydrogenase; W.B. - whole blood Cu; RBC - red blood cell Cu; Hae. - haemolysis; NE = not elevated.

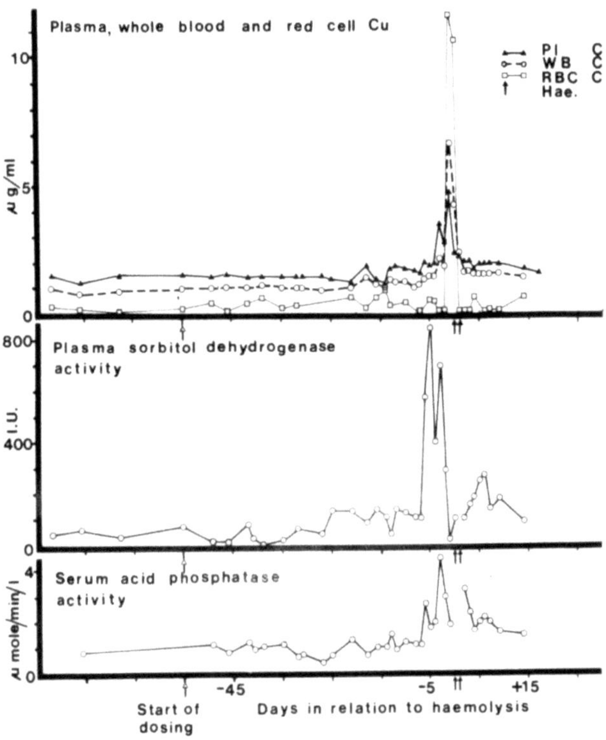

Fig (1) - Copper and Enzyme profiles in relation to Haemolysis in sheep 316. Pl.Cu = Plasma Cu, W.B. Cu = Whole Blood Cu, RBC Cu = Red Blood cell Cu. Arrows indicate days of haemolysis.

elevation was observed 3 days prior to haemolysis.

RBC Cu:- RBC Cu was calculated from the values of W.B. Cu, plasma Cu and PCV. In sheep 319 elevations of RBC Cu were observed 14,3 and 2 days prior to haemolysis and in sheep 324, 7 days prior to haemolysis. In all the other sheep the first elevations were observed 1 day prior to or on the first day of haemolysis.

It was seen that 67% and 66% of the observed elevations of AP activity during the pre and post haemolytic periods respectively were associated with elevation of SD activity. In the majority of animals the highest elevations of plasma Cu, WB Cu and RBC Cu occurred on the day of, or 1-2 days after the highest elevations of AP activity. In all but two sheep the highest elevations of RBC Cu occurred on the same day or 1 day after that of WB Cu (Table 1).

The major sources of AP are prostrate, RBC and platelets. The use of ewes and of serum from blood that was not haemolysed eliminated the possibility of AP coming from these sources. Previous studies (1,2,7) have shown that Cu accumulates in hepatic lysosomes and that morphological and biochemical changes occur in the liver during the pre-haemolytic period. The majority of the AP elevations were associated with elevations of SD activity and it was assumed that the observed elevations of AP activity during the pre-haemolytic period were due to hepatic cell necrosis. Increased activity of lysosomal enzymes has been observed in serum during hepatic necrosis in humans (8). Our results indicate that necrosis of hepatocytes may be occurring as early as 37-49 days prior to haemolysis i.e. 6-16 days after first drenching with Cu.

It was found that prior to haemolysis 85% and 75% of the elevations of plasma and WB Cu respectively were associated with elevations of AP activity indicating that these rises were due to rupture of Cu loaded hepatic lysosomes. In sheep,AP (6), and Cu (9) clear very rapidly from the blood stream. Thus the observed elevations of these values on consecutive days may not result from the necrosis of one set of cells but from the necrosis of different cells each day.

The highest elevations of AP activity, observed 1-3 days prior to haemolysis, and of Cu in plasma and WB recorded 1-2 days prior to or on the first day of haemolysis may be due to marked necrosis of hepatocytes heavily loaded with Cu and acid hydrolases. The mechanism which triggers the rapid uptake of copper by RBC, 0-24 hours prior to haemolysis, is not clear but it appears that lysis of RBC is due to reduced deformability brought about by the direct and indirect effects of Cu on the RBC membrane (10,11).

ACKNOWLEDGEMENT

The Work was supported by grants from the Australian Research Grants Committee and the Australian Wool Corporation Wool Research Trust Fund.

REFERENCES

1. Ishmael, J., Gopinath, C. and J.McC. Howell, 1971, Res. Vet. Sci. 12:358 - 366.

2. Ishmael, J., Gopinath, C. and J. McC. Howell, 1972, Res. Vet. Sci. 13:22 - 29.

3. Gooneratne, S.R., Howell, J.McC. and J.M. Gawthorne, 1981, Br. J. Nutr. (in press).

4. Ford, E.J.H., 1967, J. Comp. Path. 77:405 - 411.

5. Walter, K. and C. Schutt, 1974. In "Methods of Enzymatic Analysis" (2) (E.H.U. Bergmeryer) (Academic Press Inc., N.Y., San Francisco, London) pp 856-859.

6. Shannon, A.D., Quin, J.W. and F.C. Courtice, 1977, Res. Vet. Sci. 22:209 - 215.

7. Gooneratne, S.R., Howell, J. McC. and R.D. Cook, 1980, Am. J. Pathol. 99:429 - 450.

8. Gove, C.D., Wardle, E.N. and R. Williams, 1981, J. Clin. Pathol. 34:13 - 16.

9. Neethling, L.P., Brown, N.M.M. and P.J. DE Wet, 1968, Jl. S. Afr. Vet. Med. Ass. 39 (2):13 - 19.

10. Adams, K.F., Johnson, G. (Jr.), Hornowski, K.E. and T.H. Lineberger, 1979, Biochimica et Biophysica Acta, 550:279 - 287.

11. Soli, N.E. and I. Nafstad, 1976, Acta vet. scand. 17:316 - 327.

DISCUSSION

WOOLLIAMS (Scotland). I wanted to ask what rate of dosing you used and how you relate the results to different dosing results?

KUMARATILAKE (Australia). We gave one dose rate to all the test animals which was a 0.2% solution of copper sulphate at the rate of 15 ml per/kg body wt for 5 days a week until haemolysis.

DANKS (Australia). Could I ask whether you looked at the non-caeruloplasmin copper in the plasma when you were looking at the plasma copper. In human Wilson's disease and in acute copper poisoning of the accidental type which we sometimes see in children we see a sequence very similar to what you are proposing here. In the acute haemolytic crisis of Wilson's Disease something causes an episode of hepatic necrosis, release of copper and haemolysis, and in those circumstances we have extraordinary high levels of non-caeruloplasmin copper. Whereas in normal humans this is of the order of 5 or 10% of the total plasma copper, it may go up to 75% of the total copper in the acute phase.

KUMARATILAKE (Australia). No. We did not measure non-caeruloplasmin copper in plasma, we measured only plasma copper.

BREMNER (Scotland). I can perhaps answer Professor Danks question. We have looked at the blood from copper poisoned sheep - you certainly find the copper is not associated with total plasma, you have a normal complement, there is no great increase. Perhaps I can ask a question of the speaker? The onset of the haemolytic crisis is often associated in ewes with emphasis on stress. Have you any thoughts on how that may operate with regard to your own findings? Quite often you find under practical conditions that the crisis comes on shortly after stress such as a change in the nature of the diet or reduction of food intake or the poor sheep is chased by a dog or something like that. Do you have any thoughts?

KUMARATILAKE (Australia). There was a reduction in intake, of food, but other than that stress was minimised since the animals were housed in well ventilated rooms.

HOWELL (Australia). There is no doubt that stress will precipitate a haemolytic crisis. Very often in Liverpool the animals went into haemolysis if there had been a frost the night before. We don't get frosts in Perth but we do get heat waves and as Mr Kumaratilake says we put them in well ventilated rooms in order to minimise stress.

HUCKER (Australia). I would like to ask why SDH was the enzyme selected for indication of hepatic damage? I have noticed previous workers often used SGOT. I ask this question because in commercial housed sheep enterprises that I have studied,copper toxicity has been a problem, and SGOT has been confused by concurrent muscle changes as indicated by elevated CPK levels. In our case we chose arginase as a liver specific enzyme and we found it gave consistently good results.

HOWELL (Australia). Sorbitol dehydrogenase and arginase are good indicators of liver damage in sheep. We use sorbitol dehydrogenase activity in plasma because the assay is relatively cheap and easy to perform.

KIDNEY FUNCTION IN COPPER TOXICITY IN SHEEP

S. R. Gooneratne, J. McC. Howell and J. S. Kumaratilake

School of Veterinary Studies
Murdoch University
MURDOCH, WESTERN AUSTRALIA, 6150
AUSTRALIA

The commonest form of copper (Cu) toxicity in farm animals is chronic copper poisoning (CCP) and sheep are particularly susceptible (1). Copper accumulates in the liver and kidney during the pre-haemolytic period (2) and during the haemolytic crisis the kidneys are swollen, black and contain large amounts of copper and iron (3). Todd (4) remarked that "the appearance of the kidney is probably the most characteristic feature of" CCP. It has been considered that kidney failure and ureamia play an important part in the death of animals in the post- haemolytic period of the disease (5). Although morphological features of the kidney are well documented (2, 3, 4), information on the kidney function during CCP is lacking. The present study was carried out to assess the function of the kidney during the pre-haemolytic, haemolytic and post-haemolytic phases.

MATERIALS AND METHODS

To induce CCP fifteen, 9-12 month old Merino ewes were dosed repeatedly with a 0.2% solution of copper sulphate at the rate of 15ml/kg body weight on 5 days of the week. Five similar sheep acted as controls. Three test sheep died during the haemolytic crisis while the rest were killed at various times during and after the haemolytic crisis. Several tests of kidney function were applied. Inulin (IN) and ρ-amino-hippuric acid (PAH) clearances were used to assess glomerular filtration rate (GFR) and renal plasma flow (RPF) respectively (6). Phenosulphonphthalein (PSP) clearances were used to assess proximal tubular transport (PTT) (7). In addition plasma Cu (Pl.Cu), packed cell volume (PCV) (8) and blood urea nitrogen (BUN) (9) were measured. In test animals a maximum of 6 kidney function assessments was carried out per sheep and in controls 3 assessments were taken. The specific times at which these tests were carried out is illustrated in Table 1.

RESULTS AND DISCUSSION

The IN , PAH and PSP clearance rates of control and test sheep are summarized in Table 1. In test sheep the clearance rates of IN, PAH, and PSP remained normal until haemolysis. At haemolysis clearance rates of In, PAH and PSP were reduced indicating glomerular and tubular damage. In our previous studies we have observed the degree of haemolysis to vary markedly among individual sheep. Sheep in severe haemolysis had high BUN levels. . In the present study, in addition to BUN three other parameters (duration of haemolysis, Pl.Cu, PCV) were also measured and these were used to classify the degree of haemolysis into mild, moderate or severe (Table 2). An animal was classified as having a mild, moderate or severe haemolysis if the measurements were within the range for 3 of the 4 parameters measured. Of the 15 test sheep 4 were killed during the first 36-48 hours of haemolysis and hence the duration of haemolysis could not be accurately determined. Therefore their haemolysis was classified if 2 of the 3 remaining parameters were satisfied. Among the 15 test sheep 6 had a mild haemolysis, 4 had a moderate haemolysis and the remaining 5 had a severe haemolysis. The loss of kidney function during haemolysis could be directly related to the degree of haemolysis (Table 3). Animals in severe haemolysis exhibited marked impairment of GFR, PTT and RPF. These sheep had elevated BUN and of the 5 sheep in this category 3 died. These sheep had the greatest loss of kidney function (Table 3). In animals undergoing a mild haemolysis lasting 24 to 48 hours the clearance of IN, PAH and PSP was comparable to their pre-haemolytic values and the animals recovered. The intermediate group exhibited a moderate haemolytic crisis lasting 48-72 hours and had moderately decreased GFR, RPF and PTT. All animals in this category also recovered.

These results reinforce the importance of kidney changes in CCP in sheep and indicate that both glomerular and tubular damage are of importance. It has been postulated that the kidney damage is caused by the combined effects of haemoglobinaemia and hypercupraemia (3). The present results indicate that if the hypercupraemia and haemoglobinaemia are mild and present for less than 48 hours they do not cause marked impairment of kidney function.

However animals with severe to moderate functional impairment which recovered from the haemolytic crisis showed an improvement in kidney function with time. But 3 weeks after the cessation of haemolysis, although the renal clearances of PAH and PSP had returned to normal, the GFR was only about three quarter of the pre-haemolytic values, thus indicating the presence of residual glomerular damage.

TABLE 1: Clearance rates of inulin (IN) ρ-amino-hippuric acid (PAH) and phenosulphonphthalein (PSP) in sheep at various stages of chronic copper poisoning.

TIME OF SAMPLING	IN CLEARANCE† (mls/min)	PAH CLEARANCE† (mls/min)	PSP CLEARANCE† % excreted/ 20 mins
(1) Prior to start of expt.	72 ±3	470 ±19	55 ±1
(2) 2nd rise of SD*	73 ±3	449 ± 9	57 ±1
(3) Next rise SD*/10 days after the 2nd rise	65 ±3	437 ±14	58 ±1
(4) At haemolysis	37 ±5	349 ±56	39 ±3
(5) 1 week after cessation of haemolysis	52 ±6	383 ±57	49 ±5
(6) 3 weeks after cessation of haemolysis	51 ±5	460 ±44	60 ±5
(1) Prior to start of expt.	67 ±6	421 ±69	56 ±3
(2) 3 weeks after start of experiment	70 ±5	417 ±53	54 ±2
(3) 6 weeks after start of experiment	64 ±1	409 ±57	56 ±3

† Mean ±SE *SD - Sorbitol dehydrogenase

TABLE 2: Classification of Haemolysis

DEGREE OF HAEMOLYSIS & NO. OF SHEEP	DURATION (Hours)	PLASMA Cu (μg/ml)	PCV %	BUN - mg/dl
MILD (6)	24 - 48	< 2	> 20	< 100
MODERATE (4)	48 - 72	2 - 4	15 - 20	100 - 200
SEVERE (5)	> 72	> 4	< 15	> 200
CONTROLS (5)	-	0.56 - 1.22	26 - 38	48 - 77

TABLE 3: Inulin (IN), ρ-amino-hippuric acid (PAH) and phenosulphonphthalein (PSP) clearance rates measured at haemolysis.

DEGREE OF HAEMOLYSIS & No. of sheep	IN CLEARANCE* (mls/min)	PAH CLEARANCE* (mls/min)	PSP CLEARANCE* % excreted/ 20 mins
MILD (6)	57 ±7	567 ±73	51 ±3
MODERATE (4)	31 ±5	262 ±24	42 ±1
SEVERE - TOTAL (5)	17 ±4	158 ±24	21 ±1
SEVERE - DIED (3)	13 ±1	144 ±14	21 ±3
CONTROLS (5)	67 ±2	415 ±32	55 ±1

* Mean ± SE

CONCLUSION

In CCP kidney function remains unaltered until the haemolytic crisis.

By measuring the following parameters at haemolysis, (a) the duration of haemolysis (b) Pl. Cu (c) PCV and (d) BUN, it is possible to categorize the degree of haemolysis to mild, moderate or severe.

At haemolysis there appears to be both glomerular and tubular damage.

Kidney function was directly related to the degree of haemolysis.

Using data obtained from the kidney function tests performed on the second day of the haemolytic crisis it is possible, in most cases, to predict which animals would recover and those which would die.

REFERENCES

1. Underwood, E.J. 1977. In "Trace Element in Human and Animal Nutrition" 4th edition (Academic Press, New York, San Francisco, London) pp 56-108.

2. Howell, J.McC. 1977. In "The Veterinary Annual" Seventh issue (eds. C.S.G. Grunsell and F.W.G. Hill) (Wright-Scientechnica-Bristol) pp 70-73.

3. Gopinath, C., Hall, G.A., and J.McC. Howell, 1974, Res. Vet. Sci. 16:57 - 69.

4. Todd, J.R., 1969, Proc. Nutr. Soc. 28:189 - 198.

5. Howell, J.McC. 1978. In "Proceedings of the Third International Symposium on Trace Element Metabolism in Man and Animals" (ed. M. Kirchgessner) (Arbeitskreis Für Tierernährungsforschung, Weihenstephan) pp 536-539.

6. English, P.B., Hogan, A.E., and H.L. McDougall, 1977, Am. J. vet. Res. 38:1317 - 1322.

7. Finco, D.R., 1971, J. am. vet. med. Assoc. 159:336 - 340.

8. Gooneratne, S.R. 1979. In Ph.D. Thesis "Morphometric, electron microscopic, biochemical and pathological investigation of liver of normal and copper poisoned sheep" Murdoch University, Western Australia.

9. Sigma Technical Bulletin, No. 640, 1980.

ACKNOWLEDGEMENT

We wish to thank Mr M Gates and Miss K Burroughs for their technical assistance.

This work was supported by a grant from the Australian Wool Corporation Wool Research Trust Fund.

DISCUSSION

SUTTLE (Scotland). Why do you think kidney function has such a crucial role? Could it be that copper clearance is correlated with things like inulin clearance and this is the important feature?

GOONERATNE (Australia). I think most of the reports say that animals which die subsequent to haemolysis do so because of kidney failure. It is also reported that there is liver damage prior to haemolysis. We are investigating the changes in the kidney, if any, prior to haemolysis and we are particularly interested in the changes which occur during haemolysis.

BREMNER (Scotland). There is a massive rise in kidney copper concentration at the time of haemolysis, values going up around 200 ppm, in my experience. How quickly does that copper clear from the kidney in the animals which survive, have you any ideas.

GOONERATNE (Australia). It takes a fair while. If the animal had about 200 ppm in the kidney at haemolysis the level could be at least 150 - 175 even three weeks after haemolysis. I don't know whether the level comes back to normal at all, but there is a slight reduction.

HAYWOOD (England). Do you completely discount primary damage to the kidney by copper? Are you now saying that all the kidney damage is secondary to the effect of haemoglobin.

GOONERATNE (Australia). Maybe copper also plays a part but there are reports that if you inject copper without causing haemolysis there is very little kidney damage and also if you give substances which cause haemolysis only, without increasing copper that there is very little kidney damage and that when these two are combined you get the massive kidney damage that we find in chronic copper poisoning in sheep.

COPPER TOXICITY AND TOLERANCE IN THE RAT

S. Haywood

Department of Veterinary Pathology
University of Liverpool, England

Copper toxicity in the rat has never been fully investigated. This study* aimed to relate changes in the liver and kidney to Cu accumulation over a period of time, at both the cellular and subcellular level.

MATERIALS AND METHODS

Young male rats were fed 2000 mg/kg copper as $CuSO_4$ in the diet for 15 weeks. Groups of 4 were killed at intervals, the liver and kidneys were retained for histology and Cu analysis (1) and the blood/plasma for Cu content; bilirubin; alanine amino-transferase, alkaline phosphatase and copper oxidase (caeruloplasmin) activity (2). Subsequently the livers from another group similarly treated were homogenised and subjected to analytical subcellular fractionation after a modification of Peters' method (3). Lysosomal distribution was identified in the gradient by N acetyl-β-glucosamidase activity and Cu content spectrophotometrically.

RESULTS

I Liver and kidney copper analysis (Fig 1)

Liver Cu content rose to 3359.7 ± 1145.3 µg/g (control 17.7 ± 3.5 µg/g) at 6 weeks and thereafter fell. Kidney Cu content similarly rose to a maximum of 1447.0 ± 1049.2 µg/g (control 33.9 ± 9.9 µg/g) and then fell slightly.

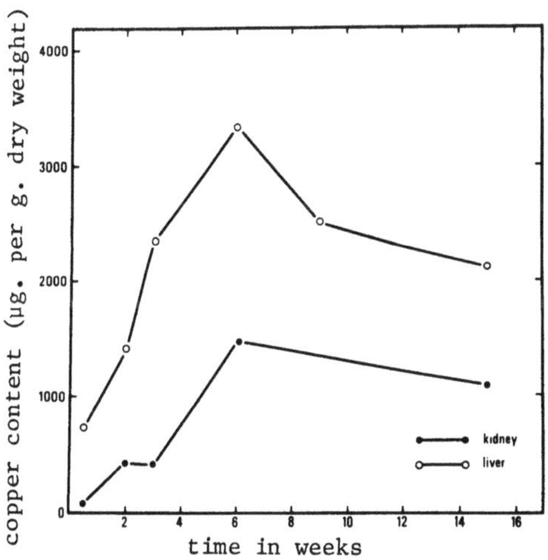

Fig. 1. Cu content of liver/kidneys of rats on Cu supplemented diet for 15 weeks.

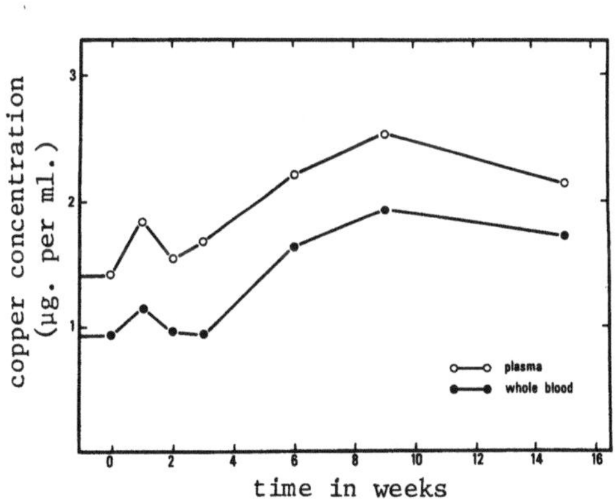

Fig. 2. Cu content of plasma and blood on high Cu diet.

II Blood/plasma analysis

The Cu content of both plasma and blood (Fig. 2) rose to their highest values of 2.5 ± 0.2 and 1.9 ± 0.3 µg/ml respectively at 9 weeks (control 1.4 ± 0.2 and 0.9 ± 0.2 µg/ml) which later fell slightly.

* This work is currently supported by The Wellcome Trust.

Plasma caeruloplasmin content rose to 43.9 ± 6.3 (control 21.9 ± 3.8) mg/100 ml at 9 weeks then fell off slightly; alanine amino transferase activity rose likewise to 143.8 ± 53.3 (control 26.9 ± 11.7) Karmen units/ml but activity was maintained. Alkaline phosphatase activity 7.6 - 13.4 Sigma units per ml was within the normal range as was bilirubin (0.2 - 0.5 mg/100 ml).

III. Histological changes

During the cumulative period the liver showed increasing signs of cellular disturbance in the periportal zone characterised by foci of cellular hypertrophy and necrosis with associated inflammatory response (Fig. 3a). At the same time there was pericanalicular Cu accumulation (Fig. 3b). By 6 weeks there were more severe widespread changes (Fig. 4a) and a loss of copper from the hepatocytes (Fig. 4b). Subsequently the liver recovered. The kidney also showed the most severe changes at 6 weeks (Fig. 5) after which tubular reconstitution occurred.

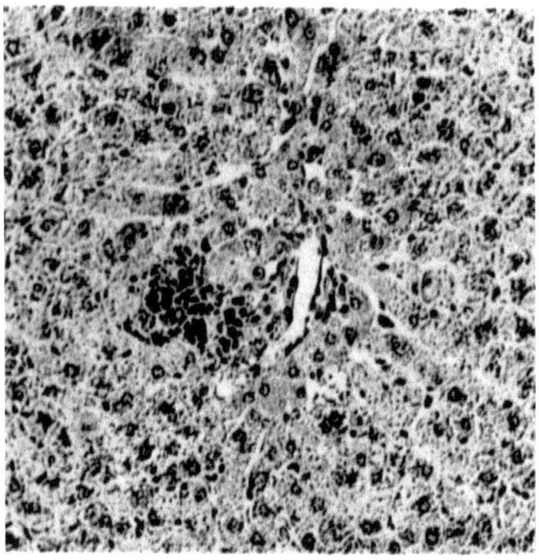

Fig. 3a. Liver of rat after 3 weeks on Cu supplemented diet. Inflammatory focus in periportal zone. H & E X 230

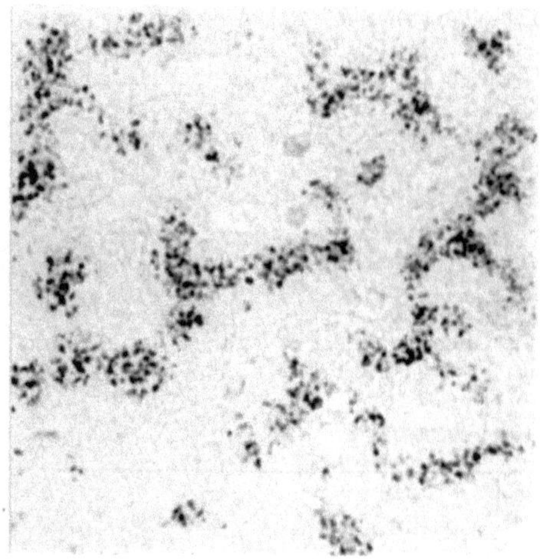

Fig. 3b. Same liver as 3a. Pericanalicular deposition of Cu granules. Rubeanic acid X 720

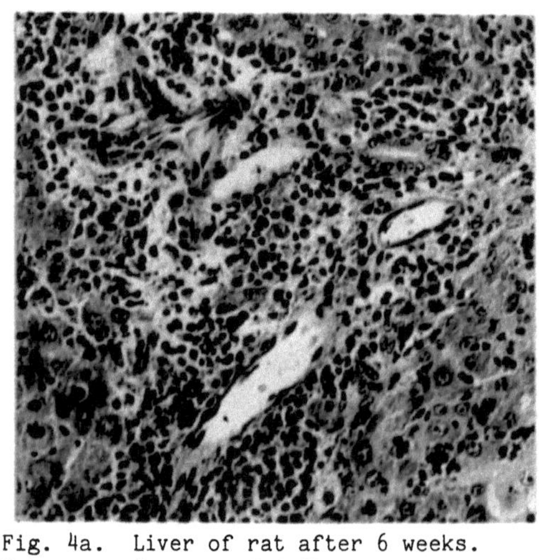

Fig. 4a. Liver of rat after 6 weeks. Extensive damage and inflammatory response periportally. H & E X 284

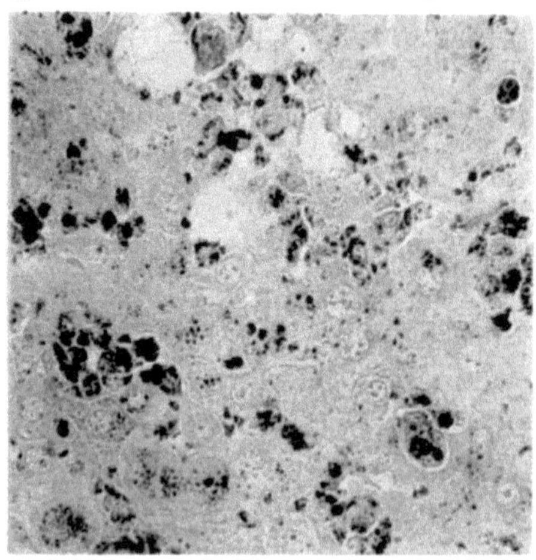

Fig. 4b. Same liver as 4a. Concentration of Cu staining granules in phagocytes. Rubeanic acid X 375

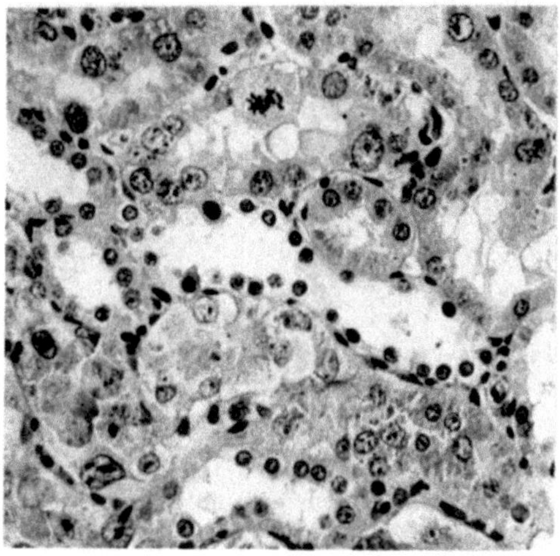

Fig. 5. Kidney of rat after 6 weeks on Cu-supplemented diet showing tubular necrosis and early regeneration.
H & E X 300

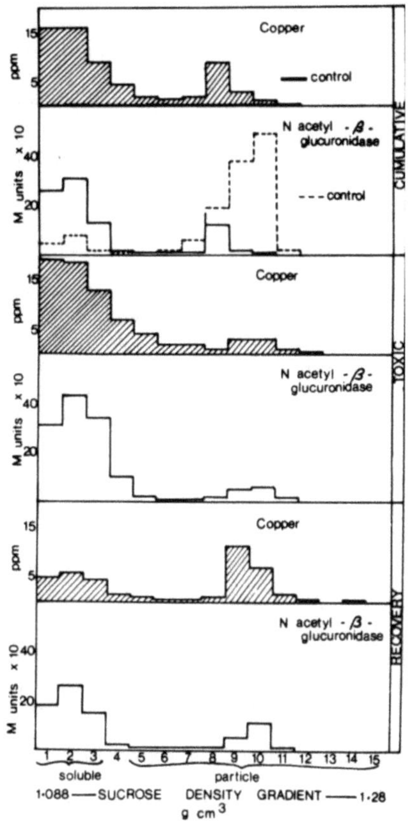

Fig. 6. Histogram to show Cu distribution in cytoplasm of liver over experimental period.

IV. Subcellular fractionation (Fig. 6)

Cu-containing lysosomes showed an increase in fragility which was greatest when the liver was fully loaded, after which they became less fragile as Cu levels fell during the recovery phase.

CONCLUSIONS

When rats are fed continuously high levels of Cu they accumulate Cu in the liver (and kidney) to toxic levels after which adaptation and recovery of damaged organs occurs and they become tolerant.

Toxicity is characterised by very high liver/kidney Cu levels and is associated with lysosomal damage and release of Cu into the cytoplasm.

Tolerance is characterised by an overall fall in tissue Cu levels and increased lysosomal stability. The reason for this is not known, but may involve blood transport of excess Cu and renal excretion.

This work demonstrates the existence of homeostatic mechanisms capable of transforming a potentially lethal heavy metal (Cu) toxicosis into a state of tolerance.

REFERENCES

1. Haywood, S. (1980) J. comp. Path. 90, 217-232.

2. Haywood, S. and Comerford, B. (1980) J. comp. Path. 90, 233-238.

3. Peters, T.J. (1977) Clin. Sci. & Mol. Med. 53, 505.

SELENIUM IN COPPER TOXICITY IN SHEEP

S.R. Gooneratne and J.McC. Howell

School of Veterinary Studies
Murdoch University
MURDOCH, WESTERN AUSTRALIA 6150
AUSTRALIA

INTRODUCTION

Chronic copper (Cu) toxicity is a well recognised nutritional hazard in sheep (1). In addition to the major changes which occur in erythrocytes, liver and kidney, a variety of other changes have been described including an increase in permeability of muscle membranes (2). Degenerative changes in muscle are known to be produced by selenium (Se) and vitamin E (Vit. E) deficiency (3). The reports on Cu-Se interactions in sheep (4, 5) and the precipitation of Se deficiency by high dietary levels of Cu in chicks (6) encouraged us to investigate the nutritional interrelationship between Cu and Se in copper poisoned sheep.

Liver Se levels have not previously been reported in sheep undergoing the haemolytic crisis of chronic copper poisoning (CCP). The present study presents the Cu and Se levels prior to, and at haemolysis and correlates these results with our previously published data on liver glutathione peroxidase (GSH-Px) levels and ultrastructural changes in muscle (2).

ANIMALS

Twenty one, 9-12 month old Merino wethers were each given 2 injections of 1ml Selvite E (68 I.U. Vit. E. and 5mg Se/ml) a week apart the last being given 1 week before the start of the experiment. To induce CCP nine sheep were dosed repeatedly with a 0.2% solution of copper sulphate at the rate of 10ml/kg body weight on 5 days of the week. Twelve similar but undosed sheep acted as controls. Six Cu dosed sheep were killed at the 'haemolytic crisis' of CCP and three others were killed at the second rise of sorbitol dehydrogenase which was approximately three weeks prior to haemolysis.

RESULTS AND DISCUSSION

Liver Cu, Se and GSH-Px levels of control and test sheep are summarized in Table 1. Significant increases in liver GSH-Px activity and Se levels were observed on Cu loading.

TABLE 1:- Copper (Cu) and selenium (Se) levels, and glutathione peroxidase (GSH-Px) activity in liver of control sheep and those killed prior to and during the haemolytic crisis. (Values given are the mean and the standard error of the mean.)

GROUP	Cu*	Se*	GSH-Px†
CONTROL (n=12)	288 ± 25	3.1 ±0.1	70 ±4
PRE-HAEMOLYTIC (n=3)	2946 ±220 (P<0.01)	4.6 ±0.4 (P<0.05)	120 ±7 (P<0.01)
HAEMOLYTIC (n=6)	4206 ±267 (P<0.01)	5.6 ±0.4 (P<0.01)	106 ±7 (P<0.01)

* *μg/g liver dry matter* † *μmoles NADPH oxidised/min/mg protein*

These results are in contrast to those reported for animals with a simple stomach by Jensen (6) who demonstrated that Cu can induce a Se deficiency in chicks and by Godwin et al (7) who were of the opinion that Se deficiency in the rat was associated with a lowered threshold for Cu toxicity.

In sheep the relationship between Se and Cu appears to be more complex. In lambs with a suboptimal Cu status Thomson & Lawson (5) observed a significant increase of liver Cu concentration after Se administration. However when lambs were supplemented with Cu this effect could not be seen. Awad et al (8) and Lee & Jones (9) found that Se had little or no effect on liver Cu concentration in sheep. In both these studies liver Cu levels were between 150-460 μg Cu/g liver DM which is within the normal range for sheep (10). Liver Se concentrations were not measured in these studies.

Published results for liver Se levels measured in lambs grazing (11) or fed normal or excess Cu (12) are also conflicting. Sivertsen et al (11) did not observe any relationship between liver Cu (approximately 2400 μg Cu/g liver DM) and Se concentration but White et al (12) found that increases in dietary Cu resulted in a significant increase in Se concentration and GSH-Px activity in livers with a maximum of 1363μg Cu/g liver DM. The latter suggested that the increased Se retention in sheep was a response to tissue damage caused by Cu accumulation. The maximum liver Cu levels observed in sheep studied by White et al (12) were less than 50% of the liver Cu of the sheep killed in the pre-haemolytic stage of the present study and their animals were possibly 4-6 weeks away from the development of a 'haemolytic crisis'. The Se levels reported here (Table 1) for control and pre-haemolytic sheep though slightly elevated compare favourably with the results of White et al (12) but the liver Se levels of the animals in haemolysis were even higher.

It has been suggested that the increase in plasma creatine kinase which occurs at haemolysis may be due to an increase in muscle membrane permeability (2, 13), and that this may have been brought about by a variety of factors including Vit. E and Se depletion in tissues (2). The present results rule out the possibility of a Se depletion in liver. We did not find a depletion of GSH-Px in the left ventricle of these sheep (2). However alterations in mitochondria were the most frequently seen ultrastructural change in the semitendinosus, triceps and cardiac muscle of sheep killed at haemolysis and cardiac muscle appeared to be the most affected (2). Similar changes in mitochondria have been described as the earliest change in muscle fibres of Vit. E deficient rabbits (14), chicks (15) and rats (16). Mitochondrial alterations were also seen in liver of sheep killed at haemolysis (17).

It is tempting to presume that a depletion of Vitamin E occurs in Cu loaded sheep.

CONCLUSIONS

In sheep accumulation of Cu in liver was associated with an increase in both Se concentration and GSH-Px activity in that organ. This was most significant in sheep undergoing a 'haemolytic crisis'.

Swelling and vacuolation of mitochondria occurred in liver and muscle of sheep during the 'haemolytic crisis'. This was most prominent in cardiac muscle. These changes resembled the earliest changes in muscle fibres of Vit. E deficient rabbits, chicks and rats.

REFERENCES

1. Howell, J.Mc.C., 1977.In"The Veterinary Annual", 7th issue (eds. C.S.G. Grunsell & F.W.G. Hill) pp 70-73.

2. Gooneratne, S.R., and J.McC. Howell, 1980, Res. Vet. Sci. 28:351-361.

3. Jenkins, K.J., and M. Hidiroglou, 1972, Can. J. An. Sci. 52:591-598.

4. Hill, M.K., Walker, S.D., and A.G. Taylor, 1968, N.Z. J. Ag. Res. 12:261-270.

5. Thompson, G.G., and B.M. Lawson, 1970, N.Z. Vet. J. 18:79-82.

6. Jensen, L.S., 1975, Proc. Soc. Exp. Biol. and Med. 149:113-116.

7. Godwin, K.O., Partick, E.J., and C.N. Fuss, 1978. In "Proceedings of the 3rd International Symposium on Trace Element Metabolism in Man and Animals", (ed. M. Kirchgessner) (Arbeitskreis für Tierernährungsforschung, Weihenstephan) pp 185-186.

8. Awad, Y.L., Ahmed, A.A., Lotfi, A.Y., and F. Fahmy, 1973, Zbl. Vet. Med. A. 20:742-747.

9. Lee, H.J., and G.B. Jones, 1976, Aust. J. Ag. Res. 27:447-452.

10. Underwood, E.J., 1977. In "Trace Elements in Human and Animal Nutrition", 4th edition, (Academic Press, N.Y.) p. 95-108.

11. Sivertsen, T., Karlsen, J.T., Norheim, G., and A. Frøslie, 1978, Acta. Vet. Scand. 19: 472-474.

12. White, C.L., Cloninger, R.W., Hansen, J.A., Hoekstra, W.G., and A.L. Pope, 1979. In "Proceedings of the Nutrition Society of Australia", 4:148-149.

13. Thompson, R.H., and J.R. Todd, 1974, Res. Vet. Sci. 16:97-99.

14. Van Vleet, J.F., Hall, B.V., and J. Simon, 1968, Am. J. Path. 52:1067-1079.

15. Cheville, N.F., 1966, Path. Vet. 3:208-225.

16. Howes, E.L.J., Price, H.M., and J.M. Blumberg, 1964, Am. J. Path. 45:599-631.

17. Gooneratne, S.R., J.McC. Howell, and R.D. Cook, 1980, Am. J. Path. 99:429-450.

ACKNOWLEDGEMENTS

We wish to thank Dr. J.M. Gawthorne for many valuable discussions and Mr. P. Young, C.S.I.R.O., Wembley for carrying out selenium estimations and the Australian Research Grants Commission for financial support.

DISCUSSION

BINNERTS (The Netherlands). Could the lesions seen in mitochondria be artifacts?

GOONERATNE (Australia). We have consistently seen mitochondrial alterations in many of the muscles of sheep undergoing the haemolytic crisis. Similar changes were also seen in hepatic mitochondria. These changes were absent in the control sheep.

LLOYD DAVIES (Australia). Do you think that there is an interaction between Cu and Se?

GOONERATNE (Australia). We have observed elevated Se levels in Cu loaded liver of sheep. Dr. Colin White has reported similar changes.

TOXIC DIETARY SELENIUM LEVELS FOR WEANLING SWINE

A. L. Moxon and D. C. Mahan

Animal Science Department, Ohio Agricultural Research and Development Center and The Ohio State University, Wooster, Ohio 44691 U.S.A.

In 1969 Patrias and Olson (1) studied the selenium content of corn samples produced in major corn producing states west of Ohio. Values from their samples ranged from .05 ppm to over .15 ppm. Mahan *et al.* (2) found an average selenium content of .020 ppm in 62 corn samples from southwestern Ohio countries with 58% of these samples containing less than .020 ppm. Thirteen soybean samples from the same area averaged .067 ppm selenium. Diets formulated in Ohio from this corn and soybean oil meal generally contain from .030 to .070 ppm selenium depending on the relative ratio of each in the mixture. Supplemental selenium added at the level of .100 ppm has been shown to improve reproductive performance of sows when fed such corn and soybean meal diets (3).

More recent research in Ohio has shown that the young pig's dietary Se requirement may be as high as .50 ppm at 2-weeks postweaning, declining to .3 ppm by 20 kg body weight and .10 ppm from 20 kg to market weight (4,5,6).

As a result of these findings and other Ohio research with postweaning pigs, increased supplemental inorganic selenium seems advisable. Consequently there was a need to establish the toxic levels of dietary selenium in the diets for young swine.

EXPERIMENTAL PROCEDURE

A total of 72 pigs weighing approximately 7.7 kg were allotted by litter and weight from lactating dams to nine lots of 8 pigs each. They were given a 20% protein corn-soybean meal diet with gradual supplemental selenium levels added as sodium selenite. The supplemental selenium concentrations varied from 0 (basal) to 40 ppm. The selenium concentrations for each treatment group are presented in Table I.

Blood samples were taken initially and at weekly intervals with serum collected and frozen for later selenium analyses. Animals were weighed and feed consumption measured at weekly intervals. Representative animals were sacrificed and tissue samples taken for selenium analyses after 37 days.

Selenium analyses were performed on the feed, serum and tissue samples using the method as described by Olson (7).

RESULTS

There was a linear reduction of feed intake at 5 ppm and above resulting in lowered animal gains (Table I). No deaths resulted from any of the high dietary selenium levels except in the 20 and 40 ppm diets in which the deaths occurred as a result of starvation.

Pigs given the 40 ppm selenium diet consumed no feed after the first week. Those pigs fed 20 or 40 ppm Se were very weak with those fed the 40 ppm selenium losing an average 1.95 kg body weight during the trial. Some pigs on the higher levels of selenium showed loss of hair (alopecia) by the 17th day. Some pigs in lots with 10 ppm selenium or more showed cracked hoofs by the end of the feeding trial.

The voluntary refusal of seleniferous diets has been previously observed in rats by Franke (8) in 1935 and in dogs by Rhian and Moxon (9) in 1943. Humans apparently lack the ability to quickly detect selenium in their foods to restrict their intake. A botanist (10) in Honduras reports consuming Sopucaia nuts (*Lecythis elliptica*) daily for a two-week period. After 4 weeks he began losing hair from his head. After 4 to 5 months his fingernails and toe nails began to split and later as the new nails grew out the old nails were shed. A sample of these nuts was analyzed and found to contain 3,300 ppm selenium on dry-matter fat-free basis.

TABLE I. Effect of Various Dietary Selenium Levels on Pig Performance

Item					Added Dietary Se (ppm)[a]				
	0	2.5	5.0	7.5	10	15	20	40	0/40[b]
No. Pigs	8	8	8	8	8	8	8	8	8
No. Deaths	0	0	0	0	0	0	2[c]	8[c]	0
Initial Wt, kg	7.89	7.72	7.63	7.67	7.81	7.85	7.67	7.63	8.04
Final Wt, kg	18.70	18.61	17.43	15.98	15.12	11.21	9.13	5.68	18.61
Daily Gain/Grams									
0- 7 days	22.7	131.7	95.3	31.8	86.3	36.3	-18.2	-63.6	163.4
7-14 "	227.0	217.9	181.6	131.7	158.9	86.3	4.5	-59.0	249.7
14-21 "	304.0	304.2	190.7	195.2	204.3	72.6	-9.1	-99.9	227.0
21-28 "	408.6	358.7	358.7	367.7	245.2	131.7	27.2	-59.0	286.0
28-37 "	449.0	417.7	454.7	354.1	272.4	118.0	49.9	--	585.7
0-37 Days	290.6	295.1	267.9	222.5	195.2	90.8	18.2	--	304.2
Daily Feed/Grams									
0- 7 Days	249.7	281.5	245.2	208.3	222.5	272.4	154.4	158.9	317.8
7-14 "	286.0	340.5	336.0	258.8	308.7	200.0	249.7	--[d]	431.3
14-21 "	517.0	581.1	485.8	417.7	390.4	399.5	72.6	--	549.3
21-28 "	717.3	681.0	603.8	535.7	472.2	354.1	200.0	--	581.1
28-37 "	821.7	762.7	790.6	667.4	540.3	354.1	286.0	--	862.6
0-37 Days	535.7	540.3	580.5	431.3	395.0	317.8	190.7	--	549.3
Average Daily Se[e] intake mg.		.130	.255	.330	.390	.480	.385	--	.120

[a]Diets were analyzed for total selenium content. The analyses confirmed the calculated levels.
[b]A choice of 2 diets was provided to this lot of pigs. Over the 37-day trial period the pigs consumed a total of only 1.135 kg of the 40.0 ppm Se diet and over 162 kg of diet without added Se.
[c]Two pigs on each of these treatments died. Pigs refused to consume the diets with high selenium content. Thus, the apparent cause of death was starvation. The 6 remaining pigs on the 40 ppm selenium diet were sacrificed because of their starved condition.
[d]No feed consumed after 1st week.
[e]Average daily intake of added dietary selenium.

In the case of pigs in Lot 9 with a choice of basal (0 selenium added) and 40 ppm selenium diet they ate from both feeders on the first day and subsequently avoided the seleniferous diet. On day 21 the locations of the 2 feeders were switched. The pigs ate some of the 40 ppm feed again the first few days in the new location and began consuming only the low-Se diet until the end of the trial. During the 37 day trial period the 8 pigs ate a total 1.135 kg of the 40 ppm diet and 160 kg from the basal (0 added selenium/feeder).

The selenium content of blood serum in the basal group was similar with an initial value of .025 and .030 ppm Se at 37 days. Serum selenium values in all other groups increased as the experiment progressed with average values over 2.0 ppm in the 10, 20 and 40 ppm groups by the end of the 37 day period. The serum selenium values for the 0/40 ppm diet group decreased from .085 ppm selenium initial value to .056 ppm on day 21. The locations of the 0 and 40 ppm feeders were switched on day 21 and on day 28 the serum selenium had risen to .353 ppm selenium. On day 37 the serum selenium was down to .160 ppm.

The selenium content of tissues collected on day 37 showed some interesting variations. Livers from pigs fed the basal diet averaged .106 ppm selenium compared with .183 ppm average in pancreatic tissue. In all groups with added inorganic selenium the liver Se concentration was more than double the content of pancreas tissue. The selenium content of *longissimus* muscle increased as dietary selenium increased with values being .059, .149 and .504 for the basal, 2.5 and 40 ppm diets. The liver Se was lower than that of the kidney cortex samples in the 0, 2.5 and 0/40 ppm diet groups; whereas in all other groups, the average selenium contents of livers was higher than from the kidney cortex. The selenium contents of tissues from the 0/40 ppm diet group was intermediate between the corresponding tissues from the 0 and 2.5 ppm selenium diet groups.

REFERENCES

1. Patrias, G. and O. E. Olson. 1969. Selenium contents of samples of corn from midwestern states. Feedstuffs 41 (No. 43) page 32.

2. Mahan, D. C., Thomas L. Smith, J. D. Latshaw, A. L. Moxon and J. H. Cline. 1973. The current dilemna of the selenium-vitamin E problem. O.A.R.D.C.

3. Mahan, D. C., L. H. Penhale, J. H. Cline, A. L. Moxon, A. W. Fetter and J. T. Yarrington. 1974. Efficacy of supplemental selenium in reproductive diets on sow and progeny performance. J. Animal Science 39:536-543.

4. Mahan, D. C. and A. L. Moxon. 1978. Effect of increasing the level of inorganic selenium supplementation in the postweaning diets of swine. J. Animal Science 46:384-390.

5. Mahan, D. C. and A. L. Moxon. Toxic dietary and injectable selenium levels for young pigs. Ohio Swine Research and Industry Report Animal Science Series 79-2, pg 25-30. O.A.R.D.C.

6. Mahan, D. C. and A. L. Moxon. 1980. Effect of dietary selenium and injectable vitamin E-selenium for weanling swine. Nutrition Reports International 21:829-836.

7. Olson, O. E. 1969. Fluorometric analysis of selenium in plants. J. Assoc. Official Agr. Chem. 52:627-634.

8. Franke, K. W. 1935. A new toxicant occurring naturally in certain samples of plant foodstuffs. XI. The effect of feeding toxic and control foodstuffs alternately. J. Nutrition 10:233-239.

9. Rhian, Morris and A. L. Moxon. 1943. Chronic selenium poisoning in dogs and its prevention by arsenic. J. Pharmacology and Expt'l. Therapeutics 78:249-264.

10. Dickson, John D. 1969. Notes on hair and nail loss after ingesting Sopucaia Nuts (*Lecythis elliptica*). Economic Botany 23:133-134.

DISCUSSION

ALLEN (UK). Dr. Moxon in your summary you refer to disorders of co-ordination in some of the piglets; can you attribute that to any central nervous system disorder?

MOXON (USA). No that was just starvation.

ALLEN (UK). From your intake data you were feeding up to 40 parts per million of Se per kd diet and you presented some very nice graphs of the increment in tissue selenium concentrations but your initial evidence on food intake suggested those pigs had not taken in selenium from the first day onwards.

MOXON (USA). They had not taken much selenium but they got enough to build up the tissue. They ate a day or two and then they stopped on the high level of selenium feed.

ALLEN (UK). But even if they took it on the first day they would still get the initial dose of something like 5 mg per kilogram.

MOXON (USA). Yes. Apparently they retain the selenium they took pretty well.

LINDER (USA). For the pigs that had the choice of 0 or 40 ppm Se, could you estimate how much selenium they actually ate? Also, do you think that pigs that have a choice would always take in adequate selenium?

MOXON (USA). Yes they would and that showed up I think in the fact that the ones eating 2½ parts per million grew better than ones that had no added selenium. There was only .6 ppm in their basal diet.

BEHNE (Germany). Why do the pigs immediately refuse the food with a high selenium content. Is there a difference in the taste or an immediate change in metabolic processes?

MOXON (USA). I do not think it is taste because they would eat at either of the feeders for a day or so. I think it takes about one day to learn what is going on and then they avoid the high selenium feed.

MINERAL TOLERANCE OF DOMESTIC ANIMALS: A NAS-NRC REVIEW

C.B. Ammerman

University of Florida
Gainesville, Florida 32611, USA

All mineral elements, whether considered to be essential or potentially toxic, can have an adverse effect upon the animal if included in the diet at excessively high levels. For essential elements, the dietary level which will just promote optimum response can be considered as the minimum requirement and this level can be variable. The level at which a mineral exhibits a toxic effect is also quite variable. There is no single requirement for a mineral by an animal and neither is there a single safe or maximum level at which a mineral can be tolerated without adverse effect. Factors which can influence both the minimum requirement and the maximum level which can be tolerated by animals include the following: 1) Biological availability of the mineral source; 2) Growth potential of the animal; 3) Animal function (growth, production of milk, etc.); 4) Previous treatment; 5) Age; 6) Environment; 7) Levels of other dietary nutrients.

There are several causes for concern about the possibility of excessive mineral intake by animals under practical conditions. Among these reasons are the following: 1) Plant tissue levels vary in mineral content; 2) Leaves and stems of forages can become contaminated with minerals; 3) Water levels vary in mineral content either from natural causes or from contamination; 4) By-product feedstuffs may contain elevated mineral levels.

Development of a NAS-NRC publication. In the early and mid 1970's, consulting nutritionists, and various industry groups and organizations in the United States repeatedly expressed concerns about the need for an organized source of information on the toxicity of mineral elements to animals. The Committee on Animal Nutrition under the National Research Council, National Academy of Sciences responded in 1976 by appointing a Subcommittee on "Mineral Toxicity in Animals". Members of the subcommittee included representatives from industry; the Bureau of Foods, and Bureau of Veterinary Medicine in the Food and Drug Administration; and Universities.

A primary responsibility of the subcommittee was that of summarizing the available literature on mineral tolerance in domestic animals and establishing "Maximum Tolerable Levels" (MTL) where data were adequate. The maximum tolerable level for a mineral was defined as "...that dietary level that, when fed for a limited period, will not impair animal performance and should not produce unsafe residues in human food derived from the animal." Where possible, these levels were generated for 36 mineral elements for cattle, sheep, swine, poultry, horses and rabbits (NRC, 1980). Brief sections were included also on essentiality, metabolism, sources and tissue levels for each element.

As stated above, there can never be just one MTL for an element or species just as there can never be only one value as a minimum requirement for an essential nutrient. There must occur under practical feeding situations a series of MTL's just as there would be a series of minumum requirements for essential mineral elements under varied feeding conditions. Thus, the MTL's should be used as guidelines and with similar judgments applied as are used with minumum requirement values.

Summary. All mineral elements, whether essential or nonessential dietary nutrients, produce adverse effects in animals if consumed at sufficiently high levels. A recent NAS-NRC publication entitled "Mineral Tolerance of Domestic Animals" lists maximum tolerable levels of dietary minerals for domestic animals for 36 mineral elements and discusses each element in relation to its potential adverse influence on the animal.

1. NRC. 1980. Mineral Tolerance of Domestic Animals. C.B. Ammerman, J.P. Fontenot, M.R.S. Fox, H.D. Hutchinson, P. Lepore, H.D. Stowe, D.J. Thompson and D.E. Ullrey (Authors). National Academy of Sciences - National Research Council, Washington, DC.

DISCUSSION

LINDER (USA). Where can this publication be purchased?

AMMERMAN (USA). From the National Academy of Sciences, 2101 Constitution Avenue, Washington DC, 20418.

SEDIVY (USA). I believe 25 ppm Cu is the level quoted in this new publication as being the lower limit at which copper toxicity occurs in sheep. Would you consider the copper content of feeds in Midwestern USA to be adequate?

AMMERMAN (USA). In your area dietary levels of copper that are below 10 - 12 ppm should be safe. We showed 25 ppm in the publication, but in other parts of the world, e.g. Scotland this concentration is regarded as being highly toxic. The availability of dietary copper appears to be affected by many factors. Good nutritional practice dictates that we do not approach the maximum tolerable level.

WIENER (Scotland). I think that genetic differences need greater emphasis. We have recently completed a copper toxicity trial with lambs given two diets, 12 or 19 ppm Cu, for 12 weeks. Texel cross breeds showed signs of impending copper poisoning but three of the other four breeds were far from this stage. On 12 ppm Cu Texels retained more copper in liver than Scottish Blackface lambs fed 19 ppm Cu. Genetic differences should be taken into account when making recommendations.

TRACE ELEMENTS AND HUMAN DISEASE

TRACE ELEMENTS IN HUMAN DISEASE, WITH PARTICULAR REFERENCE TO COPPER AND ZINC

D.M. Danks

Department of Paediatrics, University of Melbourne and Genetics Research Unit, Royal Children's Hospital Research Foundation, PARKVILLE, VICTORIA, 3052, AUSTRALIA

This will not be the type of review which summarizes the current state of knowledge. Others at this congress could tackle that task, but I could not do it justice. I plan to discuss a personal approach to the study of certain trace elements based upon a training in clinical paediatrics and in genetics, and upon an amateur interest in the role of copper and other trace elements in the body. Reference is made to recent reviews where possible rather than to original findings.

I consider that the real challenge in the next decade of trace element research is to delineate precisely the processes involved in transport, storage and utilisation of these elements in the body. We have learned something about requirements for trace elements, about their turnover and homeostasis in a general way and about their ultimate functions as components of enzymes or as activators. We know very little about the detailed events at cell membranes and within cells which determine the availability of the elements to the enzymes and protect the body against toxic accumulation.

Gene mutations provide very powerful tools for the unravelling of complex normal processes. The "experiment of nature" is beautifully designed, affecting just a single component of a complex process. Careful analysis allows precise identification of the component which is faulty, thereby defining one component of the normal process. This approach was developed to a high level of precision in studying bacterial and fungal metabolism. It also proved effective in the definition of the major components of the immune system in man (1) and has been used very successfully in studying vitamin B_{12} metabolism in man (2). In immunology the strategy involved comparison of human genetic diseases with the effects of ablation of components of the immune system of chick embryos. The strategy of the vitamin B_{12} work was more deliberate and more comparable to that used in micro-organisms.

I wish to argue that the next phase of study of trace metals should include a deliberate search for mutations affecting trace metal utilization in man and animals, plus use of tissue culture systems for genetic analysis and for induction of those mutations which are not found in the animals.

ANIMAL MUTANTS

Ideally one would wish to find, in an appropriate animal species, mutants affecting every process in the absorption, transport, storage, utilization and excretion of each trace element - a tall order. To be even more demanding it would be good to find each mutant in two or more species - a small animal with a rapid reproductive rate is good for some studies, but others require larger animals from which repeated blood or tissue samples are readily obtained. For the present mice and man are the species of choice, particularly because so many mutants have been recognized and documented already in both species (3, 4). Unfortunately most murine mutants are known by fairly crude characteristics of appearance and/or behaviour (e.g. *mottled* or *crinkled* because of coat colour variation, *quaking* or *jimpy* because of neurological disturbances) and few have been defined at a biochemical level. Several groups have been looking over these mutants for features suggestive of underlining trace element defects, but the yield has been rather disappointing to date.

Mottled mutants (brindled$^{Mo^{br}}$, blotchy$^{Mo^{br}}$, viable brindled$^{Mo^{vbr}}$) are definitely related to disturbed copper transport and will be discussed in more detail. Claims have been made for a relationship of *crinkled* (cr) and *quaking* (qk) to copper (5, 6), but others have been unable to confirm this relationship (7). *Sex-linked anaemia* (sla) involves defective intestinal transport of iron (8), with interesting similarities to the malabsorption of copper in Mo mutants.

The *pallid* (pa) mutant appears to involve a disturbance in the availability of manganese (7).

Lethal milk (lm) causes zinc deficiency and can be corrected by zinc supplementation of the lactating dam or of the pup (9).

One must ask how the search for trace element mutants could be intensified. It is not easy to suggest an efficient approach. So far selection has depended on suspicious clinical features followed by attempted amelioration of effects by supply of the element suspected or by analysis of tissues for the element. The advent of relatively simple methods of analysing for multiple trace elements may allow systematic screening by measurements in certain tissues from each mutant available - a daunting task.

Some mutants affecting trace elements are known in other animals. Many Bedlington terriers develop a recessively inherited liver cirrhosis related to chronic copper retention (10). Certain Friesian cattle develop a zinc deficiency state (parakeratosis) very similar to acrodermatitis enteropathica (11). Zinc deficiency has been claimed in Alaskan malamute chondrodysplasia, but the evidence is not very convincing (12).

HUMAN MUTANTS

Man is, of all animals, the one studied most intensively by biochemical methods. This is because humans are large, generally co-operative and tractable to blood and tissue sampling, and intensely interested in their own diseases. We know two genetic defects of copper transport (Wilson's disease and Menkes' disease) (13) one of zinc absorption (acrodermatitis enteropathica) (14) an interesting disorder affecting the molybdenum cofactor (15, 16) and many defects of iron transport and utilization. We also know many mutants affecting cobalamin metabolism, but it is not really fair to call them defects of cobalt utilization because the human body does not transport the metal as such, requiring the organic complex performed.

It is worth reflecting briefly on the method of recognition of several of these human mutants as a guide to ways in which others may be discovered. The copper defect in Menkes' syndrome was sought and found because copper deficiency seemed to link together disturbed structures of hair and of elastic fibres, as well as depigmentation (13). The coincidence of xanthinuria (xanthine oxidase deficiency) and sulphituria (sulphite oxidase deficiency) pointed to a disturbance of molybdenum availability (15) just as cobalamin deficiency provided a link between methylmalonic aciduria and homocystinuria (17).

A good knowledge of the functions of the various trace elements would make doctors and veterinarians alert to the possibility of a deficiency as the cause of symptoms and the co-incidence of two or more appropriate symptoms becomes a very strong lead requiring investigation. Training of these professionals should emphasize this knowledge.

It can be seen that the accumulation of mutants for use in the identification of steps in trace metal "handling" is slow and laborious. Next one has to find out which process is at fault in each mutant. (Incidentally, it is difficult to know what word to use to describe absorption, transport, storage, utilization and excretion of a trace element. "Metabolism" is not really an appropriate term for it implies alteration of that which is metabolized, and this is not true of trace elements).

GENETIC ANALYSIS OF METABOLISM

The power of genetic analysis of metabolism in bacteria has been mentioned. In studying bacterial auxotrophs it was possible to combine the specificity of mutations in affecting just one enzyme (indeed just one polypeptide chain is a multimeric enzyme) and the power of genetic complementation analysis. When two bacteria which appeared to have a defect in the same step in metabolism were fused together, normal metabolism was sometimes restored. Usually there proved to be two sequential enzyme functions involved in what had been considered one process. Sometimes there was only one enzyme, but it was a dimer of two different polypeptides and some active molecules were formed in the hybrid cell.

These same approaches are possible in human or animal cells provided that the functions to be studied are active in the types of cells which can be grown in tissue culture and provided the abnormal phenotype can be determined accurately. Ideally one wants a phenotypic effect which allows selection of hybrid cells, but this may not be essential.

A few words may be needed about selection of mutant cells. The penicillin selection system for bacterial auxotrophs is the prototype. Penicillin kills only those bacteria which are multiplying. Auxotrophs plated on a basal medium will not multiply and survive while penicillin kills the dividing wild type organisms. Transfer to media with additional growth factors rescues the mutant colonies. Bromodeoxy uridine (BrdU) can be used similarly in mammalian cells. It is incorporated into dividing cells which then die if exposed to UV light. Mutant cells unable to grow in the selective medium will survive. Other selection systems can be devised for specific purposes.

TISSUE CULTURE STUDIES

Of all the mutants affecting trace elements which have been mentioned above, only Menkes' syndrome and *mottled* mice have been well studied in tissue culture and shown abnormal. Both these are X chromosomal mutants and the disturbance of copper transport is so closely similar that homology of the genes concerned seems very probable (13, 18). We are currently attempting stringent tests for homology in fused cells.

The phenotype of both the human and mouse mutations in cell culture (fibroblastic and lymphoid cells from human patients; fibroblastic and renal epithelial cells from mice) comprises increased copper content, normal rate of copper isotope uptake, increased copper isotope retention, reduced copper isotope release and increased sensitivity to toxic effects of copper added to the medium (19).

It is clear that the basic defect must reside in one of the copper transport mechanisms within these cell types. However, energetic work in a number of laboratories has not yet defined the lesion. The excess copper present in the cells is attached to a metallothionein-like molecule. A mutation of metallothionein causing irreversible binding of copper is one possibility; a defect in a process required to release copper from normal metallothionein molecules is a second possibility; a third mechanism could involve failure of transport by some other intracellular ligand with secondary superinduction of metallothionein. Other possible defects could be proposed.

Three groups have independently studied copper transport in fibroblastic cells cultured from patients with Wilson's disease. Goka and Howell mentioned high copper levels in cells from one patient (20). Chan *et al.* described elevations in 7 patients (21). We observed elevated levels in each of 7 patients on at least one occasion, but could not find conditions in which this elevation was observed consistently (22). We have done further work with many conditions of culture and testing, including those used by Chan, and still predictable results elude us - a frustrating situation for we desperately want two copper mutations demonstrable in cultured cells for our research strategy. We have also looked, without success, for phenotypic effects in cell culture in Indian childhood cirrhosis (see below), in Bedlington terriers with copper retention (Camakaris & de Blaquiere, unpublished) and in *crinkled* mice (7).

No-one seems to have looked for phenotypic effects in cultured cells in the mutants involving other trace elements listed above. It would seem desirable to make such studies. *Sex-linked anaemia* of mice (sla) seems a particularly strong candidate to show a phenotype in culture because intestinal mucosal cells accumulate excessive amounts of iron just as one sees for copper in *mottled* mice (and in Menkes' syndrome).

If we are to make full use of the power of the genetic approach to the understanding of trace elements we must try to find expression of each defect in cells which can be manipulated in culture so as to use the power of genetic complementation analysis.

We may not have to wait to find every mutant we want for these studies in patients or in animals. We should be able to mimic the bacterial geneticist by producing and selecting the mutants we want in cell cultures. To illustrate the strategy, one might start with a culture of Menkes' cells (or *mottled* mouse cells) and grow them with steadily increasing levels of copper in the medium in order to select a spontaneous mutant which can tolerate the copper because it has a defect in copper uptake in addition to the original defect. Chemical mutagens might be used to increase the frequency of mutations. Alternatively one might start with normal cells grown in a high copper level which is just tolerable and use a BrdU selection system to pick up cells which are more sensitive than others to the copper hoping to find, for instance, mutants lacking metallothionein. We do not yet have any results to report from experiments of this type, but we do have considerable experience in using the BrdU selection system for mutants affecting intermediary metabolism.

Once a number of mutants affecting copper transport have been isolated then complementation analysis can be applied. At this stage one no longer needs selection techniques, but can use end points such as cell copper levels, copper isotope uptake or release.

HUMAN TRACE ELEMENT MUTANTS

In the final segment of this review I wish to discuss some of the known human mutations affecting trace elements, with particular emphasis upon what they tell us about normal processes.

<u>Wilson's disease</u> deserves first comment in deference to its priority of discovery as a

trace element disorder, dating from 1935. The main message of the struggle to understand Wilson's disease is that a defect which is confined to an inaccessible organ like the liver is very difficult to study (13). Defective biliary excretion of copper and greatly diminished synthesis of caeruloplasmin are the most fundamental disturbances in this disease. Since gene mutations usually affect only one process, this suggests that a common process underlies both biliary excretion and incorporation into caeruloplasmin. This conclusion is not supported by kinetic modelling studies which have been performed in rats and in sheep (23, 24). In both these species biliary copper appears to come from one hepatic pool and caeruloplasmin from another. However, one must not lose sight of the fact that both these processes are much less efficient in the sheep than in the rat. The fact that these species differ in the efficiency of both processes (not just one or the other) may really support the contention of one underlying process, a process which may be defective in all sheep and in those few humans with Wilson's disease. Some have suggested that the same process is defective in the normal human fetus (25). If only one could reliably demonstrate a defect in copper transport in cultured cells from patients with Wilson's disease, knowledge would surge ahead.

Some mention of other liver diseases associated with hepatic copper retention seems appropriate (13). Copper retention is observed in all patients with primary biliary cirrhosis (26) and with Indian childhood cirrhosis (27). Primary biliary cirrhosis is a disease of middle-aged women in which disturbance of bile salt excretion and pathological changes in small bile ducts are prominent. It does not show Mendelian inheritance and has some features of an auto-immune disease. Copper accumulates along with, rather than in advance of, the progress of the liver pathological changes. These features suggest a rather specific, but secondary, intereference in copper excretion in the bile.

The situation is more difficult to interpret in Indian childhood cirrhosis. This condition does show a marked familial incidence and debate over many years has not clearly decided between a major genetic factor or a familial exposure to infective or toxic agents. The degree of copper accumulation is very gross, the copper levels (4000 μg per g dry weight) observed at 4 or 5 years of age being several times those observed in much older patients with Wilson's disease. A primary defect in copper transport seems a possibility, but is far from certain. We have been unable to find any disturbance in copper levels or in uptake or release of copper isotope in cultured fibroblasts (Camakaris & de Blaquiere, unpublished).

One interesting patient with the recessively inherited condition of congenital lymph-oedema and cholestatic jaundice, first described by Aagenes showed remarkable hepatic copper retention and copper deposition in the basal ganglia with neurological symptoms (28). Unfortunately we have not had access to further cases and the Norwegian doctors who have a number of cases have not measured liver copper. This could be another genetic defect of copper excretion, but a rather specific secondary disturbance is more likely.

The neurological effects in this patient and the Kayser-Fleischer rings described in some adults with primary biliary cirrhosis support the idea that copper overflows to these tissues from the liver. This is one of the reasons for believing that the basic defect in Wilson's disease is likely to be confined to the liver.

In Menkes' syndrome the basic defect is certainly widespread in many body cells (13,18). Copper accumulates to abnormal levels in cells and yet remains unavailable to the copper enzymes. This statement can be defended most precisely from our own studies showing lysyl oxidase deficiency in fibroblastic cells despite a copper content 5 times normal (29). Many other observations are consistent with the statement, though not so critical. In the affected fetus one sees increased copper levels in all tissues except liver, yet affected babies are born with a number of quite definite consequences of the disease already apparent. After birth the basic defect in intestinal cells causes severe malabsorption of copper with secondary copper deficiency.

The symptoms and signs of Menkes' syndrome provide the best information available regarding the consequences of severe copper deficiency in humans. The prominence of arterial disease (with rupture or thrombosis of major arteries before 2 years of age) may be telling us that marginal chronic copper deficiency could play a role in other forms of arterial disease. The absence of anaemia is surprising and unexplained.

In acrodermatitis enteropathica there is certainly a defect in zinc absorption, but it is surprising to find that there is still debate whether the defect is in the mucosal absorptive process itself or in factors normally present in milk and in pancreatic secretions which assist in zinc absorption. Several patients described with normal plasma zinc levels have nonetheless responded to zinc administration (30). There may be several different defects among the cases currently classified as acrodermatitis enteropathica. Surprisingly little use has been made of intestinal mucosal biopsy in the study of this disease (31) and no reports of studies in tissue culture have been reported. The integration of human studies and studies of related animal mutants (lethal milk mouse and lethal trait A46 in Friesian cattle (11)) seems desirable.

Dominantly inherited elevation of plasma zinc (32) is a harmless genetic variant which is not yet fully explained, but may be related to increased binding to albumin. If genetic defects of zinc transport can be demonstrated in cultured cells then it will be very interesting to test interactions with plasma of patients with this variant.

Finally, a very interesting defect in molybdenum utilization has recently been reported in one patient from the Netherlands (15). Two further cases in Paris and London have been mentioned (16). This patient has the biochemical and clinical consequences of defects of two molybdo-enzymes - sulphite oxidase and xanthine dehydrogenase. No molybdenum was detected in the liver despite normal levels in the plasma. A deficiency of the molybdenum cofactor, a pterin-containing molecule required to get Mo into liver mitochondria and to achieve incorporation into the molybdo-enzymes, is proposed (16). Further study of this disease is likely to help in elucidating the nature and role of this co-factor.

The ideas presented in this paper have been greatly influenced by discussions and research collaboration with Dr. Jim Camakaris in particular and also with Miss Suzanna Herd, Mr. Jeff Mann, Mrs. Marie Phillips and Dr. Peter Royce.

REFERENCES

1. Pahwa, R.N., Pahwa, S.G., O'Reilly, R., Smithwick, E.M. and Good, R.A., 1979. In "Inborn Errors of Immunity and Phagocytosis" (ed. F. Guttler, J.W.T. and R.A. Harkness) (MTP, Lancaster) pp 3-68.
2. Fenton, W.A. and L.E. Rosenberg, 1978, Ann. Rev. Genet. 12:223-248.
3. Green, E.L., 1966. "Biology of the Laboratory Mouse" (McGraw-Hill, N.Y.) 2nd Ed.
4. McKusick, V.A., 1978. "Mendelian Inheritance in Man" (Johns Hopkins, Baltimore) 5th Ed.
5. Hurley, L.S., C.L. Keen and B. Lonnerdal. 1980. In "Biological Roles of Copper" (Ciba Symposium, No. 79) (Excerpta Medica, Amsterdam) pp 227-238.
6. Hurley, L.S., 1981, Johns Hopk. Med. J. 148:1-10.
7. Mann, J.R., J. Camakaris, J. Gillespie, B. Koellreuter, J.-M. Matthieu, P.M. Royce and D.M. Danks, 1981, Biol. Trace Element Res. (in press).
8. Bannerman, R.M. and J.A. Edwards, 1976, Brit. J. Haematol. 32:299-307.
9. Piletz, J.E. and R.E. Ganschow, 1978, Science 199:181-183.
10. Twedt, D.C. and I. Sternlieb, 1979, J. Amer. Vet. Assoc. 175:269-275.
11. Brummerstedt, E., A. Basse, T. Flagstad and E. Andresen, 1977, Amer. J. Path. 87: 725-728.
12. Brown, R.G., G.N. Hoag, M.E. Smart and L.H. Mitchell, 1978, Growth 42:1-6.
13. Danks, D.M. 1981. In "The Metabolic Basis of Inherited Disease" (eds. J.B. Stanbury, J.B. Wyngaarden and D.S. Fredrickson) (McGraw-Hill, N.Y.) (in press).
14. Hambidge, K.M., 1978, J. Hum. Nutr. 32:99-108.
15. Johnson, J.L., W.R. Waud, K.V. Rajagopalan, M. Duran, F.A. Beemer and S.K. Wadman, 1980, Proc. Nat. Acad. Sci. USA 77:3715-3719.
16. Johnson, J.L., B.E. Hainline and K.V. Rajagopalan, 1980, J. Biol. Chem. 255:1783-1786.
17. Baumgartner, R., J.C. Linnell, G.E. Gaull, C. Bauchmann, H. Wick and B. Steinmann, 1979, Helv. Paediatr. Acta 34:483-492.
18. Danks, D.M., 1977, Inorg. Perspect. Biol. Med. 1:73-100.
19. Camakaris, J., D.M. Danks, L. Ackland, E. Cartwright, P. Borger and R.G.H. Cotton,1980, Biochem. Genet. 18:117-131.
20. Goka, T.J., R.E. Stevenson, P.M. Hefferan and R.R. Howell, 1976, Proc. Nat. Acad. Sci. USA 73:604-606.
21. Chan, W.Y., W. Cushing, M.A. Cofeman and O.M. Rennert, 1980, Science 208:299-300.
22. Camakaris, J., L. Ackland and D.M. Danks, 1980, J. Inher. Metab. Dis. 3:155-158.
23. Owen, C.A. Jr. 1980. In "Biological Roles of Copper" (Ciba Symposium No. 79) (Excerpta Medica Amsterdam) pp 267-276.
24. Weber, K.M., R.C. Boston and D.D. Leaver, 1980, Aust. J. Agric. Res. 31:773-790.
25. Epstein, O. and S. Sherlock, 1981, Lancet 1:303-305.
26. Sternlieb, I., 1980, Gastroenterology 78:1615-1628.

27. Tanner, M.S., B. Portmann, A.P. Mowat, R. Williams, A.N. Pandit and C.F. Mills, 1979, Lancet 2:1203-1205.
28. Smith, A.L. and D.M. Danks, 1978, Brit. Med. J. 2:1400-1401.
29. Royce, P.M., J. Camakaris and D.M. Danks, 1980, Biochem. J. 198:579-586.
30. Krieger, I. and G.W. Evans, 1980, J. Pediat. 96:32-35.
31. Atherton, D.J., D.P.R. Muller, P.J. Aggett and J.T. Harries, 1979, Clin. Sci. 56: 505-507.
32. Smith, J.C., J.A. Zeller, E.D. Brown and S.C. Ong, 1976, Science 193:496-498.

DISCUSSION

HOWELL (Australia). I wonder if Professor Danks would care to say something more about the overflow of copper in Wilson's disease. We have produced some very high levels of copper in the liver of our copper poisoned sheep and we have got the haemolysis, the liver changes, and the kidney changes equivalent to those seen in Wilson's disease. We haven't got the dramatic elevations of CNS copper levels and we think the changes that we have described in the central nervous system are much more likely to be attributable to liver or kidney failure than to the excess of copper.

DANKS (Australia). Yes. I suspect that it may prove to be a species difference. It appears to be also the case that Bedlington Terriers do not develop brain lesions like those of Wilson's disease, although most of the other components of Wilson's disease, or of what you see in copper toxicity in sheep, have been reported in the Bedlington Terriers. We seem to have two different organisms (the dog and the sheep) which are resistant to brain effects. Conversely we have just miniscule evidence, namely our one patient with this Åagenes' syndrome indicating that in humans the brain can be susceptible to the overflow of copper even when the overflow is occurring for some reason other than for Wilson's disease. In other words, it isn't only in Wilson's disease that you get the brain involvement because in this child the symptomotology was identical to Wilson's disease, the brain pathology was just like Wilson's disease and the brain copper desposition was just like Wilson's disease. There have not been any reports of brain involvement in the other forms of human copper-accumulating liver disease to my knowledge, although overflow to cornea with the formation of Kayser-Fleischer rings has been described in primary biliary cirrhosis, and overflow to kidney has been described. Perhaps this is age specific in humans and the overflow has to start earlier in life to have this effect, although, if that is the case, there is a very long lag between the overflow and the production of symptoms. It would be very interesting to look more carefully at the basal ganglia of patients who do eventially die of primary biliary cirrhosis and the other human copper-accumulating diseases to see whether they also show a preclinical degree of the same effect.

MAY (Wales). What is known about the fate of copper incorporated into ceruloplasmin?

DANKS (Australia). I am sure there are other people in this audience who know much more about that than I do. I have always wondered where does copper go to after it has been used by any of the copper enzymes. We know that it eventually comes out in the bowel - but how on earth does it get there? What happens when ceruloplasmin breaks down? What happens when molecules of superoxide dismutase break down, or molecules of cytochrome oxidase - where does the copper go to? We talk glibbly of it being conserved and reutilised, but do we know that it is? We also talk glibbly of copper stores in various places, but I am not sure that anybody has ever proved that they operate as stores. Even the high copper level of the neonatal liver is talked of as though neonates are very clever in accumulating a lot of copper because he is going to want it later on. On the other hand it is equally possible that he just hasn't yet learnt to excrete copper in his bile or make ceruloplasmin. As soon as he does acquire these skills copper goes out of the liver and a lot of it goes out in the bile. We tend to attribute knowledge and predetermination to a lot of these processes which is not necessarily true.

WIENER (Scotland). May I be allowed to comment, since Professor Howell has already turned attention to sheep. I was, of course, very delighted with Professor Dank's model of using mutants in order to study the pathways in trace element metabolism. But I am slightly worried that some non-genetistists may conclude that genetic variation is manifested only by a normal and a mutant which is very different from the normal. So I just want to say that if the people in this room were an equivalently diverse population of sheep, I would expect the efficiency of absorption of their dietary copper to vary several-fold and that this has consequences for their Cu requirements and their responses in situations of dietary copper deficiency or dietary copper excess. In sheep of least, deficiency and excess are relative terms which would vary with the genotype. So in addition to the genetic story which Professor Danks has so ably put before us, the quantitative genetic aspect should not be forgotten and that genetic variation may be part of normal variability - at least this would be so for ruminants. In this context, I think the debate which you, Professor Danks are having with Lucille Hurley over the crinkled mouse is one which should be pursued vigorously because it may give you the clue for not just whether but how a mutant may have a different effect in different genetic backgrounds. I think the difference in your two results is an exception which you should treasure.

VERSIECK (Belgium). Professor Danks, you mentioned several genetic disorders of trace element metabolism. May I ask you to comment also on the syndrome of heritable hyperzincemia, described by Smith and co-workers in Science, some years ago.

DANKS (Australia). This is an interesting mutant without any functional consequence yet noted but which certainly is, I hope, capable of being used for interaction with other mutants in order to determine its effects. There was a familiar occurrence documented through two to three generations of an apparently dominant single gene effect upon the serum zinc level. The precise mechanism of it, to my knowledge, was not worked out in the paper and I have not seen any further publication nor identification of any further families. I certainly agree with Dr. Wiener in that one must treasure the exceptions and I think that this would be an exceptional family which may shed great light upon the normal processes of zinc transport.

If I could just briefly comment on what Dr. Wiener said, I agree; of course with him that one has the whole genetic background influencing the matter. I think there are two important, and to some extent separable, processes here. For the approach I am expounding of trying to identify components of the copper transport process, one needs to find a single gene mutations of big effect. If one can identify all of the steps of the process, I suspect one may find that much of the polygenic variation will prove attributable to minor polymorphic variation in these individual processes.

MILLS (Scotland). In the search for diagnostic indices of dietary deficiency, we have often debated the possible value that mutants of the type you are discussing might have. Clinical features of deficiency differ however. Cerebral vascular lesions appear to be rare or not as well developed in the nutritionaly Cu deficient animal as in the Menkes child. Do such differences arise from differences in transport mechanisms in the genetic defect? Will data for the mutant cell yield information that will be of greatest use for study of transport mechanisms or for detection of new roles of elements, as in the case of the molybdenum deficiency sulphite oxidase cases. In looking at the mutant you may look at a heavily accentuated defect in transport but how do we know that it reflects a step in transport in the normal subject.

DANKS (Australia). I suppose my answer to that would be that we are looking a far way off to do the thing that you'd like to do, but I see these mutants as being a way of drawing a map of the roads - identifying what roads there are! Of course the mutants will not tell you the width of the road and the speed of the traffic down each of the roads, but at least once you have found all the roads then you know where to measure the flow.

HAYWOOD (England). Could you comment more upon the absorption of copper in Wilson's disease, and the possibility of renal excretion? It seems to me, exciting as your ideas are, many diseases are multifactorial and there may be defects at many different sites. I may be quite wrong but I believe there are absorption abnormalities in Wilson's disease.

DANKS (Australia). I would like to contest the idea that genetic disease which shows a simple mendelian pattern is ever multifactorial in its basic defect. Genetic disease which shows a simple mendelian pattern is unifactorial in its basic defect. It may be multiconsequential and there may be a great pleomorphism of the consequences, but not of the basic defect. I think that we must be careful not to get these two things confused. Yes, of course Wilson's disease does have effects on many organs, but my credo as a genetisist, and there are thousands of mutations in many different organisms to support my credo, is that underneath it all there is one process at fault and all the rest flow as secondary effects because of the interaction of that fault with the physiology of the process in the body. When one understands the thing precisely, one can mark out every one of those steps like you can in some mutant haemoglobins. You can describe every one of the physiological processes which has gone wrong and why it had gone wrong. To answer the more specific points of your question, I think that to my knowledge, and we haven't personally studied the copper uptake in Wilson's dieases, there is not an increase in copper uptake, but rather a decrease in copper loss in the bile. The renal copper loss is secondary to the high level of non-ceruloplasmin copper in the plasma which is the filterable component of the plasma copper and, in the later stages of the disease, also compounded by damage to the renal tubules which is producing an aminoaciduria. The aminoaciduria is tending to carry copper with it. These are secondary consequences which come as the disease progresses.

VAN DEN HAMER (The Netherlands). I would like to ask you about the special form of Menkes' disease you mentioned. In one of your first papers you describe that in the oral copper 64 test hardly any copper 64 was detected at any time in the blood. Did you apply the same test to the new case of Menkes' disease with late onset?

DANKS (Australia). I afraid we didn't. As other people have pointed out since that original report one can measure uptake more precisely with Cu67 or by other methods. The method we used was a relatively crude method for quantifying absorption and it seemed necessary to use a more precise method if we were to compare absorption in this mild patient with that of classical cases.

VAN DEN HAMER (The Netherlands). We looked at a Menkes' patient with a very early onset, and when we gave some Cu64 orally we saw a very pronounced uptake of the Cu64. This was not like in a normal control, where you firstly see a sharp uptake peak, then a decrease and finally an appearance of Cu64 in the form of ceruloplasmin. This patient could not be distinguished clinically from any other of the five patients we looked at. Only after it was found that this patient had the slightly higher plasma copper and therefore also a slightly higher plasma ceruloplasmin level than most Menkes patients. We then looked at Cu64 uptake and found this rather pronounced difference.

DANKS (Australia). It is extremely interesting. Once again an exception that I would want to treasure, looking to see whether this patient might have a different defect of copper transport. Did he have a linked family history?

VAN DEN HAMER (The Netherlands). Yes there was.

DANKS (Australia). Bad luck. I was hoping it might be a defect of a different step controlled by a different gene or another chromosome.

AN ACQUIRED FORM OF ACRODERMATITIS ENTEROPATHICA DUE TO LONGTERM LACTOSE-FREE MILK ALIMENTATION

TAKAFUMI MORISHIMA, SHIGERU YAGI and TSUKASA TAKEMURA
DEPARTMENT OF DERMATOLOGY, NIHON UNIVERSITY SCHOOL OF MEDICINE, TOKYO, JAPAN

Acrodermatitis enteropathica (AE) is a genetically determined disorder characterized by psoriasiform to impetiginous skin lesions on the acral areas and around natural orifices, alopecia, and diarrhea. Besides these three cardinal findings, there are usually many symptoms such as stomatitis, glossitis, blepharitis, photophobia, retardation of growth, personality changes, and susceptibility to infection. In 1973, Barnes and Moynahan (1) reported that oral zinc therapy resulted in dramatic resolution of the stigma of AE, and AE is now believed to be an inherited human zinc deficiency disorder.

In 1975, Kay and Tasman-Jones (2), and Okada et al. (3) found that symptoms similar to AE developed in patients receiving intravenous hyperalimentation, and clarified that its pathogenesis was due to zinc deficiency state.

We have recently observed a 4-month-old baby, fed on a lactose-free milk for about 3 months, who developed AE-like lesions which disappeared dramatically in about 2 weeks by switching the milk to a general cow's milk formula, without administration of zinc.

REPORT OF A CASE

The patient, a boy, was born on 2nd March, 1978, of healthy parents, not consanguineous, after a full-term uneventful pregnancy. He was given cow's milk formula from the time of birth. From about 20 days after birth, diarrhea began, 5-6 times a day. Because of this persistent and intractable diarrhea, he was hospitalized on 31st March in the pediatric clinic of a private hospital and was given dietary and antibiotic therapy. Alimentation with lactose-free milk was especially successful for the treatment of diarrhea and he was discharged on 13th May. Although he no longer had diarrhea, he continued to be fed lactose-free milk. His mother noticed a moist dermatitis around his natural orifices from the end of May, and an incomplete loss of hairs from the beginning of July. The baby failed to gain weight, lost the power to ingest milk, and his total serum protein level fell to 4.2 g/dl on 3rd July from 6.8 g/dl at the time of discharge (Fig. 2). He was hospitalized in the Dermatology Department of Surugadai Nihon University Hospital on 27th July.

Physical examination revealed a poorly nourished boy. There was no diarrhea. Dermatological findings:Exudative and erosive lesions were prominent around the eyelids, mouth, and nostrils (Fig. 1-A). Numerous erosions and erythemas, covered with scales or crusts, were found from the neck to the chest (Fig. 1-C). Psoriasiform erythemas were present on the scrotum (Fig. 1-D) and perianal area, and candidiasis-like paronychia on fingers. Hair on the scalp was sparse.

These clinical findings were consistent with those of AE. To examine the nature of faeces, the baby was given FMS milk, a cow's milk formula, in place of lactose-free milk, without administration of zinc. As his ability to ingest milk was rather weak, milk was fed forcibly through a stomach tube. Rather unexpectedly, however, the skin lesions disappeared completely by 14 days after hospitalization. Fig. 1 shows the same area of the skin at the time of initial examination (A) and after 2 weeks (B). Fig. 2 shows the correlation between the clinical course and results of clinical examinations, especially those for serum zinc, alkaline phosphatase, and total protein. It should be especially noted that serum zinc, which was 17 µg/dl at the time of initial examination, increased to 63 µg/dl two weeks later. Subsequently, his growth became good, his scalp hair increased, and skin eruptions did not recur. Serum zinc remained within the normal range and he was discharged on 8th September. At the time of writing, 3 years after discharge, there has been no sign of a recurrence.

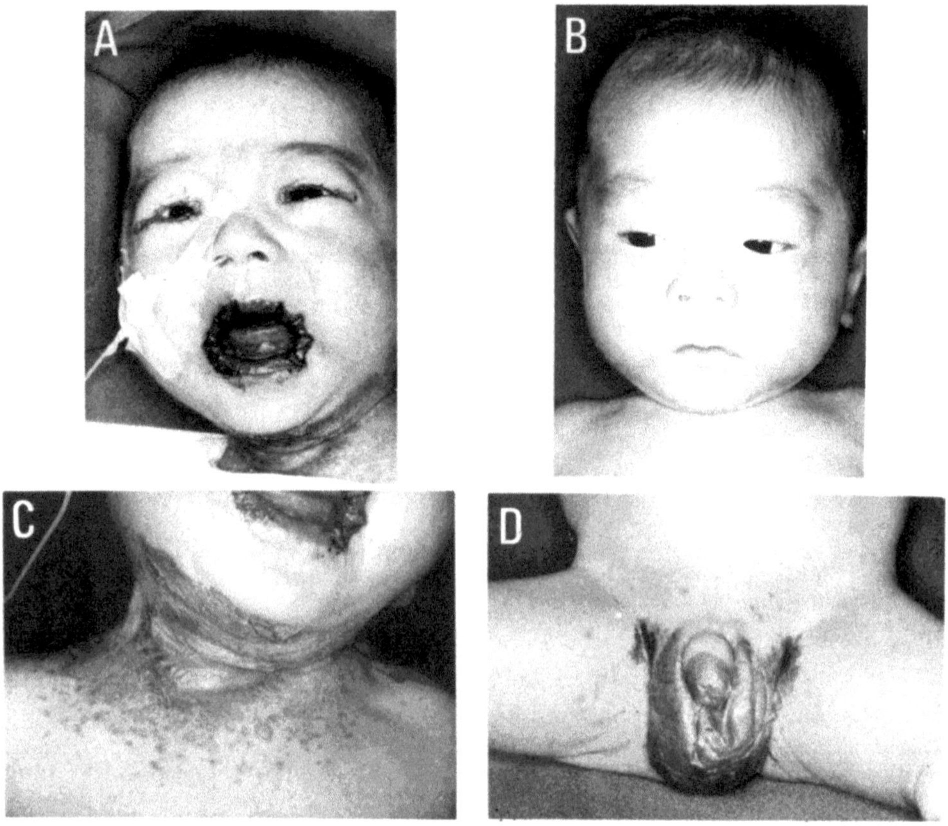

Fig. 1 Exudative and erosive lesions around eyelids, nostrils,and mouth (A),impetiginous lesions from the neck to the chest(C),and psoriasiform lesions on the scrotum(D). These lesions disappeared dramatically about 2 weeks after switching from lactose-free milk to ordinary cow's milk formula(B).

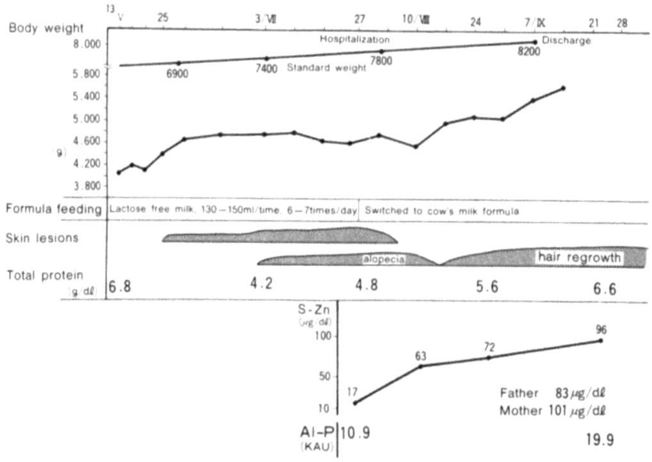

Fig. 2 Correlation between clinical course and clinical examination data.

DISCUSSION

The skin lesions seen in this baby are similar to the typical symptoms of hereditary AE but his condition can be distinguished from it by the fact that his symptoms disappeared in a short period just by switching from lactose-free milk to ordinary cow's milk formula, without zinc therapy, and that the low serum zinc level returned to the normal range. Why did zinc deficiency occur in this case? From his clinical course, we consider that his zinc deficiency was associated with the alimentation of lactose-free milk for about 3 months.

Lactose-free milk has a composition not too different from that of cow's milk formula except lactose is removed and replaced with glucose. It is widely used for infantile diarrhea, especially for the treatment of lactose intolerance. However, as is a problem for specialized milk formulae in general, hyperosmolarity and insufficiency of trace metals have been pointed out (5). Zinc content, measured in our laboratory was 0.5 mg/100 g in lactose-free milk and 1.2 mg/100 g in FMS, a cow's milk. Although zinc content is low in lactose-free milk, it seems that there have been no reports that AE-like lesions develop often in babies fed on lactose-free milk. It has been noted that all of the sugar in lactose-free milk has been replaced by glucose and this has resulted in hyperosmolarity. There have been some reports on side effects, such as development of necrotizing enterocolitis in low birth-weight infants fed on elemental formulae (4) and of hypoproteinemia fed on lactose-free milk (5).

It may be possible to summarize the pathogenesis of zinc deficiency in this case as follows: A large amount of zinc was lost because of the intractable diarrhea, and a large amount of zinc was required for the synthesis of new proteins, especially the zinc metalloenzymes, during the recovery period. Alimentation with lactose-free milk resulted in insufficient oral ingestion of zinc, and induced a further lowering of zinc in the tissues due to poor zinc absorption and hypoproteinemia caused by some pathological changes in the bowels. It was presumed that the zinc deficiency state developed by such a process and the AE-like lesions developed as a consequence.

It is considered from the above clinical experience that AE, a human zinc deficiency, should be classified into two types, hereditary and acquired. We have already reported two cases of hereditary type which were treated successfully by zinc therapy (6) (7). Typical skin lesions recurred after stopping oral ingestion of zinc sulfate in both cases. These facts suggest that zinc therapy must be continued throughout the life time of a patient in the hereditary type of AE. Therefore, it is important to discriminate between the two types of AE, hereditary and acquired. The acquired form can develop from intravenous hyperalimentation, longterm lactose-free milk alimentation or longterm penicillamine administration.

REFERENCES

1. Barnes, P.M. and E.J. Moynahan, 1973, Proc. Roy. Soc. Med. 66:327-329

2. Kay, R.G. and C. Tasman-Jones, 1975, Lancet 2:605-606

3. Okada, T., et al., 1975, The Stride of Med., 92:436-442 (in Japanese).

4. Book, L.S. et al., 1975, J. Ped., 87:602-605.

5. Iida, Y., Ped., 1978, 19:223-228 (in Japanese).

6. Morishima, T. et al., 1975, Jpn. J. Clin. Dermatol., 29:991-997 (in Japanese).

7. Morishima, T. et al., 1976, Jpn. J. Clin. Dermatol., 30:879-885 (in Japanese).

DISCUSSION

AGGETT (Scotland). I think we have been waiting for this type of case for some time. The first demonstration of zinc deficiency associated with lactose free milk was that described by Moynahan and Barnes when they discovered the role of zinc in the aetiology of acrodermatitis enteropathica. That first patient who responded to zinc supplements was, in fact, on a demineralized casein product marketed for the management of lactose tolerance. This prompted Professor Clayton at the Institute of Child Health in London to arrange balance studies looking at children being fed entirely on this product and the results which are to be published show that not only is there a negative zinc balance in these infants but there is also a negative copper balance in some instances. I think that there may well be an underlying copper problem if we keep these children on prolonged feeding with this sort of product without giving any attention at all to these sort of supplements.

ELMES (Wales). I was involved in the retrospective analysis of the case of acute zinc deficiency in a child in the days before pediatricians were aware of these problems. This child had a similar history of diarrhoea and intestinal disturbance that actually led to surgery. He was put onto a soya milk based formula and then developed the types of lesions you described. He only recovered completely when fed a beef protein hydrolysate without zinc supplementation. Now as to why your child developed this, I would like to postulate that he had a transient lactose intolerance that you very successfully treated but unfortunately you put him into a severe state of zinc depletion by your treatment and that by the time his lactose intolerance was treated he was able to absorb cow's milk again.

LONNERDAL (U.S.A). How was this lactose-free milk produced? Was it produced by using a demineralised whey product or was it by incubating whey with lactase to degrade the lactose? I think that is a very important point because if you are using a demineralised whey product you will lose all the ligands that possibly promote zinc absorption, while if you are using a lactase degraded product you will have all those ligands still there.

A CASE OF ALOPECIA IN MAN ASSOCIATED WITH ZINC DEFICIENCY

B. Momcilović, S. Handl and V. Zjacić

Institute for Medical Research and Occupational Health
and

M. Stojanovic
Department of Medicine, Clinical Hospital,
ZAGREB
YUGOSLAVIA

The essential role of zinc in human nutrition and metabolism is well documented (1,2). The normal functioning of the skin and hair appears to be closely related to the zinc status. Cutaneous manifestations of dermatitis enteropathica (3), wound healing (4,5), and gastric ulcer (6) are reported to have been cured by a zinc sulphate therapy. Moreover, a case of penicillamine-induced zinc deficiency followed by hair loss was described where the hair recovered after zinc supplementation (7). The aim of this work is to present a case of alopecia areata totailis in man associated with zinc deficiency.

PATIENT AND TREATMENT

A boy born in 1966 became gradually alopetic in 1977 without an apparent reason. He is the second of two children from a high income family. The past history revealed two febrile tetanies in infancy and measles, chickenpox, mumps, rubella and infectious mononucleosis by the age of five. Since then he has had no health problems. His dietary habits included drinking of large amounts of pasteurized cow's milk (1-2 lit/day), a lot of 'empty' calorie foods and little meat. When he had become alopetic the boy underwent a thorough medical examination which did not show organic causes of the illness. Tonsillectomy and a symptomatic polyvitamin therapy were unsuccessful. Paramedical treatments inducing reactive hyperaemia, bullae and crusts were followed by a growth of scanty hair at the top of capillitium only, which fell out shortly afterwards. The boy became depressed with transitional emotional outbursts but he was a good pupil at school.

In 1980 low serum zinc values were found (Table 1) and oral zinc supplementation therapy was immediately initiated (70 mg $ZnSO_4 \cdot 7H_2O$ three times a day with plenty of water after meals). By that time the boy was entirely alopetic and even eyebrows and eyelashes were missing (Fig. 1). During the treatment zinc and copper in serum were determined by the flameless atomic absorption spectroscopy (8) whereas iron, unsaturated iron binding capacity (UIBC), total iron binding capacity (TIBC), iron saturation, calcium, phosphorus, and magnesium were analysed by standard analytical methods in the hospital laboratory (9). Testosterone and cortisol were determined by radioimmunoassay kits (Wiena Laboratories L.M.C., Austria, and New England Nuclear, U.S.A. respectively).

Approximately two months after the commencement of treatment dark coloured hair started to appear on the scalp. At the same time eyelashes and eyebrows appeared as well as the first pubic hair. After six months of treatment the whole scalp was covered by thin hair (Fig. 2) and after one year of treatment the head was covered with hair of normal thickness and even some curling was noted (Fig. 3). Other clinical signs included increased apetite and increase in body height from initial 159 cm to 165.5 cm i.e. a two-fold increase in relation to the previous one-year period.

Changes in the clinical picture were followed by a laboratory finding of an increase of zinc and iron in serum from low to normal levels whereas copper decreased from high to moderate levels. Iron saturation and UIBC also showed signs of better iron utilization. TIBC and serum levels of calcium, phosphorus and magnesium were unaffected by zinc supplementation. Testosterone and cortisol levels were 0.160 and 25.8 μg/100 ml at the institution of treatment and 0.193 and 13.8 μg/100 ml three months later respectively i.e. within normal range.

TABLE 1

Laboratory findings during the treatment of alopecia areata

		1980		1981	
		April	October	January	May
Zinc	µg/100 ml	61	98	123	130
Copper	µg/100 ml	140	122	106	-
Iron	µg/100 ml	65	76	101	-
UIBC	µg/100 ml	275	285	255	-
TIBC	µg/100 ml	340	361	356	-
Fe sat.	%	19.1	21.1	28.3	-
Calcium	mg/100 ml	9.7	10.1	10.1	-
Phosphorus	mg/100 ml	4.0	4.2	6.0	-
Magnesium	meq/lit.	1.7	1.3	1.6	-

Normal ranges at this hospital laboratory; Zn 92-125, Cu 91-141, Fe 80-175, UIBC 140-250, TIBC 250-400, Fe sat. 25-50, Ca 9-11, P (children) 4-7, Mg 1.5-2.3.

DISCUSSION

The etiology of alopecia areata was associated with parasitic infestations, endocrine disturbances, neurotrophic disorders and focal infections to name a few without apparent success and consequently with disappointing results in symptomatic therapy (10). Our observation clearly demonstrated that at least some forms of alopecia areata are associated with trace elements deficiency, zinc deficiency in particular. In such cases simple supplementation of zinc to the diet will help to treat the illness leaving the therapy with corticosteroids and dinitrochlorbenzene (11) to other ill understood forms of alopecia areata. An apparently dissociative pattern of metabolic behaviour of zinc and iron on one side and copper metabolism on the other revealed subtle interrelationships between these three trace elements which might be used as prognostic criteria during the treatment. A close similarity was noted in the metabolic behaviour of zinc and iron (12). Whereas a dissociation marked iron and copper metabolism i.e. high copper and low iron values were observed in even such unrelated health disturbances as stress and iron anaemia (13, 14). Moreover, zinc and copper antagonism has been recently used in the treatment of Wilson's disease where long-term treatment with zinc sulphate was followed by a decrease of accumulated tissue copper (15).

ACKNOWLEDGEMENT

This work was supported by a research grant from the Scientific Research Council of the Socialistic Republic of Croatia, Yugoslavia. The authors thank Mrs. Dj.Karlić, M.D., for helpful suggestions and H. Stanković, M.Sc., for analysis of serum samples.

REFERENCES

1. Halsted, J.A., J.C.Jr. Smith, and M.I. Irwin, 1974, J. Nutr., 104:354-378.
2. Agett, P.J., and J.T. Harries, 1979, Arch. Dis. Childh., 54:909-917.
3. Sunderman, Jr. F.W., 1975, Ann. Clin. Lab. Sci., 5:132-145.
4. Haley, J.V., 1979, J. Surg. Res., 27:168-174.
5. Williams, K.J., R. Meltzer, R.A. Brown, Y. Tanaka, and R.C.J. Chiu, 1979, J. Surg. Res., 27:62-67.
6. Frommer, D.J., 1975, Med. J. Aust., 2:793-796.
7. Klinberg, W.G., A.S. Prasad, and D. Oberleas, 1976, In "Trace elements in human health and disease" (ed. A.S. Prasad) (Academic Press, N.Y.), vol. I., pp 51-65.
8. Analytical methods using the HGA graphite furnace, 1977, (Perkin-Elmer Co., Norwald) pp 1-125.

9. Wessman, N., and V.J. Pileggi, 1974, In "Clinical Chemistry" (eds. R.H. Henry, D.C. Cannon, J.W. Wintelman) pp 639-754.

10. Cajkovac, S., 1970, In "Skin diseases" (Yug. Acad. Sci. Arts, Zagreb) pp 509-513.

11. Happle, R., K. Cebulla, and K. Echternacht-Happle, 1978, Arch. Dermatol., 144:1629-1631.

12. Hawkins, T., J.M. Marks, V.M. Plummer, and M.W. Greaves, 1976, Clin. Exptl. Dermatol., 1:243-252.

13. Andel, M., V. Brodan, and E. Kuhn, 1978, Horm. Metab. Res., 10:171.

14. Schulz, J., and N.J. Smith, 1958, Am. J. Dis. Child., 95:120-125.

15. Hoogenraad, T.U., R. Koeveoet, and E.G.W.M. de Ruyter Korver, 1979, Eur. Neurol., 18:205-211.

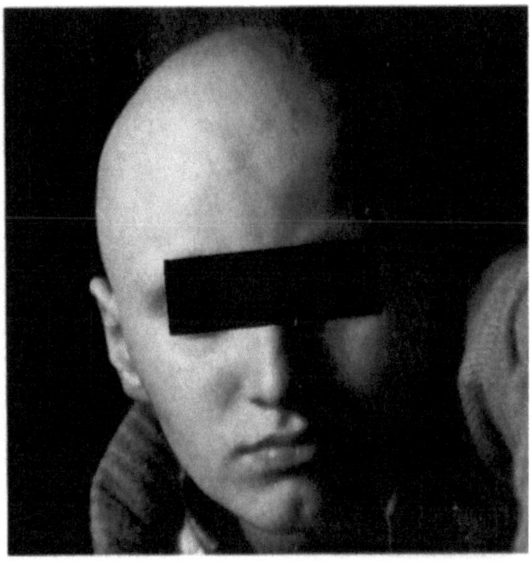

Fig. 1. Photograph taken before treatment.

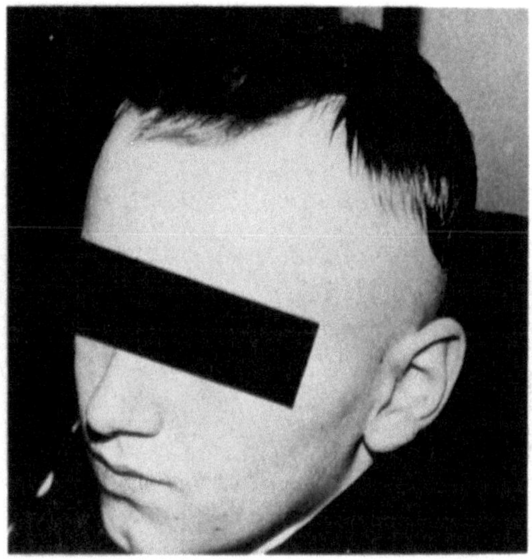

Fig. 2. Photograph taken after 6 months of treatment.

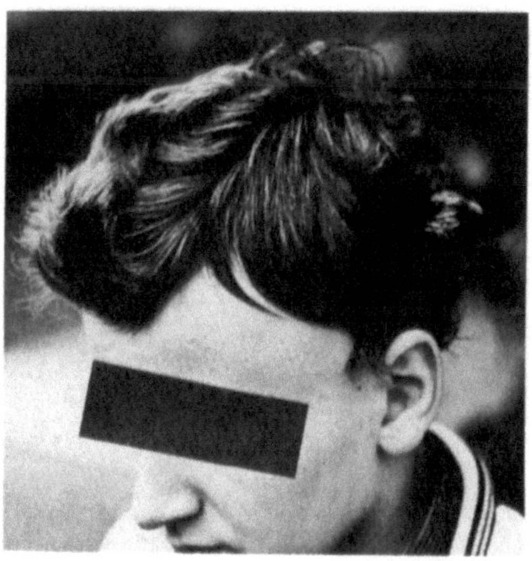

Fig. 3. Photograph taken after 1 year of treatment.

DISCUSSION

GOLDEN (Jamaica). Did this boy's dietary intake of any other nutrient e.g. protein quantity or quality, vary. Did his appetite change?

MOMCILOVIC (Yugoslavia). Since the beginning of therapy the amount of milk consumed was reduced because of it's high Ca content. His appetite greatly increased. In the two year period before treatment he grew 6.5cm and after one year of zinc supplementation he grew 6.5 cm.

ZINC AND FOLATE MALABSORPTION IN COELIAC DISEASE

S. Jameson, O. Björklund, L. Lööf, J. Ulfberg
Department of Internal Medicine
University Hospital
S-750 14 UPPSALA
SWEDEN

INTRODUCTION

Coeliac disease (CD), defined as the condition in which there is an abnormal jejunal mucosa that responds morphologically to treatment with a gluten-free diet (GFD) (1), is more common than was earlier thought. Many CD-patients only have vague or no gastrointestinal symptoms. Recurrent iron deficiency anemia and folate deficiency anemia are common findings and often the only symptoms. With an increased awareness of the disease we diagnose now many early cases and we find today in our department abnormally low serum folate levels only in 50 % of new CD-cases whereas serum-zinc regularly is abnormally low in almost all. Zinc and folate absorption are connected as it has been shown that the activity of folate-conjugase, the jejunal enzyme which divides complex folates to monofolate, is reduced, when zinc deficiency is introduced in man (2). Folate deficiency as a consequence of jejunal disease is well established (3). Folic acid absorption test is a more sensitive test than xylose test (3). Zinc absorption test is also in use and some patients with CD had markedly impaired zinc uptake following oral loading (4). Consequently we wanted to study the value of a combined oral zinc and folic acid absorption test in the diagnosis of CD.

MATERIAL AND METHODS

Seventeen consequtive patients with villous atrophy of varying degree in jejunal biopsies when investigated for possible gastrointestinal disease took part. All later on responded with normalization or improvement of mucosal changes when on a GFD. Nine women and eight men had a mean age of 44, SD 15, range 21-74 years. Twelwe age and sex matched healthy asymptomatic adults without any medication served as controls. A standard oral absorption test was used. Five mg folic acid (Folacin) in 30 ml water was given fasting in the morning. One hour later 0.2 gm zinc sulphate equivalent to 45 mg Zn (Solvezinc) which is an effervescent preparation containing citrate, was given in 30 ml water. Blood samples were drawn at 0', 60', 120', 180', and 210'. Serum-folate was analyzed microbiologically, serum-zinc with atomic absorption spectrophotometry. Five patients had been given folate treatment before the investigation. Seven patients have been retested with the oral absorption test after six months on GFD and were rebiopsied at the same time.

RESULTS

Fasting serum-zinc was abnormally low in 16/17 patients (normal range 11-20 µmol/l) and in 1/12 controls. Serum-folate was abnormally low in 7/17 patients, in 5/12 untreated and 2/5 folate treated patients (normal range 7-20 nmol/l).

The results from the absorption tests are given in the following tables.

S-folate nmol/l mean	0'	60'	120'	180'	210'	peak	S.D.	n
CD untreated	14.5	88	221	200	219	255	134	12
CD folate treated	25.6	287	552	447	381	557	185	5
CD all	17.7	150	319	273	270	344	202	17
Controls	17.6	546	668	505	408	678	176	12
t-test 17 - 12, p	n.s.	<0.001	<0.001	<0.001	<0.001	<0.001		

S-Zn µmol/l mean	0'	60'	120'	150'	peak	S.D.	n
CD untreated	8.8	17	21	18	22	6.7	12
CD folate treated	9.8	20	27	25	28	9.5	5
CD all	9.1	18	23	20	24	7.8	17
Controls	13.2	25	31	29	32	4.0	12
t-test 17-12, p	<0.001	<0.02	<0.01	<0.001	<0.005		

The increase in serum zinc 0' - peak value was in 12 untreated CD 13.6 µmol/l, S.D. 5.4, and in 12 controls 19.0, S.D. 5.1, t-test $p < 0.02$.

A positive correlation was found between peak S-Zn levels and peak S-folate in 12 untreated CD patients and 12 controls, $r = 0.64$, $p < 0.001$. Only one CD-case overlapped into control range, a man with slight villous atrophy in jejunal biopsy.

Seven patients were retested after six months on GFD. They had all normalized or improved their folate and zinc absorption curves, corresponding to normalization or improvement of mucosal changes in jejunal biopsies. One of them had earlier got folate therapy but she further increased her uptake when on GFD.

DISCUSSION

The results show that the rise in serum-zinc and serum-folate after oral pharmacological doses is a useful test of the absorptive capacity of the proximal small gut. Zinc uptake is a controlled process, the details of which are still not fully known in man. A low molecular weight zinc-binding compound has been purified from human milk and evidence is provided that this compound is citrate (5). A similar ligand is identified in human duodenal secretions after pancreatic stimulation (6). A positive correlation has been found between zinc absorption and the protein content in meals (7). An inhibition of zinc absorption was seen with nonheme iron in man (8). The role of intracellular ligands in the regulation of zinc absorption has recently been reviewed (9). Studies in mice indicate a significant role for metallothionein in zinc absorption with a positive relationship between absorption and the synthesis of metallothionein (10). Prostaglandin E_2 facilitates zinc transport across the intestinal mucosa in rats (11).

When using zinc tolerance tests plasma zinc response was a function of dosage and varied with the amount of zinc taken (12). High plasma peak zinc levels were also found when using a dose of 120 mg oysters (13). Zinc sulphate, at 220 mg dosage, gave much less serum zinc levels in patients with alcoholic cirrhosis than in healthy controls (14). Decreased absorption of zinc and iron was found in patients with iron deficiency anemia and geophagia (15). Patients treated for obesity by jejuno-ileostomy had lesser zinc uptake than healthy controls (16). Abnormally low zinc uptake has also been found in Crohn's disease (23). Patients with adult CD had impaired zinc uptake following oral loading (4). Zinc uptake by enterocytes was markedly reduced in jejunal mucosal biopsies from patients with acrodermatitis enteropathica (AE) (17). The zinc binding ligand in duodenal secretions was less efficient in zinc uptake in AE than in controls (6). Zinc deficiency is suggested to be the cause of coeliac infertility (18). GFD normalizes serum-zinc levels in CD and dermatitis herpetiformis (DH) (19).

We found a lesser zinc uptake in patients with villous atrophy of varying degree than in normal controls expressed both as absolute serum-zinc levels or total increase. There was overlapping between the groups, however. Two patients who only had slight villous changes and five who had previously been given folate treatment reached the highest levels, the latter in spite of villous atrophy. The effervescent zinc preparation we used contains citrate, which might have improved zinc uptake and concealed a more pronounced difference. Already existing deficiencies could affect and reduce protein synthesis, what possibly could explain the difference in uptake between folate treated and untreated patients. Patients with villous changes also have a reduced uptake surface area. All seven patients retested after six months on GFD showed markedly improved or normalized zinc and folate uptake, parallel to improvement of villous changes.

The complex mechanisms of intestinal absorption of folates have recently been reviewed (20). Still unresolved controversies remain. Folic-acid absorption tests have been in use since long. Patients with idiopathic steatorrhoea, many of which with villous atrophy, had all abnormal folic acid absorption (3). Malabsorption of dietary folate in adult CD and DH has also been found (21). It has also been shown that mono- and polyfolates were poorly absorbed in untreated CD, but were normally absorbed after treatment with GFD (22). Zinc depletion results in a decreased intestinal hydrolysis of pteroylpolyglutamate and it is suggested that intestinal conjugase in man is a zinc-dependent enzyme (2).

Our data show a better folate uptake in patients with villous atrophy who previously had received folate treatment compared to untreated patients. One patient further improved her folate uptake when on GFD. Totally seven patients improved or normalized folate uptake, parallel to improvement of villous changes, when on GFD. Folate deficiency seems further to impair folate and zinc absorption, irrespective of surface area.

SUMMARY

In coeliac disease there is a malabsorption of folic acid and zinc, when tested with pharmacological preparations in oral loading tests. Zinc malabsorption seems to precede folate malabsorption. Abnormally low serum zinc levels is an early and common finding in CD. Folic acid treatment increases the ability to absorb folic acid and zinc, in spite of villous atrophy. Gluten free diet normalized mucosal changes as well as folate and zinc absorption in seven patients. A combined oral zinc and folate loading test seems to discriminate patients with villous atrophy of varying degree from healthy controls.

REFERENCES

1. Booth, C.C., 1970, Br. Med. J. 3:725 - 731.

2. Tamura, T., and B. Shane, 1978, Am. J. Clin. Nutr. 31:1984 - 1987.

3. Chanarin, I., and M.C. Bennett, 1962, Br. Med. J. 1:985 - 988.

4. Clover, S.C., and P.W. Brunt, 1976, Gut 17: 813.

5. Lönnerdal, B., and A.G. Stanislowski, 1980, J. Inorg. Biochem. 12:71 - 78.

6. Casey, C.E., and K.M. Hambidge, 1979. J. Pediatr. 95:1008 - 1010.

7. Sandström, B., and B.Arvidsson, 1980, Am. J. clin. Nutr. 33:739 - 745.

9. Solomons, N.W., and R.A. Jacob, 1979, Gastroenterology 76:1252.

10. Starcher, B.C., and J.G. Glauber, 1980, J. Nutr. 110:1391 - 1397.

11. Song, M.K., and N.F. Adham, 1978, Am. J. Physiol. 234: E99 - E105.

12. Oelshlegel, F.J., and G.J. Brewer, 1977. In "Zinc Metabolism" (Eds. G.J. Brewer and A.S. Prasad) (A.R. Liss, New York) pp 299-311.

13. Solomons, N.W., and R.A. Jacobs, 1979, J. Lab. Clin. Med. 94:335 - 343.

14. Sullivan, J.F., and M.M. Jetton, 1979, J. Lab. Clin. Med. 93: 485 - 492.

15. Arcasoy, A., and A.O. Cavdar, 1978, Acta Haematol. 60:76 - 84.

16. Andersson, K.-E., and L. Bratt, 1976, Europ. J. clin. Pharmacol. 9:423 - 428.

17. Atherton, D.J., and D.P. Muller, 1979, Clin. Sci. 56:505 - 507.

18. Jameson, S., 1976, Acta Med. Scand. Suppl. 593:1 - 89.

19. Jameson, S., and G. Järnerot, 1979, Opuscula Medica 24:14 - 15.

20. Halsted, C.H., 1979, Am. J. Clin. Nutr. 32:846 - 855.

21. Hoffbrand, A.V., and A.P. Douglas, 1970, Br. Med. J. 3:85 - 89.

22. Halsted, C.H., and A.M Reisenauer, 1978, Gut 19:886 - 891.

23. McClain, C., and C. Soutor, 1980, Gastroenterology 78:272 - 279.

DISCUSSION

ELMES (Wales). Have you only studied the partial villous atrophy of coeliac disease or have you studied any other forms of partial villous atrophy? The second question is, have you any opinion as to the region of the small intestine that is most concerned with zinc absorption? You follow the serum levels over a temporal range that in a fasting patient should take you quite far down the small intestinal tract and as you have quoted you find zinc deficiency not only in coeliac disease which tends to affect the upper small intestine, but also in Crohn's disease which usually affects the lower small intestine.

JAMESON (Sweden). We find the same defects in uptake in other villous atrophy patients too; there were no responses in those patients who didn't respond to gluten free diets.

ELMES (Wales). But still suffering from coeliac disease?

JAMESON (Sweden). I wouldn't say that these patients have coeliac disease who don't respond to gluten free diets. Villous atrophy is not only due to coeliac disease.

ELMES (Wales). So you have no comments on the site of zinc absorption?

JAMESON (Sweden). No, we have seen in a few patients with duodenitis the defect in zinc uptake and we have also seen a defect in uptake in patients with gastric atrophy, but we think that these patients are early coeliacs.

VAN DEN HAMER (The Netherlands). Two questions. Firstly, could you compare your method of testing with zinc with the method using radio-zinc. With your method using cold zinc you are always looking at an increase on top of the already existing zinc. If you use radio-zinc you would possibly find a sharper and more easily detectable increase. Second question, do you have any idea why the zinc absorption is strongly increased in Wilson's disease?

JAMESON (Sweden). We haven't studied zinc uptake in Wilson's disease, but I know about one case of Wilson's disease which had been treated with zinc for a long time and got complete normalisation - that is a literature case. We have not used radio-isotopes because the only one available is zinc 65, and we won't use that in normal doses. We also do not have a whole body counter which could be used in other parts of Sweden apart from Uppsala. Zinc isotopes with short half lives are not available yet, but we will get them in one year I hope.

ANON. I think its rather simple to produce zinc-69m with a reactor: this is a short lived isotope.

ANON. I would just like to make a couple of comments based on our own experience with a similar test in Aberdeen. One is on the interpretation of your results - you have expressed your results in terms of the various peaks achieved at hourly intervals up until 3½ hours. Have in fact you interpreted it in terms of the increments over and above the base line level at the start of the test? This is a technique that we have used and it may perhaps point to something that Dr. Elmes is bringing up; we found that there was a very marked difference between coeliac and normal patients in the amount of zinc absorbed over the first three hours as interpreted by measuring the area under the graph. However, at 6 hours when one presumes that distal absorption is taking place, there was in fact not such a clear distinction at all between the coeliac patients who are on a normal diet and between the normal patients. Clearly that would imply that there was some absorption occurring in the coeliac patient somewhere, and since we know that coeliac disease is both a patchy lesion and principally a proximal intestine lesion, I think you certainly chose the right time interval to do your test.

JAMESON (Sweden). We made these tests in order to get a rapid test to discriminate between normals and abnormals which could be useful in clinical work. That is why we use these sampling times. We have calculated the areas under the curve and they show significant differences also. We have also made comparisons between peak increases and there was a significant difference between the groups in peak increase if we use the untreated patients as controls. Those patients with earlier folate treatment have a better zinc uptake and many of them are within the normal range. However there is a slower and more prolonged uptake of zinc compared with untreated patients, and the area under the curve is less in our coeliac patients than in normals.

EFFECTS OF DIPHENYLHYDANTOIN ON ZINC AND COPPER METABOLISM

R. Palm, G. Hallmans and K. Wing

Biophysical Laboratory and Departments of Neurology,
Pathology and Oral Roentgenology
University of Umeå, 901 87 UMEÅ
SWEDEN

Diphenylhydantoin, or phenytoin, has for more than 40 years been one of the main drugs in epilepsy therapy. Phenytoin has the characteristics of a chelator and, as the symptoms of phenytoin intoxication resemble those of acute zinc deficiency (Henkin *et al* 1971), it is possible that the drug may cause zinc deficiency. The serum zinc (S-Zn) concentration in patients treated with phenytoin has been reported to be decreased (Barbeau and Donaldson 1974), unchanged (Schott and Delves 1978) and increased (Pippenenger *et al* 1980). The serum copper (S-Cu) concentration has been found to be unchanged (Barbeau and Donaldson 1974) and increased (Schott and Delves 1978, Pippenenger *et al* 1980). The concentration of the copper-containing enzyme ceruloplasmin in serum (S-CP) has been found to be increased in patients on phenytoin (Taylor *et al* 1974). Very high S-Cu concentrations have also been reported during phenytoin intoxication (Schott and Delves 1978).

The purpose of the present study was to determine the zinc and copper status in epilepsy patients before and during phenytoin therapy and in patients on long-term therapy in order to test the hypothesis that phenytoin can cause a zinc deficiency.

MATERIALS AND METHODS

Zinc and copper status in eleven untreated epileptics, six men and five women, were determined immediately prior to the initiation of phenytoin therapy, during the first 72 hours of therapy and again after one month of therapy. None of the eleven patients had experienced a seizure during the last 14 days before the start of therapy. None of them had any clinical or laboratory signs of infections, liver disease or alcoholism and none was on pharmacotherapy including contraceptives. Phenytoin was given orally in doses of 300 mg daily, 200 mg at 8 a.m. and 100 mg at 6 p.m.. S-Zn, S-Cu, S-CP, S-phenytoin, S-albumin and S-α_2-macroglobulin were determined before the start of therapy and 24, 48, 72 hours and one month after the first dose of phenytoin. All samples were taken in the morning after a night's fasting and before the morning dose of phenytoin. The concentrations in eleven age- and sex-matched, healthy volunteers were used as controls. In some patients the urinary excretion of Zn and Cu and the zinc concentrations in the cerebrospinal fluid (CSF) were determined before the start of and after one month of therapy.

The S-Zn, S-Cu, S-CP and S-phenytoin concentrations were determined in twenty epileptics, ten males and ten females, on long-term therapy with phenytoin alone or in combination with other anti-epileptic drugs. The concentrations in 20 age-matched, healthy volunteers, ten males and ten females, were used as control values. All samples were taken after a night's fasting.

Flame atomic absorption spectrophotometry (AAS) (Hallmans 1978) was used to determine the zinc and copper concentrations in serum and urine and the concentrations of zinc in CSF (Palm, Hallmans and Sjöström 1981). Serum concentrations of albumin, ceruloplasmin and α_2-macroglobulin were determined by electroimmunoassay (Laurell 1972) and the S-phenytoin concentrations were measured with the EMIT-technique (Kupferberg 1978).

RESULTS

S-Zn concentrations were found to be significantly lower in untreated epileptics compared with controls ($P < 0{,}05$) (Fig. 1). At the start of phenytoin therapy slight increases of both S-Zn and S-Cu concentrations were found after 24 hours and they persisted at 48 and 72 hours. After one month S-Cu had increased even more and S-Zn had returned to pre-therapy levels. S-CP was found to be increased after one month. No changes in serum concentrations of albumin and α_2-macroglobulin were found. Urinary excretion of zinc and copper was unchanged during the study. The concentrations of zinc in CSF in untreated epileptics (0,16 $\pm$0,06 µmoles/l) did not differ from that in controls (0,14$\pm$0,03), and no change was found after one month of therapy (0,18$\pm$0,05).

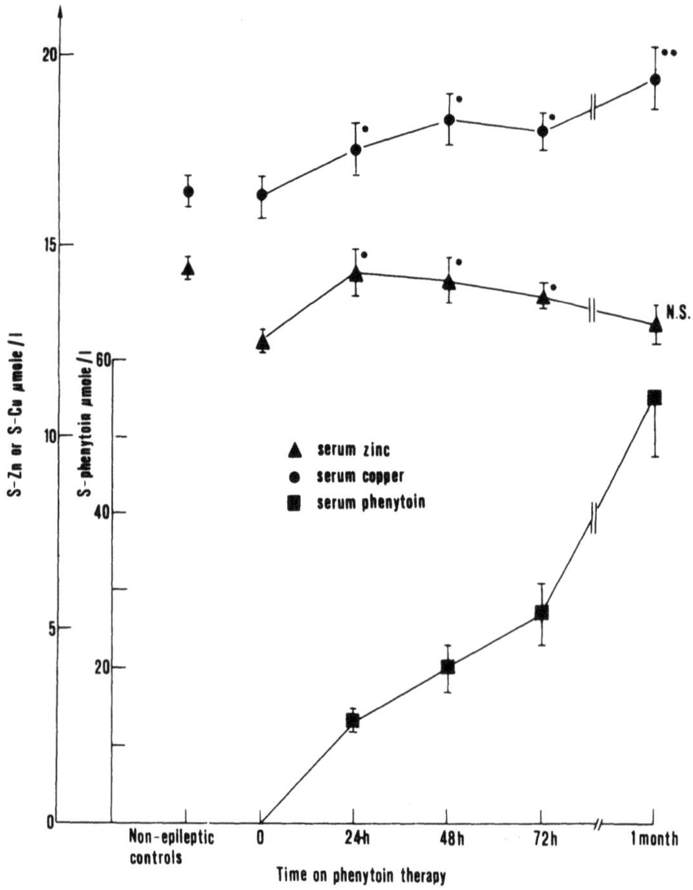

Fig.1. Serum concentrations of zinc, copper and phenytoin in eleven healthy volunteers and in eleven epileptic patients at the start of phenytoin therapy. The results are presented as the means $\pm$S.E.M.. The results of t-test comparisons with pre-therapy concentrations are indicated in the figure (* $p < 0,05$; ** $p < 0,01$).

Higher S-Cu and S-CP concentrations were found in epileptic patients on long-term phenytoin therapy than in controls of the same age and sex ($p < 0,01$), while no difference was found for the S-Zn concentrations (Table I).

	N	Mean age	S-Zn µmoles/l	S-Cu µmoles/l	S-CP g/l	S-Phenytoin µmoles/l
Male epileptics	10	37	14,6$\pm$1,1	20,5$\pm$0,7***	0,35$\pm$0,09***	60$\pm$9
Male controls	10	37	14.4$\pm$0,7	15,0$\pm$0,6	0,26$\pm$0,02	-
Female epileptics	10	42	13,6$\pm$0,6	25,2$\pm$1,9**	0,43$\pm$0,03**	51$\pm$9
Female controls	10	40	13,6$\pm$0,6	18,1$\pm$0,6	0,30$\pm$0,02	-

Table I. Serum concentrations of zinc, copper, ceruloplasmin and phenytoin in epilepsy patients and controls. The results are presented as the means $\pm$S.E.M.. The results of t-test comparisons with controls are indicated (** $p < 0,01$; *** $p < 0,001$).

DISCUSSION

The S-Zn concentration was low in the untreated epileptics. This may indicate that they had a slight zinc deficiency. They were healthy in all other respects. The S-Zn increased to near normal levels immediately after the start of phenytoin therapy but were decreased somewhat after one month's therapy. This temporary increase in S-Zn may reflect increased absorption, decreased excretion or a redistribution in the body's tissues and fluids.

Evidence for a zinc deficiency in phenytoin therapy has not been found in this study but the increased S-Cu and S-CP concentrations may be an indirect sign of zinc deficiency. It is possible that some of the S-Zn is bound to phenytoin and unavailable to the rest of the body. The possibility of a hidden zinc deficiency indicates the need for a method for determining the fraction of S-Zn bound to phenytoin.

Some pilot studies have been performed in rats with the purpose of determining whether or not phenytoin interferes with zinc metabolism. Preliminary results have demonstrated a decrease in the intestinal absorption of zinc (^{65}Zn) from single meals to which human therapeutic doses of phenytoin had been added. No other evidence for an effect of phenytoin on zinc metabolism has as yet been seen. Further studies on the effects of phenytoin using radiozinc in rats are planned.

REFERENCES

1. Henkin, R.I., P.M. Patten, P.K. Re *et al*, 1975, Arch. Neurol. 32:745-751.

2. Barbeau, A., J. Donaldson, 1974, Arch. Neurol. 30:52-58.

3. Schott, G.D. and H.T. Delves, 1978, Br. J. Clin. Pharmac. 5:279-280.

4. Pippenenger, C.E., C. Garlock, F. Fernandez *et al*, 1980, In "Advances in Epileptology: XI th Epilepsy International Symposium" (ed. R. Canger).

5. Taylor, J.D., P.M. Kahn, and T.N. Higgins, 1974, Am. J. Clin. Pathol. 61:557-558.

6. Hallmans, G., 1978, Acta Dermatoven. 58, suppl 80.

7. Palm, R., G. Hallmans, R. Sjöström, 1981, in manuscript.

8. Laurell, C.B., 1972, Scand. J. Clin. Invest. 29, suppl 124:71-82.

9. Kupferberg, H.J., 1978, In "Antiepileptic Drugs: Quantitative Analysis and Interpretation" pp 9-17.

DISCUSSION

BREMNER (Scotland). Did you follow changes in copper metabolism in the studies with rats?

PALM (Sweden). Yes, copper accumulated in liver, kidney and certain parts of the brain.

VAN DEN HAMER (The Netherlands). Were your patients hospitalised, or were they outpatients? In Holland, I know of two trials with phenytoin, one with hospitalised patients and one with outpatients. The former patients experienced a great increase in plasma copper but in the latter there was no increase at all. I wonder if the plasma copper changes are determined more by the disease than by the therapy?

PALM (Sweden). We had both hospitalised and outpatients in our study and both groups showed a similar response.

DERMALLY APPLIED COPPER SALICYLATES AS ANTI-INFLAMMATORY AGENTS.

W.R. Walker[1], S.J. Beveridge[2] and M.W. Whitehouse[3]

[1]Department of Chemistry, [2]Faculty of Medicine, University of Newcastle, New South Wales, 2308; and [3]Department of Pathology, University of Adelaide. Adelaide, S.A. 5001.

INTRODUCTION

The 'Copper Bracelet' has been shown to be of some therapeutic value to a significant number of "arthritic" people (1). A subsequent study (2) confirmed that metallic copper dissolves in human sweat and the following reaction (HL is one of the numerous acids in sweat) was suggested:

$$2Cu^{0} + 4HL + O_2 \rightarrow 2CuL_2 + 2H_2O.$$

The copper(II) complex bis(glycinato) copper(II) monohydrate has been prepared (3) by the aeration of metallic copper in an aqueous glycine solution. Labelled with ^{64}Cu it was used to perfuse intact cat skin (4).

Schubert (5) suggested that salicylate delivered copper to the cells of the body and that a copper salicylate was involved in the antipyretic action of salicylates in rats. This fact has been included in a lengthy review on copper pharmacology by Sorenson (6). Our review on copper salicylates (7) describes a copper salicylate mixture called "Permalon" that was used for the treatment of rheumatoid and other degenerative diseases as early as 1950. The water soluble preparation would be $Na_2[Cu(sal)_2]$ where H_2sal is salicylic acid. The anti-inflammatory effects of other copper complexes have been reported recently (8) and claimed to be dependent on the route of administration. The "copper approach" to anti-rheumatic drugs is discussed further in a recent comprehensive review (9).

Because of the problems associated with intraveous, oral and other types of application of copper drugs, this investigation of the *topical* application of lipophilic copper(II) complexes has been undertaken. Copper(II) salicylate complexes have been synthesised and tested because they mimic superoxide dismustase (an essential metallo-enzyme in host defence) in scavenging superoxide radicals ($\bullet O_2^-$). It has been suggested (10) that:

$$Cu(II)(Hsal)_2 + \bullet O_2^- \rightarrow Cu(I)(Hsal)_2 + O_2$$

$$Cu(I)(Hsal)_2 + 2H^+ + \bullet O_2^- \rightarrow Cu(II)(Hsal)_2 + H_2O_2.$$

It appears, therefore, that copper(I) is involved as has been postulated previously (11).

These copper salicylates also mimic caeruloplasmin, the extracellular copper-containing plasma protein, which destroys oxygen-derived free radicals (12).

The lipophilic copper(II) complexes are of the type

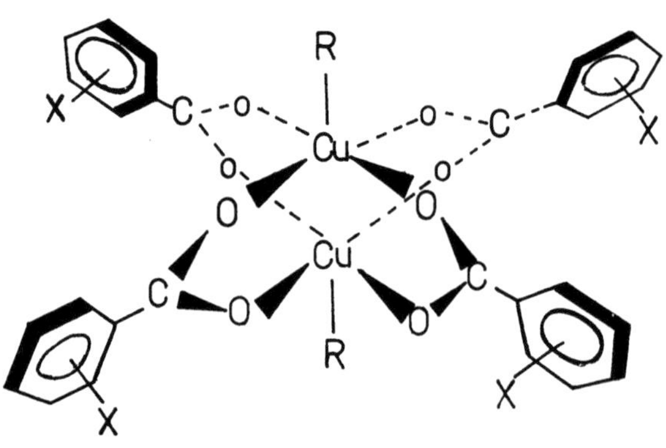

Alcusal[R]
X = OH on C_2
R = C_2H_5OH

Dermcusal[R]
X = OH on C_2
R = DMSO

RESULTS AND DISCUSSION

It has been shown that copper(II) acetate (the same dimeric structure as Alcusal[R]) can perfuse human skin (13). We have confirmed that Alcusal[R] labelled with Cu^{64} perfuses pig and rat skin. Elevated salicylate and copper levels in the plasma of a human volunteer (SJB) have been found several hours after topical application of Alcusal[R] to the forearms.

Alcusal[R]

This and related formulations have been applied to the shaved dorsal skin of rats (ca 20 cm^2) just below the neck. Full details of their anti-inflammatory activity and pharmacology have been reported (14). The abstract of this paper states: "In rats, dermal (dorsum) application of an ethanolic copper salicylate complex (ECS) in ethanol with glycerol (Alcusal[R]) effectively suppressed established polyarthritis, carrageenan-induced paw oedema and the inflammatory response to hydroxylapatite microcrystals. These responses appear to be due to some degree of synergism between the copper(II) and salicylate, i.e. not solely due to either species alone. Evidence for transdermal absorption of ^{64}Cu from ^{64}Cu-ECS is provided. Some safety aspects of this novel drug are discussed".

Properties of Topically Applied Alcusal[R] buffered with sodium acetate (pH 4.7) include

1. Percutaneous delivery of Cu(II) to Cu deficient child with a mild allelic variant of Menkes' Symdrome[a].
2. Relief of swelling/joint stiffness in patients with rheumatoid arthritis[b].
3. Suppression of established polyarthritis in rats[c]
4. Prevention of the inflammatory response in rats to hydroxyl apatite micro crystals and to soluble irritants (histamine, carrageenan)[c].
5. Minimal toxicity to inflamed rats[c].

---

[a]Case report, Department of Pediatrics, Royal Children's Hospital, Melbourne, communication from Professor D. Danks, 1980. [b]Unpublished reports from Drs. G. Crighton and C.J. Hawkins (Brisbane) and Dr. R.G. Robinson (North Sydney). [c]Ref. (14).

Analogues of Alcusal[R] (with other metals, or with other substituted benzoic acids and with fatty acids such as hexanoic acid) and salicylic acid alone (no Cu), were all less effective than Alcusal[R] when applied dermally.

Dermcusal[R]

The DMSO complex $Cu(Hsal)_2 \cdot (CH_3)_2SO$, applied to the skin of animals as a solution in DMSO-glycerol (in the presence of 3:1 molar excess of salicylic acid) has shown the following:

1. Beneficial in treating sesamoiditis, lumbar-sacral arthritis sprained check ligaments, chronic arthritis of fetlock and tendon sprain in horses[a].
2. Restoration of mobility in spontaneous canine arthritis[b].
3. Suppression of established polyarthritis in rats[c].
4. Prevention of inflammation induced by zymosan or carrageenan in rats[c].
5. Toxicity was minimal when applied at therapeutic doses to inflamed rats and dogs[b,c].

---

[a]J. Kohnke, Veterinary Director, Medical Research (Marketing) Pty. Ltd., North Ryde, Australia. [b]Personal communication from Professor B. Boettcher, Biological Sciences Department, University of Newcastle. [c]See Reference (15).

Details of experiments with Dermcusal[R] have been published in detail (15).

Direct evidence that the effects of Alcusal[R] and Dermcusal[R] are due to the Cu(II) content has been obtained. This follows:

1. Omission of the copper results in gross reduction or complete loss of pharmacological activity.

2. Significant reduction of pharmacological activity when Cu(II) chelating agents (such as citrate, D-penicillamine) were administered concurrrently by i.p. injection.

3. Effective mimicry in acute anti-inflammatory assays of CuL_2 where L = ligands which of themselves possess no anti-inflammatory activity (e.g. decanoic acid, 5-methyl-salicylic acid).

4. Inactivity of topically applied Cu(I) thio-complexes (e.g. Cu(I)-2-mercaptobenzoic acid).

Other lipophilic copper(II) preparations

Copper complexes with other anti-inflammatory drugs e.g. niflumic acid, phenylbutazone in DMSO likewise exhibit superior activity to those drugs applied alone (without Cu) to the skin.

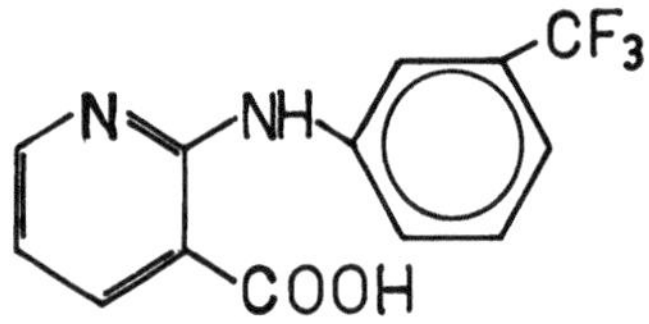

Niflumic Acid (NFA)

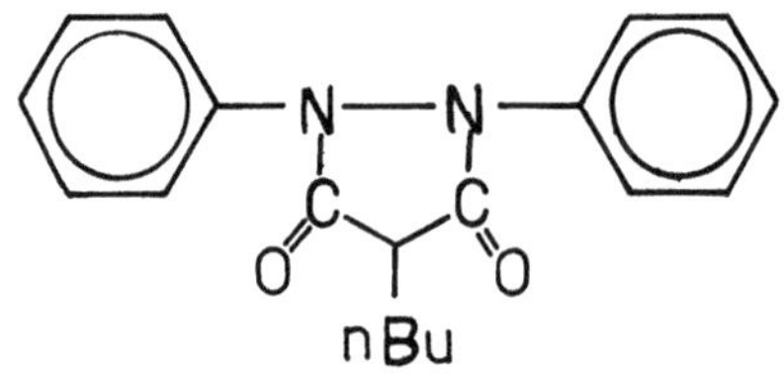

Phenylbutazone (PB)

Many other complexes prepared in DMSO exhibited acute anti-inflammatory activity (carrageenan paw oedema) but were not significantly active in suppressing adjuvant-induced arthritis. Only Cu-NFA and Cu-Pb exceeded Dermcusal in anti-inflammatory activity but Cu-NFA caused far more skin toxicity. Details of this work concerning lipophilic copper-(II) formulations and correlations between their composition and anti-inflammatory/anti-arthritic activity when applied to the skin of rats, has been reported (16). We are also investigating the complex $Cu(PB)_2$.

CONCLUSION

Topical application promises to be a superior alternative to orally ingested drugs and clinical evaluation of both Alcusal^R and Dermcusal^R would now seem justified.

REFERENCES

1. Walker, W.R. and Daphne M. Keats, 1976, Agents and Actions, 6: 454-459.
2. Walker, W.R. and B. Griffin, 1976. Search, 7: 100-101.
3. Beveridge, S.J. and W.R. Walker, 1980. Aust. J. Chem., 33: 2331-5.
4. Walker, W.R., R.R. Reeves, M. Brosnan and G.D. Coleman, 1977. Bioinorg. Chem., 7: 271-276.
5. Schubert, J., 1966. Sc. American, 214(5): 49.
6. Sorenson, J.R.J., 1976. Prog. in Med. Chem., 15: 211-260.
7. Walker, W.R. and S.J. Beveridge, 1979. Inorg. Perspect. Biol. Med., 2: 93-125.
8. Brown, D.H., W.E. Smith and J.W. Teape, 1980. J. Med. Chem., 23: 729-734.
9. Bonta, I.L., M.J. Parnham, J.E. Vincent and P.C. Bragt, 1980. Prog. in Med. Chem., 17: 186-260.
10. O'Young, C.L. and S.J. Lippard, 1980. J. Am. Chem. Soc., 102: 4920-4924.
11. Whitehouse, M.W. and W.R. Walker, 1978. Agents and Actions, 8(1-2): 85-90.
12. Goldstein, I.M., H.B. Kaplan, H.S. Edelson and G. Weissmann, 1979. J. Biol. Chem., 254: 4040-4045
13. Odinstova, N.A. (U.S.S.R.). Deposited Doc. 1976. VINITI, 2967-76, 61-3 (Russ.) Chem. Abs., 89:360.
14. Walker, W.R., S.J. Beveridge and M.W. Whitehouse, 1980. Agents and Actions, 10(1): 1-10.
15. Beveridge, S.J., W.R. Walker and M.W. Whitehouse, 1980. J. Pharm. Pharmacol., 32: 425-427.
16. Beveridge, S.J., M.W. Whitehouse and W.R. Walker, 1981. Agents and Actions, in press.

DISCUSSION

B. GOLDEN (Jamaica). How did you measure the absorption of copper in the Menke's case?

WALKER (Australia). If David Danks was here he would answer that. He wrote to me and said that the serum copper was elevated in the patient. I would assume that he used A.A. spectrophotometry.

STEVENS (Australia). Could you comment on the possible role of gold in the injection form as a treatment for rheumatoid arthritis and its possible relation to the effects of copper.

WALKER (Australia). I have published with Michael Whitehouse in Agents and Actions, a paper called 'Copper and Inflammation' and in that article we put up eight hypotheses concerning the function of copper. We mentioned that it is not unlikely that gold (I), that is the sort of gold in Nyochrysin is to reduce the copper (II) to copper (I). I have been on record as saying that copper (I) is the active form. I suggest you read that paper.

SUTTLE (Scotland). Have you any estimates of either the rates or amounts of copper that were absorbed in these dermal preparations?

WALKER (Australia). Yes indeed. We set up perfusion apparatus as described by Tregear in which you put radioactive copper on skin that is held in place over a container (like the skin of a drum). A plot of the activity of copper against time shows that for 2 or 3 hours nothing is absorbed then you get absorption occurring linearly with time. We have shown that at least one mg of the complex would go in over 24 hours. That has also been shown by a student of mine who put ALCUSAL on his forearm and after 1-4 hours the salicylic acid levels were up and the copper levels were up after 2 hours.

SUTTLE (Scotland). My interest is that when we give 50-100 milligrams of a copper complex to sheep or cattle it is flammatory and these sort of preparations might be an interesting alternative to our copper injections in the field.

WALKER (Australia). Thank you for the comment.

MOMCILOVIC (Yugoslavia). What was the level of copper in your ointment and how stable is the ointment on the skin? Does it remove easily or do you have to remove it after some time or must it peel off or what?

WALKER (Australia). Both the preparations we call ALCUSAL and DERMCUSAL contain 640 milligrams of copper per 100 ml, so that 1 ml rubbed on would give a maximum of 6.4 milligrams of copper. We think that about 5% is absorbed, hence a dose of several mls could be applied. Now when you use DMSO it is quite dramatic the way it is absorbed. Of course we plan to use this in the veterinary field.

BIOLOGICAL HALF-LIFE OF SELENITE-SELENIUM IN SOME HUMAN DISEASES

T. Westermarck[1], T. Rahola[2], M. Suomela[2] and A.-K. Kallio[3]

[1]Helsinki Central Institution for the Mentally Subnormal, SF-02400 KIRKKONUMMI, [2]Institute of Radiation Protection, and [3]the School for Handicapped Children in Helsinki, HELSINKI, FINLAND

ABSTRACT

Selenium (Se) retention and losses were measured with a whole-body counting technique in 3 patients with Batten or Spielmeyer-Sjögren or juvenile type of neuronal ceroid lipofuscinosis (NCL), 5 patients with Duchenne type of muscular dystrophy (DMD) and 5 controls, using 4.5 - 18 µCi of oral ^{75}Se-sodium selenite as a tracer. After the administration of ^{75}Se whole-body counting measurements were performed 3 - 5 times a week for the first 2 weeks and then monthly for 3 - 16 months.

The biological half-life (mean ± S.D.) was shorter in the DMD patients than in the NCL patients and the controls, 65 days ± 6 days, 98 days ± 18 days, and 101 days ± 14 days, respectively.

INTRODUCTION

In 1973 Se was shown to be one of the trace elements essential to man (Awasthi et al. 1975). Low Se levels have been observed in patients with DMD and in patients with NCL (Westermarck 1977).

The purpose of this study was to clarify whether the retention and biological half-life of Se are different in these two types of patients and in the control persons. Further more the effect of regular sodium selenite supplementation was studied in the DMD patients.

MATERIAL AND METHODS

Of the patients with DMD, 3 lived at the School for Handicapped Children in Helsinki (VH, HJ and MN) and two at home (JA and TI). The 3 patients with NCL lived at the Rinnekoti Institution. So did also 2 of the controls (TA and KH) suffering from mild cerebral palsy. The 3 other controls lived at home and one of them (AI) was the brother of (TI) one of the DMD patients.

Each subject was given an oral dose of sodium selenite containing 4.5 - 18 µCi ^{75}Se and < 0.01 - 1.8 mg stable Se. Five patients with DMD were orally given sodium selenite tablets daily starting 70 - 135 days after the beginning of the study. The daily dosage was 0.025 mg Se/kg body-weight.

The whole-body ^{75}Se activity was measured in an iron-chamber using a multi-detector linear scanner (Suomela 1967). After the administration of ^{75}Se, whole-body counting measurements were performed 3 - 5 times a week for the first 2 weeks and then monthly for 3 - 16 months. The statistical counting error was negligible for at least 150 days, but by the end of the measuring period (500 days) it had reached 25 %.

RESULTS

In the DMD patients the retention of ^{75}Se appeared to be less than in the NCL patients and the controls (Table 1). The biological half-life was shorter in the patients with DMD than in the NCL patients and the controls ($P < 0.01$) (Table 1).

The retention curves of ^{75}Se whole-body activity for a DMD patient (TI) and his healthy brother (AI) are presented in Fig. 1. The effect of the administration of a daily dose of 0.5 mg stable Se as sodium selenite is also seen in the retention curve. This curve also demonstrates that the biological half-life during Se supplementation treatment was the same as before the treatment, except the period immediately after the start of treatment. The same trend was exhibited by the other patients given Se supplementation.

Table 1. Age, sex and weight of patients and controls. Administered amount of stable selenium and ^{75}Se. Whole-body retention and biological half-life of ^{75}Se in two groups of patients and one control group.

Subject	Sex	Age years	Weight kg	Se-carrier mg	Dose µCi	Se-75 Retention % of dose	$T_{\frac{1}{2}}$ (BIOL.) days
BATTEN's							
MH	F	17	53	< 0.01	16	81	80 ± 4
HK	M	14	47	< 0.01	15	47	105 ± 4
JS	M	18	68	< 0.01	18	57	109 ± 11
Average ± 1 SD							98 ± 16
DUCHENNE's							
JA	M	22	50	1.6	12	30	66 ± 2
VH	M	15	54	0.8	7.4	48	72 ± 3
TI[a]	M	9	23	1.2	11	27	57 ± 3
HJ	M	13	44	0.7	6.5	52	65 ± 3
MN	M	16	32	0.5	4.5	39	64 ± 3
Average ± 1 SD							65 ± 6
CONTROL							
TA[b]	M	19	65	< 0.01	4.7	66	124 ± 6
LE	F	37	65	1.3	12	42	89 ± 2
KH[b]	M	14	44	< 0.01	4.6	52	101 ± 3
AI[a]	M	12	34	1.4	9.2	(42)[c]	97 ± 4
TS	F	37	60	1.0	8.7	49	94 ± 2
Average ± 1 SD							101 ± 14

[a] Brothers on the same diet [b] Patients with mild cerebral palsy [c] Approximate

DISCUSSION

In investigations of this kind, dietary intake of Se naturally influences the results. All the persons examined were on an average Finnish diet.

Biological half-life values in the controls did not differ from the values reported in healthy women in New Zealand (Thomson & Stewart 1974). Our study shows that DMD patients have an increased excretion of Se, indicating that they need Se supplementation. In contrast the NCL patients seem to have a normal total elimination of Se, although aberrant excretion of ^{75}Se in urine has been demonstrated (Westermarck et al. 1981).

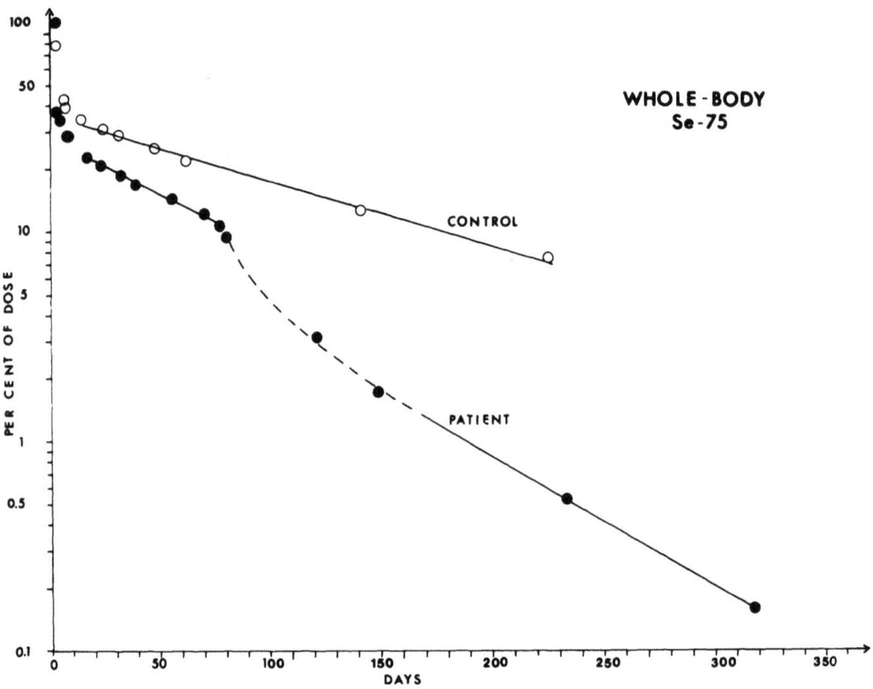

Fig. 1. Retention curve of ^{75}Se activity in the whole-body of a patient (TI) with Duchenne muscular dystrophy, and his healthy brother (AI). The patient was given sodiumselenite tablets in a daily dose of 0.025 mg Se/kg body weight beginning 70 days after the administration of the ^{75}Se activity.

ACKNOWLEDGEMENTS

This study was supported by the Rinnekoti Research Foundation.

REFERENCES

1. Awasthi, Y.C., Beutler, E. and Srivastava, S.K., 1975. Purtification and properties of human erythrocyte glutathione peroxidase. J. Biol. Chem., 250:5144-5149.

2. Suomela, M., 1967. The construction and calibration of the whole-body counter in the Institute of Radiation Physics. Institute of Radiation Physics, Report SFL-A6, Helsinki.

3. Thomson, C.D. and Stewart, R.D.H., 1974. The metabolism of ^{75}Se selenite in young women. Brit. J. Nutr., 32:47-57.

4. Westermarck, T., 1977. Selenium content of tissues in Finnish infants and adults with various diseases, and studies on the effects of selenium supplementation in neuronal ceroid lipofuscinosis patients. Acta pharmacol. et toxicol., 41:121-128.

5. Westermarck, T., Rahola, T., Suomela, M., Santavuori, P. and Simell, O., 1981. Selenium absorption and losses in Spielmeyer-Sjögren type of neuronal ceroid lipofuscinosis (Batten's disease) using total body counting for Se-75 detection. In: Armstrong, D., Koppang, N. and Rider, J.A. eds. Ceroid-Lipofuscinosis (Batten's disease), Elsevier/North-Holland Biomedical Press, Amsterdam, in press.

DISCUSSION

ANON. Why is the half life of ^{75}Se so reduced in patients with Duchenne's disease?

WESTERMARCK (Finland). We do not know.

ROBINSON (N.Z.). What was the blood selenium in your patients before you started?.

WESTERMARCK (Finland). In Batten's disease the mean was about 40 ng/ml in serum and I think it was about 50-60 ng/ml in whole blood.

ROBINSON (N.Z.). Did you measure cells separately for enzyme activity?

WESTERMARCK (Finland). Yes, the GSH-Px activity in erythrocytes was very low indeed in Batten's disease but in the patients with Duchenne's muscle dystrophy it varied. Some of these were very low too.

ANON. Are there hormonal changes connected with these diseases?

WESTERMARCK (Finland). I do not know - I do not think so.

LOMBECK (Germany). You showed us one curve of ^{75}Se activity in a patient with Duchenne's disease, where there was a break and it was no straight line. Do you think that it is just one half-lifetime you find in this long experiment with the whole body counter - couldn't there be several half-lifetimes for selenium?

WESTERMARCK (Finland). The curve changed when we had supplemented the patients with selenium.

LOMBECK (Germany). You think they were straight in the controls so that it was just one compartment?

WESTERMARCK (Finland). Yes they are straight after a few days.

SELENIUM IN BLOOD IN HEALTH AND IN CERTAIN DISEASES

Plantin,L.-O., Meurling,S.,Nycquist,O.,Strandvik,B. and Tunell,R.

Karolinska Institute Research Center, the Pediatric Clinic and the Cardiologic Department,Huddinge Hospital and Department of Pediatric Surgery Uppsala

INTRODUCTION

During the last decade interest has more and more focussed on selenium as an essential trace element and as a constituent of the enzyme glutathione peroxidase (GSHpx). In contrast to most other essential trace elements as copper,iodine and zinc the "normal" selenium levels in blood vary rather much between different geographical locations. Studies by Westermarck (12) have demonstrated that the blood Se concentrations are related to the environmental Se concentrations. The Chinese studies of the Keshan disease also demonstrate the importance of Se supply through regional food products (6). These in their turn depend on the availability of Se in the ground.Epidemiological studies by Shamberger (9) show an inverse correlation between the selenium levels and coronary diseases. There are indications that a low selenium intake is associated with greater risks for several other serious diseases as certain forms of cancer according to Jansson(5),Schrauzer(10) Broghamer(3) and for multiple sclerosis according to Ahlrot-Westerlund(1). Against this background it is a very important and serious task to establish the selenium concentration levels in the population especially in infants and children. The fact that Sweden is a low selenium region makes this still more important. We have therefore decided to analyse blood samples from healthy people and in certain diseases that might be selenium dependent. The plasma or serum Se probably reflect the current Se intake whereas the erythrocytes probably indicate the average Se intake over a longer period. We therefore decided to analyze both plasma and red cells for Se.Baglan and cow-rkers(2) suggested that the placenta could be used as an indicator of trace element exposure. Their work as well as that of Thieme and coworkers(11) seem to supprt this idea. We have also collected placentas from two different locations in Sweden for analysis of Se and other trace elements. Blood samples from two disease groups cystic fibrosis (CF) and acute myocardial infarction(AMI) have also been analyzed.

METHODS

Commercially available heparinized blood sampling tubes were analyzed for their Se content in order to check if they could be used for selenium blood sampling. The dried heparin was collected from the tubes and submitted to the analytical procedure for selenium. No Se could be detected which means that there was less than 0.2 nanogram of Se in one sampling tube. This type of tubes was therefore used to collect blood by venipuncture. The plasma and red cells were separated by centrifugation. About 0.5 g of each was transferred to quartz ampoules and freeze-dried. The tubes were sealed and the ampoules collected into batches of about 40 which were then sent for neutron irradiation for one week at a flux of $2 \cdot 10^{13}$ n/cm^{2}s in the Norwegian nuclear reactor JEEP II. This method for neutron activation analysis has been reported previously by Plantin(7). A summary is presented in fig.1. The control patients were healthy asymptomatic blood donors between 20 and 67 years of age and children admitted to the surgical clinic for minor surgical affections but otherwise healthy.The children with CF had a well established diagnosis and had been under treatment for several years. Placentas were obtained from healthy mothers giving birth to normal healthy children. From the placenta a normal macroscopically unaffected cotyledon was dissected and samples were taken from for trace element analysis. All handling of the placenta and the sampling was made in a laminar flow hood.Blood samples from patients with acute myocardial infarction (AMI) were obtained immediately after their admittance to the hospital. The samples were analyzed only after the diagnosis had been confirmed by ECG, the typical symptoms and the results of analyses of the enzymes ASAT,LD, CK tot and CKMB.

METHOD FOR SELENIUM ANALYSIS:

1. DISSOLVE THE NEUTRON IRRADIATED SAMPLE IN HOT MIXTURE OF 5 ML CONC. NITRIC ACID AND 5 ML PERCHLORIC ACID TOGETHER WITH CARRIER.
2. FILTER THROUGH QUARTZ AND WASH WITH 10 ML HYDROCHLORIC ACID.
3. EVAPORATE TO A SMALL VOLUME.
4. PRECIPITATE SELENIUM WITH 10 ML 20 % HYDROXYLAMINE HYDROCHLORIDE.
5. HEAT TO 90 C AND FILTER OFF SELENIUM.
6. DRY, WEIGH AND COUNT AT 136 AND 265 KEV.

FIG. 1. Method for selenium determination after neutron irradiation of the sample.

SELENIUM CONC. IN HEALTHY CONTROLS

AGE GROUP	TYPE	SE NG/G	NUMBER
20-39	PL	75.7	13
20-39	RC	99.6	13
40-69	PL	80.2	18
40-69	RC	94.8	16

THERE IS NO SIGNIFICANT DIFFERENCE IN SELENIUM CONC. BETWEEN AGE GROUPS OR BETWEEN SEXES.

FIG. 2. Selenium concentration in plasma and red cells in two adult age groups.

RESULTS

The selenium concentration in plasma and red cells from the adult controls are shown in fig.2. The values are subdivided into two age groups 20-39 years and 40-69 years There is however no statistical difference between the Se values either for plasma or for the red cells. Further analysis of variance shows that there is no significant difference in Se concentration either between sexes or between age groups for both plasma and red cells. Se is however higher in red cells than in plasma ($P < 0.001$) Thus all results for plasma Se for adults can be pooled to a plasma mean and S.D. of 80.5 ± 16.2 ng/g and the Se values for red cells have a pooled mean and S.D. = 98.9 ± 20.2 ng/g. The plasma values for 35 children between 4 -18 years have a mean of 67.5 ± 12.6 ng/g which is significantly different from the adult mean on the 99.9 % probability level. The results for the analyses of Se in placentas are shown in fig.3. As can be seen the Se has a higher concentration in the placenta than in blood. In fig.4 a comparison is made between the Se concentration in plasma from children with cystic fibrosis and with the control group. There is no statistical difference between the Se values in the two groups. Fig. 5 shows the results from the AMI patients as compared to controls in the same age range. The AMI patients have lower Se conc in plasma but higher Se in red cells than the controls. Statistically the difference is probably significant on the 95 % probability level

SELENIUM IN PLACENTAL TISSUE

FROM 2 DIFFERENT AREAS IN SWEDEN

	HUDDINGE	SKELLEFTEÅ
NO OF SAMPLES	12	10
SE CONC. NG/G	126	163
% STAND. DEV.	19.2	12.5

$t=3.894 \quad P<0.001$

FIG. 3. Selenium concentration in placental tissue from two different areas in Sweden.

DISCUSSION

The selenium concentration in blood plasma and erythrocytes from the healthy controls are very low in comparison with what is considered normal in the United States and many other countries. Only from certain parts of Finland (12), from China (6) and from New Zealand (8) have lower values been reported. That the placenta has a higher concentration

DIAGNOSIS	Se ng/g	S.D.	NO OF CASES
HEALTHY	67.5	12.6	35
CYST.FIBR.	70.3	23.5	9

FIG. 4. Analyses of selenium in plasma from children (age 4-18 years)

SELENIUM IN ACUTE MYOCARDIAL INFARCTION ng/g

	CONTROLS	AMI	P
PLASMA	83.7 + 17.7	70.3 +14.5	P<0.05
RED CELLS	97.5 + 25.1	116.2 +14.9	P<0.05
NO OF CASES	18	12	

FIG. 5. Comparison between Se levels in plasma and red cells in AMI patients and controls.

of Se than blood has been reported previously by Baglan(2) and is confirmed here. It is evident from fig.3 that the placentas from Skellefteå have a considerably higher Se concentration than the placentas from Huddinge (t=3.894, P< 0.001). The area around Skellefteå is known to be rich in all sorts of minerals and from the copper mines in the area selenium is obtained as a by-product when the copper is refined. Castillo and coworkers(4) reported that serum levels of Se in 250 patients with cystic fibrosis was 105 ng/ml which is lower than the U.S. mean of 150 ng/ml. They did not consider their patients as Se deficient. From fig 4 can be seen that our values for the CF patients do not differ from the controls. There seems to be no evidence that Se is an important factor for CF patients. With the epidemiological and other evidence mentioned above that Se plays a role in coronary diseases it is a little surprising to find that there are so small differences in Se concentrations in plasma and red cells between the AMI patients and the controls. There is however an indication that plasma Se is low in the acute phase of the disease. This investigation continues.

REFERENCES

1.Ahlrot-Westerlund,B. & Schalin,G.,1981,In "Geomedical Aspects in Present and Future Research"(ed. J.Låg)(Universitetsforlaget,Oslo) pp.81-114.
2.Baglan,R.J.,Brill,A.B.,Schulert,A.,Wilson,D.,Larsen,K.,Dyer,N.,Mansour,M.,Schaffner,W., Hoffman,L. and Davies,J.,1974,Environmental Research 8:64 - 70.
3.Broghamer,W.L.,McConnell,K.P. and Blotcky,A.L.,1976,Cancer 37:1384.
4.Castillo,R.,Landon,C.,Lewiston,N.,Morris,V. and Levander,O., In "Second International Symposium on Selenium in Biology and Medicine.Texas Tech.University,Lubbock 1980)In press
5.Jansson,B.,Seibert,G.B. and Speer,J.F.,1975,Cancer 36:2373.
6.Keshan Disease Research Group of the Chinese Academy of Medical Sciences,Beijing, 1979,Chinese Medical J. 92:471-476.
7.Plantin,L.-O.,In "Mineral Elements '80.A Nordic Symposium.Helsinki 1980", In press.
8.Robinson,M.F. and Thomson,C.D., In same ref as ref 4. In press
9.Shamberger,R.J., In same ref as ref no 4. In press.
10.Schrauzer,G.N.,1976,Bioinorganic Chem. 5:275.
11.Thieme,R.,Schramel,P.,Klose,B.-J. and Waidl,E.,1975,Geburtshilfe und Frauenheilkunde 35:349-353.
12.Westermarck,T.,Raunu,P.,Kirjarinta,M. and Lappalainen,L.,1977,Acta Pharm.et Toxicol. 40:465-475.

DISCUSSION

ANON. How soon after the myocardial infarction were the bloods taken?

PLANTIN (Sweden). They were taken just after the patients were admitted to hospital and only those who were confirmed with acute myocardilitis were then analysed. So it was taken in the actual stage of the disease.

ANON. Thank you. Do you think that could be a non-specific effect of trauma?

PLANTIN (Sweden). We do not know. The statistical significance is very weak and we have to continue this investigation.

BEHNE (Germany). Could you estimate how high the selenium content in the placentas would be if you subtract the amount of selenium in the residual blood.

PLANTIN (Sweden). The blood content was measured and it was around 12% of the mean. I suppose there will be a higher :alue of selenium if you subtracted blood values because blood tends to be lower in selenium than tissues.

PROHASKA (U.S.A.). There have been some reports about the low levels of erythrocyte glutathione peroxidase in multiple sclerosis patients - you mentioned something in your abstract about M.S. patients, did you measure blood and plasma selenium levels in multiple sclerosis patients, and if you did what kind of values did you get?

PLANTIN (Sweden). These investigations are underway but the data are not ready for presentation - I cannot say.

ANON. Just a small technical point, we are interested in sampling organs and I would like details of your titanium knife.

PLANTIN (Sweden). The titanium knife is made by the International Atomic Energy Agency and is distributed by them from Vienna. It is made of pure titanium sheet that is sharpened by a special device.

VERSIECK (Belgium). Have you measured serum copper or zinc levels in patients with myocardial infarction?

PLANTIN (Sweden). We have not measured these other elements. We were just concentrating on selenium this time for the reasons I mentioned in the talk.

ROBINSON (New Zealand). About cystic fibrosis; we have found no difference with controls in the few we have looked at, similarly for multiple sclerosis. We haven't had many to look at, but those we have don't appear to be different.

KESHAN DISEASE

Lianzhen Zhu

Institute of Health, Chinese Academy of Medical Sciences,
BEIJING, THE PEOPLE'S REPUBLIC OF CHINA

Keshan disease is an endemic cardiomyopathy in China. It was first recognized in 1935 in Keshan county which is located in northeast China. Since then, more and more areas have been found affected. The affected areas are focally located in a long, belt-like zone running across the country from northeast to southwest, mainly in mountainous countries. The susceptible population are children aged below 10 years old and women of childbearing age.

THE TYPES OF KESHAN DISEASE (1)

Keshan disease is classified into four clinicopathological types: acute, subacute, chronic and latent. The acute type occurs suddenly and death may occur from cardiogenic shock within a few hours. The subacute type, which commonly occurs in children, has a slower onset than the acute type, but a quicker onset than the chronic type. The cases usually have the symptoms and signs of both the acute and the chronic type, most cases have facial edema and gallop rhythm. Chronic cases may persist for a number of years with varying degrees of congestive heart failure, and with moderate or severe enlargement of the heart. Causes of the Latent type have normal heart function, but mild heart enlargement or abnormal changes in the ECG pattern.

THE ETIOLOGY OF KESHAN DISEASE

The etiology of Keshan disease is not yet clear. However, two facts indicate that selenium deficiency may play an important role in this disease. Firstly, all areas where Keshan disease is endemic are selenium deficient. In the years of 1972-1973 (2) a comparative study of human blood and hair selenium levels, whole blood GSHpx activities and selenium excretion in urine of children in affected and non-affected areas was performed. It revealed that there is an invariable association of selenium insufficiency with Keshan disease (Table 1).

Table 1 Blood and Hair Selenium Levels, Whole Blood GSHpx Activities and Selenium Excretion of Populations in Keshan Disease Affected and Non-affected Areas

Blood Se Level (ppm)	Affected Area ($\bar{X}$ ± SE)	Non-affected Area ($\bar{X}$ ± SE)
Blood Se Level (ppm)	0.021 ± 0.001(173)	0.095 ± 0.009(111)
Hair Se Level (ppm)	0.074 ± 0.001(1478)	0.359 ± 0.004(1745)
Se Excreted from 12 hrs.-night-urine(μg)	0.69 ± 0.06(9)	1.50 ± 0.04(10)
Whole Blood GSHpx Activities (unit)*	60.5 ± 0.07(63)	77.5 ± 2.1(22)

* Enzyme activity unit = μM of GSH oxidized per 8 μl whole blood in 5 minutes.

The top value of blood selenium contents tested in seven affected areas was 0.025 ppm, but the blood selenium level in the affected areas often dropped below 0.01 ppm. Examination of the extent of variation in hair selenium levels of the agricultural populations suggested that the average values of affected areas are all below 0.12 ppm, and no less than 0.2 ppm for non-affected areas. Those of the intermediate zones are all above 0.12ppm.

The selenium contents of staple foods in affected areas were also lower (Table 2).

Table 2 Selenium Contents (ppm) of Staple Foods of Keshan Disease Affected and Non-affected Areas ($\bar{X}$ ± SD)

Foods	Affected Area	Non-affected Area
Corn	0.005 ± 0.002(178)	0.036 ± 0.056(82)
Rice	0.007 ± 0.003(49)	0.024 ± 0.038(49)
Wheat	0.008 ± 0.006(116)	0.032 ± 0.020(30)
Soybean	0.010 ± 0.008(134)	0.069 ± 0.076(31)

Secondly, selenium supplementation protects children against Keshan disease. Observations were made on the use of sodium selenite in the prevention of Keshan disease in Sichuan Province during 1974-1977 (3). The difference of incidence between supplemented and unsupplemented groups was highly significant both in 1974 and in 1975. In 1976, when all the children were given sodium selenite, the rate was further decreased. In 1977, no new cases of Keshan disease were discovered among 12,749 children observed. The fatality rate of children in the unsupplemented group of 106 cases in 1974 and 1975 was 50%, but was only 5.9% in the supplemented group (Table 3).

Table 3 The Reduction of Incidence of Keshan Disease in Selenium Supplemented Children during 1974-1977

Year	Group	Number of Subjects Observed	Case			
			Number	Alive	Turn to Chronic	Death
1974	Control	3,985	54	27	2	27
	Treated	4,510	10	10	1	0
1975	Control	5,445	52	26	3	26
	Treated	6,767	7	6	0	1
1976	Control	212	1	1	1	0
	Treated	12,579	4	2	0	2
1977	Control	---				
	Treated	12,749	0	0	0	0

However, it seems unlikely that selenium is the sole factor responsible for the occurrence of Keshan disease, because some of the epidemiological characteristics, such as the heavy incidence in certain seasons and years could not be explained by a decrease in body selenium level. So besides selenium deficiency, other factors, such as nutrition status, viral infection, mycotic toxication and many others should also be considered to be relevant to the etiology of Keshan disease.

THE PREVENTION OF KESHAN DISEASE IN CHINA

Provision of sodium selenite tablets to children in affected areas is the main measure taken to prevent Keshan disease in China. The dosage for children aged 1-5 years old is 0.5 mg once a week, and for children aged 6-9 years old is 1.0 mg once a week.

Because dispensing tablets every week is time consuming, supplementation of table salt with sodium selenite to a concentration of 10-15 ppm (of sodium selenite) has been tried. Observations on residents who were treated with selenite table salt for three years and four months revealed that the salt enriched with 10 ppm of sodium selenite was enough to maintain hair selenium concentration at a level above 0.2 ppm, if an average of 15 g of the salt could be taken from meals daily by an adult.

Spraying sodium selenite solution on the leaves of crops during their preflowering stage was also tried. Selenium contents of corn and rice could be raised to a level of about 0.07 and 0.05 ppm respectively by this method. Availability of the increased selenium was similar to that occurring naturally in corn, as demonstrated by animal experiments. Furthermore, by this means the bio-availability of selenium was better than that from selenium tablets. However, the selenium recovery in the seed was very poor, probably not more than 2% of the amount applied. So some difficulty is encountered in practical application.

REFERENCES

1. Chen, Xiaoshu et al: Studies on the Relation of Selenium and Keshan Disease, 1980, Biol. Trace Element Res. 2 :91.

2. Yang, Guangqi: Relation of Selenium Deficiency to the Occurrence of Keshan Disease. Unpublished.

3. Keshan Disease Research Group of the Chinese Academy of medical Science: Observation on the Effect of Sodium Selenite in Prevention of Keshan Disease, Chinese Med. J. 92:471, 1979.

4. Yang, Guangqi: Different Ways for the Supplementation of Selenium to the Residents in Keshan Disease Areas. In press.

DISCUSSION

WHITE (Australia). Dr Zhu, considering the failure of Marion Robinson's group to observe any diseases associated with low blood selenium levels in New Zealand, do you suspect that genetic factors may be involved in Keshan disease?

ZHU (China). Keshan Disease is an endemic disease and is not due to genetic factors. From our observations, when a family moved from a non-affected area to an affected one, there was a tendency for the children of this particular family to develop Keshan Disease. But when this family left the affected area, the children gradually returned to normal health.

HETZEL (Australia). Is the disease recent or has it been known for a long time?

ZHU (China). The disease was first recognised in 1935 due to a very high fatality rate in that year. We have not collected any information regarding Keshan Disease before 1935 so it is possible that Keshan Disease may have occurred before 1935. Since 1935, more and more areas have been found to be affected by Keshan Disease. For example, it was found in Sichuan Province in the 1960's.

VALENTINE (USA). What made you suspect selenium as an aetiological agent?

ZHU (China). The aetiology of Keshan Disease is not yet clear. However, 3 facts indicate that selenium deficiency may play an important role in this disease. First, all areas where Keshan Disease is endemic were selenium deficient, as indicated by selenium-responsive diseases in livestock. Secondly, no Keshan Disease has been found in regions with high selenium content. Thirdly, selenium supplementation protected children against Keshan Disease. Other factors such as malnutrition, viral infection, mycotic toxication and many others have also been considered to be relevant to the aetiology of Keshan Disease.

HETZEL (Australia). Perhaps I could just comment that Chinese epidemiology is very advanced. A recent text has just become available in the West to indicate that the Chinese have monitored their situation to a much greater extent than we thought.

ULLREY (USA). Have estimates of Vitamin E intake in this population been made?

ZHU (China). No.

BRUCE (Sweden). Are you now supplementing all the people with selenium in these suspect areas or are you using other methods also.

ZHU (China). We only used sodium selenite tablets to supplement the population in these areas.

BRUCE (Sweden). Do you use selenium supplements in all areas or only to special families or special regions.

ZHU (China). Selenium supplements have been given to all children below 10 years old regardless of whether they have Keshan Disease or not.

POSSIBLE ROLE OF TRACE ELEMENTS IN THE AETIOLOGY OF ENDEMIC GENU VALGUM

K.A.V.R. Krishnamachari

Assistant Director, National Institute of Nutrition
(Indian Council of Medical Research)
Jamai-Osmania, Hyderabad-500007
INDIA

Endemic fluorosis is a public health problem in several states of India (1). Fluoride in large quantities are ingested by the residents of endemic areas over prolonged periods of time since in these areas, fluorides are present in soluble forms in the subsoil water which forms the source of drinking water. Fluorosis is characterised by osteosclerosis and calcification of ligaments, muscular attachments and interosseous membranes. Dental mottling is a characterstic sign of fluorosis.

Endemic genu valgum was identified in some endemic fluorosis areas in India where fluoride content of drinking water was between 3.5 and 6.0 ppm (2). Genu valgum develops between 7 and 16 years of age, with a higher prevalence in males (3). Many of these subjects are undernourished with regard to several nutrients, including calcium. Epidemiological studies indicated a higher prevalence of genu valgum among those whose staple was Sorghum Vulgare in comparison with among those whose staple was rice (3). Also, an inverse correlation was observed between the prevalence of genu valgum and copper content of drinking water in the corresponding village (4). Radiological examination of more than 80 subjects had revealed that in this disorder there is combination of osteosclerosis, osteoporosis and osteomalacia. There were distinctive features of secondary hyperparathyroidism such as thinning of the cortex, subcortical resorption, expansion of lower ends of radius and ulna, microcystic appearance of short long bones, etc. (5).

BIOCHEMICAL FINDINGS

Serum calcium, inorganic phosphorus and alkaline phosphatase levels were estimated by standard procedures in subjects of endemic genu valgum and were found to be within normal limits (1). Circulating immunoreactive parathyroid hormone levels were estimated in ten subjects of endemic genu valgum, in fifteen subjects suffering from endemic skeletal fluorosis and in normal match control subjects. The mean levels of the hormone were 5390 pg/ml and 2674 pg/ml in endemic genu valgum and endemic fluorosis respectively which were significantly higher than in the corresponding controls (6). Plasma levels of thyrocalcitonin were also estimated which revealed that in subjects with endemic genu valgum the levels were significantly lower than in matched controls (7). Urinary hydroxyproline was estimated in ten endemic genu valgum subjects and in seven age and sex matched controls. The mean values for endemic genu valgum and endemic fluorosis groups were 66.2 mg and 38.3 mg in 24 hour period for the total hydroxyproline and 5.9 mg and 3.1 mg respectively in 24 hour period for the non-dialyzable fraction. There was a statistically significant increase in both these fractions in subjects with endemic genu valgum when compared with the corresponding values obtained in matched controls. Ca^{47} turnover studies were carried out in nine subjects with endemic genu valgum and in four controls. Also the studies were conducted in subjects of skeletal fluorosis and in matched controls. The results indicated that there is evidence of increased Ca^{47} turnover, increased accretion rate and increased resorption rate in both genu valgum group and fluorosis group as compared to age matched control subjects. It was also observed that after correcting for the age, these parameters were significantly higher in subjects with genu valgum deformity than in subjects with skeletal fluorosis (8). Serum copper, zinc and magnesium were estimated in subjects with genu valgum using atomic absorption spectrophotometer. All the three minerals estimated were within normal limits in genu valgum subjects.

BONE ANALYSIS FOR TRACE MINERALS

Twenty subjects suffering from endemic genu valgum were surgically corrected by osteotomy and wedge resection appropriately by a surgeon. Surgical samples of cortical bone were aseptically collected and processed for atomic absorption spectrophotometres for trace minerals analysis. The results of analysis are presented in the table. The values are compared with the values obtained in bone samples obtained from non-fluorotic subjects residing in normal areas. An attempt was made to obtain bone samples from comparable regions in age and sex matched subjects.

Table

Group	Zn (μg/g)	Cu (μg/g)	Cr (μg/g)	Mg (μg/g)	Mn (μg/g)	F (μg/g)	P (mg/g)	Ca (mg/g)
Endemic genu valgum	81.0	1.3**	6.7	3281*	0.6**	7283**	76.5	188.0
Control subjects	96.8	3.7	4.5	4768	1.2	3190	85.7	232.0

** $P < 0.001$

* $P < 0.05$

Values are expressed on dry weight basis

It was observed that copper and manganese contents of bone were significantly lower in endemic genu valgum subjects while in this group fluoride content was significantly higher than in the control subjects. Recent studies from the National Institute of Nutrition have indicated that in endemic genu valgum the levels of 25-Hydroxy Cholecalciferol are not decreased (9). Results of earlier studies had shown that molybdenum content of foods cultivated and consumed in endemic genu valgum areas was higher than in control areas (10).

These studies indicate that apart from exposure to high amounts of environmental fluoride, the subjects of endemic genu valgum suffer from possible altered metabolic status with regard to trace minerals such as copper and manganese. It is well known that copper deficiency of a tissue like bone may be associated with osteoporotic changes. The observation that the target tissue viz. bone has low content of manganese and copper is very important in relation to chronic fluoride effect which the subjects were exposed to.

SUMMARY

Endemic genu valgum is a distinct clinical entity of fluoride toxicity and is characterised by secondary hyperparathyroidism and osteoporosis among undernourished population groups who subsist on sorghum or rice as the staple. Hormonal changes and also changes in calcium turnover were observed in these subjects. This clinical entity is entirely different from nutritional rickets. Both epidemiological and biochemical studies indicate that in these subjects there are specific differences between subjects who develop the deformity and the others. The possible role of trace minerals in relation to the aetiology of endemic genu valgum is being investigated.

ACKNOWLEDGEMENTS

I thank the Director General of Indian Council of Medical Research and Director, National Institute of Nutrition, for the travel grants and support and the organizers of the TEMA-4 for the financial assistance for my stay during the conference.

REFERENCES

1.Krishnamachari,K.A.V.R.,1978,Indian.J.Med.Res.68(Suppl):94-98.
2.Krishnamachari,K.A.V.R.,and Kamala Krishnaswamy,1973,Lancet.2:877-879.
3.Krishnamachari,K.A.V.R.,and Kamala Krishnaswamy,1974,Indian.J.Med.Res.62: 1415-1423.
4.Krishnamachari,K.A.V.R.,1976,Indian.J.Med.Res.64:284-291.
5.Krishnamachari,K.A.V.R.,and B.Sivakumar,1976,Fluoride,9:185-191.
6.Sivakumar,B.,and K.A.V.R.Krishnamachari,1976,Horm.Metab.Res.8:317-319.
7.National Institute of Nutrition, Hyderabad, Annual Report, 1978.
8.Narasinga Rao,B.S.,Krishnamachari,K.A.V.R.,and C.Vijayasarathy,1979, Br.J.Nutr.41:7-14.
9.National Institute of Nutrition, Hyderabad, Annual Report, 1980.
10.Deosthale,Y.G.,Krishnamachari,K.A.V.R.,and Bhavani Belavady,1977,Ind.J. Agric.Sci.47:333-335

HAIR CHROMIUM CONCENTRATIONS AND ARTERIOSCLEROTIC HEART DISEASE.

Michel Côté and Denis Shapcott

Faculté de Médecine
Université de Sherbrooke
Sherbrooke (Quebec) Canada J1H 5N4

INTRODUCTION

Many studies, comprehensively reviewed by Mertz[1], strongly suggest a link between chromium "deficiency" and arteriosclerotic heart disease (ASHD). The biologically active form of chromium (Cr), glucose tolerance factor, is required to maintain normal glucose metabolism. Cr supplementation has also been shown to improve coronary risk factors including glucose intolerance and elevated serum levels of cholesterol. Despite acknowledged analytical difficulties in defining an individual's chromium status, we had previously found a significally lower head hair chromium concentration (HCC) in males with documented ASHD when compared to a control group of normal individuals as defined by medical history alone. This difference was more pronounced in diabetics and younger males[2].

This study compares HCC of ASHD cases and controls with normal coronary arteriograms. It evaluates the variation after a period of three months in HCC in controls and in ASHD cases with and without Cr supplementation. Finally it analyses the relationship of HCC to serum levels of glucose, cholesterol and triglyceride in ASHD cases.

MATERIALS AND METHODS

Eighty-six males with ASHD had documented transmural myocardial infarction and/or greater than seventy-five percent narrowing of a coronary artery as demonstrated by coronary arteriography. Initially, fasting serum levels of glucose, cholesterol, triglycerides and 2 hours p.c. glucose were determined and HCC sampled (0 mo). A half were selected at ramdom and given one mg of chromium as chromic chloride daily for a period of 3 months (3 mo) when the above analyses were repeated. The eighty-six male controls had normal coronary arteries documented by coronary arteriography but some suffered other forms of heart disease. HCC was also sampled at 0 mo and 3 mo in the controls. Of the eighty-six controls, fifty-one had normal hemodynamic studies and normal coronary arteries, eighteen had valvular heart disease, thirteen cardiomyopathies and other types of cardiac abnormalities were found in four.

Hair chromium was measured by atomic absorption spectroscopy using the technique developed in our laboratory[3]. All samples were analysed simultaneously.

RESULTS

There was no significant difference in age between the normal controls (mean 52.1 ± 11.2 years) and ASHD cases (mean 55.8 ± 9.2 years).

TABLE 1 - HEAD HAIR CHROMIUM CONCENTRATION MEAN ± SD IN (NG/G) IN ASHD CASES AND NORMAL CONTROLS.

ASHD N 86	373 ± 200	$P < 0.0001$
Controls N 86	636 ± 389	

The mean HCC is significantly lower in ASHD group when compared to normals. This fact is further illustrated in Fig. 1 showing the frequency distribution of HCC in percent of ASHD cases and controls.

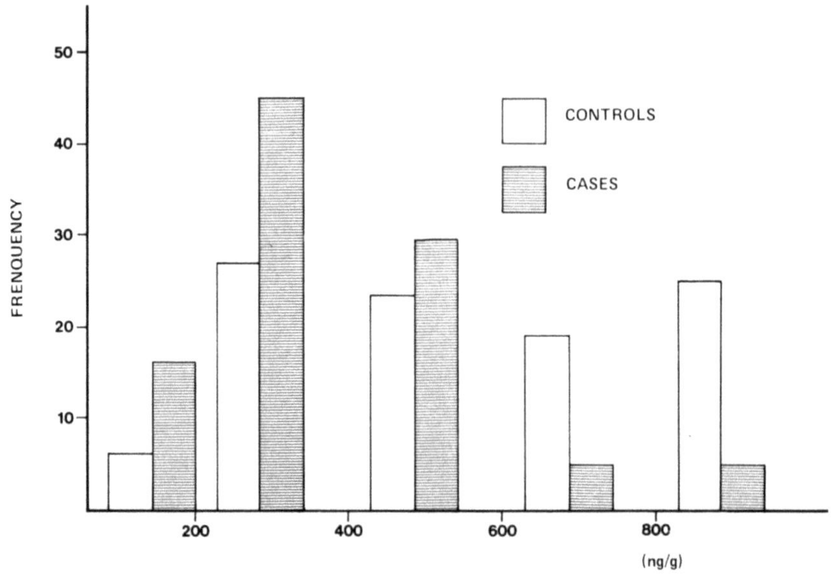

Fig. 1 - Frequency distribution in percent of ASHD cases and controls and head hair chromium concentrations (ng/g)

Fifty-six percent of ASHD cases are in the lowest quartile of HCC as determined for the normal controls (Tab. 2).

TABLE 2 - PERCENT DISTRIBUTIONS OF CASES AND CONTROLS BY HEAD HAIR CHRONIUM CONCENTRATION QUARTILES (NG/G) AS DETERMINED FOR THE CONTROLS

	Q_1 170-365	Q_2 366-485	Q_3 486-750	Q_4 751 +
ASHD	56	21	17	6
Controls	25	25	25	25

The frequency distribution of low HCC is highest in diabetics with ASHD. Seventy-five percent are found in the lowest quartile (Tab. 3)

TABLE 3 - PERCENT DISTRIBUTION OF HAIR CHROMIUM CONCENTRATION QUARTILES (NG/G) OF DIABETICS WITH ASHD

	Q_1 170-365	Q_2 366-485	Q_3 486-750	Q_4 751 +
ASHD + Diabetes N 20	75	10	10	5
Controls N 86	25	25	25	25

The median HCC was not significantly different after three months for all subjects, controls and ASHD cases including those who had received chromium supplementation. Fourty-five percent of ASHD cases with chromium supplementation did however increase their HCC by one or more quartile but twenty seven percent decreased (Tab. 4).

Thirty-two percent of controls also had changes in HCC at 3 months of two or more quartiles.

TABLE 4 - PERCENT DISTRIBUTION OF CASES WITH AND WITHOUT Cr SUPPLEMENTATION, AND CONTROLS BY CHANGES IN HAIR CHROMIUM CONCENTRATION QUARTILES AT 3 MONTHS.

	No change	Change*
Controls N 62	68	32
ASHD + Cr N 22	54.5	45.5
ASHD - Cr N 31	71	29

* Change = 2 or 3 quartiles

There were no significant differences for blood levels of glucose, 2 hours pc glucose, triglyceride or cholesterol between ASHD cases at 3 months with and without chromium supplementation.

CONCLUSION

Patients with arteriosclerotic heart disease have significantly lower HCC when compared to men with normal coronary arteries. This difference is more pronounced in cases with glucose intolerance.

Despite the high compliance of ASHD cases administered inorganic chromium chloride supplementation, hair chromium concentrations failed to reach the levels of normal controls in most cases.

Furthermore a third of normal controls and ASHD cases without chromium supplementation had noteworthy variations in HCC with time suggesting that the level of HCC alone has its limitations in defining an individual's chromium "status".

These results however continue to support our hypothesis that chromium deficiency whether from dietary deficiency or metabolic derangement is a risk factor for coronary artery disease.

REFERENCES

1. Mertz, W. (1979) Chromium in Nutrition and Metabolism (Ed. D. Shapcott, J. Hubert) (Elsevier/North-Holland Biomed. Press), pp 1-14

2. Côté, M., L. Munan et al (1979) Chromium in Nutrition and Metabolism. (Ed. D. Shapcott, J. Hubert) (Elsevier/North-Holland Biomed Press) pp. 223-228,

3. Shapcott, D. et al (1977) Clin. Biochem., 10:178-180

DISCUSSION

COYLE (Australia). Could you elaborate on you hair chromium method and whether you have tried a serum chromium assay as well?

COTÉ (Canada). The method for measuring chromium in hair was as follows: the samples were degreased in hexane, washed in triton X100, digested in nitric acid, dried and oxidised at a low temperature and measured by atomic absorption spectrometry. As to the serum levels I think Dr. Shapcott a few days ago had an extensive presentation on the method evaluating chromium status in individuals. The problem is like many trace elements in that the levels are so low in the serum and in the urine too that we don't think that the measurements are reliable.

CORNELIS (Belgium). My question also concerns your sampling technique. What kind of hair did you take, I mean hairs already fully grown, and on the point of falling out or hairs which were still firmly attached to the scalp? What part of the hair did you take?

COTÉ (Canada). I did all the sampling myself and I tried to take it from the same place on the head every time. I took several hairs and cut them at 1 inch. Bigger problems were finding enough hair and its subsequent preparation. We used normal stainless steel scissors for cutting.

CORNELIS (Belgium). Are you not afraid that the hair was contaminated in this way?

COTÉ (Canada). It is a possibility but I don't think there was any contamination.

CORNELIS (Belgium). What was your sample size? Did you take one gram of hair or just 10 micrograms to put into the atomic absorption spectrophotometer?

COTÉ (Canada). I would like to refer back to Dr. Shapcott since I did not do the analysis myself.

SHAPCOTT (Canada). We used about 50 mg of hair per sample. Anything smaller than that introduced positive errors of dilution i.e. due to large multiplication. We dissolved the hair in nitric acid, dried it, oxidized it and ashed it. This ashed sample was then dissolved in weak nitric acid and analysed directly by AAS.

CORNELIS (Belgium). I would like to suggest that contamination from scissors might be a problem with such small samples, especially when you have no controls as such.

SHAPCOTT (Canada). This is possible.

CORNELIS (Belgium). This might be an explanation for the fact that when you take very small hair samples you get much higher chromium values.

SHAPCOTT (Canada). No, we feel that it is not a blank effect but rather that there is always an analytical variation which is multiplied tremendously by a dilution factor when you take a very small amount of hair.

ABDULLA (Sweden). I wonder if you have any information concerning the dietary intake of these two groups and if the dietary levels were the same could you comment on why there was a difference in the hair levels?

SHAPCOTT (Canada). Well we honestly don't know, the possibilities are multiple. Water has certainly been looked at in some parts of the world, for example in Finland where they found a low level of chromium in the water in parts of the country which had a high incidence of coronary disease.

GUTHRIE (New Zealand). What is the minimum time required to detect an increase in hair chromium following supplementation with chromium.

SHAPCOTT (Canada). I don't know. One of the things we tried to show is that chromium sampling at one time does not appear to reflect chromium status because it varies so much. We have a study going on with medical students where we sample their hair and urine every month for chromium. We are now up to about a year and a half but we haven't analysed these results. I think we will probably find fantastic variations in individual hair chromium concentrations and I think hair is probably a bit like a tree when you are looking at th rings of the tree and finding various factors that affect the chromium deposition in the hair. We know that a lot of factors increase chromium in the serum, for example stress, heart attacks, glucose, and maybe the deposition is dependent on that kind of stress.

BROMINE DEFICIENCY AND INSOMNIA IN PATIENTS ON DIALYSIS

P.L.Oe, R.D.Vis, J.H.Meijer, F.van Langevelde, W.Allon, C.v.d.Meer and H.Verheul

Free University Renal Unit, Department of Clinical Pharmacy and Institute of Physics, de Boelelaan 1117, 1081 HV AMSTERDAM, THE NETHERLANDS.

Unusually low concentrations of bromine were found in serum from patients on chronic haemodialysis. Bromine deficiency was due to removal of bromine via the artificial kidney. Twenty-one patients who had been on chronic haemodialysis for at least one year were interviewed about aspects of their sleep. The average sleeping score during six weeks was determined to express the quality of sleep. Bromine in serum of these patients was determined according to the technique of proton-induced X-ray emission. A significant correlation was found between bromine deficiency and quality of sleep. Addition of bromine to dialysate resulted in a significant improvement of the quality of sleep. We may conclude, that there is strong evidence for an important rôle of bromine deficiency in causing insomnia of patients on chronic haemodialysis.

Introduction

A study of the transport of trace elements during haemodialysis revealed unusually low concentrations of the trace elements bromine and rubidium in serum from patients on chronic intermittent haemodialysis, due to removal of these elements via the artificial kidney (1). Subnormal blood bromine levels in patients on maintenance haemodialysis have also been reported by others (2,3,4). Postmortem analysis of gray matter of the brain from dialyzed patients revealed highly significant depressions of bromine concentration (5). However, the possible physiopathological meaning of this observation remained unknown (3). The purpose of the present study was to elucidate the possible rôle of bromine deficiency in causing insomnia of patients on maintenance haemodialysis.

Patients and methods

Twenty-one stable patients (nine male and twelve female patients), undergoing twice weekly maintenance haemodialysis for a period of at least one year, participated in the study. The average age was 48.4 years (range 21-76 years). The average duration of maintenance haemodialysis was 3.4 years (range 1.0-11.4 years). A double-blind study of 20 weeks during which sodium bromide or chloride (placebo) was added to the dialysate was done in four of these patients. Twenty-five and 50 µmol/l was added for periods of 10 weeks each. Informed consent was obtained before the double-blind trial.

For the quantification of insomnia five techniques are available (6): 1. Electroencephalography 2. Registration of movements 3. Determination of arousal threshold 4. Observation of the sleeping person 5. Interviewing about qualities of sleep. The last method was chosen because of the suitability for out-patients. Moreover, there was experience with this method in the therapeutic evaluation of hypnotics (7). The patients had to complete daily a form with four questions about their sleeping behaviour (see table I). The average score during six weeks was calculated to express the quality of sleep.

Table I. Four questions about sleep

a. How did you sleep?	3 = well
	2 = moderately
	1 = miserably
b. How long did you sleep?	
	4 = 7 to 8 hours
	3 = 5 to 6 hours
	2 = 2 to 5 hours
	1 = 0 to 2 hours
c. How long did it take you to fall asleep?	3 = short time
	2 = intermediate
	1 = long time
d. How did you feel after awaking?	
	3 = well
	2 = moderately
	1 = miserably

For the determination of bromine in serum 0.5 ml of the sample was added to 0.5 ml of a 10.0 ppm Mo solution made by dissolving 18.4 mg $(NH_4)_6Mo_7O_{24}\cdot 4\ H_2O$ in 1.0 l of distilled water. After mixing, 10 µl of the solution was freezedried on a Hostaphan backing foil (4 µm thickness) and irradiated for 10 min with 2.4 MeV protons accelerated with the AVF cyclotron of the Vrije Universiteit. For the detection of the X-rays a 100 mm^2 Si(Li) detector with a pulsed optical feedback preamplifier was used. A pile-up rejector was included in the electronics and spectra were stored on magnetic tape and analysed off-line. For details see ref. 8.

To determine the correlation between bromine concentration in serum and the quality of sleep, a rank correlation test of Kendall was applied. The test of Wilcoxon was performed to determine the significance of the effect of addition of sodium bromide or placebo (sodium chloride) on the quality of sleep.

Results and discussion

Earlier investigation showed already, that depletion of bromine might be responsible for insomnia (1). Table II shows, that the mean bromine concentration in serum of 18.8 µmol/l was markedly lower than the value considered as normal by Bowen (42 µmol/l), which was based on a compilation of numerous measurements with different techniques (9). The mean average percentage of the maximum obtainable score during six weeks was 83. Fig. 1 shows a significant correlation between bromine concentration in serum and the average percentage of the maximum obtainable score during 6 weeks ($p < 0.005$), indicating an inverse relationship between bromine deficiency and the quality of sleep in patients on maintenance haemodialysis.

Table II. Serum bromine concentration and average percentage of maximum obtainable score of 21 patients, undergoing maintenance haemodialysis

Case	Age (years)	Diagnosis	Maintenance haemodialysis (years)	Serum bromine (µmol/l)	Average sleeping score	Use of hypnotic
1. B.	32	CG	3.8	18.8	10.7	+
2. E.-S.	62	CP	1.0	18.8	12.6	
3. G.-V.	52	PK	1.2	15.0	8.3	+
4. J.	21	CG	2.2	20.0	12.9	
5. M.-D.	42	CG	2.4	20.0	12.3	+
6. P.	46	CP	11.4	17.5	7.6	+
7. R.	59	DG	3.0	18.8	7.9	+
8. S.	66	CP	4.1	18.1	11.8	+
9. V.	32	CG	1.0	17.5	11.1	
10. W.	55	CP	1.0	17.5	11.4	
11. S.	33	NAS	6.6	21.3	11.1	+
12. G.	76	CG	8.2	16.3	10.8	+
13. A.-K.	61	NAS	1.4	23.8	12.3	
14. B.-K.	37	CP	2.6	22.5	11.2	
15. B.-J.	60	AN	1.0	12.5	8.7	+
16. M.	62	AN	8.4	8.8	9.0	+
17. W.-H.	66	NAS	1.8	11.3	10.2	
18. G.-T.	50	MM	1.3	13.8	10.6	
19. B.	52	DG	1.6	21.3	12.8	
20. K.	23	CG	5.8	22.5	12.6	
21. L.	29	HN	1.0	38.8	11.8	
Mean ± S.D.	48.4 ± 15.7		3.4 ± 3.0	18.8 ± 6.0	10.8 ± 1.7	

CG = chronic glomerulonephritis; CP = chronic pyelonephritis; PK = polycystic kidneys; DG = diabetic glomerulosclerosis; NAS = nephroangiosclerosis; AN = analgesic nephropathy; MM = renal damage due to multiple myeloma; HN = hereditary nephritis.

Table III. The effect of addition of sodium bromide or placebo (sodium chloride) to dialysate on sleeping score of 4 patients on maintenance dialysis (maximum obtainable score = 13)

Case	Observation period	Addition	Trial period
3.	8.4 ± 0.6	bromide	9.4 ± 0.5
6.	7.4 ± 0.7	chloride	7.5 ± 0.5
15.	8.4 ± 1.5	bromide	10.5 ± 1.0
16.	8.3 ± 1.5	chloride	7.4 ± 0.7

Table III shows the result of a double-blind trial in four patients on maintenance haemodialysis. The addition of sodium bromide to the dialysate effectuated in patients 3 and 15 a significant improvement of sleeping score ($p<0.005$ and $p<0.05$ respectively). Concomitantly their serum bromine concentrations increased to the expected values. The addition of sodium chloride to dialysate in patients 6 and 16 did not improve the sleeping score at all. The serum bromine concentrations of these patients remained unchanged.

The significant correlation between bromine deficiency and the quality of sleep and the significant effect of the addition of sodium bromide on sleeping score strongly suggest, that bromine deficiency plays an important rôle in causing insomnia of patients on chronic intermittent haemodialysis. However, further investigations are necessary to get a definite answer about the physiopathological meaning of bromine deficiency.

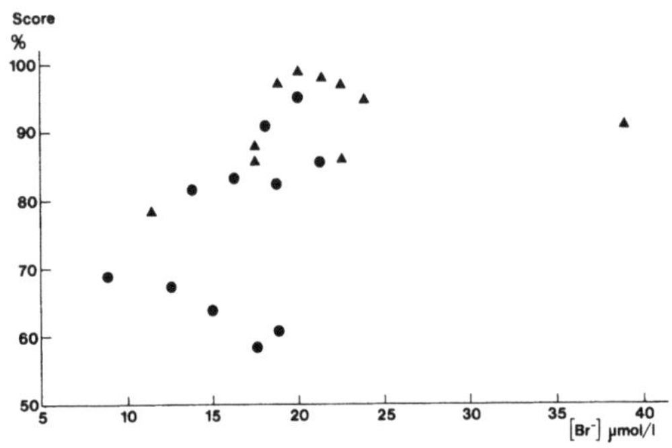

Fig. 1. Correlation between bromine deficiency and quality of sleep. Closed circles: users of hypnotics.

References

1. Vis, R.D., P.L.Oe, and H.Verheul, 1979, Artificial Organs 3 (supplement): 44-48.
2. Alfrey, A.C., H.Rudolph, and W.R.Smythe, 1975, Kidney int. suppl.2: 85-89.
3. Cornelis, R., S.Ringoir, N.Lameire, L.Mees, and J.Hoste, 1979, Mineral Electrolyte Metab. 2: 186-192.
4. Tsukamoto, Y., S.Iwanami, and F.Marumo, 1980, Nephron 26: 174-179.
5. Alfrey, A.C., and W.R.Smythe, 1978. In "Proceedings of the Eleventh Contractors Conference, Artificial Kidney Program. Bethesda, Md.: National Institutes of Health, DHEW" p.137.
6. Zelvelder, W.G., 1971. In Thesis "Therapeutic evaluation of hypnotics" (Rijksuniversiteit, Utrecht) pp 10-14.
7. Oswald, I., 1968, Pharmacological reviews 20: 277.
8. Vis, R.D., 1977. In Thesis "On the use of an ion-accelerator for trace analysis" (Vrije Universiteit, Amsterdam) pp 25-65.
9. Bowen, H.J.M., 1974, J. Radioanal. Chem. 19: 215-226.

DISCUSSION

VERSIECK (Belgium). Dr. Oe, you have shown us a slide with bromine, rubidium and strontium values in normal subjects and patients on haemodialysis. Were these values on serum?

OE (The Netherlands). Yes, on serum. If I remember, the rubidium value was 2.3 μg/ml.

VERSIECK (Belgium). This seems to be an extremely high value. May I ask you if you have analysed biological reference materials in order to check the accuracy of your method for that element.

OE (The Netherlands). My colleague (Dr. Vis) has done them.

GARNYS (Australia). Have you any idea of the variation of normal bromine values between urban and city populations? I am not sure what the situation is in Holland, but here atmospheric bromine levels in cities are very much higher than in urban areas.

OE (The Netherlands). I am just wondering why you ask this question because we are not aware that there may be a difference.

GARNYS (Australia). One of the main additives in petrol, particularly a lead additive is as a bromine compound and the bromine level in city environments is very much higher. There are not any data available on the serum bromine levels in city dwellers. We seem to have a technique and I am interest in serum and I was just wondering whether you looked at country patients versus city patients?

OE (The Netherlands). We are interested in your question. Thanks for your valuable suggestion.

HORVATH (U.S.A.). I notice that your rubidium values are also decreased and yet you did not test the effect of rubidium on sleep. I wonder if you are considering studying the effect of rubidium or other elements as well as bromine?

OE (The Netherlands). Bromine is selected because there are several results from the literature suggesting that bromine deficiency may have a relationship with insomnia. For example, fifty years ago doctors prescribed bromide for sleep, before bromide was replaced by barbiturates. Experimental data in cats demonstrated a bromine-containing substance in the cerebrospinal fluid and when you injected this substance intravenously the "encephalé isolé" cat preparation, the time course of the "rolling eye movement" - sleep increasing effect and the "not-rolling eye movement" - sleep decreasing effect on the electro-encephalogram coincided with the rising phase of appearance of this substance in the cerebrospinal fluid. There are not data available from the literature that rubidium or other elements can be related to the phenomenon of sleep.

TRACE ELEMENT PATTERNS IN BLOOD OF PATIENTS WITH RENAL FAILURE

R.Cornelis[a], S.Ringoir[b], L.Mees[a], N.Lameire[b], B.Wallaeys[a] and J.Hoste[a]

[a] Institute for Nuclear Sciences, Laboratory for Analytical Chemistry

[b] Division of Nephrology of the Department of Medicine, University Hospital

[a] Proeftuinstraat 86, B-9000 GENT

BELGIUM

One of the kidneys' major tasks consists in eliminating waste products. The complete or partial loss of the kidney function deeply affects the organism and will be reflected in the element patterns of the blood.

The major part of our research dealt with trace elements in blood, serum and red blood cells of end-stage kidney patients, either on hemodialysis or on continuous ambulatory peritoneal dialysis (CAPD).

CHRONIC HEMODIALYSIS PATIENTS

Apart from metabolic changes inherent to kidney failure, trace elements in chronic hemodialysis patients are also liable to be subject to excessive losses or to accumulation through the very permeable dialysis membranes. The concentrations both at the start and at the end of treatment, for ten patients can be summarized as following :

- Serum Cs is higher than normal before dialysis, but drops to about 50% of the original value. The red blood cell Cs remains normal (1).
- The serum Rb-content is normal before dialysis, but lower after, then only yielding 50% of the original value. The red blood cell Rb is outspokenly below normal as a result of the treatment. Both alkali-metals behave notably analogous : there is a similar fall in concentration in serum and there is constancy in red blood cells (1). The dialysate contains less K than normally present in blood, to alleviate the body of superfluous K. A low Rb-content could theoretically be a consequence of the K-depletion.
- Serum Se is always below normal, whereas red blood cell Se stays at its normal level (1). There is no influence noticeable after a single dialysis session. Se is an essential element. The significance of this sub-normal serum Se-value for the chronic dialysis patient remains unknown.
- The serum Zn-values are slightly lower than normal, both at the start and at the end of the treatment. Serum Zn is a little enriched during the dialysis. The red blood cell Zn is slightly, nevertheless statistically, higher than normal, and remains practically unchanged (1).
- At the start of the dialysis, the serum Fe-content is normal in eight out of the ten patients, and even rises slightly above normal towards the end. The two other patients had parenteral Fe administered during the treatment and the serum Fe-content rose respectively three and seven times above its previous level. The red blood cell Fe-content of the first eight patients is normal and remains so. The other two also show an unaltered red blood cell Fe, i.e. respectively normal and subnormal.
- The Co-data of serum and of red blood cells are abnormally high (4 to 150 times higher than normal) and they alter course, but in a dissimilar way, as a result of the dialysis process. Too high Co-contents may induce degenerative heart troubles. The impact of these Co-contents is difficult to assess. Maybe the quality of the blood samples was not to the mark. True enough, they were obtained without intervention of a stainless steel needle, through direct sucking of the blood out of the external blood line into the syringe. Hidden contamination hazards might still exist.
- Br-levels in the blood of these chronic hemodialysis patients appear to be below normal and to decrease even further during dialysis. It is possible that their bromine-stores may even become depleted (2).

CAPD : CONTINUOUS AMBULATORY PERITONEAL DIALYSIS PATIENTS

Rather recently we started to investigate eleven trace elements in serum and red blood cells of patients who were since more than one year on continuous ambulatory peritoneal dialysis (CAPD). The procedure consisted of two sampling steps : 1. at the time fresh dialysate was introduced into the body; 2. six hours later, when the dialysate needed renewal. The fresh as well as the redundant dialysate were sampled as well. The analytical technique employed was neutron activation analysis. The determination of some elements required radiochemical separations (Mn and Cu : oxine extractions (6) ; Cr : distillation as CrO_2Cl_2). All the other elements could be determined by instrumental neutron activation analysis (2)(4).

TABLE 1 : a - Br,Co,Cr,Cs,Cu,Fe,Hg,Mn,Rb,Se and Zn contents in serum and red blood cells of a CAPD patient, as well as in the dialysate, at the start and at the end of a 6 h treatment

b - Cu,Mn and Zn in CAPD patients (b-1 and b-2)

Case	Element (unit)	SERUM (ml)				RED BLOOD CELLS (g)				DIALYSATE (ml)		
		Normal mean ± s.d.	1. start	2. end	*ratio 2/1*	Normal mean ± s.d.	1. start	2. end	*ratio 2/1*	1.start	2. end	*ratio 2/1*
a man, 45y CAPD 18 months	Br (μg)	4.87±2.02 (7)	0.94	0.92	*0.98*	4.14°	0.55	0.55	*1.00*	0.384	0.939	*2.44*
	Co (ng)	0.108±0.060 (4)	0.422	0.329	*0.78*					0.069	0.158	*2.29*
	Cr (ng)	0.160±0.083 (4)	3.63	3.60	*0.99*					1.66	0.74	*0.45*
	Cs (ng)	0.74±0.25 (3)	1.21	1.10	*0.91*	4.82±2.10 (3)	5.95	5.84	*0.98*	0.005	0.93	*186*
	Cu (μg)	1.04±0.24 (6)	0.87	0.85	*0.98*	0.66±0.08 (6)	0.91	0.81	*0.89*	0.011	0.024	*2.18*
	Fe (μg)	1.63±0.43 (3)	1.13	0.79	*0.70*	1025±136 (3)	1148	1082	*0.94*	N.D.	N.D.	
	Hg (ng)	2.2±0.9 (8)	10.6	7.6	*0.72*	16.6±2.2 (8)	21.4	22.4	*1.05*	N.D.	N.D.	
	Mn (ng)	0.57±0.13 (6)	0.64	0.46	*0.72*	15.0±4.9 (6)	14.9	13.2	*0.89*	0.53	0.26	*0.49*
	Rb (μg)	0.173±0.036 (3)	0.181	0.170	*0.94*	4.28±0.98 (3)	3.55	3.23	*0.91*	0.002	0.142	*71*
	Se (μg)	0.130±0.021 (3)	0.109	0.089	*0.82*	0.163±0.030 (3)	0.146	0.136	*0.93*	0.002	0.005	*2.50*
	Zn (μg)	1.13±0.20 (3)	1.07	0.72	*0.67*	11.1±1.8 (3)	15.1	13.5	*0.89*	0.045	0.019	*0.42*
b*-1	Cu (μg)	1.07±0.24 (6)	1.43	1.28	*0.89*	0.66±0.08 (6)	0.53	0.44	*0.83*	0.011	0.021	*1.91*
b+-2			1.71	1.74	*1.02*		0.73	0.67	*0.92*	0.012	0.024	*2.00*
b-1	Mn (ng)	0.57±0.13 (6)	0.60	0.70	*1.17*	15.0±4.9 (6)	5.75	5.69	*0.99*	0.45	0.18	*0.40*
b-2			0.45	0.61	*1.35*		10.71	11.74	*1.10*	0.35	0.19	*0.54*
b-1	Zn (μg)	1.13±0.20 (3)	0.63	0.59	*0.94*	11.1±1.8 (3)	10.9	10.2	*0.94*			
b-2			0.75	0.74	*0.99*		12.5	12.4	*0.99*			

N.D. not detected ; b*-1 man, 53 y, CAPD 12 months ; b+-2 woman, 66 y, CAPD 17 months

° calculated value, from the blood and serum contents, assuming a hematocrit of 0.45

The evaluation of the concentrations (duplicate analyses) of Br, Co, Cr, Cs, Cu, Fe, Hg, Mn, Rb, Se and Zn in serum and red blood cells, as well as in the dialysate, of one CAPD patient (case a), can be read in table 1, part a. The following comments can be made :

- The Br-levels both in serum and red blood cells, are below the normal range but do not change as a consequence of this peritoneal dialysis session. The increase in the Br-level of the dialysate shows that some Br leaves the body through the peritoneum.
- Co in serum is nearly normal, be it a little too high. The content decreases about 22% and this goes together with a rise in the Co-content of the withdrawn dialysate.
- The Cr-values in the sera are about 23 times higher than the normal mean and remain unaltered through the treatment. The concentration of Cr in the dialysate is exceptionally high, i.e. 10 times the normal Cr-content of serum. It is the only trace element present in the dialysate in a concentration exceeding that of the normal mean serum level. Some Cr of the dialysate seems to be absorbed into the body.
- The Cs-content of serum falls within the higher range and decreases a little. The Cs red blood cell content is normal. Cs in the fresh dialysate appears to be extremely low (5 pg/ml) but it increases nearly 200 times to the end.
- The Cu-content of serum is normal and it remains so. The Cu-values of the red blood cells are within the normal range, be it on the higher side. Cu decreases about 10%. The original Cu-concentration of the fresh dialysate is about 100 times lower than the normal mean serum Cu value and it doubles after use.
- The serum Fe-values are normal but exhibit a 30% decrease in this particular case. The red blood cell values are normal. Fe is not detected in the dialysate.
- Hg in serum seems abnormally high in comparison with the given literature value. It decreases by about 28% to the end of this peritoneal dialysis session. The Hg content of the red blood cells is on the high side but remains unchanged. No Hg is detectable in the dialysates.
- Mn in the sera and the red blood cells is normal and it shows a decrease in both matrices. The original Mn-content of the dialysate is comparable to that of serum but it decreases to half of its original value.
- Rb in serum and red blood cells is normal and decreases only a little. Rb is hardly detectable in fresh dialysate (2 ng/ml) but increases seventy times after use.
- The Se-values of serum and red blood cells fall within the normal range but decrease to the end. Se is very low in the dialysate but appears to double after use.
- The Zn content in serum is normal at the start but diminishes for about 33%. Both Zn-values in the red blood cells are higher than normal; there appears to be a 9% decrease. The Zn-content of the dialysate is very low and is more than halved at the end of the treatment.

Results for Cu, Mn and Zn in two other CAPD patients are also available (to be read in table 1 - part b).

- Cu in serum is high in these two patients, but still within the normal range. The red blood cell values are originally within the normal range but go down at the end of the treatment (case b-1 falls below the normal range). It is noteworthy to mention that the Cu-contents of the 3 dialysates show nearly identical values as well fresh as after use (they become twice the original value).
- The Mn contents of the sera vary between the normal boundaries. The red blood cell concentrations of case b-1 are on the lower side. The Mn concentrations of the dialysates nearly halved and this is similar to case 1.
- The serum Zn-values of case b-1 are below normal, its red blood cells are normal. In case b-2, the serum Zn value is normal and the red blood cell Zn close to the upper limit.

We plan to extend this study to at least ten CAPD patients in order to test and eventually confirm the present findings. The final results will allow us to decide about normality, excess or loss of these important trace elements in patients who are since more than one year on CAPD treatment.

REFERENCES

1. Cornelis, R., L.Mees, S.Ringoir and J.Hoste, 1979, Mineral Electrolyte Metab. 2:88-93
2. Cornelis, R., S. Ringoir, N.Lameire, L.Mees and J.Hoste, 1979, Mineral Electrolyte Metab. 2:186-192
3. Versieck, J., J.Hoste, F.Barbier, H.Michels and J.De Rudder, 1977, Clin.Chem. 23: 1301-1305
4. Versieck, J., J.Hoste, F.Barbier, H.Steyaert, J.De Rudder and H.Michels, 1978, Clin.Chem. 24:303-308
5. Thiers, R.E., J.F.Williams and J.H.Yoe, 1955, Anal.Chem. 27:1725-1731
6. Versieck, J., A.Speecke, J.Hoste and F.Barbier, 1974, Clin.Chem. 20:1141-1145
7. Versieck, J. and R.Cornelis, 1980, Anal.Chim.Acta 116:217-254
8. Magos, L., 1972, J.AOAC, 55:966-971

DISCUSSION

BEHNE (Germany). During hemodialysis the water content in the blood can change quite considerably. Did you get the same changes for the contents of the protein-bound trace elements when you related them to the dry weight or to the protein content of the serum.

CORNELIS (Belgium). The elimination of water from the body is one of the purposes of the hemodialysis treatment and accounts for 2 up to 7 kg per session. Our values are expressed per millilitre serum and are not correlated to the protein content, although this was quantified as well.

SHAPCOTT (Canada). The salts which are used in these dialysis procedures are not chemically pure. Would you care to speculate on how much change is produced simply by having excessive amounts of trace elements in the actual fluid? Do you think this is a significant factor or not?

CORNELIS (Belgium). I have shown you the concentrations of the various elements in the dialysate used for the CAPD patients and these concentrations were all comparable to the normal serum value with the exception of chromium, which was ten times higher.

SHAPCOTT (Canada). Yes but with the hemodialysis.

CORNELIS (Belgium). With the hemodialysis we did not systematically analyse the dialysate because it is a continuous flow system. We found that the cobalt content was below the nanogram per ml range. Nevertheless as so many litres are passing through the artificial kidney, there could occur a transfer of trace elements through the membrane into the blood and then absorption by the proteins. This might have been the case with cobalt.

SERUM TRACE ELEMENT LEVELS IN HEPATOBILIARY DISEASES

J. Versieck, J. Hoste, F. Barbier, L. Vanballenberghe, J. De Rudder, R. Cornelis

Department of Internal Medicine, Division of Gastroenterology; and Institute for Nuclear Sciences, Laboratory for Analytical Chemistry; University of Ghent; Ghent, Belgium

INTRODUCTION

The serum copper and zinc concentrations of healthy subjects are now generally accepted to be about 1 μg/ml (1). On the contrary, the serum manganese and molybdenum levels are surrounded by controversy (1). For manganese, the reported values vary from 0.54 ± 0.16 ng/ml (2) up to 34.3 ± 7.0 ng/ml (3) and, for molybdenum, from 0.59 ± 0.23 ng/ml (4) up to 257 ± 205 ng/ml (5).

Results of analyses of biological reference materials have conclusively shown that part of the controversy is certainly attributable to analytical inaccuracies. However, there are solid reasons to suppose that another source of errors should be added. Indeed, strong evidence has accumulated that the high values are largely due to unsuspected contamination of the samples during their collection and manipulation (1).

Using neutron activation analysis, we developed methods for the determination of the 4 elements. Stringent precautions were applied to avoid extraneous additions. In this publication, we summarize our observations in normal subjects and in patients with hepatobiliary diseases.

MATERIALS AND METHODS

We collected our samples with a polypropylene catheter (Intranule 110 16, Vygon) in high-purity quartz tubes (Spectrosil, Thermal Syndicate Limited), cleaned with extreme care (rinsed with redistilled water, soaked for about 2 days in 30 % hydrogen peroxide, rinsed with quartz-bidistilled water, boiled for about 4 hours in a mixture of equal volumes of Suprapur nitric and sulphuric acid (E. Merck), boiled for another 4 hours in quartz-bidistilled water, rinsed again and, finally, steam-cleaned for about 6 hours with quartz-bidistilled water). To avoid airborne contamination, we kept these tubes carefully closed as much as possible with individual, thoroughly cleaned Teflon stoppers (Du Pont Company) and carried them to and from the laboratory in an air-tight polyethylene box. Before irradiation, we handled, lyophilized, and prepared our samples in a dust-free room as detailed elsewhere (6). For the determination of manganese, copper, and zinc we transferred about 120 mg of lyophilized serum into meticulously cleaned polyethylene containers (Kartell SpA), for the determination of molybdenum about 1600 mg.

After irradiation at a flux of $\sim 1.8 \cdot 10^{12}$ $n \cdot cm^{-2} \cdot s^{-1}$ for about 4 hours (determination of manganese, copper, and zinc) or for about 7 hours (determination of molybdenum) in the Thetis Reactor of Ghent University, the samples were mineralized, processed, and measured as described in previous publications (7, 8).

Routine liver function tests were carried out with established methods. Statistical tests were performed with standard formulas.

We checked the accuracy and precision of our techniques by analysing biological reference materials. We obtained the following results (mean ± standard deviation) - for manganese, in Bowen's kale: 14.08 ± 0.23 μg/g and in NBS bovine liver: 9.7 ± 0.3 μg/g; for copper, in Bowen's kale: 4.439 ± 0.164 μg/g and in NBS bovine liver: 180.6 ± 6.0 μg/g; for zinc, in Bowen's kale: 31.44 ± 1.37 μg/g and in NBS bovine liver: 129 ± 7 μg/g; and, finally, for molybdenum, in Bowen's kale: 2.14 ± 0.08 μg/g and in NBS bovine liver: 3.42 ± 0.11 μg/g.

RESULTS

In healthy subjects, we determined the following levels (mean ± standard deviation) - for manganese: 0.57 ± 0.13 ng/ml; copper: 1.07 ± 0.24 μg/ml; zinc: 0.94 ± 0.13 μg/ml; and molybdenum: 0.55 ± 0.21 ng/ml.

In patients with acute viral hepatitis, we found - for manganese: 2.32 ± 0.96 ng/ml; copper: 1.43 ± 0.36 μg/ml; zinc: 0.90 ± 0.16 μg/ml; and molybdenum: 3.10 ± 1.46 ng/ml. When compared to the levels in healthy subjects, the manganese and molybdenum values are markedly increased whereas the copper value is slightly increased. In 6 patients, the serum manganese concentration was determined both during the initial and the subsiding phase of the disease. It was found to normalize in all patients (9). In 13 patients, the serum molybdenum concentration was determined 2 to 4 times during the initial and the subsiding phase of the infection. It was also observed to return toward normal, in parallel with the routine liver function tests (10). When the results of the initial phase and the subsiding phase were combined, significant positive correlations were observed between the serum molybdenum level and the serum bilirubin (r = 0.432), alkaline phosphatase (r = 0.511), aspartate transaminase (r = 0.710) and alanine transaminase (r = 0.683) (10).

In patients with chronic persistent hepatitis, we observed - for manganese: 0.58 ± 0.13 ng/ml; copper: 1.04 ± 0.17 μg/ml; zinc: 1.02 ± 0.16 μg/ml; and molybdenum: 0.61 ± 0.15 ng/ml. These figures do not differ significantly from the normal.

In patients with chronic active hepatitis and postnecrotic cirrhosis, we obtained - for manganese: 0.84 ± 0.24 ng/ml and 0.84 ± 0.20 ng/ml; copper: 1.25 ± 0.31 μg/ml and 1.09 ± 0.30 μg/ml; zinc: 0.79 ± 0.20 μg/ml and 0.81 ± 0.33 μg/ml; and molybdenum: 0.97 ± 0.49 ng/ml and 1.01 ± 0.50 ng/ml. When compared to the levels in healthy subjects, the manganese and molybdenum values are significantly increased whereas the zinc values are decreased. When the results of the patients with acute hepatitis, chronic persistent hepatitis, chronic active hepatitis, and postnecrotic cirrhosis were grouped, significant positive correlations were observed between the serum manganese level and the serum bilirubin (r = 0.430), aspartate transaminase (after double logarithmic transformation, r = 0.945) and alanine transaminase (after double logarithmic transformation, r = 0.870) (9).

In patients with alcoholic liver disease, the results were as follows - manganese: 0.65 ± 0.18 ng/ml; copper: 1.24 ± 0.34 μg/ml; zinc: 0.76 ± 0.10 μg/ml (significantly decreased); and molybdenum: 1.32 ± 0.56 ng/ml (increased). The lowest serum zinc concentration (0.61 μg/ml) was observed in a woman with decompensated cirrhosis accompanied by renal failure, whereas the highest 3 serum molybdenum concentrations (2.59, 1.87, and 1.85 ng/ml) were seen in a women 1 ½ months after a bout of alcoholic hepatitis, and in a man and a woman on admission because of alcoholic hepatitis, with deep jaundice (serum bilirubin 47.3 and 13.3 mg/dl, respectively). In the man, the value decreased to 1.12 ng/ml after 3 weeks hospitalization when the serum bilirubin had fallen to 9.4 mg/dl.

In patients with liver metastases, we found - for manganese: 1.08 ± 0.50 ng/ml (significantly raised); copper: 1.88 ± 0.24 μg/ml (raised); zinc: 0.77 ± 0.20 μg/ml (lowered); and molybdenum: 1.40 ± 0.39 ng/ml (raised). A significant negative correlation was found between the serum zinc concentration and the serum bilirubin (r = -0.619) and alkaline phosphatase (r = -0.645), a significant positive correlation between the serum molybdenum concentration and the serum alkaline phosphatase (r = 0.615) (9, 10). In 15 patients who were only symptomatically treated, the serum molybdenum concentration turned out to be closely related to the survival time (10).

In patients with gallstones, we observed - for manganese: 0.59 ± 0.20 ng/ml; copper: 1.48 ± 0.28 μg/ml (significantly increased); zinc: 0.90 ± 0.20 μg/ml; and molybdenum: 1.28 ± 0.38 ng/ml (increased). Though the mean serum zinc level was normal, 2 patients with multiple stones in the common bile duct and a bout of cholangitis were observed to show a definite low level, namely 0.53 and 0.67 μg/ml. In general, the patients examined for manganese, copper, and zinc showed a less severe clinical picture than those examined for molybdenum. Indeed, whereas some of the first series were asymptomatic, all of the second series showed complicating obstruction and infection or concomitant pancreatitis. One patient with cholecystitis was re-examined 1 month after cholecystectomy: the serum molybdenum level was found to have decreased from 1.60 ng/ml to 1.10 ng/ml.

Finally, in patients with extrahepatic biliary obstruction, due to a benign postoperative stricture or to a tumor of the biliary tract or the head of the pancreas with extrinsic pressure on the bile duct, we obtained - for manganese: 0.97 ± 0.39 ng/ml; copper: 2.45 ± 0.59 μg/ml; zinc: 0.91 ± 0.29 μg/ml; and molybdenum: 1.62 ± 0.70 ng/ml. When compared to the levels in healthy subjects, the manganese, copper, and molybdenum values are significantly raised. It is interesting to note that, whereas the serum copper level was increased in all patients, the serum manganese and molybdenum levels were only increased in 2 of the 5 and in 10 of the

13 patients, respectively. One of the patients having a normal serum molybdenum level was re-examined 15 days after admission when the bilirubinemia had gone up from 12.4 to 44.2 mg/dl: the value was found to have increased from 0.83 to 1.43 ng/ml.

DISCUSSION

The levels of manganese we found in serum of healthy subjects are nearly identical with those published by other workers taking adequate precautions to avoid sample contamination (1). The levels of molybdenum are lower than reported by other investigators, with the exception of those obtained by Kasperek and colleagues (4) in supernatant plasma. We thoroughly checked the accuracy and precision of our technique by determining the concentration of the element in biological reference materials. There are solid reasons to suppose that several published values may have been subject to methodological inaccuracies or errors caused by unsuspected extraneous additions to the samples.

Our observations on copper and zinc in serum of patients with hepatobiliary diseases, namely a slightly but significantly increased copper level in acute hepatitis, a significantly decreased zinc level in patients with chronic active hepatitis, postnecrotic cirrhosis, and alcoholic liver disease, a significantly increased copper and decreased zinc in patients with liver metastases, and a significantly increased copper in patients with gallstones and extrahepatic biliary obstruction, agree with those described by several other investigators.

As far as we know, no reliable data on manganese and molybdenum in patients with hepatobiliary diseases have been previously reported by other investigators. We found the mean value of these 2 elements to be significantly increased in patients with acute viral hepatitis, chronic active hepatitis, postnecrotic liver cirrhosis, liver metastases, and extrahepatic biliary obstruction, whereas the mean value of molybdenum was also increased in patients with alcoholic liver disease and symptomatic gallstone disease.

The mechanisms responsible for the observed abnormalities are far from established. The liver is known to be a very important organ in the economy of several trace elements. The potential clinical significance of the occuring phenomena is only beginning to emerge (11, 12).

ACKNOWLEDGEMENTS

We thank Y. Odent for her assistance during the preparation of this manuscript. This work was supported by the Fonds voor Geneeskundig Wetenschappelijk Onderzoek and the Interuniversitair Instituut voor Kernwetenschappen.

REFERENCES

1. Versieck, J., and R. Cornelis, 1980, Anal. Chim. Acta, 116:217-254.
2. Heydorn, K., and K. Nørgård, 1973, J. Radioanal. Chem. 15:683-693.
3. Danys, J., and M. Kušleikaitė, 1971, Z. Inn. Med. 26:718-721.
4. Kasperek, K., G.V. Iyengar, J. Kiem, H. Borberg, and L.E. Feinendegen, 1979, Clin. Chem. 25:711-715.
5. de la Cruz, B. 1973. In "Trace Elements in Relation to Cardiovascular Diseases" (International Atomic Energy Agency, Vienna) pp 19-27.
6. Speecke, A., J. Hoste, and J. Versieck, 1976. In "Accuracy in Trace Analysis: Sampling, Sample Handling, Analysis - Vol. I" (ed. P. LaFleur) (U.S. Department of Commerce. National Bureau of Standards, Washington) pp 299-310.
7. Versieck, J., A. Speecke, J. Hoste, and F. Barbier, 1973, Z. Klin. Chem. Klin. Biochem. 11:193-196.
8. Versieck, J., J. Hoste, F. Barbier, L. Vanballenberghe, J. De Rudder, and R. Cornelis, 1978, Clin. Chim. Acta, 87:135-140.
9. Versieck, J., F. Barbier, A. Speecke, and J. Hoste, 1974, Clin. Chem. 20:1141-1145.
10. Versieck, J., J. Hoste, L. Vanballenberghe, F. Barbier, R. Cornelis, and I. Waelput, 1981, J. Lab. Clin. Med. 97: 535-544.
11. Halsted, J.A., and J.C. Smith, 1974, Gastroenterology, 67:193-194.
12. Weismann, K., E. Christensen, and V. Dreyer, 1979, Acta Med. Scand. 205:361-366.

DISCUSSION

SALTMAN (U.S.A.). That is a beautiful analytical study. We have been doing similar studies on serum trace elements but in our case, looking at bone diseases (bone dysfunction - osteoperosis, slow healing of broken bones and in some cases slow wound healing - post surgically) and note a quite different, but I think meaningful pattern that is beginning to emerge. We have about five different patients that have slow bone growth and healing, and those same people have no detectable manganese and low copper - about half of normal - and low zinc - about half of normal - and an elevated calcium level. Have you looked at any bone abnormalities in your studies?

VERSIECK (Belgium). In our laboratory we have not examined patients with bone disease. Personally, I focused my attention on patients with hepatobiliary diseases. Furthermore, we have done a study in patients with myocardial infarction and as you heard yesterday, Dr. Cornelis has done studies on patients with renal failure. However, I was surprised, if I understand you correctly, that you have seen patients with no detectable manganese in their serum. What is the detection level of your method?

SALTMAN (U.S.A.). We can detect down to the levels which you are detecting; about 0.5 ppb. We have compared serum from these unusual patients with a series of normals and we just cannot detect any manganese in their serum. As far as I know, there is only one other case. That was reported several years ago but was never confirmed. We now have five different cases that we believe are true zero manganese or non detectable manganese in serum. We are doing this with a graphite furnace and we find that when we begin to add increasing amounts of known manganese to the serum that have zero manganese we can detect it, so it is not a matrix effect - it is truly an effect of an altered metabolism. The low manganese in several cases also corresponds with low copper and zinc and high calcium. I hope that maybe some of your people will make a friend in orthopaedics and maybe do some of their cases.

VERSIECK (Belgium). Thank you for this very interesting comment.

K. KRISHNAMACHARI (India). In India we have the problem of osteoporosis occurring past the age of 30-35 along with cases of liver disease. It becomes very difficult for us to attribute the serum changes in manganese to the osteolitic bone or to the liver diseases. Have you compared your manganese values with something like gamma GT which is very specific for the liver necrosis, and also could you explain how long these changes take place after the liver repair has taken place?

VERSIECK (Belgium). We do not consider the serum abnormalities, for instance for manganese, as very specific for liver disease, but that needs further studies. In answer to your question on establishing the diagnosis of liver disease, we rely on the routine liver function test. The purpose of this communication was not to recommend these abnormalities as a liver function test. We looked for abnormalities of the metabolism of trace elements in liver diseases to have an insight into what is happening in these situations, not for diagnostic purposes.

TRACE ELEMENTS IN LIVERS FROM A S.I.D.S. SURVEY

P.E. Wilson

Government Chemical Laboratories
30 Plain St., PERTH WESTERN AUSTRALIA 6000

During 1973-1975 a survey of Sudden Infant Death Syndrome cases was carried out by Dr. J. Hilton a forensic pathologist with the State Health Laboratories of the Western Australian Public Health Department.

Samples of liver from this survey were analysed for selenium and subsequently for calcium, magnesium, copper, iron, manganese, zinc, cadmium and lead at the Government Chemical Laboratories.

METHOD

About 8-10 g of wet tissue, accurately weighed, was digested in nitric-perchloric acids then made to 25 mL with 6M HCl. A 5 mL aliquot was taken for the determination of calcium, magnesium, copper, iron, manganese and zinc by A.A.S. using an air acetylene flame. Manganese and low copper were determined in the 6M HCl solution against standards with a matched acid background. Calcium, magnesium, high copper, iron, manganese and zinc were determined in suitably diluted solutions containing 1500 µg/mL strontium. Standards were prepared at the same acid concentration as the test solutions and contained the same level of strontium.

Cadmium and lead were determined on the remaining solution by solvent extraction of the A.P.D.C. complexes formed at pH 4.0, after preliminary extraction of the 6M HCl solution with MIBK to remove iron and with oxime at pH 4.0 to remove copper and zinc. A venturi sampler[1] was used for sample aspiration where solution volume was limited.

To control the analyses dried bovine liver, standardised by replicated analyses against NBS bovine liver, was used at two weights in each sample batch.

To guard against contamination random blanks were used, vigorous cleaning of equipment was maintained and reagents were purified or selected by test before use.

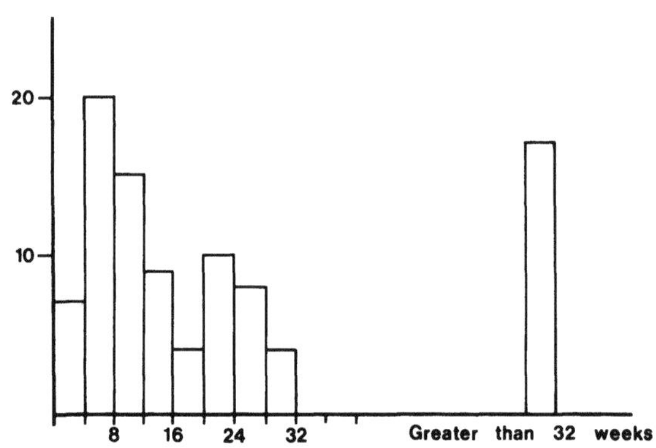

FIGURE 1 Age distribution from Survey

In Fig. 1 the age distribution for 94 of 116 cases analysed is shown. This distribution is essentially similar to those published by Beckwith[2] and Finkle et al[3] as being typical for SIDS.

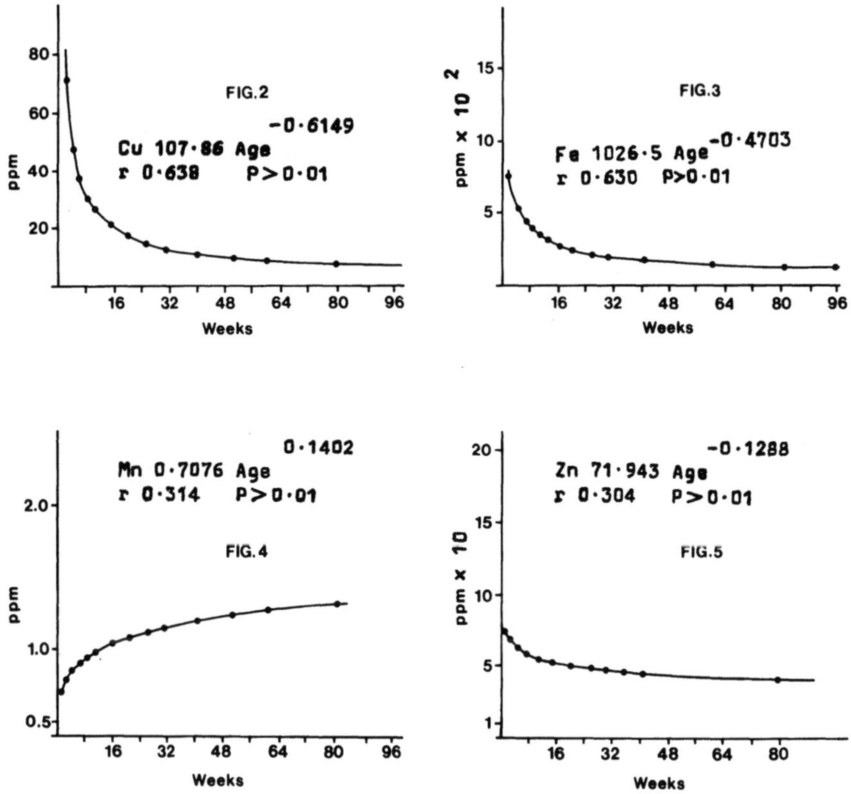

FIGURES 2-5 Regression on age, regression equation, correlation coefficient and significance for copper, iron, manganese and zinc

DISCUSSION

Plots of the analytical data were made and indicated that values for copper, iron, manganese and zinc were related to age. Figs 2-5 show regression on age, regression equation, correlation coefficient and significance for these elements. A linear regression Mn = 0.00931 Age + 0.9257; r = 0.431 fits the manganese data better but the shown regression was used for evaluation of interelement effects.

Regression on age was calculated for all elements so that interactions could be evaluated. To check the relationships found the data were divided into two sets, based on the fit of values to the copper age regression. Values more than twice or less than half the calculated value were placed in a 'non-fit' category and totalled 31 samples. Regression on these sub populations were determined and showed improved age correlations in the 'fit' set for copper, iron, manganese and zinc with improvement in copper-iron and copper-manganese correlation. Iron-calcium, and iron-zinc correlations showed improvement but iron-manganese was not improved. All age correlations were reduced in the 'non fit' set with the exception of manganese. Manganese-iron and manganese calcium correlations were improved. Based on these results conclusions drawn for the full data set were :-

(1) Copper and iron are well correlated with age.
(2) Manganese and zinc are reasonably correlated with age.
(3) Calcium is poorly correlated with age.
(4) Magnesium is not correlated with age.
(5) Iron-manganese were correlated over and above age correlation. The correlation of manganese with age may reflect this.
(6) Magnesium is anomalous, it is not correlated with age or iron values but appears to be slightly correlated with manganese, copper and zinc.

Mean values with standard deviation were calculated for all elements and are shown in Table 1 for those elements not correlated with age. These parameters are not shown for the age-correlated elements since they appear to have little meaning. This table also shows comparison with other published data, Lapin et al[4] and Iyengar et al[5] for SIDS and adults livers respectively.

TABLE 1. LEVELS IN LIVER TISSUE μg/g WET TISSUE

mean; (range), {No.}

	W.A. SURVEY		LITERATURE VALUES LAPIN 1976[4]		IYENGAR 1978[5]
	SIDS	CONTROL	SIDS	CONTROL	ADULT
Ca	39.8 (22-70) {105}	29.0 (19-43) {6}	- -	- -	86.3 (54-133) {39}
Mg	152.9 (120-209) {106}	160 (150-190) {6}	200 - {13}	233 - {14}	215 (172-240) {30}
Se	0.24 (0.10-0.54) {119}	0.25 (0.20-0.35) {6}	0.99 - {13}	0.93 - {14}	0.38 (0.097-0.68) {38}
Pb	0.42 (0.10-1.3) {106}	0.39 (0.15-0.81) {6}	- -	- -	1.28 (0.81-2.3) {102}

CADMIUM

The cadmium values for 104 samples determined were all low indicating little exposure to this element. The range of values found is presented in Table 2.

TABLE 2. CADMIUM VALUES

μg/g wet tissue	No. of Samples
0.04 - 0.1	18
0.02 - 0.04	11
0.02 - less than 0.01	75

A full data printout for the elements calcium, magnesium, copper, iron, manganese, zinc, selenium and lead is available if required.

ACKNOWLEDGEMENTS

I thank Dr. J. Hilton, W.A. State Health Laboratories for permission to analyse samples from his SIDS Survey for the additional elements. I express my grateful thanks to Z.E. SPADEK, D. TRANTHIM-FRYER, A. ROWLEY and B. CODLING for their expert help.

REFERENCES

1. Wilson, P.E. 1979 Atomic Absorption Newsletter 18:115.
2. Beckwith, B. 1975 "The Sudden Infant Death Syndrome" U.S. Dept. of Health, Education & Welfare.
3. Finkle, B.S., McCloskey, K.L., Kopjak, Ladislav and Carroll, J.M. 1979 J. For. Sci. 24:775-789.
4. Lapin, C.A., Morrow, G. III., Chavail, M., Belkie, D.P. and Fisher, R.S. 1976 J. of Ped. 89:607-608.
5. Iyengar, G.V., Kolimer, W.E., Bowen, H.J.M., 1978 "The Elemental Composition of Human Tissues and Body Fluids" Verlag Chemie, Weinheim, New York.

DISCUSSION

ANON. What is the distribution of minerals within your sudden death syndrome samples and what might be expected under normal circumstances.

WILSON (Australia). Unfortunately, I couldn't find very much in the literature on the normal liver values - that is why I made the comparison with adult values and why I thought it might be useful to publish this data. I suspect that the levels are normal, or appear to be normal but I don't honestly know - I would appreciate any comments on this from anyone else.

IYENGAR (Germany). As far as I know there is no literature information on that - I have been keeping a close watch especially as I am preparing a second compilation now. I am trying to bring in more infant tissue data and the only person who has supplied some information is in a book by Hamilton. I don't know whether you saw his book on Chemical Elements in Man which was published last year, but in this he made a few comments, although not many results.

WILSON (Australia). Yes, the data is very scant.

PERVAN (Australia). Recently in the United States there has been a postulation that low ascorbic acid levels may be contributing towards S.I.D.S. They in fact used some earlier figures from the Eastern States of Australia, where large doses of ascorbic acid lowered the infant mortality in aboriginal children. Can you comment on this?

WILSON (Australia). I am afraid this is not my field.

READING (Australia). Has hair analysis been done in any SIDS children for zinc or other trace elements?

WILSON (Australia). No, with the tissue being available, liver was selected. We did look at blood, kidney, the odd spleen and things like this. However, the reason for reporting liver values is because the liver was the most abundant and available sample and seemed to present the one which had the least problems as far as trying to analyse and get consistent results.

READING (Austraia). Please could you clarify, was zinc very significant in the findings in these SIDS children? Was the zinc low?

WILSON (Australia). The zinc values that I have shown in the liver don't appear to be low by comparison with adult data - they are low but they don't seem to be deficient or anything like that, but I don't really know of any data that shows this.

READING (Australia). I looked at one family where the first patient was a near cot death and the third one was a cot death and the fourth one was a near cot death, and this family was very low in vitamin B6, not B1 or vitamin C, but I didn't look at zinc at that time. Zinc is certainly very important in B6 metabolism and it might now be interesting to see what the zinc levels are in this family, especially if the zinc levels are low too. B6/zinc is important for GABA synthesis. There is a 3-fold increase in GABA during sleep.

WILSON (Australia). Yes, we did the odd cobalt on these and except to say that they are very very low as you expect in humans, I have got no other comment.

ANON. Do you think that expressing your results per gram wet weight is the most appropriate way to express them when you have an organ which has a variable amount of fat? About 50% of the liver is extracellular fluid, it has got a lot of blood in it and to express results of this nature in terms of per gram wet weight instead of maybe per unit of DNA or some other measurement might be more appropriate. We have had a lot of problems in deciding what is the best way to express the results of trace elements in even a very simple thing like a red cell. In malnourished children we reach quite different conclusions if we express, for instance the zinc in a red cell per the number of cells or per gram of haemoglobin or per volume of cell. I would have thought that in such a variable organ such as the liver to express the results per gram wet weight is maybe to err.

WILSON (Australia). I come from a traditional sort of animal background and we have always expressed them as either wet weight or dry weight and I suppose I am biased by that particular approach and I take note of your comments. Thank you.

TRACE ELEMENT INTERACTIONS

PREDICTING THE EFFECTS OF MOLYBDENUM AND SULPHUR CONCENTRATIONS ON THE ABSORBABILITY OF COPPER IN GRASS AND FORAGE CROPS TO RUMINANTS

N.F. Suttle

Moredun Research Institute
408 Gilmerton Road, Edinburgh, EH17 7JH.

INTRODUCTION

Small increases in the Mo and S concentrations of semi-purified diets caused up to four-fold reductions in the absorption of Cu(^{A}Cu) by sheep (1). Furthermore, 90% of the variation in ^{A}Cu in subsequent experiments was accounted for by an equation which allowed for the independent effect of S and the interaction between Mo and S (2). In the absence of any information on natural foodstuffs, the equation for semi-purified diets has been used by the Agricultural Research Council (ARC,3) as a provisional means of allowing for the effects of Mo and S on the Cu requirements of ruminants and by others to predict the ^{A}Cu in herbage (4) and brassicas (5) grazed by cattle; such extrapolations may be unjustified. The ^{A}Cu in natural foodstuffs was therefore examined.

MATERIALS AND METHODS

INDOOR EXPERIMENTS

Groups of 4-6 Scottish Blackface ewes, 4-8 years old, were given 0.55-0.7kgDM/d as silage, hay and fresh leafy or root brassicas. All foodstuffs were grown locally and were offered once daily for 21d after a gradual transition over 3d from the semi-purified diet previously used to induce hypocupraemia. Deionised drinking water was freely available.

GRAZING EXPERIMENTS

Similar groups of animals grazed plots of permanent pasture whose Mo and S contents had been altered by the administration of Na_2MoO_4, as a foliar spray, or $(NH_4)_2SO_4$, as crystaline solid, to the closely grazed sward in March. Subsequent growth was grazed for 21-28 days in June/July and September/October. In the first year pasture was conserved in summer as hay from plots treated with Na_2MoO_4 only. Estimates of herbage intake were obtained from data for faecal DM output and the digestibility (D) of herbage DM. Frequent faeces collections were made, using harnesses fitted with nylon gauze bags. Samples of herbage were obtained on 2-3 occasions. 'D' values were estimated from the ratio of Cu concentrations in the faeces and herbage i.e. using Cu as an indigestible marker and also by an in vitro method. The Cu concentration method gave consistently higher values (74.1 ± 2.8 v 62.3 ± 5.0%) for summer pasture and was taken as the preferred estimate, reflecting selective grazing. For autumn pasture the difference varied from 0.3-12.7% and was positively correlated with faecal ash content, which ranged from 16.0-25% compared with a uniform of 14.1 ± 0.3% in summer. It was assumed that the ingestion of soil led to overestimates of digestibility by the Cu method and that the in vitro method gave the preferred estimate in autumn when pasture was relatively scarce.

MEASUREMENT OF ^{A}Cu

The technique was the repletion method used to evaluate the ^{A}Cu in semi-purified diets. The change in plasma Cu (P,mg/l) after 21d in the initially hypocupraemic ewes was converted to the amount of Cu (mg/d) absorbed by the equation $2.135\ A = P + 0.262$ which obtained when Cu was given by continuous intravenous infusion (6).

ANALYTICAL TECHNIQUES

The methods used for determining Cu, Mo and S have been described previously (7).

RESULTS

CHANGES IN PLASMA Cu

The principle differences within and between foodstuffs in their ability to replete the hypocupraemic ewes are illustrated in fig. 1. A small increase in Mo concentration in grazed herbage significantly reduced the response in plasma Cu in summer (fig. 1a): the effect was less marked but still significant when pasture was conserved as hay (fig. 1b).

Fig. 1. Changes in plasma Cu in hypocupraemic ewes repleted with (a) fresh herbage or (b) hay with (●,▲) and without (○,△) Mo fertilization (c) on permanent pasture in summer (□) and autumn (■) and (d) silage (▽), swedes (▼) or kale (X).

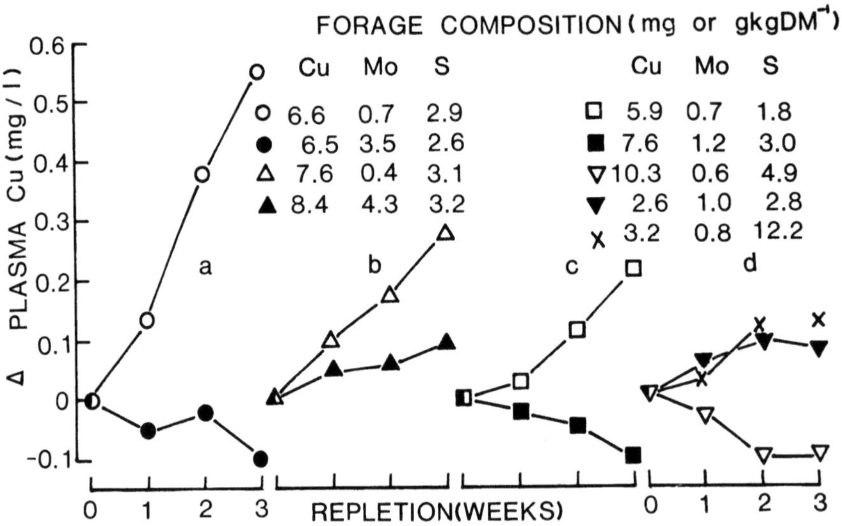

Seasonal changes were evident in that autumn regrowth invariably produced smaller responses in plasma Cu than the same pasture grazed in summer (fig. 1c). Differences between crops were also prominent; some silages failed to replete the ewes despite containing high concentrations of Cu whereas root and leafy brassicas containing less Cu allowed appreciable increases in plasma Cu (fig. 1d).

EFFICIENCY OF Cu ABSOPRTION

The estimates of ^{A}Cu as a percentage of intake are summarised in Table 1.

Table 1. Absorbability of Cu (%) in relation to the composition of natural foodstuffs

Crop	No of estimates	Mean A ± s.d.	Range Mo (mg/kgDM)	S (g/kgDM)	Cu (mg/kgDM)
Summer pasture	10	2.3 ± 1.0	0.6-3.5	1.75-4.4	6.7-8.3
Autumn pasture	8	1.2 ± 0.8	0.5-6.1	2.9-4.7	6.3-10.4
Hay	8	7.2 ± 3.4	0.3-4.3	1.1-3.9	2.2-8.7
Silage	7	4.9 ± 3.2	0.3-1.3	2.1-4.9	6.3-10.4
Leafy brassicas	5	12.8 ± 3.2	0.4-1.1	7.3-12.2	2.7-6.0
Root brassicas	2	6.7 ± 0.9	1.0	2.8-3.9	1.8-3.6

Fresh herbage was the least absorbable Cu source: values were on average 52% lower in autumn than in summer and the maximum value recorded was 4.1%. Cu in hay was absorbed 3 times more efficiently on average than that in summer pasture. The ^{A}Cu in silage varied widely between the extremes found in fresh grass and hay. The leafy and root brassica crops on the other hand contained Cu which was consistently well absorbed. Significant proportions of the variations in ^{A}Cu within forages were accounted for by relationships with Mo and S concentrations and those generally giving the best fits to the data are given in Table 2. They indicate widely differing quantitative relationships, each dissimilar from that obtaining in semi-purified diets. Only autumn pasture failed to give a significant relationship because ^{A}Cu was uniformly low.

It should be noted that the silages and brassicas studied in the experiments did not include materials of elevated Mo content and the relationship for these crops may change when this variable is 'stretched'.

Table 2. Multiple regression equation relating efficiency of absorption (A,%) to concentrations of Mo(mg/kgDM) and S (g/kgDM) in the diet.

	Constant	Mo	S	Mo X S	R	d.f.
Brassicas	0.042	-	-	0.0096**	0.84	5
Silage	0.174	-0.067**	-0.0243**	-	0.96	4
Summer pasture	0.075	-0.0303*	-0.0134*	0.0083	0.76	6
Hay	0.075	0.0784**	-	-0.0271**	0.89	5
Semi-purified diet (A)	0.066	-	-0.0086***	-0.00061***	0.83	29
Semi-purified diet ($^{A}\log_{10}$)	-1.157	-	-0.0074***	-0.00137***	0.93	29

DISCUSSION

These first estimates of ^{A}Cu in natural foodstuffs indicate wide variation within and between foodstuffs, due to the effects of Mo and S. There was, however, no correlation between observed ^{A}Cu and that predicted from equations for semi-purified diets (2) and there seems little doubt that specific equations will be required for each forage examined. Interrelationships may be governed by the rate or extent to which Mo and S are solubilised or degraded from a particular foodstuff during digestion in the rumen since the Cu x Mo x S interaction is initiated there (1). The presence of non-degradable forms of S such as methyl-cysteine sulphoxide and thioglycosides in brassica crops (5) should result in less impairment of ^{A}Cu than would be predicted from total S contents alone. The high ^{A}Cu values in brassicas may also partly reflect metabolic disturbances (5) which stimulate caeruloplasmin synthesis giving overestimates of ^{A}Cu: the repletion technique assumes systemic effects on Cu metabolism to be absent. The rapid degradation of silage N in the rumen might favour the antagonism and explain the low ^{A}Cu's recorded in some silages: the converse may apply with hay. It remains difficult to explain, however, why the Mo and S in some herbages and silages should have more marked inhibitory effects on ^{A}Cu than those of MoO_4^{2-} and SO_4^{2-} added as soluble salts to semi-purified diets.

The data support the principle, incorporated into recent estimates of Cu requirements (3), that variations in ^{A}Cu must be allowed for. The value provisionally suggested for sheep on 'antagonist-free' diets (6%) would however appear to be an over-estimate as far as the grazing Blackface ewe is concerned and her requirements may be considerably higher than those indicated by ARC, particularly in autumn. It seems unlikely that errors in estimating the DM intake of the grazing animal or conducting a short-term study following a change of diet would alone be responsible for a three-fold under-estimate of ^{A}Cu. The low ^{A}Cu in autumn pasture may be attributable in part to the increased ingestion of soil (8).

REFERENCES

1. Suttle, N.F. (1978). Proc. Symp. Sulphur in Forages, Wexford, Ireland. Ed. J.C. Brogan. An foras Toluntais, Dublin, p. 197.

2. Suttle, N.F. and McLauchlan, M. (1976). Proc. Nutr. Soc., 35, 22A.

3. Agricultural Research Council (1980). The Nutrient Requirements of Livestock Commonwealth Agricultural Bureaux, Farnham Royal, p. 230.

4. Givens, D.I. and Hopkins, J.R. (1978). J. agric. Sci., Camb., 91, 13.

5. Barry, T.N., McDonald, R.C. and Reid, T.C. (1981). J. agric. Sci., Camb., 96, 257.

6. Suttle, N.F. (1974). Br. J. Nutr., 32, 395.

7. Suttle, N.F. (1981). Vet. Rec., 108. (in press).

8. Suttle, N.F., Thornton, I. and Alloway, B.J. (1975). J. agric. Sci., Camb., 84, 249.

DISCUSSION

LLOYD DAVIES (Australia). In estimating intakes on pasture there is always a possibility of high errors. A 10-20% variation in estimation of the intake might bring the results from pasture-fed sheep closer to those from animals given semi-purified diets.

SUTTLE (Scotland). The digestibility data are given in more detail in the full paper. The copper ratio method gave us a 12% higher digestibility than the in vitro method for summer pasture and we concluded that it provided the more accurate estimate of digestibility allowing for herbage selection. In autumn the situation was complicated in that in certain samples there was again a difference and as the faecal ash content went up the discrepancy between the copper ratio and the in vitro method increased. Soil ingestion may have contributed to the low absorption of copper in the autumn pasture and certainly made the copper ratio method a little bit suspect but if we use either one or the other of the two methods it really doesn't alter the picture greatly.

FRASER (New Zealand). In view of the fact that hypocupremia is more common in cattle than in sheep, at least in New Zealand, have you any thoughts of doing similar work in cattle?

SUTTLE (Scotland). We have compared the responses of cattle and sheep on a semi purified diet. We found that for given molybdenum-sulphur combinations the absorption of copper is about 1% lower in cattle than in sheep. I hesitate to comment on relative responses at pasture.

EFFECTS OF THIOANIONS OF MOLYBDENUM AND TUNGSTEN UPON COPPER METABOLISM

C. F. Mills, I. Bremner, B. W. Young & N. T. Davies

Rowett Research Institute, Bucksburn, Aberdeen AB2 9SB, Scotland

Studies of the mechanism of action of Mo and S as antagonists of Cu metabolism in ruminants have led to the suggestion that the oxythio or tetrathio analogues of molybdate rather than the MoO_4^{2-} anion may play important roles by inhibiting either the absorption of Cu or its utilisation subsequent to absorption (1,2).

However, evidence for the intraruminal synthesis of MoS_4^{2-} from diets containing MoO_4^{2-} and reducible sources of S is scanty. It is confined to the spectrophotometric demonstration of the presence of the MoS_4^{2-} ion in rumen contents or suspensions of rumen microorganisms exposed to Mo concentrations considerably exceeding those known to inhibit the utilization of Cu by ruminants (1,2). From studies of kinetic aspects of the reaction between MoO_4^{2-} and S^{2-} in inorganic media it has in fact been suggested that while the reaction MoO_4^{2-} - MoO_3S^{2-} - MoO_2S^{2-} - $MoOS_3^{2-}$ may proceed readily at the pH and S^{2-} concentrations commonly found in the rumen, complete substitution to MoS_4^{2-} is less likely (3).

The effectiveness of low dietary concentrations of Mo (>2 mg Mo/kg) as MoS_4^{2-} upon the absorption and metabolism of Cu by rats has been demonstrated previously (4,5). In view of the above suggestions we have now compared the effects of MoS_4^{2-} with those of the oxythiomolybdates, $MoO_2S_2^{2-}$ and $MoOS_3^{2-}$ and have also investigated the influence of the corresponding di- and tetrathiotungstate analogues upon copper metabolism in rats.

In our previous studies of the action of MoS_4^{2-} upon Cu metabolism of rats we have noted repeatedly that growth arrest, defects in melanogenesis in hair and inhibition of caeruloplasmin synthesis occur more rapidly than could probably be accounted for merely by inhibition of Cu absorption. Except for blood plasma in which the (molar) Cu/Mo ratio of treated rats was consistently 2:1, the absence of clear stoichiometric relationships between the Cu and Mo contents of tissues in which the proportion of Cu in the cytosol and the activity of Cu-enzymes were depressed has led us to suspect that at least some of the systemic consequences of MoS_4^{2-} upon Cu metabolism were indirect. Speculation as to the possible origin of such effects led us to re-examine evidence (6) that dietary Mo virtually doubles the biological half life of sulphide within the rumen of the sheep; we concluded that stoichiometric conversion of ingested Mo to MoS_4^{2-} could not account for the magnitude of this effect. If, in contrast, the formation of derivatives such as MoS_4^{2-} inhibited S^{2-} oxidation by rumen and other tissues it could well account for both the above sets of observations. As a preliminary to a more detailed study of the effects of MoS_4^{2-} upon S^{2-} oxidation we have thus investigated whether this Mo source influences the response of rats to low dietary concentrations of S^{2-}.

Effects of oxythio- and tetrathiomolybdates and their W analogues on Cu metabolism

1. Absorption of ^{64}Cu. Rats weighing 150-175 g were offered for one week a semi-synthetic diet providing 3 mg Cu/kg and the Mo or W supplements indicated in Table 1. They were trained to consume a 1 g allocation of diet rapidly before the remainder of each daily supply was given. On day 8, 5 µCi ^{64}Cu was given in 1 g diet and carcass retention of ^{64}Cu was determined 4 h later using procedures described previously. Results are given in Table 1. Both MoS_4^{2-} and WS_4^{2-} strongly inhibited ^{64}Cu absorption but, in contrast, their oxythioanalogues, $MoO_2S_2^{2-}$, $WO_2S_2^{2-}$ and $MoOS_3^{2-}$ had no significant effect. The tetrathio anions of Mo and W strongly enhanced the retention of ^{64}Cu by blood (principally by plasma) at the expense of liver and kidney. Similar but less marked anomalies in ^{64}Cu distribution occurred when the oxythio derivatives were given and it was evident that this effect was directly proportional to the degree of thio-substitution of the Mo or W source.

TABLE 1. Effects of Thio- and Oxythio-derivatives of Mo and W on Cu Absorption and Distribution.

Treatment		Control	$MoO_2S_2^{2-}$	$MoOS_3^{2-}$	MoS_4^{2-}	WS_4^{2-}
Mo or W concentration (mg/kg diet)		0	6	8	6	32
Absorption of ^{64}Cu (relative to control*)		1.00	0.88	1.22	0.20	0.28
Proportion of absorbed ^{64}Cu in	Liver	0.28	0.18	0.10	0.09	0.09
	Kidneys	0.17	0.11	0.05	0.03	0.03
	Plasma	0.13	0.25	0.37	0.57	0.58

*control rats absorbed 33 ± 8% of the ^{64}Cu dose.

2. Effects on Cu distribution and utilisation. Rats weighing 40-50 g were offered for 5-6 weeks the above diet supplemented with di-, tri- or tetrathio anions of Mo and W. Diets provided 2-8 mg Mo/kg or 4-32 mg W/kg. None of the oxythio anions of W or Mo inhibited growth, produced anaemia or clinical signs of Cu deficiency or depressed caeruloplasmin activity. In contrast, WS_4^{2-} inhibited growth and, like MoS_4^{2-}, induced all the above defects unless dietary Cu was increased (Table 2).

TABLE 2. Effects of Thio- and Oxythio- Derivatives of Mo and W on Growth, Haematological Status and Plasma Caeruloplasmin in Rats.

		% of control				
Treatment	Control	$MoO_2S_2^{2-}$	$MoOS_3^{2-}$	MoS_4^{2-}	$WO_2S_2^{2-}$	WS_4^{2-}
Mo or W concentration (mg/kg diet)	0	6	8	6	32	16
Weight of rats (g)	265±37	105	107	53	96	49
Blood haemoglobin (g/l)	135±2	106	99	43	97	43
Plasma caeruloplasmin (U/l)	24.5±3.1	158	98	<4	147	28

Values for Hb and caeruloplasmin of rats receiving Mo or W-containing diets are given as % of controls. All rats receiving 32 mg W as WS_4/kg and some rats receiving 6 mg Mo as MoS_4/kg, died.

Although most Mo and W derivatives affected liver and kidney Cu, only the tetrathio compounds decreased the solubility of Cu in the cytosol (Table 3).

Irrespective of their clinical effects or potency as inhibitors of caeruloplasmin or cytochrome oxidase all thioanions enhanced the Cu (and Mo) content of the plasma albumin fraction and this increment was TCA-insoluble. Such effects are not induced in rats given MoO_4^{2-} and we thus conclude that the failure of the oxythioanions to induce clinical and biochemical signs of Cu deficiency results, not from their destruction in the g.i. tract but from a stringent requirement for complete thiosubstitution for inhibition of Cu utilisation.

3. Synergistic relationships between MoS_4^{2-} and S^{2-}. The influence of MoS_4^{2-} upon the tolerance of rats given low dietary concentrations of S^{2-} has been investigated. Secondary effects on Cu metabolism were assessed. Synergistic interactions between MoS_4^{2-} and S^{2-} were evident with respect to inhibition of growth, induction of achromotrichia, anaemia, cardiomegaly and diarrhoea, depression of liver Cu and enhancement of liver Fe. While many such effects were probably attributable to the secondary induction of Cu deficiency, the appearance of sulphaemoglobin in rats given (MoS_4^{2-} + S^{2-}), but not S^{2-} alone, suggested that oxidation of S^{2-} was also inhibited. The possibility that dietary S^{2-} may enhance the stability of MoS_4^{2-} within the g.i. tract is not precluded.

TABLE 3. Effects of Thio- and Oxythio-Derivatives of Mo and W on Plasma and Tissue Copper

Treatment	Control	$MoO_2S_2^{2-}$	$MoOS_3^{2-}$	MoS_4^{2-}	$WO_2S_2^{2-}$	WS_4^{2-}
Mo or W concentration (mg/kg diet)	0	6	8	6	32	16
Plasma Cu (μg/ml)						
Total	0.74	2.75	2.35	1.32	6.58	1.50
TCA-soluble	0.74	0.23	0.10	0.05	1.49	0.03
Liver Cu						
Total (μg/g wet wt)	3.85	3.94	3.99	2.10	4.89	3.12
% in cytosol	48	48	-	-	55	15
Kidney Cu						
Total (μg/g wet wt)	5.82	9.41	9.01	8.10	6.37	13.8
% in cytosol	52	45	32	24	57	29

THE THIOMOLYBDATE HYPOTHESIS: AN APPRAISAL

The following observations support the concept that sulphido-derivatives of Mo may play a role in the Mo-S antagonism of Cu metabolism in ruminants (i) the MoO_4^{2-} ion *per se* has no inhibitory effect on Cu metabolism, (ii) although partially thio-substituted derivatives of MoO_4^{2-} are relatively ineffective antagonists of Cu, MoS_4^{2-} strongly inhibits Cu absorption and its systemic effects extend to a depression of caeruloplasmin and cytochrome oxidase activity, to growth failure, diarrhoea and other clinical manifestations preventable by Cu, (iii) additional evidence indicates that MoS_4^{2-} may inhibit the normally rapid oxidation of S^{2-} - a situation likely to influence, markedly, the absorption and tissue utilisation of Cu.

We now appreciate that the effects of thio- or oxythiomolybdates (or their W analogues) upon plasma Cu content and distribution, especially the enhanced proportion of total Cu associated with albumin in a form insoluble in TCA, neither directly reflect their capacity to induce clinical and biochemical signs of Cu deficiency nor to affect Cu absorption. Only the fully substituted thio-derivatives exhibit all these properties; the oxythio anions merely cause anomalies in plasma and tissue Cu content and distribution that probably have little or no pathological relevance. While MoS_4^{2-} is a more effective inhibitor of Cu absorption even than S^{2-}, two lines of evidence suggest that the presence of free MoS_4^{2-} in the gastrointestinal tract of ruminants is unlikely to account for the effects of relatively low dietary concentrations of Mo (eg. 1-10 mg Mo/kg) upon Cu metabolism in these species. Firstly, the free ion is only detectable by spectrophotometric examination of rumen contents when Mo concentration is relatively high. Secondly, ruminants given low dietary concentrations of Mo rarely exhibit the anomalies in plasma Cu distribution that invariably result when rats absorb $MoO_nS_{4-n}^{2-}$. The effect of derivatives of higher molecular weight which, while retaining a strong affinity for Cu, are not absorbed is thus being studied.

The inhibitory effect of the MoS_4^{2-} moiety upon Cu absorption may well reflect its relatively greater stability within the gastrointestinal tract than that of S^{2-}. It is also evident that both MoS_4^{2-} and $MoOS_3^{2-}$ readily react with a variety of biologically relevant, Fe-containing, complexes to form binuclear derivatives within which they act as bridging ligands (7, Simpson and Mills, unpublished). Terminal S atoms remaining free in such bridges retain a strong affinity for Cu and react to yield highly insoluble products. The relevance of such Mo-S derivatives to the inhibition of Cu absorption is being investigated.

REFERENCES

1. Dick, A.T., D.W. Dewey and J.M. Gawthorne, 1975, J. Agric. Sci. 85:567-568.
2. Mills, C.F. and I. Bremner, 1980. In "Molybdenum and Molybdenum-containing Enzymes" (ed. M. Coughlan) (Pergamon Press, Oxford) pp 519-542.
3. Clarke, N.J. and S.H. Laurie, 1979, J. Inorg. Biochem. 12:37-43.
4,5. Mills, C.F., T.T. El-Gallad, I. Bremner and G. Wenham, 1981a,b, J. Inorg. Biochem. (in press).
6. Gawthorne, J.M. and C.J. Nader, 1976, Br. J. Nutr. 35:11-23.
7. Mitchell, P.C.H. and D.A. Parker, 1976, J. Chem. Soc. (Dalton) 1976, 1821.

DISCUSSION

GAWTHORNE (Australia). What do you think is the mechanism of the synergism between thiomolybdate and sulphide? Does thiomolybdate inhibit sulphide oxidation in the tissues, or does sulphide maintain thiomolybdate in a particular form that interacts with copper metabolism?

MILLS (Scotland). I think that both possibilities exist and at the moment we cannot decide between them. However, sulph-haemoglobin appears when tetrathiomolybdate is given together with sulphide, but not when sulphide is given alone, indicating that sulphide is not being oxidised properly in the presence of tetrathiomolybdate.

SUTTLE (Scotland). The physico-chemical evidence that Clark and Lawrie have presented in favour of a trithio- rather than tetrathiomolybdate ignores the fact that 90% of the molybdenum in the rumen is in the solid phase and when you take that into account it shifts the ratio of sulphide: to thiomolybdate. As far as the tetrathiomolybdate is concerned if it is given to sheep at a relatively low level of about 4 parts per million it will reduce copper absorption without having any systemic affects. The problem to my mind is that it is no worse than molybdenate itself. If tetrathiomolybdate is the active principle there has to be a quantitive conversion of molybdate to thiomolybdate.

MILLS (Scotland). I generally agree with your comments, but with one exception. When we fed tetrathiomolybdate at those concentrations we saw some systemic effects.

EFFECTS OF THIOMOLYBDATE IN RATS FED A HIGH COPPER DIET

J.M. GAWTHORNE, J.McC. HOWELL and R.S. WYBURN

School of Veterinary Studies, Murdoch University,
Murdoch, Western Australia 6150

Thiomolybdates have been postulated to be the active components in the interaction between copper, molybdenum and sulphur in ruminants (1). They may be synthesised in the rumen by microorganisms acting on dietary molybdenum and sulphur. The compounds have the potential to cause a reduction in the absorption of copper from the intestinal tract and to lower the availability of copper at metabolic sites of action. Although formation of thiomolybdates by washed suspensions of rumen microorganisms can be demonstrated in vitro, the concentrations that are likely to be generated in vivo are below the detection limits of spectrophotometric techniques (1,2). Thus the existence of thiomolybdates in ruminants and the involvement of these compounds in antagonising copper metabolism continues to rest on circumstantial evidence. The effects of thiomolybdate in rats comprises some of the evidence.

Mills et al, (2) reported that administration of thiomolybdates to rats fed a low copper diet (3 ppm) mimicked many of the effects of a high sulphate and molybdate diet in ruminants. The changes were reversed when additional copper was given. In the present paper we confirm and extend these findings and those of Spence et al, (3) and discuss the results of administering thiomolybdate to rats given a diet containing approximately 80 times more copper than that used by these workers.

EXPERIMENTAL

Weanling PVG/c rats similar in appearance to Hooded Wistar rats were fed a commercial rat cube diet. The diet contained an unusually high concentration of copper (250 ppm) but clinical or histological signs of copper toxicity or adverse effects on growth or breeding performance had not occurred in rats fed this diet for extended periods of time. Ammonium tetrathiomolybdate (TM) was dissolved in drinking water at a range of concentrations up to 0.25 mg/ml. Control animals received ammonium chloride in the water at the same molar concentration of ammonium ions as for 0.25 mg TM/ml. In one experiment control animals were pair-fed to animals given 0.25 mg TM/ml. Animals were killed after 8 weeks on these treatments.

With the exception of some reduction in caeruloplasmin oxidase activity and cytochrome oxidase activity there were no discernable effects of administering thiomolybdate at concentrations up to 0.15 mg/ml.

However, marked changes occurred when 0.2 - 0.25 mg TM/ml was given. Growth rate was impaired principally because of reduced food and water intake, but a 10 - 25% decrease in body weight persisted relative to pair-fed controls. The bodyweight of thiomolybdate-treated animals averaged 159 g at 8 weeks compared to 250 g for control animals. Greying of black hair was detectable at 14 - 18 days after commencing the treatment and by 6 weeks greying and loss of hair was obvious in all treated animals. Caeruloplasmin oxidase activity of control animals increased rapidly after weaning and approached adult levels in about 3 weeks. In animals given 0.2 mg TM/ml the post-weaning rise in caeruloplasmin activity was less than half that of control animals and by 8 weeks the activity had declined to zero. The relationship between caeruloplasmin oxidase activity and thiomolybdate concentration in drinking water was sigmoidal, decreasing from 120 IU in control animals to zero in animals given 0.25 mg TM/ml. Cytochrome oxidase activity in brain mitochondria was reduced by approximately 40% in animals receiving 0.2 or 0.25 TM/ml. A normocytic normochromic anaemia in which haemoglobin concentrations were reduced from 14.8 g/dl to 10.7 g/dl was observed. Bone abnormalities similar to those described by Spence et al, (3) were induced. The changes were most significant in the femurs and tibias in which epiphyseal plates were disrupted and widened, evidence of new periosteal bone formation was seen in growth plate regions and at other sites presumed to be tendon insertions.

The extent of the changes were so great in some animals that spontaneous fractures occurred and movement was restricted. The other tissues of thiomolybdate-treated animals appeared normal, except for testes which were smaller, less tense and a darker shade of pink than those of control animals. Histological sections of testes and epididymis revealed that thiomolybdate had caused a loss of spermatids and a cessation of spermatogenesis. No histological abnormalities were found by light microscopy in tissues other than testes and bone. Changes in the intestinal tract such as those described by Fell et al, (4) were not observed. At no time was diarrhoea induced in animals given thiomolybdate.

Thiomolybdate increased the concentration of copper and iron in liver,testes and femur but did not alter the concentrations of zinc, except in femur (Table 1).

Table 1 - Effect of thiomolybdate on tissue levels of Cu, Fe and Zn in rats.

TREATMENT	LIVER			TESTES			FEMUR		
	Cu	Fe	Zn	Cu	Fe	Zn	Cu	Fe	Zn
Pair-fed controls	4.7	176.8	35.3	1.5	11.3	33.6	1.3	45.8	210.0
0.25 mg TM/ml	27.5	643.4	32.0	46.0	33.0	28.3	26.6	69.4	118.2

(Thiomolybdate given in drinking water. Values are μg/g fresh wt. 6 rats/group)

Subcellular fractionation of liver revealed that thiomolybdate had markedly increased the copper content in all fractions except cytosol (Table 2).

Whereas approximately 60% of the copper in the liver of control rats was extracted by tris-acetate buffer containing 1% mercaptoethanol, less than 25% of the copper from animals given 0.25 mg TM/ml was extracted. Only 6% of copper in the kidneys of thiomolybdate-treated rats was solubilised. Gel filtration of the extracts showed that thiomolybdate had produced a distortion in the pattern of copper-containing fractions in which the proportion of high molecular weight forms (> 74,000 MW) was increased, and low molecular weight forms, including metallothioneins (10,000 MW) was decreased.

Table 2 - The effect of thiomolybdate on the subcellular distribution of copper in rat liver.

TREATMENT	SUB-CELLULAR FRACTION				
	Nuclear	Heavy Mitochondrial	Light Mitochondrial	Microsomal	Cytosol
Control	0.5	0.9	1.1	0.2	2.4
0.25 mg TM/ml	3.3	10.2	3.1	1.4	2.8

(Thiomolybdate given in drinking water. Values are μg/g fresh wt. 4 rats/group)

We have used bovine serum albumin (BSA) as a model protein to investigate the effects of interaction with thiomolybdate and copper. BSA formed a brownish-yellow complex with thiomolybdate and this complex had a high affinity for copper. The thiomolybdate-treated protein was more slowly digested by pronase, a proteolytic enzyme. Copper had no additional effect on the slower rate of digestion.

DISCUSSION

These results confirm that in rats thiomolybdate can induce signs that are characteristic of copper deficiency (including low activity of the copper-containing enzymes cytochrome oxidase and caeruloplasmin oxidase). Despite the high concentration of copper in the diet, if enough thiomolybdate was given in the drinking water, clinical signs were induced. Paradoxically these signs occurred in the presence of very high concentrations of copper in the liver, testes and bone. We therefore conclude that copper was trapped in these tissues in unavailable forms.

The interaction of thiomolybdate with body tissues and fluids is poorly understood.

Inorganic molybdate and hydrogen sulphide are known to react with the amino acids cysteine and histidine to form dimers linked by Mo_2S_2 bridges (5,6,7). There is also evidence that thiomolybdate reacts with plasma proteins and that the products of the reaction have high affinity for copper. The copper complexed in this way precipitates with the protein when TCA is added (1,2). We have found that thiomolybdate-treated BSA binds copper and is more slowly degraded by a proteolytic enzyme (pronase) than untreated albumin. Taken together, these observations raise the possibility that thiomolybdate may interact not only with plasma proteins, but also with some proteins inside cells. The products of the interaction may bind copper, resist proteolytic degradation, and accumulate. The shift we have observed towards high MW forms of copper in extracts of liver and kidney may be an early stage in the interaction, and subsequent stages may involve the uptake of these altered proteins by lysosomes. If true, this would explain the observed increase in the copper content of particulate sub-cellular fractions that contain lysosomes and the low availability of copper for the synthesis of copper-containing enzymes.

The following scheme summarises our hypothesis for the effects of absorbed TM on copper metabolism.

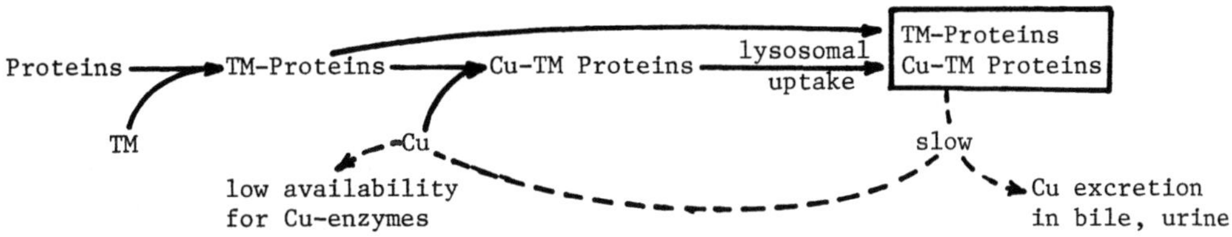

In addition, the absorption of copper and TM from the gastrointestinal tract may be reduced by the formation of Cu-TM-protein complexes that are poorly digested and absorbed. Conditions for such interactions would be particularly favourable in adult ruminants because of the long retention time, sulphide concentration and neutral pH of the rumen.

ACKNOWLEDGEMENT

We are grateful to Mr.M McKenzie and Mr.K Chong for care of the animals and technical assistance with the trace element analyses, and Mr. M Gates for assistance with preparation of the pathology specimens.

REFERENCES

1. Dick, A.T., D.W. Dewey and J.M. Gawthorne, 1975. J. Agric. Sci. Camb. 85:567-568.
2. Mills, C.F., I. Bremner, T.T. El-Gallad, A.C. Dalgarno and B.W. Young. 1978. "Proceedings of the Third International Symposium on Trace Element Metabolism in Man and Animals" (ed. M. Kirchgessner) (Arbertzkreis fur Tierenahrungs-forschung Weichenstephan), pp. 150-158.
3. Spence, J.A., N.F. Suttle, G. Wenham, T.T. El-Gallad and I. Bremner, 1980. J. Comp. Path. 90:139-153.
4. Fell, B.F., D. Dinsdale and T.D. El-Gallad, 1979. J. Comp. Path. 89:495-514.
5. Kay, A. and P.C.H. Mitchell, 1970. J. Chem. Soc. (A):2421-2428.
6. Spivack, B. and Z. Dori, 1970. Chem. Comm. 1716-1717.
7. Spivack, B. and Z. Dori, 1973. J. Chem. Soc., Dalton Trans. 1077-1080.

DISCUSSION

MILLS (Scotland). We have found very similar effects of thiomolybdates on the solubility of copper in liver cytosols. It is interesting that the proportion of copper in cytosol is inversely proportion to the degree of substitution of sulphur in the thiomolybdate - as more sulphur is put in, the solubility declines. Did you attempt to do your pronase experiment anaerobically - I wonder if this is a variable that may influence the results?

GAWTHORNE (Australia). Yes, this may be a variable. We did not do the experiments anaerobically, but proceeded as rapidly as possible once the TM-protein was made.

ULLREY (USA). If copper was fulfilling the function of an oxidase assisting in the mobilisation of iron for haemoglobin synthesis would you care to speculate about the reason for the normocytic normochromic anaemia as opposed to a microcytic hypochromic anaemia?

GAWTHORNE (Australia). I have no explanation for it.

SUTTLE (Scotland). Could you please restate for me that part of your species comparison in which you were suggesting a difference in the time for absorption of thiomolybdates in rats compared with ruminants.

GAWTHORNE (Australia). We are postulating that the reaction of thiomolybdates with protein may be a slow process. In the ruminant there is time for this to happen because of the slow turnover through the rumen. This may also be the reason why the molybdenum in rumen contents is associated with a particulate fraction containing microorganisms (i.e. in the form of thiomolybdate-protein complexes). However, in the rat the thiomolybdate goes straight into the stomach and the faster rate of passage leaves less time for interaction with proteins.

SUTTLE (Scotland). But why shouldn't that reaction take place in the rat intestine?

GAWTHORNE (Australia). It probably does take place to some extent in the intestines, but the overall time for the reaction would be less in rats compared with ruminants because of the larger retention time of the latter. We are speculating that there may be a greater percentage of the copper in an absorbable form in the rat, compared with ruminants because of the difference in rates of passage of digesta.

BREMNER (Scotland). Are you sure that the lesions you get are not a form of molybdenosis rather than a form of copper deficiency?

GAWTHORNE (Australia). I am reasonably sure it is copper deficiency because a similar pattern of effects, including low activity of the copper-dependent enzymes has been observed by us at 250 ppm Cu in the diet, and your group at 3 ppm Cu. If it was a molybdenosis I would not have expected these signs to appear in your experiments where small quantities of thiomolybdate were given.

BREMNER (Scotland). Yes, but the high level of copper in your diet may only have reduced molybdenum absorption. There is also the problem that the lesions are more severe than is generally found in copper deficiency induced with purified diets. By giving extra copper to thiomolybdate-treated rats we may think that we are relieving copper deficiency symptoms, but perhaps we are merely blocking the absorption of thiomolybdate?

GAWTHORNE (Australia). Yes, that is a possibility. However, during a dietary deficiency of copper we are simply providing less copper to metabolic sites, but by treatment with thiomolybdates we are placing a copper antagonist right at the metabolic sites, and are probably pulling copper away from proteins. This may be why the lesions are more severe.

THE EFFECT OF INJECTING SHEEP WITH THIOMOLYBDATES

A.J. Gordon[1] & J.L. Hill[2]

CSIRO[1], Division of Animal Production, P.O. Box 239, BLACKTOWN, N.S.W., 2148
CSIRO[2], Division of Animal Production, Private Bag, P.O., WEMBLEY, W.A., 6014

Interest has centred in this laboratory on factors which affect the physical properties of wool fibres. Copper deficiency has been shown to cause both a reduction in the tensile strength of wool fibres (Marston, 1946; Palmer, 1949) and the production of abnormally crimped and (in black sheep) depigmented fibres (Dick, 1954; Lee, 1956). It is thought to act as a catalyst in the oxidation of thiol groups to disulphide groups in the keratinization of the wool fibre (Marston, 1946) but details of its role have not been investigated. Thiomolybdates may be useful in a systematic approach to such a study because a single i/v injection of thiomolybdate causes an immediate, albeit transient alteration in the distribution of plasma Cu (Dick *et al.*, 1975). It was decided to determine if repeated injections of thiomolybdate (i) affected the properties of growing wool, (ii) caused longer term changes in Cu metabolism as indicated by changes in the concentration of Cu in liver as well as plasma and (iii) altered the concentrations of other elements (e.g. Zn and Fe) whose metabolism is interrelated with that of Cu. An additional treatment of feed intake level was included because it was uncertain what effect this factor had on the liver and plasma levels of the elements examined and because this was likely to influence the sulphur content of the wool and hence the keratinization process.

MATERIALS & METHODS

1. Ammonium tetrathiomolybdate (ATM). This was prepared by bubbling H_2S through a solution of ammonium molybdate (10 g) dissolved in ammonium hydroxide (106 ml, d = 0.88) and water (24 ml) for several hours. The orange crystals were filtered, washed with ethanol, dried and stored at -4°C. An aqueous solution exhibited absorption maxima at 467, 317, 241 and 194 nm.

2. Black sheep. Three adult Merinos were given twice daily injections (s/c or i/p) of ATM in 0.9% saline at the rate of 1 mg ATM/kg.d for 5 (2 sheep) or 10 (1 sheep) days. Another was given an i/v infusion of ATM (1 mg/kg) for 27 h. The wool was subsequently examined for signs of depigmentation. Fibre diameter in the depigmented regions was compared with that of the adjacent black wool, using the C.S.I.R.O. Fibre-Fineness-Distribution-Analyser (Lynch and Michie, 1976).

3. White sheep. Twenty four adult Merino ewes, individually penned, were assigned to two feed intake levels (300 g/d, n = 12; or 1200 g/d, approximately *ad lib*, n = 12) of lucerne chaff-oats (60%-40%) pellets which contained *c.* 4 ppm Cu and 15 ppm Zn (dry matter basis). Twice daily injections (i/p) of ATM were commenced when the mean live weights of the sheep on each intake level were 23.6 kg or 41.9 kg respectively and this regimen continued for 17 days. Four sheep on each intake received either 0, 1 or 2 mg/kg.d ATM in 0.9% saline (20 mg/ml). Dyebands were applied at the start of the regimen. Liver samples were taken by biopsy, 7 days prior to commencing ATM dosing and on the last day of the regimen. In these samples Cu, Zn and Fe were determined by atomic absorption spectroscopy and Mo by the colourimetric method of Bingley (1963). Blood plasma samples were collected on the sixteenth day of the regimen and deproteinized with TCA, centrifuged, and Cu, Zn and Mo in the supernatant determined by atomic absorption spectroscopy. The trace element concentration was analyzed by analysis of variance, assuming a completely randomized design, with feed intake and ATM level as the treatments. With liver, the values before dosing were used as covariates to adjust the values after dosing. Wool was clipped from mid-side patches (100 cm^2) at approximately fortnightly intervals before, during and after ATM dosing. The scoured, conditioned clippings were weighed to determine the effects on wool growth rate. Mid-side staples were taken 140 days after the last day of dosing and measured for staple strength (Gordon and Donnelly, 1979). Fibre diameter was measured in adjacent sections 2 mm in length, of selected staples (Lynch and Michie, 1976), one section of which, having an abnormal crimp near the dyeband was judged to correspond to the wool grown during the injection period.

RESULTS & DISCUSSION

In all four black sheep, ATM administration caused the appearance of a white band in the wool, the width of which (< 0.5 to 2 mm) was proportional to the duration of the treatment. The width of this band was consistent with normal wool growth rates in the injection period and clipping and weighing wool from mid-side patches of the white sheep confirmed that ATM did not suppress wool growth. Although the fibre diameter in the depigmented region was 1-3 μm finer than in the adjacent black wool this may reflect a reduction in the fibre volume occupied by pigment granules because there was no reduction of fibre diameter in the white sheep. The band width in the black sheep infused with ATM for 27 h (< 0.5 mm) was consistent with an effect over just 1-2 cell cycles in the follicle bulb since one cycle is estimated at 15-21 h (Black and Reis, 1979). A narrow (2-3 mm) band of abnormal crimp was evident near the dyeband level in white wool after ATM injection. It was no more evident in sheep on one intake level than another suggesting that the crimping process may be controlled at the level of low S protein synthesis since this would be less affected by intake than high S protein synthesis. The staples appeared to break mainly at the affected region when extended but their strengths were no different to that of the control staples. There was a reduction (P < 0.05) in liver Cu in these sheep (Table 1). Comparing the covariate adjusted values after dosing, the sheep given ATM contained on average 30% less copper than the controls (P < 0.01) but there was no difference between the two levels of ATM.

TABLE 1. Liver contents (ppm DM±sem) at Two Feed Intake Levels of Mo, Cu and Fe Before and After Injecting Sheep for 17 d with Thiomolybdate (ATM). (Values in brackets are after covariate adjustment.)

Feed Intake (g/d)	N	ATM (mg/ kg.d)	Mo Before	Mo After		Cu Before	Cu After		Fe Before	Fe After
300	4	0	2.6†±0.1	5.5±1.9	(4.6)	873†±113	790±106	(539)	1570†±150	1325±160
300	4	1	1.1 ±0.3	64.0±5.5	(59.2)	564 ± 61	408± 50	(403)	1295 ±163	1220±156
300	4	2	2.1 ±0.5	65.3±9.0	(62.1)	559 ±186	387±139	(388)	1333 ±118	1045± 79
1200	4	0	2.4 ±0.5	3.6±0.5	(3.5)	564 ± 85	644±120	(626)	503 ± 94	608±139
1200	4	1	3.5 ±0.3	53.0±7.1	(53.3)	752 ±103	560± 86	(420)	353 ± 52	440± 57
1200	4	2	4.0 ±0.3	43.5±3.7	(44.8)	500 ± 49	408± 68	(446)	355 ± 76	503± 47
Intake effect‡			**	NS	(NS)	NS	NS	(NS)	**	**
ATM effect‡			NS	**	(**)	NS	*	(**)	NS	NS
Intake x ATM‡			**	NS	(NS)	NS	NS	(NS)	NS	NS

† Mean of 3 animals; ‡ From analysis of variance; *, P < 0.05; **, P < 0.01; NS, not significant.

Sheep on Cu deficient intakes can have liver Cu values as low as 2-10 ppm (Marston *et al.*, 1948). Such low values may be necessary before fibre strength is reduced. Subsequent experiments using larger doses of ATM (4 and 5 mg/kg.d) have not caused much further reduction in liver Cu than the *c.* 30% decline observed in the present work, despite causing concentrations of liver Mo (> 200 ppm) which were much higher than those presently reported (44-65 ppm, Table 1). This suggests that only a portion of the total liver Cu (*c.* 30%) is liable to interference by ATM although the length of the dosing period and frequency of ATM dosing may influence this.

In contrast to Cu and Mo neither Fe nor Zn in the liver was affected by ATM dosing but with Fe there was a marked effect (P < 0.01) of intake (Table 1), the underfed sheep having greater concentrations of Fe than those fed *ad lib.* Thus, care should be taken in interpreting liver Fe levels for grazing sheep with no records of feed intake, such as those reported by Marston *et al.* (1948) and Lee (1956) in studies relating to copper deficiency. Some *ad lib* sheep before dosing also had more liver Mo than those on 300 g/d but the difference between intakes was not as striking as for Fe. There was no effect of either intake or ATM dosing on liver Zn, the mean value for all sheep before and after dosing being 137 ± 3 ppm.

TABLE 2. TCA-Soluble Plasma Contents (ppm plasma ± sem) at Two Feed Intake Levels of Mo, Cu and Zn After Injecting Sheep for 16 d with Thiomolybdate (ATM)

Feed Intake (g/d)	N	ATM (mg/kg.d) 0	1	2	‡Effect of: Intake	ATM
			Mo			
300	4	0.24 ± 0.12	0.99 ± 0.17	1.41 ± 0.25	NS	**
1200	4	0.18 ± 0.14	0.77 ± 0.20	2.18 ± 0.84		
			Cu			
300	4	1.02 ± 0.15	0.68 ± 0.04	0.70 ± 0.04	**	*
1200	4	1.14 ± 0.04	1.13 ± 0.16	0.94 ± 0.07		
			Zn			
300	4	0.99 ± 0.05	1.17 ± 0.09	1.07 ± 0.05	**	*
1200	4	1.15 ± 0.05	1.31 ± 0.08	1.51 ± 0.14		

‡ Legend - see Table 1. There were no interactions between intake and ATM.

In contrast to liver, TCA-soluble plasma levels of Zn (Table 2) were increased by ATM and were related ($r = 0.6$, $n = 24$) to Mo levels (Table 2) in this fraction. This rise in Zn may be attributed to the fall in Cu (Table 2) since these elements tend to be antagonistic. The increase in TCA-soluble plasma Zn after repeated injections of ATM, contrasts with the effect of a single i/v injection of ATM, which although causing a temporary reduction in TCA-soluble plasma Cu to almost zero levels does not result in a concomitant temporary rise in TCA-soluble Zn (A.J. Gordon, unpublished data). The effects of ATM on Mo, Cu and Zn were, as expected, most marked at the highest level of ATM but this was not statistically significant. Again, in contrast to the liver, the *ad lib* fed sheep, when compared to the underfed sheep had higher values for Cu and Zn but not Mo.

In conclusion, the reduction in the concentrations of Cu in both the liver and the TCA-soluble fraction of plasma is consistent with the hypothesis that the primary effect of ATM is to bind copper. The lack of effect of ATM on liver Zn suggests that the rise in the TCA-soluble plasma Zn concentration is probably secondary to the effects on Cu. The implications for wool growth of these alterations in Cu and Zn metabolism have yet to be elucidated.

REFERENCES

1. Bingley, J.B. 1963. J. Agric. and Food Chem. 11:130-131.

2. Black, J.L., and P.J. Reis, 1979, In "Physiological and Environmental Limitations to Wool Growth", Proceedings of a National Workshop, Leura, N.S.W., Australia (eds J.L. Black and P.J. Reis) (U.N.E. Publishing Unit) pp 267-294.

3. Dick, A.T. 1954. Aust. Vet. J. 30:196-202.

4. Dick, A.T., Dewey, D.W., and J.M. Gawthorne, 1975. J. Agric. Sci., Camb., 85:567-568.

5. Gordon, A.J., and J.B. Donnelly, 1979. Aust. J. Agric. Res. 30:949-963.

6. Lee, H.J. 1956. J. Agric. Sci. 47:218-224.

7. Lynch, L.J., and N.A. Michie, 1976. Textile Res. J. 46:653-660.

8. Marston, H.R. 1946. In Proceedings of a Symposium on Fibrous Proteins (Society of Dyers and Colorists, Leeds) pp 207-214.

9. Marston, H.R., Lee, H.J., and I.W. McDonald, 1948. J. Agric. Sci. 38:222-228.

10. Palmer, R.C. 1949. J. Agric. Sci. 39:265-273.

DISCUSSION

GAWTHORNE (Australia). We gave black sheep tetrathiomolybdate in the drinking water and found that in contrast to intravenous administration it was difficult to induce white bands in the wool. A slow greying of the wool occurred. Have you tried oral administration of thiomolybdates?

GORDON (Australia). I have not given thiomolybdates in food or water because I was concerned about oxidation. When these compounds are exposed to air significant colour changes occur although in solution there is no change in the absorption spectrum. I always make up a fresh solution for each injection and keep the solutions anaerobic during infusions.

THE EFFECT OF COPPER AND MOLYBDENUM ON ^{75}Se-SELENOMETHIONINE METABOLISM IN SHEEP

C.L. White[1], W.G. Hoekstra and A.L. Pope
School of Agriculture and Life Sciences, University of Wisconsin,
MADISON 53706

[1] Currently, CSIRO Division of Animal Production,
PRIVATE BAG, P.O., WEMBLEY, WESTERN AUSTRALIA 6014
AUSTRALIA

High levels of copper have induced selenium-responsive diseases in chickens (Jensen, 1975) and rats (Godwin et al., 1978). Jensen suggested that copper may form insoluble complexes with selenite in the gut thereby reducing selenium availability. However, recent studies with copper-loaded sheep have suggested that copper supplements lead to an increase in apparent selenium retention (White et al. (1979), Gooneratne and Howell (1981)). The following experiment was designed to investigate the effect of copper supplements on selenium absorption, retention and tissue distribution in sheep.

Copper absorption is influenced by sulphur and molybdenum intakes through the formation of copper thiomolybdates (Dick et al. 1975). Selenium can also react with molybdenum to form selenomolybdates, compounds analogous to those formed with sulphur (Mueller and Diemann, 1968). A further aim of this experiment was to investigate the effect of combinations of dietary copper and molybdenum on selenium metabolism in sheep.

MATERIALS AND METHODS

Twelve Hampshire-Suffolk ram lambs were divided at weaning into four groups of three and offered diets containing either; 3.9 mg Cu and 1.2 mg Mo per kg diet (basal); basal plus 10 mg Cu/kg (Cu); basal plus 10 mg Mo/kg (Mo); or basal plus 10 mg Cu and 10 mg Mo/kg (Cu+Mo). The basal diet contained 1.6 g S/kg and 0.07 mg Se/kg, and copper and molybdenum were added to this diet as $CuSO_4.5H_2O$ and Na_2MoO_4, respectively, in the amounts described above. The lambs were offered 883 g dry diet per head per day which was all consumed after a 7 day adjustment period. Selenium was administered in gelatin capsules containing 0.17 mg DL-selenomethionine and $216x10^6$ dpm ^{75}Se L-selenomethionine (specific activity >3.3 μ Ci/ng).

Urine and faeces samples were collected over 7 days, and blood samples over 13 days, after which the animals were slaughtered and tissues were weighed and homogenized (polytron) for counting and analysis for selenium (Hoffman et al., 1968), copper (Heckman, 1971) and glutathione peroxidase (GSHpx) (Paglia and Valentine, 1967)).

RESULTS AND DISCUSSION

The level of copper or molybdenum had no significant effect on ^{75}Se excretion or retention. Seven day cumulative excretion in faeces accounted for between 27 and 35% of the dose (Figure 1A), while urinary excretion accounted for 8 to 12% (Figure 1B). Total retention (mean±SEM) after 7 days (% of dose) was 63.8±0.5 for basal, 53.4±7.3 for Cu, 58.1±7.7 for Mo and 56.3±3.7 for Cu+Mo. Differences were not significant; hence it is unlikely that copper or molybdenum at the levels used formed seleno complexes in sufficient quantities in the gut to significantly alter selenium availability.

Percent dry matter digestibility was not affected by dietary treatment and ranged from 77.2±1.8 to 80.2±0.2 for the Mo and basal sheep, respectively. Haematocrit values ranged from a minimum of 31±2.5 to a maximum of 36±0.7 for the Mo and Cu+Mo sheep, respectively. Liver copper concentrations (mg/kg fresh weight) were 77±11 (basal), 256±7(Cu), 69±2 (Mo) and 316±33 (Cu+Mo).

Blood ^{75}Se reached a plateau 24 hours after dosing, and remained at this level for 13 days (Figure 1C). Plasma activity peaked at 24 hrs after dosing, but subsequently declined over the 13 day measurement period. Packed cells, continued to accumulate ^{75}Se and this accounted for the whole blood plateau. Blood counts at 1, 7 and 13 days revealed no significant treatment effects when expressed on a total blood volume basis (calculated as 6% body weight).

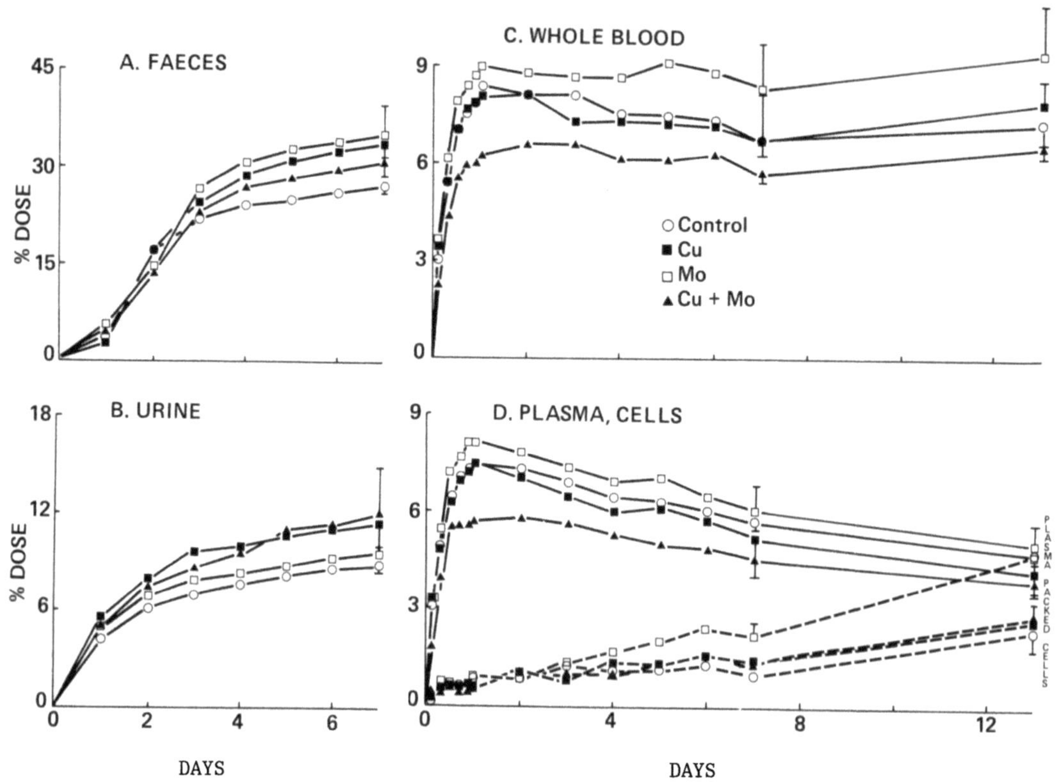

Figure 1. Appearance of ^{75}Se in faeces, urine, blood, plasma and packed cells. A,B, Cumulative excretion of ^{75}Se in faeces and urine, respectively. C, appearance of ^{75}Se in whole blood. D, appearance of ^{75}Se in plasma and packed cells. Results are expressed as a percentage of the dose, and blood volume was estimated as 6% liveweight.

The kidney contained the highest concentration of ^{75}Se and Se, while heart muscle contained the highest GSHpx activity (Figure 2). Testes contained higher GSHpx activities than liver or kidney, but had comparable concentrations of ^{75}Se to that of liver. It is apparent that the relative proportions of ^{75}Se to Se and GSHpx reflect the properties of the tissue in terms of whether selenium is being stored or utilized in forms other than GSHpx. This close similarity between GSHpx:Se and GSHpx to ^{75}Se in all tissues except heart indicate that ^{75}Se-selenomethionine was behaving in a manner closely analagous to that of endogenous selenium. For heart, it would seem that relatively more ^{75}Se from oral selenomethionine is incorporated into GSHpx within a given time, than Se from endogenous sources.

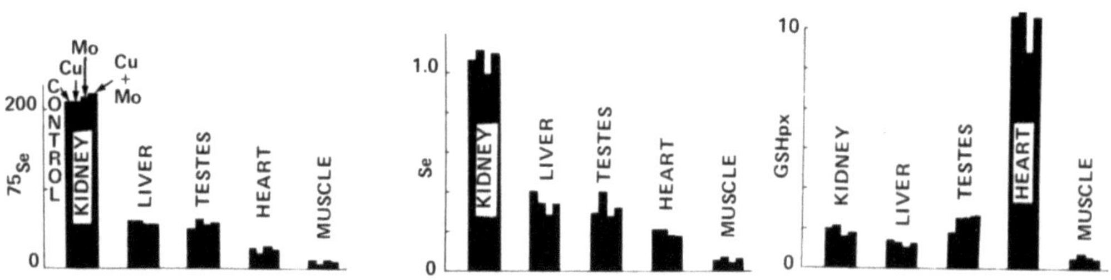

Figure 2. Comparative tissue levels of ^{75}Se (cpm/g fresh), Se (mg/kg fresh) and GSHpx (μM GSH oxid./min/g fresh weight).

SUMMARY

Increasing dietary copper did not reduce either apparent absorption or biological potency of selenium. With added molybdate, there was a trend toward higher levels of ^{75}Se in blood, but not in ^{75}Se apparent absorption or retention. Supplements of molybdenum and copper in combination had no effect on urinary or faecal excretion of ^{75}Se and it is unlikely that these trace metals are responsible for the unpredictable outbreaks of selenium-responsive white muscle disease.

ACKNOWLEDGEMENTS

The authors appreciate the technical assistance from Mr. J. Leiberman and Mr. T. Cadwalader, and the financial support for C.L.W. provided by the Wisconsin Alumini Research Fund.

LITERATURE CITED

1. Dick, A.T., D.W. Dewey and J.M. Gawthorne, 1975, J. Agric. Sci. Cam. 85:567 - 568.

2. Godwin, K.O., E.J. Partick and C.N. Fuss, 1978, In "Trace Element Metabolism in Man and Animals" (ed. M. Kirchgessner) pp 185-187.

3. Gooneratne, S.R., and J.McC. Howell, 1981, This volume.

4. Heckman, M., 1971, J. Assoc. Offic. Anal. Chem. 54:666.

5. Hoffman, J., R.J. Westerby and M. Hidiroglou, 1968. J. A. O.A.C. 51:1039 - 1042.

6. Jensen, L., 1975, Proc. Soc. Exp. Biol. Med. 149:113 - 116.

7. Mueller, A., and E. Diemann, 1968, Z. Naturforsch. B. 23:1605 - 6.

8. Paglia, D.E., and W.M. Valentine, 1967, J. Lab. Clin. Med. 70:156 - 169.

9. Underwood, E.J., 1976, In "Trace Elements in Human and Animal Nutrition" (4th Ed.). (Academic Press, New York) pp 324.

10. White, C.L., R.W. Cloninger, J.A. Hansen, W.G. Hoekstra and A.L. Pope, 1979, Proc. Nutr. Soc. Aust. 4:148.

DISCUSSION

MOIR (Australia). In view of the higher excretion of ^{75}Se in faeces compared with urine do you think that the selenomethionine passed through the rumen unchanged or was it converted to some other complex?

WHITE (Australia). We do not have data on the forms of selenium passing out of the rumen, but based on previous studies where we have fed selenomethionine, a very large proportion of it ends up in wool, so my own thoughts are that a large proportion is getting through the rumen as selenomethionine.

WEIGAND (Germany). You have said that you did not find significant differences between treatments. If you repeated the experiment with a larger number of animals perhaps the differences would be significant.

WHITE (Australia). Yes, there were large variations in two of the groups that influenced the degree of statistical significance. We repeated the experiment with ewes and lambs and we did not find the same results as we got in the first experiment. So at this stage I do not know what is happening. It is possible that there was a contamination of selenium which resulted in the high GSHpx and selenium values in the weaning result and by the time we did the ^{75}Se experiment these differences had disappeared. We analysed the diets for selenium and there did not appear to be any differences.

ZINC-INDUCED COPPER DEFICIENCY IN SWINE

G.M. Hill, P.K. Ku, E.R. Miller, D.E. Ullrey
H.D. Stowe, T.A. Losty and B.L. O'Dell

Michigan State University, EAST LANSING, MI
and
University of Missouri, COLUMBIA, MO
UNITED STATES OF AMERICA

INTRODUCTION

The antagonistic relationship of high levels of dietary zinc on tissue copper has been reported in several species. Because of the high concentration of tissue copper in the neonate compared to the adult pig, it is difficult to produce a copper deficiency with a semi-purified diet that produces normal growth with copper supplementation. Unpublished work from this laboratory showed that the liver of neonates from dams fed 5000 ppm zinc contained less than 4 ppm copper (wet basis). Thus, it was hypothesized that these neonates would become deficient in copper if their diet was not supplemented and would serve as useful models to study copper deficiency in the pig.

METHODS

Gilts weighing approximately 30 kg were fed a corn-soybean meal diet meeting all known dietary requirements and supplemented with 5000 ppm zinc from feed grade zinc oxide. They were managed in the same manner as other sows in the herd. When their second litter was farrowed, their pigs were allowed to nurse until 3 to 5 days of age when they were weaned and placed in individual stainless steel pens. All pigs were fed the basal diet (Table 1) and deionized-distilled water ad libitum for a one week adjustment period and were then allotted to dietary treatments based on sex and litter. Pigs were weighed and bled initially and weekly until termination of the experiment (5 weeks). The dietary treatments were 0, 5 and 10 ppm added copper as copper sulfate.

A one way analysis of variance and the Bonferroni t test were used to find differences between treatment means.

RESULTS AND DISCUSSION

After 14 days, pigs receiving the 0 ppm supplemented diet weighed significantly less ($P<0.5$) and had reduced hemoglobin, hematocrit, and serum copper concentrations and no detectable caeruloplasmin activity (Table 2). These trends continued throughout the study. There were no statistically significant differences between the values observed for the pigs receiving 5 or 10 ppm copper. Three of the five pigs receiving the unsupplemented diet died. All pigs in this group appeared unthrifty, had an abnormal stance and hyperventilated when handled. Similar clinical signs have been previously described (1,2,3).

At the end of the study, the pigs were killed and tissues collected. Lysyl oxidase assays were performed using soluble elastin, insoluble elastin and soluble collagen preparations from the aorta as substrates. The activities were calculated on the basis of dpm released per g of extracted protein and dpm released per mg of aortic tissue. The effect of the dietary levels of copper on lysyl oxidase activity was most apparent with the soluble elastin substrate (Table 3). The activity of the unsupplemented group was 16.4% of that of the 5 ppm treatment group ($P<.05$). A depression of lysyl oxidase activity in the copper-deficient chick has been previously reported (4,5).

Cytochromes a and a_3 are together called cytochrome C oxidase, and the copper that it contains is believed to transfer electrons from the a_3 heme to oxygen (6). Cytochrome C oxidase activity was depressed significantly in the heart ($P<.01$) and liver ($P<.05$) of pigs from 0 ppm treatment (Table 4). The depression of this enzyme and the reduced hemoglobin of the unsupplemented group may account for the hyperventilation which occurred when the animals were stressed.

The copper concentration in the liver, heart, kidney and pancreas (Table 5) was significantly higher in animals receiving either 5 or 10 ppm supplementary copper, but liver stores appeared to be somewhat lower than in normally-fed pigs. Zinc concentrations in the heart and kidney of the unsupplemented pigs (Table 4) were reduced ($P<.05$).

Assessment of the parameters measured in this study indicates that it is possible to produce a markedly copper-deficient pig using the offspring of sows fed 5000 ppm zinc. These parameters include previously unreported aortic lysyl oxidase and hepatic and cardiac cytochrome C oxidase activities, which are much reduced in the copper-deficient pigs and which support our previous conclusion (7) that the dietary copper requirement of the pig is about 6 ppm.

LITERATURE CITED

1. Teague, H.S. and L.E. Carpenter. 1951. J. Nutr. 43:389-399.
2. Lahey, M.E., C.J. Gubler, M.S. Chase, G.E. Cartwright and M.M. Wintrobe. 1952. Blood, J. Hematol. 7:1053-1074.
3. Myers, Gretchen. 1966. "A Copper Deficiency in Young Growing Pig". M.S. Thesis, Purdue University, W. Lafayette, IN.
4. Siegel, R.C., S.R. Pinnell and G.R. Martin. 1970. Biochem. 9:4486-4492.
5. Harris, E.D., W.A. Gonnerman, J.E. Savage and B.L. O'Dell. 1974. Biochim Biophys. Acta 341:332-344.
6. Lehninger, A.L. 1977. "Biochemistry". Worth Publishers Inc., NY.
7. Okonkwo, A.C., P.K. Ku, E.R. Miller, K.K. Keahey and D.E. Ullrey. 1979. J. Nutr. 109:939-948.

Table 1. Basal diet

Ingredient	% of diet
Dried skimmed milk	60.0
Glucose	15.4
Corn starch	11.2
Corn oil	5.0
α-Cellulose	3.0
Salt, iodized	.5
Trace mineral premix[a]	.2
$CaHPO_4 \cdot 2H_2O$	.8
Vitamin premix[b]	1.1
Selenium premix[c]	1.4
Chromium premix[d]	1.4

[a]Contained 4.4% ZnO, 37.3% $FeSO_4 \cdot 7H_2O$, .62% $MnSO_4 \cdot H_2O$, 3% $CoCO_3$, 40.2% $MgCO_3$, 14.5% corn starch.

[b]Supplied per kg of diet: 110 mg ascorbic acid, .22 mg biotin, 22 mg Ca pantothenate, 1000 mg choline chloride, 4.4 mg folic acid, 8.8 mg para-amino benzoic acid, 3.3 mg pyridoxine HCl, 8.8 mg riboflavin, 2.2 mg 2-methyl-1, 4-naphthoquinone, 110 mg myo-inositol, 22 IU all-rac-α-tocopheryl acetate, 33 mg nicotinic acid, 6.6 mg thiamin HCl, 4400 IU retinyl palmitate, 440 IU cholecalciferol, 20 μg cyanocobalamin.

[c]Provided .1 mg selenium (from Na_2SeO_3) per kg diet.

[d]Provided 1 mg chromium (from $CrCl_3 \cdot 6H_2O$) per kg diet.

Table 2. Weight, caeruloplasmin activity and concentration of serum Cu, hemoglobin and hematocrit in pigs

Days on Test	Diet Cu, ppm		
	0	5	10
Weight, kg			
0	2.62	2.69	2.47
14	4.81[a]	6.56[b]	6.01[ab]
35	8.40[a]	16.90[b]	15.54[b]
Serum Cu, g/dl			
0	12	11	9
35	11[a]	213[b]	190[b]
Caeruloplasmin, ΔOD/min/ml			
0	.004	.005	.008
35	.000[a]	.499[b]	.474[b]
Hemoglobin, g/dl			
0	8.9	9.6	9.1
35	5.5[a]	12.2[b]	12.0[b]
Hematocrit, %			
0	28.6	30.5	28.4
35	21.2[a]	37.9[b]	37.9[b]

[ab]Difference superscripts on the same line, designate a significant difference ($P<.05$).

Table 3. Activity of aorta lysyl oxidase of pigs

Substrate	Diet,Cu ppm		
	0	5	10
	dpm/μg extracted protein		
Insoluble elastin	7.2[a]	19.9[b]	14.8[ab]
Soluble elastin	16.9[a]	118.6[b]	93.3[b]
Soluble collagen	2.0[a]	5.2[b]	4.0[b]
		dpm/mg tissue	
Insoluble elastin	78.2	62.4	63.8
Soluble elastin	12.9[a]	418.2[b]	330.9[b]
Soluble collagen	11.2	16.1	15.1

[ab]See footnote, Table 2.

Table 4. Cytochrome C oxidase in heart and liver of pigs (ΔOD/min/mg protein)

	Diet Cu, ppm		
Tissue	0	5	10
Heart	.53[a]	3.81[b]	3.72[b]
Liver	.35[a]	.91[b]	.95[b]

[ab]See footnote, Table 2.

Table 5. Copper and zinc concentration in the liver, heart, kidney and pancreas (ppm, wet)

	Diet Cu, ppm					
Tissue	0		5		10	
	Cu	Zn	Cu	Zn	Cu	Zn
Liver	.28[a]	73	15.90[b]	127	15.88[b]	103
Heart	.17[a]	8[a]	2.59[b]	11[ab]	2.48[b]	12[b]
Kidney	1.95[a]	14[a]	5.31[b]	19[b]	3.98[ab]	15[a]
Pancreas	.33[a]	35	1.07[b]	45	1.21[c]	40

[abc]See footnote, Table 2.

DISCUSSION

SCHWARZ (Germany). Were there any fertility problems in the sows fed the high zinc diet?

ULLREY (USA). Not until the second pregnancy when sows fed the high zinc diet produced fewer pigs per litter.

SCHWARZ (Germany). What copper concentration would you recommend for practical diets?

ULLREY (USA). Our evidence suggests that 10 ppm Cu should be adequate for reproducing sows given practical diets.

EFFECT OF DIETARY ZINC AND CADMIUM ON SELENIUM AND CADMIUM AVAILABILITY TO RATS FED LETTUCE LEAVES WITH VARYING SELENIUM LEVELS

Ross M. Welch and William A. House
U.S. Department of Agriculture
U.S. Plant, Soil and Nutrition Laboratory
ITHACA, NEW YORK 14853 U.S.A.

Previously (1,2), we assessed the bioavailability of Cd and Se in intrinsically labelled lettuce leaves fed in single meals to Zn-deficient rats. We now report the absorption, retention and biological half-life of ^{109}Cd and ^{75}Se in Zn-depleted and in Zn-adequate rats fed labelled lettuce leaves that differed in Se content. The leaves were fed in single meals to rats maintained on diets with and without supplemental Cd. Dietary Cd levels were varied because preconditioning of animals to Cd reportedly affects Cd metabolism (3).

METHODS

Lettuce plants (*Lactuca sativa*) were grown in nutrient solutions that contained 0.113 μg Cd/ml as ^{109}Cd labelled $CdSO_4$ and either 0.022 or 0.109 μg Se/ml as ^{75}Se labelled Na_2SeO_3; Cd level was reduced 50% after 47 days of plant growth. Additional plants were grown without added Cd or Se. Leaves harvested from plants grown for 68 days were processed and analyzed (2) for total and radioactive Cd and Se. Respective concentrations of Cd and Se in leaves averaged 47.5 and 1.4 μg/g dry wt in plants (n=8) grown in the low-Se solution, and 62.4 and 7.0 μg/g dry wt in plants (n=8) grown in the high-Se solution. Cd and Se levels in plants grown without added Cd or Se were <0.05 and 0.01 μg/g dry wt, respectively. The leaves were incorporated into meals fed to male, Sprague-Dawley rats.

Weanling rats (n=48) were purchased and were fed a commercial diet (2) for five days. All rats were housed individually (4) and had free access to deionized water. The rats were then allotted to four groups of 12 each. During a 13-day adaptation period, the rats were fed a basal diet (5) supplemented with either 0 or 37 μg Zn/g of diet as $ZnCO_3$ and either 0 or 10 μg Cd/g of diet as $CdSO_4$. On day 14, the rats were fasted for 16 hours, and rats in each group of 12 were allotted by weight to two groups of six each. The average body weight of Zn-depleted rats was 74 g, and that of Zn-supplemented rats was 147 grams. Each of the eight groups of rats was assigned to a different treatment (table 1). Treatments consisted of feeding meals that contained 1.0 g of labelled lettuce leaves mixed with 1.0 g of diet and 0.5 g of sucrose. Each treatment contained the appropriate diet fed during the adaptation period. Lettuce leaves used in the meals provided about 34 μg of Cd and about either 1 or 6 μg of Se; labeled and unlabelled lettuce leaves were mixed to adjust the levels of Cd and Se.

Table 1. Dietary levels of Zn, Cd, and Se, and amounts of Zn, Cd and Se consumed by rats fed single meals that contained intrinsically labeled lettuce leaves

Treatment number	Dietary levels			Consumed in meal[1,2]		
	Zn	Cd	Se	Zn	Cd	Se
		μg/g			μg	
1	<0.5	<0.05	0.6	33.3[a]	33.9[a]	1.7[a]
2	<0.5	10.2	0.6	32.5[a]	43.0[b]	1.7[a]
3	<0.5	<0.05	0.6	32.9[a]	31.7[a]	6.4[b]
4	<0.5	10.2	0.6	32.6[a]	40.4[b]	6.4[b]
5	37.3	<0.05	0.6	69.9[b]	33.6[a]	1.7[a]
6	37.3	10.2	0.6	68.7[b]	43.7[b]	1.7[a]
7	37.3	<0.05	0.6	72.7[b]	30.8[a]	6.5[b]
8	37.3	10.2	0.6	73.3[b]	40.5[b]	6.5[b]
Pooled standard error of mean				1.3	0.3	0.03

[1]Values are means (n=6); means in same column not followed by a common letter differ significantly (P<0.05).
[2]Includes amount contributed by diet and lettuce leaves.

The meals were offered to the rats for four hours, and then were replaced with the diets fed during the adaptation period. Table 1 shows the amounts of Zn, Cd and Se consumed by the rats when fed the radioactively labelled meals. Activities of ^{109}Cd and ^{75}Se in the rats were determined (2) simultaneously immediately after the feeding period and at about 24-hour intervals thereafter for 13 days. Absorption, retention and biological half-life of ^{109}Cd and ^{75}Se were determined (6). Data were evaluated by analysis of variance, and statements of significance are based on a probability level of 95%.

RESULTS AND DISCUSSION

Dietary Zn inhibited absorption and retention of Cd by rats fed low-Se meals (table 2). Accordingly, Zn-adequate rats absorbed only about one-third as much ^{109}Cd as Zn-deficient rats when the Se level in the meal was low. We reported previously (1) that increasing Zn levels in lettuce leaves fed in single meals to Zn-deficient rats depressed the absorption of ^{109}Cd from those meals. Others reported inhibition of Cd absorption by Zn in experiments with Japanese quail (7), in in vitro studies using rat intestine (8), and in studies using rat hepatocytes (9).

Absorption of ^{109}Cd by Zn-depleted rats fed lettuce leaves increased from 3.2 to 8.3% of the dose in response to increased Se levels in the leaves (2). Similarly, our present data (table 2) show that both Zn-depleted and Zn-adequate rats absorbed and retained more ^{109}Cd from high-Se than from low-Se lettuce; the response to Se was greater in Zn-adequate than in Zn-depleted rats. Moreover, there was a significant interaction between Se and Zn on ^{109}Cd absorption; consequently, Zn supplementation reduced Cd absorption in rats fed low-Se meals but not in rats fed high-Se meals.

We suggest that some Cd in low-Se lettuce was present as inorganic Cd^{2+} ions, and that increased levels of Se in the leaves affected either the form or distribution of Cd in the leaf tissues. Therefore, Cd in high-Se lettuce may be present as a Se-Cd organic complex. The latter speculation stems from the fact that Cd has an affinity for sulfhydryl groups and because Se often follows the biological pathway of sulfur. Dietary Zn could reduce the availability to rats of Cd in low-Se lettuce if inorganic Zn and Cd compete for absorption sites (mechanisms) in the alimentary tract. On the other hand, dietary Zn may not inhibit Cd absorption from high-Se lettuce because a portion of the Cd in the leaf tissue may be present as a highly available Se-Cd complex that does not compete with Zn.

Metallothionein (MTN) is involved in metabolism of both Zn and Cd (10). The increased biological half-life of ^{109}Cd in Zn-supplemented rats compared to Zn-depleted rats (table 2) may be related to synthesis of MTN in intestinal mucosa. Incorporation of ^{109}Cd into MTN in intestinal mucosa is lower in Zn-depleted than in Zn-adequate rats (11).

Initially, as indicated by others (10), we assumed that increased intake of dietary Cd would induce MTN synthesis in the rats, and would, therefore, affect ^{109}Cd metabolism. Contrary to our expectations, supplementary dietary Cd did not affect absorption, retention or biological half-life of ^{109}Cd in rats fed intrinsically labelled lettuce leaves (table 2). Possibly, the low bioavailability of $CdSO_4$(1) coupled with a moderate dietary Cd level fed for a relatively short period of time resulted in too little absorption of Cd to markedly affect MTN synthesis in the rats.

Retention and biological half-life of ^{75}Se in the rats increased with increased intake of Zn (table 2). Most of the Se in the lettuce leaves was probably in the form of organic complexes. Plants grown on selenate supplements are known to synthesize selenocysteine and incorporate this Se analog of cysteine into plant proteins (12). In rats, either increased incorporation of selenoamino acids into proteins or inhibition of Se excretion by Zn could account for the increased retention and biological half-life of radioselenium in rats fed supplementary Zn. Incorporation of selenocysteine and selenomethionine into skin protein of rats is considerably lower in Zn-deficient than in Zn-adequate rats (13). Additionally, if Zn stimulated MTN synthesis, ^{75}Se as selenocysteine may have been incorporated into MTN, a protein that has a high percentage of cysteine in its polypeptide chains.

In summary, availability to rats of ^{109}Cd and ^{75}Se in intrinsically labelled lettuce was assessed by a "whole body" radioassay method. Dietary Zn decreased Cd availability when Se levels were low; however, Se stimulated Cd absorption, and a high intake of Se nullified the protective effect of Zn on Cd availability. We speculated that the forms of Cd and Se in lettuce leaves influenced the availability of these elements to rats.

Table 2. Absorption, retention and biological half-life of radiocadmium and radioselenium in rats fed meals that contained intrinsically labelled lettuce leaves[1]

Treatment				Absorption		Retention[2]		Biological Half-life	
No.	Zn	Cd	Se	^{109}Cd	^{75}Se	^{109}Cd	^{75}Se	^{109}Cd	^{75}Se
				% of dose		% of dose		hours	
1	-	-	-	5.0[a]	48.7[a]	1.2[a]	15.4[c]	170[ab]	181[a]
2	-	+	-	4.7[a]	46.1[a]	1.3[a]	14.8[bc]	171[bc]	184[a]
3	-	-	+	7.9[b]	40.2[b]	2.2[b]	13.4[ab]	169[ab]	189[a]
4	-	+	+	8.0[b]	38.1[b]	2.1[b]	12.6[a]	140[b]	188[a]
5	+	-	-	1.6[c]	45.5[a]	0.5[c]	17.1[de]	201[ab]	213[b]
6	+	+	-	1.7[c]	47.6[a]	0.5[c]	18.3[e]	183[ab]	215[b]
7	+	-	+	8.1[a]	40.4[b]	2.8[d]	16.0[cd]	215[a]	225[bc]
8	+	+	+	8.3[a]	37.3[b]	2.7[d]	15.1[c]	221[a]	229[c]
Standard error of mean				0.5	1.5	0.1	0.5	19	4

[1]Values are means (n=6); means in same column not followed by a common letter differ significantly ($P<0.05$).

[2]Retention at 306 hours after eating radioactively labelled meal.

REFERENCES

1. Welch, R.M., W.A. House, and D.R. Van Campen, 1978, Nutr.Rep.Int. 17:35-42.

2. Welch, R.M., and W.A. House, 1980, Nutr.Rep.Int. 21:135-145.

3. Parizek, J. 1975. In "Physiological and Biochemical Aspects of Heavy Elements in Our Environment" (eds. J.P.W. Houtman and C.J.A. van der Hamer) (Delft University Press) pp 37-48.

4. Welch, R.M., W.A. House, and W.H. Allaway, 1974, J. Nutr. 104:733-740.

5. Welch, R.M., W.A. House, and D.R. Van Campen, 1977, J. Nutr. 107:929-933.

6. Van Campen, D.R., and W.A. House, 1974, J. Nutr. 104:84-90.

7. Jacobs, R.M., A.O. Lee Jones, M.N. Morra, and M.R. Spivey Fox, 1978, Fed. Proc. 37:405 (abstract).

8. Kingsley, B.S., and J.M. Frazier, 1979, Am. J. Physiol. 236:C139-C143.

9. Lyall, V., A. Mahmood, and R. Nath, 1979, Indian J. Biochem. Biophys. 16:80-83.

10. Bremner, I. 1978, World Rev. Nutr. Diet. 32:165-197.

11. Hahn, C.J., and G.W. Evans, 1975, Am. J. Physiol. 228:1020-1023.

12. Brown, T.A., and A. Shrift, 1980, Plant Physiol. 66:758-761.

13. McConnell, K.P., J.M. Hsu, J.L. Herrman, and W.L. Anthony, 1974, Proc. Soc. Exp. Biol. Med. 145:970-974.

DISCUSSION

SCHWARZ (Germany). What was the selenium concentration in the basal diet?

WELCH (USA). The basal diet contained about 0.5 ppm Se. The lettuce leaves were given in addition to this diet to provide an extra 1 μg or 6 μg Se as a single meal in the low and high treatments respectively.

SUTTLE (Scotland). Your figures for selenium absorption of 40 - 45% are lower than what we are accustomed to in studies with selenite. Would you care to comment?

WELCH (USA). We find a similar percentage absorption for sodium selenite in our system. The lettuce leaf selenium is slightly more available, but not by much.

BEHNE (Germany). For what period of time after administration of ^{75}Se did you follow the biological half-life?

WELCH (USA). The radioactivity in the rats was counted each day for 13 days.

SUNDE (Scotland). What was the molecular weight of the selenium-cadmium protein you discovered during gel chromatography?

WELCH (USA). We initially thought that the MW was between 5,000 - 10,000 daltons, but we later discovered that the selenocysteine-cadmium complex also chromatographs in this position.

ZINC AND VITAMIN A INTERRELATIONSHIP IN METABOLISM

S.P.ARORA

DIVISION OF DAIRY CATTLE NUTRITION AND PHYSIOLOGY
NATIONAL DAIRY RESEARCH INSTITUTE
KARNAL,HARYANA, 132001
INDIA

INTRODUCTION

Dietary necessity of zinc has been well demonstrated in different species of livestock and human beings because of its widespread deficiency in feeds. Zinc has been recognized as a cofactor or constituent of several enzymes and alcohol dehydrogenase is one to which it is linked alongwith diphosphopyridine nucleotide (1). Skin keratinization has been reported as a result of dietary zinc deficiency which appeared to be similar to hyperkeratosis recorded due to vitamin A deficiency. Therefore certain experiments were conducted to study the influence of zinc on the tissue utilization of vitamin A.

INDUCTION OF NIGHT BLINDNESS SYNDROMES

In a preliminary study, thirty-two growing lambs were randomly distributed into two groups. The lambs of group 1 were fed 1 kg basal diet per head per day containing 23 ppm zinc and 10,000 IU vitamin A; and those in group 2 were supplemented 48 ppm zinc over and above the level in 1 kg basal diet. On day 24, two lambs from group 1 exhibited symptoms of night blindness. They were changed over to the zinc supplemented diet and a single dose of 1,000,000 IU vitamin A was injected intramuscularly. Levels of vitamin A in blood sera were estimated from time to time. After a period of 17 days, lamb No.5817 showed recovery and its serum vitamin A level was 40.0 µg as compared to 23.3 µg per 100 ml earlier. Similarly lamb No.5814 showed a maximum level of 26.0 µg per 100 ml during sickness and recovered after a period of 27 days. In a subsequent experiment, these 2 lambs with 2 other lambs were fed a basal diet containing 23 ppm zinc and 10,000 IU of vitamin A per kg of ration. Another set of 4 lambs in group II were fed the same basal diet plus vitamin A and supplemented with 50 ppm zinc. The blood serum vitamin A level was significantly less in group I (37.1 µg/100 ml) as compared to the level (48.5 µg/100 ml) in group II (Fig.1)(2).

SKIN KERATINIZATION AND VITAMIN A LEVELS

In an another experiment, ten lambs were depleted of zinc and vitamin A for 40 days and divided into two groups. The lambs in group 1 were fed a basal diet consisting of concentrate mixture containing 25 ppm and hay containing 30 ppm zinc. The lambs in group 2 were supplemented in addition with 100 mg zinc. All the five lambs of low zinc group showed the symptoms of alopecia and parakeratosis after 4 weeks which was rated in severity from 2 to 4 out of a total score of 5. The lambs were treated with 100 mg $ZnCl_2$ in solution form and they recovered in about another 4 weeks (3). Later after their recovery, vitamin A was administered to each lamb in two trials as shown in Table 1. The trials were initiated when the mean serum vitamin A values dropped below 20 µg per

FIG. 1

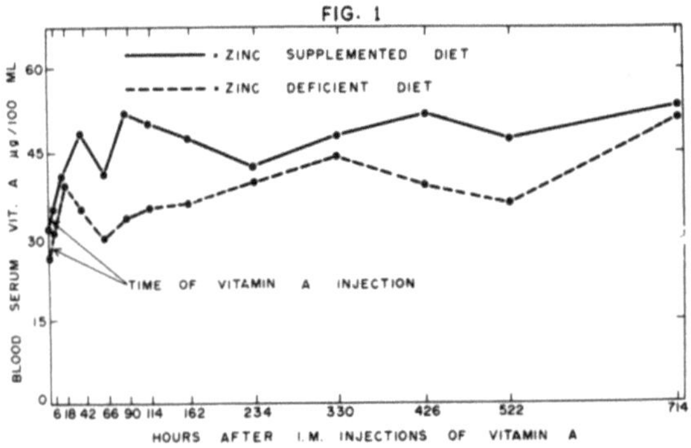

100 ml in each group. Data from both the trials demonstrated that zinc supplementation had significantly increased the vitamin A pool size in the blood serum (3).

In an investigation, 4 groups of piglets suffering from parakeratosis were given different lines of therapeutic treatments except one control group. The piglets of group 1 given oral dose of zinc treatment @ 50 mg per day showed recovery in 18 days time. The animals of group 2 treated with a single injection of 100,000 IU vitamin A plus 50 mg oral zinc daily per animal also recovered within the same period. The piglets of group 3 given vitamin B complex treatment did not show resolution as in control group 4(4).

ALCOHOL DEHYDROGENASE ACTIVITY AND RETINOL BINDING PROTEIN(RBP)STATUS

To understand further the relationship between zinc and its effect on vitamin A utilization,alcohol dehydrogenase activity was estimated in the small intestine,liver and retina of rats fed zinc deficient diet and then compared with rats fed zinc-supplemented diet. Thirteen albino rats in group1 received semi-purified diet containing 6 ppm zinc and those in group 2 received 46 ppm. After 4 weeks of experimental feeding, the rats were sacrificed and their intestinal, liver and retinal tissues were dissected for estimation of alcohol dehydrogenase activity. The activity was 2.75 times less in the retina, 1.26 times less in the intestine and 1.59 times less in the liver of group 1 than in group 2 (Table 2). Within the same individual, the enzyme activity decrease was not uniform in all the different tissues which conformed to the occurrence of lesions shown either as night blindness or skin keratinization (5).

An experiment was also conducted to estimate RBP in blood serum of 42 rats on zinc deficient and zinc supplemented diets. The rats in group 1 were fed conventional diet containing 19 ppm zinc and the other groups 2 and 3 received a purified diet containing 5 ppm zinc for a period of 5 weeks. When the rats started showing the symptoms of alopecia in the latter two groups, the rats in group 3 were given 45 ppm zinc in the diet. Two weeks after feeding, the rats were sacrificed. The mean values of RBP as shown in Table 3 were significantly different ($P \angle .05$) between groups (6).

DIETARY ZINC LEVEL AND B-CAROTENE CONVERSION

This study was undertaken to investigate the requirement of dietary zinc which would be necessary to convert β-carotene to vitamin A and latter's mobilization from liver into blood. Nine bucks divided into 3 groups were zinc and vitamin A depleted by feeding a purified diet containing 3 ppm zinc and wheat straw containing 16 ppm zinc, till the vitamin A levels were less than 10 µg per 100 ml serum. A 100 mg dose of β-carotene was administered orally to each animal for 10 days and then stopped, but zinc supplementation at the levels of zero, 20 and 40 ppm were continued in groups 1, 2 and 3 respectively upto 25 days. Blood samples were collected on days 2,4,6,8,10,15,18 and 25 for vitamin A analysis in serum. Serum vitamin A levels increased significantly in group 3 as compared to groups 1 and 2 showing thereby that dietary zinc requirement might be more than 50 ppm in goats (Table 4)(7).

A similar study undertaken in calves, given zero, 25 and 50 ppm zinc over the basal level in groups 1 through 3 and daily dose of 300 mg β-carotene for 10 days, revealed that there was a significant increase in the level of serum vitamin A in calves of group 3 than in groups 1 and 2(Table 5). Concommitantly β-carotene level was less in group 3 than groups 1 and 2 (8).

TABLE 1

VITAMIN A LEVELS IN BLOOD SERUM OF LAMBS (µg/100 ml SERUM)

VIT.A DOSE	BEFORE ADMIN.		DAY 2		DAY 4		DAY 10	
	CNL	+Zn	CNL	+Zn	CNL	+Zn	CNL	+Zn
Single injection (300,000 IU)	15.98	15.23	34.88	55.50	14.60	25.85	18.75	26.75
Single injection (300,000 IU)+ 2000 IU Daily	15.85	13.75	20.88	58.75	47.63	77.06	23.63	30.13

TABLE 2

ALCOHOL DEHYDROGENASE ACTIVITY (UNITS/mg PROTEIN/MIN.)

NO. OF RATS	RETINA		INTESTINE		LIVER	
	CNL	Zn	CNL	Zn	CNL	Zn
13	0.43	1.14	0.096	0.121	0.125	0.199
RATIO	2.75		1.26		1.59	

TABLE 3

RETINOL BINDING PROTEIN (μg/ml SERUM) IN RATS

GROUP	I (19ppm Zn)	44.97
GROUP	II (5ppm Zn)	36.83
GROUP	III (45ppm Zn)	39.94

TABLE 4

VITAMIN A LEVELS IN GOATS (μg/100 ml SERUM)

Zn TREATED	INITIAL	DAY				
		2	6	10	18	25
CNL Gr.I	7.60	18.40	19.75	19.35	12.70	9.05
20 ppm Gr.II	8.55	18.65	21.75	23.25	13.70	8.50
40 ppm Gr.III	7.10	26.10	31.70	31.60	15.85	9.50

TABLE 5

BLOOD SERUM VITAMIN A LEVELS IN CALVES (μg/100 ml)

Zn TREATED	DAY						
	0	2	4	6	8	18	25
35PPM	17.24	27.72	29.60	34.96	35.84	24.52	23.84
60PPM	16.60	29.00	32.96	36.48	37.40	33.68	32.48
80PPM	16.88	33.76	39.76	43.36	48.28	37.60	35.44

CONCLUSIONS

Series of experiments conducted to elucidate the influence of zinc on various aspects of vitamin A metabolism in body tissues, reveal that dietary zinc around 80 ppm may be necessary for vitamin A mobilization and utilization. Further the results suggest that zinc and vitamin A therapy together may be necessary for treatment of deficiency syndromes (8).

-o-o-o-

REFERENCES

1. Vallee, B.L., 1959, Physiol. Rev. 39:443.
2. Arora, S.P., 1969, Tissue distribution of administered Zn-65 and possible effect of dietary zinc on vitamin A metabolism. Ph.D.Thesis, Univ. of Illi. Urbana.
3. Saraswat, R.C. and Arora, S.P., 1972, Indian J. Anim. Sci.,42:358-62.
4. Gupta, B.S., S.P.Arora, V.S.Richaria, S.K.Ranjhan, B.S.Dhodapkar, J.L.Vegad and J.H.Byers, 1972, Indian J. Anim. Hlth.,11:65-68.
5. Prasad, C.S. and S.P.Arora, 1978, Indian J. Anim. Sci.,48:582-84.
6. Prasad, C.S. and S.P.Arora, 1979, Indian J. Dairy Sci.,32:275-79.
7. Chhabra, A., S.P.Arora and Jai Kishan, 1980, Indian J. Anim. Sci.,50:879-81.
8. Chhabra, A. and S.P.Arora, 1981, Indian Sci. Congr. Proc.P.19-20.
9. Smith, J.C.Jr., E.G.McDaniel, F.F.Fan and J.A.Holsted,1973, Science, 181:954-55.

.....

INFLUENCE OF Zn STATUS ON Pb METABOLISM IN RATS

H.Schenkel, H.-J.Lantzsch, S.Scheuermann

Institute of Animal Nutrition
University of Hohenheim
PO BOX 700562, 7000 STUTTGART 70
FEDERAL REPUBLIC OF GERMANY

It is well known that moderate to higher Zn concentrations in the diet were able to alleviate some toxic effects of ingested lead (7,8). Low dietary intake of Zn has been found to increase Pb absorption in some experiments with rats (3, 4, 5, 7). In the present experiment, it was investigated, if there is also an effect of dietary Zn on parameters of Pb metabolism at a even lower dietary Pb concentration. Furthermore, the effect of a preceding Zn depletion was investigated.

MATERIAL AND METHODS

An experiment with 4 groups of growing male rats (Sprague Dawley, 6 animals per group) was carried out. The rats received semisynthetic diets based on ovalbumin and glucose (6) during 20 days. The diets of group I and III contained 6 mg Zn/kg and of group II and IV 14.9 mg Zn/kg (dry matter basis), respectively. All diets contained about 10 mg Pb/kg as Pb nitrate. The rats of group I and II were pretreated by a 10 day Zn depletion (0,7 mg Zn/kg without Pb supplementation) immediately before the beginning of the experiment. After an 8 day balance period the animals received an i.p. injection of physiological saline and 250 mg $CaNa_2EDTA$/kg body weight on the following day. Two days after $CaNa_2EDTA$ application the rats were killed and Zn and Pb content in liver, kidney, femur and remaining body was determined. The activity of delta-aminolevulinic acid dehydratase in red blood cells was also investigated.

RESULTS AND DISCUSSION

As could be seen in table 1, there was a significant effect of the pretreatment on apparent Pb absorption and Pb balance. Pb balance of animals in group I and II was only 47% and 45% of the values in group III and IV, respectively. The same effect could be shown with respect to the Pb content of whole body. There was also a tendency to higher Pb contents in the unpretreated groups. Furthermore, a significant effect of dietary Zn on Pb content in whole body could be found. The values for group I and III exceed those of group II and IV by 32% and 31%, respectively. However, marked differences in Pb concentrations of the organs investigated could be stated. In contrast to the Pb content of whole bodies in livers of pretreated animals higher amounts of Pb were found. The Pb concentration in kidneys of rats in group I and III were higher than in group II and IV, whereas no pretreatment effect could be seen. However, in femur the Pb concentration of the pretreated animals was only 61% and 46% of the unpretreated animals, respectively. Consequent to these findings a marked influence on Pb distribution in liver and femur by the pretreatment and in liver by the dietary zinc level could be stated. In the unpretreated animals a higher percentage of whole body Pb could be found in the femur. In liver of unpretreated animals the percentage was lower.

The activity of delta-aminolevulinic acid dehydratase in red blood cells depended as well on dietary Zn intake (1, 2, 5) as on the preceding Zn depletion. Compared with the enzyme activities of rats, which under the same experimental conditions received a diet with 15mg Zn/kg unsupplemented with Pb, enzyme values of rats of group I to IV reached only 48%, 56%, 45% and 70%, respectively (Schenkel et al., in preparation).

The increased urinary Pb excretion after i.p. $CaNa_2EDTA$ application paralleled with the whole body Pb content. This was in good agreement with the diagnostic and therapeutic use of this chelating agent in Pb toxicity (8).

So among the parameters investigated the best correlation could be found between urine Pb after $CaNa_2EDTA$ application and whole body Pb content ($r=0,711$, $p<0,001$). Because the different pretreatment effects on Pb concentration in liver and femur only the Pb concentration in kidney was significantly correlated with the whole body content in all experimental groups ($r= 0,635$, $p<0,01$). The negative correlation between the activity of delta-aminolevulinic acid dehydratase in red blood cells and whole body Pb was not statistically significant.

Table 1: Effect of different pretreatment (Zn depletion) and dietary Zn intake on parameters of Pb metabolism (mean ± standard deviation)

Group		I	II	III	IV	sign.effects ($p < 0,01$) pretreatment	dietary Zn	pretr x diet.Zn
Pretreatment		with	with	without	without			
dietary Zn (mg/kg)		6	15	6	15			
dietary Pb (mg/kg)		10	10	10	10			
Pb balance	(mcg/d)	8.37	11.33	18.05	25.18	xx		
		±8.85	±7.13	±8.29	±8.91			
	(%) [1]	7.11	8.97	14.85	19.68	xx		
		±7.50	±5.64	±6.83	±6.94			
apparent absorption	(%) [1]	9.66	11.42	16.74	21.53	xx		
		±7.45	±6.06	±7.04	±6.74			
whole body Pb	(mcg)	138.7	113.1	165.3	126.2		xx	
		±35.9	±14.7	±21.8	±27.8			
liver Pb	(mcg/g) [2]	0.44	0.61	0.36	0.31	xx		xx
		±0.03	±0.12	±0.09	±0.05			
	(%) [3]	2.32	3.52	1.54	1.78	xx	xx	
		±0.56	±0.73	±0.49	±0.19			
kidney Pb	(mcg/g) [2]	1.92	1.47	2.05	1.43		xx	
		±0.33	±0.10	±0.26	±0.24			
	(%) [3]	2.42	2.26	2.29	2.13			
		±0.60	±0.34	±0.35	±0.41			
Femur Pb	(mcg/g) [2]	15.86	17.00	34.23	27.57	xx		
		±4.83	±4.43	±5.60	±6.58			
	(%) [3]	7.41	9.50	14.41	15.34	xx		
		±2.07	±1.72	±2.68	±3.94			
d-ALAD	(U/l)	2.78	3.61	2.90	4.57		xx	
		±0.76	±0.73	±0.44	±2.99			
urine Pb after $CaNa_2EDTA$ appl.	(mcg/d)	21.59	15.61	25.45	19.40		xx	
		±4.22	±2.55	±5.25	±2.99			

1) in per cent of dietary intake
2) fresh weight
3) in percent of whole body content

SUMMARY

In an experiment with growing rats marked effects of dietary zinc intake and a preceding Zn depletionperiod on Pb concentrations in whole body, liver, kidney and femur and the activity of delta-aminolevulinic acid dehydratase in red blood cells could be stated. There was a close relationship between the whole body Pb burden and the urinary Pb excretion after an i.p. application of $CaNa_2EDTA$.

LITERATURE

1. Abdulla, M.,Svensson, S., Mathur, A., and K.Wallenius, 1978. In "Trace Element Metabolism in Man and Animals-3" (ed. M.Kirchgessner) (ATW, Weihenstephan) pp 64-67

2. Abdulla, M., Svensson, S., Haeger-Aronsen, B., Mathur, A.,and K.Wallenius, 1978, Enzyme 23:170-175

3. Cerklewski, F.L., and R.M.Forbes, 1976, J.Nutr. 106:689-696

4. El-Gazzar, R.M., Finelli, V.N., Boiano, J., and H.G.Petering, 1978, Toxicol. Letters 1:227-234

5. Finelli, V.N., Klauder, D.S., Karaffa, M.A., and H.G.Petering, 1975, Biochem. Biophys. Res. Commun. 65:303-311

6. Lantzsch, H.-J., and K.H.Menke, 1973, Zschr.Tierphysiol. Tierernährg.Futtermittelkde 32:129-143

7. "Mineral Tolerance of Domestic Animals" (ed. National Research Council) (National Academy Press, Washington) pp 256-276

8. Tsuchiya, K., 1979. In "Handbook on the Toxicology of Metals" (ed. L.Friberg, G.F.Nordberg and V.B.Vouk) (Elsevier, Amsterdam) pp. 451-484

DISCUSSION

ANON. Do you think that EDTA might have preferentially removed lead from some tissues compared with others?

SCHENKEL (Germany). The quantity of urinary lead excreted after $CaNa_2EDTA$ administration was nearly 10% of the body content. We cannot state that there was a preferential removal from some tissues because the animals were not killed at the time of collecting the urine, but two days after the EDTA administration.

^{64}Cu METABOLISM IN INFANT RATS FED MILK ENRICHED WITH ZINC, IRON AND COPPER

B. Momčilović

Institute for Medical Research and Occupational Health,
158 M.Pijade
Zagreb
Yugoslavia

Milk appears to be an attractive vehicle for trace element fortification because it is widely consumed by infants and children (1). Indeed, better indices of growth and haematological status were observed in infants fed milk enriched with either zinc, iron or copper (2-4). However, the effect of unilateral fortification is limited by both Liebig's Law of the minimum and Voisin's Law of the maximum (5). Therefore unilateral fortification of food products could be either uneconomical if overabundant elements are wasted or harmful if the body is not capable of excreting excessive mineral load or both (6). The aim of this experiment was to investigate the effect of milk enriched with physiological doses of zinc, iron and copper (7) added either separately or in combination, on ^{64}Cu metabolism in infant rats.

METHODS AND RESULTS

The experiment was performed on 88 six-day-old infant rats in litters of six. Each littermate was randomly assigned to one of the eight available (8) dietary treatments: A) milk (pasteurized cow's milk containing 3.5 µg Zn/ml, 0.5 µg Fe/ml and 0.05 µg Cu/ml (9); B) milk + Zn (12 µg Zn/ml from $ZnSO_4.7\ H_2O$); C) milk + Fe (25 µg Fe/ml from $FeSO_4.7\ H_2O$); D) milk + Cu (5 µg Cu/ml from $CuSO_4.5\ H_2O$); E) milk + Zn + Fe; F) milk + Zn + Cu; G) milk + Fe + Cu; and H) milk + Zn + Fe + Cu. The animals were artificially fed 0.45 ml of the respective diet, marked with 74 kBq 64-Cu/rat, with a dropper for a period of 8 h (10). The isotope was in an almost carrier-free form (Radiochemical Centre, Amersham, England). The radioactivity of 64-Cu in the whole body and liver, kidneys, brain and femur was determined 4 days later, in a single-channel twin crystal or a well-type scintillation count respectively (Nuclear Chicago, U.S.A.) (11). The data were classified simultaneously in two ways, according to the effect of copper, and other trace element supplementation (12). The significance of the difference among the groups was determined by Duncan's test (13).

Whole body and liver 64-Cu retention were not affected by dietary treatments (Table 1). However, significant distributional dose-dependent effects of copper on 64-Cu metabolism were observed. 64-Cu retention in the kidneys and the brain was lower in infant rats fed milk enriched with copper than in those fed diets to which copper was not added ($P<0.01$ for diets A,B,C,E vs diets D,F,G,H in both organs). An antagonistic distributional effect of zinc supplementation on 64-Cu deposition in the femur was observed. 64-Cu retention in the femur of zinc fed animals was lower than in animals fed milk without zinc ($P<0.05$ for diets A,C,D,G vs diets B,E,F,H). Whole body weight and body weight gain were not affected by the type of dietary treatment.

DISCUSSION

The observed high 64-Cu whole body retention in infants as compared to adult rats (14) revealed an age specific phenomenon most probably associated with both milk diet (15) and pinocytotic capacity of the gastrointestinal tract in infant animals (16). Moreover, 64-Cu absorption appears independent of the physiological level of stable copper in the diet indicating an inadequate homeostatic control of copper absorption at an early period of life. The observed overall distributional effect of copper vs non-copper containing diets in lowering 64-Cu retention in the brain and kidneys of infant rats could be of pharmacokinetic importance as it may be a sign of approaching a physiological threshold in the complex fortification of milk with trace elements. Moreover, the distributional effect of zinc vs non-zinc diets on 64-Cu deposition in the femur revealed a subtle antagonism between zinc and copper in the bone which is a target for zinc deposition (17). Zinc may alter copper metabolism by competing for various binding sites within a specific metabolic system, including intestinal absorption (18). However, this effect appears to be dose dependent as Murthy and co-workers (19) found no interaction between zinc and

TABLE 1.

The effect of milk fortified with either zinc, iron, or copper on 64-Cu retention and tissue distribution in infant rats.

	Milk**	^{64}Cu (% dose)* Milk+Zn***	Milk+Fe****	Milk+Zn+Fe
Whole body				
Control	74.9±2.6	74.0±3.0	75.8±2.7	72.2±1.8
*****Copper	73.8±3.1	69.5±0.8	71.2±1.5	76.6±3.5
Liver				
Control	27.6±3.1	26.6±2.6	25.6±2.6	26.0±1.9
Copper	26.6±2.8	21.4±1.3	26.0±1.9	22.5±2.5
Kidneys				
Control[1]	$0.42\pm0.02^{b,c}$	0.51 ± 0.03^{a}	$0.48\pm0.02^{a,b}$	$0.43\pm0.01^{b,c}$
Copper[2]	$0.47\pm0.02^{a,b}$	$0.43\pm0.02^{b,c}$	0.39 ± 0.02^{c}	$0.44\pm0.02^{b,c}$
Brain				
Control[1]	0.58 ± 0.02^{b}	0.71 ± 0.05^{a}	0.71 ± 0.03^{a}	$0.61\pm0.02^{a,b}$
Copper[2]	$0.62\pm0.03^{a,b}$	0.59 ± 0.04^{b}	0.58 ± 0.03^{b}	0.59 ± 0.02^{b}
Femur				
Control	0.058 ± 0.004^{c}	$0.066\pm0.005^{a-c}$	0.077 ± 0.005^{a}	0.062 ± 0.004^{b}
Copper	$0.075\pm0.004^{a,b}$	$0.056\pm0.005^{b,c}$	0.060 ± 0.003^{c}	0.059 ± 0.005^{c}

*Mean (X_{11})±S.E. **Pasteurized cow's milk (9) ***Containing added 12 μg Zn/ml from $ZnSO_4.7H_2O$ **** Containing added 25 μg Fe/ml from $FeSO_4.7H_2O$ *****Containing added 5 μg Cu/ml from $CuSO_4.5H_2O$. [1-2] P<0.05. [a-c]Means bearing various superscripts for the same tissue differ significantly (P<0.05).

copper by using doses of up to 40 ppm Zn and 2 ppm Cu in the diet whereas Alfaro and Heaton (20) stated that small amounts of copper even facilitated zinc absorption. Alexander and co-workers (21) suggested complex fortification of infant food with zinc, iron and copper but this and earlier experiments revealed a different pattern of 64-Cu 59-Fe and 65-Zn whole body and tissue distribution in infant rats fed fortified milk (22-24). Therefore it is reasonable to assume that more pharmacokinetic data on trace element nutrition at an infant period of life are necessary as an uncritical, polypragmatic fortification of food products with trace elements may not be innocuous (25).

ACKNOWLEDGEMENT

This work was supported by a research grant from the Scientific Research Council of the Socialistic Republic of Croatia, Yugoslavia. The author thanks Mrs. M. Landeka and M. Buben for their technical assistance and Mrs. N. Breber, Dj. Breski and Mr. C.Tominac for their care and handling of the animals.

REFERENCES

1. Carmichael, D., J. Christopher, J. Hegenauer, and P. Saltman, 1975, Am.J.Clin.Nutr., 28:487-493.

2. Cordano, A., 1974, Ann. Nestle, No. 33:2-16.

3. Walrawens, P.A., and K.M. Hambidge, 1976, Am.J.Clin.Nutr., 29:1114-1121.

4. Lammi, A.T., V.A. Lovrić, and W.S. Hemphill, 1975, Med.J.Aust., 1:441-443.

5. Schutte, K.H., 1964, In "The biology of the trace elements" (Crosby Lockwood Son., London) p 44.

6. Underwood, E.J., 1976, In "Trace elements in human health and disease" (ed. A.S. Prasad)(Academic Press N.Y.) pp 269-279.

7. National Research Council, 1972, In "Nutrient requirements of laboratory animals". (Nat.Acad.Sci., Washington D.C) p 64.

8. Momčilović, B., 1979, Period.Biol., 81: 559-565.

9. Momčilović, B., and D. Kello, 1977, Nutr.Rep.Intl., 15:651-657.

10. Momčilović, B., and I. Rabar, 1979, Period. Biol., 81:27-31.

11. Comar, C.L. 1955, In "Radioisotopes in biology and agriculture" (McGraw Hill, N.Y.) p 176.

12. Kamitian, G., 1972, In "Calculator assisted statistics series No.133" Wang Lab. Inc., Tewksbury) pp 20-25.

13. Duncan, D.B., 1955, Biometrics, 11:1-42.

14. Owen, Jr. C.A., 1965, Am.J.Physiol., 209:900-904.

15. Momčilović, B., and K. Kostial, 1974, Environ.Res., 8:214-220.

16. Martinsson, K., and L. Jonsson, 1976, Zbl.Vet.Med.A,23:277-282.

17. Momčilović, B., Belonje, B., Giroux, A., Shah, B.G., 1975, Nutr.Rep.Intl., 12:197-203.

18. Evans, G.W., 1973, Physiol.Rev., 53:535-570.

19. Murthy, L., L.M. Klevay, and H.G. Petering, 1974, J.Nutr., 104:1458-1465.

20. Alfaro, B., and F.W. Heaton, 1974, Br.J.Nutr., 32:435-445.

21. Alexander, F.W., B.E. Clayton, and H.T. Delves, c974, Quart.J.Med., 43:89-111.

22. Momčilović, B., 1978, In "Proceedings of the Third International Symposium on Trace Element Metabolism in Animals" Ed. M. Kirchgessner, Arbeitskreis fur Tierernahrungsfurschung Weihenstaphan p.448.

23. Momčilović, B., 1981, this Symposium.

24. Keen, C.L. and L.S. Hurley, 1980, Mech.Aging Develop., 13:161-176.

25. Morrison, A.B. and J.A. Campbell, 1963, Canad.Med.Ass.J., 88:523-527.

DISCUSSION

GRACE (New Zealand). In one of your first slides when you were outlining your experimental set-up, you listed the concentration of the copper, zinc and iron you administered. Did that include what was in the milk?

MOMČILOVIĆ (Yugoslavia). No, those concentrations were added to the milk.

BREMNER (Scotland). Could you give us some information as to how you assess the absorption of Copper ^{64}. Was this done by whole body counting and if so did you remove the gastro-intestinal tract?

MOMČILOVIĆ (Yugoslavia). Yes the whole body absorption was assessed in the whole body count. We did not present the results where the gastro-intestinal tract was removed so it does not affect the overall picture.

^{59}Fe METABOLISM IN INFANT RATS FED MILK ENRICHED WITH ZINC, IRON AND COPPER

B. Momčilović

Institute for Medical Research and Occupational Health,
158 M. Pijade
Zagreb
Yugoslavia

Large segments of the world population are in a less than satisfactory nutritional status with respect to zinc, iron and copper and the relevance to man of these data needs to be ascertained (1). Infants and children appear especially vulnerable to the deficit of either zinc, iron or copper; better indices of growth and haematological status were observed in infants fed milk enriched with one of these trace elements (2-4). However, the nutritional physiology of trace elements is rarely confined to a single element in isolation form the others because of frequent interactions at the absorption and cellular level which can affect minimum requirements and influence maximum tolerance (5). The aim of this experiment was to investigate the effect of milk enriched with physiological doses of zinc, iron and copper (6) added either separately or in combination on 59-Fe metabolism in infant rats.

METHODS AND RESULTS

The experiment was performed on 144 six-day-old infant rats in litters of six. Each littermate was randomly assigned (7) to one of the eight available treatments: A) milk (pasteurized cow's milk containing 3.5 µg Zn/ml, 0.5 µg Fe/ml and 0.05 µg Cu/ml) (8); B) milk + Zn (12 µg Zn/ml from $ZnSO_4.7\ H_2O$); C) milk + Fe (25 µg Fe/ml from $FeSO_4.7\ H_2O$); D) milk + Cu (5 µg Cu/ml from $CuSO_4.5\ H_2O$); E) milk + Zn + Fe; F) milk + Zn + Cu; G) milk + Fe + Cu; and H) milk + Zn + Fe + Cu. The animals were artifically fed 0.45 ml of the respective diet labelled with 3.7 kBq 59-Fe/rat with a dropper for a period of 8 h (9). The isotope was almost carrier-free (Radiochemical Centre, Amersham, England). The 59-Fe radioactivity in the whole body and liver, kidneys, brain and femur was determined 4 days later in a single-channel twin crystal or a well-type scintillation counter NaI(Tl) respectively (Nuclear Chicago, U.S.A.) (10). The data were classified simultaneously in two ways according to the effect of iron and other trace element supplementation (11). The significance of the difference among the groups was determined by Duncan's test (12).

Overall whole body retention of 59-Fe was decreased in infant rats fed zinc-containing diets (diets B,E,F,H vs diets A,C,D,G)($P<0.05$) whereas the retention of radioactive iron was not appreciably affected in animals fed iron vs non-iron and copper vs non-copper containing diets respectively (Table 1). However, overall 59-Fe retention was decreased in the kidney of animals fed iron-containing diets (diets C,E,G,H vs diets A,B,D,F)($P<0.05$) whereas in increase in 59-Fe retention was observed in the kidneys of infant rats fed copper containing diets (diets D,F,G,H vs diets A,B,C,E)($P<0.01$). Both iron and copper decreased 59-Fe retention in the brain and femur of infant rats ($P<0.01$ for both elements and tissues). 59-Fe retention in the liver appeared unaffected by any of the treatments.

DISCUSSION

The results of this experiment indicate a high age specific whole body and tissue 59-Fe retention apparently related to the high iron requirements in the infant period of life (13,14). Iron homeostasis is unique in that unlike other trace elements it is regulated primarily by absorption and not by excretion (15). However, the homeostatic control mechanism of absorption is apparently lacking in infant rats at the physiological level of milk fortification as reflected by high 59-Fe retention irrespective of the level of stable iron in the diet. Zinc interaction with iron as revealed by lower 59-Fe retention in the whole body of infant rats fed zinc vs non-zinc-containing diets supports in part the view of Mahloudij and co-workers (16) that a simultaneous administration of iron and zinc may impair their utilization. However, in a previous experiment on adult rats it was demonstrated that the specificity of zinc binding ligands was impaired in the presence of adequate amounts of both zinc and iron (17).

The profound meaning of subtle redistributional changes in the tissue of infant rats induced either by an increased level of dietary iron or by interaction of copper vs non-copper containing diets with 59-Fe retention remains to be elucidated. Iron absorption was increased in animals fed copper deficient diet but the effect was less significant when copper was fed to iron deficient rats (18). Moreover 59-Fe tissue distribution pattern in infant rats is different from that of 65-Zn (19) and 64-Cu (20). This indicates the importance of a pharmacokinetic approach to complex trace element fortification as any change in food density (21) might affect the time course of action, distribution, metabolism and excretion of trace elements.

TABLE 1.

The effect of milk fortified with either zinc, iron or copper on 59-Fe retention and tissue distribution in infant rats

	Milk**	^{59}Fe (% dose)* Milk+Zn***	Milk+Cu****	Milk+Zn+Cu
Whole body				
Control	94.0±2.4[a,b]	92.8±2.9[a-d]	92.3±1.8[b-d]	91.4±3.4[d]
*****Iron	91.6±2.1[c,d]	93.4±3.3[a-c]	94.5±2.3[a]	91.6±2.0[c,d]
Liver				
Control	7.8±3.0[c]	9.9±2.8[a-c]	10.0±2.8[a-c]	8.2±1.6[a-c]
Iron	10.5±3.7[a,b]	9.7±3.0[a-c]	10.7±6.1[a]	8.1±2.2[b,c]
Kidneys				
Control[1]	1.26±0.36[b]	1.53±0.41[a]	1.58±0.27[a]	1.62±0.41[a]
Iron[2]	1.37±0.24[a,b]	1.17±0.25[b]	1.40±0.28[a,b]	1.58±0.27[a]
Brain				
Control[3]	1.20±0.14[a,b]	1.29±0.15[a]	1.20±0.17[a,b]	1.07±0.12[d]
Iron[4]	1.13±0.08[c,d]	1.13±0.08[c,d]	1.17±0.06[b,c]	1.06±0.11[d]
Femur				
Control[3]	0.255±0.038[b]	0.301±0.085[a]	0.194±0.031[c,d]	0.209±0.046[c,d]
Iron[4]	0.200±0.039[c,d]	0.190±0.043[d]	0.227±0.050[b,c]	0.216±0.040[c,d]

*Mean (X_{18})±S.E. **Pasteurized cow's milk (8) ***Containing added 12μg Zn/ml from $ZnSO_4 \cdot 7H_2O$ ***Containing added 5 μg Cu/ml from $CuSO_4 \cdot 5H_2O$ *****Containing added 25 μg Fe/ml from $FeSO_4 \cdot 7H_2O$ [1,2]$P<0.05$ [3,4]$P<0.01$ [a-d]Means bearing various superscripts for the same tissue differ significantly ($P<0.01$).

ACKNOWLEDGEMENT

This work was supported by a research grant from the Scientific Research Council of the Socialistic Republic of Croatia, Yugoslavia. The author thanks Mrs. M. Landeka for her technical assistance and Mrs. M. Buben, N. Breber, K. Pribić, M. Vnucec and S.Tomasić for their care and handling of animals.

REFERENCES

1. Becking, G.C., 1976, Fed. Proc., 35:2480-2485.

2. Cordano, A., 1974, Ann. Nestle, No. 33:2-16

3. Walrawens, P.A., and K.M. Hambidge, 1976, Am.J.Clin.Nutr., 29:1114-1121.

4. Lammi, A.T., V.A. Lovric, and W.S. Hemphill, 1975, Med.J.Aust., 1:441-443.

5. Underwood, E.J., 1976, In "Trace elements in human health and disease" (ed. A.S. Prasad)(Academic Press, N.Y.) pp 269-279.

6. National Research Council, 1972, In "Nutrient requirements of laboratory animals" (Nat.Acad.Sci.,Washington D.C.) p 64.

7. Momčilović, B., 1979, Period.Biol., 81:559-565.

8. Momčilović, B., and D. Kello, 1977, Nutr.Rep.Intil.,15:651-657.

9. Momčilović, B., and I. Rabar, 1979, Peiord.Biol., 81:27-31.

10. Comar, C.L., 1955, In "Radioisotopes in biology and agriculture" (McGraw Hill, N.Y.) p 176.

11. Kamitian, G., 1972, In "Calculator assisted statistics series No.133 (Wang Lab., Towksbury) pp 20-25.

12. Duncan, D.B., 1955, Biometrics, 11:1-42.

13. Ezekiel, E. 1967, J.Lab.Clin.Med., 70:138-149.

14. Gallagher, N.D., R. Mason, and K.F. Foley, 1973, Gastroenterology, 64:438-444.

15. Linder, M.C., and H.N. Munro, 1977, Fed. Proc., 36: 2017-2023.

16. Mahloudij, M., J.G. Reinhold, M. Haghshenass, H.A. Ronaghy, M.R. Spivey-Fox, and J.A. Halsted, 1975, Am.J.Clin.Nutr., 28:721-725.

17. Momčilović, B., and D. Kello, 1979, Nutr.Rep.Intl.,20:429-436.

18. Owen, Jr.C.A. 1973, Am.J.Physiol., 224-514-518.

19. Momčilović, B., 1978, In "Proceedings of the Third International Symposium on Trace Element Metabolism in Animals" p 448. Ed. M. Kirchgessner, Arbeitskreis fur Tierernahrungsfurschung Weihenstaphan.

20. Momčilović, B., 1981, this Symposium.

21. Wretlind, A., 1977, Nutr. Metab., 21:210-214.

INFLUENCE OF ORAL ZINC INTAKE ON OTHER ELEMENTS

M. Abdulla and S. Svensson

Unit for Community Care Sciences at Dalby and Research
Department 2, University Hospital, S-221 85 Lund, Sweden

SUMMARY

Although systemic and local administration of zinc has been in practice for many decades, little is known regarding the influence of oral zinc intake on the absorption and distribution in various compartments of other essential and toxic elements with similar chemical properties and electronic configuration. We have investigated the influence of oral zinc intake in healthy adults on the changes in the concentrations of calcium, magnesium, zinc, copper and lead in blood. Three separate studies have been carried out. In the first study seven adults consumed 135 mg zinc/day for 12 weeks and five subjects of the same age received a placebo tablet for the same period. In the second study, six adults consumed 90 mg zinc/day for 12 weeks and finally in the third study, seven adults consumed 45 mg zinc/day for the same interval. In the last two studies six age-matched subjects served as controls. Blood samples were collected prior to, during and after the experimental period. On an average, 10 - 12 samplings were made in each study. Concentrations of zinc, copper, magnesium, calcium and lead were determined by atomic absorption spectrophotometry. We also measured the activities of the zinc-dependent enzymes, delta-aminolevulinic acid dehydratase and alkaline phosphatase in erythrocytes and granulocytes respectively. Results showed that plasma as well as the erythrocyte zinc levels were increased after two weeks of zinc intake, reached maximum levels after six weeks and remained almost constant thereafter. Alongwith the rise of zinc in plasma and whole blood, we noted an increase in the activities of the two enzymes mentioned above. Copper concentrations in the plasma started to decrease after two weeks of zinc intake and reached 50 % below the normal levels after six weeks. Copper:zinc quotient in plasma became extremely low towards the end of the experimental period. The above-mentioned changes became normal after four weeks of termination of the zinc intake. There were no changes in copper levels in whole blood throughout the experimental period. The concentration of iron, calcium and magnesium did not alter during the period of zinc intake. Lead levels in whole blood was reduced by 20 - 30 % of the normal values after six weeks of zinc intake. Our results so far indicate that prolonged oral zinc intake may interfere with the absorption and distribution of copper and lead. Oral zinc therapy in doses used in our study does not influence the absorption of other elements. Zinc therapy seems to be effective against plumbism.

INTRODUCTION

Although zinc is traditionally classified as a trace element in plant and animal nutrition, next to iron, it is the most common trace element found in the human body. An average healthy adult human body contains 2 - 2.5 g of zinc which is half that of iron and about 15 - 25 times as that of copper (1). High concentrations of zinc are found in bone, skin, prostata and hair. Zinc is present in at least 80 metallo-enzymes that have been identified and is involved in a wide range of cellular activities including nucleic acid and protein synthesis. Zinc has attracted much interest in clinical and nutritional research since the beginning of 1960's when Prasad et al. (2) described cases of dwarfism in the middle-eastern countries associated with zinc deficiency. Zinc deficiency hypogonadism associated with dwarfism, hepatosplenomegaly and geophagia is a well-established syndrome today. In recent years, results of several investigations indicate that impaired wound healing and zinc deficiency are correlated (3). Zinc also plays a key role in the treatment of acrodermatitis enteropathica (4). Several reports from different parts of the world indicate that zinc deficiency is associated with protein-calorie malnutrition and it is fairly common among growing children, pregnant women and elderly in developing countries. Zinc deficiency has been found to have devastating effect on the reproductive functions of several animal species, including man (5). Recent studies conducted in our laboratories and elsewhere indicate that zinc has some protective effect against the toxicity of heavy metals like lead and cadmium (6). Zinc is also known to play an active role in carcinogenesis (7).

In spite of the fact that local administration of zinc had been in practice since historical times, the clinical application of systemic zinc therapy became popular only recently. Oral zinc therapy has been found to have effect on several clinical conditions.

Despite the extensive work done in the near past regarding the in vivo effect of zinc, little is known regarding the influence of oral zinc on the absorption and distribution in various body compartments of other essential and toxic elements with similar chemical properties and electronic configuration. The present study has been done to investigate the influence or oral zinc intake on the concentrations of calcium, magnesium, iron, copper and lead in blood of healthy human adults.

MATERIAL AND METHODS

Three separate studies have been done. In the first study, which was conducted as a single blind one, seven healthy adults (average age 25) consumed 200 mg zinc sulphate ($ZnSO_4$, 7 H_2O, corresponding to 45 mg zinc) thrice a day for 12 weeks and five subjects of the same average age consumed a placebo tablet thrice a day for the same interval. In the second study, six adults consumed 200 mg zinc sulphate twice daily and finally in the third study, seven adults consumed 200 mg zinc sulphate daily for 12 weeks. In the last two studies, six age-matched subjects without zinc intake served as controls. Blood samples were collected prior to, during and after the experimental period. On an average, 10 - 12 samplings were made within a period of 3 - 4 months from the start. The samples were collected in specially-washed plastic tubes containing 0.1 ml heparin. Each time 20 - 50 ml of venous blood was drawn. The activities of delta-aminolevulinic acid dehydratase (ALA-D) in erythrocytes and alkaline phosphatase in granulocytes were measured as described elsewhere (6, 8). For the determination of zinc and copper in whole blood, 5 ml of venous blood was diluted with suitable quantities of deionised water. The magnesium concentration in whole blood was determined after the dilution of the samples with 1 % lanthanum chloride. All elements studied in the present investigation were analysed by atomic absorption spectrophotometry.

RESULTS AND DISCUSSION

In the first study with high oral zinc load, we observed that zinc levels in the plasma started to increase after 2 - 4 weeks from the start and reached maximum levels after 6 - 8 weeks. In the placebo group, there was no increase in plasma zinc throughout the experimental period. Figure 1 shows part of the results. In the second and third studies, similar trends were observed, although the increases in the plasma levels were not so high as that was observed in the first study. In the last two studies, we also measured the erythrocyte zinc and found a 15 % increase after 4 weeks. The concentrations of iron, magnesium and calcium did not alter throughout the experimental period in all groups. We did not measure the concentration of copper and lead in the first study. In the last two studies, we followed the copper concentrations in plasma throughout the experimental period and the results showed that the copper levels in plasma started to decrease after two weeks of zinc intake and the copper:zinc ratio in the plasma became significantly low ($p < 0.001$) during the final stage of zinc intake. Similar trend was also found in the third study. In the last study, we also measured the lead levels in whole blood and the results showed that there was a 20 - 30 % decrease in lead levels after four weeks of zinc intake alongwith an increase of zinc in erythrocytes.

The results concerning the activities of the enzymes measured in erythrocytes and granulocytes are presented elsewhere (8). The results in short indicated that alongwith an increase of zinc in blood (after 2 - 4 weeks of zinc intake), the activities of both enzymes mentioned earlier also increased. These increases, however, returned to original levels after the termination of zinc intake.

We also observed a temporary increase in blood sugar after 2 - 4 weeks of zinc intake in the second and third studies. A glucose tolerance test showed, however, no abnormalities in carbohydrate metabolism. The blood sugar levels returned to normal levels after the termination of the zinc therapy. Similarly, we found that most of the subjects participating in the last two studies developed an intolerance to ethyl alcohol which remained for some time even after the termination of zinc intake. Clinically, there were no other complaints from the participants in all groups except for occasional nausea and abdominal distress.

In recent years, both oral and parenteral administration of zinc in humans and several animal species have been found to improve several clinical conditions such as retarded growth, abnormalities in reproduction and impaired hair growth associated with zinc deficiency. The beneficial effect of zinc against heavy metal poisoning has attracted considerable interest recently. Little is known regarding the mechanism(s) involved in the absorption and distribution of zinc. The interaction between zinc and other divalent elements with similar chemical properties and electronic configuration seems to be of great importance in understanding the beneficial and damaging aspects of zinc therapy. Our results indicate that zinc may influence the distribution of copper and lead both at the cellular and intestinal levels. Zinc absorption and transfer to the vascular system are related to the synthesis of metallothionine-like proteins in the enterocytes (9). Metallothionine-like proteins may also bind metals other than zinc and this may explain partly the reductions of copper and lead found in plasma and blood after zinc intake. When the oral load of zinc is high, the absorption of the above metals may be reduced alongwith an increase in fecal excretion. It is likely that the absorption of elements other than copper and lead also may be influenced by zinc therapy for longer periods than three months. This aspect ought to be kept in mind in clinical trials of oral zinc therapy for longer periods. The temporary increase of blood sugar and alcohol intolerance during oral zinc therapy are two other factors which need further clarification. Such studies are in progress in our laboratory at present.

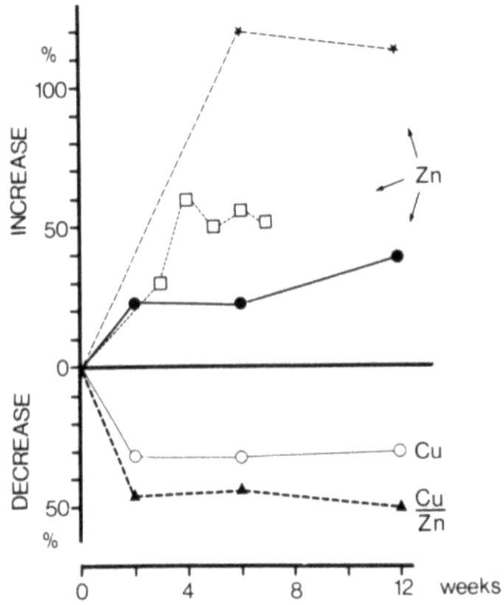

Fig. 1. Plasma levels of zinc and copper during oral zinc intake

* 135 mg/day
□ 90 mg/day
● 45 mg/day

REFERENCES

1. Underwood, E.J. 1971. In "Trace Elements in Human and Animal Nutrition" (Academic Press Inc., N.Y.) pp 208-252

2. Prasad, A.S., A. Miale, Z. Farid, H.H. Sandstead, and A.R. Schulert. 1963. J. Lab. Clin. Med. 61:537-549

3. Hallböök, T., and E. Lanner. 1967. Lancet II:780-782

4. Moynahan, E.J. 1975. Proc. R. Sco. Med. 68:276

5. Jameson, S. 1976. Acta Med Scand. No. 37, suppl. 593, pp 5-20

6. Abdulla, M. 1980. In "Mechanisms of Toxicity and Hazard Evaluation" (ed. Holmstedt, Lauwerys, Mercier and Roberfroid) (Elsevier/North Holland Biomedical Press) pp 599-602

7. Abdulla, M., A. Björklund, A. Mathur, I. Qvist, C. Tropé, and K. Wallenius. 1981. In "Recent Advances in Clinical Nutrition" (ed. A. Howard and I. MacLean) In press

8. Abdulla, M., and S. Svensson. 1979. Scand. J. Clin. Lab. Invest. 39:31-36

9. Cousins, R.J. 1979. Nutr. Rev. 37:4, 97-103

DISCUSSION

LLOYD DAVIES (Australia). One of the effects of zinc administration in animals is to reduce copper availability. Does this also occur in humans?

ABDULLA (Sweden). I presume so, although only plasma copper was investigated in our experiments. The copper concentration in the plasma of some subjects was as low as 50% of normal values. This might be important because zinc therapy is being used so widely today - for treatment of impotency or skin lesions.

INFLUENCE OF DIETARY Ca ON Cd METABOLISM IN PIGS

H.Schenkel, B.Krehl

Institute of Animal Nutrition
University of Hohenheim
PO BOX 700562, 7000 STUTTGART 70
FEDERAL REPUBLIC OF GERMANY

In a number of experiments it could be found that on low dietary Ca intake there was an increase in Cd incorporation in some tissues. Nearly all investigations were carried out with rats as experimental animals using diets with very low Ca and/or high Cd concentrations. Due to different experimental conditions not all of the results were in accordance (2 - 14). There was a lack of information with respect to farm animals. Therefore the present experiment was conducted with fattening pigs using dietary Ca- and Cd concentrations, which can occur under practical conditions.

MATERIAL AND METHODS

Three groups of pigs were fed diets based on barley and soybean meal. The barley was cultivated on a contaminated area and contained 0.88 mg/kg (DM). The Cd concentration in the diets was 0.66 mg/kg. Due to different Ca carbonate supplementation in the diets of group I to III 0.48%, 0.73% and 1.06% Ca were determined. The animals received the diets during the whole fattening period (30-100 kg body weight). After slaughtering Cd-concentrations were determined in liver, bile, kidney, pancreas, spleen, duodenum, brain, musc. longissimus, heart, tongue, rib and incisors.

RESULTS AND DISCUSSION

The Cd concentrations in the analyzed tissue and body fluids varied between 9.22 mg/kg dry weight in a kidney sample and 9 mcg/kg dry weight in a brain sample (table 1).

Table 1: Cd concentrations (mg/kg dry weight) in different tissues of pigs fed a barley-soybean meal diet containing 0.66 mg/kg

	Kidney	Duodenum	Liver	Pancreas	Bile	Spleen
mean	4.41	1.30	0.63	0.24	0.12	0.10
standard deviation	1.49	0.44	0.18	0.06	0.04	0.05
range	2.39-9.22	0.58-2.36	0.32-1.19	0.04-0.39	0.03-0.26	0.04-0.21

	M.long.	Incisor	Heart	Tongue	Brain	Rib
mean	0.066	0.061	0.043	0.039	0.024	0.017
standard deviation	0.041	0.049	0.018	0.017	0.013	0.008
range	0.014-0.18	0.019-0.27	0.021-0.12	0.020-0.080	0.009-0.080	0.012-0.073

The mean Cd concentration in the dry matter of kidney and duodenum was higher and in liver nearly equal than dietary Cd concentration, whereas in the remaining tissues Cd concentrations were lower than dietary concentration.

The different Ca intake has an marked effect on Cd concentration in kidney, liver and spleen (table 2).

Table 2: Cd concentration in kidney, liver, spleen and pancreas after different Ca intake (mg Cd/kg dry weight; mean $\pm$ standard deviation)

Group	dietary Ca (%)	Kidney	Liver	Spleen	Pancreas
I	0.48	5.63 ± 1.66^{a}	0.756 ± 0.184^{a}	0.134 ± 0.042^{a}	0.264 ± 0.058^{a}
II	0.73	4.24 ± 1.22^{b}	0.626 ± 0.184^{ab}	0.106 ± 0.050^{a}	0.217 ± 0.069^{a}
III	1.06	3.47 ± 0.65^{c}	0.525 ± 0.109^{b}	0.073 ± 0.031^{b}	0.232 ± 0.041^{a}

groups sharing not the same subscript were significantly ($p < 0,05$) different

The Cd concentrations in kidney, liver and spleen of the animals in group III was only at 62%, 59%, and 54% of the values in group I, respectively. A similar tendency could also be stated for Cd concentration in pancreas. Consequently significant correlations could be found between Cd concentrations in these organs (table 3).

Table 3: Coefficients of correlation of Cd-concentration in liver, kidney, pancreas, and spleen

	liver	pancreas	spleen
kidney	0.669***	0.382*	0.591***
liver		0.600***	0.660***
pancreas			-0.164

* $p < 0,05$; *** $p < 0,001$

It cannot be concluded from the present experiment whether this effect was caused by an increased Cd absorption or a decrease in Cd excretion. Cousins (1,14) mentioned an increased Cd binding to the Ca binding protein in intestinal mucosa and in consequence an increased Cd absorption in Ca deficiency.

The mean Cd concentration in bile was 0.12 ± 0.05 mg/kg, 0.09 ± 0.03 mg/kg, and 0.14 mg/kg dry weight in group I, II and III, respectively. On an average of the 3 groups a ratio for Cd concentration in bile and liver of 0.19 could be found. In comparison the ratios for Cu, Mg, Mn, Fe, and Zn were 0.98, 0.82, 0.61, 0.06 and 0.04. Assuming an excretion of 1500 ml bile per day it could be calculated that the quantity of Cd excreted via bile was equivalent to 0.8-1.3% of daily Cd intake in the 3 groups. Corresponding values for Cu, Mg, Mn, Fe and Zn were 4.3%, 1.7%, 0.5%, 0.8% and 0.3%, respectively.

SUMMARY

It could be shown in an experiment with fattening pigs that at low level Cd intake even a relatively small variation in Ca intake caused different Cd concentrations in kidney, liver and spleen.Highest concentrations in these organs could be stated when a diet with moderate Ca deficiency was fed. Bile seems to be an essential way of Cd excretion, although no direct relationship between Cd-concentration in liver and bile could be found in the present experiment.

LITERATURE

1. Cousins, R.J., 1979, Environm. Health Perspect, 28:131-136
2. Gergely, A., and E.Morava, 1979, In "Kadmium-Symposium" (Wiss. Beitr. F.Schiller-Univ., Jena) pp 131-135
3. Hamilton, D.L., and W.M.Smith, 1978, Environm. Res. 15:175-184
4. Itokawa, Y., Abe, T., and S.Tanaka, 1973, Arch. Environm. Health, 26:241-244
5. Itokawa, Y., Abe, T., Tabei, R., and S.Tanaka, 1974, Arch. Environm. Health, 28:149-154
6. Itokawa, Y., Nishino, K., Takashima, M., Nakata, R., Kaito, H., Okamoto, E., Daijo, K., and J.Kawamura, 1978, Environm. Res., 15:206-217
7. Kello, D., 1979, Arch. Environm. Health, 34:30-33
8. Larsson, S.E., and M.Piscator, 1971, Israel. J.Med.Sci., 7:495-501
9. Piscator, M., and S.E.Larsson, 1974, In "Trace Element Metabolism in Man and Animals-2" (ed. W.G.Hoekstra, J.W.Suttle, H.E.Ganther, W.Mertz) (University Park Press, Baltimore) pp 687-689
10. Pond, W.G., and E.Walker jr., 1975, Proc.Soc.Exp.Biol.Med., 148:665-668
11. Sasser, L.B., Callaham, R.J., and G.E.Jarboe, 1978, in "Trace Element Metabolism in Man and Animals-3" (ed. M.Kirchgessner) (ATW, Weihenstephan) pp 562-565
12. Washko, P.W., and R.J.Cousins, 1975, Nutr.Rep.Int., 11:113-127
13. Washko, P.W., and R.J.Cousins, 1976, J.Toxicol.Environm.Health, 1:1055-1066
14. Washko, P.W., and R.J.Cousins, 1977, J.Nutr., 107:920-928

ACKNOWLEDGEMENT

This investigation was supported by the Ministerium für Ernährung, Landwirtschaft, Umwelt und Forsten, Baden-Württemberg, F.R.G.

Ca, Mg AND Ca/P RATIO OF RAT MOLARS AFTER Cd SUPPLY

G.R. Wesenberg and G. Fosse

Institute of Anatomy
Årstadveien 19, N-5000 BERGEN
NORWAY

Cadmium is a toxic element of increasing use in industry, and the danger of Cd pollution from the waste products of these industries are imminent (1, 2). Cadmium has as its primary target organs kidney cortex, liver and other soft tissues (3-6), but Cd can to some lesser extent accumulate in bone and teeth (5, 7).

We have previuosly reported that the Cd content of rat molars was significantly correlated with total Cd intake and with kidney cortex Cd content (5, 6).

Cadmium has been reported to interfere with Ca metabolism (8), but there are few reports on Cd's effect on Mg or P metabolism. Cd has also been reported to increase caries susceptibility (9-11) and to decrease enamel solubility (12) in experimental animals.

The aim of this study was therefore to observe whether Cd supply, in addition to Cd accumulation, resulted in a change in the normal Ca and Mg contents of rat molars. We also wanted to see if the Cd administration had an effect on the Ca/P ratio in enamel and dentin of the molars.

MATERIALS AND METHODS

Ten groups of inbred Wistar rats were exposed to graded doses from 2.5 to 25 ppm of $CdCl_2$ in the drinking water from conception until sacrifice at the age of ten weeks. The increment between groups was always 2.5 ppm. In addition one group served as a control group and was given tap water throughout the experiment. Thus there were altogether eleven groups. In utero and during sucking the Cd was given through the dams to ensure a continuous exposure, even if a moderate placental and mammary barrier to Cd exists (13-15).

All six molars from the right half of the upper and lower jaws of ten animals in each group were washed in 0.01N HCl, rinsed in distilled water and dried at 45°C for 2 hours. The molars of the left half of the upper and lower jaws of the same animals were pooled for each group, and the enamel and dentin for each pooled sample was separated by the aceton-bromoform flotation method (16). The weighed samples of whole molars as well as separated enamel and dentin were dissolved in 2N HCl and stored at 80°C for two days. The Cd concentration in this solution was chelated by APDC-saturated xylene (7). Prior to the chelating 400 µl of the original solution was removed from each sample, and the Ca and Mg concentrations were determined in appropriate dilutions.

Cd as well as Ca and Mg were determined by AAS. The 228.8, 422.7 and 285.2 nm absorption lines were used for Cd, Ca and Mg determinations, respectively.

Two upper and two lower molars from each of eight additional animals, four in the control group and four in the 22.5 ppm Cd group, were planed transversely, mounted on aluminium stubs, carbon coated and subjected to x-ray microanalyses on a Philips PSEM 500 unit with a Kevex 5000A microprobe. All spectra were obtained with the same magnification, counting time and spot size, and with constant distance and angle between the specimen surface and the detector.

X-ray spectra from four different areas in the enamel and two different areas in the dentin of each molar were obtained.

All results were treated statistically by means of Student's t-tests and linear regression.

RESULTS AND DISCUSSION

The Cd and Ca levels were both significantly elevated on the 0.1 percent level for Cd doses above 2.5 ppm; the Mg concentrations were significantly lowered on the 0.1 percent level for doses below 12.5 ppm, and in the 20 and 25 ppm groups (Fig. 1).

There was a significant positive correlation ($p < 0.001$) between Cd and Ca, and between Ca and Mg in whole molars. The correlation tests were carried out using the individual values in each group through all groups.

As there is only one value for enamel and dentin for each Cd dose, the results only indicate that the increase of Ca was mainly in enamel, while the decrease of Mg was mainly in dentin (Fig. 2).

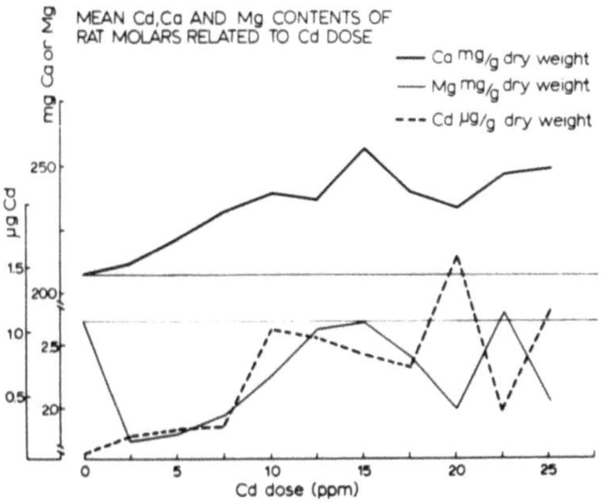

Fig. 1.: Mean Cd, Ca and Mg contents of whole molars of Cd loaded rats as found by AAS.

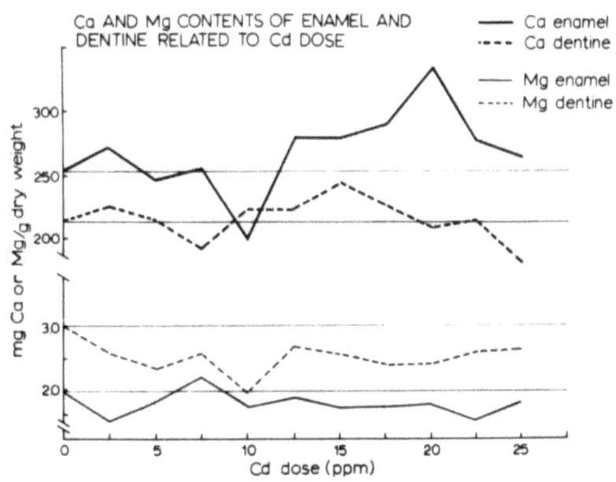

Fig. 2.: Ca and Mg contents of separated enamel and dentin in molars of Cd loaded rats as found by AAS.

Neither in enamel nor in dentin could Cd or Mg be detected from the background noise. The Ca and P values were expressed as the areas delineated by the K_α peaks of their emission curves, and <u>not</u> as absolute values.

There was no significant difference between the Ca/P ratios of controls and experimentals neither in enamel nor in dentin.

Our experimental design did not enable us to state whether the observed increase of Ca and decrease of Mg was a result of increased Ca uptake and thus a decreased Mg uptake, or a result of redistribution of Ca and Mg from other tissues.

Jenkins (17) has stated that wide variations in the Ca/P ratio in the diet have only a small effect on the Ca/P ratio of bone and teeth, but a marked effect on their CO_3 content. Sobel et al. (18) have suggested that animals with an enamel of high CO_3 content is more liable to caries than that of control animals. It may therefore be possible that Cd altered the mineralization of the enamel and dentin, and thus caused a higher susceptibility to caries.

In conclusion this study showed:

1. That graded doses of Cd in the drinking water of rats resulted in a increase of Ca and a decrease of Mg in the molars.

2. The increase of Ca seemed to be in enamel, while the decrease of Mg seemed to be in dentin.

3. Cd supply of 22.5 ppm had no effect on the Ca/P ratio neither in enamel nor in dentin. Hence phosphorous content must also be raised in enamel.

ACKNOWLEDGEMENTS

This study was supported by the Norwegian Research Council for Science and the Humanities and the Borregaard Research Fund.

REFERENCES

1. Fleischer, M., A.F. Sarofim, D.W. Fasset, P. Hammond, H.T. Shacklette, I.C.T. Nisbet and S. Epstein, 1974, Environ. Health Perspect. 7:253 -323.

2. Yost, K.J., 1979, Environ. Health Perspect. 28:5 - 16.

3. Friberg, L., and E. Odeblad, 1957, Acta Path. 41:96 - 98.

4. Gunn, S.A., and T.C. Gould, 1957, Proc. Soc. Exp. Biol. Med. 96:820 - 823.

5. Wesenberg, G.B.R., G. Fosse, N.-P.B. Justesen and P. Rasmussen, 1979, Intern. J. Environ. Stud. 14:223 - 230.

6. Wesenberg, G.B.R., G. Fosse, P. Rasmussen and N.-P.B. Justesen, 1981, Intern. J. Environ. Stud. 16:147 - 155.

7. Fosse, G., and N.-P.B. Justesen, 1977, Intern. J. Environ. Stud. 11:17 - 27.

8. Bremner, I., 1978, Wld. Rev. Nutr. Diet. 32:165 - 197.

9. Ginn, J.T., and J.F. Volker, 1944, Proc. Soc. Exp. Biol. Med. 57:189 - 191.

10. Leicester, H.M., 1946, J. Dent. Res. 25:337 - 340.

11. Wisotzky, J., and J.W. Hein, 1958, J.A.D.A. 57:796 - 800.

12. Manly, R.S., and B.G. Bibby, 1949, J. Dent. Res. 28:160 - 171.

13. Lauwerys, R., J.P. Buchet, H. Roels and G. Humbermont, 1978, Environ. Res. 15:278 - 289.

14. Lucis, O.J., R. Lucis and Z. Sharkh, 1972, Arch. Environ. Health. 25:14 - 22.

15. Neathery, M.W., W.J. Miller, R.P. Gentry, P.E. Stake and D.M. Blackmon, 1974, J. Dairy Sci. 57:1177 - 1183.

16. Manly, R.S., and H.C. Hodge, 1939, J. Dent. Res. 18:133 - 141.

17. Jenkins, G.N., 1978, In "The Physiology and Biochemistry of the mouth" (Blackwell Scientific Publications, Oxford, London, Edinburgh, Melbourn) pp 241-242.

18. Sobel, A.E., J.H. Shaw, A. Hanok and S. Nobel, 1960, J. Dent. Res. 39:462 - 472.

DISCUSSION

ANON. Did you take just one position on the enamel or did you scan across the width of the enamel to look for differences?

FOSSE (Norway). We looked at four areas on each tooth and scanned three spots within each area.

INTERACTION BETWEEN NICKEL, COPPER AND IRON IN RATS

F.H. Nielsen and T.J. Zimmerman

USDA, SEA, Grand Forks Human Nutrition Research Center
and
Department of Biochemistry, University of North Dakota
Grand Forks, North Dakota, 58202
United States of America

Hill and Matrone (1) used a coordination chemistry approach, drawing freely from valence bond, molecular orbital, and ligand field theories, to develop the hypothesis that ions with similar valence shell electronic structures would competitively interact in biological systems. According to that hypothesis, nickel would interact with either copper or iron, because nickel could have similar chemical properties as those elements, thus forming similar complexes. Because copper and iron form different types of complexes, any interaction between them would most likely be noncompetitive, or occur where cationic properties, but not complex structure, are important. The preceding suggests that there are complex relationships among nickel, copper and iron in biological systems which we decided to confirm by using factorially arranged experiments.

MATERIALS AND METHODS

Female weanling Sprague-Dawley rats were weighed individually upon arrival and housed three per all-plastic cage (2) located inside a laminar-flow rack. The rats were assigned to groups of six in fully-crossed, three factor, two by three by three experiments with a completely randomized factorial arrangement of treatments. The treatments were the supplementation of the basal diet with iron [as a mixture of ferric and ferrous sulfates (3)] at 15 and 45 µg/g, nickel ($NiCl_2 \cdot 3H_2O$) at 0, 5 and 50 µg/g and copper ($CuSO_4 \cdot 5H_2O$) at either 0, 0.5 and 5 µg/g (exp. 1), or 0, 0.25, and 12 µg/g (exp. 2). A gram of basal diet (4) contained in exp. 1 approximately 16 ng of nickel, 2.3 µg of iron and 0.47 µg of copper, and in exp. 2, 20 ng of nickel, 1.3 µg of iron and 0.39 µg of copper as determined by atomic absorption spectrometric methods (4). Fresh food and deionized water in plastic cups were provided each day. Room temperature was maintained at 23°. Automatically controlled room lighting provided 12 hours of light and 12 hours of darkness.

The rats were fed their respective diets for 11 weeks in exp. 1 and 8 weeks in exp. 2, weighed, then decapitated subsequent to ether anesthesia and cardiac exsanguination. The liver, heart, one kidney and spleen were removed, blotted dry, and weighed immediately. Microhematocrits were determined on heart blood. The following parameters were determined by using described methods (3,4): hemoglobin; plasma alkaline phosphatase activity, total lipids, lipid phosphorous, cholesterol, copper, iron, nickel and zinc; liver total lipids, lipid phosphorous, cholesterol, copper, iron, manganese and zinc; heart copper, iron, manganese, nickel and zinc; and ratios of heart wt./body wt., kidney wt./body wt., liver wt./body wt. and spleen wt./body wt. The data were treated by a three-way analysis of variance (5). For brevity, only the hematocrit, heart wt./body wt. ratio, liver copper and liver iron data are presented here.

RESULTS AND DISCUSSION

Our findings showed the complexity of the relationships among dietary nickel, copper and iron. Nickel interacted with copper and this interaction was influenced by iron. The first sign of a nickel-copper interaction occurred during the last week of each experiment. In exp. 1, three rats fed the diet without additional copper and with 50 µg of nickel/g began to lose weight - two in the group fed 15 µg of iron/g, and one in the group fed 45 µg of iron/g of diet. In exp. 2, six rats, three in each group described above, began to lose weight and two died (one from each group) from heart rupture before the experiment was completed. In severely copper-deficient rats fed the dietary supplements of 0 and 5 µg/g of nickel, none died or lost weight.

Data from the two experiments were similar except when the diet contained no supplemental copper and 15 µg of iron/g. The rats fed that diet were more anemic in exp. 2 than in exp. 1 - possibly because the copper-iron deficiency was more severe in exp. 2 than in exp. 1. In copper deficiency, the liver content of copper is depressed,

Effect in Rats of Copper, Iron, Nickel and Their Interaction on Hematocrit, Heart Size, and Liver Content of Copper and Iron

Treatment[1]				Experiment 1				Experiment 2			
Cu		Fe	Ni		Heart wt.[2]	Liver[3]			Heart wt.[2]	Liver[3]	
Exp. 1	Exp. 2			Hct.	ratio	Cu	Fe	Hct.	ratio	Cu	Fe
μg/g	μg/g	μg/g		%		μg/g	μg/g	%		μg/g	μg/g
0.0	0.0	15	0	36.2	0.499	2.04	300	17.3	0.902	1.34	538
0.0	0.0	15	5	28.4	0.636	2.25	540	17.9	0.719	1.48	571
0.0	0.0	15	50	30.0	0.913	2.01	592	20.1	0.748	1.49	633
0.5	0.25	15	0	40.3	0.373	7.08	514	29.7	0.508	2.83	406
0.5	0.25	15	5	40.4	0.378	6.65	393	30.2	0.532	3.17	457
0.5	0.25	15	50	40.7	0.362	6.68	454	28.9	0.613	2.62	558
5.0	12.0	15	0	40.6	0.330	13.69	440	38.4	0.396	26.49	128
5.0	12.0	15	5	41.2	0.331	13.67	457	38.1	0.389	21.52	124
5.0	12.0	15	50	41.1	0.343	13.25	273	39.1	0.391	18.60	138
0.0	0.0	45	0	36.8	0.497	1.78	533	37.4	0.501	2.04	689
0.0	0.0	45	5	37.8	0.530	2.05	809	36.5	0.550	2.13	862
0.0	0.0	45	50	33.2	0.596	2.06	506	29.0	0.678	1.90	971
0.5	0.25	45	0	40.7	0.363	6.78	603	40.6	0.412	4.51	919
0.5	0.25	45	5	41.8	0.366	6.98	700	41.2	0.426	3.76	1194
0.5	0.25	45	50	41.2	0.358	6.50	709	40.3	0.461	3.85	968
5.0	12.0	45	0	40.7	0.324	13.62	604	41.7	0.360	15.14	600
5.0	12.0	45	5	41.8	0.334	12.93	504	41.6	0.372	14.96	618
5.0	12.0	45	50	41.8	0.325	13.23	542	41.1	0.367	14.74	564

Analysis of Variance - P Values

Copper effect	0.0001	0.0001	0.0001	NS	0.0001	0.0001	0.0001	0.0001
Iron effect	0.0005	0.02	NS	0.0001	0.0001	0.0001	0.0001	0.0001
Nickel effect	NS	0.01	NS	NS	0.04	NS	0.03	0.004
Nickel x copper	0.003	0.002	NS	0.02	NS	NS	0.03	0.04
Nickel x iron	0.03	NS	NS	NS	0.0002	0.02	NS	NS
Copper x iron	0.006	0.04	NS	NS	0.0001	0.0001	0.0001	0.0001
Nickel x copper x iron	NS	NS	NS	0.03	0.0001	0.0007	0.03	NS

[1]Amounts of Ni (nickel chloride), Fe (mixture of 40% ferrous sulfate and 60% ferric sulfate), and Cu (copper sulfate) supplemented to the diet.

[2]Grams of heart/100 grams of body weight.

[3]Dry weight basis.

and iron is elevated. Those signs of copper deficiency were apparently more severe in exp. 2 than in exp. 1 when rats were fed no supplemental copper and inadequate iron (see table). Also, in rats fed adequate copper and 15 μg of iron/g of diet, the liver iron content was apparently lower in exp. 2 than in exp. 1. The apparent differences in iron and copper status of rats presumably account for the statistical differences between the two experiments.

Our findings suggest that nickel and copper are antagonistic. Demonstration of this antagonism, however, depended upon the levels of copper, iron and nickel in the diet. If copper-, or copper plus iron-, deprivation was not too severe or mild, signs of copper deficiency were more severe with than without nickel; that effect was greater when dietary nickel was 50 μg/g rather than 5 μg/g. Support for this statement included the data shown in the table for hematocrit and heart wt./body wt. ratio from groups fed the following dietary supplements: 0 copper and either 15 or 45 μg of iron/g in exp. 1; 0 copper and 45 μg of iron/g, or 0.25 μg of copper and 15 μg of iron/g in exp. 2.

On the other hand, if the copper-, or copper plus iron-, deprivation was too severe, or quite mild, neither 5 nor 50 μg of nickel/g of diet consistently antagonized copper deficiency signs. In exp. 2, nickel supplementation of the very anemic rats fed diets containing 0 copper and 15 μg of iron/g did not intensify the copper deficiency signs of depressed hematocrit and elevated heart wt./body wt. ratio. Also, those copper deficiency signs were not induced, or intensified, in rats fed borderline copper-deficient diets, such as those containing supplements of 0.5 μg of copper and 15 or 45 μg of iron/g in exp. 1, and 0.25 μg of copper and 45 μg of iron/g in exp. 2.

Some indication of the nature of the antagonism between nickel and copper were obtained from the liver copper data. Nickel supplementation did not depress the level of copper in liver of copper-deficient rats. This indicates that nickel did not exacerbate copper deficiency signs by interfering with the absorption of copper. The antagonism between copper and nickel was probably due to the isomorphous replacement of copper by nickel at various functional sites which interfered with some biological processes.

The mechanism through which iron influenced the interaction between nickel and copper is unclear. A possible explanation could be that iron deprivation inhibited copper absorption. That possibility seems questionable because, when dietary copper was adequate, iron deprivation elevated the level of copper in the liver. Perhaps, copper deficiency was affected by iron deprivation through an effect on the level of nickel in tissues. Iron-deprived rats exhibited elevated absorption of nickel (6). In exp. 1, iron deprivation elevated the level of nickel in heart. Thus, iron deprivation could have had an effect on copper deficiency similar to that of elevated dietary nickel. However, further analyses of tissues are needed to confirm this suggestion.

It is evident from the findings that to show an antagonism between copper and nickel, proper dietary levels of these elements must be used. This antagonism can be attenuated by altering dietary iron. Thus, in factorially arranged experiments, by carefully choosing the dietary levels of the three elements, it should be possible to show a nickel-copper interaction, nickel-copper-iron interaction, or no interaction.

REFERENCES

1. Hill, C.H., and G. Matrone, 1970, Fed. Proc., Fed. Amer. Soc. Exp. Biol. 29: 1474-1481.

2. Nielsen, F.H., and B. Bailey, 1979, Lab. Anim. Sci. 29: 502-506.

3. Nielsen, F.H., 1980, Biol. Trace Element Res. 2: 199-210.

4. Nielsen, F.H., T.R. Shuler, T.J. Zimmerman, M.E. Collings, and E.O. Uthus, 1979, Biol. Trace Element Res. 1: 325-335.

5. Scheffé, H., 1959, "The Analysis of Variance" (John Wiley, NY) pp 90-137.

6. Ragan, H.A., 1978, Pac. Northwest Lab. Annu. Rep. DOE Assist. Secr. Environ. PNL-2500-Pt. 1, pp. 1.9-1.10.

DISCUSSION

PAYNTER (Australia). Did you measure the activity of any copper-dependent enzymes?

NIELSEN (USA). We only measured caeruloplasmin oxidase activity. It was depressed in copper deficiency but was not affected by the interaction between copper and nickel. That is, at 0, 0.25 and 0.5 ppm Cu caeruloplasmin activity was very depressed but the lack of iron or high levels of nickel did not exacerbate this sign of copper deficiency.

BINNERTS (The Netherlands). Did you observe whether high levels of copper induced a nickel deficiency?

NIELSEN (USA). The highest level of copper used in this experiment was 12 ppm, and there was no induction of nickel deficiency. We have not tried higher levels of copper.

INTERACTION BETWEEN BORON AND CHOLECALCIFEROL IN THE CHICK

C.D. Hunt and F.H. Nielsen

Department of Anatomy, University of North Dakota
and
USDA, SEA, Grand Forks Human Nutrition Research Center
Grand Forks, North Dakota, 58202
United States of America

Boron is recognized as essential for normal growth and development of higher plants (1). However, to date, conclusive evidence showing a physiological role for boron in animals has not been described. Several attempts to induce a boron deficiency in rats were unsuccessful, although the diets used apparently contained only 155-163 ng of boron/g (2-4). Skinner and McHargue (5) reported that supplemental dietary boron was beneficial to potassium-deficient rats. That finding was not confirmed by Follis (6) who used a different diet with an unknown boron content, and different levels of boron supplementation.

Borate, or its derivatives, competitively inhibit *in vitro* oxidoreductase enzymes which require pyridine or flavin nucleotides, and serine protease enzymes. Examples of the oxidoreductase enzymes inhibited by borate include yeast alcohol dehydrogenase (7), aldehyde dehydrogenase (8), xanthine oxidase (9), and cytochrome b_5 reductase (10). Smith and Johnson (11) suggested that borate inhibited these enzymes by competing for the NAD, or flavin, co-factor. Borate apparently complexes with ribosyl hydroxy groups of NAD (12). Examples of serine protease enzymes inhibited by borate include chymotrypsin (13), subtilisin (14) and protease 3 from *Streptomyces griseus* (15). Borate and boronic acid derivatives apparently inhibited these enzymes by forming transition state analogues (14). When injected IP into rats fed diets apparently containing less than 1 ng of boron/g, boron stimulated hepatic RNA synthesis (16).

Although the effects of boron *in vitro*, or upon IP injection, were probably pharmacological, they support the possibility that boron has physiological action. Boron meets some of the criteria for essentiality (17) including: It is a low molecular weight element which forms chelates and thus is chemically suitable for biological function. It is ubiquitous on the earth's crust and in sea water, therefore, it has been generally available to animals during their evolution. It is present in significant quantity in animals. It is toxic to animals orally only at relatively high levels. Homeostatic mechanisms for boron in animals are implied by serum levels, excretion rates, and lack of excessive accumulation. The major criterion for essentiality in animals not fulfilled by boron is the demonstration that its deficiency reproducibly impairs a biological function from optimal to suboptimal.

In some of our early studies on arsenic nutrition and metabolism, chicks exhibited poor growth and leg abnormalities even though all known essential nutrients were added to the diet in sufficient quantity. We designed studies to determine the shortcoming of the diet and found that boron supplementation stimulated growth. Subsequent to those studies, we found that the diet contained inadequate cholecalciferol because the indicated potency of the cholecalciferol supplement was apparently not true. After correcting for the impotent cholecalciferol, we found that boron did not markedly stimulate growth. This suggested an interaction between boron and cholecalciferol. The present studies were done to confirm that interaction.

MATERIALS AND METHODS

Day-old cockerel chicks were weighed individually upon arrival and housed in all plastic environmental chambers (18). The chicks were assigned to groups of 25 in fully-crossed, two factor, two by two experiments with a completely randomized factorial arrangement of treatments. The treatments were the supplementation of the basal diet (containing 0.28 µg of boron/g) with boron (as boric acid) at 0 and 3 µg/g, cholecalciferol at either 250 and 2500 IU/kg (Exps. 1 and 2), or 125 and 2500 IU/kg (Exp. 3). Fresh food and deionized water in plastic containers were provided each day. Environmental temperature was maintained at approximately 28° throughout the experiments. The chicks were provided 24 hours of light daily by using fluorescent lighting filtered through acrylic plastic.

The chicks were fed their respective diets for 28 to 32 days, weighed, and decapitated subsequent to cardiac exsanguination with a heparin-coated needle and syringe. The liver, spleen and heart were removed, blotted dry and weighed immediately. Microhematocrits were determined on heart blood. Hemoglobin and plasma cholesterol, alkaline phosphatase activity, and uric acid were determined by our usual methods (19). For brevity, only selected data from experiment 3 are presented. The data were treated by two-way analysis of variance (20).

RESULTS AND DISCUSSION

Our findings confirmed an interrelationship between boron and cholecalciferol (see table). The interrelationship was a reflection of contrasting effects of dietary boron on cholecalciferol-deficient or -adequate chicks. For example, in experiment 3, growth was 38% greater in boron-supplemented than boron-deprived chicks fed the cholecalciferol-deficient diet. When the diet contained adequate cholecalciferol, growth was only 11% greater in boron-supplemented than boron-deprived chicks. Similarly, plasma alkaline phosphatase activity and heart wt./body wt. ratio were apparently unaffected by dietary boron in cholecalciferol-adequate chicks. However, the apparently cholecalciferol-deficiency-induced elevation in plasma alkaline phosphatase activity and heart wt./body wt. ratio was greater in boron-deprived than boron-supplemented chicks. Also, rachitic long bones were found in 17 of 21 in boron-deprived, and in 9 of 22 in boron-supplemented, chicks fed the cholecalciferol-deficient diet. The growth, plasma alkaline phosphatase activity, heart wt./body wt. ratio and bone findings suggest that boron performs the

Effect in Chicks of Boron, Cholecalciferol and Their Interaction on Selected Parameters (Experiment 3)

Treatment[1] B	Treatment[1] Cholecalciferol	Body Weight[2]	Heart wt. Ratio[3]	Hct.	Plasma alk. Phosphatase Act.	Plasma Uric Acid	Spleen wt. Ratio[3]
mg/kg	IU/kg	g		%	units[4]	mg %	
0	125	561	0.729	34.1	4.65	1.55	0.133
3	125	775	0.644	32.8	2.93	1.28	0.143
0	2500	896	0.605	33.2	1.77	2.02	0.111
3	2500	994	0.621	32.3	1.75	1.89	0.116
Analysis of Variance - P Values							
Cholecalciferol effect		0.0001	0.0002	NS	0.0001	0.0001	0.0001
Boron effect		0.0002	NS	0.06	0.0004	0.05	NS
Boron x cholecalciferol		NS	0.01	NS	0.0005	NS	NS

[1]Amounts of boron (boric acid) and cholecalciferol supplemented to the diet.

[2]Weight at age 32 days.

[3]Grams of tissue/100 grams of body weight.

[4]Units were μmoles of p-nitrophenyl phosphate split/min/ml of plasma.

function of cholecalciferol at some metabolic site or is necessary for the efficient metabolism of cholecalciferol. On the other hand, cholecalciferol deficiency may be enhancing some of the signs of boron deficiency. In experiment 3, boron deficiency apparently elevated hematocrits and uric acid, and significantly depressed growth. The effect of dietary boron on these parameters was more marked in cholecalciferol-deficient than cholecalciferol-adequate chicks. Perhaps cholecalciferol deficiency enhances the need for boron. Thus, in the present studies, the low-boron diet with adequate cholecalciferol may be marginally deficient, but in the absence of adequate cholecalciferol, moderately deficient, in boron.

The findings do not establish an exact locus for the interrelationship between boron and cholecalciferol. Most likely, the interrelationship involves calcium and/or phosphorous metabolism. In addition to our findings, other evidence to support that suggestion was reported by Elsair et al. (21). They found that boron alleviated the negative effects of fluoride toxicity on calcium-phosphorous metabolism. For example, boron corrected the fluoride-induced negative calcium and phosphorous balances caused by depressed intestinal absorption of calcium and phosphorous, and depressed renal reabsorption of phosphorous. Boron also partially alleviated the fluoride-induced secondary hyperparathyroidism signs of hypercalcemia, hypophosphatemia, and depressed renal absorption of phosphorous.

The findings of the present study indicate a new area of research has been identified. Furthermore, elucidation of the interrelationship between boron and cholecalciferol may establish boron as an essential nutrient for animals and humans.

REFERENCES

1. Warrington, K., 1926, Ann. Bot. 40: 27-42.
2. Hove, E., C.A. Elvehjem, and E.B. Hart, 1939, Am. J. Physiol. 127: 689-701.
3. Orent-Keiles, E., 1941, Proc. Soc. Exp. Biol. Med. 44: 199-202.
4. Teresi, J.D., E. Hove, C.A. Elvehjem, and E.B. Hart, 1944, Am. J. Physiol. 140: 513-518.
5. Skinner, J.T., and J.S. McHargue, 1945, Am. J. Physiol. 143: 385-390.
6. Follis, R.H., 1947, Am. J. Physiol. 150: 520-522.
7. Weser, U., 1968, Z. Physiol. Chem. 349: 1479-1482.
8. Deitrich, R.A., 1967, Arch. Biochem. Biophys. 119: 253-263.
9. Roush, A., and E.R. Norris, 1950, Arch. Biochem. 29: 344-347.
10. Strittmatter, P., 1964, J. Biol. Chem. 239: 3043-3050.
11. Smith, K.W., and S.L. Johnson, 1976, Biochemistry 15: 560-565.
12. Johnson, S.L., and K.W. Smith, 1976, Biochemistry 15: 553-559.
13. Berezin, I.V., G. Ya. Kolomiitseva, A.V. Levashev, and K. Martinek, 1967, Mol. Biol. 1: 67-74.
14. Lindquist, R.N., and C. Terry, 1974, Arch. Biochem. Biophys. 160: 135-144.
15. Bauer, C.-A., and G. Pettersson, 1974, Eur. J. Biochem. 45: 473-477.
16. Weser, U., 1967, Proc. Soc. Exp. Biol. Med. 126: 669-671.
17. Nielsen, F.H., 1974, In "Trace Element Metabolism in Animals-2" (ed. W.G. Hoekstra, J.W. Suttie, H.E. Ganther, and W. Mertz) (University Park Press, Baltimore) pp 381-395.
18. Hunt, C.D., 1979, In "The Effect of Dietary Vanadium on ^{48}V Metabolism and Proximal Tibial Growth Plate Morphology in the Chick", Ph.D. Thesis, University of North Dakota, Grand Forks.
19. Nielsen, F.H., 1980, J. Nutr. 110: 965-973.
20. Scheffé, H., 1959, "The Analysis of Variance" (John Wiley, NY) pp 90-137.
21. Elsair, J., R. Merad, R. Denine, M. Reggabi, S. Benali, M. Azzouz, K. Khelfat, and M. Tabet Aoul, 1980, Flouride 13: 30-38.

DISCUSSION

HILL (UK). The recommended levels of 100 - 200 I.U. cholecalciferol per kilogram diet are not very different from the lower of your two treatment concentrations. The very high levels you used are not found in commercial diets. Is it possible that in practical diets the boron that is present is reducing the need for cholecalciferol?

NIELSEN (USA). The reason that we used the very high level of cholecalciferol was to be sure that the chickens would not suffer from cholecalciferol deficiency. We were concerned about the potency of some of our preparations. The results with 250 IU of low-potency cholecalciferol in the first two experiments were similar to those obtained with 125 IU of potent cholecalciferol in the last experiment. The usual recommended cholecalciferol allowance for growing chickens is 200 IU/kg diet. Although I am not ready to say that boron is essential for cholecalciferol metabolism, these results suggest that this might be true.

FUNCTIONS OF TRACE ELEMENTS

TRACE ELEMENTS AND IMMUNOCOMPETENCE

Kathleen M. Nauss and Paul M. Newberne

Laboratory of Nutritional Pathology
Massachusetts Institute of Technology
Cambridge, Massachusetts 02139

INTRODUCTION

The interrelationships between general nutrition, mineral elements and the immune system are complex and poorly understood. There has been a sharp upturn however in interest and in research activity in recent years relative to nutrition and immunocompetence. There is now available sufficient data to clearly indicate important functional roles for many nutrients, and, in particular, some of the trace elements in the development and maintenance of the immune system. Space limitations do not permit a review with in-depth coverage; those elements about which more is known will be emphasized. These will include the essential elements iron, zinc, magnesium, selenium, copper, manganese and iodine. The heavy metals lead, cadmium and mercury will be briefly discussed because of the implications that these, either directly or indirectly, can have profound influences on immunocompetence.

Nutritionists and immunologists, historically parochial in their scientific endeavors, are now bringing to bear the expertise of each discipline in collaborative research destined to elucidate important effects and mechanisms about nutrition and immunocompetence. The following pages will detail selected aspects of what is now known about trace elements and host resistance to disease.

IRON

Iron status and infection in infancy has been the subject of excellent recent reviews (1-3). It was more than fifty years ago when MacKay (4) reported that infants given iron had fifty percent fewer respiratory and gastrointestinal infections compared to those who did not receive dietary iron supplements. These observations supported the hypothesis that iron deficiency predisposes to increased susceptibility to infection. Immunocompetence has been studied in human patients and in animals with varying degrees of iron deficiency anemia (5-7). Reports of both human and animal studies provide conflicting data.

The major sickle cell hemoglobinopathies are associated with a high incidence of infection. These include bacterial septicemia, Salmonella osteomyelitis and severe pneumonia. A possible explanation suggested for the increased susceptibility to infections is hyperferremia, (to be discussed later), a result of chronic severe hemolytic states associated with these diseases (8). On the other hand, significant iron loss is not a feature of sickle cell anemia in the first few years of life. Although hemolysis has its onset in early infancy (9) iron absorption is probably increased in sickle cell anemia because of the erythroid marrow hyperplasia. Part of the increased absorption can be balanced by iron loss through hemosiderinuria, often observed in these patients.

Other investigators have reported that iron deficiency is not necessarily associated with increased susceptibility to infection (10,11). One report (12) has indicated that individuals with iron deficiency anemia experienced less susceptibility to infections than patients with other types of anemia. Fletcher et al (13) found lymphocyte numbers decreased in iron-deficiency and Jarvis and Jacobs (14) observed vacuolar changes in mitochondria of 90% of the lymphocytes of iron deficient patients. In another study the lower percentage of T-lymphocytes in iron-deficient children was corrected by iron therapy over a 4-week period (15,16).

There have been no reported clinical abnormalities in serum immunoglobulin concentrations (17,19) but iron deficiency results in a decrease in antibody synthesis after immunization with tetanus toxoid (20). Cell-mediated immunity is reported to be diminished in iron deficiency by some investigators (13,15,18,19,21,22,23) but reported not to be affected by others (24-26). Such conflicting results may be because of differences in nutritional status, infection, inadequate controls, timing of the studies, methods, folate status, iron therapy and others. In general, clinical studies have been less well designed and controlled, compared to animal investigations although the latter are few in number. Even so, data from them are conflicting.

Iron-deficient pigs with experimental anemia suffered a very high mortality from E. coli endotoxin (27). Iron deficient rats had significantly decreased antibody production when vaccinated with tetanous toxoid (20). Iron deficient rats challenged with Streptococcus pneumoniae had a 1000 fold increase in susceptibility to infection, compared to control rats (28). Baggs and Miller found that rats born to iron deficient mothers suffered increased morbidity and mortality, compared to controls, when exposed to Salmonella typhimurium (29, 30). These investigators concluded that there was a decreased number of myeloperoxidase-containing phagocytes in the infected gastrointestinal tract and this predisposed to more severe disease. There were a number of critical factors not considered in the latter studies: the assay for myeloperoxidase would not differentiate the enzyme specific for neutrophils, eosinophils and monocytes from non-specific substances containing peroxidase activity; morphologic studies were not reported, to identify the cells in the gut nor were other body defenses measured, leaving the results open to question.

Bacteria require adequate quantities of iron to reach their full potential for growth, replication and production and release of some of the exotoxins. It has been shown that, under controlled conditions, iron facilitates growth and multiplication of microorganisms (8). An hypothesis states that increased availability of iron (unbound) increases the virulence and capacity of bacteria to invade the host. This concept is supported by a number of lines of argument. For example, the total concentration of iron in biologic fluids (plasma, gastrointestinal secretions, milk) far exceeds the minimum requirement for bacterial growth. This would appear to argue against a growth-promoting effect of excess iron. However, iron-binding proteins (transferrin, lactoferrin, gastroferrin) are present in these fluids and they are capable of binding iron with sufficient affinity to reduce the free iron available to bacterial systems to a very small amount.

In vivo studies (8) have shown that, as the serum iron level increases and transferrin becomes more saturated, the amount of free iron in equilibrium with the transferrin-iron complex becomes sufficiently high to allow bacterial growth. In addition, some microbial systems can apparently extract iron from free hemoglobin. Serum iron however appears to be unavailable for bacteria under normal conditions because it is tightly bound to transferrin and other binding proteins. Furthermore, serum from mammalian species is normally bacteriostatic; if ionic iron is added in concentrations sufficient to saturate 60-80% of available transferrin, the bacteriostatic effect is abolished (31). Thus, it appears that, while excess free iron may enhance infectious disease, mechanisms are available to withhold iron from invading microorganisms (2,32,33). In cases where protein synthesis is diminished or where protein is not available, such as in PCM, binding of iron may not be possible because binding proteins are not present in adequate supply (12,34-36).

Under other conditions body iron stores are increased but bacterial infections do not appear to be increased. These include hemachromatosis where there is no unusual susceptibility to infection in children or adults (37,38). There does appear to be a problem where iron-binding proteins are in short supply and unbound iron may be in excess (12,35,36,39).

The data referred to above point out the difficulties in interpreting currently available information about the effects of iron on the immune system and host defense; it is equivocal.

ZINC

Overt zinc deficiency was first observed and reported among rural inhabitants of the Middle East in the early 1960's (40). Although there has been little evidence for nutritional deprivation, children in the United States have exhibited zinc deficiency (41). These observations were somewhat surprising because zinc is ubiquitous in food, water, and in the general environment. Zinc forms strong chelates with many ligands and often occurs in nature in the form of complexes. Some complexes are readily absorbed, others are poorly absorbed (42). These factors complicate the assessment of zinc in mammalian systems since dietary concentrations provide no information on bioavailability. There is now a considerable body of knowledge and a burgeoning interest in the influence of zinc on the immune system; the accumulated data to date indicate that this trace element has a profound effect on host defense mechanisms relative to infectious disease.

Hereditary zinc deficiency in humans (acrodermatitis enteropathica) and in domestic animals (43,44) is associated with a decrease in cell mediated immunity which is reversed by zinc therapy. Some aspects of zinc deficiency do not fit the usual pattern; for example, the synthesis of plasma acute phase glycoproteins is not compromised by zinc deficiency (45). On the other hand it appears that most host defenses against infectious diseases are diminished by zinc deficiency. Zinc deficient animals and children exhibit atrophy of the thymolymphatic system, depressed cell-mediated immunity and increased susceptibility to infection (46,47).

Children with protein energy malnutrition have atrophy of the thymolymphatic system and immunological changes similar to those associated with zinc deficiency (48). Children suffering from these kinds of immunologic deficits have decreased liver concentrations of zinc (49-51). Acrodermatitis enteropathica, a genetic disorder of zinc absorption, referred to above is associated with severe skin lesions, CNS disturbances, gastrointestinal malfunction and above normal susceptibility to infections (52).

A number of investigators have observed an effect of zinc on in vivo and in vitro systems related to host metabolic responses to inflammation (53,54). Among the roles suggested for zinc is a potentiation of fever and temperature-dependent host defenses.

Zinc deficiency has a profound effect on the anatomy of the lymphoid tissues of experimental animals (44,55-57). The deficiency results in hypoplasia of the thymus, spleen, lymph nodes, Peyers patches and other intestinal lymphoid tissues. Both T- and B-cell populations are under the influence of zinc deficiency but the T-cell areas of the nodes are most severely affected (58). Zinc supplementation reverses the hypoplasia in a relatively short period of a few weeks. A significant clinical study was conducted by Golden et al (59) using chest X-rays to demonstrate thymic atrophy. Eight children with coexisting zinc deficiency and protein energy malnutrition had an increased thymus size and plasma zinc levels within 10 days following supplementation with 2 mg/kg/day of zinc acetate.

A number of derangements have been noted in zinc deficient immune systems of human patients and in experimental animal models (55,60-64). There is a depression in T-helper lymphocyte functions, diminished activity of natural killer cells, decreased T-lymphocyte killer activities and depressed cytotoxic activity of NK cells. A depression in immunoglobulin concentrations in zinc deficiency has been reported (62) in human patients and a disordered profile of Ig distribution in mice (57). In general, mitogen stimulation of lymphocytes in zinc deficiency is depressed.

The thymic atrophy associated with zinc deficiency is accompanied by decreased output of thymopoietin and other measurable humoral factors (62,65,66).

From the brief resume of zinc and the thymolymphatic system it is clear that this trace element is of signal importance to the immune system and host defense mechanisms. While effects of deficiency are profound, reversal on supplementation appears to take place in a remarkably short period.

MAGNESIUM

There are no reported changes in the immune system of human subjects relative to magnesium deficiency. There are, however, significant modifications in the immune system of magnesium deficient rats and other animal species. This is particularly the case in chronic studies. A selected representative sample of such observations are reported in the next few paragraphs.

A magnesium deficient diet, fed to rats, resulted in a leucocytosis including increases in the number of neutrophils, mononuclear cells and, especially, eosinophils (67-70). Magnesium deficient rats also exhibited dermal hyperemia as well as the white cell increases (68,71). There was a degranulation of mast cells, increased urinary histamine, and a progressive change in WBC to leukemia (68,71). Others reported an abnormal distribution and a decreased granularity of mast cells in magnesium deficient rats (70) but no measurable change in histamine excretion. Battifora et al (69) confirmed that myelogenous leukemia occurred in magnesium deficient rats; cells from these rats could be transplanted to syngeneic animals during early postnatal life. Morever, Sprague-Dawley rats developed malignant lymphoma on a magnesium deficient diet after six months (72). There was also an appearance of spontaneous malignant lymphomas in magnesium deficient rats and an increased susceptibility to transplanted lymphoma cells (73).

Immunoglobulin concentrations in the serum of magnesium-deficient rats (74) and mice were depressed (75). There is disagreement about the capacity of magnesium-deficient animals to respond favorably to immunization (74-77).

In contrast to the usual history of nutritional deficiency atrophy of the thymus gland with many nutrients, in magnesium deficient rodents, thymic hyperplasia may be the rule (76), despite the fact that the deprived animals may become more susceptible to parasitic infections.

SELENIUM

There are no reports of selenium affecting the immune functions of humans although ^{75}Se is taken up by lymphocytes utilizing a sulfhydryl-requiring mechanism (78). In other studies it was demonstrated that either a deficiency or an excess of selenium reduced the development of HA antibody titer (79-81). Contrary to these effects, diets containing 1.25-2.25 ppm selenium were associated with an increase in primary immune response to sheep RBC reflected by increased numbers of spleen plaque-forming cells and higher IgM and IgG titers of HA antibodies.

The addition of 2.5 μg/ml of selenium to the drinking water potentiated vaccination of mice with P. berghei (82).

While studies relative to the effects of selenium are few in number, they indicate that selenium does have a profound effect on the immunocompetence of animals; effects on man are lacking.

COPPER

In spontaneous and controlled experiments with Salmonella typhimurium infection in rats it has been shown that copper deficiency results in high mortality (83). Furthermore, lymphocytes of copper deficient rats exhibit less mitogenic stimulation than supplemented controls (84). Ionic copper has also been shown to enhance the phagocytic activity and oxygen uptake by the granulocytes of the dog (85). While data are inadequate it appears that this trace element affects the immune system of mammals.

IODINE

Doses of iodine as high as 1.2 mg/day to calves resulted in diminished antibody response to vaccination with Brucella or Leptospira. In vitro studies revealed a reduction in lymphocytic response to stimulation with PHA and decreased phagocytic activity of peripheral blood neutrophils (86). Iodine appears to be involved with a capacity for neutrophils to destroy bacteria following ingestion through the production of myeloperoxidase (87). It has been postulated that the thyroid gland may be providing the necessary iodide to the neutrophils by way of the thyroid hormone. Human neutrophils generate both T_3 and T_4 under controlled conditions (88). Further, the deiodination of T_4 is significantly enhanced in the peripheral leukocytes of septic monkeys (89). Peroxide and the superoxide radical are important to such reactions (90).

HEAVY METALS AND IMMUNE RESPONSE

LEAD

When lambs were fed solubilized lead acetate, daily for 12 weeks, in graded doses up to 16 mg/kg serum levels reached 99 μg/dl in the highest dose group. In spite of these levels there were no detectable effects on serum IgM or IgG or on HA antibody titer responses to immunization with Serratia marcescens (91). However, mice (92) or rabbits (93) given lead acetate had a decreased number of plaque-forming cells with SRBC and depressed neutralizing antibody responses to pseudorabies vaccination respectively. Other investigators have found that mice given lead in the drinking water for 15 weeks were no different from controls relative to stimulation of peripheral blood lymphocytes by Con-A or LPS; stimulation by PHA was significantly decreased (94).

CADMIUM

Dietary cadmium at non-toxic doses increased resistance of chickens to S. gallinarium but toxic levels (100 ppm) reduced resistance under similar conditions. The latter was easily reversed (95,96). Chvapil et al found no effect of moderate concentrations of cadmium on phagocytic activity of dog neutrophils in vitro (97) but others (98) have reported reduced phagocytic activity of neutrophils and macrophages from mice.

MERCURY

There are sparse data on mercury and immunocompetence. Hill (96) has observed lowered resistance of chickens to S. gallinarium, at 500 ppm exposure, and rabbits exhibited diminished synthesis of neutralizing antibodies to pseudorabies vaccination (93).

REFERENCES

1. O'Brien, D. (ed.) 1979. "Pediatric Nutrition Handbook," Chapter 37. Amer. Acad. Pediat. (Evanston, Ill.).

2. Bullen, J.J., H.J. Rogers, and E. Griffiths, 1978, Curr. Top. Microbiol. Immunol. 80: 1-30.

3. Chandra, R.K., G. Woodford, and P. Hyam, 1977. In "Iron Metabolism" (ed. H. Kies) Elsevier, (Amsterdam) pp. 249-268.

4. MacKay, H.M., 1928, Arch. Diseases of Children 3:117-122.

5. Andelman, M.B. and B.R. Sered, 1966, Am. J. Diseases of Child. 111:45-54.

6. Salmit, P., P. Hanninen, and T. Pelttonen, 1963, Acta. Paediat. Scan. 140:114-120.

7. Shaw, R. and W.O. Robertson, 1964, Ohio Med. J. 60:45-52.

8. Weinberg, E.D. 1978, Microbiol. Rev. 42:45-66.

9. Suskind, R.M. (ed.) 1977. In "Malnutrition and the Immune Response, (Raven Press, New York).

10. Burman, D. 1972, Arch. Dis. Child. 47:361-370.

11. James, J.A., and M. Combs, 1960, Pediatr. 26:368-375.

12. Masame, A.E.J., J.M. Muingi, and G.B.R. Swai, 1974, Lancet 2:314-316.

13. Fletcher, J., J. Mather, M.J. Lewis, and G. Whiting, 1975, J. Infect. Dis. 131:44-50.

14. Jarvis, J.H., and A. Jacobs, 1974, J. Clin. Pathol. 27:973-979.

15. Bhaskaram, C., and V. Reddy, 1975, Brit. Med. J. 3:522.

16. Bhaskaram, C., P. Siva, and K.A. Krishnamachari, 1977, Lancet 1:1000.

17. Chandra, R.K., and A.K. Saraya, 1975, J. Pediatr. 86:899-902.

18. MacDougall, L.G., R. Anderson, G.M. McNab, and J. Katz, 1975, J. Pediatr. 86:833-843.

19. Sawitsky, B., R. Kauter, and A. Sawitsky, 1976, Am. J. Med. Sci. 272:153-160.

20. Nalder, B.N., A.W. Mahoney, R. Ramakrishnan, and D.G. Hendricks, 1972, J. Nutr. 102: 535-542.

21. Jacobs, A., and D.H.M. Joynson, 1972, J. Nutr. 102:535-542.

22. Joynson, D.H.M., A. Jacobs, W.D. Murray, and A.E. Dolby, 1972, Lancet 2:1058-1059.

23. Srikantia, S.G., P. Siva, C. Bhaskaram, and K.A. Krishnamachari, 1976, Lancet 1:1307-1309.

24. Gross, R.L., J.V.O. Reid, P.M. Newberne, et al, 1975, Am. J. Clin. Nutr. 28:225-232.

25. Kulapongs, P., V. Vithayasai, R. Suskind, and R.E. Olson, 1974, Lancet 2:689-691.

26. Suskind, R., P. Kulapongs, V. Vithayasai, and R.E. Olsen, 1977. In "Malnutrition and The Immune Response, Raven Press (New York) pp. 387-393.

27. Osborne, J.C., and J.W. Davis, 1968, J. Am. Vet. Med. Assoc. 152:1630-1632.

28. Hegsted, D.M., E.S. Murray, and K.J. Welsh, 1975, Proc. Soc. Exp. Biol. Med.

29. Baggs, R.B., and S.A. Miller, 1973, J. Nutr. 103:1554-1560.

30. Baggs, R.B., and S.A. Miller, 1974, J. Infect. Diseases 130:409-418.

31. Bullen, J.J., H.J. Rogers, and L. Leigh, 1972, Brit. Med. J. 1:59.

32. Weinberg, E.D., 1975, JAMA 231:39-41.

33. Bullen, J.J., and J.A. Armstrong, 1979, Immunology 36:781-791.

34. McFarlane, H., S. Reddy, K.J. Adcock et al, 1970, Brit. Med. J. 4:268-270.

35. Murray, M.J., and A.B. Murray, 1978, Am. J. Clin. Nutr. 31:700.

36. Murray, M.J., A.B. Murray, M.B. Murray, and C.J. Murray, 1978, Brit. Med. J. 2:1113-1115.

37. Finch, S.C., and C.A. Finch, 1955, Medicine 34:381-386.

38. Perkins, I.W.S., C.R.B. Blackburn, and R.W. Beale, 1965, Amer. J. Med. 39:118-123.

39. McFarlane, H., S. Reddy, K.J. Adcock et al, 1970, Brit. Med. J. 4:268-270.

40. Prasad, A.S., A. Miale, Z. Farid, H.H. Sandstead, and A.R. Schebert, 1975, J. Lab. Clin. Med. 61:537-549.

41. Hambidge, K.M., and P.A. Walravens, 1976. In "Trace Elements in Human Health and Disease. Vol. 1, Zinc and Copper," Acad. Press (New York) pp. 21-31.

42. O'Dell, B.L., 1969, Am. J. Clin. Nutr. 22:1315-1322.

43. Nelder, K.H., and A.M. Hambidge, 1975, New Eng. J. Med. 292:879-882.

44. Brummerstedt, B., A. Basse, T. Flagstad, and E. Andresen, 1974, Am. J. Path. 87:725-728.

45. Pekarek, R.S., and M.C. Powanda, 1976, J. Nutr. 106:905-912.

46. Fraker, P.J., S.N. Haas, Luecke, R.W., 1977, J. Nutr. 107:1889-1895.

47. Kroneman, J., G.J.W. Van de Mey, A. Helder, and B.L. Cent, 1975, Vet. Med. 22:201.

48. Edelman, R. 1977. In "Malnutrition and the Immune Response" (ed. R. Suskind)(Raven Press, N.Y.) p. 47.

49. Warren, P.J., J.D. Hansen, and B.J. Leorman, 1969, Proc. Nutr. Soc. 28:6a.

50. Cheek, D.B., P.E. Hill, A. Cordano, and G.G. Graham, 1970, Pediatr. Res. 4:135-142.

51. Kahlili, M., A. Kabiel, A. El-Kahteb et al, 1974, Am. J. Clin. Nutr. 27:260-269.

52. Moynahan, E.J., and P.M. Barnes, 1973, Lancet 1:676.

53. Mapes, C., P.T. Bailey, C.F. Matson, E.C. Hever, and P. Sobocinski, 1978, J. of Cellular Physiol. 95:115-120.

54. Chvapil, M., L. Stankova, Z. Bartos, T. Cox, and Nichols, W., 1979, J. Reticuloendothel. Soc. 25:345-354.

55. Fernandes, G., M. Nair, K. Onoe et al, 1979, Proc. Nat. Acad. Sci. USA 76:457-461.

56. Fraker, P.J., P. DePasquale-Jardien, C.M. Zwickl, and R.W. Luecke, 1978, Proc. Nat. Acad. Sci. USA 75:5660-5664.

57. Beach, R.S., M.E. Gershwin, R.K. Makishima, and L.S. Hurley, 1980, J. Nutr. 110:805-815.

58. Tanaka, T., G. Fernandes, C. Tsao et al, 1978, Fed. Proc. 37:931.

59. Golden, M.H.N., A.A. Jackson and B.E. Golden, 1979, Lancet 2:1057-1059.

60. DePasquale-Jardien, P., and P.J. Fraker, 1979, J. Nutr. 109:1847-1855.

61. Chandra, R.K., and B. Au, 1980, Am. J. Clin. Nutr. 33:736-738.

62. Cunningham-Rundles, C., S. Cunningham-Rundles, and J. Garafolu et al, 1979, Fed. Proc. 38:1222.

63. Gross, R.L., N. Osdin, L. Fong, and P.M. Newberne, 1979, Am. J. Clin. Nutr. 32:1260-1265.

64. Pekarek, R.S., H.H. Sandstead, R.A. Jacob, and D.F. Barcome, 1979, Am. J. Clin. Nutr. 32:1466-1471.

65. Good, R.A., A. West, and G. Fernandes, 1980, Fed. Proc. 39:3098-3104.

66. Imata, T., G.S. Incefy, T. Tanaka et al, 1978, Fed. Proc. 37:1827.

67. Kashiwa, H.K., and G.F. Hungerford, 1958, Proc. Soc. Exp. Biol. Med. 99:441-443.

68. McCreary, P.A., H.A. Battifora, G. Laing, and G.M. Haas, 1966, Proc. Soc. Exp. Biol. Med. 121:1130-1133.

69. Battifora, H.A., P.A. McCreary, and B.M. Hahneman et al, 1968, Arch. Path. 86:610-620.

70. Bois, P., 1963, Brit. J. Exp. Path. 44:151-155.

71. McCreary, P.A., H.A. Battifora, and B.M. Hahneman et al, 1967, Blood 29:683-690.

72. Haas, G.M., G.H. Laing, P.A. McCreary, and R.M. Galt, 1978, Clin. Res. 26:710a.

73. McCreary, P., G. Laing, and G. Haas, 1973, Am. J. Pathol. 70:89-90.

74. Alcock, N.N., and M.E. Shils, 1974, Proc. Exp. Biol. Med. 145:855-858.

75. Elin, R.J., 1975, Proc. Exp. Biol. Med. 148:260-624.

76. Haas, G.M., P.A. McCreary, G.H. Laing, and R.M. Galt, 1980. In "Magnesium in Health and Disease" (eds M. Cantin and M. Seelig) (Spectrum, Jamaica, N.Y.) pp. 185-200.

77. McCoy, J.H., and M.A. Kenney, 1975, J. Nutr. 105:791-797.

78. Porter, E.K., J.A. Karle, and A. Shraft, 1979, J. Nutr. 109:1901-1908.

79. Spallholz, J.E., J.L. Martin, M.L. Gerlock, and R.H. Heinzerling, 1973, Proc. Exp. Biol. Med. 143:685-689.

80. Spallholz, J.E., J.L. Martin, M.L. Gerlock, and R.H. Heinzerling, 1973, Infect. Immun. 8:841-842.

81. Spallholz, J.E., J.L. Martin, M.L. Gerlock, and R.H. Heinzerling, 1975, Proc. Exp. Biol. Med. 148:37-40.

82. Desowitz, R.S., and J.W. Barnwell, 1980, Infect. Immun. 27:87-89.

83. Newberne, P.M., R.B. Wilson, and G. Williams, 1970, Brit. J. Exp. Path. 51:229-235.

84. Newberne, P.M., and R. Gross, 1977, unpublished.

85. Chvapil, M., L. Stankova, and C. Zukoski, 1977, J. Lab. Clin. Med. 89:135-146.

86. Haggard, D.L., H.D. Stowe, G.H. Conner, and D.W. Johnson, 1980, Am. J. Vet. Res. 44:539-543.

87. Klebanoff, S.J., and W.L. Green, 1973, J. Clin. Invest. 52:60-62.

88. Woeber, K.A., 1978, J. Clin. Invest. 62:577-584.

89. DeRubertis, F.R., 1974, J. Lab. Clin. Med. 83:902-910.

90. Woeber, K.A., 1976, Endocrin. 99:887-890.

91. Hoffman, L.J., W.B. Buck, and L.Y. Quinn, 1980, Am. J. Vet. Res. 41:331-337.

92. Koller, L.D., and S. Kovacic, 1974, Nature 250:148-150.

93. Koller, L.D., 1973, Am. J. Vet. Res. 34:1457-1458.

94. Nerlan, B.A., and L. Taddeini, 1979, Clin. Res. 27:223A.

95. Hill, C.H., 1979, Fed. Proc. 38:2129-2133.

96. Hill, C.H., 1980, J. Nutr. 110:433-436.

97. Loose, L.D., J.B. Silkworth, and D.W. Simpson, 1978, Infect. Immun. 22:378-381.

98. Loose, L.D., J.B. Silkworth, and D. Warrington, 1977, Biochem. Biophys. Res. Commun. 79:326-332.

DISCUSSION

PRASAD (USA). I would like to make a comment - as you know, Dr Giblet reported nucleoside phosphorylase deficiency in the T cell in human disorders of the congenital type and ADA deficiency in the B cell. So because of this lead we have looked into the activity of nucleoside phosphorylase in the red cells and muscle of animals and it appears that zinc is regulating the effect of phosphorylase but not of ADA. Indeed in our human model and chronic uremics we find that the B cell is not affected at all, but the T cells are. My question is, you found some B cell effect, could this be related to T helper cell function?

NEWBERNE (USA). Yes, I think there is evidence of that - not very much reported on it but it does seem to affect the helper cells and the subsets.

READING (Australia). Dr Horrobin talks about prostaglandins and the importance of zinc for prostaglandin E1 series synthesis which has an immuno potentiator effect on T lymphocytes. He also mentions manganese being very important in these pathways and thus manganese seems to have a role as an immuno potentiator also. In some of the recent literature on complement metabolism magnesium seems to be very important, and I think that is worth mentioning.

NEWBERNE (USA). Do you mean effects on the complement system?

READING (Australia). On the complement system - magnesium has a role. There is one more thing, when you see low iron in coeliac disease and the low complements C3 and C4 it is hard to know whether this is related to the iron deficiency or whether it is just suppression by the gluten. Quite often it is a mixed picture - you can have iron deficiency with absolutely normal complements. So iron may not suppress the C3, it may be the food allergy which is upsetting the complement.

NEWBERNE (USA). Does gluten have a direct affect on the complement system?

READING (Australia). I can't answer. I know it is associated, but I am not sure how.

PROHASKA (USA). I would like to point out that between species there is a very great deal of variation even in closely related species like two different species of rodents. In the copper deficient mice we get splenomeglia and when we look at the transformation of lymphocytes in vitro we find that it is T cell rather than B cell specific - it seems to be just opposite in the rat.

NEWBERNE (USA). I think that is a very important point. You can't generalise from one species. If you are going to try and relate it to man the work must be done in man.

NEWNHAM (Australia). Recent work in Sydney has shown that diseases such as asthma, psonasis and rheumatoid arthritis, are often associated with inadequate secretions from the pancreas and high pH in the duodenum. However, by giving alkali and pancreatic enzymes, those diseases can often be helped. I have also had considerable success in treating rheumatoid arthritis with boron, another trace element which is not often investigated. Any comment?

NEWBERNE (USA). If you are right, what do you think is the mechanism of the boron influence on inflammatory cells?

NEWNHAM (Australia). Boron I think is involved in RNA and DNA production. Cartilage ligament has a poor blood supply and suffers first when boron is inadequate. Boron passes through the body quickly, half of it is lost in 12 hrs but on its way through there must be an RNA-ase or similar enzyme involved.

WILLIAMS (Wales). The previous questioner introduced rheumatoid arthritis into the discussion. The copper dependance of rheumatoid arthritis has been extensively investigated recently and it has been shown that copper deficiency in animals can lead to an increased incidence of rheumatoid arthritis. In your opinion, is it a failure in the immune response mechanism of the animals that produces the inflammation associated with the arthritis?

NEWBERNE (USA). I think it might be a combination of things, more than just a simple effect on the immune system. In copper deficiency there is a defect in connective tissue synthesis and maintenance. It is a very interesting field and there is insufficient information on the interaction of copper with the immune system.

SEGAL (Australia). I would like to ask whether there is any evidence that people suffering from Menkes disease also have arthritis?

WILLIAMS (Wales). There is a small incidence of childhood arthritis but in general it is a disease of older people. Menkes disease patients are unlikely to reach old age.

READING (Australia). I see copper deficiency quite often in chronic neurasthenics (patients with chronic lack of energy - most are adults), most of these don't have arthritis.

GOLDEN (Jamaica). There are very wide differences between species in response to particular infective organisms. The infectious spectrum that we see in protein-energy malnutrition is very interesting. There are certain infections which are very uncommon in an immuno-suppressed host, particularly diseases like meningitis, but extremely uncommon in protein-energy malnutrition compared to the well nourished. Would you like to speculate about the reasons for this?

NEWBERNE (USA). I am glad you brought this up because it is a point of extreme importance. The infective organism is going to have a very distinct effect on the outcome of the interaction between the host and the organism. A part of that interaction is the ability of the organism to get iron for instance. *Pseudomonas* secretes a very strong iron binding material. Although this is not the whole answer, I think the variation between different kinds of infective organisms in secreting siderophores must have some effect.

SUTTLE (Scotland). We have been studying the susceptibility of copper deficient mice to infection. In three experiments, we have found a reduced LD50 in copper deficient mice when challenged with *Pasteurella haemolytica* in hog mucin.

NEWBERNE (USA). Years ago we did studies in beagle dogs. We fattened them up and exposed them to either a viral or bacterial infection. Invariably the fat dogs were much more sensitive to infection. Recently we were able to do a little more work, and found that in some way this is connected with the prostaglandin status. The zinc and iron levels were deranged as well as the concentrations of some steroids.

BRAIN MICROTUBULE ASSEMBLY IN ZINC-DEFICIENCY

John E. Hesketh

Rowett Research Institute, BUCKSBURN, ABERDEEN AB2 9SB, SCOTLAND, U.K.

Microtubules consist of linear arrays of the major protein monomer tubulin together with smaller amounts of microtubule-associated proteins. The *in vitro* assembly of tubulin into microtubular structures is influenced by protein factors, inhibited by Ca^{++} and promoted by Mg^{++} (1). Zinc ions have been found to stimulate tubulin assembly *in vitro* leading to the formation of several types of unusual supramolecular assemblies (2,3) and high zinc concentrations have been shown to induce tubulin aggregates, similar to those found *in vitro*, in dorsal rat ganglion cultures (4). Despite an accumulation of evidence showing an interaction of zinc with tubulin, it is not yet established whether tubulin assembly is regulated by metal ions such as zinc, *in vivo*. However, certain cytotoxic effects of colchicine, a drug known to interact with tubulin, are exacerbated by dietary induced zinc deficiency (5) suggesting that manipulation of the zinc levels *in vivo* may alter microtubule function. In the present experiments we have examined the ability of brain tubulin to reassemble *in vitro* and have compared such reassembly in brain extracts from zinc-deficient and zinc-supplemented animals.

Male Large White x Landrace pigs, weighing 20-30 kg were housed singly in wooden pens and allowed free access to tap water which contained 0.5 g/kg Zn. The zinc deficient diet contained per 400 kg air-dry weight: maize 300 kg, soya 80 kg, $CaCO_3$ 12 kg, $CaHPO_4$ 8 kg, $MnSO_4$ 113.6 g, $FeSO_4$ 139.2 g, $CoCl_2$ 1.6 g, NaCl 2 kg, $CuSO_4$ 31.4 g, KIO_3 660 mg and 36 g vitamin supplement. The zinc content of the diet was 22-25 g/kg, the phytate content 0.70-0.88% giving a molar ratio of between 32:1 and 35:1. The control diet was supplemented with 264 g $ZnSO_4$/400 kg diet. Initially, animals were fed according to body weight but when the food-intake of the zinc deficient animals started to decline the controls were fed the average daily intake/kg body weight of the zinc-deficient group. Animals were killed, in pairs, by bleeding from the neck vessels after anaesthesia with sodium pentobarbitone. Male Hooded Lister weanling rats of the Rowett Institute strain, weighing 40-50 g, were individually housed and fed a semi-synthetic diet based on sucrose, egg-albumen and arachis oil with vitamin and mineral supplements (6). The zinc deficient diet contained 0.7 mg/kg Zn and the control diet was supplemented to give a Zn content of 40 mg/kg. Controls were pair-fed. Animals were killed by cervical dislocation.

Pig cerebellum or whole rat brain was homogenised in 2 vols 10 mM sodium phosphate buffer pH 7.0, 10 mM $MgCl_2$, 1 mM guanosine triphosphate (GTP). The homogenate was kept at 0-4°C for 30 min to allow full depolymerization of microtubules and then centrifuged at 100,000 g for 30 min at 0-4°C. Aliquots of supernatant fluid were taken for zinc analysis, measurement of colchicine-binding and assessment of microtubule assembly by turbidometry (3) following addition of 1.5 mM GTP. Turbidity was monitored at 5 or 10 min intervals during incubation for 1 hour at 37°C and samples of the supernatant from rat brain were taken at the end of the incubation for examination by electron microscopy; samples were stained with 1% uranyl acetate for 10 secs, washed in distilled water and air-dried. The remaining supernatant fluid containing reassembled microtubules was centrifuged at 100,000 g x 30 min at 20-25°C to pellet microtubular protein. The free sulphydryl group content of this pelleted material was estimated using Ellman's reagent (7) and alkaline phosphatase measured by the liberation of nitrophenol from p-nitrophenol phosphate at pH 10.4 (8).

Dietary zinc deprivation in the pig led to a fall in trichloroacetic acid-soluble plasma zinc after 2 weeks and the levels remained low (0.2-0.4 µg/ml) during the subsequent 7 weeks of the experiment. After 4-6 weeks of treatment the animals deprived of dietary zinc showed various clinical signs of zinc deficiency (9), namely parakeratosis of the skin, loss of appetite and retardation of growth as shown by reduced weight gain. In the rat zinc deficiency was associated with reduced weight gain and cycling of food intake (9). High speed supernatants prepared from cerebral cortex of zinc-deficient pigs were found to contain less zinc than those from control animals: the zinc content (µg/ml) of preparations from the deficient group was 1.21 $\pm$ 0.04 (s.e.m.), n = 4 compared to 1.36 $\pm$ 0.03 (s.e.m.), n = 4 for preparations from the control group. This difference was less marked in the rat preparations where the zinc content (μg/ml) of high speed supernatants was 2.15 ± 0.05 (s.e.m.) for the deficient group and 2.24 ± 0.05 (s.e.m.) for the supplemented animals.

The polymerization of tubulin in high-speed brain supernatant fluids was measured turbidometrically following the addition of GTP. Pigs were studied in pairs, one control and one zinc-deficient; in each experiment, the zinc-deficient animal showed a reduced rate of increase in turbidity compared to the control preparations and the difference in turbidity increase was statistically significant ($p < 0.02$) when analysed by a paired 't' test. The effect was seen primarily over the first 30 min suggesting a reduction in the rate of assembly or a longer lag phase. In a second experiment, using rats, zinc deficiency was again found to be associated with a reduced rate of tubulin reassembly (Table 1). Electron microscopy showed such supernatants to contain many microtubules, suggesting that the observed turbidity changes do indeed reflect assembly of microtubules. In both the rat and the pig zinc deficiency had no effect on the tubulin content, as judged by colchicine-binding activity, of high-speed brain supernatant fluids (Table 1). It was thus not possible to account for differences in assembly by differences in tubulin content.

TABLE 1. Microtubule reassembly in high-speed supernatants from brains of rats fed a zinc-deficient diet and from zinc-supplemented controls

	Zinc-deficient	Zinc-supplemented
Rate of reassembly (change in absorbance/hr)	0.224 ± 0.021* (6)	0.289 ± 0.023 (6)
Supernatant colchicine-binding activity (10^6 cpm/ml)	1.35 ± 0.14 (6)	1.27 ± 0.08 (6)
Microtubule free SH group content (mol SH x 10^{-7}/mg protein)	0.56 ± 0.06** (5)	0.78 ± 0.04 (5)
Microtubule alkaline phosphatase (change in absorbance/hr/mg protein)	0.69 ± 0.16 (5)	0.95 ± 0.13 (5)

Values shown are means ± s.e.m. with number of animals in parentheses. Groups were compared using a Students 't' test, * $p < 0.05$, ** $p < 0.02$ compared to zinc-supplemented controls.

The free sulphydryl group content of microtubular protein pelleted from zinc-deficient rat brain supernatant fluid (following incubation for 1 hour with GTP) was lower ($p < 0.02$) than that of microtubule protein from control preparations (Table 1). The mean microtubule alkaline phosphatase activity was lower in the zinc-deficient animals but the effect was not statistically significant.

The rate of assembly in both rat and pig preparations was considerably slower than that found with purified microtubule or tubulin preparations and slightly slower than, but comparable to, that found previously with brain high-speed supernatants (10). No EGTA was added to assembly buffers in the present experiments since it would have disturbed the zinc equilibrium and the subsequent lack of control of the calcium concentration would be expected to slow microtubule assembly (1). The slow rate of reaction made it not possible to determine the effect of assembly at equilibrium. The observed turbidity changes suggest that the reduction in assembly is attributable to a lower rate of assembly; alternatively it could be due to a lower stability of formed microtubules. The present experiments show that restriction of zinc availability to an animal causes an impairment in the ability of its brain tubulin to assemble under in vitro conditions; this shows that small variations in zinc concentrations near to the physiological range (20-30 μM) can influence tubulin assembly and it strongly suggests that zinc has some function in the assembly of microtubules, in vivo. Several authors (2,3) have shown that high zinc (and other metal) concentrations in vitro cause formation of sheets of tubulin polymers rather than microtubules. Recently, however, it has been shown (11) that the product of in vitro microtubule assembly in the presence of 25-40 μM $ZnCl_2$ is a function of the tubulin concentration; at low cation/tubulin ratios microtubules are formed. It may be that such an effect seen at low cation/tubulin ratios and the present observed effect in extracts from zinc-deficient animals reflect a specific physiological interaction of zinc with microtubules during the polymerization process while the sheet polymers formed at higher zinc concentrations reflect a non-specific, non-physiological interaction. Purified microtubular preparations will assemble in the presence of EGTA, a known Zn chelater and in the absence of added zinc (4) suggesting either that there is no absolute requirement for zinc or that there is some zinc very tightly bound to tubulin or microtubule associated proteins. Measurement of microtubular zinc content has given equivocal results (12). The observed difference in microtubule assembly between zinc-deficient and control animals could be explained either in terms of differences in the concentration of free zinc or in the amount bound to microtubular proteins.

Zinc has been shown to influence the phosphatase activity of microtubule preparations (8), and in addition several reports have suggested that certain sulphydryl groups are important in microtubule assembly (13,14). The finding that the free sulphydryl group content of crude microtubule protein is reduced in preparations from zinc-deficient rats compared with zinc-supplemented controls is compatible with the concept that the reduced ability of tubulin from brains of zinc-deficient animals to reassemble is associated with a reduction in the number of free sulphydryl groups in tubulin or microtubule-associated proteins.

REFERENCES

1. Weisenberg, R. C., 1972, Science 177:1104-5.
2. Larsson, H., Wallin, M., and A. Edstrom, 1976, Exp. Cell. Res. 100:104-110.
3. Gaskin, F., and Y. Kress, 1977, J. Biol. Chem. 252:6918-6924.
4. Gaskin, F., Kress, Y., Brosnan, C., and M. Bornstein, 1978, Neurosci. 3:1117-1128.
5. Dinsdale, D., and R. B. Williams, 1977, Brit. J. Nutr. 37:135-142.
6. Williams, R. B., and C. F. Mills, 1970, Brit. J. Nutr. 24:989-1003.
7. Ellman, G. L., 1959, Arch. Biochem. Biophys. 82:70-77.
8. Larsson, H., Wallin, M., and A. Edstrom, 1979, J. Neurochem. 32:155-161.
9. Chesters, J. K., 1978, World Rev. Nutr. Diet. 32:135-164.
10. Fellous, A., Francon, J., Virion, A., and J. Nunez, 1975, FEBS Letts. 57:5-8.
11. Haskins, K. M., Zombola, R. R., Boling, J. M., Lee, Y. C., and R. C. Himes, 1980, Biochem. Biophys. Res. Comm. 95:1703-1709.
12. Wallin, M., Deinum, J., Rindby, A., and C. Lagercrantz, 1980, Biochem. Biophys. Acta 623:368-375.
13. Wallin, M., Larsson, H., and A. Edstrom, 1979, Exp. Cell. Res. 107:219-225.
14. Timasheff, S. N., and L. M. Grisham, 1980, Ann. Rev. Biochem. 49:565-591.

DISCUSSION

NEWBERNE (USA). You comment on zinc deficiency having an effect on protein synthesis or tubulin synthesis as opposed to assembly. Is there a difference there, is it affecting protein synthesis as well as assembly.

HESKETH (Scotland). I measured the amount of tubulin in the supernatant and I couldn't find any difference, so I don't think it is affecting the synthesis.

ELMES (Wales). Can I take it that you are offering me a biochemical explanation of the abnormal fibrillar bodies we have seen on electron microscopy of the rat zinc deficient intestine and also in some cases of human coeliac disease in which there is sometimes an associated zinc deficiency?

HESKETH (Scotland). You may be right. I looked at cilia from the bronchus of certain zinc deficient animals to look at the more structural situation and I was'nt able to find any difference in cilia from zinc deficient animals. I think that may be explained by the fact that it is a structural situation where there may be lower turnover. In any situation where there is a high turnover of microtubules restriction of zinc could be important.

SMITH (Australia). Does adding back zinc at physiological concentrations to the zinc deficient extracts of brain tubulin restore the rate of tubulin assembly?

HESKETH (Scotland). I haven't tried those experiments.

SMITH (Australia). Was the initial concentration of tubulin in the assay the same too.

HESKETH (Scotland). Yes.

AUGHEY (Scotland). The obvious site for tubulin change and interchange is the seminiferous epithelium where there is a tremendous turnover of cells and the production of cells which have long flagelli. Have you looked at these cells in zinc deficient boar for example?

HESKETH (Scotland). No.

O'DELL (USA). In this very interesting observation, the sulphydryl groups in particular were decreased. What was the concentration of zinc in your phosphate buffers etc. in the system that you used for the assembly?

HESKETH (Scotland). It was negligible.

O'DELL (USA). So you are saying that the state of the sulphydryl groups present before you started the assembly may be the answer?

HESKETH (Scotland). The sulphydryl groups were measured in the microtubules after assembly, so the difference in sulphydryl groups was seen after assembly had taken place i.e. in the reduced amount of microtubules.

O'DELL (USA). Could you look at sulphydryl group content before assembly?

HESKETH (Scotland). I haven't tried it - it might be possible. Unfortunately most of the methods of preparing tubulin involve its rather characteristic ability to re-assemble and disassemble in the warm and in the cold. So methods of purification would involve some assembly.

ETHANOL AND HEPATIC SUPEROXIDE DISMUTASE IN RATS

I.E. Dreosti, I.R. Record, R.A. Buckley, S.J. Manuel and F.J. Fraser
Division of Human Nutrition, CSIRO
Adelaide, South Australia 5000
Australia

The mammalian metalloenzyme superoxide dismutase (EC 1.15.1.1) exists in two forms, one in which manganese is the prosthetic group (Mn-SOD) and the other which contains copper and zinc (Cu,Zn-SOD). Both enzymes are important in protecting biomembranes against oxidative damage by naturally-occurring superoxide radicals. Although similar in function the forms can be distinguished, as Cu,Zn-SOD is inhibited by cyanide at concentrations above 1 mM (1).

Intracellular location of the enzyme appears to differ. In rat liver, most Cu,Zn-SOD activity is found in the cytosol although some occurs in the mitochondrial intermembranal space. Mn-SOD is generally (2) but not always (3,4) found in the mitochondrial matrix space. Both enzymes are influenced to some extent by the nutritional intake of the prosthetic cation. Insufficient dietary copper lowers Cu,Zn-SOD in many species (5-8) as does manganese depletion with respect to Mn-SOD (4,8). Both forms of the enzyme appear to be inducible and respond rapidly to increased levels of superoxide radicals, although the magnitude of the change may vary (9,10).

Increased levels of SOD associated with alcoholism were first reported in fetal rat livers following maternal intoxication during pregnancy (11) and subsequently in adult rat livers after a single dose of ethanol (12) and in erythrocytes from human alcoholics (13). Such changes probably reflect a protective response to superoxide anions arising from the metabolism of ethanol by microsomal NADPH oxidase, NADPH-cytochrome C reductase (EC 1.6.2.4) and aldehyde dehydrogenase (EC 1.2.1.5) (12). In human alcoholics Cu,Zn-SOD is considered responsible for the adaptive changes seen in erythrocyte SOD (13), but nothing is known concerning the response in rat liver to alcohol intoxication. However, since prolonged ingestion of ethanol in rats is known to cause accumulation of hepatic manganese (14) the possibility was studied that Mn-SOD may be induced by alcohol in rat livers.

METHODS

Two groups of five adult female rats (200 g) of the Hooded Wistar strain were fed a stock colony diet (W. Charlick, Adelaide, Australia) for 12 months. One group received aqueous ethanol (20%, ad libitum) as their drinking fluid while controls were given water. At termination, tissues were assayed for copper, manganese and zinc by atomic absorption spectrophotometry (Perkin-Elmer 5000,) and SOD was measured using the method of Misra and Fridovich (15).

RESULTS AND DISCUSSION

Copper levels were reduced in the plasma, liver and kidneys of alcohol-treated rats while manganese decreased in the plasma and hair. On the other hand, manganese accumulated in the liver and kidneys and zinc rose slightly in the plasma (Table 1).

TABLE 1. Effect of Alcohol Consumption on Tissue Trace Element Levels in Rats

Tissue		Concentration of Element (μg/g wet wt)			Moisture (%)
		Copper	Manganese	Zinc	
Plasma	C*	2.01±0.1	0.16±0.03	1.2±0.04	-
		p<0.005	p<0.01	p<0.01	
	A†	1.7±0.05	0.08±0.01	1.35±0.03	-
Liver	C	4.0±0.1	1.9±0.1	24.3±0.8	70.4±0.2
		p<0.0005	p<0.0005		
	A	3.4±0.02	2.5±0.1	23.9±0.3	71.1±0.1
Kidney	C	25.4±1.3	0.46±0.01	27.9±1.3	74.7±0.5
		p<0.005	p<0.05		
	A	18.7±1.4	0.62±0.08	27.2±0.7	76.6±0.5
Hair	C	12.4±0.2	1.5±0.2	176±2	10.6±0.5
			p<0.005		
	A	13.4±1.2	0.5±0.04	185±22	9.4±0.6
Bone	C	2.4±0.2	1.2±0.1	174±6	38.5±0.4
	A	2.5±0.1	1.3±0.1	195±11	37.0±2.1

*C = Control
†A = Alcohol-treated

The data confirm the report by Barak et al. (14) that alcohol increases the level of hepatic manganese in rats and are in accord with the views of Dreosti et al. (16) that in rats, in the absence of liver or kidney damage, plasma zinc is not lowered by the ingestion of ethanol.

Total hepatic SOD activity increased in the alcohol-treated rats, possibly in response to a greater release of superoxide radicals accompanying the metabolism of ethanol (12). CN-insensitive SOD (Mn-SOD) appeared to be largely responsible for the overall increase in enzyme activity (Table 2).

Table 2. Effect of Alcohol Consumption on Hepatic Superoxide Dismutase in Rats

Treatment	Superoxide Dismutase Activity (units/mg protein)		
	Total SOD	CN^--insensitive SOD*	$\frac{CN^-\text{-insensitive SOD}}{\text{Total SOD}} \times 100$
Control*	7.1±0.6	2.8±0.4	39±2
	0.1> p>0.05	p<0.01	p<0.005
Alcohol-treated	9.1±1.1	5.2±0.7	57±4

* 2 mM CN^-

In general, the present data are consistent with the findings of Stevens et al. (9) concerning the form of SOD induced by oxygen exposure, but differ from other workers who attribute most of the inducible SOD activity to the Cu,Zn-enzyme (10,13). Possibly, both forms of SOD are responsive to superoxide radicals and the species of enzyme which predominates depends largely on the availability of the prosthetic cation and on the intracellular metabolism of particular tissue concerned.

In the present study, low copper and high manganese levels in the livers of the alcohol-treated rats may have provided conditions favourable for greater synthesis of Mn-SOD. Alone, however, hepatic manganese does not stimulate Mn-SOD activity (4) so that it seems probable that the alcohol response requires not only the appropriate metal ions but also the presence of superoxide radicals.

SUMMARY

In rats, ingestion of 20% ethanol for 12 months led to an increased concentration of manganese in the liver accompanied by reduced levels of copper. Total SOD activity was elevated reflecting a marked increase in CN^--insensitive, Mn-SOD.

REFERENCES

1. Fridovich, I., 1975, Ann. Rev. Biochem. 44: 147-159.

2. Taylor, D.D., 1975, Biochem. J. 147: 493-504.

3. McCord, J., Boyle, J.A., Day, E.D., Rizzolo, L.J. and M.L. Salin. 1977. In "Superoxide and Superoxide Dismutases" (eds. A.M. Michelson, J.M. McCord and I. Fridovich) (Academic, N.Y.) pp. 129-138.

4. De Rosa, G., Keen, C.L., Leach, R.M. and L.S. Hurley, 1980, J. Nutr. 110: 795-804.

5. Williams, D.M., Lynch, R.E., Lee, G.R. and G.E. Cartwright, 1975, Proc. Soc. Exp. Biol. Med. 149: 534-536.

6. Bettger, W.J., Savage, J.E. and B.L. O'Dell, 1979, Nutr. Rep. Int. 19: 893-900.

7. Paynter, D.I., Moir, R.J. and E.J. Underwood, 1979, J. Nutr. 109: 1570-1576.

8. Paynter, D.I., 1980, J. Nutr. 110: 437-447.

9. Stevens, J.B. and A.D. Autor, 1977, Lab. Invest. 37: 470-478.

10. Utsumi, K., Yoshioka, T., Yamanaka, N. and T. Nakazawa, 1977, FEBS Letters, 79: 1-3.

11. Dreosti, I.E. and I.R. Record, 1979, Br. J. Nutr. 41: 399-402.

12. Valenzuela, A., Fernandez, N., Fernandez, V., Ugarte, G. and L.A. Videla, 1980, FEBS Letters, 111: 11-13.

13. Del Villano, B.C., Tischfield, J.A., Schacter, L.P., Stilwil, D. and S.I. Miller, 1979, Alcoholism: Clin. Exp. Res. 3: 291-296.

14. Barak, A.J., Beckenhauer, H.C. and F.J. Kerrigan, 1967, Gut, 8: 454-457.

15. Misra, H.P. and I. Fridovich, 1977, Arch. Biochem. Biophys. 181: 308-312.

16. Dreosti, I.E., Buckley, R.A. and I.R. Record, 1980, Biol. Trace Element Res. 2: 31-39.

DISCUSSION

AGGETT (Scotland). I understand that alcohol is principally metabolised in the microsomal fraction and that the manganese SOD is in the mitochondria. What is the mechanism of induction of the enzyme at this location, in a different compartment to the site of alcohol metabolism?

DREOSTI (Australia). There are two possibilities. One is that in the rat there may be a cytosolic Mn-SOD in addition to the mitochondrial one and that it is this enzyme which is induced. The other possibility is that superoxide radicals from alcohol metabolism might cross the mitochondrial membranes, damage them, and cause the induction.

PROHASKA (USA). Despite the mitochondrial location of Mn-SOD it is synthesised on cytoplasmic ribosomes in proximity to the site of alcohol metabolism.

PAYNTER (Australia). Are you sure that a substance as transitory as superoxide can induce superoxide dismutase? Perhaps the changes in superoxide dismutase activity are secondary to changes in liver lipids, especially since it is a membrane-localised enzyme.

DREOSTI (Australia). Many superoxide generating systems have been tried and observed to induce superoxide dismutase activity, so it seems likely that superoxide is one of the inducers. There is some question whether Mn-SOD is intramitochondrial or membrane localised in the rat.

HOWELL (Australia). In your abstract you state that there were no histological changes in liver. Did you look for an increase in collagen or look for changes in the endoplasmic reticulum content under electron microscopy?

DREOSTI (Australia). I'm sorry I can't comment on that because the histology was done by one of my colleagues.

PROGRESS OF NICKEL METABOLISM AND NUTRITION RESEARCH

M. Kirchgeßner, Dora A. Roth-Maier, and A. Schnegg

Institut für Ernährungsphysiologie
Technische Universität München
D-8050 FREISING-WEIHENSTEPHAN
FED. REP. GERMANY

As effects of Ni deficiency on metabolism we reported in TEMA 3 reduced blood parameters (erythrocytes, hematocrit and Hb) as consequence of decreased iron absorption in Ni deficiency, and also numerous changes in enzyme activities (13). These experiments were continued, whereby in Ni deficiency, apart from the reduction in the activity of 6 various liver dehydrogenases (G6P-DH, ICDH, GlDH, LDH, MDH and α-HBDH), mainly a decrease of activities of enzymes and concentrations of metabolites of glucose degradation was observed (5). The activity of hepatic G6P-DH of 30-day-old rats is diminished by 82 % after Ni depletion (2). Even at a dietary supply of 100 ppm Fe, activity was lowered by 56 %. Especially, also activity of pancreas α-amylase was so affected that its activity decreased in Ni deficiency by 57 % (4). Thus, the reduced activity of LDH, which had diminished by 72 % (11) may be explained, too. Accordingly, glucose and glycogen contents in the organism were lowered up to 90 %, if only 50 ppb Ni are fed. In the same way serum ATP concentration of Ni depleted rats fell by 25 % compared to well supplied control animals (12).

The surprising proportionate reduction of activities of enzymes involved in glucose catabolism could be a direct consequence of the decreased degradation rate of starch by α-amylase. This leads to an energetic deficit that results in the lack of glucose and glycogen in liver. Also the lowered activity of enzymes involved in amino acid metabolism may be connected with this reduced glucose supply. Likewise, the retarded growth in Ni deficiency may thus be explained. Assuming a 50 % reduced starch hydrolysis as a consequence of depressed α-amylase activity in Ni deficiency, the evoked lower energy supply corresponds exactly to the lower weight gains that were found in Ni deficiency (6).

TABLE 1. Live weight and enzyme activities of rats after different dietary Ni supply

Diet ppm Ni	Weight g	Liver G6P-DH mU/mg prot.	Liver LDH U/mg prot.	Pancreas α-Amylase U/mg prot.
0.060	79.0[a]	102.5[a]	2.19[a]	202[a]
	±14.9	± 37.5	±0.15	± 51
0.150	84.0[a]	199.8[b]	2.48[b]	204[a]
	±14.1	± 37.8	±0.31	± 41
20	81.0[a]	229.7[b]	3.01[c]	353[b]
	±10.4	± 30.4	±0.26	± 99

A further trial was conducted to investigate, if changes in enzyme activities will occur in suboptimal Ni supply, too (7). Whereas weight gains are not influenced at a dietary supply of 60 and 150 ppb Ni, respectively, activity of α-amylase was reduced by 43 per cent at 60 ppb, and dependent on this, G6P-DH was lowered by 56 per cent and LDH by 27 per cent (Table 1). Also, at a supply of 150 ppb Ni, α-amylase was still diminished. This means that optimum requirement is higher than the minimum one. Thereby it may also be concluded from the here presented results that for the growth of rats a dietary Ni content of 150 or even 60 ppb seems to be sufficient. The enzyme activities, however, may demonstrate that this range is a suboptimal supply. The optimal Ni requirement, however, is significantly higher. This became also evident in balance trials with nongravid and gravid sows with different Ni supply. For a zero-balance, which indicates adequate requirements, a daily dietary supply of 1.4 mg Ni or 0.7 ppm is required (9, Table 2).

Table 2. Total Ni balance and anabolism of gravid versus nongravid sows in response to varying dietary trace element supply (21st day of gravidity until birth), per animal

Group	1	2	3	4
Physiol. stage	nongravid		gravid	
Trace element supply	low	high	low	high
Ni intake, mg	139.4	386.7	151.4	410.2
Fecal excretion, mg	128.7	353.2	137.0	325.6
Urinary excretion, mg	5.2	5.4	5.5	5.4
Balance, mg	4.5	28.1	8.9	79.2
Balance, %	3	7	6	19
Excretion in gest. products, mg	-	-	0.6	1.2
Anabolism, mg	-	-	8.3	78.0
Anabolism, %			5	19

Essentiality of Ni may be derived also from experiments concerning pregnancy anabolism of rats and sows, in which Ni was accreted in the organism during pregnancy besides the fetuses and the other gestational products (14, 10, 9). The results in table 2 show that during gravidity in the well supplied group an anabolism of Ni occurs, which amounted to 19 % of the Ni intake or was about 0.87 mg daily. Therewith it is established that, also in the case of Ni, homeostatic regulation is changed by gravidity. Adult animals, namely, show in the case of homeostatically regulated trace elements approximately zero-balance within certain limits, even after high intakes.

Excretion of Ni at parturition via conception products is very low and attains only 0.4 % of the total fecal and urinary excretion or averages about 4 % of the total balance. As peculiarity of Ni we found in the experiments with sows, that Ni concentration in dry matter is nearly 25 times higher in amniotic fluid than in fetuses, whereas Cu and Zn were distributed in both products to a similar extent (9). Thus, Ni is excreted via amniotic fluid in the high proportion of 25 % of the total excretion of fetuses and amniotic fluid, whereas Cu and Zn content in amniotic fluid is only 2 and 0.5 % of the total content, respectively (Fig. 1). This corresponds nearly to the distribution in dry matter excretion. In especially conducted trials with rats, these findings were confirmed (Fig. 1). Comparing the concentrations and the excretions in the individual conception products with total excretion it is indicated here, too, that Ni reacts differently from the other trace elements in amniotic fluid and also in uterus (Table 3). Similar results were obtained for the distribution of Ni and other trace elements in egg white and yolk, whereby Ni concentration was higher in egg white.

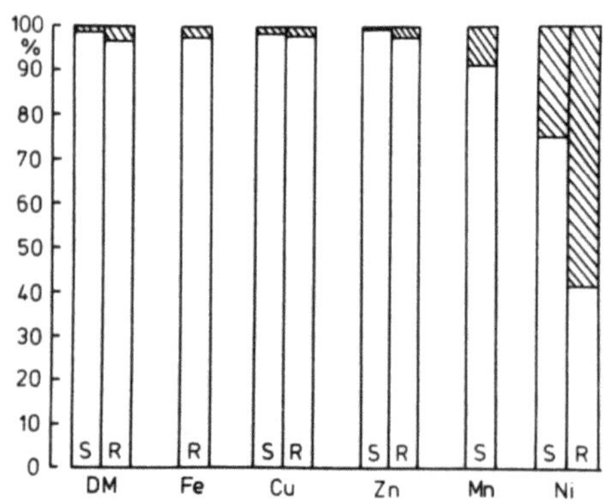

Fig. 1: Distribution of dry matter (DM) and total trace element content in fetuses and amniotic fluid of sows (S) and rats (R)

TABLE 3. Relative trace element concentration and content in gestational products of rats

	Fetuses (= 100)	Amniotic fluid	Placenta	Uterus
Concentration in dry matter				
Ni	100	3400	114	240
Fe	100	200	160	50
Cu	100	60	120	60
Zn	100	90	99	70
Mn	100	-	30	30
Total content				
Ni	100	140	50	110
Fe	100	3	30	8
Cu	100	2	40	20
Zn	100	2	20	17
Mn	100	-	6	5

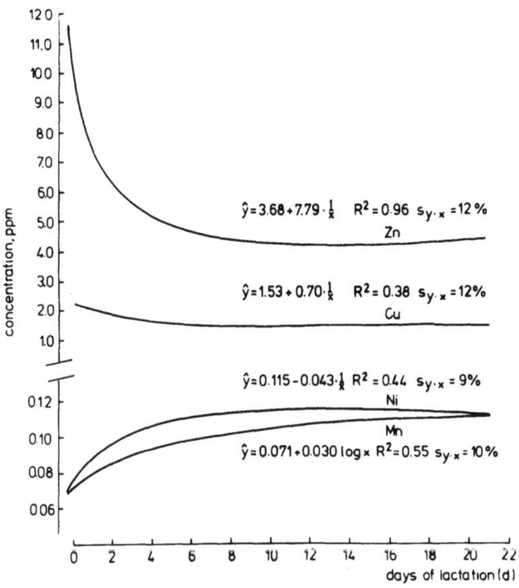

Fig 2 Concentration of Ni, Cu, Zn, and Mn in sow milk during lactation ($x = d+1$)

Also regarding the concentrations of trace elements in the milk of rats (3) and sows (8) Ni reacts inversely, compared to copper and zinc (Fig. 2). Ni concentration is low in first colostrum with an increase during lactation. The mean Ni concentration in normal sow milk is 0.11 ppm and a dietary Ni supplementation beyond the optimal requirement does not influence the Ni content of the sow milk. In rat milk Ni content increases from 0.3 ppm in colostrum to 1.5 ppm. It is possible that in both species there is a relationship between these low Ni concentration in colostrum and the high Ni concentration in amniotic fluid. In contrast, in first cow colostrum, Ni concentration of 0.1 ppm was higher than in normal cow milk, which contains 0.02 ppm (1).

SUMMARY

In various experiments with rats and sows further evidence was furnished for essentiality of Ni: (1) reduction in the concentration of glucose metabolites, enzyme activities of glucose catabolism, and mainly of pancreas α-amylase after Ni depletion but also after suboptimal dietary supply; (2) elevated Ni accretion during gravidity; (3) distribution of Ni in fetuses, amniotic fluid and milk compared with other trace elements.

REFERENCES

1. Kirchgeßner, M., 1959, Z. Tierphysiol., Tierernährg. u. Futtermittelkde. 14: 270-277
2. Kirchgeßner, M., and A. Schnegg, 1976, Bioinorganic Chem. 6: 155-161
3. Kirchgeßner, M., and A. Schnegg, 1976, Arch. Tierernährung 26: 773-776
4. Kirchgeßner, M., and A. Schnegg, 1979, Nutr. Metabol. 23: 62-64
5. Kirchgeßner, M., and Schnegg, A. 1980 a. In "Nickel in the Environment" (ed. J. O. Nriagu) (John Wiley Sons, New York) p. 635-652
6. Kirchgeßner, M., and Schnegg, A. 1980 b. In "Nickel, 3. Spurenelement-Symposium" (ed. M. Anke et al.)(Leipzig, Jena) p. 23-26
7. Kirchgeßner, M., and A. Schnegg, 1981, Nutr. Metabol. (in press)
8. Kirchgeßner, M., Dora A. Roth-Maier, and R. Spörl, 1980, Z. Tierphysiol., Tierernährg. u. Futtermittelkde. 44: 233-238
9. Kirchgeßner, M., Dora A. Roth-Maier, and R. Spörl, 1981, Arch. Tierernährung 31: 21-34
10. Kirchgeßner, M., R. Spörl, and Dora A. Roth-Maier, 1980, Z. Tierphysiol., Tierernährg. u. Futtermittelkde. 44: 98 - 111
11. Schnegg, A., and M. Kirchgeßner, 1977, Z. Tierphysiol., Tierernährg. u. Futtermittelkde. 38: 200 - 205
12. Schnegg, A., and M. Kirchgeßner, 1977, Z. Tierphysiol., Tierernährg. u. Futtermittelkde. 39: 247 - 251
13. Schnegg, A., and Kirchgeßner, M., 1978. In "Trace Element Metabolism in Man and Animals- 3" (ed. M. Kirchgeßner) (ATW, Freising-Weihenstephan) p. 236-243
14. Spörl, R., and M. Kirchgeßner, 1977, Z. Tierphysiol., Tierernährg. u. Futtermittelkde. 38: 205 - 210.

DISCUSSSION

BRATTER (Germany). Did you investigate very carefully all possibilities of nickel contamination during sampling of the amniotic fluid? If contamination can be excluded can you comment on the reasons why nickel is concentrated in amniotic fluid? As far as I know the pool of amniotic fluid has a good exchange rate and I am wondering why nickel is concentrated in contrast to the other essential trace elements.

KIRCHGESSNER (Germany). I think we can exclude contamination problems. We get similar results for sows and rats, and in chicken eggs the concentration of nickel is higher in the white than the yolk. I am convinced it is a real effect, and we have yet to investigate the mechanism.

ULLREY (USA). Did you say that 0.7ppm nickel is the optimum for sows?

KIRCHGESSNER (Germany). Based on balance studies we find this is the requirement for gravid or non-gravid sows.

ULLREY (USA). Would you please give a brief description of the ingredients used in the diets for sows?

KIRCHGESSNER (Germany). The diet contains ingredients such as oats, barley, straw, cellulose and skim milk powder. The basal diet contains 0.63mg Ni/kg.

MAMMALIAN SYNTHESIS AND DEGRADATION OF SELENOCYSTEINE

Kenji SODA, Takeshi NAKAMURA, Hidehiko TANAKA*, and Nobuyoshi ESAKI

Laboratory of Microbial Biochemistry, Institute for Chemical Research, Uji, Kyoto-Fu 611 and *Laboratory of Biochemistry, Kyoto College of Pharmacy, Kyoto 607, JAPAN

INTRODUCTION

Various selenium-containing amino acids occur in nature, and most are physiologically important. Several mammalian and microbial proteins, in particular enzymes, contain selenium amino acids as a prosthetic group. The selenium moiety of bacterial glycine reductase, bacterial formate dehydrogenase, and glutathione peroxidase of rat liver has been identified as selenocysteine (1). Although selenocysteine is assumed to be synthesized from selenomethionine by transsulfurylation through selenocystathionine and also from selenide by the serine sulfhydrylase reaction in mammalian tissues, no evidence has been obtained. In the present paper, we describe the enzymatic synthesis and degradation of selenocysteine in mammalian tissues.

MATERIALS AND METHODS

Cystathionine β-synthase (EC 4.2.1.22) was purified from rat liver according to the method of Kimura and Nakagawa (2). Crystalline cystathionine γ-lyase (EC 4.4.1.1) of rat liver, which was prepared by the method of Matsuo and Greenberg (3), was kindly supplied by Dr. Morino. Alanine dehydrogenase (EC 1.4.1.1) (4) and leucine dehydrogenase (EC 1.4.1.9) (5) were purified and crystallized from cell-free extracts of *Bacillus sphaericus* (IFO 3525) and *Bacillus megaterium* (ICR 1390), respectively. Crystalline D-amino acid oxidase was prepared from pig kidney (6). DL-Selenomethionine and DL-selenocystine were purchased from Sigma, L-homocysteine, L-cystathionine, dithiothreitol and heptafluorobutyric acid (HFB) from Nakarai Chemicals, Kyoto, Japan, glutathione, NADH, NADPH were from Kohjin, Tokyo, Japan, and aluminum selenide from Alpha Division-Ventron. *Se*-Benzyl-L-selenohomocysteine was synthesized from dibenzyldiselenide and α-amino-γ-bromobutyric acid. L-Selenohomocysteine and L-selenocystathionine were synthesized from *Se*-benzyl-L-selenohomocysteine. Selenol amino acids were prepared from the corresponding diselenides by reduction with dithiothreitol in a stream of N_2. H_2Se was prepared from aluminum selenide and water. The other chemicals were analytical grade reagents.

The α,γ- and α,β-elimination reactions catalyzed by cystathionine γ-lyase were followed by determination of α-keto acids formed with 3-methyl-2-benzothiazolone hydrazone HCl (MBTH) (7). The standard assay system contained 3 μmol of L-cystathionine (or L-selenocystathionine), 50 μmol of potassium phosphate buffer (pH 7.5), 5 nmol of pyridoxal-P and 1.5 μg of enzyme in a final volume of 0.5 ml. The reaction was performed at 37°C for 10 min. The β-replacement reaction catalyzed by cystathionine β-synthase was measured by determination of the amino acid formed with a Hitachi high performance amino acid analyzer (model 835) on a 4 X 150 mm column. Cysteine and cystathionine were determined also by the acid ninhydrin method. The incubation mixture consisted of 15 μmol of L-serine, 30 μmol of DL-homocysteine, 10 nmol of pyridoxal-P, 100 μmol of Tricine-NaOH buffer (pH 8.6) and enzyme in a final volume of 1 ml. Reactions were conducted at 37°C for 30 min in sealed tubes in which air was displaced with N_2.

Selenocysteine reductase activity was measured by determination of H_2Se formed with lead acetate. The standard reaction mixture contained 2 μmol of DL-selenocysteine, 5 μmol of dithiothreitol, 0.01 μmol of pyridoxal-P, 0.02 mg of bovine serum albumin, 60 μmol of Tricine-NaOH buffer (pH 8.5) and enzyme in a final volume of 0.5 ml. After incubation at 37°C for 20 min, 3.5 ml of 5 mM lead acetate solution was added to the reaction mixture. Lead acetate solution was added to the reaction mixture prior to enzyme in a blank. Turbidity of yellowish brown colloidal PbSe was measured at 400 nm within 15 min, because it diminishes at a rate of 0.3% per min. An apparent molar turbidity coefficient of colloidal PbSe at 400 nm was 1.18×10^4 on the basis of determination of H_2Se with bis(5-carboxy-4-nitrophenyl)disulfide (DTNB) ($\varepsilon = 1.40 \times 10^4$ at pH 7.4) (15).

Gas chromatograph-mass spectrometry was performed with a Shimadzu-LKB 9000 gas chromatograph-mass spectrometer. Gas chromatography was run on a glass column (3 mm X 1 m) packed with 3% SE 52 on Chromosorb W (60-80 mesh). Helium was used as a carrier gas at a flow rate of 30 ml/min. The column was programmed from 100 to 270°C at a rate of 5°C/min.

RESULTS AND DISCUSSION

SYNTHESIS OF SELENOCYSTATHIONINE WITH CYSTATHIONINE β-SYNTHASE

The ability of cystathionine β-synthase to catalyze the formation of selenocystathionine from L-serine and L-selenohomocysteine was examined. The reaction mixture consisted of 1.0 μmol of L-serine, 1.5 μmol of L-selenohomocysteine, 3.8 μmol of dithiothreitol, 50.0 μmol of Tricine-NaOH buffer (pH 8.6), 5.0 nmol of pyridoxal-P and enzyme in a final volume of 0.5 ml. The boiled enzyme was substituted for the native enzyme in a blank. After incubation at 37°C for 30 min in a sealed tube in which air was replaced by N_2, the reaction was stopped by addition of 0.1 ml of 50% trichloroacetic acid. After centrifugation, an aliquot of the supernatant solution was subjected to an amino acid analyzer. The product emerged at the same retention time (47.0 min) as the authentic selenocystathionine under the conditions used. Their identity was further confirmed by gas chromatograph-mass spectrometry. The HFB-*n*-propyl ester of the amino acid formed enzymatically, whose retention time was 29.5 min was identical gas chromatographically with the same derivative of authentic selenocystathionine. The mass spectrum of the derivative of the product showed the characteristic selenium isotope pattern, and coincided with that of authentic selenocystathionine. Thus, the product was identified as selenocystathionine. The Km values for the substrates were determined as follows: for selenocystathionine synthesis, L-serine, 0.3 mM and L-selenohomocysteine, 2.5 mM; for cystathionine synthesis, L-serine, 1.2 mM and L-homocysteine, 1.2 mM. The ratio of Vmax value for the selenocystathionine synthesis to that for the cystathionine was 0.69. L-Homocysteine competes with L-selenohomocysteine to inhibit the selenocystathionine synthesis in the presence of 5 mM L-serine, and *vice versa*. The apparent Ki values for L-selenohomocysteine and L-homocysteine were calculated to be 3.4 and 0.85 mM, respectively. The values are substantially consistent with the Km values shown above. This shows that both cystathionine and selenocystathionine syntheses are carried out at the same active site.

ELIMINATION OF SELENOCYSTATHIONINE BY CYSTATHIONINE γ-LYASE

Cystathionine γ-lyase as well as cystathionine β-synthase functions in transsulfurylation pathway of sulfur amino acid metabolism. Cavallini et al. (8) reported that the enzyme catalyzes also α,β-elimination of L-cystine. However, we have found that the α,β-elimination of L-cystathionine proceeds much more slowly (<3%) than the α,γ-elimination. When L-selenocystathionine was used as a substrate, the enzyme catalyzed both α,γ- and α,β-elimination to yield α-ketobutyrate, pyruvate, selenohomocysteine, selenocysteine, and ammonia. Selenohomocysteine and selenocysteine were identified by amino acid analysis and gas chromatograph-mass spectrometry after derivatization as described above. Pyruvate and α-ketobutyrate were converted enzymatically to alanine and α-aminobutyrate, followed by determination with an amino acid analyzer. The ratio of the velocity of α,β-elimination to that of α,γ-elimination of selenocystathionine was 0.5. The Vmax values for α,γ-elimination of selenocystathionine and cystathionine were calculated to be 19.3 and 6.3 μmol/mg/min, respectively. The Km values were closely similar: 1.7 mM for L-selenocystathionine and 1.3 mM for L-cystathionine. Cystathionine γ-lyase can eliminate further the following amino acids formed from selenocystathionine and cystathionine; selenohomocysteine, homocysteine, selenocysteine, and cysteine. Selenium amino acids are decomposed 2.5 to 3 times faster than the corresponding sulfur analogs.

SYNTHESIS OF SELENOCYSTEINE BY COUPLING OF CYSTATHIONINE β-SYNTHASE AND CYSTATHIONINE γ-LYASE REACTIONS

We examined whether selenocysteine is synthesized from selenohomocysteine and serine through selenocystathionine by coupling of cystathionine β-synthase and cystathionine γ-lyase reactions. The reaction system (0.5 ml) contained 5 nmol of pyridoxal-P, 40 μmol of Tricine-NaOH buffer (pH 8.0), 1.5 μmol of L-serine, 1.5 μmol of L-selenohomocysteine, 10 μmol of dithiothreitol, 0.06 unit of cystathionine β-synthase, and 0.12 unit of cystathionine γ-lyase. The mixture was incubated at 37°C for 10, 20, and 30 min under N_2. Selenocysteine was identified by amino acid analysis after conversion to *Se*-ethylselenocysteine. The synthesis of selenocysteine proceeded at a rate of 3 nmol/min. We observed the formation of selenocysteine also with rat liver homogenate (4 mg as protein), which contained 0.02 unit of cystathionine β-synthase and 0.08 unit of cystathionine γ-lyase. The reaction rate was 0.3

nmol/min under the conditions.

POSSIBILITY OF SELENOCYSTEINE FORMATION BY SERINE SULFHYDRYLASE REACTION OF CYSTATHIONINE β-SYNTHASE

Braunstein et al. (9) reported that cystathionine β-synthase of chicken liver catalyzes synthesis of cysteine from L-serine and hydrogen sulfide. We have found that the rat liver enzyme also catalyzes the reaction at a rate of 12% of cystathionine synthesis. However, attempts were made to synthesize selenocysteine from L-serine and H_2Se by the enzyme without success. The enzyme was not inactivated when incubated at 37°C for 30 min in the reaction mixture containing 20 mM H_2Se. We could not observe the α,β-elimination of selenocysteine with the enzyme.

ENZYMATIC REDUCTION OF SELENOCYSTEINE

When DL-selenocysteine was incubated with a rat liver homogenate, the formation of L-alanine and H_2Se was observed. H_2Se was identified with DTNB and lead acetate. Alanine was indetified by amino acid analysis (retention time, 46.0 min), and gas chromatography of the HFB *n*-propyl ester (retention time, 5.0 min). Chirality of alanine and unreacted selenocysteine were determined to be L and D with L-alanine dehydrogenase and D-amino acid oxidase, respectively. Redox balance between the substrate and the products is probably kept by oxidation-reduction of dithiothreitol which is contained in the reaction mixture to prepare selenocysteine from selenocystine as follows. Various thiols can substitute for dithiothreitol

$$\text{selenocysteine} + DTT_{red} \rightarrow \text{alanine} + H_2Se + DTT_{ox}$$

(DTT_{red}, dithiothreitol; DTT_{ox}, *trans*-4,5-dihydroxy-1,2-dithiane)

in the reaction mixture. The relative velocity of H_2Se formation was as follows: dithiothreitol, 100; 2,3-dimercapto-1-propanol, 68; 2-mercaptoethanol, 65; L-cysteine, 10; L-homocysteine, 45; and glutathione, 35%. These results show the occurrence of a new enzyme that catalyzes reductive cleavage of selenocysteine into alanine and H_2Se, and we termed the enzyme as selenocysteine reductase.

SELENOCYSTEINE REDUCTASE ACTIVITY OF SEVERAL MAMMALIAN TISSUES

The selenocysteine reductase activity was found in various mammalian tissues such as liver, kidney, pancreas, adrenal, heart, lung, testis, brain, thymus, spleen, and muscle of rat, dog, mouse, pig and some other mammals. The specific activities of livers and kidneys were generally higher than those of other tissues. However, no activity occurred in rat blood and rat fat. We have chosen pig liver for the purpose of purification of the enzyme in view of easy availableness and abundance of enzyme activity.

PURIFICATION AND PROPERTIES OF SELENOCYSTEINE REDUCTASE FROM PIG LIVER

The pig liver enzyme was purified to homogeneity. The molecular weight of the enzyme was estimated to be approximately 93,000 by the Sephadex G-200 gel filtration method. A molecular weight of 85,000 ± 3,000 was obtained also by the sedimentation equilibrium method. The enzyme showed the maximum reactivity at about pH 9.0. The enzyme was found to act specifically on selenocysteine to produce alanine and H_2Se. The following compounds were inert; DL-selenocystine, L-cystine, L-cysteine, L-serine, β-chloro-L-alanine, L-cysteine sulfinate, *S*-methyl-L-cysteine, *Se*-ethyl-DL-selenocysteine, L-selenohomocysteine, L-homocysteine, selenocystamine, selenocysteamine. L-Cysteine was found to behave as a competitive inhibitor against DL-selenocysteine. Selenocysteine reductase, which has an absorption maximum at 420 nm, contains pyridoxal-P as a coenzyme.

REFERENCES

(1) Stadtman, T. C., 1980, Ann. Rev. Biochem. 49:93-110. (2) Kimura, H., and Nakagawa, H., 1971, J. Biochem. 69:711-723. (3) Matsuo, Y., and Greenberg, D. M., 1958, J. Biol. Chem. 230: 545-560. (4) Ohshima, T., and Soda, K., 1979, Eur. J. Biochem. 100:29-39. (5) Ohshima, T., Misono, H., and Soda, K., 1978, J. Biol. Chem. 253:5719-5725. (6) Jenkins, W. T., Behlen, A., and Rogers, P. I., 1979, Anal. Biochem. 94:105-108. (7) Soda, K. 1968, Anal. Biochem. 25:228-235. (8) Cavallini, D., Mondovi, B., DeMarco, C., and Sciscia-Santoro, A., 1962, Arch. Biochem. Biophys. 96:456-457. (9) Braunstein, A. E., Goryachenkova, E. V., Tolosa, E. A., Willhardt, I. H., and Yefremova, L. L., 1971, Biochim. Biophys. Acta 242:247-260.

DISCUSSION

SUNDE (Scotland). Have you checked to see whether the product of the reductase is alanine or is it dehydroalanine? What is the role of the thiols in the reaction?

SODA (Japan). The product is certainly L-alanine, which was identified by amino acid analysis, gas-chromatography mass-spectrometry (as N-heptafluoroputyryl-non-propyl ester) and reaction with L-alanine dehydrogenase. Free dehydroalanine is extramely labile; it is decomposed spontaneously to pyruvate and NH_3.

SUNDE (Scotland). What is the role of the reducing equivalents (DTT) in the reaction?

SODA (Japan). Now, we are investigating the mechanism of this enzyme reaction, including a role of a reducing agent, but I suppose as follows. Selenocysteine binds to the enzyme-bound pyridoxal 5'-phosphate to form a Schiff base (an aldimine intermediate). Then α-proton of the bound selenocystine is removed to yield a ketimine intermediate: In this intermediate, selenium is removed as an element, and a proton from a reducing agent is attached to β-C(δ-CH_2-) to form an aminoacrylate intermediate, which is then tautomerized into an aldimine intermediate of alanine through a ketimine intermediate of pyruvate. This aldimine complex is hydrolyzed to alanine by the pyrydoxal phosphate enzyme.

SUNDE (Scotland). To date, do you have evidence at all for the direct formation of selenocysteine from selenide or inorganic selenium, as opposed to using O-Acetyl compounds or Selenohomocysteine or something like that?

SODA (Japan). The formation of selenocysteine from O-acetylserine and hydrogen selenide may be possible, but we have no information about this pathway.

PROGRESS IN RESEARCH ON NEWER TRACE ELEMENTS: THE METABOLISM OF VANADIUM AS INVESTIGATED BY NUCLEAR AND RADIOCHEMICAL TECHNIQUES

E. Sabbioni and E. Marafante
Radiochemistry and Nuclear Chemistry Division
Commission of the European Communities
Joint Research Centre - Ispra Establishment
21020 Ispra (Varese) - Italy

INTRODUCTION

The discovery of vanadium as an essential nutritional requirement in experimental animals (1) and its potential impact on the environment and humans as a result of coal and oil combustion for electrical energy production (2) has greatly increased the research being carried out during the past decade on effects and metabolism of this element. The inhibition of a selected class of enzyme in vitro such as phosphohydrolases is a well known type of biochemical effect of vanadium which has led to speculative proposals about a possible regulatory role of the element in mammalian tissues (3). Metabolic studies such as timed distribution of tri-, tetra- and pentavanadate ions indicate that vanadium can interact with cellular components (4,5). These interactions govern the pathways of vanadium in the body leading to a possible existence of more than one pool of the element in the cell.

The aim of this work, carried out by nuclear and radiochemical techniques, is to identify the different pools of vanadium in the rat and to correlate them to the levels of the element in the tissues of exposed and unexposed animals.

MATERIALS AND METHODS

^{48}V of high specific radioactivity (> 37 Mbq/μg of V) was prepared by proton activation of a Ti target in the cyclotron of Milan University (6,7). The concentration of vanadium in rat tissues was determined by neutron activation analysis (NAA) in a special facility of the ESSOR reactor of the JRC-Ispra in a thermal neutron flux of 2×10^{14} neutrons $cm^{-2} sec^{-1}$ in order to determine the induced short lived ^{52}V ($T \frac{1}{2} = 3.77$ min).

Male albino rats of COBS strain were used. The sacrifice, removal of tissues, preparation of intracellular fractions, gel filtration of total cytosol on Sephadex G-150 and of heated cytosol on Sepharose 6B followed by immunoprecipitation of ferritin as well as partial purification of Fe haemoproteins were carried out as previously described (4,5,8).

RESULTS AND DISCUSSIONS

Table 1 reports the results of the in vivo timed distribution of ^{48}V between the liver cytosol components of rats i.p. injected with 10 μg of $^{48}VO^{2+}$/rat as determined by gel filtration. Three pools I, II and III of vanadium corresponding to three emerging peaks in the eluate are present.

Calibration of the column with ^{59}Fe labelled ferritin and transferrin and immuno-precipitation of ferritin from the three ^{48}V-containing peaks emerging after gel filtration on Sepharose 6B of the heated cytosol (8) suggest that pool I and pool II represent V-bound ferritin (V-fer) and V-bound transferrin (V-transf). The nature of the third pool of vanadium corresponding to V-bound low molecular weight components (V-LMWC) was not identified. In comparison to the results at 1 day, those at 9 and 21 day post-injection periods show only the presence in the eluate of V-fer and V-transf while the V-LMWC disappeared completely.

Nine days after the administration of vanadium, partial purification of heart myoglobin, liver mitochondrial and microsomal cytochromes b, b5 and c, ferriporphyrin and RBC haemoglobin (Table 2) show no significant incorporation of ^{48}V into these proteins suggesting that only non-heme Fe-containing proteins would be able to bind vanadium in vivo while Fe hemoproteins would not associate with the element.

Table 1 - Different pools of vanadium in rat liver cytosol as determined by Sephadex G150 chromatography

Position of the ^{48}V emerging peaks (mol. wt.)	Pool of vanadium	^{48}V incorporated (% of radioactivity in the eluate)		
		1d	9d	21d
> 200.000	I (V-ferritin)	4.8	14.8	37
70.000 - 80.000	II (V-transferrin)	66	85.2	63
< 5000	III (V-low mol. wt comp.)	29.2	N.D. *	N.D. *

* N.D. not detected

Table 2 - Incorporation of ^{48}V into Fe haemoproteins of the rat

Partially purified Fe-hemoprotein	Total fraction from which purification occurred	^{48}V incorporated (% of radioactivity in the total fraction
Myoglobin	Heart homogenate	0.45
Cytochrome b5	Liver mitochondria	0.13
Cytochrome c		< 0.13
Ferriporphyrin		~ 0.12
Cytochrome b5	Liver microsomes	< 0.3
RBC cytosol	Total blood	5

Table 3 - Different pools of vanadium in rat plasma as identified by gel filtration on Sephadex G150

Time	^{48}V incorporated (% of radioactivity in 1 ml of plasma)		
	Not recovered from Sephadex	Transferrin	LMWC
0.5 h	96.8	2.9	0.3
8 h	90.9	8.9	0.21
1 d	81.6	18.3	0.12
4 d	15.8	84.2	–
8 d	5.7	94.3	–
21 d	5	95	–

Table 3 reports the results of the timed distribution of vanadium between the blood components after i.v. injection of 10 μg of $^{48}V^{5+}$/rat. At any time the bulk of blood vanadium (> 90%) is present in the plasma and less than 10% in the red blood cell (RBC). At 0.5 h, gel filtration of plasma on Sephadex G-150 shows the existence of two pools of vanadium. The first major pool corresponds to vanadium associated with transferrin as confirmed also by ion exchange chromatography on DEAE-Sephadex and polyacrylamide gel electrophoresis (5). The second pool corresponds to an emerging peak in the region of the chromatogram identical to that of peak III from gel filtration of the liver cytosol (V-LMWC). After 4 days this latter type of vanadium disappeared completely. Up to 1 day the V-transf and V-LMWC, however, represent only a small part of the total vanadium in the plasma because the greater part (> 80%) is apparently present in another diffusible form not recovered from gel filtration (V-unSeph).

Table 4 summarizes the type of pools of vanadium, including the V-bound lactoferrin isolated in vitro in this lab. (9), identified by ^{48}V radiotracer experiments while Table 5 reports the data on the distribution of ^{48}V in rat tissues, 24 h after a single i.p. injection of 0.1 μg of $^{48}V^{5+}$ / rat as well as the concentrations of vanadium in the tissues of normal rats as determined by NAA.

Some conclusions can be drawn by relating the identified different pools of vanadium with the measured levels of vanadium in the tissues. When introduced into the body, the greater part of the vanadium is excreted or deposited and stored in the bone. Only a small part of the element is able to interact with proteins which would be responsible for its retention. On the other hand, the normal amounts of vanadium in rat tissues are very low, of the order of a few ppb, suggesting that these amounts may correspond to the small part of V-bound protein retained. If this is correct, any variation of the V-bound protein may significantly influence the level in the tissues of vanadium. This conclusion is of particular concern with respect to the significance of environmental exposure to vanadium. Vanadium is poorly absorbed via the gastrointestinal tract (about 1%) while the degree of absorption via inhalation is significantly higher

Table 4 - Different pools of vanadium in the rat

Identified pool	Fraction in which the pool was identified and remarks
Pool I: V-ferritin (V-fer)	Liver cytosol (Sepharose 6B and immunoprecipitation). Recovered from Sephadex G-150. No dialyzable.
Pool II: V-transferrin (V-transf)	Plasma (also of rabbits, pigeon and man) and liver cytosol. Recovered from Sephadex G-150. No dialyzable. Biological half life: $T_{\frac{1}{2}} = 8$ d.
Pool III: V-low mol. wt components (V-LMWC)	Plasma as well as liver and kidney cytosol. Recovered from Sephadex G-150. Dialyzable.
Pool IV: unknown (V-unSeph)	Plasma. Not recovered from Sephadex G-150. Ultrafiltrable.
Pool V: V-lactoferrin (V-lact)	Isolated in vitro from bovine milk Recovered from Sephadex G-150. No dialyzable.

Table 5 - Vanadium content in unexposed rats and in animals exposed i.p. to 0.1 μg of $^{48}V^{5+}$/rat and sacrificed 48 h later

Tissue	Unexposed animals (ppb wt.w.) [by NAA]	Exposed animals (% of the dose injected) [by ^{48}V radiotracer]
Femur	15.5	1.2
Liver	6.8	5.8
Lung	2.6	0.1
Spleen	2.7	0.24
Testicles	2.2	0.34
Kidney	1	3.6
Brain	0.3	0.004
Blood	0.9 *	0.15 *

* % dose/ml of blood

and represents the most important route of exposure for man (10). In this laboratory, some similarities have been observed in the metabolic pathways of vanadium when injected or intracheally instilled into rats so that the exposure of man to current levels of vanadium in air may increase the low amounts of vanadium-bound protein influencing significantly the low normal levels of the element in the tissues. In this connection, the finding that the element can be exchanged in vitro from transferrin to ferritin (11) may be of particular significance. This mechanism may determine the extent of the concentration of vanadium in V-transf and V-fer pools which would ultimately determine the level of the element in the tissues.

REFERENCES

1. Frieden, E., 1972, Sci. Am.227:52-59.
2. Goetz, L. and Sabbioni, E., 1980, Rep. EUR 6998/2.
3. Cantley, L.C., Josephson, L., Warner, R., Yanagisawa, M., Lechene, C. and Guidotti, G., J. Biol. Chem., 1977, 252:7421-7423.
4. Sabbioni, E. and Marafante, E., Bioinorg. Chem., 1978, 9:389-407.
5. Sabbioni, E., Marafante, E., Amantini, L., Ubertalli, L. and Birattari, C., Bioinorg. Chem., 1978, 8:503-515.
6. Sabbioni, E. and Girardi, F., Sci. Total Environ., 1977, 7, 145-179.
7. Sabbioni, E., Marafante, E., Goetz, L. and Birattari, C., Radiochem. Radioanal. Letters, 1977, 31:39-46.
8. Sabbioni, E. and Marafante, E., J. Toxicol. Environ. Health, 1981, in press.
9. Sabbioni, E. and Rade, J., Toxicol. Letters, 1980, 5:381-387.
10. Pepin, G., Bouley, G. and Boudene, C., C.R. Acad. Sc. Paris, 1977, 285D:451-454.
11. Sabbioni, E., Rade, J. and Bertolero, F., J. Inorg. Biochem., 1980, 12:307-315.

PECULIARITY OF HEART AND MUSCLE FERRITINS

M.C. Linder, G.M. Nagel and M. Roboz

Department of Chemistry, California State University
FULLERTON, CALIFORNIA 92634 USA

INTRODUCTION

Ferritin is a large protein found in all mammalian cells, but especially in liver, spleen and bone marrow, and is considered the main site for iron storage. When iron is administered, there is an increased rate of synthesis of ferritin subunits by a mechanism which involves the mobilization of existing mRNA (1), and this results in an accumulation of ferritin able to sequester additional iron. In iron deficiency, ferritin levels are decreased in response to a lesser need for iron storage.

Different forms of ferritin are found in different tissues, comprised of variable amounts of subunits of two sizes: 19,000 and 20-21,000 daltons (2) The number of different polypeptide chains in each size class and the number of genetic cistrons involved in defining ferritin structure is not known, although recent data indicate some processing of ferritin subunits occurs after synthesis (3). This may account for some of the subunit differences observed.

The ferritin most studied, structurally, is that of horse spleen. It has 24 subunits arranged in a roughly spherical shell, with 432 symmetry, as determined by X-ray diffraction (4). It contains 6 channels through which iron can enter and exit. The complete amino acid sequence and X-ray fine structure are currently being completed.

In contrast to other tissues, heart and other muscles contain two major forms of ferritin which may be separated in polyacrylamide gel electrophoresis (PAGE)(1). Concentrations of these two ferritins, henceforth designated the "slow" and "fast" forms on the basis of their electrophoretic migration in alkaline pH, vary with the type of muscle, age, sex and with isoproterenol treatment (1), at least in the rat. Current studies in our laboratory indicate that the "fast" form is located in the myocytes, while the other is present in the non-muscle cell population (5). The observation that hearts of male and female rats contain the same total ferritin concentration but different proportions of the two forms prompted us to investigate whether one might be the structural precursor of the other. Using ^{3}H-leucine, we showed that both ferritins incorporated radioactivity in parallel, thus ruling this out. Both also incorporated ^{59}Fe in parallel, with the "fast" form showing a more rapid iron uptake in both heart and diaphragm muscle (6).

Analysis of heart and other ferritins by quantitative PAGE, in which they are subjected to differential sieving in a series of gels with decreasing pore size (1), indicated that the majority of ferritins had apparent molecular weights in the range of 420-490,000 but that both of the muscle ferritins were larger and/or more asymmetric (1). This technique has the drawback that it cannot distinguish between effects on migration due to size and asymmetry, both for the substance under analysis and the standards employed. For this reason, a thorough study of size and shape by independent means was indicated. A summary of the results (below) is mainly taken from Linder, et al (7).

RESULTS AND DISCUSSION

Ferritins from rat heart and diaphragm and from human heart, were isolated, and their size and shape analysed by gel filtration, sedimentation velocity and sedimentation equilibrium, in comparison with ferritins isolated from rat liver and horse spleen. Using Ultrogel AcA 34 (LKB) and standard proteins (thyroglobulin; 690,000 daltons, Stokes' radius 81 Å ; rat liver ferritin, 495,000 and 68.5; aspartate transcarbamylase, 310,000 and 60.5; aldolase, 160,000 and 48), apparent molecular weights of 528,000-540,000 were obtained for the "fast" muscle ferritin, and 503,000 for the "slow" species. Stokes' radii were 72 and 69 Å for the two forms, respectively.

These ferritins were then analysed by sedimentation velocity, both before and after removal of their iron. The rat muscle samples contained both the "fast" and "slow" forms, as these were impossible to separate fully and only 1-3 mg could be obtained from 40 iron-treated rats. In the case of the human heart, all but 1 of 6 preparations contained only the "fast" ferritin characteristic of muscle cells. The iron-containing (holo) ferritins sedimented as a broad boundary, with high sedimentation coefficients (Table I), reflecting a heterogeneity of iron content. Upon iron removal, clear-cut boundaries were observed, with sedimentation coefficients near 17S for rat liver and horse spleen,

Table I: Sedimentation Velocity of Heart and Other Tissue Ferritins

Ferritin Source	(Fe:Protein) ug/ug	No. of Preps.	Sedimentation Coefficient (Mean±SD;No.) $s_{20,w}$	
Holoferritins:				
Rat Liver	0.15, 0.09	2	60.0S ± 2.2 (3)	-
Horse Spleen	0.21	1	75.7S, 78.9S (2)	-
Rat Heart & Diaphr.	0.09	1	58.0S, 61.7S (2)	5.7S ± 0.4 (8)
Human Heart	0.11	1	57.1S, 62.1S (2)	2.6S, 2.6S (2)
Apoferritins:				
Rat Liver	<0.005	2	17.1S ± 0.6 (4)	-
Horse Spleen	<0.005	2	17.4S, 17.2S (2)	-
Rat Heart & Diaphr.	<0.005	3	18.5S ± 0.4 (4)*	2.9S, 3.3S (2)
Human Heart	<0.005	2	18.2S ± 0.5 (3)**	2.3S, 2.6S (2)

*p<0.001 for difference from rat liver **p<0.05 for difference from rat liver

but with significantly higher values for preparations of muscle ferritins, containing the "fast" or mixtures of "fast" and "slow" forms. In the case of the muscle preparations, there also were variable amounts of slow-sedimenting material. This represents subunits (see below) which associated with the complete ferritins during their purification, through heat and acid treatment, ammonium sulfate fractionation, iron removal and repeated gel filtration (1).

The same kinds of preparations were analysed by sedimentation equilibrium. Those containing a single form of ferritin gave a very linear plot of ln Δy vs. r^2. Plots for preparations with both the "slow" and "fast" muscle ferritins (rat heart and diaphragm)

Table II: Molecular Weights of Apoferritins Obtained by Sedimentation Equilibrium

Ferritin Source	No. of Preps.	Form	Molecular Weight (Mean ± SD; No.)
Rat Liver	2	-	495,000 ± 2000 (5)
Horse Spleen	2	-	501,000 ± 18000 (3)
Rat Heart and Diaphragm	2	"Slow"	489,000 ± 11000 (3)
		"Fast"	678,000 ± 9000 (4)*
Human Heart	1	"Fast"	756,000, 756,000 (2)

*p<0.001 for difference from rat liver and horse spleen apoferritins

gave two distinct successive slopes which were used to get two molecular weights. Only those muscle ferritin preparations not containing slow-sedimenting material were used. With this technique, we confirmed that the molecular weight of horse spleen apoferritin is in the range shown by other researchers (8,9). Rat liver ferritin gave a similar value, and so did the "slow" muscle ferritin. The only peculiar ferritin appeared to be the "fast" form characteristic of muscle cells. Not only was it larger but almost twice as large as other ferritins. Apart from the possibility of species differences, the value obtained for human heart (which is greater than that for the rat muscle) should be considered more accurate, in that it was determined in the absence of the "slow" form.

A reanalysis of the various ferritins by SDS electrophoresis to assess the size of the subunits present confirmed that, as shown previously (1), all contained subunits in the range of 19,000-23,000 daltons. For "slow" heart ferritin and ferritins from horse spleen and rat liver, this is consistent with a protein shell of 24 subunits within which the iron oxyhydroxide may crystallize and deposit (1). However, since the "fast" muscle ferritin is larger it must contain many more subunits (Table III), possibly as many as 38-40. On the basis of the molecular weights and sedimentation coefficients determined, axial ratios were calculated (Table III). The data suggest that, ferritins, the "fast" muscle form is much more asymmetrical than other ferritins. This would explain why despite its greater molecular weight it sediments at a rate close to that of the other ferritins. Obviously, these data are not compatible with a structure of the kind envisaged for horse spleen ferritin (4).

Table III: Summary: Size and Shape of Ferritins From Muscle and Other Tissues

	Horse Spleen or Rat Liver	Muscle Ferritins "Slow"	Muscle Ferritins "Fast"
Molecular Weight	495,000 - 500,000	490,000	700,000 - 750,000
Number of Subunits	24	24	34 - 38
Stokes' Radius (Å)	68.5	69	72
Axial Ratio (a/b) (95% confidence limits)	1.0 - 3.6	(1.0 - 3.6)*	5.3 - 7.0

*uncertain due to sedimentation coefficient

The muscle ferritin preparations contained varying amounts of slow-sedimenting 2.75 material (Table I) which associated itself with ferritin throughout its purification, and was not visible as an extra band in electrophoresis. It could only be removed from the heavy ferritin by preparative centrifugation at 105,000 x g, in a discontinuous sucrose gradient (10). This material is immunoprecipitable with ferritin antibody. Further work has confirmed it usually contains about 3% iron by weight, (5.7S), and has a molecular weight of 95,000 (10). Upon iron removal, its sedimentation coefficient drops to 2.7S, with a molecular weight of 38,000 by sedimentation equilibrium or gel filtration consistent with ferritin subunit dimers. Such subunit material can be deliberately produced from horse spleen apoferritin by dissociation in 67% acetic acid, and dialysis into glycine buffer, pH 3.0 (11). However, in the case of horse spleen ferritin this material readily reassociates into apoferritin at physiological pH. Thus, muscles appear to contain variable amounts of semi-free ferritin subunits, mostly in dimeric form. These may be as yet unprocessed subunit structures on their way to becoming new ferritin, or products of partial ferritin breakdown not seen in other tissues.

REFERENCES

1. Munro, H.N. and M.C. Linder, 1978, Physiol.Revs. 58:317-396.
2. Drysdale, J.W., P. Arosio, T. Adelman, J.T. Hazard and D. Brooke, 1975. In "Proteins of Iron Storage and Transport in Biochemistry and Medicine" (ed. R.R. Crichton)(North Holland, Amsterdam) pp 359-366.
3. Conlon-Hollingshead, C. and H.N. Munro, 1980, Fed. Proc. 39:2154.
4. Banyard, S.H., D.K. Stammers, and P.M. Harrison, 1978, Nature 271:282-284.
5. Linder, M.C., M. Roboz, M.J. McKown, R. Zak, and W.M. Pardridge, 1981, manuscript submitted.
6. LaCross, D. and M.C. Linder, 1980, Biochim. Biophys. Acta 663:45-55.
7. Linder, M.C., G.M. Nagel, M. Roboz, and D.M. Hungerford, Jr., 1981, J.Biol.Chem., in press.
8. Bjork, I. and W.W. Fish, 1971, Bioch. 10:2844-2848.
9. Crichton, R.R., R. Eason, A. Barclay and C.F.A. Bryce, 1973, Bioch.J. 131: 855-857.
10. Linder, M.C., G.M. Nagel and M. Roboz, 1981, manuscript in preparation.
11. Harrison, P.M., and D.W. Gregory, 1968, Nature 220:560-578.

ACKNOWLEDGEMENTS

Supported by USPHS Grant No. HL 24337.

METALLOTHIONEIN

THE NATURE AND FUNCTION OF METALLOTHIONEIN

Ian Bremner

Rowett Research Institute
BUCKSBURN, ABERDEEN
SCOTLAND

Metallothionein was first isolated and characterized as a cadmium- and zinc-binding protein from horse kidneys over 20 years ago (1). Although this event initially attracted little attention, subsequent concern about environmental pollution with heavy metals and the suggestion that this protein may play a role in their detoxification has led to a major increase in interest in its occurrence. This interest has been further stimulated by the realization that the protein is also involved in the metabolism of the essential trace elements, copper and zinc. Metallothionein has now been isolated from a wide range of animal tissues, including liver, kidneys and intestinal mucosa. It can be produced by several cell lines when they are cultured in cadmium- and zinc-containing media and has been found in fish, in marine invertebrates like crabs and mussels, in plants and in microorganisms. Since this ubiquitous protein is therefore of considerable antiquity in terms of evolutionary development, it seems likely that it plays an important role in some fundamental biological process. This suggestion will be discussed in this paper.

NATURE OF METALLOTHIONEIN

The most characteristic features of metallothionein are its low molecular weight (about 6500), its high metal content (7-10%) and its abnormal amino acid composition. Particularly noteworthy are the high cysteine content (20 out of the 61 amino acid residues in the polypeptide chain) and the absence of aromatic amino acids. Two isomers of metallothionein have been isolated from most tissues but these differ only slightly in amino acid composition and other properties. Neither the source nor the isomeric form of metallothionein affect the distribution of the cysteinyl residues along the peptide chain, although there are minor changes in the occurrence and position of other residues (2). There is always a high preponderance of Cys-x-Cys sequences, where x is a residue other than cysteine. Serine and lysine residues are usually in close juxtaposition to the cysteinyl residues and may contribute to the maintainance of the highly ordered tertiary structure of the metalloprotein (3). The fixed distribution of the cysteinyl residues is probably related to the high metal-binding capacity of the protein, since spectroscopic studies indicate that all the cysteinyl residues are involved in metal-binding.

The molar ratio of cysteine residues to metal atoms is close to 3.0 in the cadmium- and zinc-containing proteins but only 1.5-2.0 in the copper-rich proteins, implying some difference in the binding of this metal. ^{1}H-NMR studies (3) confirm that no other amino acids are involved in cadmium- and zinc-binding, which could suggest that metal-trimercaptide complexes are present. However, ^{113}Cd NMR (4) and EXAFS measurements (5) on cadmium- and zinc-metallothioneins from rabbit and sheep liver, show that each metal atom has a similar environment and has 4 sulphur atoms arranged tetrahedrally around it. In the case of the zinc protein, the metal-sulphur bonds have a length of <u>ca</u> 2.28 Å. It is apparent, therefore, that some of the cysteinyl sulphur atoms must serve as bridging ligands between different metal atoms, perhaps in a ferrodoxin- or adamantane-type structure. However, unequivocal elucidation of the structure of the protein must await the preparation of metallothionein crystals and their examination by x-ray crystallographic procedures.

Little is known of the strength of the metal-sulphur bonds in different metallothioneins but it has been suggested that the stability constants of the cadmium- and zinc-proteins are around 10^{-15} and 10^{-11} respectively. Copper is bound even more firmly to the protein, in the form of Cu(I), and only dissociates at pH values <0.5. These differences in binding strength are of some significance in explaining the involvement of metallothionein in the interactions between these metals.

SYNTHESIS OF METALLOTHIONEIN

The metal-free protein has never been isolated from any biological source and it is believed that metallothionein always occurs as the fully-substituted metal derivative. Indeed, most metabolic studies on the protein have had to rely on the measurement of the metal content of particular chromatographic fractions when estimates are required of the metallothionein concentration in samples. It is not surprising therefore that the search for the function of metallothionein should have been concentrated on its possible role in metal metabolism.

Parenteral and dietary administration of cadmium, zinc, copper, mercury, silver or gold have frequently been found to increase the amount of that metal present as metallothionein in the liver, kidneys and/or intestinal mucosa in several species. This is generally accompanied by increased incorporation of either zinc or copper into the protein, as a secondary metal. Since all these metals also stimulate the incorporation of [^{35}S]-cysteine into the protein, they do not simply bind to pre-existing thionein but induce its synthesis de novo. This is confirmed by the inhibitory effect of prior administration of cycloheximide, actinomycin D and cordycepin on the metal-induced synthesis of metallothionein, which indicates that the metals must affect the transcription of poly(A)-containing messenger RNA for the protein (6).

Further proof of this has been provided by the isolation of increased quantities of metallothionein m-RNA from the liver of zinc-injected rats and by the subsequent translation of this in a cell-free protein-synthesizing system employing wheat-germ extract. It is significant that there is a good correlation between the metallothionein m-RNA concentration and the rate of synthesis of the protein in the liver of zinc-injected rats, with maxima being attained about 6 hours after zinc injection and basal levels being restored after 18-24 hours (6).

Increased production of translatable metallothionein m-RNA has also been noted when primary or transformed cells, such as liver parenchymal, kidney, Chinese hamster ovary (CHO) and HeLa cells, are cultured in media containing added zinc or cadmium. Moreover, these studies on cultured cells have shown that zinc-metallothionein synthesis can also be stimulated by the addition of dexamethasone and other glucocorticoids to the medium (6). This is consistent with the increased concentrations of hepatic zinc-metallothionein which occurs in chicks injected with hydrocortisone and in chicks and rats subjected to different types of stress (7,8,9). Thus, restriction of food intake, infection, endotoxaemia, laporotomy and imposition of stress by such means as exercise or change in environmental temperature can greatly increase hepatic zinc-metallothionein concentrations and increase [^{35}S]-cysteine incorporation into the protein. These effects which are probably mediated by the concomitant changes in the levels of circulating glucocorticoids, clearly demonstrate that metallothionein synthesis can be induced under physiological conditions by means other than the administration of large amounts of a particular metal.

It could be argued that the increased metallothionein production in the liver of the stressed animals and in the hormone-treated cells is merely a consequence of the associated increased cellular uptake of zinc, since increased tissue zinc levels almost invariably result in metallothionein synthesis. However, recent studies with cultured HeLa cells indicate that dexamethasone is indeed a primary inducer of metallothionein synthesis and can operate even when zinc uptake by the cell is inhibited by cycloheximide (10).

NATURAL OCCURRENCE OF METALLOTHIONEIN

The increase in hepatic metallothionein concentrations in response to starvation (7) provided some of the earliest evidence that production of this protein could be influenced by physiological state. Hitherto, most investigations had concentrated on the effects of administration of non-essential metals, such as cadmium and mercury, on its synthesis. Much emphasis was placed on the virtual absence of metallothionein from the liver of the laboratory rat, since it appeared that some special stimulus was essential for induction of synthesis of the protein.

Subsequent studies have clearly demonstrated that this is not so. Even in normal rats, considerable amounts of (copper,zinc)-thionein are present in the kidneys, especially in females, and there is some evidence that this is influenced by female sex hormones (11).

Zinc-metallothionein is present in the cytosolic fraction of human livers (12) and (copper, zinc)-metallothioneins occur naturally in the liver of several species of domestic animals (13). The most convincing evidence that metallothionein is synthesized under normal physiological conditions is the occurrence of large amounts of the zinc-protein in the liver of foetal and neonatal animals. Maximum concentrations are attained at about 80 days gestational age in the foetal lamb (14) but at about 2 days after birth in the rat (15). This difference between species is probably a reflection of the differences in the maturity of the foetuses at birth and may indicate a particular need for zinc at a certain stage of development.

The concentration of copper bound to metallothionein in the liver cytosol also increases during foetal and neonatal development and is highest at about 2 weeks of age in the rat (15). However, a second pool of copper metallothionein has recently been identified in the liver of neonatal (and perhaps adult) animals (16). This is situated in the particulate fraction of the liver, perhaps in the lysosomes, and can only be extracted on disruption of membranes under reducing conditions. This second metallothionein fraction probably corresponds to neonatal hepatic mitochondrocuprein and may therefore be present in polymeric form.

Marked changes occur in the tissue concentrations of copper and zinc present as metallothionein as the copper and zinc status of growing and adult animals is varied (7,13). Thus, no copper or zinc occur in this form in copper- and zinc-deficient animals, respectively, whereas a major proportion, 40-75%, of the additional metal is generally bound to this protein when tissue copper and zinc concentrations are increased. One notable exception to this is the virtual absence of copper-metallothionein from the liver of zinc-deficient animals regardless of their liver copper content. This is not a result of impaired induction of metallothionein synthesis in zinc-deficiency, since this occurs at a normal rate when zinc-deficient rats are injected with copper. Instead, it appears to be a reflection of an increased rate of degradation of copper-metallothionein in these animals (17).

DEGRADATION OF METALLOTHIONEIN

Little is known of the processes involved in the degradation of metallothionein or of the ultimate fate of the metal which is incorporated into the protein. This is arguably one of the main reasons for the uncertainty about the role of the protein in metal metabolism. Several estimates have been made, using pulse labelling techniques, of the biological half-life of different metallothioneins in the liver and kidneys of the rat. The cadmium-containing proteins generally have a half-life of 3-5 days, whereas the zinc- and copper-proteins have half-lives of about 20 and 12 hours, respectively (6,17,18). The nature of the bound metals is clearly a major determinant of the biological stability of the protein and even a slight increase in the relative copper content can increase the turnover rate of the protein (18).

It is generally assumed that metallothioneins are degraded within the lysosomes, although there is as yet no direct evidence that this is the case. Both cadmium- and zinc-metallothioneins are readily hydrolysed when incubated with rat liver lysosomal extracts, the rate of degradation increasing in the order cadmium- $<$ zinc- $<$ apo-protein (6). This is in accord with the findings in vivo, and can be related to the chemical stability of the metalloprotein complexes. The greater dissociation of zinc than of cadmium from the protein at the pH within the lysosomes could explain the greater susceptibility of zinc-metallothionein to proteolysis. However, the findings on the degradation of the copper-protein are not consistent with this view. This is chemically the most stable of the common metallothioneins and no significant dissociation of copper would be expected at the pH within the lysosome. Studies on the degradation of copper-metallothionein by rat liver lysosomal extracts in vitro confirm that the protein is not readily hydrolysed (19), yet the biological half-life of the protein in vivo is less than that of the zinc- and cadmium-derivatives. The reason for this apparent anomaly has not been established, but it could indicate that processes other than lysosomal degradation of the protein are involved in the mobilization of copper from the protein. These could include chelation of the copper by other ligands, oxidation of the Cu(I) to Cu(II), or perhaps formation of aggregated forms of the protein within the lysosomes.

The various forms of metallothionein differ in one other important respect with regard to their degradation. Thus, when metallothionein synthesis has been induced by copper or zinc, the subsequent removal of the metals from the protein occurs at about the same rate as the protein is degraded (6,17). In contrast, the rate of removal of cadmium from the hepatic and renal proteins is extremely slow, with half-lives of several months, despite a biological half-life of the protein moiety of only 3-5 days (18). Clearly, the liberated cadmium is not readily excreted from the tissue and is able to induce further synthesis of the protein, whereas both copper and zinc are available for other metabolic processes and do not cause further metallothionein synthesis.

METALLOTHIONEIN AND HEAVY METAL DETOXICATION

The incorporation of cadmium into metallothionein is commonly regarded as the principal means whereby the intracellular toxicity of the metal is reduced. Several studies have shown, both in vivo and in vitro, that cadmium is less toxic when bound to metallothionein than when present in ionic form or as weak metal-protein complexes (20). For example, cadmium-metallothionein is a less potent inhibitor of certain enzymes than is an equivalent molar concentration of ionic cadmium. Dose-response investigations on chronic cadmium toxicity indicate that clinical lesions only appear when the capacity to incorporate the metal into metallothionein is saturated. The protective effect of the prior injection of a small dose of cadmium or zinc against the effects of a subsequent larger, and normally toxic dose of cadmium has been attributed to 'priming' of the metallothionein-synthesizing system, although this does not explain all facets of this protective effect.

The most convincing evidence that metallothionein may be involved in the detoxication of cadmium is perhaps the finding that the development of cadmium resistance in some cell lines, as with CHO or human skin epithelial cells, is associated with an increase in their capacity to produce metallothionein (21). It is surprising in some respects that such a mechanism for detoxifying cadmium should apparently have evolved since itrequires continual synthesis of a short-lived protein. It might have been expected, at least on teleological grounds, that more effective control would have been achieved by either limiting the uptake or promoting the excretion of the metal. However, it is possible that cadmium, and other heavy metals, are utilizing a metal-binding system which originally evolved to regulate the metabolism of copper and zinc.

METALLOTHIONEIN AND HEPATIC COPPER AND ZINC METABOLISM

Several roles have been proposed for metallothionein in the control of copper and zinc metabolism, although none of these has, as yet, been unequivocally proven. The increased production of the protein when liver zinc concentrations are increased above a critical level, the block in hepatic zinc uptake when protein synthesis is inhibited by actinomycin D and the short biological half-life of the protein have suggested an involvement in the hepatic uptake and temporary storage or cellular detoxication of the metal. Metallothionein-bound zinc can be used for metabolic purposes, as is indicated by the growth response and low faecal zinc excretion by zinc-deficient rats after a large proportion of an injected dose of zinc was transiently incorporated into the hepatic protein (19). However significant excretion of zinc from hepatic metallothionein may also occur in other circumstances, as when rats are transferred from high-zinc to zinc-deficient rats (22).

The elevated zinc-metallothionein content in the liver of foetal and neonatal animals has also been assumed to represent a storage reserve for zinc (14,15). Indeed, about half of the total body zinc in foetal lambs is in this form at 80-90 days gestational age. However, as this pool of zinc is only capable of supplying 5-10% of the subsequent demand of the lamb foetus for zinc up to the time of birth, its storage capacity is somewhat limited. An alternative suggestion is that zinc-metallothionein in foetal and neonatal liver acts as a specific reservoir of zinc for the supply of the metal to particular cytosolic components in the liver during this period of rapid development (15). It is of interest in this regard that zinc-requiring apo-enzymes, such as alkaline phosphatase and carbonic anhydrase, can be activated in vitro by zinc-metallothionein (23). This may occur by direct transfer of zinc between the proteins, although it would seem that there would be strong stereochemical constraints on such a process.

It has yet to be shown that such a transfer occurs in vivo but it is nevertheless significant that a similar role as a metal donor has been proposed for copper-metallothionein from *Neurospora crassa*, since it was found that this protein reactivated apotyrosinase in vitro more effectively than did free ionic copper (24). This was not the only function ascribed to copper-metallothionein, however, as it also appeared to be involved in the storage and cellular detoxification of the metal when the organism was grown in copper-supplemented medium.

Some insight into the possible roles of metallothionein in hepatic copper metabolism has been gained by relating the distribution of copper in the liver of different species to their tolerance to increased exposure to copper (13). Thus in the pig, which is highly tolerant of copper, a high proportion of the hepatic copper is present in the cytosol as metallothionein. Copper is rapidly excreted in the bile of these animals and it is known that the copper-protein has a short biological half-life. In contrast, only a small proportion of liver copper in the sheep occurs as metallothionein and even this is dependent on the liver zinc content. Biliary copper excretion in these animals occurs at a low rate and it is probable that this is responsible for the ready accumulation of copper in these animals and for the eventual development of copper toxicosis. There is also a difference in the association of copper within the particulate fraction in the liver of these animals. Preliminary studies (19) indicate that much of the copper in this fraction of copper-loaded pigs is also present as metallothionein whereas no such additional metallothionein-pool occurs in sheep liver. It seems therefore that one role of the protein could be to temporarily store copper in a relatively non-toxic form, from which it can be subsequently excreted or used for a variety of metabolic purposes. However, proof of this hypothesis must await the discovery of the relationship between the two pools of copper-metallothionein in the liver and the precise intracellular source of biliary copper.

METALLOTHIONEIN AND THE ABSORPTION OF ZINC AND COPPER

Cousins (25) has suggested that metallothionein may also play an important role in the homeostatic control of intestinal zinc absorption, since there is often an inverse relationship between binding of zinc to mucosal metallothionein and the efficiency of absorption of the metal. This implies that the protein serves as an inducible zinc-binding compartment in the mucosa, which can diminish the transfer of intestinal zinc to the plasma. While there is considerable evidence to support this view, much of it was unfortunately obtained in experiments where the zinc status of rats was increased by zinc injection or by feeding diets with very high zinc content. There is therefore some doubt as to the importance of this regulatory process at physiological zinc intakes, especially as mucosal metallothionein concentrations in the rat are only significantly increased at dietary zinc contents of about 900 mg/kg (26). Moreover, variations in the efficiency of zinc absorption have been recorded in rats over a narrow range of dietary zinc intakes, without any increase in mucosal zinc uptake (27). Metallothionein may therefore only be involved in controlling zinc absorption at near-toxic zinc intakes, when other control mechanisms are saturated, or alternatively may be involved in zinc excretion rather than absorption processes.

Mucosal metallothionein may also function, on occasion, in the control of the intestinal absorption of copper. Thus, the decreased copper absorption, and the ultimate development of copper deficiency, in zinc-supplemented rats has been attributed to the induction of mucosal metallothionein synthesis by zinc and to the subsequent incorporation of copper into the protein (26). This is consistent with the higher affinity of copper for binding sites on the protein. Significantly, lower levels of dietary zinc (150 and 450 mg/kg) which do not increase metallothionein concentrations in the mucosa have no effect on copper absorption. It appears, therefore, that copper which is bound to the newly-synthesized metallothionein in the intestine is not readily transferred to the plasma.

An inverse relationship between the efficiency of copper absorption and the binding of copper to intestinal metallothionein has also been observed in oestrogen-treated rats (28) and in mottled mice with genetically-induced copper deficiency (19). This could again indicate an involvement of metallothionein in the control of copper absorption but it cannot be excluded that the accumulation of copper-metallothionein in the mucosa of these animals is the result and not the cause of decreased copper absorption and of impaired transport of copper to the plasma. The accumulation of mucosal copper-metallothionein, as also occurs in neonatal rats in which pinocytotic uptake of copper has taken place (29), may simply be another illustration of induced synthesis of this protein when intracellular concentrations of copper or of other metals are increased. There is no convincing evidence at present that metallothionein is involved in normal homeostatic control of copper absorption in response to changes in dietary copper supply (26).

CONCLUSION

Substantial progress has been made in recent years in the determination of the structure of metallothionein and in the elucidation of the mechanisms involved in its synthesis. However the physiological function of this highly-conserved protein is still largely a matter of conjecture. Roles have been postulated for the protein in the control of the absorption and in the hepatic storage or detoxication of certain metals. However, the widespread occurrence of metallothionein in microorganisms, plants and throughout the animal kingdom, its presence in practically all animal tissues and the ability of so many metals to induce its synthesis may indicate that its primary function is not necessarily concerned with such specific processes. Rather, it may simply be that production of metallothionein is a general response to increased intracellular concentrations of certain metals, perhaps as a cellular detoxication mechanism. In the case of copper and zinc, this is merely a short-term response and the subsequent fate of the metal (and therefore the apparent function of the protein) will be largely determined by the particular needs of the animal.

REFERENCES

1. Kagi, J.H.R. and B.L. Vallee, 1960, J. Biol. Chem. 235:3460-3465.
2. Kissling, M.M. and J.H.R. Kagi, 1979. In "Metallothionein" (ed. J.H.R. Kagi and M. Nordberg) (Birkhauser, Basel) pp 145-151.
3. Vasak, M., A. Galdes, H.A.O. Hill, J.H.R. Kagi, I. Bremner and B.W. Young, 1980, Biochemistry. 19:416-425.
4. Otvos, J.D. and I.M. Armitage, 1980, Proc. Nat. Acad. Sci. USA. 77:7094-7098.
5. Garner, C.D., I. Bremner, S.S. Hasnain and J. Bordas (unpublished observations).
6. Cousins, R.J., 1979, In "Metallothionein" (ed. J.H.R. Kagi and M. Nordberg) (Birkhauser, Basel) pp 293-301.
7. Bremner, I. and N.T. Davies, 1975, Biochem. J. 149:733-738.
8. Oh, S.H., J.T. Deagen, P.D. Whanger and P.H. Weswig, 1978, Am. J. Physiol. 234:E282-E285.
9. Sas, B. and I. Bremner, 1979, J. Inorg. Biochem. 11:67-76.
10. Karin, M. and H.R. Herschman, 1981, Eur. J. Biochem. 113:267-272.
11. Bremner, I., R.B. Williams and B.W. Young, 1981, J. Inorg. Biochem. (in press).
12. Buhler, R.H.O. and J.H.R. Kagi, 1974, FEBS Lett. 39:229-234.
13. Bremner, I., 1979. In "Metallothionein" (ed. J.H.R. Kagi and M. Nordberg) (Birkhauser, Basel) pp 273-280.
14. Bremner, I., R.B. Williams and B.W. Young, 1977, Br. J. Nutr. 38:87-92.
15. Mason, R., A. Bakka, G.P. Samarawickrama and M. Webb, 1981, Br. J. Nutr. 45:375-390.
16. Riordan, J.R. and V. Richards, 1980, J. Biol. Chem. 255:5380-5383.
17. Bremner, I., W.G. Hoekstra, N.T. Davies and B.W. Young, 1978, Biochem. J., 174:883-892.
18. Cain, K. and D.E. Holt, 1979, Chem.-Biol. Interact. 28:91-106.
19. Bremner, I., R.K. Mehra, B.W. Young and N.T. Davies, unpublished results.
20. Webb, M., 1979, In "The chemistry, biochemistry and biology of cadmium" (ed. M. Webb) (Elsevier/North Holland, Amsterdam) pp 195-266.
21. Hildebrand, C.E., R.A. Tobey, E.W. Campbell and M.D. Enger, 1979, Exp. Cell Res. 124:237-246.
22. Chen, R.W., E.J. Vasey and P.D. Whanger, 1977, J. Nutr. 107:805-813.
23. Udom, A.O. and F.O. Brady, 1980, Biochem. J. 187:329-335.
24. Lerch, K., 1980, Nature, 284:368-370.
25. Cousins, R.J., 1979, Am. J. Clin. Nutr. 32:339-345.
26. Hall, A.C., B.W. Young and I. Bremner, 1979, J. Inorg. Biochem. 11:57-66.
27. Jackson, M.J., D.A. Jones and R.H.T. Edwards, Br. J. Nutr, in press.
28. Cohen, D.I., B. Illowsky and M.C. Linder, 1979, Am. J. Physiol. 236:E309-E315.
29. Johnson, W.T. and G.W. Evans, 1980, Proc. Soc. Exp. Biol. Med. 165:496-501.

DISCUSSION

CAMAKARIS (Australia). Do you think that metals such as mercury and silver induce metallothioneins indirectly by displacing zinc from the protein and that zinc is the primary inducer?

BREMNER (Scotland). It is probably true that the effect of certain metals on metallothionein induction can be classified as a stress-related phenomenon. In such conditions zinc accumulates in the liver and may be the primary cause of the metallothionein increase as you suggest. I cannot say that silver is one of these metals. There is, of course, no doubt that many metals induce metallothionein synthesis in their own right.

CAMAKARIS (Australia). Is there any evidence of polymerised metallothionein in lysosomes of animals other than sheep?

BREMNER (Scotland). I don't know if one can be sure it is present in the liver of adult sheep. We have looked at the accumulation of copper in the lysosomes of copper poisoned sheep. Copper and sulphur accumulate but we do not know if these elements are present as metallothionein. The best source of polymerised, or lysosomal metallothionein is the neonatal animal. Dr. Riordon in Canada has isolated it from human fetal liver.

SALTMAN (USA). Would you care to speculate on why iron has absolutely no affect on this protein and maybe tell us about differences between non-heme proteins and metallothionein that would explain why some metals bind to one and not the other.

BREMNER (Scotland). This has been a continuing puzzle to many people interested in chelation chemistry - at the moment there is no satisfactory explanation. Lead will not induce synthesis of metallothionein, but under the right conditions purified metallothionein will bind lead in vitro. I don't think iron is bound in these circumstances.

DREOSTI (Australia). Most of the evidence indicates that metallothionein is a short-term store for zinc and copper, is there any evidence that it is involved with long-term storage as well?

BREMNER (Scotland). I think of metallothionein as a type of 'sink' that gets the free metal ions out of the way quickly to prevent toxic effects. What happens to the metals thereafter will depend on the metabolic requirements at the time. For instance, when we injected zinc deficient rats with zinc a lot of the zinc was bound to liver metallothionein and subsequently very little of it was excreted in faeces - it was used in the growth of the animals. In contrast, when Dr. Whanger gave rats zinc in the diet over a long period, zinc also accumulated in liver metallothionein, but subsequently this zinc was excreted - apparently because the animals tissues were replete. I do not know if metallothionein can act as a long-term store for metals.

WHANGER (USA). We should keep in mind the species differences in metallothionein turnover. In rats and chickens the half life of zinc metallothionein is one or two days but in sheep and cattle it is between twenty two and twenty five days.

SUTTLE (Scotland). I was intrigued by your observations on the species differences between ruminants and non-ruminants. I wonder if I could extend the argument a little bit and suggest that they originate in differences in the efficiencies of copper and zinc absorption between the two groups. When a ruminant is fed as a ruminant its efficiency of Cu absorption is exceedingly low and it does not need assistance to cope with copper. With zinc on the other hand the efficiency of absorption in ruminants and non-ruminants appears to be essentially the same and therefore we have a metallothionein system which seems to be zinc driven.

BREMNER (Scotland). I should explain that some of the ruminants we worked with were veal calves, fed on a liquid milk substitute. They did not have a functional rumen so maybe the metal uptake was much the same as that in non-ruminant animals.

ELMES (Wales). A very wide range of copper and zinc values have been reported for carcinomas. I notice that you mentioned that protein synthesis inhibitors like cycloheximide inhibit the formation of metallothionein. I would like to suggest that interpretation of copper and zinc data would be assisted if any previous treatment with such inhibitors was noted and published.

COPPER-METALLOTHIONEIN METABOLISM IN THE KIDNEY OF BRINDLED MICE

Hans W. Prins & Cornelis J.A. Van den Hamer

Interuniversity Reactor Institute
Mekelweg 15, 2629 JB Delft, The Netherlands

INTRODUCTION

Brindled is a semi-dominant, X-linked mutation in mice [1], which is homologous to that of Menkes' disease in man [2]. Both mutations are characterized by an abnormal distribution of copper over the tissues. In various tissues, e.g., kidney, copper is accumulated in the cytoplasm as copper-metallothionein [3]. Other tissues, e.g., brain, do not receive adequate amounts of copper, which results in pathological features of Cu-deficiency [4]. Since the metallothionein fraction (MT-fraction) is the cytoplasmic Cu-binding protein fraction, which is predominantly affected by the mutation [5], we studied simultaneously the metabolism of metal and protein moiety of copper-metallothionein (Cu-MT) in this fraction.

MATERIALS AND METHODS

The L-[35S]cystine was obtained from New England Nuclear and diluted with L-cystine to a specific activity of 2 Ci/mmole. The 64Cu, prepared at our institute, had a specific activity of 50 or 4800 mCi/mmole; the latter was used in experiments which lasted for more than 48 hours. The 64Cu and 35S radioactivity was simultaneously determined in a liquid scintillation counter (Packard Instruments, model 2650). The Cu-concentrations were determined by flameless AAS (Perkin Elmer 373-HGA 76B).

The mice were intraperitoneally dosed with 10 μCi L-[35S]cystine and 4 hours later with 4 μg 64Cu. The moment of the Cu-injection is designated as zero hour. The liver and kidney were homogenized in 2 vol. (v/w) of 10 mM Tris-acetate (pH=7.4), and centrifuged at 120,000 x g for 1 hour, and the supernatant was fractionated on Sephadex G-75 columns (1.8 x 65 cm).

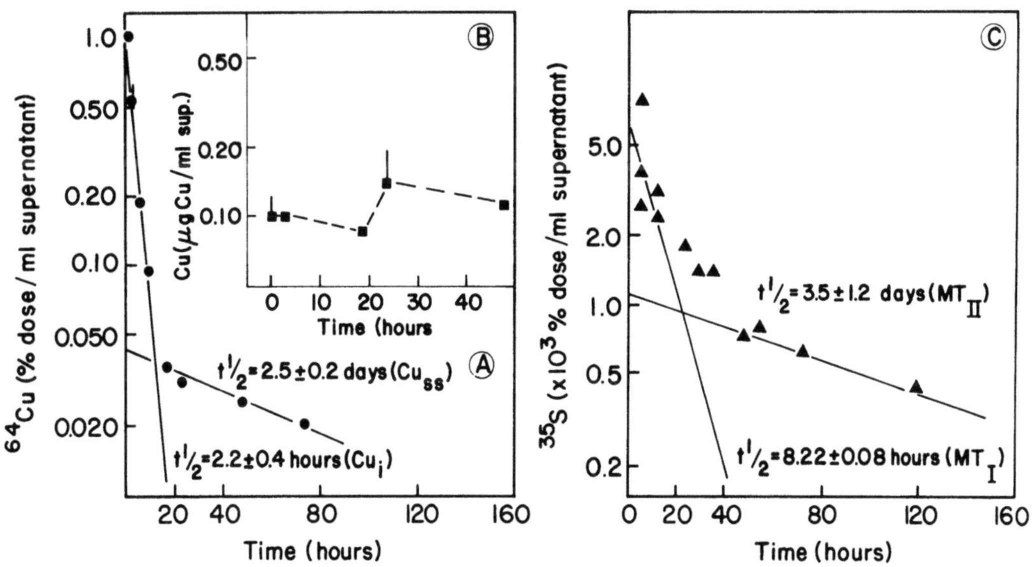

Fig. 1. Metabolism of radioactive copper (64Cu, fig. 1A), stable copper (Cu, fig. 1B) and the protein moiety of metallothionein (MT) in the renal MT-fraction of normal Brindled mice. Prior to the administration of the 64Cu dose, MT was in vivo labeled with L-[35S]cystine.

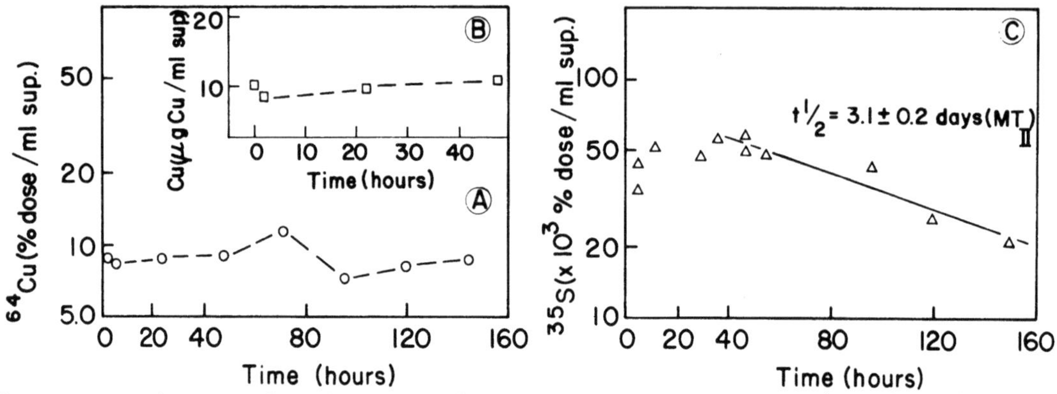

Fig. 2. Metabolism of radioactive copper (64Cu, fig. 2A), stable copper (Cu, fig. 2B) and the protein moiety of metallothionein (MT) in the renal MT-fraction of heterozygous Brindled mice. Prior to the administration of the 64Cu dose, MT was in vivo labeled with L-[35S]-cystine.

The 35S-labeled fractions containing MT were corrected for the contribution of 35S-labeled non-MT proteins [6]. The isolation of the MT-fraction was performed under anaerobic conditions at 4°.

RESULTS AND DISCUSSION

Fig. 1 shows the turnover of 64Cu and Cu in the renal MT-fraction of normal Brindled mice. In the MT-fraction, copper is bound exclusively to MT [3], of which the protein moiety is catabolized with two different half-lives, viz., MT_I: 8.22 ± 0.08 hours; MT_{II}: 3.5 ± 1.2 days. The major part of the injected 64Cu (Cu_i) is excreted with a short half-life (2.2 ± 0.4 hours). Since there is no concomitant catabolism of 35S-MT, it is concluded that Cu_i is only temporarily bound to MT and is excreted from the cytosol, prior to the catabolism of MT.

A minor part of the 64Cu is excreted with a longer half-life (2.5 ± 0.2 days), which is similar to that of MT_{II}. This fraction of the injected 64Cu is not incorporated into apo-MT, but became and "intrinsic constituent" of MT_{II} after exchange with stable copper [5]. It therefore represents the turnover of copper in the MT-fraction under steady state conditions (Cu_{ss}). The turnover rate is 0.028 μg Cu/ml supernatant per day. Since the half-lives of Cu_{ss} and MT_{II} are not significantly different ($P > 0.25$), it is assumed that both are simultaneously catabolized.

The metabolism of MT_I is not accompanied by the turnover of 64Cu. This could indicate that copper is not bound to MT_I under normal conditions and that MT_I probably represents native MT, which consists of Zn-MT. Similar half-lives of Cu,Zn-MT are only found when the animals are exposed to high copper loads which resulted in the induction of the MT-synthesis [6]. In our experiments the copper dose did not induce the MT synthesis, whereas the simultaneous occurrence of MT_I and MT_{II} is also observed in liver and kidney of Brindled mice which were not injected with Cu [7]. In the liver the Cu-metabolism in the MT-fraction is qualitatively similar to that of the normal kidney [5].

Fig. 2 shows the turnover of 64Cu and Cu in the MT-fraction of mutated Brindled mice. In comparison with normal mice, the Cu-concentration is a hundredfold and the initial 64Cu-uptake is tenfold increased. There is no excretion of 64Cu from the MT-fraction observed in the first six days post injection. Furthermore, the incorporation of L-[35S]cystine is tenfold increased and MT is only degraded with the longer half-life, viz., MT_{II}: 3.1 ± 0.2 days. From these experiments it is concluded that the excretion of both Cu_i and Cu_{ss} from the cytosol is impaired by the Brindled mutation. Since it is unlikely that a mutation results in two independent aberrations of Cu-metabolism, these results indicate that in the normal kidney Cu_i and Cu_{ss} use the same excretion pathway. The absence of MT_I indicates that, among others, the metal status of a tissue determines the MT_I to MT_{II} ratio and herewith whether MT functions as a metal transport protein (MT_I) or as a metal storage protein (MT_{II}).

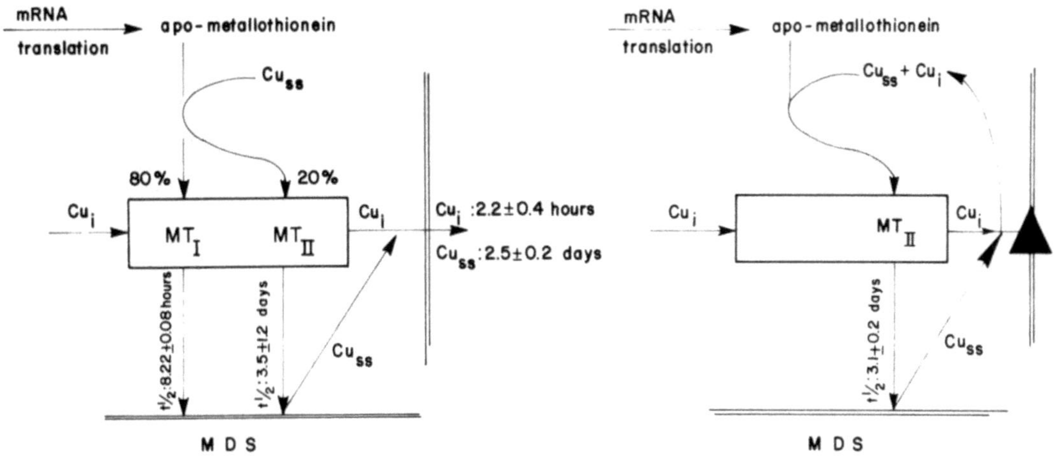

Fig. 3. Models for the metabolism of Cu_{ss}, Cu_i, and the protein moiety of metallothionein in the MT-fraction of the kidney of normal (Fig. 3A) and heterozygous (Fig. 3B) Brindled mice. MDS = metallothionein degrading system.

MODELS

Fig. 3A shows a model for the Cu-metabolism in the renal MT-fraction of normal Brindled mice. Cu_{ss} is incorporated in apo-MT, which results in the formation of MT_{II}. MT_I represents native Zn-MT. Both Cu_{ss} and MT_{II} are simultaneously degraded but are excreted from the cytosol through different pathways. Since hardly anything is known about the MT proteolysis in vivo, the compartment in the models in which MT is degraded is not differentiated [7] and referred to as MT degrading system (MDS). After injection of small amounts of copper, the metal (Cu_i) is only temporarily bound to MT_I and/or MT_{II} and is excreted prior to the catabolism of the protein moiety. Cu_i and Cu_{ss} are excreted through the same pathway.

Fig. 3B shows a model for copper metabolism in the renal MT-fraction of heterozygous Brindled mice. From the absence of the half-lives of either Cu_i and Cu_{ss}, it is concluded that their excretion from the cytosol is impaired and that the Brindled mutation results in an impaired renal reabsorption of copper. To prevent the occurrence of toxic intracellular Cu-concentrations, Cu is recycled in MT. This results in an increased synthesis of MT (Fig. 1C and 2C) Cu_i and Cu_{ss} are incorporated in MT_{II}, the MT-population with the lowest turnover. This situation is comparable to that described for renal cadmium metabolism in the rat [8].

In conclusion, in the kidney of normal Brindled mice, two populations of MT can be distinguished, viz., MT_I (metal transport) and MT_{II} (metal storage). Under normal steady state conditions, Cu_{ss} is bound to MT_{II} and both are simultaneously degraded. MT_I represents probably native Zn-MT. The Brindled mutation results in an impaired reabsorption of copper. The Cu-concentration increases and all apo-MT molecules are used to sequester copper and results in the formation of Cu-MT_{II}; the protein moiety of MT_I is undetectable.

REFERENCES

1. Hunt, D.M., 1974, Nature 249:52 - 853.
2. Danks, D.M., 1977, Inorg. Perspec. Biol. Med. 1:73 - 100.
3. Prins, H.W., and C.J.A. Van den Hamer, 1980, J. Nutr. 110:151 - 157.
4. Yajima, K. and K. Suzuki, 1979, J. Neuropath. Exp. Neur. 38:35 - 46.
5. Prins, H.W., 1981, Thesis TH Delft.
6. Bremner, I., W.G. Hoekstra, N.T. Davies, and B.W. Young, 1978, Biochem. J. 174:883 - 892.
7. Prins, H.W., and C.J.A. Van den Hamer, 1981, Life Sci., in press.
8. Feldman, S.L., K.S. Squibb, and R.J. Cousins, 1978, J. Toxicol. Environm. Health 4:805 - 813.

DISCUSSION

DANKS (Australia). How do you explain the lack of availability of the copper to the copper enzymes in many tissues if the defect is a fault in the excretory process as your model suggests?

VAN DEN HAMER (The Netherlands). The distribution of copper between the storage proteins and the enzymes will depend on the relative stability constants. I think that the copper available to the enzymes will be a function of the total intracellular copper and the half-life of the enzyme. Copper proteins and copper enzymes with long half-lives will get enough copper, at least to give a minimum level of enzyme activity, while those with short half-lives will not.

CAMAKARIS (Australia). Two questions. Firstly, if you were to load kidneys from normal mice with the same amount of copper as kidneys from mutant mice would you expect similar results? Secondly, what do you think determines whether copper will enter the pool of rapid turnover rather than that of slow-turnover?

VAN DEN HAMER (The Netherlands). In answer to your first question, I don't think the experiment can be done because the copper level in kidneys from mutant mice is about 100 times higher than normal, and to attain such concentrations in normal mice all sorts of changes would occur, associated with copper poisoning. With regard to your second question, when you look at the fraction that has a half-life of 3 days in the mutant mouse it is almost completely loaded with copper. We found 7 copper atoms : 1 zinc atom and a trace of cadmium. In the normal mouse it is very likely that the fraction with 3 day half-life is also loaded with copper, but that with 8 hour half-life is not. We find 100 times more copper and ^{35}S in the fraction with 3 day half-life.

METALLOTHIONEIN m-RNA IN NEONATAL AND ADULT RAT LIVERS

J. Mercer, *T. Stevenson, *J. Camakaris, I. Lazdins and D. M. Danks

Genetics Research Unit Royal Children's Hospital Research Foundation, Melbourne, and *Department of Genetics, University of Melbourne, MELBOURNE, AUSTRALIA

Metallothioneins (MT) are low molecular weight proteins which are probably involved in Cu and Zn homeostasis and also in the detoxification of heavy metals(1). They have an unusual amino acid composition with about one third of the residues being cysteine whilst leucine and aromatic amino acids are absent. Clearly an understanding of the control of MT synthesis is important in elucidating the role of MT in normal metabolism and in pathological states.

Neonatal rats have high concentrations of Cu and Zn in their livers and this is associated with high levels of MT(2). Adult rat livers have relatively low levels of MT, but these levels can be increased by administration of Cu, Zn or Cd. There is good evidence for the induction of MT by Cd and Zn in adult rat liver being due to increased levels of translatable MT mRNA(3,4). Winge et al(5) claim that Cu induces a protein in rat liver termed "Cu-chelatin" whilst Bremner and Young(6) claim that Cu induces MT.

METHODS

Total RNA was prepared by a modification of the guanidine hydrochloride procedure of Deeley et al(7). This RNA was used to direct translation in a wheat germ cell free system using ^{3}H leucine, ^{35}S cysteine or ^{3}H serine as a radio-labelled amino acid. The translation products were carboxyamidomethylated to prevent aggregation, and analysed on 17.5% acrylamide slab gels containing SDS. The radioactive polypeptides were detected by fluorography.

Authentic Zn induced MT was prepared from rat liver according to the method of Ryden and Deutsch(8) and carboxyamidomethylated using ^{14}C iodoacetamide.

RESULTS AND DISCUSSION

MT-mRNA in neonatal rat liver

Wong and Klaassen(2) demonstrated that in neonatal rat livers MT levels are high, peaking at about 4 days after birth and declining rapidly thereafter. Using oligo (dT) cellulose poly A^{+} neonatal rat liver RNA was prepared and subsequently fractionated on sucrose gradients. Fractions were used to direct translation. Analysis on SDS gels revealed that a band which labelled heavily with cysteine, did not label with leucine, and had an apparent molecular weight of 14,000 was enriched in the 9S region of the gradient (fractions 6 to 9 in Figure 1). This band also labelled heavily with serine and comigrated with authentic Zn induced MT. The properties of this band are consistent with it being MT.

Total RNA was also prepared from livers of rats aged -1, 0 (day of birth), 1, 2, 3, 5 and 14 days, and its translation products were analysed. RNA from all these ages directed translation of a protein identifiable as MT using the criteria described above. It was found that the amount of translatable MT mRNA did not show any significant decline between 5 and 14 days after birth. There was however much more translatable MT mRNA in neonatal rat liver when compared to adult uninduced rat liver, and these levels did not appear to drop concomitantly with the drop in MT concentration which occurs between 4 and 14 days(2). Therefore some control mechanism other than at transcription would need to be operating to explain this drop.

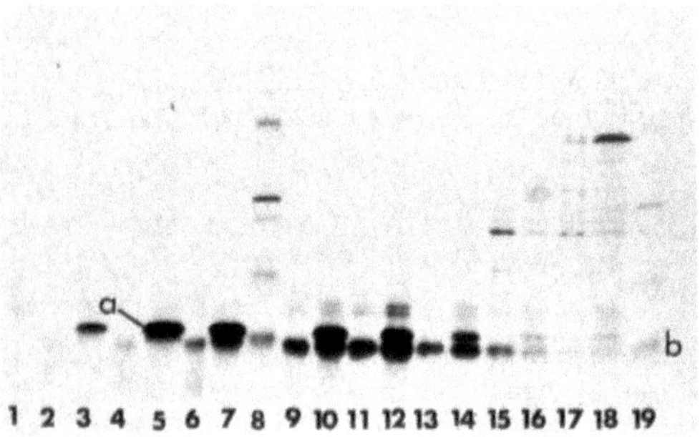

Figure 1 SDS gel of translations of neonatal rat liver RNA sucrose gradient fractions (f) using ^{3}H leucine or ^{35}S cysteine labelling. (1) no added RNA, cys; (2) f5, leu; (3) f5, cys; (4) f6, leu; (5) f6, cys; (6) f7, leu; (7) f7, cys; (8) markers; (9) f8, leu; (10) f8, cys; (11) f9, leu; (12) f9, cys; (13) f10,leu; (14) f10, cys; (15) f11, leu; (16) f11, cys; (17) f12, leu; (18) f12, cys; (19) markers (a) translated MT band, cys labelled not leu labelled; (b) 14,300 dalton marker

Cu induced MT mRNA in adult rat liver

Wistar and Sprague-Dawley male rats were injected intraperitoneally with copper chloride, and five hours later total RNA was extracted from the liver and used to direct translation. Figure 2 shows that two bands were increased following Cu injection. These bands labelled heavily with cysteine but did not label with leucine. They had an apparent molecular weight of about 14,000. The less intense band which has a slightly higher mobility comigrated with authentic radio-labelled carboxyamidomethylated Zn induced MT. However it was found that non-radioactively labelled carboxyamidomethylated authentic Zn induced MT comigrated with the more intense band. Further work is needed to clarify these anomalies in migration. The Cu induced bands also comigrated with MT translated by neonatal mRNA and Zn induced mRNA. On the basis of all these properties the cysteine rich band(s) enharced by Cu administration is likely to be MT.

MT-mRNA from Cu induced livers had a sedimentation coefficient of 9S which is the same as that of MT mRNA from neonatal rat liver and Cd induced rat liver(3). The levels of translatable MT mRNA were increased 5-fold five hours after Cu administration which compares with 2-fold in the case of Zn induction and 4-fold in the case of Cd induction(3).

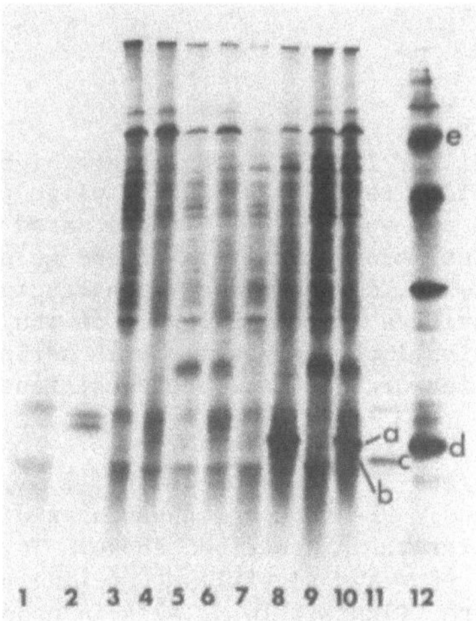

Figure 2 SDS gels of total RNA translations using ^{3}H leucine or ^{35}S cysteine.
(1) no added RNA, leu; (2) no added RNA, cys; (3) untreated rat, leu; (4) untreated rat, cys; (5) saline injected rat, leu; (6) saline injected rat, cys; (7) Cu injected Sprague-Dawley rat, leu; (8) Cu injected Sprague-Dawley rat, cys; (9) Cu injected Wistar rat, leu; (10) Cu injected Wistar rat, cys; (11) authentic Zn induced MT; (12) molecular weight markers
(a) translated MT band most intensely cys labelled, not labelled with leu;
(b) translated MT band less intensely cys labelled, not labelled with leu;
(c) authentic Zn induced MT;
(d) 14,300 dalton marker - ^{14}C methylated lysozyme;
(e) 69,000 dalton marker - ^{14}C methylated bovine serum albumin.

References

1. Kägi, J.H.R., and M. Nordberg, 1979. In "Metallothionein" (Birkhäuser Verlag) pp 46-116.

2. Wong, K., and C. D. Klaassen, 1979, J. Biol. Chem. 254: 12399-12403.

3. Anderson, R.D., and U. Weser, 1978, Biochem. J. 175: 841-852.

4. Shapiro, S.G., Squibb, K.S., Markowitz, L.A., and R. J. Cousins, 1978, Biochem. J. 175: 833-840.

5. Winge, D.R., Premakumar, R., Wiley, R.D., and K. V. Rajagopalan, 1975, Arch. Biochem. Biophys. 170: 253-266.

6. Bremner, I., and B. W. Young, 1976, Biochem. J. 157: 517-520.

7. Deeley, R.G., Gordon, J.I., Burns, A.T.H., Mullinix, K.P., Binastein, M., and R. F. Goldberger, 1977, J. Biol. Chem. 252: 8310-8319.

8. Ryden, L., and H. F. Deutsch, 1978, J. Biol. Chem. 253: 519-524.

DISCUSSION

LINDER (USA). I am interested in the persistence of mRNA for metallothionein during a period of decline in the amount of metallothionein. This is reminiscent of what happens with ferritin messenger. In situations where there is little ferritin synthesis the ferritin messenger persists and there may be a decline in ferritin concentration. Have you considered that the mRNA may be on RNP particles rather than on polyribosomes.

CAMAKARIS (Australia). No we haven't looked at that. It is an interesting suggestion.

WHANGER (USA). When we feed copper to rats at a high level in the diet rather than inject it we don't observe much copper in metallothionein but find it in a protein of about 30,000 MW. Have you ever used dietary copper to see if there is an increase in translatable mRNA?

CAMAKARIS (Australia). No we have only used intraperitoneal injections.

LIVER COPPER-BINDING PROTEINS IN DISEASE

Cornelis J.A. Van den Hamer and A.R. Janssens

Interuniversity Reactor Institute
Mekelweg 15, 2629 JB Delft, The Netherlands

INTRODUCTION

Most of the copper in the liver is present in the cytoplasm of the hepatocyte [1] and distributed over a number of proteins. One of these proteins has been identified as metallothionein. Such a distribution has been found in a number of species, e.g., rat, rabbit, mouse, sheep and cow and also in man.

Since 1948 it is known that Cu accumulates in the liver of patients with Wilson's disease [2]. Later on, a number of other liver disorders were identified as being associated with excessive storage of Cu in the liver, viz., primary biliary cirrhosis (PBC), Indian childhood cirrhosis and various other diseases involving cholestasis (for review see [3]). Because the Cu accumulation depends on, among other things, the duration and extent of the cholestasis, rather mild Cu elevations are usually found in case of most extrahepatic biliary obstructions.

In the course of a study of the Cu-metabolism of PBC [4] the distribution of the Cu in the liver of such patients was studied and compared with that of controls and of patients with other liver diseases.

MATERIALS AND METHODS

Most liver samples, particularly those of controls, were obtained at autopsy. They were taken within 6 hours after death, quickly frozen and stored at -80° C. Occasionally samples were obtained surgically while three were the result of a liver transplantation. The diagnosis of PBC was based on a number of criteria, among which the presence of antimitochrondrial antibodies in the plasma. Neither of the controls had at the time of death or in the past a disorder known to influence the liver Cu.

Total liver-Cu was determined by destructive NAA [5] or flameless AAS after destruction with HNO_3 and H_2O_2 and was expressed as μg Cu/g wet liver.

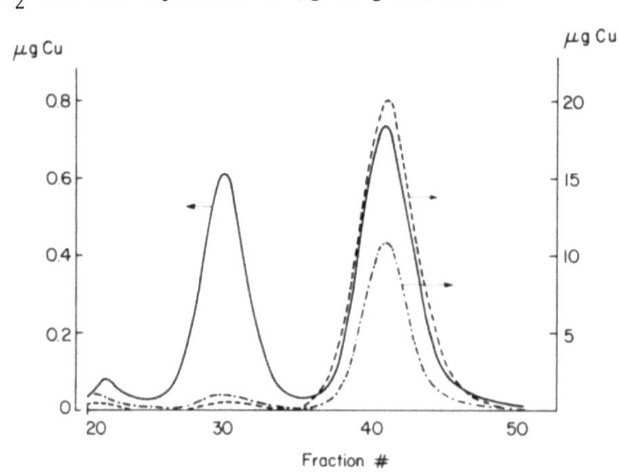

Fig. 1. Sephadex G-75 separation of the cytoplasmic Cu-binding fractions of a normal liver (———), a liver of a patient with PBC (-----) and one of a patient with a syndrome of Alagille (-.-.-.-). The Cu in the chromatographic fractions was calculated as if the cytoplasm from 1 g liver were passed through the column. Note the difference in ordinate between the normal liver and the livers of both patients.

The liver samples were homogenized in 10mM Tris-acetate buffer (pH = 7.4; passed through a short Chelex-100 column) and centrifuged for one hour at 120,000 x g. The supernatant was passed through a Sephadex G-75 column (1.8 x 70 cm) under exclusion of oxygen at 4° C. The fractions were analysed for Cu by flameless AAS. Three Cu-containing fractions were found with molecular weights of >70,000, 30,000, and 10,000 daltons, resp. (Fig. 1). The last two will be indicated as "30,000 MW fractions" and "10,000 MW fractions", resp. In two cases (one PBC liver and a liver from a patient with the syndrome of Alagille) the 10,000 MW fractions was further fractionated using a Sephadex A-25 column and a Tris-acetate gradient of 10-300 mM (pH = 7.4). Acid washed glassware was used throughout.

RESULTS AND DISCUSSION

Of the three fractions found after Sephadex G-75 chromatography the 30,000 MW fraction is probably superoxide dismutase. There is little doubt that the 10,000 MW fraction is metallothionein; its behavior on Sephadex A-25 chromatography and its typical UV-spectrum [6], studied in two samples also point in this direction.

	Liver-Cu (μg/g)	R-10/30	Cu in 30,000 MW fraction (μg/g liver)
Normals n = 9	4.9 ± 1.8	1.3 ± 0.5	2.0 ± 1.0
"Normals" n = 2	21, 34	7, 11	3.4, 3,1
PBC n = 8	52 ± 38	9.5 ± 7	2.1 ± 0.76
Other forms of cirrhosis			
Juvenile	11	5.2	1.4
Alcohol	15	13.2	0.9
Alagille	194	19.3	4.5
Wilson*	183	> 3.2	---
Children			
Cu-Deficiency 1 y old	8.7	2.7	1.0
Menkes 2 y old	3.5	0.7	1.3
Normal 2 y old	5.0	1.6	1.3

Table 1. Liver-Cu, ratio of the Cu in the 10,000 and the 30,000 MW fractions (R-10/30) and the Cu in the 30,000 MW fraction of various livers.
* The 10,000 and 30,000 MW fractions were poorly separated. Therefore the R-10/30 could only be estimated while the Cu in the 30,000 MW fraction could not be calculated at all.

Eleven liver samples were received from patients who had died of an accident or a disease without known involvement of the liver and were therefore labeled normal. On analysis most of them had indeed a normal Cu-content (range 2.4 - 7.5 μg/g; see also Table 1). However, two of these supposedly normal livers ("Normals" in Table 1) were found to have elevated Cu-levels (21 and 34 μg/g, resp.), well above the range of the other normals. They were therefore excluded from the computation of the average of the normals. Since, with the exception of three samples, the amount of Cu present in the 30,000 MW fraction of all livers (normal and pathologic) analyzed was - consistent with its designation as superoxide dismutase - fairly consistent (1.96 ± 0.85 μg/g; n = 27) the Cu in the 10,000 MW fractions was expressed as the ratio of this Cu to that in the 30,000 fraction (R-10/30), in this way avoiding the uncertainty introduced by the incomplete homogenization of some of the samples. In the normal livers the Cu in the 30,000 MW fraction and in the 10,000 MW fraction were about equal: the R-10/30 was in average 1.26. The two livers excluded from the normal group had significantly higher R-10/30's (7 and 11, resp.).

It should be noted that the average liver-Cu of our group of normals is lower than the values usually quoted in the clinical literature [7, 8]. Even one of the samples excluded from our control group and showing a greatly increased R-10/30 (7) would be regarded as normal on basis of these literature values.

In case of PBC the liver-Cu is usually, though not always much increased [9]. Such elevated liver-Cu is the result of an increase in both particulate and cytoplasmic Cu. The increase in cytoplasmic Cu is almost exclusively caused by an increase of the Cu associated with the 10,000 MW fraction, as indicated by the increase in R-10/30. Histologically - using antibodies against metallothionein coupled with peroxidase - the metallothionein was found predominantly at the periphery of the liver lobules [10].

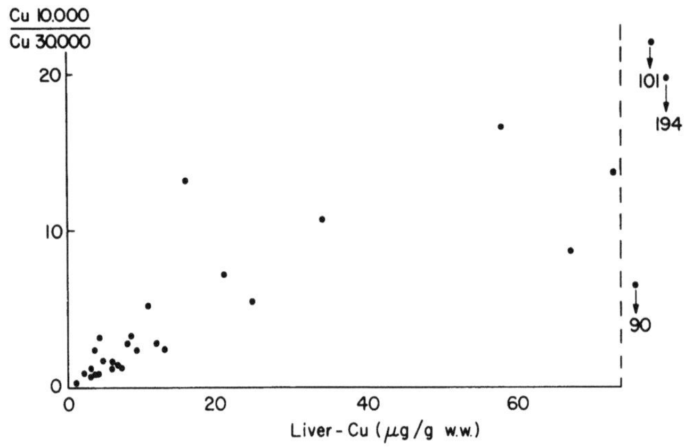

Fig. 2. The ratio of the Cu in the 10,000 and the 30,000 MW fractions (R-10/30) as function of the total liver-Cu.

The increase in R-10/30 is clearly not a feature specific for PBC as shown by the livers from patients with other forms of cirrhosis and also by the values obtained in case of two supposedly normal livers ("Normals" in Table 1) mentioned above. Such an increase in R-10/30 seems to occur as soon as the total liver-Cu starts to increase, regardless of the cause. Although there is quite a lot of scatter, a clear correlation was found between the total liver-Cu and the R-10/30 (Fig. 2). In fact, elevated R-10/30 values were seen in case of livers with only mildly increased Cu-content or with a Cu-content that would have been regarded as normal. Typical examples are two livers with other forms of cirrhosis than PBC (Table 1). Another example is that of one of the patients with PBC. Her disease was diagnosed in an early stage on basis of a positive antimitochondrial antibody test. She had a liver-Cu of 12 μg/g which would not have been pathognostic. Yet, the R-10/30 of her liver was clearly elevated (2.8).

Summarizing one can conclude that a) the normal human liver-Cu is 5 ± 2 μg/g; b) the Cu in the cytoplasmic 30,000 MW fraction is roughly constant (about 2 μg/g); c) in normal livers about equal amounts of Cu are present in the 30,000 MW fraction and bound to metallothionein; d) when the total liver-Cu increases the Cu in the metallothionein increases more clearly; and e) in liver disease the increase in cytoplasmic liver-Cu is caused by an increase in metallothionein-Cu.

REFERENCES

1. Bloomer, L.C., and G.R. Lee, 1978. In "Metals and the Liver" (ed. L.W. Powell) (M. Dekker, N.Y.) pp 179-239.
2. Cumings, J.N, 1948. In "Heavy Metals and the Brain" (Thomas, Springfield) pp 3-71.
3. Smallwood, R.A., 1978. In "Metals and the Liver" (ed. L.W. Powell) (M. Dekker, N.Y.) pp 313-330.
4. Janssens, A.R., and C.J.A. Van den Hamer, 1980, Proc. 15th Meeting Eur. Ass. Study Liver, Belgrade
5. Tjioe, P.S., J.J.M. De Goeij, and J.P.W. Houtman, 1977, J. Radioanal. Chem. 37:511-522.
6. Kägi, J.H.R., and M. Nordberg, 1979, Experientia Supplementum 34.
7. Smallwood, R.A., H.A. Williams, V.M. Rosenoer, and S. Sherlock, 1968, Lancet 2:1310-1313.
8. Evans, J., S. Newman, and S. Sherlock, 1978, Gastroenterology 75:875-878.
9. Nooijen, J.L., C.J.A. Van den Hamer, J.P.W. Houtman, and S.W. Schalm, 1981, Clin. Chim. Acta, in press.
10. Janssens, A.R., D.J. Ruiter, R. Bosman, M.J. Van Noord, and C.J.A. Van den Hamer, 1981, Proc. Dutch Soc. Gastroenterol., Veldhoven.

DISCUSSION

ELMES (Wales). What technique did you use to histochemically demonstrate the copper in the liver? Do you think that this copper you were demonstrating really shows the cellular distribution of metallothionein?

VAN DEN HAMER (The Netherlands). It was an immunological stain for metallothionein and not a stain for copper.

ELMES (Wales). We tried some immunofluorescent techniques which were successful for detecting rat liver metallothionein, but we haven't used your type of technique.

VAN DEN HAMER (The Netherlands). Well, there shouldn't be any difference but I think I told you we could not detect the normal levels, we only observed a positive stain in the case of the elevated levels of MT associated with primary biliary cirrhosis.

ELMES (Wales). I was just wanting to confirm that this was not the rubeanic acid technique for copper.

VERSIECK (Belgium). Could you tell us briefly what is an alagille's syndrome?

VAN DEN HAMER (The Netherlands). Some kind of a liver disease involving cirrhosis but I am not familiar with the clinical features.

VERSIECK (Belgium). Is there anything known about the idiopathogenesis or about the aetiology of the disease?

VAN DEN HAMER (The Netherlands). Yes, but I think Professor Danks would be better to answer this.

DANKS (Australia). Alagille's sydrome is a disorder which is characterised by deficiency of intrahepatic bile ducts - the existence of abnormalities of the pulmonary artery - an unusual facial appearance and sometimes some other less constant features. It commonly occurs as an autosomal dominant inherited condition although there are other families in which you see isolated cases which seem very similar but may not be identical. We have looked at a number of patients with that condition and we have been very puzzled by extremely variable increases of liver copper. We have looked repeatedly at liver copper in patients with complete atresia of the extrahepatic bile ducts and are also quite puzzled by the extraordinary rate of accumulation of copper in these patients who have no bile duct system at all. Something very peculiar goes on, they certainly all develop increased copper in due course but there was a great variation from one patient to another in the tempo of this occurring. I think it is a very nice idea to look at these specific fractions, which we have not done, and the idea of an immunological type assay is very appealing indeed.

MORISHIMA (Japan). You have demonstrated an increase in the copper metallothionein fraction in primary cirrhosis suggesting that probably there is a denovo biosynthesis of the protein. Did you measure the biosynthesis using the incorporation of labelled cysteine to see if it is a real biosynthesis, or have you only seen the incorporation of copper?

VAN DEN HAMER (The Netherlands). I think you can measure it but I don't think that the people in the hospital will approve injecting the patients with ^{35}S cysteine.

MORISHIMA (Japan). Yes, but for example to use other methods to measure the protein.

VAN DEN HAMER (The Netherlands). Well, we can measure the total amount of protein but that doesn't tell you too much about the synthesis; it only shows that you have an elevated level but says nothing about the rate of synthesis.

COPPER-METALLOTHIONEIN METABOLISM IN RAT HEPATOMA CELLS CULTURED IN VITRO

Johannes L. Nooijen & Cornelis J.A. Van den Hamer

Interuniversity Reactor Institute
Mekelweg 15, 2629 JB Delft, The Netherlands

INTRODUCTION

Information about trace element metabolism in malignancies in general is scarce. Some trace elements are known to be carcinogenic [1], while others are thought to have a protective effect against certain types of cancer [2]. Most reports concerning trace elements and malignancies concern Cu and Zn. Increased ceruloplasmin levels in plasma have been reported [3], both in clinical and animal studies, but little is known about the Cu metabolism of tumor cells at the cellular level. In cancerous tissue of various origin , including primary liver tumor tissue, increased Cu concentrations are reported [4,5].
In the present study the Cu metabolism of a rat hepatoma cell line cultured in vitro will be described.

MATERIALS AND METHODS

The hepatoma cell line under investigation (HTC cells) was grown as a monolayer culture in Ham's F10 medium supplemented with 12% foetal calf serum (FCS). All studies with FCS supplemented to the medium were performed with cells in exponential growth phase. The 64Cu prepared at our institute had a specific activity of 50 or 4800 mCi/mmole. The latter was used in release experiments. The L-[35S]cystine (spec. act. 300 Ci/mmole) was obtained from New England Nuclear, and diluted with L-cystine to 2 Ci/mmole. The 64Cu activity in the total cell was measured by gamma ray counting (Packard Instruments, model 5120). The 64Cu, 65Zn, and 35S activity in Sephadex G-75 fractions was measured by liquid scintillation counting (Packard Instruments, model 2650).

After incubation with the various compounds cells were harvested, homogenized in 10 mM Tris-acetate, pH 7.4, under anaerobic conditions, centrifuged at 120,000 g for 1 h at 4 oC., and the supernatant was fractionated on a Sephadex G-75 column (2x70cm).
In release experiments the radioactive medium was removed, cells were washed several times and reincubated with fresh medium.

RESULTS AND DISCUSSION

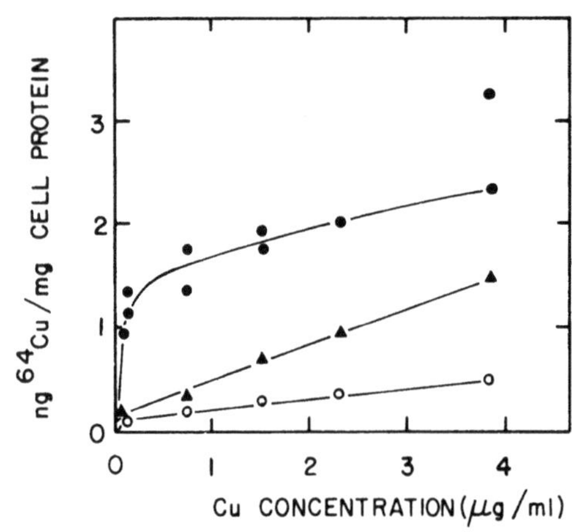

Fig. 1. Effect of 64Cu concentration on 64Cu accumulation by HTC cells during a 2 hours contact period (•), and the momentarily binding in the absence (▲), and presence (o) of FCS.

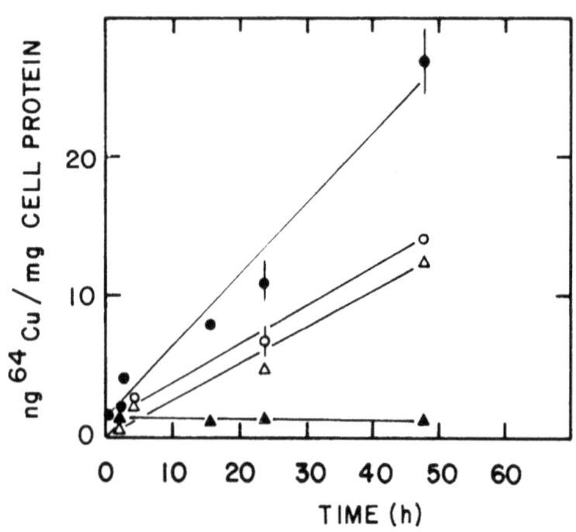

Fig. 2. 64Cu accumulation by HTC cells after being cultivated with 1.5 μg 64Cu/ml for various lengths of time. 64Cu content of total cell (•), cytosol (o), 30,000 MW fraction (Δ) and MT (▲).

Fig. 1 shows that for the momentarily binding of 64Cu to the cell surface of these cells a competition exists between binding sites on proteins present in the medium and those on the cell membrane. Furthermore it appeared that the process of Cu accumulation by these cells was saturable above a Cu concentration in the medium of 0.25 μg/ml, and also temperature dependent. Fig. 2 shows that linearly with time these growing cells continue to accumulate 64Cu for at least 48 hours. Much of the 64Cu is actually transported across the cell membrane and recovered in the cytosol. Shortly after applying the 64Cu to the cells (within 2 hours) most of the 64Cu in the cytosol is bound to a protein with molecular weight (MW) of 10,000 which could be characterized as metallothionein (MT). After about 3 hours a steady state level of 64Cu bound to MT was reached. The amount of 64Cu bound to the 30,000 MW Cu-binding protein, as suggested by various authors possibly superoxide dismutase, linearly increased with time, and ultimately contained the majority of the 64Cu in the cytosol. Only a minor fraction was bound by proteins with high MW.

During a 24 hour contact with 10 times more 64Cu in the medium (15.4 μg/ml) it seems likely that some other uptake process gets involved, since 4 times more 64Cu was accumulated under these conditions, while at a concentration of 1.54 μg/ml the uptake process was already saturated. Most of the 64Cu accumulated under these conditions of high Cu concentration in the medium has been recovered in the cytosol bound to MT.

Fig. 3 shows the incorporation of 35S from L-[35S]cystine into MT under conditions of varying Cu concentration in the medium. Between 24 and 48 hours after applying this L-[35S]cystine to the medium a steady state level is reached in experiment a and b. From this steady state level the concentration of MT can then be calculated assuming 20 cysteine residues per MT [6]. In experiment c an increased synthesis of MT could be demonstrated after 48 hours.

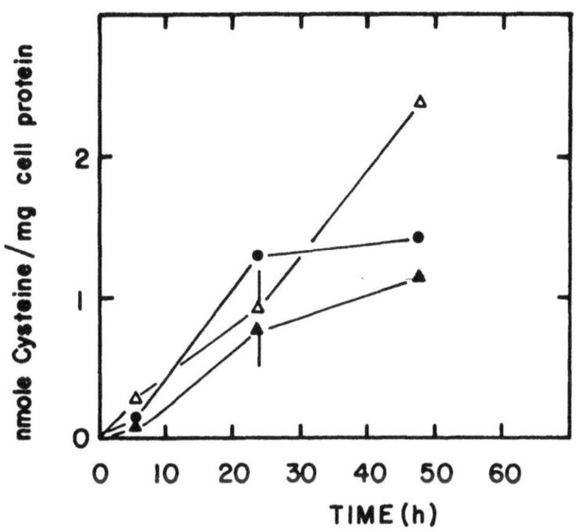

Fig. 3. Incorporation of 35S from L-[35S]cystine into MT after being cultivated for various lengths of time in the presence and absence of Cu. a. no Cu (●), b. 1.54 μg Cu/ml (▲), c. 15.4 μg Cu/ml (Δ).

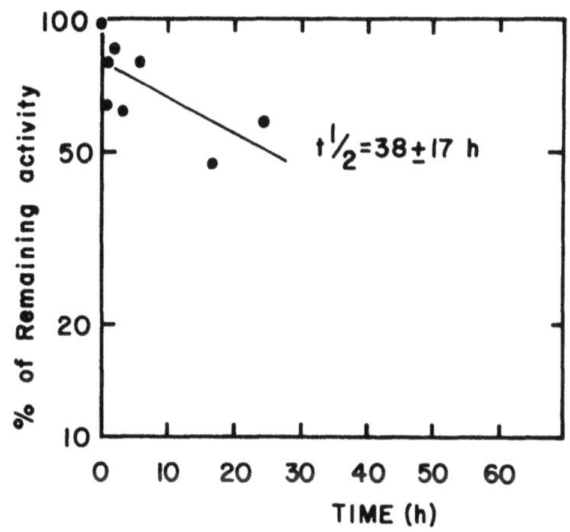

Fig. 4. Biological half life of MT synthesized by the HTC cells after a 24 hour contact with L-[35S]cystine without additional Cu in the medium.

From the concentration of MT (calculated from Fig. 3) and the amount of 64Cu bound to it one can calculate that in case of a 24 hour contact with 1.54 μg 64Cu/ml medium the ratio Cu/MT is only 0.5; in case of 15.4 μg 64Cu/ml this ratio is 7.8. This indicates that in case of 1.54 μg 64Cu/ml only a small fraction of the available Cu-binding sites on MT (10[6]) are occupied. The cells therefore do not need to synthesize extra MT to bind this Cu. 24 hours after applying 15.4 μg 64Cu/ml to the cells, MT is almost fully loaded with 64Cu.

From release experiments a biological half life ($t_{\frac{1}{2}}$) for the protein moiety of MT of 38±17 h was measured (Fig. 4). This is only slightly higher than the $t_{\frac{1}{2}}$ of Zn-thionein in liver of normal rats, which has been reported by others as between 10 and 30 hours [6]. From release experiments after a 24 hour contact with L-[35S]cystine and 1.54 μg 64Cu/ml a $t_{\frac{1}{2}}$ of

64Cu bound to MT of 2.6±0.4 h and for MT of 57±9 h was observed, which last value was not significantly different from 38±17 h as measured for MT without 64Cu bound to it (Fig. 5). The 64Cu released by MT probably is excreted by the cells. Two mechanisms for degradation of the MT-64Cu complex are possible; the 64Cu is released either with or without simultaneous degradation of the protein. In the first model this means that the 64Cu is not randomly distributed over the available MT molecules (in a ratio of 0.5 Cu/MT), but is preferentially bound to a small fraction of these. A rapid degradation of this small fraction of MT will therefore not be measured since the majority will be catabolized with the t½ of 57±9h.

When the cells were exposed to 15.4 μg 64Cu/ml medium for 24 hours before removing the 64Cu-containing medium (Fig. 6) again the t½ for the 64Cu bound to MT was very much smaller (4.7±0.5 h) than observed for the protein moiety of MT (33± 3 h), that was degraded with the same t½ as measured under conditions without additional 64Cu in the medium. Since under these conditions of high Cu concentration in the medium the MT was substantially loaded with 64Cu (7.8 Cu/MT), this indicates that Cu is bound to MT and transferred very rapidly without simultaneous degradation of the MT itself. The MT partly or totally released from its 64Cu apparently becomes available again to bind newly accumulated 64Cu. Identical half lives of the metals (Cu, Zn) bound to MT and the protein were only found when rats were exposed to high Cu loads which resulted in the induction of MT synthesis [6].

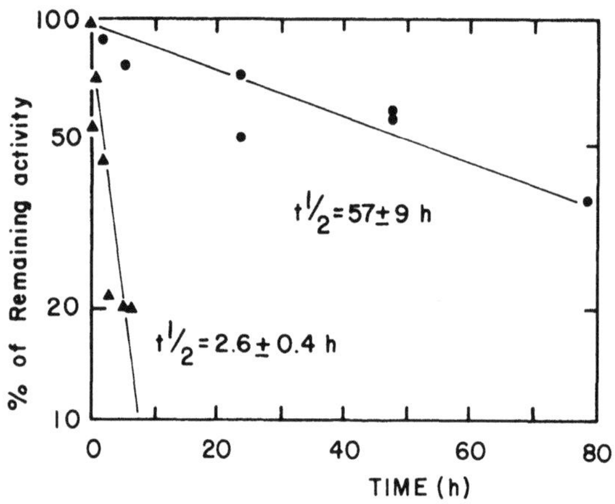

Fig. 5. Biological half life of 64Cu bound to MT (▲) and the protein moiety of MT (●) in the HTC cells as measured after a 24 hour contact with L-[35S]cystine and 1.54 μg 64Cu/ml.

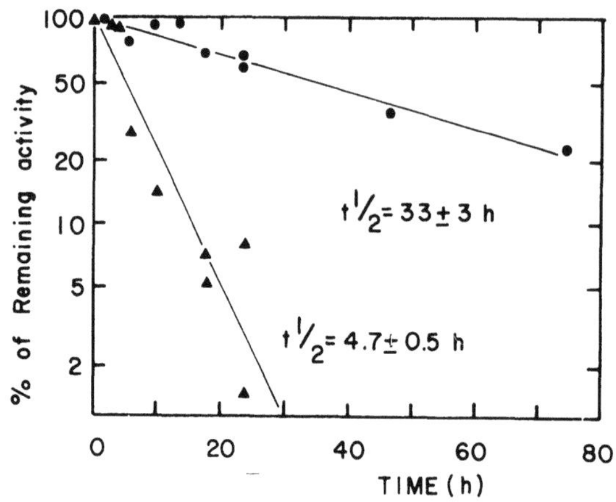

Fig. 6. Biological half life of 64Cu bound to MT (▲) and the protein moiety of MT (●) in the HTC cells as measured after a 24 hour contact with L-[35S]cystine and 15.4 μg 64Cu/ml.

In conclusion in these hepatoma cells the MT plays an important role in initial binding and regulation of Cu. Furthermore it could be demonstrated that the MT is not simultaneously degraded with the release of the Cu bound to it, but can be reused again.

Comparison of the Cu metabolism of these hepatoma cells with a primary culture of rat liver parenchymal cells showed less binding sites on the cell membrane of the hepatoma cells, thus resulting in a decreased uptake. Furthermore the accumulated Cu was less easily released from these cells, probably explaining the often found increased Cu levels in hepatoma tissue.

REFERENCES

1. Fürst, A., and R.T. Haro, 1969, Prog. Exp. Tumor Res. 12: 103.
2. Schrauzer, G.N., J.E. McGinness, and K. Kuehn, 1980, Carcinogenesis, 1:199-201.
3. Sunderman, F.W., E.A. Trudeau, E. Horak, J.M. Mitchell, and P.R. Allpas, 1979, Ann. Clin. Lab. Sci. 9:60-67.
4. Mulay, I.L., R. Roy, B.E. Knox, N.H. Suhr, and W.E. Delaney, 1971, J. Natl. Cancer Inst., 47:1-13.
5. Arnold, M., and D. Sasse, 1961, Cancer Res., 21:761-766.
6. Nordberg, M., and Y. Kojima, 1979. In "Metallothionein" (eds. J.H.R. Kägi and M. Nordberg) (Birkhäuser Verlag Basel, Boston, Stuttgart), pp 41-124.

DISCUSSION

LINDER (USA). Do you have any idea how fast the ^{35}S-cysteine is being recycled in tissue culture? I would expect the label to be rapidly recycled and therefore you might have a discrepancy between copper 64 half lives and sulphur 35 half lives.

VAN DEN HAMER (The Netherlands). I think there will be some re-incorporation but I do not think it will be too much. First of all the excess sulphur is usually excreted as sulphate, and secondly you are dealing here with growing cells that need an expanding amount of amino acids for the protein synthesis.

BREMNER (Scotland). Did you find incorporation of zinc into your copper induced metallothionein?

VAN DEN HAMER (The Netherlands). Yes.

SMITH (Australia). Is there any information on the dissociation constant of copper metallothionein? This would indicate the degree of affinity a ligand would require to take the copper off.

BREMNER (Scotland). I have not seen figures for copper metallothionein, but for the cadmium and zinc metallothioneins the stability constants are of the order of 10^{-16} and 10^{-11}. In the case of copper the binding is very tight and the cation is only released at pH's below 1.

VAN DEN HAMER (The Netherlands). I think that this implies that whatever takes over the copper has one or more -SH groups, or perhaps the fast turnover copper of half-life about 2h is more loosely bound than steady-state copper.

EFFECTS OF DIETARY SULFUR ON TISSUE METALLOTHIONEIN IN RATS

P.D. Whanger, S.H. Oh and J.T. Deagen

Department of Agricultural Chemistry, Oregon State University
CORVALLIS, OREGON 97337 USA

Since Metallothionein (MT) is known to contain about one-third of its amino acids as cysteine (1), dietary sulfur would be expected to influence its level in animal tissues. Thus, the present studies were conducted to investigate the influence of dietary sulfur upon the levels of MT in rat tissues.

TORULA YEAST DIETS

Since torula yeast has a low sulfur content, this protein was used in diets for rats in the first experiment. This basal diet is similar to that previously described (2), except that corn oil was substituted for stripped lard and vitamin E was included in the vitamin mix. It contained torula yeast (30%), sucrose (51.5%), solka floc (7.5%), corn oil (5%), minerals (5%) and vitamins (1%) and 100 ppm zinc as zinc acetate was added to this diet for all rats. A total of 24 male weanling rats were fed either the basal diet (12 rats) or this plus an addition of 1% DL-methionine (12 rats). After 4, 8, 16, and 28 days, 3 rats fed each of the two diets were killed and the zinc content determined in livers and kidneys. The amount of zinc in tissue MT was determined as previously reported (3).

As shown in figure 1, the amount of zinc in the whole liver as well as in hepatic MT was greater in rats fed the basal diet than in those fed the basal diet plus 1.0% methionine. The zinc content in hepatic MT was highest in rats at the beginning of the experiment and gradually decreased throughout the remainder of the study. This is because the zinc in hepatic MT is higher in newborn than in older rats (4). Since these results with sulfur were opposite than expected, another study was conducted to investigate this apparent paradox.

CRYSTALLINE AMINO ACID DIETS

The basal diet used in this study contained glucose (70%), crystalline amino acids (19%) (omitting the sulfur amino acids), vegetable oil (5%), salts (3%) and vitamins (1%) as previously described (5). Zinc as zinc acetate was added to all diets at level of 1,000 ppm. A total of 36 male weanling rats were divided into 4 groups and fed this diet with additions of either 0.2, 0.5, 1.0 or 1.4% DL-methionine. After 4, 8, 16 and 28 days 3 rats from each dietary group were killed, and the amount of zinc in hepatic and renal MT determined as indicated above. Three rats were killed at the beginning of the experiment and the amount of zinc determined in renal and hepatic MT to establish initial values.

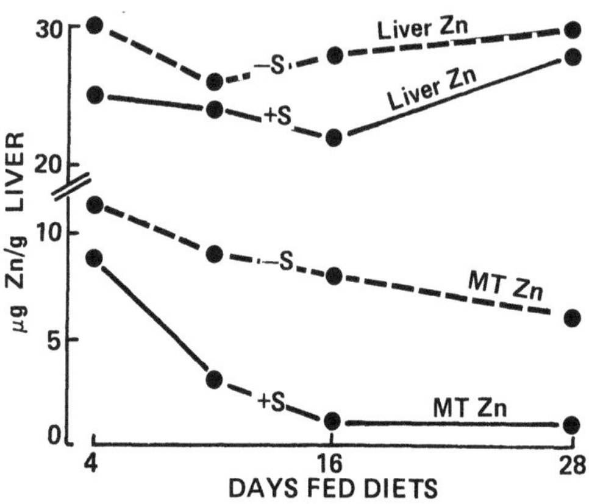

Figure 1. Effects of dietary sulfur upon liver and MT zinc levels in rats fed torula yeast based diets.

As shown in figure 2, zinc accumulated to the greatest extent in hepatic and renal MT of rats fed the diet with 1.0% methionine and lowest for rats fed the diet with 0.2% methionine. Although the levels were higher after 8 and 16 days in rats fed the diet with 1.4% methionine, the amount of zinc in MT was similar in rats fed the diet with

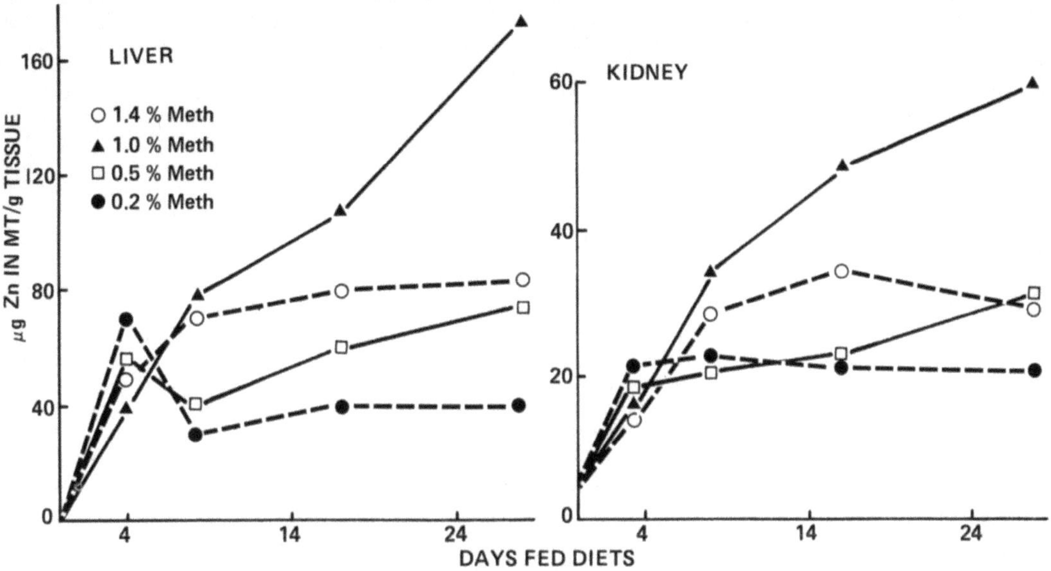

Figure 2. Effects of dietary sulfur upon hepatic and renal MT levels in rats fed crystalline amino acid based diets.

either 0.5 or 1.4% methionine at the end of the experiment. These results indicate that excess sulfur appears to decrease MT levels, offering a possible explanation for the results obtained with the torula yeast diets.

TABLE 1. Amino acid composition of zinc - containing proteins in rat liver and kidney.

Amino Acids	Liver*		Kidney*	
	MT1	MT2	MT1	MT2
Lysine	18.7	17.2	17.7	8.2
Aspartate	9.2	8.5	9.2	10.2
Threonine	4.2	3.8	4.3	2.8
Serine	8.7	8.6	8.7	8.6
Glutamate	7.5	10.6	7.6	12.8
Proline	3.9	3.4	3.9	4.9
Glycine	4.5	4.1	4.5	3.5
Alanine	6.6	6.1	6.6	11.3
Cysteine**	30.5	30.8	31.5	32.7
Valine	2.8	2.7	2.8	1.8
Methionine**	2.1	1.8	2.1	2.2
Isoleucine	0.5	0.5	0.4	0.3
Leucine	0.8	1.5	0.5	0.5
Tyrosine	trace	0.4	0.1	0.1
Phenylalanine	trace	trace	0.1	0.1
Histidine	trace	trace	trace	trace

* Residues are expressed as mole %.
** Determined respectively as cysteic acid and methionine sulfone.

In order to further investigate the effects of sulfur, the half-life of hepatic and renal MT was determined in rats fed the diets with either 0.2 or 1.0% methionine by using ^{14}C-cystine in pulse-dosed experiments(6). The half-life of MT was found to be very similar in livers of sulfur deficient (1.3 days) and supplemented (1.6 days) rats, but the half-life in the kidney was found to be almost twice as long in sulfur deficient rats (4.3 days) as compared to sulfur supplemented ones (2.7 days). Hence, the sulfur status of animals appears to influence the synthesis of MT and some effects are tissue specific.

For confirmatory purposes, the zinc binding proteins considered to be MT in liver and kidney were purified with respect to zinc by gel filtration, ion exchange resin, and hydroxylapatite chromatography similar to that previously described (7) from rats fed the diet with 1.0% methionine. The amino acid content of the purified proteins is shown in table 1. Two species of MT were found in both liver and kidney. The analysis indicated a high cysteine content (up to 32.7%) and a low content of aromatic amino acids, which is characteristic of MT (1). Other workers have shown zinc is bound to MT in the liver (8,9). Although it was logical to conclude that zinc was also bound to renal MT, indirect evidence can sometimes be misleading. For example, the cadmium-binding protein in tissues of the American oyster (10) and the low molecular weight cadmium-containing protein in rat testes (11) elute at the identical position on gel filtration as MT, but the amino acid analyses of the purified proteins indicate that they are definitely not this protein. Thus, the only sure way to identify a metal-binding protein as MT is to do an amino acid analysis on the purified protein.

SUMMARY

Two experiments were conducted to investigate the influence of dietary sulfur on metallothionein (MT) levels in tissues of rats. Addition of 1.0% methionine to a torula yeast based diet in the first experiment caused a lower MT level and zinc content in liver than in those fed the diet without methionine. In the second experiment levels of 0.2, 0.5, 1.0 and 1.4% methionine were added to a crystalline amino acid based diet for rats. MT levels were highest in liver and kidney of rats fed the diet with 1.0% methionine and lowest in rats fed the diet with 0.2% methionine. Sulfur deficiency had no effect on half-life of MT in liver, but resulted in a significantly longer one in the kidney. Amino acid analysis on the purified zinc proteins considered to be MT indicated a cysteine content up to 32.7% for both the liver and kidney proteins.

ACKNOWLEDGEMENTS - This work was supported by N.I.H. grant number AM 19285.

REFERENCES

1. Kagi, J.H.R. and M. Nordberg. 1979. In "Metallothionein". Birkhauser Verlag, Basel, Boston, Stuttgart, pp 41-124.

2. Schwarz, K. and C.M. Foltz. 1958. J. Biol. Chem. 233:245-251.

3. Chen, R.W., E.J. Vasey and P.D. Whanger. 1977. J. Nutr. 107:805-813.

4. Oh, S.H. and P.D. Whanger. 1979. Amer. J. Physiol. 237:E18-E22.

5. Rose, W.C. and T.R. Wood. 1941. J. Biol. Chem. 141:381-389.

6. Oh, S.H., J.T. Deagen, P.D. Whanger and P.H. Weswig. 1978. Bioinorg. Chem. 8:245-254.

7. Chen. R.W. and P.D. Whanger. 1980. Biochem. Med. 24:71-81.

8. Bremner, I. and N.T. Davies. 1975. Biochem. J. 149:733-738.

9. Wong, K.L. and C.D. Klaassen. 1979. J. Biol. Chem. 254:12399-12403.

10. Ridlington, J.W. and B.A. Fowler. 1979. Chem. Biol. Inter. 25:127-138.

11. Deagen, J.T., C.L. Holcomb and S.H. Oh. 1980. Fed. Proc. 39:622 (abs).

DISCUSSION

LINDER (USA). Did you look at the effects of dietary methionine on total protein as well as metallothionein in the tissues?

WHANGER (USA). No, unfortunately we did not.

HURLEY (USA). Did the salt mixture that you used in the diet contain all the essential trace elements? Also, do you think that the free amino acids might have altered the absorption of various trace elements and thereby influenced your observations?

WHANGER (USA). The salt mix was one that we use routinely in selenium deficiency studies. It does not contain zinc or selenium, but all other elements are present in adequate amounts. In answer to your second question, the amino acid mixture could have affected absorption of the trace elements and to avoid this we would need to use a protein with low content of sulfur amino acids - some work for the future.

PROHASKA (USA). Did you say that the cytosol fraction of testes does not contain metallothionein?

WHANGER (USA). A low molecular weight, cadmium-binding protein elutes in the same position as metallothionein on gel filtration, but the amino acid analysis of the protein indicates that it is not a metallothionein. It contains only 3 or 4% cysteine.

MILLS (Scotland). At 15 ppm Zn your crystalline amino acid diet was marginal in zinc for the requirements of rats. Is there a possibility that the catabolism of body proteins released zinc and stimulated metallothionein synthesis?

WHANGER (USA). In the torula yeast diet we added 100 ppm Zn and observed a decline in the level of metallothionein in tissues as the animals grew older. In the crystalline amino acid diet we added 1000 ppm Zn and this seemed to counteract the decline, in fact there was an increase in the level of metallothionein in tissues as the animals grew older.

ASPECTS OF TRACE ELEMENT ANALYSIS

PROBLEMS IN TRACE ELEMENT ANALYSIS OF BIOLOGICAL SYSTEMS

G.V.Iyengar

Institute of Medicine, Juelich Nuclear Research Center
Post Box 1913, D-517 Juelich, F.R.Germany

Despite the enormous advances in trace element analysis reproducibly reliable data for biological systems are still rare. Of course, general problems exist relating to specimen collection, preparatory techniques, measurement of the analytical signals, recording, calculation and interpretation of the results. In addition, there is, on one hand, a normal range of biological variation for a given species and for the site of residence. On the other, there are alterations in composition of biological material before and during the period of analysis depending on what techniques are applied. This introduces complexity compared with static inorganic systems. Thus, numerous pitfalls and shortcomings may lead to erroneous conclusions. Firm baseline reference values for the elemental concentrations need to be established in order to detect their alterations during disease conditions. The purpose of this review is to identify the sources contributing to deviations from normal values. Leaving aside disease conditions, the remaining sources broadly fall into two groups: presampling factors and analytical errors. However, in this report special attention shall be paid to presampling factors, the impact of which on the analytical results has not been fully recognised.

PRESAMPLING FACTORS

Procurement of normal human tissues for elemental analysis is faced with certain inherent difficulties since only a few samples such as blood, excretory products and occasional biopsy samples are obtainable from living subjects while the remaining ones are sought at autopsy. Irrespective of the mode of collection the samples are subject to influences from biological variations, postmortal changes and certain inadvertent errors. Moreover, there are differences in trace element concentrations between specific segments of an organ.

Biological variations

These may be divided into genetic factors, long and short-term physiological influences and seasonal changes.

Genetic predisposition of human subjects belonging to different racial groups is an ideal example in this context. However, because of overriding influences from environmental and other factors, changes of elemental concentrations in body components directly attributable to genetic factors in man are somewhat difficult to identify. Yet, a distinct case of heritable anomaly leading to extremely high concentrations of plasma Zn has been reported in 5 out of 7 members of one family, and 2 out of 3 second generation individuals (55). Under usual conditions these values would have been disregarded as those arising from sample contamination. In animals, changes arising from differences in breed are well recognised. Examples are composition of milk from different breeds (25), Cu in blood of sheep (18, 73) and several trace elements in mice strains (56).

Long-term physiological influences include ageing, sex, geographical and environmental factors, diet and pregnancy. These involve many in vivo factors that may influence the analytical result. The definition of long and short-term influences is somewhat arbitrary since factors such as diet and smoking often overlap. For example, habitual diet may be regarded as a long term effect while a single meal may be looked upon as a short term effect.

The age of the examined individuals is an important factor in elemental analysis since shifts in elemental concentrations occur as a function of age. Zn in serum and Fe in brain are two good examples in this context. As a function of age lowest values of serum Zn are found in babies under one year, while the concentration increases to reach a peak around 2o to 4o years and starts to fall after the age of 5o (36). In brain, irrespective of the site of tissue, Fe concentration continuously increases up to the age of 5 reaching a plateau between the age of 1o to 2o years and then levels off (53). Other examples include accumulation of Cd in kidney and Pb in bone throughout infancy to old age, elevated Cu and decreased Zn levels in plasma of infants, elevated Cu and Zn, and reduced Co concentration in fetal liver (66) and high concentration of Cr in hair of new born infants (17).

The best example of differences due to sex is that of Fe in liver (67). It has been shown that men have higher concentration of this element than women. The recorded concentrations on dry weight basis are 913 and 7oo micro g per g for males and females, respectively. Serum protein concentrations of men is significantly higher than in women (63). Therefore, differences in the concentrations of protein bound metals should be anticipated. Other examples include excretion of urinary Cd in Japanese men and women (64) and Zn concentrations in hair of infants (14). Sex related differences between the contents of Zn and Cd in liver and Zn, Cu and Cd in kidney have been reported for rats (46).

The influence of geographic and environmental factors has been established in several cases. Prolonged inhalation of polluted air from industrial and urban areas containing specific contaminants such as Al, Cd, Fe, Pb and Si, occupational exposure to a number of elements such as Be, Cr, Hg, Mn, Ni, Sb and Si may change the composition of certain tissues, e.g. lung, hair and skin. Dramatic accumulation of various trace elements in particular Sc was found in lung tissue from people living in Duisburg due to the environmental pollution from the steel furnaces, operating in this area (49). A classic example is the concentration of food and water in wide areas of Japan causing excessive body burdens through accumulation of Cd. Furthermore, consumption of certain foods known to contain high concentrations of Hg and Cd, e.g. sea foods, prolonged high or low intake of certain elements through water and foods grown on certain types of soils, are known to affect tissue picture, e.g. consistently low values for Se in blood of New Zealanders (15,7o), high values of As in blood of Thaiwanese (21) and of American women (35), abnormal concentrations of Pb in blood (9) and milk (13), Hg in milk (12) and regional differences for Cu and Zn in plasma (41) have been reported. Smoking has been correlated with elevation of Cd in blood (62) and kidney (44). Cd may also be released from domestic water pipes by soft acidic water (54).

Diet is a very important factor being the primary source of input for several trace elements. Absorption of elements into blood is influenced by dietary components. The best example is that of serum Zn which varies directly with the concentration and bioavailability of Zn in the dietary regimen. However, information is scanty on the dynamic processes occuring in the body following the absorption of elements from the diets, e.g. rate of clearance from blood. If an element is absorbed and cleared from blood relatively slowly this is useful in ascertaining the individual's exposure to that element, but presents problems for defining normal values for reference purpose. On the other hand, a rapid clearance in a few hours would suggest an effective homeostatic control mechanism, and therefore, under fasting conditions relatively consistent normal values may be expected, if analytical errors are controlled. In general, elements which are poorly absorbed into blood, e.g. Cr, Fe, Mn, Ni and Sn may be expected to pose less fluctuations, from diet point of view. Unfortunately, studies related to the residence times of trace elements in blood are restricted to a few. F has been studied quite extensively and in several species (2, 57, 58, 61). The data generally confirm a rapid but transient increase in plasma F following dietary and oral dose of soluble Flouride. So plasma values obtained after overnight fasting indicate true levels. In contrast, when elemental and foods rich in I are fed, the concentration of I in blood and other tissues are elevated and remain so for several weeks (5, 39). This aspect of iodine is discussed further under seasonal changes and inadvertent errors. Other comparable examples with dose-response relationships have been discussed for Zn in humans (11) and rats (5o), for Cd in lambs (8) and Pb in rats (3, 47).

Pregnancy induces several physiological and biochemical changes of which the increase in plasma volume by as much as 5o per cent and a decrease in plasma protein up to 3o per cent due to reduction of albumin concentration, are of interest from trace element point of view. These changes essentially account for the shifts observed in electrolytes and Zn, all of which are reported to be consistently low during pregnancy. The reverse tendency of elevated Cu in serum in pregnant women is well documented. However, biochemical changes brought about by pregnancy in animals do not always conform with those of human, and species specific differences should be anticipated (66).

Short-term physiological influences can be regarded as those which affect the system over short periods. Thus circadian rhythm, recent food, posture and stress can be grouped under this

Circadian rhythm or the diurnal variation has been examined only for a few trace elements such as Cu, F, Fe and Zn. Few studies have reported significant changes in Zn concentration of serum depending upon the time of the day (19,47,65). However, there is no unified picture concerning the time of the day at which peak concentration is reached, suggesting that more intricate factors are involved in controlling its level. Variations in the Fe concentration of plasma evidently exist in normals because high levels are generally found during the night and early morning (63, 75). A short-term Fe rhythm in children showing variations up to 4o per cent in two consecutive estimations performed within 3o minutes during the morning has also been reported (71). Cu and F do not undergo any cyclical variation (61).

The influence of recent and specific foods is also an important factor. Amino acids present in foods may chelate some elements in the blood and remove them from the stream. This appears to be the case with Zn which drops to less than 7o per cent of its initial fasting value within 3 to 6 hours following a large meal (37). Severe Zn deficient diet has been shown to reduce normal serum Zn value in rats by 4o per cent within 24 hours (74). Recent dietary intakes may also act as internal contaminants for certain elements as illustrated for As. Following ingestion of a specific food (fillet of plaice), As levels in blood serum rose by more than an order of magnitude. Even after 12 hours its level did not return to fasting value (2o). Under usual conditions (ie. overnight fasting) these values would have been accepted as normal values.

Posture and stress too may contribute to shifts in normal values since changes from an upright to a recumbent position affect the concentrations of plasma proteins. This has been shown to induce alterations in the concentrations of Zn and Se (34). Venous occlusion such as is used prior to venepuncture has been reported to increase plasma protein up to 3o per cent with a concomitant increase in Zn values (69).

In human beings, the impact of seasonal changes brought about by climatic conditions, e.g. summer, may be envisaged through recreational activities, differences in diet due to the availability of fresh fruits and vegetables, and changes in excretory pattern because of excess sweating. The last factor cautions against directly comparing analytical results of urine samples (be it concentration or 24 hour clearance) collected during different climatic seasons. An important impact of seasonal changes is seen in farm animals such as dairy cows in the western world. In winter the animals are usually confined to their stalls consuming standard foods. Whereas, in summer, they go out into the fields. Depending upon the soil composition of the fields, seasonal variation may occur in milk for elements such as Cu, F, I, Mn, Mo and Se (25).

Postmortal changes

It is well established in the realm of pathology that dead tissues are subject to severe postmortal changes and that rapidly metabolizing organs such as liver, spleen, kidney and heart are severely affected by cell swelling, imbibition and autolysis. Cell swelling and imbibition produce changes in organ volume due to fluid influx and expulsion while autolysis accounts for the actual tissue degeneration. These changes can introduce great variations in elemental concentration and content. Yet, an objective assessment of these factors have been done only recently (27). Retaining rat livers inside intact carcasses for prolonged periods of time under different conditions induced fluctuations in weight and elemental composition. Both concentration and content of electrolytes and trace elements Cu, Fe, Mn, Rb and Zn varied up to 5o per cent with respect to control values. Ca showed very high values extending to a factor of 6 due to mobilization from residual food in stomach (28). These observations suggest that there is a need for standardising sampling and storage conditions for autopsy samples aimed for elemental concentration studies.

Inadvertent errors

These may be defined as factors deceptively present in the sample thus falsifying the results. These errors, as the definition itself suggests, are difficult to detect and the analyst has little or no control on them. A few examples are presented which arise from medication, hemolysis, subclinical conditions, critically small sample weights and other miscellaneous situations.

Certain types of medication e.g. chelation therapy are recognised for their role in upsetting the trace element balance. However, there are a few other baffling situations which have received less attention. For example, prior exposure to I containing drugs or X-ray contrast media generally elevate the tissue I concentrations (7). Increases ranging from 3 to 26 times that of control values were measured by these investigators for extrathyroidal tissues several weeks after the treatment. Elevated levels were seen in adipose tissue even after 2 years post-treatment. In blood serum the biological halflife for ingested iodine is quoted to be 2.5 weeks (6o). The wide spread use of I containing drugs (some as Prophylaxis) and X-ray contrast media signal a formidable source of internal source of contamination for this element.Careful evaluation of case history is necessa y to minimise such errors. Yet another example is seen in pregnant women taking antinausea drugs and sleeping pills. These drugs have been associated with elevated values of Cu, Mn and Zn in hair and placenta.

Hemolysis is another example in this context. Normal plasma contains much less hemoglobin in relation to serum which may contain 1o to 2o mg per 1oo ml, an equivalent of 35o to 7oo ng of Fe per ml. Below a level of 3o mg of hemoglobin per 1oo ml of serum, naked eye cannot distinguish the presence of hemolysis. Above this level it is noticeble depending upon the intrinsic colour of the plasma. This means that Fe concentration of up to 1000 ng per ml resulting from hemoglobin can be present, but virtually unnoticed. Colorimetric methods commonly used in the clinics for the determination of Fe in serum measure exclusively transferrin Fe, hence hemolysis in this case is of no consequence. Whereas, using modern methods such as atomic absorption and neutron activation, total Fe is measured yielding high values (26, 38).

Subclinical conditions involve marginal deficiencies and excesses arising from altered input, differences in absorption and unidentified errors of metal transport. Borderline values obtained in such cases cannot be distinguished from the normal range which is still rather wide for several elements, and therefore not well defined. Consistently high or low borderline values should in such cases be viewed with suspicion.

Weighing critically small dry biological samples (e.g. $<$ 10 mg) may contribute considerably to inadvertent errors because of variability of the moisture content on the sample container and absorption of the moisture by the sample itself. Sizeable unaccountable errors are also possible because of inconsistency in the wetness of the tissue (1).

Differences between specific segments of an organ

Significant differences in trace element concentrations between specific segments of an organ such as kidney (44,45), brain (22) and bone (10) have been illustrated. For example, sectioning of kidney into pure cortex and pure medulla is necessary since metal levels (e.g. Cd and Zn) differ significantly between tissue sections (45). Careful sectioning of hair to account for the distance from the scalp is another example of site dependency of elemental composition.

ANALYTICAL ERRORS

Analytical errors begin with sampling and sample preparation and persist through all the remaining steps such as measurement of the analytical signals, recording and calculation of the results.

Sampling is one of the critical steps in trace element analysis. A knowledge of the numerous factors and a strict adherence to all the requirements are essential if the results are to be reliable. There are some basic hazards in trace element analysis which arise from contamination of the sample by the elements, loss of trace elements and change in mean composition with respect to the elements to be analysed. Contamination of the sample may be caused by the environment, the actual sampling operation and the operating personnel. Trace elements may be lost due to the overlooked differences in the chemical state of the element, by volatilization during drying and ashing, absorption on container walls and handling tools, and also by sputtering and spraying in the case of liquids. The changes in the mean trace element composition of the sample may result from the physical or chemical changes and redistribution of the elements. The main sampling operations for solid tissues include the selection of an optimal sampling method, the sampling site and preparation of the sampling equipment. Liquid samples require additional treatment such as the use of stabilizer and phase separation. It is also necessary to obtain timed collections for samples like urine and faeces. Preservation of both solid and liquid samples is necessary during transport to the laboratory. These are discussed elsewhere (51, 52).

Sample preparation of biological material is necessary in order to prepare the samples for physical measurement and it is the responsibility of the analyst. Some essential fundamental requirements include a well defined analytical programme, prior knowledge of the approximate elemental composition of the sample matrix, reasonably clean conditions to suit the low concentration level analytical work and adequate precaution against the basic hazards as already discussed. For solid samples the sample preparation steps may include particle size reduction by fragmentation, drying, homogenisation followed by homogeneity test, etc. For liquid samples which are generally suspensions or emulsions, the following steps may be necessary; homogenisation, phase separation, preconcentration and subsampling. Final preservation and storage will be required if further analysis is not done immediately. Storage of dried samples at 2 to 4 degrees centigrade and gamma-ray irradiation reduce the rate of microbial and chemical decay. Exclusion of air dust and humidity is mandatory. In practice, since contamination of the sample with containers and handling tools cannot be avoided, the approach of controlled contamination using accessories made of quartz, teflon, plastic and certain high purity metals, e.g. Ti, must be adopted (29-33, 52).

A substantial part of the errors associated with measurement of the analytical signals may be overcome by the use of proper analytical methods (23) and quality control materials. However, it should be recognised that analysis of certified or standard reference materials (SRM or CRM) does not usually test the whole analytical procedure from beginning to the end since analytical quality control samples are usually handled differently in some way, and may therefore not be exposed to such errors, faced by real samples during sampling and sample preparation. In the absence of a suitable reference sample or matrix, validity of the analytical method should be proved by two different methods, preferably based on different principles, e.g. atomic absorption and neutron activation. It may be mentioned here that there is a lack of certified reference materials especially for the elements at ppb level. For some elements such as F and Si, there are no certified reference materials while only a few are available for elements such as I, Mo, Ni, Sn and V, among others (48). As an overall conseqence of this situation literature data reveal serious inconsistencies and a substantial part of the analytical data are in error(24, 26, 68). Mn, Co, Cr etc. in blood serum (6) and Cr in foods (42) are typical examples of this situation.

Errors arising from data handling mav involve keypunch errors, failure to consider minimum detectable limits and clerical mistakes.

Errors may also arise during interpretation of data because of choosing inappropriate statistical treatment and improper bases for expressing results. Choice of fresh, dry or ash weights, may, depending upon the matrix, present problems. It has been reported that concentration of Pb in ash within a rib generally decreases as the distance from the costochondral junction increases, but the opposite trend is revealed in dry and wet bone (1o). Similarly, choosing whole blood as basis, the activity of the selenoenzyme glutathione peroxidase was found to be lower in pregnant rats, but based on hemoglobin which represents erythrocytes, there was no difference any more (4). Discrepancies have also been reported for changes in trace element content of uterus because of alterations in water content under estrogen influence (4o). Inadequately matched controls e.g. age and sex, can also contribute to errors. For example, Fe concentration in livers of women before and after menopause differ significantly (67). Cu value in blood of subjects who are on the verge of puberty is subject to fluctuation depending upon the level of estrogen in the blood. Some implications of this problem related to Cu have been discussed (59).

CONCLUSIONS

1. There is a need to systematically evaluate the presampling factors associated with different types of biological samples, e.g. rate of clearance from blood of absorbed elements.

2. It is also necessary to evaluate critically some of the existing sampling and sample preparation techniques for various biological materials in order to recommend standard procedures and to identify suitable equipment. Only then will the intercomparison and utilization of analytical results from various sources be valid for wider application such as nutritional and physiological studies covering various population groups.

3. Since trace element determinations are commonly reported for samples such as blood, serum, urine, hair, nail and bone, efforts should be directed to prepare suitable standard reference materials for these.

4. Thorough documentation of the personal history of the subjects is an invaluable aid in assessing reference values.

5. It is imperative that analysts should be associated with trace element investgations already at the planning stage and not just from the moment a sample arrives at the laboratory.

LITERATURE CITED

1. Adrian, W.J. and M.L.Stevans, 1979, J.Wildlife Dis. 15:125-26
2. Armstrong, W., L.Singer, J.Ensinck and C.Rich, 1964, J.Clin.Invest. 43:555-56
3. Barltrop, D. and H.E.Khoo, 1975, In "Trace Substances in Environmental Health", (Ed. D.D.Hemphill), Uni. of Missouri, Columbia, Mo., 9:369-76
4. Behne, D., R. von Berswordt-Wallrabe, W.Elger and S.W.Walters, 1978, Experientia 34:986-87
5. Carr, E.A. and D.S.Riggs, 1953, Biochem.J. 54:217-25
6. Cornelis, R. and J.Versieck, 198o, In "Trace Element Analytical Chemistry in Medicine and Biology" , (Ed. P.Brätter) (de Gruyter, Berlin), pp 587-6oo
7. Costa, A., O.B.Testori and M.Migliardi, 1978, J.Endocrn. 1:221-25
8. Doyle, J.J., W.H.Pfander, S.E.Grebing and J.O.Pierce, 1972, in Ref 3, 6:181-86
9. Elwood, P.C., A.S.Leger and M.Morton, 1976, Lancet I:1295-96
1o. Esterby, S.R., 1976, Ph.D.Thesis, University of Waterloo, Canada
11. Freeland-Graves, J.H., L.M.Ebangit and P.J.Hendrikson, 198o, Amer.J.Clin.Nutr. 33:1757-66
12. Galster, W.A., 1976, Environ.Health Persp. 15:135-45
13. Ghelberg, N.W., I.Ruckert and H.Strans, 1972, Igena, Hygiene (Bucharest), 21:17-25
14. Gibson, R.S. and M.S.De Wolfe, 198o, Nutr. Rep. Int. 21:341-49
15. Griffith, N.M. and C.D.Thompson, 1974, N.Z.J.Med. 8o:199-2o5
16. Guillard, O., A.Pirion, J.Bombert and D.Reiss, 1979, Biomed. 31:193-94

17. Hambidge, K.M. and J.D.Baum, 1972, Amer.J.Clin.Nutr. 25:376-79
18. Hayter, S., G.Wiener and A.C.Field, 1973, Anim.Prod. 16:261-66
19. Hetland, O. and E,Brubakk, 1973, Scand. J.Clin.Lab.Invest. 32:225-26
2o. Heydorn, K., E.Damsgaard., Larson, N.A. and B.Neilson, 1978, IAEA-SM-227, pp 129-42
21. Heydorn, K., 197o, Clin.Chim.Acta 28:349-57
22. Höck,A., U.Demmel, H.Schicha, K.Kasperek and L.E.Feinendegen, 1975, Brain 98:49-64
23. IAEA Technical Report 197, (Vienna), 198o
24. Iyengar, G.V., 1981, Sci.Total Environ. 19:(Editorial article, in press)
25. Iyengar, G.V., 1981, The Elemental Composition of Human and Animal Milk: A Review, IAEA Technical Report, (Vienna, in press)
26. Iyengar, G.V., W.E.Kollmer and H.J.M.Bowen, 1978, The Elemental Composition of Human Tissues and Body Fluids, (Verlag Chemie) (Weinheim, FRG)
27. Iyengar, G.V. 198o, Sci.Total Environ. 15:217-36
28. Iyengar, G.V. 1981, J.Path. 134: (in press)
29. Iyengar, G.V. and B.Sansoni, 198o, In IAEA Technical Report 197, pp 73-1o1
3o. Iyengar, G.V. and K.Kasperek, 1977, J.Radioanal. Chem. 39:3o1-16
31. Iyengar, G.V., K.Kasperek and L.E.Feinendegen, 1978, Phys.Med.Biol. 23:66-76
32. Iyengar, G.V., K.Kasperek and L.E.Feinendegen, 1978, Sci.Total Environ. 1o:1-16
33. Iyengar, G.V., K.Kasperek and L.E.Feinendegen, 198o, Analyst 1o5:794-8o
34. Jürgensen, H. and D.Behne, 1977, J.Radioanal.Chem. 37:372-82
35. Kagey, B.T., J.E.Bumgarner and J.P.Creson, 1977, Ref. 3 11:252-56
36. Kasperek, K.,L.Feinendegen, I.Lombeck and H.Bremer, 1977, Eur. J.Pediat. 126:199-2o2
37. Kasperek, K., H.Schicha, V.Siller, L.Feinendegen and A.Höck, 1972, IAEA-SM-157, pp 517-26
38. Kasperek, K., J.Kiem, G.V.Iyengar and L.E.Feinendegen, 1981, Sci.Total Environ. 17:133-43
39. Kiltgaard, H.M., H.W.Dirks, W.R.Garlick and S.B.Banker, 1952, Endocrinol. 5o:17o-73
4o. Kollmer, W.E., P.Schramel and G.V.Iyengar, 1974, Endokrinologie 64:115-2o
41. Kubota, I., U.A.Lazar and F.Losee, 1968, Arch.Environ.Health 16:788-93
42. Kumpulainen, J., 198o, Report EKT-553, University of Helsinki, Helsinki, Finland
43. Lifschitz, M.D. and R.I.Henkin, 1971, J.Appl.Physiol. 31:89-94
44. Le Baron, G.J., 1977, Ph.D.Thesis, University of Waterloo, Canada
45. Livingstone, H.D., 1971, in Ref 3, 5:399-412
46. Murthy, L., D.P.Rice and H.G.Petering, 1977, Proceedings TEMA-3, (Ed.M.Kirchgessner) (Weihenstephan, Munich), pp 557-6o
47. Mylrole, A.A., L.Moore and U.Erobbogbo, 1977, Toxicol. Appl.Pharm. 41:361-67
48. Parr, R.M., 198o, in Ref. 6, pp 631-55
49. Persigehl, M., Kasperek, K., H.J.Klein and L.E.Feinendegen, 1976, Beitr.Path. 157:26o-68
5o. Rice, D.P., L.Murthy, T.Shirley, I.E.Menday and H.J.Petering, 1973, Ref.3, 7:3o5-11
51. Sansoni, B. and G.V.Iyengar, 1978, IAEA-SM-227, pp 3-5
52. Sansoni, B. and G.V.Iyengar, 198o, In Ref. 29, pp 57-71
53. Schicha, H., K.Kasperek, L.Feinendegen, V.Siller and H.Klein, 1971,Beitr.Path.142:268-74
54. Schroeder, H.A. and A.P.Nason, 1971, Clin.Chem. 17:461-7o
55. Smith, J.C., Z.A.Zeller, E.D.Bronner and S.C.Ong, 1976, Science 193 :496-98
56. Siegers, M., H.J.Heiniger, K.Kasperek, G.V.Iyengar and L.E.Feinendegen,1977, Ref.46,pp188
57. Simon,J. and J.W.Suttie, 1968, J.Nutr. 94: 511-15
58. Singer, L. and W.D.Armstrong, 196o, J.Appl.Physiol. 15:5o8-1o
59. Sorenson, J.R.J., 1978, Inorg. Persp. Biol.Med. 2:1-2o
6o. Sönkesen, P.H., R.Ekins, H.Stevans, E.Williams and J.D.N.Nabarro, 1968 Lancet II:426
61. Spencer, H.,D.Osis and I.E.Waitrowski, 1974, In Ref. 3, 8:299-3o4
62. Stoeppler, M., 1979, Vortragsveranstaltung UMWELTCHEMICALIEN der AGF, Bonn, pp.19
63. Stengle, J.M. and A.L.Schade, 1957, Brit.J.Haematol. 3:117-2o
64. Suzuki, S. and T.Taguchi, 197o, Ind. Health 8:115o-55
65. Tonitou, Y., C.Tonitou, A.Bogan, H.Beck and D.Reenberg, 1978, Clin.Chim.Acta 87:35-4o
66. Underwood, E.J., 1977, Trace Elements in Human and Animal Nutrition, (Academic)(London)
67. Van Eijk, H.G., W.E.Wittink, G.Boss and J.P.Goossens, 1974, Clin.Chim Acta 5o:275-8o
68. Versieck, J. and Cornelis, R., 198o, Anal.Chim Acta 116:217-54
69. Walker, W.E., I.Bone, B.H.Taylor and J.Kellehner, 1979, Int.Nat.J.Vit.Res. 49:413-18
7o. Watkinson, J.H., 1974, N.Z.Med.J. 8o:2o2-5
71. Werkman,H.P., J.M.F.Trijbels and E.D.A.M.Schretten, 1974, Clin. Chim.Acta 53:65-68
72. White, D.H. and M.T.Finley, 1978, In Ref 3, 12:22o-23
73. Wiener, G., and A.L.Field, 1969, J.Comp.Path. 79:7-1o
74. Wilkins, P.J., P.C.Gray and I.E.Dreosti, 1972, Brit. J.Nutr. 27:113-15
75. Wiltinks, W.F., T.Kruithof, C.Mol, G.Bos and K.Van Eijk, 1973, Clin.Chim Acta 49:99-1o4

DISCUSSION

ANON. I am a little bit confused with respect to your terminology. If you carry out an accurate analyses, then the results describe correctly a certain state of the organism and the parameters which are involved. In my opinion, these parameters cannot be regarded as possibilities of errors of analysis.

IYENGAR (Germany). I agree with you. That is why they were grouped separately. Those which are true and should be taken into consideration were put as pre-sampling factors. We should be aware of it before we deal with the results - that is what I meant, that is why I never said it was an analytical error.

WEIGAND (Germany). In your post mortem changes you had some decrease in the dry weight of your samples when they were stored at -25°. Could you give us an explanation for that?

IYENGAR (Germany). I suspect it is due to blood losses during freezing and thawing.

JUDSON (Australia). You touched on the use of reference material in trace element analysis. We are also interested in this problem and recently we sent out a questionnaire to about 50 trace element laboratories in Australia and New Zealand. Of those we sent out we got about 43 replies, and in those replies we found that about 60% of the trace element laboratories routinely use a reference material in their analytical work, either an external or internal reference material. Only about 30% used it on occasions, and about 10% never used a quality control material. I was wondering whether you also experienced this in other parts of the world.

IYENGAR (Germany). I am quite sure it must be a general problem, I think that there are only a few truly disciplined analytical groups which use reference materials as a matter of practice.

GARNYS (Australia). On the point of freezing, you could possible take note of some of the experiences using water where a preferential precipitation of certain materials forms in the final salt mush during the freezing process. We have experienced great difficulty in resolublising silicon for example, in frozen water. I think studies of what happens to frozen samples needs to be undertaken.

VERSIECK (Belgium). I wish to give a brief comment on biological reference materials. Generally the materials we have at present are excellent for checking the accuracy and precision of our analytical methods. But, as the concentration in these materials is frequently considerably higher than in real samples, it is mostly impossible to check out, for instance, our sample handling procedures. In my opiniou, what we really need for the future is a biological reference material with trace element levels that are very close to the concentrations in our real samples.

IYENGAR (Germany). Sure I agree with you, as you know the International Atomic Energy Agency is interested in this problem and depending on finances will produce a reference serum.

ELMES (Wales). Could I make a comment about this word sub clinical about which there has been a little controversy in the last day or two. I have had a conventional medical training and I hope my conventionally medically trained colleagues will agree with me that if you have a sub-clinical condition there is something definitely abnormal with your physiology but it hasn't disturbed your state of well-being enough for you to need to consult a doctor.

IYENGAR (Germany). I accept that analogy.

MILLS (Scotland). I think the correct definition of the word 'clinical' is literally something you can see by the bedside.

A NOVEL METHOD BASED ON STABLE ISOTOPES FOR THE STUDY OF SELENIUM NUTRITION IN MAN

Morteza Janghorbani, Merrill J. Christensen,
Ara Nahapetian, Bill T. G. Ting, and Vernon R. Young
Nuclear Reactor Laboratory, Department of Nutrition and Food
Science, and Clinical Research Center
Massachusetts Institute of Technology
Cambridge, MA 02139 U.S.A.

INTRODUCTION

Studies of selenium nutrition and metabolism in man have so far been hindered experimentally in part due to lack of suitable tracers. The only radioisotope of selenium available for these studies is ^{75}Se, which possesses a relatively long physical half life (120 days). In contrast, the element selenium consists of six naturally occurring stable isotopes whose natural abundances are given in Table 1.

Table 1: Stable Isotopes of Selenium

Isotope	Nat. Abun. at .%	Degree of Enrichment and Price for Available Enriched Preparations*	
		Enrichment at .%	Price $/mg
^{74}Se	0.9	55.91	92.35
^{76}Se	9.0	96.06	5.80
^{77}Se	7.5	91.77	10.35
^{78}Se	23.5	97.27	4.35
^{80}Se	50.	99.45	0.90
^{82}Se	9.0	96.81	29.35

*Data from Oak Ridge National Laboratory, Oak Ridge, TN

For instance, the stable isotope ^{74}Se naturally occurs at 0.9 atomic % abundance, is commercially available as enriched preparation with atomic abundance of 55.91% and at the cost of $92.35/mg. The availability of so many stable isotopes of selenium indicates the potential, not only for supplanting the use of radioselenium (^{75}Se) in human subjects, but also for conducting human nutrition experiments based on simultaneous multiple tagging, preparation and processing of labeled foods under normal food preparation and processing conditions, administration of enriched preparations to humans under field conditions, etc. However, stable isotopes of selenium have not yet been reported in human nutrition and metabolic studies, partly due to the lack of available measurement methodology. In this communication, we explore some of the basic issues related to the use of these isotopes in human experimentation.

EXPERIMENTAL

A. Analytical Chemistry of Stable Isotopes of Selenium. The method of Radiochemical Neutron Activation Analysis (RNAA) was applied to the analysis of human fecal, RBC, plasma, and urine samples. The method has been described in detail elsewhere (1) and is given briefly in Figure 1.

Figure 1: Outline of RNAA Procedure for the Measurement of ^{74}Se, ^{76}Se, and ^{80}Se in Biological Materials (1)

1. *Sample spiked with ^{75}Se tracer; wet ashed with conc. $HNO_3 + H_2O_2$.*

2. *Reduce Se to selenite with conc. HCl; precipitate with APCD; dissolve precipitate in conc. HNO_3; split if ^{80}Se is also to be measured; measure chemical yield for each split.*

(^{74}Se and ^{76}Se)

3a. *Irradiate at $\phi_{th} = 5 \times 10^{12} ncm^{-2}s^{-1}$ for 30 sec.; measure ^{77m}Se activity with high-resolution gamma spectrometry.*

4a. *Reirradiate at $\phi_{th} = 5 \times 10^{12} ncm^{-2}s^{-1}$ for 50-100 h; measure ^{75}Se activity with high-resolution gamma spectrometry.*

(^{80}Se)

3b. *Irradiate at 5×10^{13} $ncm^{-2}s^{-1}$ for 10 min.*

4b. *Remove ^{64}Cu activity by solvent extraction with dithizone.*

5b. *Reduce Se to Se° and measure ^{81m}Se activity with high-resolution gamma spectrometry.*

B. Human Enrichment Studies. In order to test the potential and feasibility of using stable isotopes of Se in human subjects, a single dose of ^{74}Se containing 54.4 μg of the isotope was given to each of four adult subjects with breakfast meal. Subsequently, samples of feces, blood, and urine were collected. The details of these experiments will be described elsewhere.

RESULTS AND DISCUSSION

A. Accuracy and Precision of Isotope Measurements. The method of RNAA employed here was developed specifically for the measurement of stable isotopes ^{74}Se, ^{76}Se, and ^{80}Se in matrices relevant to human nutrition and metabolism of selenium; viz. feces, plasma, RBC, and urine. Accuracy and precision of this method for the measurement of unenriched samples was investigated in each matrix as well as in NBS SRM 1577 (Bovine Liver) whose Se content is certified at 1.1 ± 0.1. The results from these studies have been presented previously (1) and indicate that overall these measurements can be routinely performed with accuracy and precisions of 10% or better. Furthermore, under favorable experimental conditions it may be possible to improve these measurements to approach the precision and accuracy of 1% although this is neither routinely achievable nor is it necessary for the majority of human experiments (1).

B. Isotope Enrichment in Human Matrices. The fundamental issue regarding the feasibility of stable isotope use in human experimentation is in relation to the practically achieved degree of isotopic enrichment subsequent to oral/I.V. administration of physiologically relevant dose(s) of these isotopes. If the achieved enrichment in the medium of interest (feces, plasma, RBC, urine) is sufficient to permit accurate measurements, then the method is suitable for the particular dosing conditions used or with regard to tracing the label in the specific medium of interest. We have explored this issue by a single oral dose of ^{74}Se (54.4 μg) in adults whose daily Se intake was about 175. μg Se and had plasma and RBC selenium levels of approximately 0.16 and 0.30 μg/ml resp. reflecting sufficient body selenium status (2). Typical mass isotopic ratios of $^{74}Se/^{76}Se$ in different media at various times post ingestion are given in Table 2 for one subject. The data presented in this table show that measurable enrichment of ^{74}Se occurred in all samples examined. The lowest degree of enrichment is in plasma and RBC, as expected, but even in these media significant enrichment took place. Thus it appears from these data that sufficient enrichment of ^{74}Se occurs in all matrices of interest subsequent to a single oral dose of 54.4 μg of the isotope and that this should allow studies of absorption, urinary excretion, net retention, and plasma-RBC turnover of the isotope to be carried out in human subjects.

Table 2: Measured Mass Isotope Ratio for $^{74}Se/^{76}Se$ in One Subject Ingesting a Single Dose of 54.4 μg ^{74}Se with Breakfast Meal

Time, post-ingestion, hours	Measured Mass Isotope Ratio,* $^{74}Se/^{76}Se$			
	Feces	RBC	Plasma	Urine
0	-	-	0.098	-
6.	-	.126	.203	.757
24. - 48.	0.742	-	-	.306
72. - 96.	0.778	-	-	-
96.	-	.137	.155	-

*Unenriched accepted ratio = 0.0945

C. Applications in Human Nutrition and Metabolism. Several nutritional and metabolic issues of selenium nutriture can now be explored directly in human subjects with this novel method. These are:

1. Bioavailability Studies. Little is known about the availability of selenium from different foods and dietary factors affecting this in man (3). The stable isotope approach now permits these studies to be carried out using the method of fecal isotope

monitoring (4) and the extrinsic tag approach. However, it now appears that selenium from composite meals may not form a single exchangeable pool prior to intestinal absorption (5) although definitive data have not yet been published. This indicates the importance of establishing the validity of the extrinsic tag approach which requires availability of foods intrinsically labeled with stable isotopes of selenium. Recently, this has been accomplished with poultry (6) and the issue of exchangeability of selenium pools is now being explored. Once different foods have been labeled with stable isotopes of selenium, the issue of dietary availability can be explored directly.

2. Retention Studies. It is now presumed that homeostasis of selenium is exerted mainly through regulation of urinary excretion (7). The stable isotope method permits determination of net retention of the ingested isotope as well as direct measurement of the extent of urinary excretion and thus allows investigation of the significance of this pathway in the regulation of selenium status in man. Since the method can measure both isotopic selenium as well as the native selenium of the sample while distinguishing excretion of the label from native body selenium, it is now possible to investigate equilibration dynamics of the ingested dose with the body pool of selenium by following changes in stable isotope enrichment in the blood and urine pools.

3. Selenium Body Turnover. Numerous studies dealing with the issues of body pool size of selenium under various conditions of selenium status, dynamics of appearance and disappearance of the ingested (or injected) isotope in the plasma and cellular elements of blood in health and disease, and so on can now be explored.

4. Selenium Nutriture of Special Population Groups. Administration of stable isotopes of selenium avoids all issues of concern relative to radioselenium and thus is an invaluable tool especially in population groups at risk of radiation exposure. Thus, important unexplored issues related to nutriture of Se in infants, adolescents, and women of all ages and child bearing status can be pursued with this approach.

REFERENCES

1. Janghorbani, M., B. T. G. Ting, and V. R. Young, 1981, submitted to Am. J. Clin. Nutr.

2. Rea, H. M., C. D. Thomson, D. R. Campbell, and M. F. Robinson, 1979, Br. J. Nutr. 42:201-208.

3. Thomson, C. D. and M. J. Robinson, 1980, Am. J. Clin. Nutr. 33:303-323.

4. Janghorbani, M., B. T. G. Ting, and V. R. Young, 1980, Clin. Chim. Acta 108:9-24.

5. Richold, M., M. F. Robinson, and R. D. H. Stewart, 1977, Br. J. Nutr. 38:19-29.

6. Janghorbani, M., M. J. Christensen, F. H. Steinke, and V. R. Young, 1981, J. Nutr., in press.

7. Burk, R. F., 1976. In "Trace Elements in Human Health and Disease", vol. II (ed. A. S. Prasad) (Academic Press, New York) pp 105-128.

DISCUSSION

BRATTER (Germany). I think a certain drawback of this method is the price. As far as I know one gram of highly enriched ^{74}Se costs about $10,000, is this right?

JANGHORBANI (U.S.A). Yes that is correct. If you had mentioned the price when I gave the talk with zinc it would have been a lot more dramatic. The selenium-74 is available at various enrichments, the one I have shown on the first table, cost $175 a mg. Now a single experiment in one of those periods requires 50 microgram of selenium-74. That indicates that one mg. of selenium-74 can be used for twenty subject periods and costs $175, so the isotopic cost for selenium is simply a very trivial component of a metabolic experiment. The cost is not as trivial for other stable isotopes such as calcium, magnesium, zinc and some others though for those isotopes also, we have done calculations and published that the overall cost of isotopes even for the most expensive, namely calcium-46, amounts to about 5 to 10% of the overall cost of conducting a human metabolic experiment. We consider that is not very important; for selenium it is certainly a very trivial thing.

LEVANDER (U.S.A.). One thing that does bother me a little bit is that you are adding ^{74}selenite, in these experiments is that correct?

JANGHORHANI (U.S.A). Selenium isotope was given as a single dose of selenite replacing the selenium supplement and it was given in water prior to the breakfast meal.

LEVANDER (U.S.A.). This was a 50 μg dose and you were giving about another 50 μg from selenium in the diet as other forms?

JANGHORBANI (U.S.A). On the days of isotope administration the mineral supplement that normally provided about 100 μg total selenium was replaced so that the selenium intake did not vary on the days of administration as compared to the rest of the experimental period. An important point, it is a very good point you raised Dr. Levander, is that these experiments were conducted in the United States with subjects whose selenium status would be considered sufficient by today's standard, the way we understand these and that of course is very much different from areas of the world such as New Zealand where selenium nutrition is considered an important problem. What these experiments are to show as a feasibility study is that in the worst case the procedure is adequate. For example, in subjects whose selenium status is high, their plasma selenium is high, their faecal output is high, their urinary output is high to the extent of about 10 times that reported by Dr. Robinson and people from New Zealand. If this procedure proves adequate under these conditions then in subjects and areas where selenium status is lower the procedure would work even better. That is part of my selling job.

SERUM VERSUS PLASMA FOR TRACE METAL ANALYSIS

Nancy W. Alcock

Department of Medicine
Memorial Sloan-Kettering Cancer Center
1275 York Avenue, New York, NEW YORK, 10021
U. S. A.

INTRODUCTION

Trace metal analysis in blood requires special attention to specimen collection and handling, since potential contamination is a problem and also, redistribution of the element in question between the various blood fractions must be avoided. Hence it is frequently suggested that serum zinc is significantly higher in plasma than in serum, (1) due to either release of zinc from platelets or to invisible hemolysis. The potential for contamination from anticoagulants must always be considered. In measuring the manganese concentration in red blood cells and plasma from a selected group of patients variable and elevated levels in plasma were found, while the levels in red cells were constant and similar to those reported in the literature (2). These findings suggested that contamination had occurred and the anticoagulant used, sodium heparin was a possible source. Hence, a comparison between heparinized blood and a sample collected without anticoagulant was made. In addition, various sodium heparin preparations were analyzed for both manganese and chromium.

MATERIALS AND METHODS

Instrument: A Perkin-Elmer Spectrophotometer Model 5000 was used with a Graphite Furnace, Model 500 with Maximum Power Temperature Sensor, Auto Sampler Model 40, Recorder Model 56 and Printer Model 10.

Reagents: Distilled water shown to contain no measurable manganese or chromium, Manganese Concentrated Standard was prepared by dissolving manganese powder (Fisher Scientific Catalogue Number M-78) in approximately 100 ml. of 5% nitric acid (v/v). The volume was made to 1 liter with distilled water.

Triton-X 100 (Fisher Scientific, Catalogue # CS-282-4M). Appropriate dilutions were made with distilled water.

Working Standards were prepared by serial dilution in water of the concentrated manganese standard. 1,2 and 3 ppb manganese standards were prepared by additions of 20, 40 and 60 µl of a 50 ppb Mn standard to either water or 0.05% Triton-X to make a total volume of 1.0 ml. Standards were similarly added to serum or plasma diluted 1/3 with 0.1% Triton-X.

Chromium Working Standards of 1, 2 and 3 ppb were prepared using a reference standard containing 1000 ppm chromium (Fisher Catalogue # SO-C-192) and the same procedure as that used for the manganese working standards.

Sodium Heparin: The strengths of heparin solutions - bovine and porcine - and their source are shown in Tables 1 and 2.

Table 1. Mn. and Cr. concentration in Bovine Heparin. Each result is from analysis of a separate vial. Vials for each strength were from the same batch.

		BOVINE Na - HEPARIN		
#	UNITS /VOL	MANUFAC-TURER *	Mn (ppb)	Cr (ppb)
1	100/ml	ES	5; 4	1; 1
2	1000/ml	ES	8; 17; 15	36; 39; 44
3	1000/ml	ES-R	210; 230; 245; 365	19; 30; 59; 30
4	5000/ml	ES	29; 37; 46	148; 210; 148
5	5000/ml	ES-R	62; 52;49	73; 45; 39

* ES - ELKINS-SINN
ES-R-ELKINS-SINN FOR ROBINS

Table II. Manganese and Chromium concentration in Porcine Heparin. Each result is from analysis of a separate vial.

	PORCINE Na - HEPARIN (a)			
#	UNITS/VOL	MANUFAC-TURER (b)	Mn	Cr
1	10,000/10ml	ES	12	15
2	50,000/10ml	ES	480	185
3	50,000/10ml	0	24	43

(a) 0.7 - 0.9% NaCl
(b) ES - ELKINS-SINN
0 - ORGANON

PROCEDURES

Programs used for manganese and chromium are shown in Tables 3 and 4 respectively.

Table 3. PROGRAM FOR Mn.

STEP	1	2	3	4
T^o C	120 (approx.)	950	1900	2700
Ramp (Sec.)	10	5	0	1
Hold (Sec.)	40	45	5	3
Int. Gas (ml/min) -Argon	300	300	10	300

λ = 279.5 nm.

Table 4. PROGRAM FOR Cr.

STEP	1	2	3	4
T^o C	120	1200	2300	2700
Ramp (Sec.)	10	20	0	1
Hold (Sec.)	40	20	5	3
Int. Gas (ml/min) -Argon	300	300	10	300

λ = 357.9 nm.

BLOOD COLLECTION

Blood was collected in a plastic syringe (Becton-Dickinson #3604)as follows:
1) in the absence of anticoagulant for serum preparation and
2) after drawing a minimum amount of sodium heparin (1000 units/ml) via a 21 gauge needle into the lower half of the syringe and then expelling as much as possible by repeated action of the syringe plunger. A minimum amount of heparin solution remained. Serum samples were also prepared from blood collected from 1 subject using both a 21 guage needle and a "Butterfly" infusion line (Becton Dickinson #4492). The procedure for the treatment of blood collected in the absence of anticoagulant is shown in Figure 1. Heparinized specimens were treated in the same manner except that they were centrifuged without delay after blood collection.

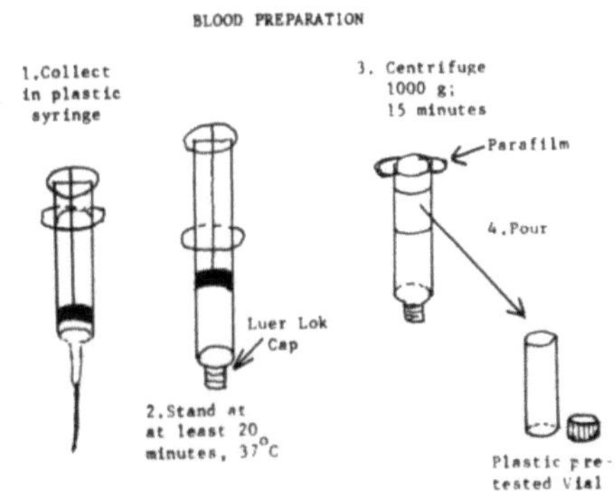

Figure 1. Procedure for separation of serum from blood.

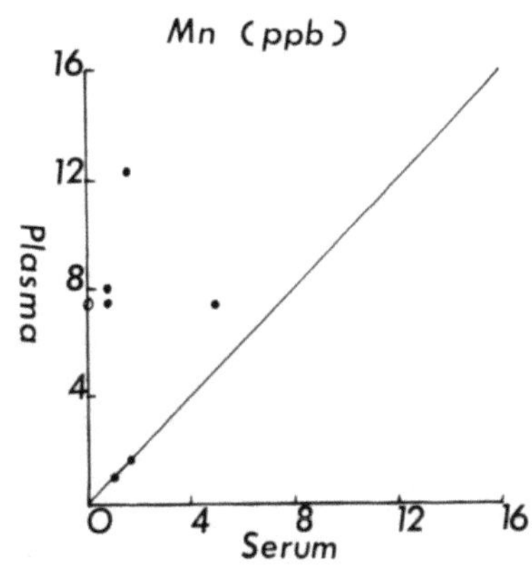

Figure 2. Mn concentration in serum versus plasma • ; Mn concentration in water collected in a heparinized syringe versus that from an untreated syringe o.

In addition to the above procedures an aliquot of serum was drawn into a 10 ml plastic syringe via a 21 gauge needle and left for 1 hour with periodic mixing. Analysis for manganese was compared with that of the original sample.

ANALYSIS

Serum and plasma were analyzed after diluting 1/3 with 0.1% Triton-X using the method of standard additions. Sodium Heparin solutions were analyzed after appropriate dilution with distilled water and absorbance compared with that from aqueous standards for manganese and chromium. The programs used for the analyses are shown in Tables 3 and 4.

RESULTS

No difference was found between the manganese concentration of serum from blood collected via the 21 gauge needle, the "Butterfly" line nor the serum left in contact with the syringe for 1 hour. The manganese concentrations from 6 plasma-serum pairs collected from 6 subjects at random are shown in Figure 2 above. In 4 of the 6 the plasma manganese concentration is greater than that in serum, while there was no difference in the remaining 2. Five of the 6 subjects had manganese concentrations in serum which were 1 - 2 ppb while the sixth had a level of 4.4 ppb. A sample of distilled water collected in a 10 ml syringe in the same manner is heparinized blood contained 7.8 ppb manganese compared with a non detectable level in similarly treated water in the absence of heparin. The concentration of manganese in various preparations of bovine sodium heparin ranged from 4 - 365 ppb (Table 1). In 3 preparations of porcine sodium heparin analyzed the manganese concentration was 12 - 480 ppb (Table 2). The range of chromium concentration in the same bovine and porcine sodium heparin was 1 - 210 and 15 - 185 ppb respectively (Tables 1 and 2).

DISCUSSION

Sodium heparin used as anticoagulant was probably the source of apparent elevation of plasma manganese measured in a group of patients. The analysis of a number of vials of sodium heparin for both manganese and chromium indicates that within batches there was considerable variability. The magnitude of contamination together with the variability between batches of different strengths preclude sodium heparin as an anticoagulant when collecting blood for trace metal analysis. Personal experience has indicated that if serum is carefully separated from blood there is no difference between its zinc concentration and that of plasma from samples collected at the same time. Since zinc is present in red blood cells in approximately 10 times the concentration in plasma, and platelets could lose zinc to the serum, the method of collection and treatment described suggest that there is no redistribution of zinc during preparation. The samples therefore appear to be suitable for trace metal analysis in general.

REFERENCES

1. Smith, J.C., Butrimovitz, G.P. and W.C. Purdy, 1979, Clin. Chem. 25: 1487-1491.

2. Cotzias, G.C., Miller, S.T., and J. Edwards, 1966, J. Lab. Clin. Med. 67: 836-843.

DISCUSSION

VERSEICK (Belgium). Just two comments. The first is that the risks you face when using heparin as an anticoagulant are well known. It was, if I remember, first published by Bowen in 1962 or so. The second is: in your abstract I see that your normal values for manganese are 1 - 2 ppb. I believe that there are good reasons to suppose that you still have underestimated the risk of contamination when doing manganese analyses. There are now several people who are working under rigorously controlled conditions who find a manganese level of 0.5 or 0.6 ppb. Manganese is one of the elements that is best homeostatically controlled. One of the investigators who has shown that very well is Cotzias. He was also from New York.

ALCOCK (USA). I am aware of these studies and I am aware of the normal values. I agree that it is very difficult to know whether one has completely eliminated contamination, I am also aware of another group working with a graphite furnace in the United States who does have similar values to the ones which I have.

CAN SKIN CELLS BE USED TO ASSAY ZINC STATUS IN ANIMALS

J. Apgar and R. M. Welch
U.S. Department of Agriculture
U.S. Plant, Soil and Nutrition Laboratory
ITHACA, NEW YORK 14853 U.S.A.

Zinc (Zn) concentration in serum or plasma is currently the most widely used criterion of Zn status; but it is affected by factors other than Zn status and is not particularly sensitive to marginal Zn deficiency. Other parameters have been suggested (1), but so far none have been shown to be of greater value than plasma or serum Zn concentration. Since skin lesions are one of the most consistent signs of Zn deficiency in humans as well as animals, an assay using skin parameters might be more specific and perhaps more sensitive than other types of assays.

One problem in designing an assay based on skin is the variation in the site which is affected. What is needed is a site that is affected consistently. In the rat the esophagus has been shown to be rapidly and consistently affected by Zn deficiency both in the young rat (2) and in the adult (unpublished data). The esophagus is not an easy tissue to sample; but various sites in the mouth are also affected in the rat (3), and sampling of oral tissues would be possible. To explore the possibility of using oral tissues as a basis for an assay, we decided to look first at the effects of Zn deficiency on the esophagus of other animals to determine whether this tissue in other species is equally sensitive to Zn status. We chose to look at the esophagus first because parakeratosis occurs throughout the esophagus in the rat whereas the affected sites in the mouth are more variable. We have so far examined the esophagus of Zn-deficient guinea pigs and mice.

Thirty young guinea pigs were given a low-Zn agar diet with casein as the protein source. (Both agar and casein were extracted with ethylenediaminetetraacetic acid to reduce the Zn concentration.) Zn concentration of the diet was <1 ppm. The Zn-adequate animals received 25 ppm Zn incorporated in the diet and were also given drinking water supplemented with Zn at the same level. Supplementation with cabbage did not increase weight gain in the Zn-supplemented animals as has been reported with another diet (4). Restricted-intake controls were individually paired with a Zn-deficient guinea pig and fed a reduced amount of diet to keep weight gain as close as possible to that of the Zn-deficient animal. Weight gain during the experiment is given in Fig. 1. Half the guinea pigs were killed at 3 weeks; the remainder at 6 weeks. Unless otherwise noted, values were not significantly different (P<.01) at 3 and 6 weeks and have been combined. Values (means ± SEM) not followed by the same superscript are significantly different (P<.01) by Duncan's multiple range test. Serum Zn concentration was 0.5±.02[a], 0.8±.04[b], and 1.3±.07[c] µg/ml in Zn-deficient (-Zn), Zn-supplemented-restricted intake (+ZnRI), and Zn-supplemented-ad libitum-fed (+Zn) animals, respectively. Serum alkaline phosphatase activity was 2.5±.2[a], 7.2±.5[b], and 6.7±.2[b] Sigma units/ml in the same groups. Femur Zn concentration was 66±2[a], 142±4[b], and 119±3[c] µg/g dry weight. Testicular weight was reduced by food restriction in the guinea pig although it is not in the rat (2). Testicular weights were 92±25[a], 87±17[a], and 356±22[b] mg/100 g body weight for -Zn, +ZnRI, and +Zn animals week 6.

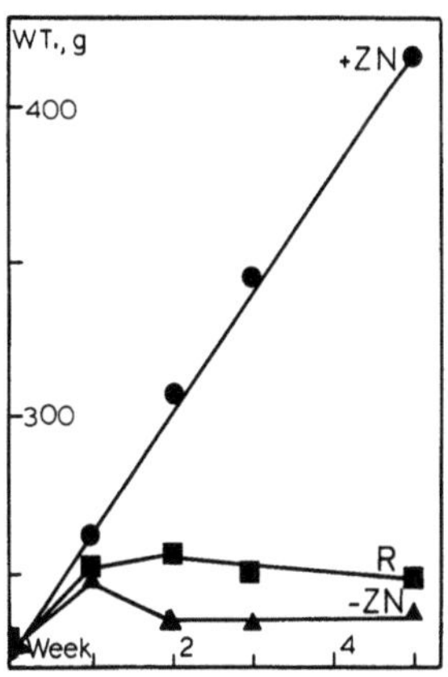

Fig. 1. Weight gain of guinea pigs given low Zn diet. (▲) no added Zn; (●)+25 ppm Zn, ad libitum food intake; (■)25 ppm Zn, restricted food intake.

At the end of 6 weeks, the -Zn animals stood in a peculiar arched posture and had difficulty righting themselves if placed on their sides. Both conditions seemed to be related to a very thick and crusty ventral skin which also prevented spreading the forelegs completely apart. In spite of this, parakeratosis of the esophagus was found in only half the deficient animals and was not particularly severe.

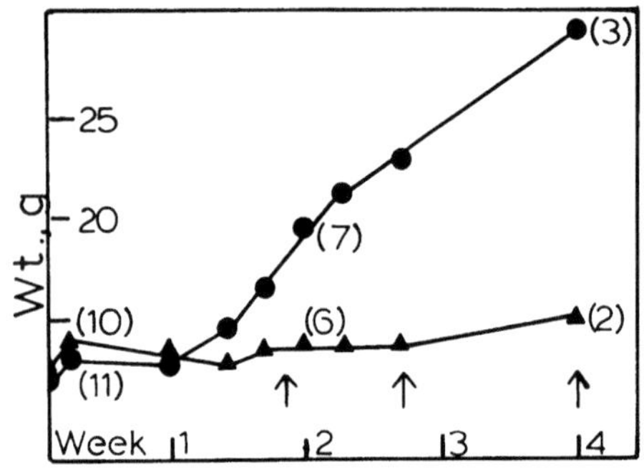

Fig. 2. Weight gain of mice given low Zn diet with (●) or without (▲) Zn supplementation.

The growth curve for 20 mice given either the agar-casein diet used in the guinea pig experiment or a low Zn egg albumen diet is shown in Fig. 2. Half of the mice were supplemented with Zn; restricted-intake controls were not included. Animals were killed at 2, 3, and 4 weeks. No parakeratosis was found in the esophagus at any interval.

The esophagus therefore is not equally sensitive to Zn status in all species, which makes the use of oral tissues as a basis for an assay of Zn status less promising. On the other hand, the report that salivary sediments in humans were affected by low Zn intake (5) suggests that human oral tissues may respond to Zn status.

It is possible, however, that the peculiar sensitivity of the rat esophagus to Zn status can be used in another way, specifically to look for a factor(s) in serum which is related to growth and differentiation of esophageal cells and which is affected by Zn status. The mucosal layer of the esophagus is similar to the epidermal layer of skin, and procedures used for isolation and growth of epidermal cells _in vitro_ can also be applied to esophageal cells. We have begun work on a system for studying esophageal cells _in vitro_ and have some preliminary data on the effects of Zn deficiency _in vivo_ on cells in this system.

Cells were obtained from esophagi that were split longitudinally, sterilized in 70% ethanol, stripped of the muscle layer, and incubated overnight in 0.2% trypsin in serum-free medium (Dulbecco's modified Eagle's medium [DME]). The mucosal layer was removed and stirred 20 minutes in DME with 5% calf serum. The medium was filtered and centrifuged. In some experiments the mucosal tissue remaining on the filter was dried and weighed; in a few cases the submucosal layer was also dried and weighed. Resuspended cells were plated in DME with 20% rat serum that had been heated to 55° for 30 minutes. Approximately 48 hours after plating ^{3}H-thymidine was added to the dishes. Approximately 20 hours later, cells were either harvested for determination of radioactivity in acid-precipitable material or fixed for radioautography.

Rats from which the esophagi were obtained were given a low-Zn egg-albumen diet. Zn-supplemented animals received 100 ppm Zn in the drinking water. Restricted-intake controls were individually paired with a -Zn rat and fed the diet at a reduced level in order to maintain weight gains at the same level as in the -Zn rats. Esophagi from 5 rats per treatment were combined to obtain sufficient cells. Statistical analysis is not presented since the data is from pooled samples.

Table 1. Effect of Zn deficiency on mucosal and submucosal weight of esophagus of young male rats and on _in vitro_ ^{3}H-thymidine incorporation by mucosal cells

Week	Cell Source	Body Weight	Mucosal Weight	Submucosal Weight	^{3}H-thymidine Incorporation
		g	mg	mg	cpm
1	+Zn	158	1.9	-	17,800
	+ZnRI	133	2.0	-	10,200
	-Zn	129	3.1	-	26,700
2	+Zn	214	2.1	-	25,600
	+ZnRI	152	2.6	-	26,700
	-Zn	149	4.6	-	148,900
3	+Zn	257	2.6	7.9	21,300
	+ZnRI	149	2.2	5.9	98,200
	-Zn	144	5.1	6.4	153,000

The data in Table 1 show that the mucosal layer of the -Zn animals was heavier than that of either +Zn or +ZnRI animals by the end of the first week. There was little difference in the weights of the submucosal layers as is evident from the weights week 4. Preparations of cells from -Zn animals also incorporated more ^{3}H-thymidine in vitro than did those from +Zn animals, indicating that there was an increase in proliferating cells in the -Zn animals as well as an increase in the differentiated cell layer. The greater incorporation of ^{3}H-thymidine by preparations of cells from -Zn animals was consistent and was observed in adult males that had been given a deficient diet for 2 months and in younger (weanling) males after 1, 2, and 3 weeks of treatment.

Table 2. Effect of serum from +Zn, +ZnRI, and -Zn animals on in vitro ^{3}H-thymidine incorporation by mucosal cells from +Zn animals.

Serum Source	^{3}H-thymidine Incorporation cpm		
	Week 1	Week 2	Week 3
+Zn	11,500	23,400	9,100
+ZnRI	1,000	24,000	10,500
-Zn	7,900	27,100	27,300

The results in Table 1 were obtained by culturing cells from the different treatments in the presence of the same serum, in this case from +Zn animals. In Table 2 the effects of serums from the different treatments on thymidine incorporation by cells from the same source (+Zn animals) are compared. The effects of the different treatments on the ability of the serums to support thymidine incorporation were less consistent than were the effects of the treatments on the cell source. As shown in Table 2, incorporation was greater in the presence of serum from +Zn animals week 1, almost the same for the different serums week 2, and greater with serum from -Zn animals week 3. The pattern was not always the same in other experiments, which suggests that the variation was not due to Zn deficiency alone. Some of the variation may be due to differences in low molecular weight nutrients in the serum that result from differences in food intake. This difference should be reduced with improvements in the medium. Requirements for growing epidermal cells in cultures are not as well delineated as those for other cell types, but improvements are appearing regularly. As we and others learn more about culturing these cells, we are confident that it will be possible to identify factors related to their growth that are affected by Zn status.

REFERENCES

1. Solomons, N.W., 1979, Am. J. Clin. Nutr. 32:856 - 871.

2. Diamond, I., H. Swenerton, and L.S. Hurley, 1971, J. Nutr. 101:77 - 84.

3. Chen, S., J. Meyer, and S. Gerson, 1975, Pharmacol. Ther. Dent. 2:83 - 95.

4. Knehans, A.W., R.L. Kincaid, W.O. Regan, and B.L. O'Dell, 1979, J. Nutr. 109:418 - 425.

5. Freeland-Graves, J.H., P.J. Hendrickson, M.L. Ebangit, and J.Y. Snowden, 1981, Am. J. Clin. Nutr. 34:312 - 327.

X-RAY MICROANALYSIS OF INTRACELLULAR ZINC IN HUMAN INTESTINAL MUCOSA

John Gwyn Jones and Margaret E Elmes

Department of Pathology
University Hospital of Wales
CARDIFF CF4 4XN
WALES UK

Previously the analysis of tissue zinc has been done on whole tissues and localization techniques have mainly been histochemical using dithizone. Human jejunal mucosa had a zinc content of 97 and 99 µg/g dry weight in control subjects (Hambidge et al 1978). Dithizone-reactive Paneth cells were found in rat intestine when fixed in a non aqueous fixative (Elmes, 1976) but no dithizone reactive cells were found in human intestinal mucosa (Elmes and Gwyn Jones, 1981).

X-ray microanalysis was used to detect and measure intracellular zinc in human intestinal mucosa using samples of histologically normal jejunum and ileum resected either during partial gastrectomy (4 patients) or intestinal resection (3 patients). The tissues were fixed using a modified glutaraldehyde technique, sectioned and ultrathin sections observed and analysed in an AEI EMMA 4 transmission electron microscope with a KEVEX solid state detector linked to a TEKTRONIX multichannel analyser. Two cells per section of the following types were analysed; enterocytes and goblet cells on the lower half of the villus and Paneth and undifferentiated stem cells at the base of the crypts. An accelerating voltage of 80 Kv, beam current of 0.1 µA and counting time of 40 secs were used for each analysis and 5 sections from different resin blocks per patient analysed. Mean whole cell zinc concentration was determined for each cell type and mean Paneth cell granule and cytoplasmic zinc concentration measured for each patient and shown in Tables 1, 2 and 3.

Table 1. Mean whole cell zinc concentrations (ppm) in the Jejunum.

Patient	Paneth	Goblet	Stem	Entero	Patient mean
1	4,700	9,700	10,100	9,300	8,500
2	10,000	15,300	19,700	16,400	15,400
3	4,200	5,300	8,000	8,200	6,400
4	2,400	4,700	4,100	5,300	4,100
Jejunal mean	5,300	8,800	11,000	9,800	

In the jejunum analysis of variance showed a significant difference both between patients and also between cell types (F ratios; patients 39.36 d.f. 3/9; cell types = 8.78 d.f. 3/9; $p < 0.01$ in both cases). A multiple range test showed that Paneth cells had a significantly lower zinc concentration than the other three cell types and there was no significant difference between stem cells, enterocytes and goblet cells.

Table 2. Mean whole cell zinc concentrations (ppm) in the Ileum.

Patient	Paneth	Goblet	Stem	Entero	Patient mean
5	6,600	9,900	11,800	13,200	10,400
6	3,400	6,500	8,700	11,000	7,400
7	9,400	6,800	9,100	13,200	9,600
Ileal mean	7,000	7,700	9,900	12,500	

Table 2 shows that in the ileum there was a significant difference between cell types ($p < 0.05$) but no significant difference between patients (F ratios, patients = 3.86, d.f. 2/6; cell types = 8.35, d.f. 3/6). The highest mean zinc concentrations were found in the enterocytes and apart from Patient 7 Paneth cells had the lowest zinc levels. A multiple range test showed a significant difference between enterocytes and between goblet and Paneth cells considered together.

The mean enterocyte zinc concentration was higher in the ileum than the jejunum but statistically significant differences could not be demonstrated in each cell type between the jejunum and ileum.

Table 3. Mean Paneth cell granule and cytoplasm zinc concentration (ppm).

	JEJUNUM			ILEUM	
Patient	Granules	Cyto	Patient	Granules	Cyto
1	8,200	5,400	5	6,800	7,800
2	8,600	8,900	6	3,800	6,400
3	2,400	3,100	7	2,900	5,600
4	3,300	1,900			
Jejunal mean	5,600	4,800	Ileal mean	4,500	6,600
S.E.	800	800	S.E.	1,200	600

Table 3 shows mean zinc concentrations in Paneth cell granules and cytoplasm. Jejunal granule zinc was higher than the cytoplasmic zinc and also higher than granule zinc in the ileum (not statistically significant).

In the ileum the cytoplasmic zinc concentration was significantly higher than the granule zinc concentration ($p < 0.05$).

All mucosal cells analysed contained significant amounts of zinc at levels 60 - 100 times greater than those found in whole tissue analysis in spite of being dithizone negative. Within each cell type a wide variation in levels between cells was found that was maximum in the goblet cells. This is probably a measure of both biological variation and zinc retention during processing. Paneth cells generally had lower mean zinc levels than the other cell types analysed, the enterocytes had the highest levels especially in the ileum. Animal experiments have indicated that the ileum has the greatest capacity for zinc absorption (Antonson, Barak and Vanderhof, 1979) and we would suggest that these high zinc values are probably due to both free zinc and metallothionein in the enterocyte as this technique may not detect zinc within a large molecule such as a metalloenzyme.

Stem cells had mean zinc levels of the same order in jejunum and ileum. Goblet cells showed a wide variation in zinc levels within a patient that may be associated with different proportions of zinc rich mucigen granules and zinc low mucin mass in the cells analyzed. The abnormal goblet and Paneth cells found in the zinc deficiency due to Acrodermatitis enteropathica may indicate an essential role for zinc in maintenance of granule stability.

Whole Paneth cell zinc and Paneth cell granule zinc were of the same order in all tissues sampled but the cytoplasmic zinc was higher than granule zinc in the ileum only.

It is evident that the dithizone reaction shows a very small proportion of intracellular zinc and this zinc is probably water soluble and lightly bound.

SUMMARY

X-ray microanalysis was used to detect and measure intracellular zinc in human intestinal mucosa. The highest levels were found in enterocytes, especially in the ileum and stem cells. Goblet cells showed a wide variation that may be associated with their secretory status. Paneth cells showed the lowest levels and in the ileum the cytoplasmic zinc was higher than the granule zinc.

Histochemical methods evidently demonstrate a very small proportion of intracellular zinc present in a tissue.

REFERENCES

1. Hambidge, K M, Neldner K H, Walravens, P A, Weston W L, Silverman A, Sabol J S, Brown R M. 1978. In "Zinc and Copper in Clinical Medicine", Spectrum Publications Inc., NY, pp 81 -98.

2. Elmes M E, 1976. J Path 118: 183 - 191.

3. Elmes M E, Gwyn Jones J, 1981. Histochem J 13: 335 - 337.

4. Antonson D L, Barak A J, Vanderhoof J A, 1979. J Nutr 109: 142 - 147.

DISCUSSION

HOWELL (Australia). Dr. Elmes, when you were describing the technique you told us how you would get the instrument to read the content of an organelle and later on you talked about whole cell values; did you widen the beam or do it as a sum of values?

ELMES (Wales). Yes you open the beam out.

SALTMAN (U.S.A.). You showed us the picture of the slide where you got a zinc stain from the rat. Why do you think the human tissue failed to stain?

ELMES (Wales). There are slight differences in the granule structure between rats and man and there is a very marked physiochemical difference in that you cannot use the same fixation on rat Paneth cells that you can on human cells because the cores just fall out. That is why Dr. Gwyn Jones had to develop a new fixation technique. So there is a physiochemical difference in the construction of the granules.

INDEXES

LIST OF PARTICIPANTS

M. Abdulla, Research Department 2, E Block 4th Floor, University Hospital, S-221 85 Lund, Sweden

I.K. Adams, 10910 E Tan Que Verde Road, Tucson, Arizona 85715, U.S.A.

P.J. Aggett, Department Physiology, Marischal College, University Aberdeen, Aberdeen Scotland, AB9 1AS, U.K.

N. Alcock, Memorial Sloan-Kettering Cancer Center, 1275 York Avenue, New York, 10021, U.S.A.

J. Allen, Department of Agriculture, Jarrah Road, South Perth 6151, Western Australia.

J.D. Allen, Veterinary School, Murdoch University, Murdoch 6150, Western Australia.

W.M. Allen, Department of Functional Pathology, Institute for Research on Animal Diseases, Compton Nr Newbury, Berkshire RG16 ONN, U.K.

C.B. Ammerman, Nutrition Laboratory, University Florida, Gainesville, Florida 32611, U.S.A.

S.P. Arora, Division of Dairy Cattle Nutrition & Physiology, National Dairy Research Institute, Haryana 132001, India.

J.R. Arthur, The Rowett Research Institute, Greenburn Road, Bucksburn, Aberdeen AB2 9SB, Scotland, U.K.

B.E. Aughey, Department Veterinary Anatomy, Veterinary School, Bearsden Road, Glasgow, Scotland G61 1OH, U.K.

D. Behne, Hahn Meitner Institute für Kernforschung, 1000 Berlin 39, Federal Republic of Germany.

R.R. Bell, School of Community Health, Western Australia Institute of Technology, Hayman Road, Bentley 6102, Western Australia.

W.T. Binnerts, Department of Animal Physiology, Agriculture University, Wageningen, The Netherlands.

C.J. Blaxland, Post Office Box 90, Brookvale 2100, New South Wales, Australia.

P. Brätter, Hahn Meitner Institute für Kernforschung, 1000 Berlin 39, Federal Republic of Germany.

A.C. Bray, University of Tasmania, GPO Box 252C, Hobart 7002, Tasmania, Australia.

I. Bremner, The Rowett Research Institute, Greenburn Road, Bucksburn, Aberdeen AB2 9SB, Scotland, U.K.

J. Brooks, I.C.I. Australia Operations Pty Ltd, PO Box 347, Dandenong 3175, Victoria, Australia.

A. Bruce, Swedish National Food Administration, Box 622, S-75126, Uppsala, Sweden.

W. Bussler, Lentzeallee 55-57, 1000 Berlin 33, Federal Republic of Germany.

J. Camakaris, Department Genetics, Melbourne University, Parkville 3052, Victoria, Australia.

W. V. de A. Camargo, Institute Biologico, Caixa Postal 7119, 94014 Sao Paulo, SP. Brazil.

J.R. Cannon, Department Organic Chemistry, University of Western Australia, Nedlands 6009, Western Australia.

B.S. Chandler, CSIRO, Division Animal Production, Private Mail Bag, Post Office Wembley 6014, Western Australia.

H.M. Chapman, School of Veterinary Studies, Murdoch University, Murdoch 6150, Western Australia.

D.A. Christensen, Department of Animal and Poultry Science, Univeristy of Saskatchewan, Saskatoon, Sask, 57N OWO, Canada.

J. Christley, Department of Anatomy, Univerisity of Sydney, Sydney 2006, New South Wales, Australia.

M. Copland, Department of Agriculture, Jarrah Road, South Perth 6151, Western Australia.

R. Cornelis, Institute for Nuclear Sciences, Proeftuin-Straat 86, Gent, 9000, Belgium.

N.D. Costa, School of Veterinary Studies, Murdoch University, Murdoch 6150, Western Australia.

M. Côté, Centre Hospitalier Universitaire, Sherbrooke, P.Q, Canada. J1H 5N4.

P. Coyle, Division of Clinical Chemistry, Institute of Medical & Veterinary Science, Box 14 Rundle Street Post Office, Adelaide 5000, South Australia.

F.J. Cumming, Biological & Health Sciences, Deakin University, Geelong 3217, Victoria, Australia.

D.H. Curnow, Clinical Biochemistry Department, Queen Elizabeth II Medical Centre, Nedlands 6009, Western Australia.

D.M. Danks, Royal Childrens' Hospital, Parkville 3052, Victoria, Australia.

W.J. Davies, Department of Agriculture, Port Lincoln 5606, South Australia.

J.R. DeLaeter, School of Applied Science, Western Australia Institute of Technology, Hayman Road, Bentley 6102, Western Australia.

R. De Souza, Smith Kline & French Labs Ltd, Mundells, Welwyn Garden City, Harts, England, AL7 1EY, U.K.

D.W. Dewey, 91 Cross Road, Magill 5072, Adelaide, South Australia.

M.J. Dibley, 29 Union Street, Subiaco 6008, Western Australia.

J. Dickson, Department of Agriculture, Jarrah Road, South Perth 6151, Western Australia.

I.E. Dreosti, Division of Human Nutrition, CSIRO, Kintore Avenue, Adelaide 5000, South Australia.

J.S. Edwardly, Center for the Application of Isotopes & Radiation, National Atomic Energy Agency, Post Office Box 2 Kebayoran Lama, Jakarta, Selatan, Indonesia.

S.R. Edwards, Clinical Biochemistry Department, Queen Elizabeth II Medical Centre, Nedlands 6009, Western Australia.

M.E. Elmes, Department Pathology, University Hospital of Wales, Heath Park, Cardiff, Wales, CF4 4XN, U.K.

H. Fels, Animal Production Division, Department of Agriculture, Jarrah Road, South Perth, 6151, Western Australia.

J. Fitzherbert, 3 Urumga Parade, Wollongong 2500, New South Wales, Australia.

G. Fosse, Institute of Anatomy, University of Bergen, Arstadveien 19, 5000 Bergen, Norway.

A.J. Fraser, Ruakura Animal Health Laboratory, Private Bag, Hamilton, New Zealand.

V.P. Garnys, A.M.D.E.L., Post Office Box 114, Eastwood 5063, South Australia.

R.J. Gartner, Animal Research Institute, Yeerongpilly 4105, Queensland, Australia.

G.A. Garton, The Rowett Research Institute, Geenburn Road, Bucksburn, Aberdeen AB2 9SB, Scotland, U.K.

J.M. Gawthorne, School of Veterinary Studies, Division Veterinary Biology, Murdoch University, Murdoch 6150, Western Australia.

B.E. Golden, Tropical Metabolism Research Unit, Kingston 7, Jamaica.

M.H.N. Golden, Tropical Metabolism Research Unit, Kingston 7, Jamaica.

R. Gooneratne, School of Veterinary Studies, Division Veterinary Biology, Murdoch University, Murdoch 6150, Western Australia.

A.J. Gordon, CSIRO, Division Animal Production, Post Office Box 239, Blacktown 2148, New South Wales, Australia.

N.D. Grace, Applied Biochemistry Division, D.S.I.R., Palmerston North, New Zealand.

R.D. Graham, Department Agronomy, WAITE Agricultural Research Institute, Glen Osmond 5064, South Australia.

J.L. Greger, 34 Merlham Drive, Madison, Wisconsin 53705, U.S.A.

B.E. Guthrie, Nutrition Department, University of Otago, Post Office Box 56, Dunedin, New Zealand.

C.G. Halpin, "Attwood" Veterinary Research Laboratory, Mickleham Road, Westmeadows 3047, Victoria, Australia.

R.J. Hannam, Department of Agriculture, Northfield Laboratories, GPO Box 1671, Adelaide 5001, Australia.

I. Harding-Barlow, 3717 Laguna Avenue, Palo Alto, CA 94306, U.S.A.

B.F. Harland, Food & Drug Administration, Division of Nutrition, HFF-268, 200 "C" St. S.W., Washington D.C. 20204, U.S.A.

I.M. Harris, 41 St. Margaret's Road, Tottenham, London N17 6TY, U.K.

S. Haywood, Department Veterinary Pathology, University of Liverpool, Post Office Box 147, Liverpool, Merseyside L69 3BX, U.K.

P.J. Healy, Veterinary Research Station, Glenfield 2167, New South Wales, Australia.

S.M. Herd, Genetics Research Unit, Royal Children's Hospital, Flemington Road, Parkville 3052, Victoria, Australia.

J.E. Hesketh, The Rowett Research Institute, Greenburn Road, Bucksburn, Aberdeen AB2 9SB, Scotland, U.K.

B.S. Hetzel, CSIRO, Division Human Nutrition, Kintore Avenue, Adelaide 5000, Australia.

J.L. Hill, CSIRO, Division Animal Production, Private Mail Bag, Post Office, Wembley 6014, Western Australia.

R. Hill, The Royal Veterinary College, Boltons Park, Potters Bar, Herts EN6 1NB, U.K.

D.J. Horvath, Agriculture Science Building, WO UAO University, Morgantown, West Virginia 26506, U.S.A.

A.L. Howarth, Post Office Box 90, Brookvale 2100, New South Wales, Australia.

J. McC. Howell, School of Veterinary Studies, Division Veterinary Biology, Murdoch University, Murdoch 6150, Western Australia.

M.J. Hoxey, 28 Sovereign Avenue, Willetton 6155, Western Australia.

D.A. Hucker, Regional Veterinary Laboratory, Department Agriculture, Post Office Box 406, Hamilton 3300, Victoria, Australia.

J.F. Hudman, Murdoch University, Murdoch 6150, Western Australia.

D.R. Hudson, CSIRO, Division Mineralogy, Private Mail Bag, Post Office, Wembley 6014, Western Australia.

L.S. Hurley, Department Nutrition, University of California, Davis, California 95616, U.S.A.

R.I. Hutagalung, Department Animal Sciences, Faculty Veterinary Medicine & Animal Science Universiti, Pertanian Malaysia, Serdang, Selangor.

V. Iyengar, Nuclear Research Center Julich, Institute of Medicine, D 517 Julich, Federal Republic of Germany.

S. Jameson, Department Internal Medicine, University Hospital, S-75014, Uppsala, Sweden.

M. Janghorbani, Nuclear Reactor Laboratory, M.I.T., 138 Albany Street, Cambridge MA, 02139, U.S.A.

N.E. Johnson, 105 Quarterdeck Drive, Madison WI 53705, U.S.A.

W.H. Johnson, 2430 Hospital Lane, Room 55, Redding CA 96001, U.S.A.

G.J. Judson, Institute of Medical & Veterinary Science, Box 14, Rundle Street Post Office, Adelaide 5001, South Australia.

K. Katoh, Ishikawa, Prefecture College of Agriculture, Suematsu, Nonoichi-Machi, Ishikawa-Gun, Ishikawa Prefecture, Japan.

C.L. Keen, Department Nutrition, University of California, Davis, California 95616, U.S.A.

M. Kirchgessner, Institut für Ernährungsphysiologie, D-805 Freising-Weihenstephan, Federal Republic of Germany.

W.E. Kollmer, Gesellschaft für Strahlen-Und, Umweltforschung, Neuherberg D-8042, Federal Republic of Germany.

K.A.V.R. Krishnamachari, National Institute of Nutrition, Indian Council of Medical Research, Hyderabad 500007, India .

R. Krishnamachari, National Institute of Nutrition, Indian Council of Medical Research, Hyderabad 500007, India.

J. Kubota, Soil Investigations NETSC, US Plant & Soil Nutrition Laboratory, Tower Road, Ithaca 14853, U.S.A.

J.S. Kumaratilake, School of Veterinary Studies, Division Veterinary Biology, Murdoch University, Murdoch 6150, Western Australia.

H. Lane, The University of Texas, Health Science Center at Houston, School of Allied Health Sciences, Post Office Box 20036, Houston Texas 77025, U.S.A.

J.P. Langlands, CSIRO, Private Bag, Armidale 2350, New South Wales, Australia.

H.J. Lantzsch, Institut of Animal Nutrition, University of Hohenheim, Emil-Wolff-Str 8, D-700 Stuttgart 70, Federal Republic of Germany.

O.A. Levander, USDA Human Nutrition Center, Beltsville, Maryland 20705, U.S.A.

G. Lewis, M.A.F.F. Central Veterinary Laboratory, New Haw Weybridge, Surrey K15 3NB, U.K.

M.C. Linder, Department of Chemistry, California State University, Fullerton, California 93634, U.S.A.

J.R. Lindsay, CSIRO, Animal Production, Private Mail Bag, Post Office, Wembley 6014, Western Australia.

B. Lino, Australian Wool Corporation, Post Office Box 4021 GPO, Sydney 2001, Australia.

H. Lloyd Davies, School of Wool & Pastoral Science, University of New South Wales, Post Office Box 1, Kensington 2035, New South Wales, Australia.

I. Lombeck, University Children's Hospital, 4000 Dusseldorf, Federal Republic of Germany.

J.F. Loneragan, School Environmental & Life Sciences, Murdoch University, Murdoch 6150, Western Australia.

B. Lönnerdal, Department Nutrition, University California, Davis, California 95616, U.S.A.

R.S.W. Lugg, Department Health & Medical Services, Curtin House, 7th Floor 60 Beaufort Street, Perth 6000, Western Australia.

U. Mack, 19 Enid Street, Bracken Ridge 4017, Queensland, Australia.

J.R. Mann, Genetics Research Unit, Royal Children's Hospital, Flemington Road, Parkville 3052, Victoria, Australia.

D.C. Martens, Virginia Polytechnic Institute & State University, College of Agriculture & Life Sciences, Department of Agronomy, Blacksburg Virginia 24061, U.S.A.

D.G. Masters, Department Animal Science & Production, University Western Australia, Nedlands 6009, Western Australia.

H.G. Masters, Department of Agriculture, Jarrah Road, South Perth 6151, Western Australia.

P.M. May, Department Chemistry, UWIST, Cardiff CF1 3NU, U.K.

C.F. Mills, The Rowett Research Institute, Greenburn Road, Bucksburn, Aberdeen AB2 9SB Scotland, U.K.

R.J. Moir, Department Animal Science & Production, University of Western Australia, Nedlands 6009, Western Australia.

B. Momčilović, Institute for Medical Research & Occupational Health, M. Pijade 158 P.O.B. 291, YU-41001 Zagreb, Yugoslavia.

T. Morishima, Department of Dermatology, School Medicine, Nihon University, 1-8-13 Kanda-Surugadai, Chiyoda-KU, Tokyo 101, Japan.

A.L. Moxon, 2133 Eddy Road, Wooster, Ohio 44691, U.S.A.

J.E. Musk, 23 Peet Crescent, Trigg 6020, Western Australia.

A. MacPherson, West of Scotland Agricultural College, Auchincruive, Ayr KA6 5HW, Scotland, U.K.

L.R. McDowell, Animal Science Department, Livestock Pavilion, University of Florida, Gainesville, Florida 32611, U.S.A.

J.D. McFarlane, Post Office Box 618, Naracoorte 5271, South Australia.

J.M. McKenzie, Nutrition Department, University of Otago, Post Office Box 56, Dunedin, New Zealand.

D. McMaster, Department of Medicine, Institute of Clinical Science, Grosvenor Road, Belfast Northern Ireland.

M.E. Nairn, School of Veterinary Studies, Murdoch University, Murdoch 6150, Western Australia.

P. Newberne, Massachusetts Institute of Technology, 50 Ames Street, E18-611, Cambridge, Massachusetts, U.S.A.

R.E. Newnham, Post Office Box 61, Thornbury 3071, Victoria, Australia.

K.A. Nicholson, 9 Terama Court, Glen Waverley 3150, Victoria, Australia.

E.J. Nielsen, USDA, SEA, GFHNRC, Box 7166, University Station, Grand Forks 58202, North Dakota, U.S.A.

F.H. Nielsen, USDA, SEA, GFHNRC, Box 7166, University Station, Grand Forks 58202, North Dakota, U.S.A.

D. Oberleas, 2916 Edinburgh Court, Lexington, Kentucky 40503, U.S.A.

B.H. O'Connor, Department of Physics, Western Australian Institute of Technology, Hayman Road, Bentley 6102, Western Australia.

B. O'Dell, Department Biochemistry, 322 Chemistry Building, University Missouri, Columbia MO 65211, U.S.A.

P.L. Oe, v.d. Veerelaan 67, 1181 Ra Amstelveen, The Netherlands.

J.E. Oldfield, Department of Animal Science, Oregon State University, Corvallis, Oregon 97331, U.S.A.

J. Pallauf, Institut für Tierernäehrung, Universität Giessen, Senckenbergstrasse 5, D-63000 Giessen, Federal Republic of Germany.

R. Palm, Department of Neurology, University Hospital, S-901 85 Umeå, Sweden.

A.J. Parker, School of Mathematical & Physical Sciences, Murdoch University, Murdoch 6150, Western Australia.

D.I. Paynter, "Attwood" Veterinary Research Laboratory, Mickleham Road, Westmeadows 3047, Victoria, Australia.

R. Peet, Department of Agriculture, Jarrah Road, South Perth 6151, Western Australia.

M. Perlman, Neonatal Unit, Hadassah University Hospital, Ein Kerem, P.O.B. 12,000, Jerusalem 91 120, Israel.

Z. Pervan, 6 The Avenue, Midland 6056, Western Australia.

D.W. Peter, CSIRO, Private Mail Bag, Post Office,Wembley 6014, Western Australia.

S.M. Peters, Golden Poultry Farming Industries, Kooree Grange, Heath Road, Leppington 2171, New South Wales, Australia.

D.S. Petterson, Department of Agriculture, Jarrah Road, South Perth 6151, Western Australia.

A.M. Phillips, Genetics Department, University of Melbourne, Melbourne 3000, Victoria, Australia.

G.D. Phillips, Intensive Care Unit, Flinders Medical Centre, Bedford Park 5042, South Australia.

A.P. Pillay, 362 Havenside Drive, Chatsworth, Durban 4092, South Africa.

L.O. Plantin, Karolinska Institute Research Center, Huddinge Hospital, Huddinge S-141 86, Sweden.

B.J. Potter, CSIRO, Division of Human Nutrition, Kintore Avenue, Adelaide 5000, South Australia.

A.S. Prasad, 4710 Cove Road, Orchard Lake MI 48033, U.S.A.

J.R. Prohaska, University of Minnesota Duluth, Department of Biochemistry, Duluth, Minnesota 55812, U.S.A.

A. Purcell, Department of Agriculture, Jarrah Road, South Perth 6151, Western Australia.

B. Radjagukguk, Department of Social Science, Faculty of Agriculture, Gadjah Mada University, Sekip, Yogyakarta, Indonesia.

C.M. Reading, Post Office Box 587, Dee Why 2099, New South Wales, Australia.

G.A. Reed, I.C.I. Australia Operations Pty Ltd, GPO Box 4311, Melbourne 3001, Victoria, Australia.

P.G. Reeves, Department Biochemistry, 322 Chemistry Building, University of Missouri, Columbia MO 65211, U.S.A.

A.M. Reichlmayr-Lais, Institut für Ernährungsphysiologie, D-805 Freising-Weihenstephan, Federal Republic of Germany.

D.J. Reuter, Box 1671 G.P.O., Adelaide 5001, South Australia.

M.F. Robinson, Nutrition Department, University of Otago, Post Office Box 56, Dunedin New Zealand.

H.P. Roth, Institut für Ernährungsphysiologie, D-805 Freising-Weihenstephan, Federal Republic of Germany.

B.E. Ruksan, Secretaria De Agricultura Y Ganaderia, Instituto Nacional De Tecnologia Agropecuaria, Centre De Investigaciones En Ciencias Veterinarias, (1712) Castelar (Pcia. Bs. As.) Buenos Aires, Argentina.

S.V. Rumball, Department Chemistry/Biochemistry/Biophysics, Massey University, Palmerston North, New Zealand.

E. Sabbioni, Radiochemistry & Nuclear Chemistry Division, C.C.R. Euratom, 21020 Ispra, Varese 21020, Italia.

D.P. Saltman, Department Biology, University California, San Diego, La Jolla, California 92093, U.S.A.

G.P. Savage, Biochemistry Department, Lincoln College, Canterbury, New Zealand.

H. Schenkel, Institute für Tierernährung, Universitat Hohenheim, D-7000 Stuttgart 70, Federal Republic of Germany.

F.J. Schwarz, Institut für Ernährungsphysiologie, D-805 Freising-Weihenstephan, Federal Republic of Germany.

R. Scott, Department of Urology, Royal Infirmary, Glasgow G4 OSF, Scotland, U.K.

W.E. Sedivy, 2140 East 19th, Fremont, Nabraska 68025, U.S.A.

D. Shapcott, Department Pediatrie, Centre Hospitalier, Universitaire Sherbrooke, Quebec IH 5N4, Canada.

A.J. Sinclair, Veterinary Research Institute, Park Drive, Parkville 3052, Victoria, Australia.

R.M. Smith, CSIRO, Division of Human Nutrition, Adelaide 5000, South Australia.

K. Soda, Institute for Chemical Research, Kyoto University, Gokasho UJI, Kyoto-FU 611, Japan.

J.T. Spickett, School of Community Health, Western Australian Institute of Technology, Hayman Road, Bentley 6102, Western Australia.

L. Steel, Department of Medicine, Institute of Clinical Science, Grosvenor Road, Belfast BT12 6BJ, Northern Ireland.

P. Steele, Animal Health Laboratory, Department of Agriculture, Jarrah Road, South Perth 6151, Western Australia.

B.J. Stevens, c/- Department of Applied Biology, R.M.I.T., Melbourne 3000, Victoria, Australia.

R.A. Sunde, The Rowett Research Institute, Greenburn Road, Bucksburn, Aberdeen AB2 9SB, Scotland, U.K.

J.W. Suttie, Department of Biochemistry, University of Wisconsin-Madison, Wisconsin, Madison 53706, U.S.A.

N.F. Suttle, Moredun Research Institute, 408 Gilmerton Road, Edinburgh EH17 7JH, Scotland, U.K.

Y. Takagi, 2-11-16 Takeshirodai, Sakai, Osaka 590-01, Japan.

I.L.H. Tengrup, Department of Surgery, Malmo General Hospital, 21401 Malmo, Sweden.

C.D. Thomson, Nutrition Department, University of Otago, Post Office Box 56, Dunedin, New Zealand.

D.E. Ullrey, Department Animal Science, Michigan State University, East Lansing, Michigan 48824, U.S.A.

J. Valentine, School of Public Health, UCLA, Los Angeles, California 90024, U.S.A.

C.J.A. Van Den Hamer, Hendrinaland 138, The Hague, 2591 TM, The Netherlands.

J.M. Versieck, Department Internal Medicine, Division of Gastroenterology, Academisch Ziekenhuis, De Pintelaan 135, B-9000 Ghent, Belgium.

C. Walker, U.S. Plant, Soil & Nutrition Laboratory, Cornell University, Ithaca, New York 14853, U.S.A.

W.R. Walker, University of Newcastle, Newcastle 2308, New South Wales, Australia.

K.M. Weber, CSIRO, Animal Health Division, Private Bag No. 1, Parkville 3052, Victoria, Australia.

W.S. Webster, Department of Anatomy, University of Sydney, Sydney 2006, New South Wales, Australia.

E. Weigand, Institut für Ernährungsphysiologie, D-805 Freising-Weihenstephan, Federal Republic of Germany.

R.M. Welch, U.S. Plant, Soil & Nutrition Laboratory, Tower Road, Ithaca 14853, U.S.A.

T.W. Westermarck, Helsinki Central Institute for the Mentally Subnormal, Kirkkonummi, SF-02400, Finland.

P.D. Whanger, Department Agricultural Chemistry, Corvallis, Oregon 97330, U.S.A.

A.M. Wheeler, Post Office Box 217, Richmond 2753, New South Wales, Australia.

C.L. White, CSIRO, Division Animal Production, Private Mail Bag Post Office, Wembley 6014, Western Australia.

G. Wiener, A.R.C. Animal Breeding Research Organisation, West Mains Road, Edinburgh EH9 3JQ, U.K.

D.R. Williams, Department of Chemistry, University of Wales Institute of Sciences & Technology, Cardiff, S. Glamorgan CF1 3NU, U.K.

J.D. Williams, Post Office Box 488, Lismore 2480, New South Wales, Australia.

P.E. Wilson, Government Chemical Laboratories, 30 Plain Street, Perth 6000, Western Australia.

R.G. Wilson, Clinical Chemistry Department, Princess Margaret Hospital, GPO Box D184, Perth 6001, Western Australia.

C.T. Wong, Department of Physiology, Faculty of Medicine, National University of Singapore, 0316 Singapore.

J.A. Woolliams, 17 Crossland Crescent, Peebles, Peebleshire EH 45 8 LF, U.K.

S. Yagi, Department of Dermatology, School of Medicine, Nihon University, 1-8-13 Surugadai-Kanda, Chiyoda-Ku, Tokyo 101, Japan.

N.G. Yates, CSIRO P3T, Post Office Box 123, Bogor, West Java, Indonesia.

L. Zhu, Institute of Health, Nan-Wei Road, Bejing, China.

AUTHOR AND DISCUSSION PARTICIPANTS INDEX

Note: Page numbers refer to the first page of each contribution.

Page numbers underlined indicates authorship.

Note: Page numbers refer to the first page of each contribution.

Page numbers underlined indicates authorship.

Note: Page numbers refer to the first page of each contribution.

Page numbers underlined indicates authorship.

Note: Page numbers refer to the first page of each contribution.

Page numbers underlined indicates authorship.

Note: Page numbers refer to the first page of each contribution.

Page numbers underlined indicates authorship.

SUBJECT INDEX

Note: Page numbers refer to the first page of each contribution.

Note: Page numbers refer to the first page of each contribution.

Note: Page numbers refer to the first page of each contribution.

Note: Page numbers refer to the first page of each contribution.

Note: Page numbers refer to the first page of each contribution.

<u>Note</u>: Page numbers refer to the first page of each contribution.

Note: Page numbers refer to the first page of each contribution.

Note: Page numbers refer to the first page of each contribution.

Note: Page numbers refer to the first page of each contribution.

Note: Page numbers refer to the first page of each contribution.

Note: Page numbers refer to the first page of each contribution.

Note: Page numbers refer to the first page of each contribution.

Note: Page numbers refer to the first page of each contribution.

MIX
Papier aus verantwortungsvollen Quellen
Paper from responsible sources
FSC® C105338

If you have any concerns about our products, you can contact us on
ProductSafety@springernature.com

In case Publisher is established outside the EU, the EU authorized representative is:
Springer Nature Customer Service Center GmbH
Europaplatz 3, 69115 Heidelberg, Germany

Printed by Libri Plureos GmbH
in Hamburg, Germany